Dr. Lynn Holmes

HOLT

ENVIRONMENTAL SCIENCE

ANNOTATED TEACHER'S EDITION

Karen Arms

This book was printed with soy-based ink on acid-free recycled content paper, containing MORE THAN 10% POSTCONSUMER WASTE.

HOLT, RINEHART AND WINSTON
Harcourt Brace & Company
Austin • New York • Orlando • Atlanta • San Francisco • Boston • Dallas • Toronto • London

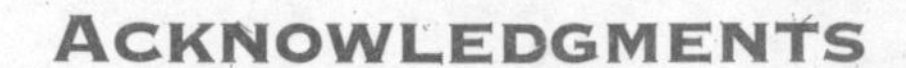

Writers

Claudia Fowler
University Laboratory School
Baton Rouge, Louisiana

Natalie Goldstein
Environmental Science Writer
Newfield, New York

Lynn Diebolt-Lewis
Olivet High School
Olivet, Michigan

Jacquelyn Jarzem, Ph.D.
Austin Community College
Austin, Texas

Stephanie Lanoue
Lago Vista High School
Lago Vista, Texas

Michael Lubich
Mapletown High School
Greensboro, Pennsylvania

Patricia Merkord
Austin High School
Austin, Texas

Carol Quay
Heritage Conservancy
Doylestown, Pennsylvania

William Quay
Science Writer
Kintnersville, Pennsylvania

Richard Shippee
Chairman: Life Science Department
Vincennes University
Vincennes, Indiana

Carol Wagner
Pflugerville High School
Pflugerville, Texas

Printed in the United States of America
ISBN 0-03-095400-2
2 3 4 5 6 7 8 9 032 97 96 95

Reviewers

Hugh Allen
American High School
Miami-Dade Community College
Miami, Florida

Jocelyn Baker
Windsor Forest High School
Savannah, Georgia

Albert Barr
Spring High School
Spring, Texas

Margaret A. Brumsted
Dartmouth High School
Dartmouth, Massachusetts

Dana Bruton
J. L. McCullough High School
The Woodlands, Texas

Lynette Celano
St. Gregory Catholic School
Phoenix, Arizona

Elizabeth Corum
Johnson High School
Savannah, Georgia

Jim Cramer
Brookfield East High School
Brookfield, Wisconsin

Pamela Gomillion
Puckett High School
Puckett, Mississippi

Wes Halverson
LBJ Science Academy
Austin, Texas

Janis Lariviere
Westlake High School
Alternative Learning Center
Austin, Texas

Clifford Lerner
Keene High School
Keene, New Hampshire

Lynn Diebolt-Lewis
Olivet High School
Olivet, Michigan

Sheila Lightbourne
Choctawhatchee High School
Fort Walton Beach, Florida

Alicia Moody
Choctawhatchee High School
Fort Walton Beach, Florida

George Newberry
Sparrows Point High School
Baltimore, Maryland

Marie Rediess
Algonac High School
Algonac, Michigan

Chester Sadonick
Bay High School
Bay Village, Ohio

Denice Sandefur
Nucla High School
Nucla, Colorado

Richard Shippee
Chairman: Life Science Department
Vincennes University
Vincennes, Indiana

Eva Silverfine
Ecologist
San Marcos, Texas

John M. Trimble
Corona Del Sol High School
Tempe, Arizona

David C. Tucker
Mt. Baker High School
Deming, Wyoming

Anne Tweed
Eaglecrest High School
Aurora, Colorado

Celia Wesenberg
Ponderosa High School
Shingle Springs, California

Anne Whitford
Elk Grove High School
Elk Grove, California

Paul W. Wilson
Owasso High School
Owasso, Oklahoma

CONTENTS

To the Teacher

Holt Environmental Science has been designed to provide you and your students with a balanced approach to the diverse study of our environment. The emphasis in this program is the study of science and the development of thinking and decision-making skills. Our goal is to provide students with the science background they need to analyze for themselves many of the issues concerning our environment.

The field of environmental science also offers a rare occasion to apply many different sciences to extend our knowledge of both our world and its inhabitants. Within what other course of study do you have the opportunity to discuss topics that range from a fragile desert biome to the debate over CFCs and the ozone layer? Perhaps it is this diverse context and its clear application that makes environmental science inherently interesting to students.

No Easy Solutions to Difficult Problems
Being extraordinarily complex, environmental problems often defy easy answers and quick solutions. Using present knowledge and technology, we are only marginally successful at predicting even the weather. It is not hard to understand how difficult it is to predict with any accuracy the outcomes of human actions that affect habitats and whole ecosystems. Although a great deal is known about the environment, there is still much to discover, particularly about how species interact with one another within complex ecosystems. There is also a real urgency to learn all we can before the environment is damaged beyond the point where informed decisions could prevent irreparable damage.

The *Holt Environmental Science* program can do much to improve your students' awareness of the environmental problems we are facing now and will be facing in the future. On an almost daily basis the media provides us with reports about environmental problems and debates. When people do not understand the science behind these problems, it is likely that the media will wield entirely too much influence over the outcome of these debates.

Awareness Is Half the Battle
This book is not designed to convert students into environmental activists. However, by studying this book, students will add significant depth to their understanding of the environment. It is hoped that this understanding will lead to an increased level of appreciation and that students will treat the environment accordingly.

Humans will continue to multiply and occupy more and more areas of the globe. And in so doing, they will interfere with most of our environment's complex natural systems. Many of the adverse consequences of that interference cannot be predicted. Others, however, are already anticipated and could be avoided. Unfortunately, environmental knowledge is not widespread. For example, you may know that dumping motor oil on the ground can contaminate enormous amounts of groundwater. Other people, however, may not be so well informed. Many people in the United States and throughout the world are still dumping oil, often because they do not realize the adverse effects of such an action.

Such environmental ignorance is very harmful and particularly disappointing considering our level of knowledge. While it is true that we don't have all the answers, it is also true that we know a great deal about our environment. It is important that we share this information. The more environmentally literate we are, the fewer careless mistakes we will make that will negatively affect our planet and all of its inhabitants.

Components of the Program

Pupil's Edition

Holt Environmental Science is a unique science textbook designed exclusively for high school courses in environmental science. This text not only engages students in the process of science, but also explores the real-world applications of this important and growing discipline. Through readings and explorations, students learn how science serves to further our understanding of the environment. At the same time, a variety of features provide students with insights into various environmental issues that stem from human activities. Exposure to such issues gives students valuable practice in applying science knowledge and skills within a real-world context. As a result, students learn more science and are better able to apply their knowledge. Students' decision-making and problem-solving skills also improve, and they are more likely to build on their knowledge as environmental issues are debated in the media.

Annotated Teacher's Edition

This Annotated Teacher's Edition contains all the information you need to teach either a semester or a full-year course in environmental science. In addition to relevant articles on subjects such as portfolios, thematic instruction, concept mapping, and safety, the teacher's commentary provides a wide range of teaching suggestions, background information, and planning aids. An Answer Key and on-page annotations provide you with all the answers you will need throughout your instruction.

For more detailed information about this Annotated Teacher's Edition, see the following page.

Teaching Transparencies

The *Holt Environmental Science Teaching Transparencies* consists of 50 four-color overhead transparencies highlighting important concepts and processes covered in the textbook. In addition, each transparency is accompanied by a blackline-master worksheet, which includes a variety of questions about the graphic. An Answer Key is provided.

Laboratory and Field Guide

The *Holt Environmental Science Laboratory and Field Guide* includes more than 30 additional investigations to support your instruction. A variety of different types of investigations are provided, including wet labs, scenario-based investigations, field investigations, issue-based activities, and student-designed experiments. Investigation work sheets are also included for the investigations in the Pupil's Edition.

Laboratory and Field Guide, Teacher's Edition

This Teacher's Edition provides overprinted, on-page answers for all questions and problems contained in the Laboratory and Field Guide. Teaching strategies and ideas for variations and extensions to the investigations are also provided.

Chapter Tests with Answer Key

The *Holt Environmental Science Chapter Tests* are blackline-master tests that provide a range of assessment options. Each test consists of objective, short-answer, critical-thinking, application, and alternative-assessment questions. An Answer Key is provided for convenient grading.

STS Science Forums

Videodiscovery's *STS Science Forums* is a unique videodisc package consisting of two double-sided videodiscs and a variety of print components, including teacher's guides and student workbooks. The *STS Science Forums* provides 12 scenarios that focus on different environmental issues, such as chemicals in drinking water, nuclear waste disposal, use of pesticides, and the impact of fossil fuels on the greenhouse effect. Each scenario begins with a proposal or problem statement. Students form lobbying groups—such as consumers, big business, and environmentalists—and analyze the issues. As "lobbyists," students build persuasive positions using the discs as research tools. In the process, they gain valuable experience in making informed and well-reasoned decisions.

Using the Teacher's Commentary

The ***Holt Environmental Science Annotated Teacher's Edition*** will help you create a rich and productive course in environmental science, regardless of your particular class requirements. A wide variety of teaching strategies is provided to accommodate a variety of teaching styles and preferences, as well as differences in student maturity and interests. All information provided in the Teacher's Commentary is arranged in an easy-to-use, logical format.

Planning Guides

A Planning Guide begins each chapter of Teacher's Commentary. This guide, in chart form, provides a section-by-section listing of all the components and features of the textbook as well as ancillaries such as Chapter Tests, Teaching Transparencies, and STS Science Forums. Each Planning Guide provides pacing information for four different teaching situations: 1/2 year course for average students, 1/2 year course for advanced students, 1 year course for average students, and 1 year course for advanced students.

Materials Organizer

A Materials Organizer follows each Planning Guide to provide a list of materials needed to complete the chapter investigation. Information about advance preparation makes planning and preparing for the investigation easy and efficient.

Chapter Overview

A Chapter Overview is included to give you a quick summary of each section of the chapter. This handy overview helps you focus your planning relative to the main points of each section.

Resources

A bibliography of relevant resources, categorized by application for teachers and students and by media type, is provided. An internal reference section, called Getting Involved, is also included. This section identifies pertinent EcoSkills that invite student participation. This section also references environmental organizations that students who are particularly interested in the chapter's subject matter can contact on their own.

Using the Quotation

Each Pupil's Edition chapter begins with a thought-provoking quotation that relates to the environment. These quotations provide an excellent opportunity to explore student perceptions, attitudes, and opinions. The strategies provided under Using the Quotation will help you use these quotations to their fullest.

Using the EcoLog

Each Pupil's Edition chapter also begins with an EcoLog box, which contains several questions that relate to the chapter. Students should answer these questions in their EcoLog prior to starting the chapter. In answering these questions, students will reveal their prior knowledge about the subject of the chapter.

Students' answers to the EcoLog questions are not intended to be graded. Instead, they are to be revisited and revised by the students at the end of the chapter (as directed on each Chapter Highlights page). By revising their answers, students will have the opportunity to dispel any misconceptions that they may have had and to see firsthand how their knowledge has grown. Only after the EcoLogs have been revised should they be used as an assessment tool.

Section-by-Section Commentary

The Teacher's Commentary for each section contains a wide array of ideas, information, and strategies organized within the following five headings: Focus, Motivator, Teaching Strategies, Closure, and Teaching Strategies for Investigation.

Focus

The Focus provides a succinct overview of what the students are to learn to successfully complete the section.

Motivator

Getting students interested in a topic is often half the battle in extending their knowledge. The Motivator describes, in detail, an idea for generating student interest at the onset of each section. Ideas may include high-interest demonstrations; short, cooperative learning activities; thought-provoking discussion topics; and role-playing scenarios.

Teaching Strategies

A variety of teaching strategies are provided to accommodate a wide range of teacher and student needs. Each section includes a selection of strategies from the following list.

Background—Interesting background information that provides added depth to the section information

Reinforcing Ideas—Strategy for ensuring that students understand an important concept

Demonstration—Procedure for illustrating a concept using a teacher demonstration

Using the Map, Using the Diagram, etc.—Strategies for strengthening process skills by using maps, diagrams, and other visual components of the section

Using the Case Study, Using the Points of View—Strategies for how and when to best use each of the features included in the chapter

Discussing the Issue—Information about relevant controversial issues that lend themselves to further discussion within the section

Debate—Topic and guidelines for organizing a student debate on a relevant issue

Safety Alert—Special notes or concerns about safety

Cooperative Learning Activity—Instructions for organizing a cooperative group activity relating to the section

Using the Theme of . . . —Specific focus question (and answer) that fosters thematic connections

Multicultural Connection—Information or activity that heightens multicultural awareness and extends knowledge of other cultures

Cross-Discipline Connection—Strategy for encouraging students to think beyond subject-area boundaries and explore relationships between science and a host of other disciplines such as math, history, and geography

Meeting Individual Needs—Strategy for adapting instruction to better fit the needs of gifted students, learners having difficulty, or learners with limited English proficiency

Notable Quote—Additional quotation to initiate thinking and contemplation

Research Activity—Longer-term activity that requires research

Real-World Connection—Strategy that relates information in the section to students' personal lives and immediate surroundings

Using the Transparency—Strategy for using a Teaching Transparency to reinforce a concept

Closure

The Closure information provides a variety of assessment strategies as well as methods for reteaching difficult concepts that students may not have fully grasped. In addition, extension information is provided for students who would like to go beyond the scope of the textbook. Each Closure contains the following information.

Section Review—Quick reference to location of answers in Answer Key

Alternative Assessment—Detailed instructions for an alternative method of assessing the information presented in the section

Reteaching—Strategy for using alternative modalities for reteaching important concepts

Extension—Higher-level activity for extending information beyond the boundaries of the section

Teaching Strategies for Investigation

The Teaching Strategies for Investigation provide valuable background, preparation, safety, and extension information related to the chapter investigation.

Portfolio

A Portfolio icon has been added to selected strategies throughout the Teacher's Commentary. The icon identifies the strategy as being one that could produce a finished product suitable for a portfolio. You may want to suggest this to your students or even make certain selections mandatory. For more information on portfolios, see pages T13–T14.

Answer Key

A complete Answer Key, tabbed for quick access, provides answers for nearly all of the questions asked within the textbook. The only exceptions are questions asked within captions, and multiple-choice questions and concept maps within the Chapter Reviews. These questions are answered with on-page annotations. In addition, on-page annotations provide page references so that all answers can be found quickly and easily in the Answer Key.

MASTER PLANNING GUIDE

Holt Environmental Science is a flexible textbook that can be used in either a full-year or 1 semester course for students' varying abilities. The Master Planning Guide shown below outlines four alternative classroom situations: 1/2 year course for average students, 1/2 year course for advanced students, 1 year course for average students, and 1 year course for advanced students. The number of days specified for each chapter are only general recommendations.

1/2 YEAR COURSE

Chapter	Average													Days	Advanced													Days
1	•	•	•		•						•	•		7–10	•	•	•		•	•					•	•		7–10
2	•	•	•		•							•		7–9	•	•	•		•							•		6–7
3	•	•	•				•				•	•		9–10	•	•	•		•		•				•	•		8–10
4	•	•	•	•								•		7–9	•	•	•	•	•							•		7–9
5	•	•			•						•	•		7–9	•	•			•						•	•		6–8
6	•						•					•		3–4	•	•					•					•		3–4
7			•	•								•		4–6			•	•	•						•	•		3–5
8														0														
9														0	•	•	•		•							•		6–7
10	•	•	•		•		•					•		9–11	•	•	•		•		•					•		8–10
11	•	•	•		•						•	•		9–11	•	•	•		•						•	•		8–10
12	•	•	•		•							•		7–9	•	•	•		•						•	•		7–9
13														0														
14						•								1–2						•								1
Section/Feature	1	2	3	4	C	M	P	E	S	F	I	T	V	70–90	1	2	3	4	C	M	P	E	S	F	I	T	V	70–90

1 YEAR COURSE

Chapter	Average													Days	Advanced													Days
1	•	•	•		•	•				•	•	•		15–16	•	•	•		•	•				•	•	•	•	15–16
2	•	•	•		•	•					•	•		15–16	•	•	•		•	•			•	•	•	•		15–16
3	•	•	•		•		•			•	•	•		16–18	•	•	•		•		•	•		•	•	•		16–18
4	•	•	•	•	•		•			•	•	•		17–19	•	•	•	•	•		•	•		•	•	•		17–19
5	•	•			•	•		•	•	•	•	•	•	14–16	•	•	•		•	•		•	•	•	•	•	•	14–16
6	•						•				•	•		6–7	•	•	•		•		•			•	•	•	•	6–7
7	•	•	•	•	•						•	•	•	8–10	•	•	•	•	•	•		•		•	•	•	•	8–10
8	•	•				•		•	•		•	•		5–6	•	•	•		•	•		•	•	•	•	•		5–6
9	•	•	•		•		•		•		•	•	•	8–10	•	•	•		•		•		•	•	•	•	•	8–10
10	•	•	•		•		•	•			•	•		16–17	•	•	•		•		•	•	•	•	•	•		16–17
11	•	•	•		•	•			•	•	•	•		17–18	•	•	•		•	•			•	•	•	•	•	17–18
12	•	•	•		•		•	•	•		•	•		14–16	•	•	•		•		•	•	•	•	•	•	•	14–16
13	•	•									•	•		5–6	•	•	•		•		•			•	•	•		5–6
14	•	•			•	•					•			4–5	•	•			•	•			•	•	•			4–5
Section/Feature	1	2	3	4	C	M	P	E	S	F	I	T	V	160–180	1	2	3	4	C	M	P	E	S	F	I	T	V	160–180

C = Case Study, **M** = Making a Difference, **P** = Points of View, **E** = Environmental Careers, **S** = EcoSkills, **F** = Field Activity, **I** = Investigation, **T** = Teaching Transparency, **V** = Videodisc: STS Science Forum

CHAPTER PLANNING GUIDES

For chapter-specific planning information, see the Planning Guides provided for each chapter in the Teacher's Commentary. Each Planning Guide contains a complete listing, by section, of all the components and features of the chapter in order of recommended use. For your convenience, the chapter Planning Guides provide the same course-length and ability-level options as the Master Planning Guide and also include more specific pacing information.

Meeting Individual Needs

Many classrooms today include students with a wide range of academic and linguistic skills. Reaching every student in these circumstances can be a challenge for the teacher. Yet, with appropriate teaching strategies and educational materials, the heterogeneous classroom can be an especially fertile learning environment. The following information will help you use *Holt Environmental Science* to meet the individual needs of diverse students.

- **Gifted Learners**

Teaching gifted students is often very rewarding. At the same time, however, these students present unique challenges. The difficulty of teaching gifted students often lies in keeping them interested, motivated, and challenged. Gifted learners who are inadequately challenged may become bored, withdrawn, or openly disruptive. Far too often, gifted students are neglected on the grounds that they need little attention or guidance to achieve at a satisfactory level.

Holt Environmental Science examines the many facets of complex environmental issues. This encourages gifted learners to gain a broad perspective of issues before taking a position. The end results are students who understand many points of view. Also, *Holt Environmental Science* emphasizes creative problem solving and decision making in activities that are suitable for your most gifted students. Because there is no single right solution to many environmental issues, students' opinions can reflect their individual abilities as they extend these open-ended activities to fit their interests and talents. To challenge gifted students to the fullest, you should utilize the Extension strategies offered within each chapter.

In addition, the Teaching Strategies contained in this Annotated Teacher's Edition include suggestions geared specifically toward gifted learners. Look for the label *Meeting Individual Needs: Gifted Learners.* Other strategies, such as Cross-Discipline Connection, Using the Points of View, and Research Activity, may also be most suitable for gifted students.

- **Learners Having Difficulty**

For any number of reasons, some students may have difficulty understanding basic concepts or may perform poorly in the average classroom environment. While *Holt Environmental Science* is engaging and interesting throughout, some individuals may require special attention in order to grasp the essential concepts of environmental science.

You might have to adjust the pace at which you provide learning goals. You may find it beneficial to use relevant, concrete examples in the classroom that relate to the daily life of the student. You may also have to reinforce a lesson a number of times. For these students, you should emphasize hands-on laboratory investigations and cooperative learning activities. These same techniques are often successful in working with students who simply have a general low interest in schoolwork. In addition, the many informative and attractive graphics reduce the possibility of students growing bored, while helping to establish the concepts of environmental science.

In the Teaching Strategies for each chapter, you will find activities and teaching suggestions geared specifically toward students having difficulty. Look for the label *Meeting Individual Needs: Learners Having Difficulty.* Other strategies, such as Real-World Connections, Reinforcing Ideas, Cooperative Learning Activity, Demonstration, and Reteaching, may also be suitable for these students.

- **Second-Language Learners**

Holt Environmental Science employs a multitude of striking visuals to support the science content. You can use these photographs, illustrations, and diagrams to particular advantage with students who are second-language learners. The feature Using the Photograph offers additional visual strategies that are well suited to these students.

Also, *Holt Environmental Science* offers activities and teaching suggestions geared specifically toward second-language learners. Look for the label *Meeting Individual Needs: Second-Language Learners.* Other strategies, including Cooperative Learning Activity and Alternative Assessment, are also suitable for students not proficient in the English language.

You can use *Holt Environmental Science* to effectively meet the needs of learners of all levels and backgrounds. By following the teaching suggestions within this Annotated Teacher's Edition, such as Motivators, Portfolios, and Multicultural Connections, as well as the specific helps for meeting the special needs listed here, you can select a variety of appropriate and suitable options for the differing ability levels of your students.

SPECIALLY CHALLENGED STUDENTS

The instruction of heterogeneous classes has become a prominent aspect of today's schools. State and federal laws mandate that all students are to have access to the least restrictive learning environment possible. Thus, many students—those with physical, mental, or behavioral disabilities—are being mainstreamed into the regular classroom. In order to accommodate them, you should try to make your classroom as accessible as possible with regard to the needs of your particular students.

- **Physically Impaired Students**
  Make your classroom as easy to move about in as possible by removing any obvious barriers. If any of your students use a wheelchair, make sure the aisles are wide enough to accommodate the chair. Make sure that the student can reach any equipment he or she needs. You may want to use a mobile demonstration table so that it can be moved to different areas of the room for maximum utility.

- **Visually Impaired Students**
  Seat students with marginal vision near the front of the room to maximize their view of both you and the chalkboard. Allow students who are completely blind to become familiar with the classroom layout before the first class begins, and promptly inform them of any changes you make. Whenever possible, provide blind students with Braille or taped versions of all printed materials, or let them use hand-held devices that convert written text into speech.

- **Students With Hearing Difficulties**
  If you have partially hearing-impaired students in your class, remember to always face them while speaking. Arrange seats in a circle or semicircle to facilitate lip reading so that hearing-impaired students can see you and the other students. Speak in simple, direct language, and avoid digressions or sudden changes in topic. During class discussions, periodically summarize what other students are saying, and repeat students' questions before answering them. Use visual media such as filmstrips, transparencies, and close-captioned films when appropriate.
  A student who is completely deaf may require a sign language interpreter. If so, let the student and the interpreter determine the most convenient seating arrangement. When asking the student a question, be sure to look at the student, not the interpreter.

- **Students With Speech Impediments**
  Mainstreaming speech-impaired students is generally not very difficult. Patience is essential, however. For example, resist the temptation to finish sentences for a student who stutters. Also pay attention to nonverbal cues such as facial expression and body language. You need not leave the speech-impaired student out of the normal classroom discussions. For example, you may call on a speech-impaired student to answer a question and then allow the student to write out his or her response on the chalkboard or overhead projector. Be supportive and encouraging, and use multisensory materials to create a more comfortable learning environment.

- **Students With Behavioral Disorders**
  Behavioral disorders are emotional or behavioral disturbances that hinder a student's overall functioning. The behaviorally impaired student may exhibit any of a variety of behaviors, ranging from extreme aggression to complete passivity. Obviously, no single teaching strategy can accommodate all behavioral disorders. As a general rule, try to be fair and consistent, yet flexible, in your dealings with these students, and make sure that you clearly state rules and expectations. Reinforce desirable behavior, and ignore or mildly admonish undesirable behavior.

Students with physical, mental, or emotional difficulties can benefit from a partner who is willing to lend assistance. You may wish to enlist the aid of other students in the classroom to make it possible for specially challenged students to engage in the same activities as other students. Encourage members of your class to assist physically impaired students. Peer tutors, student volunteers who work with specially challenged students on specific assignments and review material, can also be effective.

Using Portfolios

The study of science has always provided an excellent opportunity for students to practice a wide variety of skills, including making careful and accurate observations, logging data, writing hypotheses, designing and performing experiments, interpreting data, and sharing conclusions—just to name a few. Given this variety, it has always been a challenge to effectively assess student mastery of these skills. Traditional assessment tools such as tests, work sheets, and quizzes lend themselves to assessing some of these skills, but they are not very accurate in assessing others. After reading this information, you may find that there is an appropriate place for portfolios in your teaching of environmental science.

Portfolios: What, Why, and How

Definitions and descriptions vary as to what constitutes a portfolio. Very simply, a portfolio is a collection of work that is done by the student during the course of the year. Usually, the students themselves have a say as to what goes into the portfolio, but selections should represent the objectives outlined by the curriculum. The purpose of the portfolio is to represent the student's mastery of skills and knowledge within the subject area. This may sound simple enough, and, in fact, it can be very simple. On the other hand, portfolios can be elaborate collections of work that are limited only by the teacher's and students' imaginations.

Your initial decision to use portfolios in your assessment leads to a host of other decisions that must be made. However, before you launch into making these decisions, it is a good idea to seek input from other teachers as well as from your school administration. You may even want to discuss your ideas with parents and other members of your community. This will make it easier to get feedback later as to the impact portfolios are having on student learning and attitudes.

You will also need to plan a system for organizing and managing the portfolios. You will need to establish guidelines for what types of information will be admitted into the portfolios, how and by whom the selections will be made, and when materials can be added or revised. These decisions will be based on your individual preferences and on the level and attitudes of your students. The following guidelines may help you in developing your plans.

- Allow students to select for themselves the sample materials that best represent their level of understanding and mastery. Although you may require that certain projects or materials be placed in the portfolios, it is advised that the students play an active role in selecting their best work. This gives students ownership and encourages them to take increasing responsibility for the quantity and quality of their work.
- Allow students to revise the selections in their portfolios at any time. The portfolios should evolve as your students' skills improve.
- Identify well in advance when you will be assessing the portfolios. Students should be given plenty of warning before the portfolios are collected for assessment.
- Determine in advance the criteria you will use for grading the portfolios. Share this criteria with the students up front. You may want to make a criteria checklist that each student can place inside his or her portfolio. This checklist would provide both you and your students with a ready reference to the grading criteria.
- You may want to keep the portfolios in the classroom or in some other area, allowing frequent but controlled access to them. This will decrease the chances of portfolios being lost. This may require a significant amount of room, depending on the nature of the portfolios.
- Make available examples of high-quality portfolios so that students can see examples of excellent work. This, of course, may be possible only after you have used portfolio assessment for at least one course.

What Goes In?

As mentioned earlier, portfolios are limited only by the imaginations of the teacher and the students. Almost any kind of work can be included in a portfolio. As for the physical nature of a portfolio, it could be a folder, a three-ring binder, or even a simple box. The main criterion for the physical makeup of a portfolio is that it safely, neatly, and conveniently stores the student's work samples. The following are some examples of appropriate additions to a student portfolio. Additional specific suggestions are provided in the Teaching Strategies for each chapter of

Holt Environmental Science. They are identified with this icon.

- Experiment write-ups and data
- Displays, such as posters, dioramas, and models
- Creative writing, such as poems, lyrics, short stories, and letters
- Videotapes, cassette tapes, photographs, artwork, or other media
- Group project results
- Computer printouts or disks
- Book reports
- Class assignments
- Personal opinions about issues
- Projects that show critical-thinking and decision-making skills
- Records of observations and data collections
- Observations about field trips
- Commentary about newspaper or magazine articles
- Items that show self-assessment and reflection over time
- Brainstorming results
- Concept maps

Remember that portfolios should be considered works-in-progress and could contain various stages of a project as it is being completed. This is a departure from a "create, hand-in, and forget" mentality.

The selections of portfolio items must also align with the objectives of your course. The portfolios should represent the skills and knowledge that you are expecting your students to acquire. To accomplish this, it is necessary for you to establish guidelines for selections in advance. You may want to require that your students' portfolios include one of each of the following items.

- Results of an experiment, highlighting steps of a scientific method
- Log entries describing direct observations
- Informed personal opinion about a local issue
- Informed personal opinion about a global issue
- Example of decision-making process with regard to the environment
- Report about some aspect of the biotic environment
- Report about some aspect of the abiotic environment
- Example of persuasive communication related to the environment
- Explanation of a scientific process
- Illustration of a natural cycle

The guidelines given here are only examples. You will want to create guidelines that fit your own goals and the capabilities of your students. Be as vague as possible in identifying the medium for each item, thus providing your students with as much creative freedom as possible.

Evaluation

Because portfolios are not static, but rather ever-changing and growing bodies of work, it is recommended that they receive periodic reviews. The best way to handle this is to review each portfolio on your own and then schedule a private portfolio conference with each student. These conferences do not have to be long; they just need to allow time for discussion about the status and quality of the portfolios. The following guidelines will be helpful as you have your portfolio conferences.

- Be positive. Try to provide as much reinforcement as possible. You do not want this conference to be construed as a negative experience.
- Encourage students to reflect on their portfolios and use real examples as they discuss their progress. This student input can provide you with a wealth of information about the student's strengths and weaknesses as well as his or her attitudes.
- Encourage student self-evaluation. Encourage each student to analyze his or her own strengths and weaknesses. As students evaluate their own work, you can get very accurate insights into possible misconceptions or unrealistic notions about your expectations for portfolios.
- Encourage students to discuss any problems or concerns and also to share any success stories or favorite works. Again, this can help you determine what areas you may need to concentrate on as you continue your instruction.

Pros and Cons

Portfolios have many advantages, but they also create some challenges. Portfolios provide a very flexible assessment tool that allows you to evaluate a variety of skills. In addition, the work selected for the portfolios represents, for the most part, work that has added to the students' understanding of the subject matter. In other words, in doing the portfolio samples, students are engaged in the process of learning. This is not the case with most tests and quizzes. Most tests and quizzes take time away from learning rather than adding to it. Portfolios help empower students to take an active role in both their education and evaluation. Portfolios also help you focus on the strengths of your students, as well as diagnose their weaknesses.

On the other hand, portfolios do take time to organize and implement. They require careful planning up front and considerable time to evaluate. They may also require storage space to keep them safe and organized. Nonetheless, teachers who have invested the time have found portfolios to be excellent tools in both encouraging and assessing student performance.

Environmental Issues

Your course on environmental science provides wonderful opportunities not only to teach science, but also to show the connections between science, technology, and society. By spending time relating science to environmental issues of the day, you will be helping students apply their science knowledge. This application of knowledge will help your students become more informed citizens as well as better decision makers. In addition, you will find that by punctuating your science instruction with relative and pertinent environmental issues, you will help keep your students motivated and involved. There is nothing like a good debate to stimulate class involvement and further learning.

Two Points of View

Any time there are two differing points of view about a single topic, you have the seed of a debate, or an issue. Environmental issues abound as both human populations and technology grow. Just considering our increase in population, it is easy to see that more land and other resources must be utilized for direct human existence. And although we have made great strides in controlling the environment for our own needs, much of the technology created to do so has been accompanied by negative side effects and consequences, particularly related to our natural environment.

Almost daily, the media cover ongoing debates concerning our relationship with the environment. At any given time, it is easy to locate articles about global, national, and local environmental issues—all of which could be of real interest to students.

- **Global Issues** These are issues that transcend geographic and political boundaries. Issues such as global warming, tropical-forest devastation, ocean dumping, and population growth affect our environment on a global level.
- **National Issues** These are issues that do not normally affect other nations but that do extend beyond regional and state boundaries. Issues related to our national air and water quality, forest and wetland management, species protection, recycling, and federal environmental policies provide excellent opportunities to explore national concerns.
- **Local Issues** These are issues that relate directly to you and your students because they are the issues being debated in your local community. These issues probably have the most potential for involving students in further exploration because they are so relevant to the people in the community. Local issues may include debates on where to locate a landfill, power plant, highway, or manufacturing plant; how to handle local pollution problems; where to develop land for population growth; whether or not to invest in public transportation systems; how to protect nativc wildlife; and on what environmental programs to spend tax money.

Issues in *Holt Environmental Science*

Holt Environmental Science contains a variety of options for weaving relevant environmental issues into your instruction. Where appropriate, information about environmental issues is incorporated into the running text. In addition, detailed information related to issues is provided in high-interest features and in this Annotated Teacher's Edition.

- **Points of View** Seven two-page *Points of View* features have been included in the Pupil's Edition. For a listing of where they are located, see page xiii in the table of contents of the Pupil's Edition. Each *Points of View* feature covers a single issue from two different perspectives. For example, the controversy surrounding the old-growth forests in the Pacific Northwest is presented from differing points of view: the point of view of environmentalists and the point of view of loggers. Each point of view is given equal time and is presented with equal intensity. These features are ideal for motivating students to debate issues and explore various attitudes related to the environment.
- **Case Studies** Although not all of the 19 *Case Studies* included in the *Pupil's Edition* discuss issues, many explore interesting environmental controversies. For example, the *Case Study* on pages 68–69 discusses the

controversy surrounding the policy of allowing fires to burn naturally in national parks such as Yellowstone. The *Case Study* on pages 356–357 discusses the controversy surrounding the harvesting of whales. Use these and other *Case Studies* to give real-life context to the concepts presented in the Pupil's Edition. All of the *Case Studies* are located where their introduction is most pertinent.

- **Teacher's Commentary** The Teacher's Commentary provided in this Annotated Teacher's Edition also provides valuable suggestions for focusing students' attention on local issues.

Such strategies help you personalize your course and make it more relevant than would be possible using only global or national issues.

A Final Word About Issues

Environmental issues are the perfect vehicle for helping students practice their decision-making skills. For this to be successful, it is important that you not take a side by presenting information in a biased manner. Challenge your students to evaluate the information and draw their own conclusions.

Regardless of the outcome, always press your students to back up their opinions with sound data and information.

Thematic Instruction

Because science has traditionally been divided into distinct areas, such as biology, chemistry, and physics, students are often led to believe that nature is similarly arranged. Too often, students come to view the areas of science as being totally distinct and separate. Students fail to recognize that certain underlying principles, or themes, unite all science disciplines.

Environmental science is an ideal course for showing thematic connections between a variety of science disciplines. In learning about our interactions with both the biotic and abiotic environments, you naturally cover content that is most often delegated to the biology, earth science, or physical science courses. For example, in teaching about air pollution, you may cover the following topics:

- Wind currents and temperature inversions, both of which are normally covered in earth science
- Molecular makeup of air and air pollutants, which are normally covered in physical science
- The possible effects of pollutants on living organisms, which is normally covered in biology

By addressing these different topics in the context of air pollution, students receive a more coherent and realistic picture of what science is all about—a method of helping us understand our surroundings. And in learning how these different topics relate to one another, students begin to recognize thematic connections.

Using Themes in Your Class

By incorporating themes into your instruction of environmental science, you can illustrate that seemingly different processes, structures, and systems have underlying similarities. Although there are many different themes that could be used, *Holt Environmental Science* emphasizes the following themes:

- Energy
- Evolution
- Interacting Systems
- Stability
- Patterns of Change
- Scale and Structure

Based on these themes, two thematic focus questions have been provided in this Annotated Teacher's Edition for each chapter. Look for the label *Using the Theme of. . .* in the Teaching Strategies.

Use the focus questions to promote class discussion. In doing so, you will help your students explore thematic relationships and help promote conceptual understanding. The focus questions will also help students integrate facts and ideas and will provide a meaningful context for class discussions. Such discussions could lead to even further thematic development.

The following table identifies the location of each thematic focus question within this Annotated Teacher's Edition. Use this table to organize your thematic instruction and to preview the questions.

Reviewers

Hugh Allen
American High School
Miami-Dade Community College
Miami, Florida

Jocelyn Baker
Windsor Forest High School
Savannah, Georgia

Albert Barr
Spring High School
Spring, Texas

Margaret A. Brumsted
Dartmouth High School
Dartmouth, Massachusetts

Dana Bruton
J. L. McCullough High School
The Woodlands, Texas

Lynette Celano
St. Gregory Catholic School
Phoenix, Arizona

Elizabeth Corum
Johnson High School
Savannah, Georgia

Jim Cramer
Brookfield East High School
Brookfield, Wisconsin

Pamela Gomillion
Puckett High School
Puckett, Mississippi

Wes Halverson
LBJ Science Academy
Austin, Texas

Janis Lariviere
Westlake High School
Alternative Learning Center
Austin, Texas

Clifford Lerner
Keene High School
Keene, New Hampshire

Lynn Diebolt-Lewis
Olivet High School
Olivet, Michigan

Sheila Lightbourne
Choctawhatchee High School
Fort Walton Beach, Florida

Alicia Moody
Choctawhatchee High School
Fort Walton Beach, Florida

George Newberry
Sparrows Point High School
Baltimore, Maryland

Marie Rediess
Algonac High School
Algonac, Michigan

Chester Sadonick
Bay High School
Bay Village, Ohio

Denice Sandefur
Nucla High School
Nucla, Colorado

Richard Shippee
Chairman: Life Science Department
Vincennes University
Vincennes, Indiana

Eva Silverfine
Ecologist
San Marcos, Texas

John M. Trimble
Corona Del Sol High School
Tempe, Arizona

David C. Tucker
Mt. Baker High School
Deming, Wyoming

Anne Tweed
Eaglecrest High School
Aurora, Colorado

Celia Wesenberg
Ponderosa High School
Shingle Springs, California

Anne Whitford
Elk Grove High School
Elk Grove, California

Paul W. Wilson
Owasso High School
Owasso, Oklahoma

CONTENTS

Teacher's Commentary

includes a selection of strategies from the following list.

Background—Interesting background information that provides added depth to the section information

Reinforcing Ideas—Strategy for ensuring that students understand an important concept

Demonstration—Procedure for illustrating a concept using a teacher demonstration

Using the Map, Using the Diagram, etc.—Strategies for strengthening process skills by using maps, diagrams, and other visual components of the section

Using the Case Study, Using the Points of View—Strategies for how and when to best use each of the features included in the chapter

Discussing the Issue—Information about relevant controversial issues that lend themselves to further discussion within the section

Debate—Topic and guidelines for organizing a student debate on a relevant issue

Safety Alert—Special notes or concerns about safety

Cooperative Learning Activity—Instructions for organizing a cooperative group activity relating to the section

Using the Theme of . . . —Specific focus question (and answer) that fosters thematic connections

Multicultural Connection—Information or activity that heightens multicultural awareness and extends knowledge of other cultures

Cross-Discipline Connection—Strategy for encouraging students to think beyond subject-area boundaries and explore relationships between science and a host of other disciplines such as math, history, and geography

Meeting Individual Needs—Strategy for adapting instruction to better fit the needs of gifted students, learners having difficulty, or learners with limited English proficiency

Notable Quote—Additional quotation to initiate thinking and contemplation

Research Activity—Longer-term activity that requires research

Real-World Connection—Strategy that relates information in the section to students' personal lives and immediate surroundings

Using the Transparency—Strategy for using a Teaching Transparency to reinforce a concept

Closure

The Closure information provides a variety of assessment strategies as well as methods for reteaching difficult concepts that students may not have fully grasped. In addition, extension information is provided for students who would like to go beyond the scope of the textbook. Each Closure contains the following information.

Section Review—Quick reference to location of answers in Answer Key

Alternative Assessment—Detailed instructions for an alternative method of assessing the information presented in the section

Reteaching—Strategy for using alternative modalities for reteaching important concepts

Extension—Higher-level activity for extending information beyond the boundaries of the section

Teaching Strategies for Investigation

The Teaching Strategies for Investigation provide valuable background, preparation, safety, and extension information related to the chapter investigation.

Portfolio

A Portfolio icon has been added to selected strategies throughout the Teacher's Commentary. The icon identifies the strategy as being one that could produce a finished product suitable for a portfolio. You may want to suggest this to your students or even make certain selections mandatory. For more information on portfolios, see pages T13–T14.

Answer Key

A complete Answer Key, tabbed for quick access, provides answers for nearly all of the questions asked within the textbook. The only exceptions are questions asked within captions, and multiple-choice questions and concept maps within the Chapter Reviews. These questions are answered with on-page annotations. In addition, on-page annotations provide page references so that all answers can be found quickly and easily in the Answer Key.

Master Planning Guide

Holt Environmental Science is a flexible textbook that can be used in either a full-year or 1 semester course for students' varying abilities. The Master Planning Guide shown below outlines four alternative classroom situations: 1/2 year course for average students, 1/2 year course for advanced students, 1 year course for average students, and 1 year course for advanced students. The number of days specified for each chapter are only general recommendations.

1/2 YEAR COURSE

Chapter	Average													Days	Advanced													Days
1	•	•	•		•						•	•		7–10	•	•	•		•	•					•	•		7–10
2	•	•	•		•							•		7–9	•	•	•		•							•		6–7
3	•	•	•				•				•	•		9–10	•	•	•		•		•				•	•		8–10
4	•	•	•	•								•		7–9	•	•	•	•	•							•		7–9
5	•	•			•						•	•		7–9	•	•			•						•	•		6–8
6	•						•					•		3–4	•	•					•					•		3–4
7			•	•								•		4–6			•	•	•						•	•		3–5
8														0														
9														0	•	•	•		•							•		6–7
10	•	•	•		•		•					•		9–11	•	•	•		•		•					•		8–10
11	•	•	•		•						•	•		9–11	•	•	•		•						•	•		8–10
12	•	•	•		•							•		7–9	•	•	•		•						•	•		7–9
13														0														
14						•								1–2						•								1
Section/ Feature	1	2	3	4	C	M	P	E	S	F	I	T	V	70–90	1	2	3	4	C	M	P	E	S	F	I	T	V	70–90

1 YEAR COURSE

Chapter	Average													Days	Advanced													Days
1	•	•	•		•	•				•	•	•		15–16	•	•	•		•	•				•	•	•	•	15–16
2	•	•	•		•	•					•	•		15–16	•	•	•		•	•			•	•	•	•		15–16
3	•	•	•		•		•			•	•	•		16–18	•	•	•		•		•	•		•	•	•		16–18
4	•	•	•	•	•		•			•	•	•		17–19	•	•	•	•	•		•	•		•	•	•		17–19
5	•	•			•	•		•	•	•	•	•	•	14–16	•	•	•		•	•		•	•	•	•	•	•	14–16
6	•						•				•	•		6–7	•	•	•		•		•			•	•	•	•	6–7
7	•	•	•	•	•						•	•	•	8–10	•	•	•	•	•	•		•		•	•	•	•	8–10
8	•	•				•		•	•		•	•		5–6	•	•	•		•	•		•	•	•	•	•		5–6
9	•	•	•		•		•		•		•	•	•	8–10	•	•	•		•		•		•	•	•	•	•	8–10
10	•	•	•		•		•	•			•	•		16–17	•	•	•		•		•	•	•	•	•	•		16–17
11	•	•	•		•	•			•	•	•	•		17–18	•	•	•		•	•			•	•	•	•	•	17–18
12	•	•	•		•		•	•	•		•	•		14–16	•	•	•		•		•	•	•	•	•	•	•	14–16
13	•	•									•	•		5–6	•	•	•		•		•			•	•	•		5–6
14	•	•			•	•					•			4–5	•	•			•	•			•	•	•			4–5
Section/ Feature	1	2	3	4	C	M	P	E	S	F	I	T	V	160–180	1	2	3	4	C	M	P	E	S	F	I	T	V	160–180

C = Case Study, **M** = Making a Difference, **P** = Points of View, **E** = Environmental Careers, **S** = EcoSkills,
F = Field Activity, **I** = Investigation, **T** = Teaching Transparency, **V** = Videodisc: STS Science Forum

Chapter Planning Guides

For chapter-specific planning information, see the Planning Guides provided for each chapter in the Teacher's Commentary. Each Planning Guide contains a complete listing, by section, of all the components and features of the chapter in order of recommended use. For your convenience, the chapter Planning Guides provide the same course-length and ability-level options as the Master Planning Guide and also include more specific pacing information.

Using Portfolios

The study of science has always provided an excellent opportunity for students to practice a wide variety of skills, including making careful and accurate observations, logging data, writing hypotheses, designing and performing experiments, interpreting data, and sharing conclusions—just to name a few. Given this variety, it has always been a challenge to effectively assess student mastery of these skills. Traditional assessment tools such as tests, work sheets, and quizzes lend themselves to assessing some of these skills, but they are not very accurate in assessing others. After reading this information, you may find that there is an appropriate place for portfolios in your teaching of environmental science.

Portfolios: What, Why, and How

Definitions and descriptions vary as to what constitutes a portfolio. Very simply, a portfolio is a collection of work that is done by the student during the course of the year. Usually, the students themselves have a say as to what goes into the portfolio, but selections should represent the objectives outlined by the curriculum. The purpose of the portfolio is to represent the student's mastery of skills and knowledge within the subject area. This may sound simple enough, and, in fact, it can be very simple. On the other hand, portfolios can be elaborate collections of work that are limited only by the teacher's and students' imaginations.

Your initial decision to use portfolios in your assessment leads to a host of other decisions that must be made. However, before you launch into making these decisions, it is a good idea to seek input from other teachers as well as from your school administration. You may even want to discuss your ideas with parents and other members of your community. This will make it easier to get feedback later as to the impact portfolios are having on student learning and attitudes.

You will also need to plan a system for organizing and managing the portfolios. You will need to establish guidelines for what types of information will be admitted into the portfolios, how and by whom the selections will be made, and when materials can be added or revised. These decisions will be based on your individual preferences and on the level and attitudes of your students. The following guidelines may help you in developing your plans.

- Allow students to select for themselves the sample materials that best represent their level of understanding and mastery. Although you may require that certain projects or materials be placed in the portfolios, it is advised that the students play an active role in selecting their best work. This gives students ownership and encourages them to take increasing responsibility for the quantity and quality of their work.
- Allow students to revise the selections in their portfolios at any time. The portfolios should evolve as your students' skills improve.
- Identify well in advance when you will be assessing the portfolios. Students should be given plenty of warning before the portfolios are collected for assessment.
- Determine in advance the criteria you will use for grading the portfolios. Share this criteria with the students up front. You may want to make a criteria checklist that each student can place inside his or her portfolio. This checklist would provide both you and your students with a ready reference to the grading criteria.
- You may want to keep the portfolios in the classroom or in some other area, allowing frequent but controlled access to them. This will decrease the chances of portfolios being lost. This may require a significant amount of room, depending on the nature of the portfolios.
- Make available examples of high-quality portfolios so that students can see examples of excellent work. This, of course, may be possible only after you have used portfolio assessment for at least one course.

What Goes In?

As mentioned earlier, portfolios are limited only by the imaginations of the teacher and the students. Almost any kind of work can be included in a portfolio. As for the physical nature of a portfolio, it could be a folder, a three-ring binder, or even a simple box. The main criterion for the physical makeup of a portfolio is that it safely, neatly, and conveniently stores the student's work samples. The following are some examples of appropriate additions to a student portfolio. Additional specific suggestions are provided in the Teaching Strategies for each chapter of

Holt Environmental Science. They are identified with this icon.

- Experiment write-ups and data
- Displays, such as posters, dioramas, and models
- Creative writing, such as poems, lyrics, short stories, and letters
- Videotapes, cassette tapes, photographs, artwork, or other media
- Group project results
- Computer printouts or disks
- Book reports
- Class assignments
- Personal opinions about issues
- Projects that show critical-thinking and decision-making skills
- Records of observations and data collections
- Observations about field trips
- Commentary about newspaper or magazine articles
- Items that show self-assessment and reflection over time
- Brainstorming results
- Concept maps

Remember that portfolios should be considered works-in-progress and could contain various stages of a project as it is being completed. This is a departure from a "create, hand-in, and forget" mentality.

The selections of portfolio items must also align with the objectives of your course. The portfolios should represent the skills and knowledge that you are expecting your students to acquire. To accomplish this, it is necessary for you to establish guidelines for selections in advance. You may want to require that your students' portfolios include one of each of the following items.

- Results of an experiment, highlighting steps of a scientific method
- Log entries describing direct observations
- Informed personal opinion about a local issue
- Informed personal opinion about a global issue
- Example of decision-making process with regard to the environment
- Report about some aspect of the biotic environment
- Report about some aspect of the abiotic environment
- Example of persuasive communication related to the environment
- Explanation of a scientific process
- Illustration of a natural cycle

The guidelines given here are only examples. You will want to create guidelines that fit your own goals and the capabilities of your students. Be as vague as possible in identifying the medium for each item, thus providing your students with as much creative freedom as possible.

Evaluation

Because portfolios are not static, but rather ever-changing and growing bodies of work, it is recommended that they receive periodic reviews. The best way to handle this is to review each portfolio on your own and then schedule a private portfolio conference with each student. These conferences do not have to be long; they just need to allow time for discussion about the status and quality of the portfolios. The following guidelines will be helpful as you have your portfolio conferences.

- Be positive. Try to provide as much reinforcement as possible. You do not want this conference to be construed as a negative experience.
- Encourage students to reflect on their portfolios and use real examples as they discuss their progress. This student input can provide you with a wealth of information about the student's strengths and weaknesses as well as his or her attitudes.
- Encourage student self-evaluation. Encourage each student to analyze his or her own strengths and weaknesses. As students evaluate their own work, you can get very accurate insights into possible misconceptions or unrealistic notions about your expectations for portfolios.
- Encourage students to discuss any problems or concerns and also to share any success stories or favorite works. Again, this can help you determine what areas you may need to concentrate on as you continue your instruction.

Pros and Cons

Portfolios have many advantages, but they also create some challenges. Portfolios provide a very flexible assessment tool that allows you to evaluate a variety of skills. In addition, the work selected for the portfolios represents, for the most part, work that has added to the students' understanding of the subject matter. In other words, in doing the portfolio samples, students are engaged in the process of learning. This is not the case with most tests and quizzes. Most tests and quizzes take time away from learning rather than adding to it. Portfolios help empower students to take an active role in both their education and evaluation. Portfolios also help you focus on the strengths of your students, as well as diagnose their weaknesses.

On the other hand, portfolios do take time to organize and implement. They require careful planning up front and considerable time to evaluate. They may also require storage space to keep them safe and organized. Nonetheless, teachers who have invested the time have found portfolios to be excellent tools in both encouraging and assessing student performance.

THEMATIC CONNECTIONS						
	Energy	Evolution	Interacting Systems	Stability	Patterns of Change	Scale and Structure
Chapter 1			1.3, p. T32			
Chapter 2	2.1, p. T36			2.2, p. T38	2.3, p. T40	
Chapter 3					3.3, p. T47	3.2, p. T45
Chapter 4			4.2, p. T54	4.1, p. T51		
Chapter 5		5.1, p. T61	5.3, p. T66			5.2, p. T65
Chapter 6	6.1, p. T70		6.1, p. T69			
Chapter 7	7.2, p. T79		7.4, p. T84			7.1, p. T78
Chapter 8	8.2, p. T90					8.1, p. T87
Chapter 9		9.3, p. T99	9.2, p. T98			
Chapter 10	10.1, p. T105	10.3, p. T108	10.3, p. T108		10.1, p. T105	
Chapter 11	11.2, p. T115				11.1, p. T113	
Chapter 12	12.3, p. T125					12.2, p. T123
Chapter 13				13.1, p. T130		
Chapter 14					14.1, p. T138	

As a teacher, you can employ themes as an organizational tool to reinforce student understanding of concepts rather than merely rote memorization of facts. By discussing thematic focus questions with your class, you can lead your students to a deeper understanding of scientific principles. The following descriptions will give you added insight into the six themes selected for *Holt Environmental Science.*

Energy
Energy is the ability to put matter into motion. It is what makes the universe and everything in it dynamic and ever changing. The flow of energy from the abiotic world to the biotic world and from organism to organism helps determine how living things interact within their environments. Living things need energy for maintenance, growth, and reproduction. In the physical world, energy takes a variety of forms, including wave energy, sound energy, and light energy. The study of dynamic systems in any field of science requires an understanding of energy—its origins, how it flows through systems, how it is converted from one form to another, and how it is conserved. Energy provides the basis for all interactions, whether chemical, biological, or physical. Energy thematically connects all disciplines, and its use and misuse by humans is especially highlighted in the study of environmental science.

Evolution
Evolution is a natural progression from simple to more complex. Evolutionary change is not simply an alteration, but rather a progression of such alterations over the continuum of time. The entire universe and all of its inhabitants can thus be studied from the viewpoint of evolutionary and historical change. An evolutionary theme can be used to help students understand why things exist as they do today by investigating how they existed in the past. The evolution of life resulted from a long line of living organisms adapting to diverse and dynamic environments. Evolution is also responsible for the vast diversity of organisms that inhabit those environments.

Interacting Systems
The world consists of systems, both large and small. Interactions occur between and among both inanimate and animate systems. Through this theme, students quickly see how all life-forms (including humans) depend both on other life-forms and on their abiotic environment. Interactions also exist between various parts of the inanimate world. This theme also helps students see how changes within environmental systems can have ripple effects—effects that may not be identified immediately but may have very pronounced effects. For example, when students study food chains, they should come to the conclusion that if even one link in a food chain is damaged, the entire chain may be at risk. You might extend this to include abiotic factors such as soil composition, erosion, and water and air quality. The influences of society, pollution, and even natural disasters can all be discussed under the theme of Interacting Systems.

Stability
Stability refers to the constancy or permanence within any system. It can refer to the way in which systems as a whole do not change, or it can refer to a state of static equilibrium in which the forces causing change within a system are balanced. For example, a stable population is one in which births and immigration are balanced by an equal amount of death and emigration. Stability also relates to the predictability of change within natural processes. As students learn about fragile ecosystems, they will come to understand the importance of stability in terms of preservation and conservation. In a dynamic world, there are always areas where stability is required in order for the system to be maintained.

Patterns of Change
The natural world is characterized by patterns of change. Change through time can be analyzed in connection to its rates and patterns. Change can be predictable or unpredictable, regular or irregular. It can occur in cyclical or linear patterns and can affect both simple and complex systems. Evolution is one pattern of change. Others include

growth, urbanization, erosion, bioaccumulation, and a host of other factors that increase or decrease over time. Knowing about different patterns of change provides students with a basis for understanding natural systems and their underlying mechanisms. This understanding will help them predict what might happen in the future, a skill highly important in studying and maintaining natural systems.

Scale and Structure

This theme provides a means for understanding the nature of matter and systems within the natural world. Structure defines how matter fits together. Scale determines the level at which this structure is examined. Using this theme, you can lead your students to an understanding of the underlying framework that gives order to matter, from the most basic forms to the most complex. Since all matter, both organic and inorganic, shares certain similarities in structure and materials, students will come to realize that all living things are made of the same materials as the rest of the universe. As complex collections of the building blocks we call atoms, every living creature—human beings included—is subject to all of the physical principles that apply to molecules, mountains, and stars.

CONCEPT MAPPING

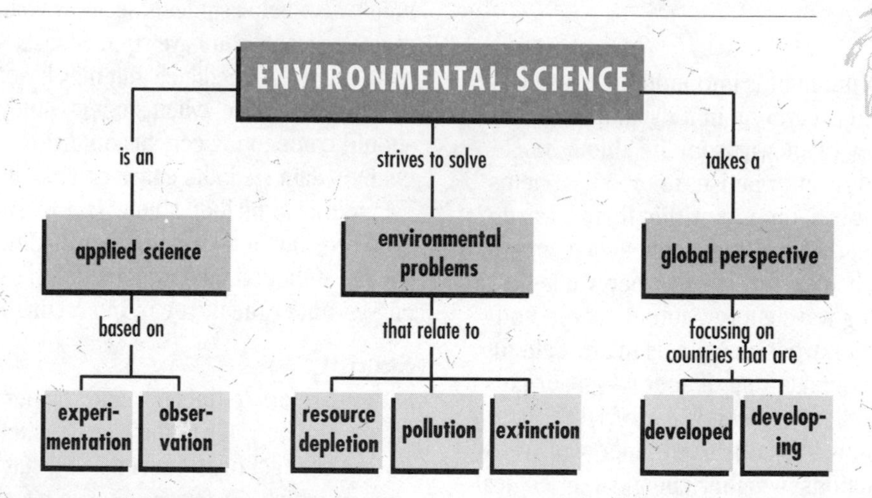

Concept mapping is a learning technique that helps students understand concepts in a unique way. Ideas are expressed as written words or phrases, and lines are drawn between related ideas. On these lines, a short linking word or phrase is provided that explains the relationship between the ideas. These connections link the concepts, and the resulting "map" shows how the concepts relate to each other. Because concept maps are visual representations, students actually see the connections between concepts and ideas.

USING CONCEPT MAPS

In *Holt Environmental Science,* concept maps are used in three ways:

1. In the Section Reviews of several chapters, students are asked to create concept maps based on what they have learned. Also, throughout this Annotated Teacher's Edition you are presented with opportunities to have your students create concept maps as a method of further reinforcing and assessing their learning.
2. On each Chapter Highlights page, a completed concept map is provided as a visual summary of the chapter.
3. A partially completed concept map is given in each Chapter Review so that students can check their understanding of the chapter content by providing the missing items.

Concept mapping has valuable potential for evaluating your students' comprehension and your own teaching focus. Because students must closely examine their existing ideas and perceptions about the subject in order to do concept mapping, their maps will show you any errors or misconceptions in their understanding. If the concept

To the Teacher

Holt Environmental Science has been designed to provide you and your students with a balanced approach to the diverse study of our environment. The emphasis in this program is the study of science and the development of thinking and decision-making skills. Our goal is to provide students with the science background they need to analyze for themselves many of the issues concerning our environment.

The field of environmental science also offers a rare occasion to apply many different sciences to extend our knowledge of both our world and its inhabitants. Within what other course of study do you have the opportunity to discuss topics that range from a fragile desert biome to the debate over CFCs and the ozone layer? Perhaps it is this diverse context and its clear application that makes environmental science inherently interesting to students.

No Easy Solutions to Difficult Problems

Being extraordinarily complex, environmental problems often defy easy answers and quick solutions. Using present knowledge and technology, we are only marginally successful at predicting even the weather. It is not hard to understand how difficult it is to predict with any accuracy the outcomes of human actions that affect habitats and whole ecosystems. Although a great deal is known about the environment, there is still much to discover, particularly about how species interact with one another within complex ecosystems. There is also a real urgency to learn all we can before the environment is damaged beyond the point where informed decisions could prevent irreparable damage.

The *Holt Environmental Science* program can do much to improve your students' awareness of the environmental problems we are facing now and will be facing in the future. On an almost daily basis the media provides us with reports about environmental problems and debates. When people do not understand the science behind these problems, it is likely that the media will wield entirely too much influence over the outcome of these debates.

Awareness Is Half the Battle

This book is not designed to convert students into environmental activists. However, by studying this book, students will add significant depth to their understanding of the environment. It is hoped that this understanding will lead to an increased level of appreciation and that students will treat the environment accordingly.

Humans will continue to multiply and occupy more and more areas of the globe. And in so doing, they will interfere with most of our environment's complex natural systems. Many of the adverse consequences of that interference cannot be predicted. Others, however, are already anticipated and could be avoided. Unfortunately, environmental knowledge is not widespread. For example, you may know that dumping motor oil on the ground can contaminate enormous amounts of groundwater. Other people, however, may not be so well informed. Many people in the United States and throughout the world are still dumping oil, often because they do not realize the adverse effects of such an action.

Such environmental ignorance is very harmful and particularly disappointing considering our level of knowledge. While it is true that we don't have all the answers, it is also true that we know a great deal about our environment. It is important that we share this information. The more environmentally literate we are, the fewer careless mistakes we will make that will negatively affect our planet and all of its inhabitants.

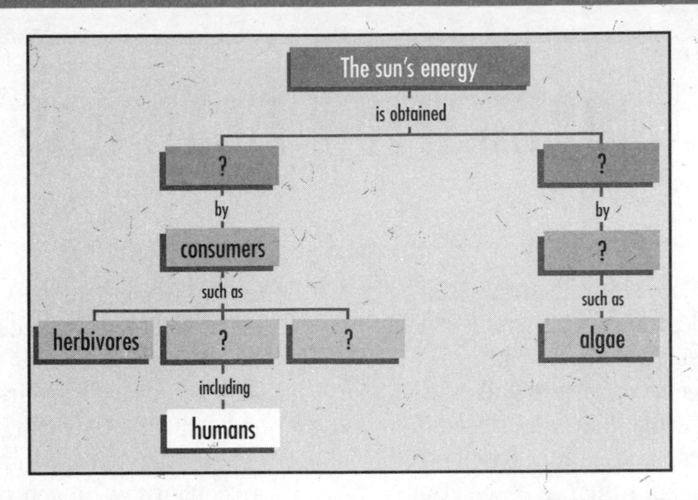

Each Chapter Review includes a partially completed concept map for students to complete.

maps indicate that several students have similar weaknesses in their comprehension of a subject, you can rethink and clarify your teaching of that material.

It is important to emphasize to your students that there is no single correct way to draw a concept map. A concept map can take almost any form as long as it is logically arranged. The better the student's understanding of the topic, the more detailed and cross-linked the map should be.

Evaluating Concept Maps

When you evaluate your students' concept maps, you should keep in mind that a good concept map has the following features:

- The map starts with one general concept and works down to more specific ideas.
- Each concept appears only once.
- All concepts are linked to at least one other concept.
- Linking words and phrases explain the relationships between concepts.
- Cross-linkages are shown where appropriate.
- More than one path is shown.
- Examples are included where appropriate.

If properly used, concept maps can increase comprehension and improve retention of new information and even of information that students are already familiar with. These benefits make concept mapping an important addition to your students' learning strategies.

How to Make a Concept Map

The steps involved in making a concept map are outlined below. Instructions are provided for the student on page xvii of the Pupil's Edition.

1. **Determine the main ideas or concepts.**
   Make a list of the main concepts or topics in the material for which you are making a concept map.
2. **Place the concepts in order from the most general to the most specific.**
   Put the most general concept at the top of your concept map. Then look at the other concepts and determine how they relate to the most general one. On your map, arrange the concepts in order from general to specific, keeping in mind that some concepts may be equally specific. When all of the concepts are positioned, draw a circle or box around each one.
3. **Determine the relationships between the concepts.**
   On the map, connect all of the related concepts with lines. On each connecting line, write a linking word or short phrase that explains the relationship.

Cooperative Learning

Cooperative learning is a learning technique in which students work in heterogeneous groups and earn recognition, rewards, and sometimes grades based on the academic performance of the group as a whole. With cooperative learning, you can orchestrate certain discussions, investigations, research projects, or other activities to foster group participation.

Benefits of Cooperative Learning

One of the benefits of cooperative learning is that it models the real-world scientific experience in which scientists work together, not in isolation, to solve difficult problems. Students working in groups learn about the joys as well as the frustrations involved in scientific inquiry. With cooperative learning, the classroom becomes a fertile environment for ideas and novel solutions. Another benefit of cooperative learning is that students sharpen social skills and develop a sense of confidence in their own abilities. Students also channel their energy into constructive tasks while satisfying their fundamental need for social interaction.

Components of Cooperative Learning

Establishing cooperative learning in the classroom requires you to relinquish some control. Through cooperative learning, students themselves become responsible for building their knowledge. Working in groups to probe and investigate ideas, answer questions, and draw conclusions about observations allows students to discover and discuss concepts in their own language. When students learn through cooperation, the knowledge they derive becomes their own; they are less dependent on you as a source of knowledge. Cooperative learning helps students become self-sufficient, self-directed, and lifelong learners.

To successfully integrate cooperative learning into your classroom, consider the following important components.

- **Group Size** Although group size will vary depending on the activity, the optimum group size for cooperative learning is three or four students but may be as large as six. For students unaccustomed to this learning style, keep the group size to about three students.
- **Group Goals** Students need to understand what is expected of them. Identify the group goal, whether it is to master specific objectives or to create a product such as a chart, a report, or an illustration. Identify and explain the specific cooperative skills required for each activity, and emphasize basic skills such as working within an appropriate noise level, staying with the group, encouraging participation by each member, and generally treating each member of the group with respect.
- **Individual Accountability** Each group member should have some specific responsibility that contributes to the learning of all group members. At the same time, each group member should reach a certain minimum level of mastery. Suggest that groups evaluate their performance after they finish an activity. Encourage students to offer improvements for problems without criticizing individuals.
- **Positive Interdependence** A learning activity becomes cooperative only when everyone realizes that *no* group member can be successful unless *all* group members are successful. This encourages positive interdependence—students working with one another to achieve group goals. Interdependence may be facilitated in a number of ways. You might assign each student to a different meaningful role or allow students to do this themselves. By supplying only one set of instructions or materials, students will be compelled to work together. Different students in the group may also be responsible for specific information by becoming an "expert" on some topic. Each member would then have to share that information with the other group members or with other groups within the class.

Grouping Strategies

A cooperative-learning group will require a variety of roles—leader, researcher, materials manager, presenter—in order to accomplish its task. Assigning roles is very important, especially at first. Consider the behavior patterns of your students, and assign roles that will complement those patterns. For example, if you have a student who is always moving around the classroom, assign that person the important role of timekeeper. It will be his or her job to sit in one spot, watch the time, and keep the group apprised of the time elapsed and the time remaining. This role is suitable for a student who tends to slow the group down or for a student who is easily distracted.

Similarly, you might assign a shy student the role of recorder. Emphasize that an important function of this role is to listen carefully to others, ask questions, and log findings. You could assign a student who tends to belittle others the role of encourager. When an encourager speaks, it must be only to say something positive. The role of summarizer may be ideal for a student who is being left out of academic work and who needs more practice at cognitive thinking.

Group leader is a role that certain students will always want and other students will always avoid. Be careful not to stereotype your students, and encourage those who may not be "born leaders." Often, those students you least suspect will surprise you by becoming the best leaders. You may also find that students prone to disruptive behavior perform well and cooperate enthusiastically when placed in the role of "leader" or "supervisor." This role builds self-esteem. However, you should give shy or disruptive students time to become familiar with cooperative learning before placing them in a position of leadership.

Cooperative learning is an effective tool for meeting the individual needs of all your students. It builds relationships among students as they interact with each other as individuals with a common goal. You may be surprised at just how sophisticated your students will become once they feel they are "in charge" of their learning.

Multicultural Instruction

Multicultural instruction serves to ensure that all students have the chance to learn, to succeed, and to become whomever they would like to become—regardless of race, gender, socioeconomic background, or disability. Multicultural instruction affirms the positive nature of this country's wide diversity by helping students develop an open mind, a positive self-image, and a realistic understanding of the world around them.

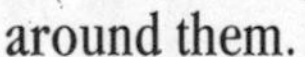

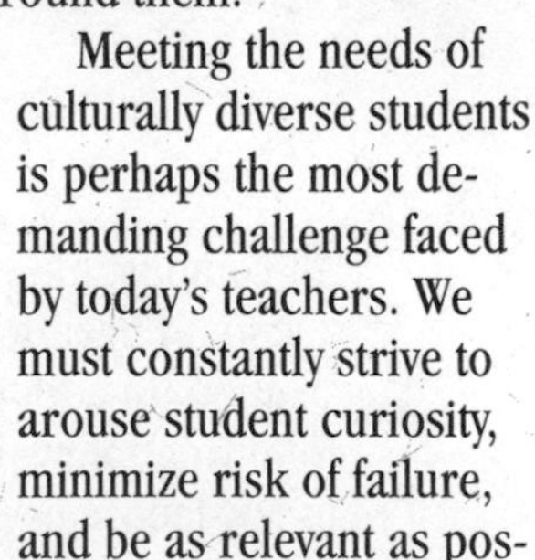

Meeting the needs of culturally diverse students is perhaps the most demanding challenge faced by today's teachers. We must constantly strive to arouse student curiosity, minimize risk of failure, and be as relevant as possible to the needs and interests of individual students. The more culturally relevant we make our science programs, the better we will serve our changing class populations.

Strategies for Promoting Multiculturalism

Let your students help you incorporate multicultural learning into your classroom by allowing them the freedom to express their feelings and attitudes during your classes. This is very easy to do in environmental science, in which a variety of environmental issues promote discussion and the formation of opinions. By personalizing your instruction in this manner, you will find that your students are more curious about the world around them and are more enthusiastic about gaining new knowledge.

Likewise, it is very important to be sensitive to the cultural identities of your students, as well as of any people you may be studying in the course of your environmental instruction. You must view these cultural identities on an equal footing. Having a healthy respect for individual cultures will go a long way in helping each of your students achieve and maintain a positive self-image.

The following strategies are provided as a menu of ideas that may help you create a positive multicultural environment.

- Recognize and convey to students that all languages are equally valid. Learning English, however, increases the range of opportunities available to the students.
- Draw special attention to the diversity of role models in *Holt Environmental Science.* Note particularly features such as *Making a Difference* and *Environmental Careers.* These features highlight the contributions of people from a variety of ethnic and cultural backgrounds. The power of such role models should not be underestimated. Positive role models can create interest and motivation and may even influence a student's pursuit of a career.
- Use cooperative learning to diversify student groups. You will find that students become more open-minded

and develop more positive, accepting, and supportive relationships with their peers. Labels concerning ethnicity, gender, ability, social class, and handicaps cease to exist. For more detailed information about cooperative learning, see pages T20–T21.

- Take every opportunity to relate environmental science to personal experiences. Invite your students to discuss any of their own experiences that may apply. You may discover some very relevant connections and analogies, and the learning process may become more interactive and personalized for the class. Likewise, this process might also add to the richness of the class by highlighting the cultural differences among your students.
- Allow students to select independent projects relevant to their own culture or cultural interests. These projects should permit students to create new, positive avenues of self-expression from their own experiences. In addition, these activities promote the development of positive attitudes toward general academics, social interactions, and the study of science.

MULTICULTURAL INSTRUCTION AND *HOLT ENVIRONMENTAL SCIENCE*

Holt Environmental Science is designed to serve the multiethnic and multicultural classrooms of today. The content includes events, concepts, and issues from diverse ethnic and cultural perspectives. As students work through the textbook, they will understand that science is a human endeavor that has been advanced by the contributions of many cultures and ethnic groups. This attitude is reinforced directly by features such as *Making a Difference* and *Environmental Careers,* which highlight the positive contributions of a wide variety of people.

To add even more depth to your multicultural instruction, multicultural teaching strategies have been included in the Teacher's Commentary of this Annotated Teacher's Edition. Look for the label *Multicultural Connection.* These strategies provide information and ideas on how to add interest and relevance to your instruction by focusing on different aspects of cultural diversity.

SAFETY GUIDELINES FOR TEACHERS

SAFETY IN THE LABORATORY

Laboratory investigations provide rich educational experiences for students. Yet safety is always a major concern in the science laboratory. The safety information in this Annotated Teacher's Edition is intended to help you organize your laboratory program, but it cannot anticipate all the safety issues of every teaching situation. Many states have school laboratory safety regulations that address such topics as storage of combustible materials and eye and body protection. Check with your state department of education. In addition to local regulations, the following safety precautions should be routine in any laboratory.

- Review the safety symbols found on page xvi and the general safety guidelines featured on pages 409–412 of this text with your students.
- Post laboratory rules in a conspicuous place in the laboratory.
- Always perform an experiment yourself before asking students to do so.
- Before the class begins an experiment, review safety rules and demonstrate safety procedures.

- Never permit students to work in your laboratory without your direct supervision or to conduct unauthorized experiments. Lock your laboratory and storage room when you are not present.
- Provide labeled disposal containers for glass (sharps) and waste chemicals.
- Allow no food or beverages in the laboratory.
- When an experiment has been completed, insist that students clean their work areas, wash and store all materials and equipment, turn off all water, gas, and electrical appliances, and wash their hands.
- Note the location of each piece of safety equipment in your laboratory, learn to use it correctly, and teach students to do the same.
- Post emergency telephone numbers, including those of the school nurse, nearest poison control center, fire department, hospital, and ambulance, in a conspicuous place near the telephone.

Working With Animals

When live animals are introduced into the laboratory for observation and experimentation, two safety standards must be used. The safety of the students is one objective and the welfare of the animal is another. The following guidelines address both of these safety concerns.

- Before introducing animals into the laboratory, make complete plans for their care and feeding, including over weekends and school holidays.
- Be sure that all mammals used in a school laboratory have been inoculated for rabies unless they have been purchased from a reliable biological supply house.
- The following animals should never be brought into the classroom: wild birds and mammals, snapping turtles, poisonous snakes, and insects that may be carriers of disease.
- Students should not bring their pets to the classroom without approval from an administrator.
- Warn students not to tease animals or subject them to unnecessary handling.

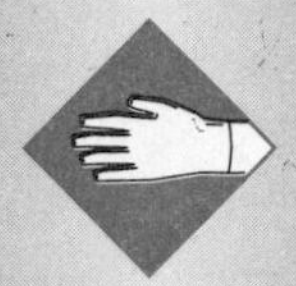

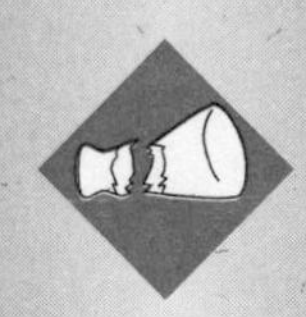

- Experiments with animals should not involve the use of drugs, toxic products, anesthetics, surgery, carcinogens, or radiation.
- Make certain that any student who is scratched or bitten by an animal receives immediate attention from a school nurse.
- Remind students to wash their hands thoroughly after handling animals.
- Never allow students to examine "road-kills" or nonpreserved specimens. Doing so increases the risk of infection.
- Wear protective gloves and splash-proof safety goggles at all times when handling preserving fluids or preserved specimens.

Working With Plants

The safe use of plants in the laboratory requires precautions taken on behalf of the plants as well as the students.

- Before introducing plants to the laboratory, make complete plans for their care and maintenance during the school week as well as on weekends and holidays.
- Check reference books to determine whether any plant you wish to bring to class has toxic effects. If so, caution students and post warnings where the plant will be located.
- Caution students never to put any part of a plant in their mouth or near their eyes or to rub plant parts or sap on their skin.
- Students should always wash their hands thoroughly after handling plants.

Working With Microorganisms

- Pathogenic bacteria should never be used.
- Petri dishes containing bacterial cultures should be sealed with tape.
- Before washing petri dishes, the cultures should be killed by heating the dishes in an autoclave or pressure cooker or by applying alcohol or a strong disinfectant.

- Wire loops used to transfer microorganisms should be flamed before and after the organisms are transferred.

Working With Chemicals

- Label student reagent containers with the name of each substance and its hazard class (flammable, corrosive, etc.).
- Dispose of hazardous waste chemicals according to federal, state, and local regulations. Refer to the Material Safety Data Sheet for recommended disposal procedures.
- Remove all sources of flames, sparks, and heat from the laboratory when any flammable material is being used.
- Chemicals should be stored in a well-ventilated area that is kept locked at all times. Flammable chemicals should be stored in a fire-resistant cabinet. Never store such chemicals in a refrigerator unless the refrigerator is specifically marked as explosion proof.
- Chemical storage areas should be kept clean, orderly, and well lighted.
- Wearing contact lenses for cosmetic reasons should be prohibited in the laboratory because liquids can be drawn up under the lenses by capillary action and thereby come into direct contact with the eyeball. Students who must wear contact lenses prescribed by a physician should wear eye-cup ANSI Z87.1 approved safety goggles.

Safety in the Field

Field work is a meaningful and relevant addition to an environmental science course. To help students view field work as a learning experience and to minimize horseplay that can lead to accidents or injury, be sure to give students clear goals for the trip. You might have students create a list of questions to answer during the trip. Also, observe the following precautions.

- Visit outdoor sites in advance, and note potential safety problems such as bodies of water, poisonous plants, venomous insects and snakes, and areas where falls might occur.
- Visit with park rangers or site administrators, and ask what precautions should be taken during the field trip.
- Discuss necessary safety measures with students, and warn them of any hazards you have discovered.
- Caution students to dress in a manner that will keep them comfortable, warm, and dry. Shoes should be sturdy, cover the entire foot, and be appropriate to the terrain. Long pants should be worn rather than shorts or skirts.
- On a trip to any body of water, include an adult skilled in water safety and CPR.
- Pack a first-aid kit, and consider including insect repellent.
- Caution students not to touch or pick wildflowers or other plants without your approval.
- Caution students not to touch wild animals without your approval. Also, warn students not to approach any wild animal that could bite, scratch, or otherwise cause injury.
- Have students travel in pairs and assume responsibility for the safety of their partners.
- Instruct students to report any injury immediately.

MASTER MATERIALS LIST

Material (Chapter No.)	Quantity per Group
acid precipitation, artificial (6)	50 mL
acid solutions of several concentrations* (6)	about 25 mL each
antifreeze* (5)	a small container
beakers (1)	1
250 mL (or glass jars) (5)	5
small (6)	2
binoculars† (4)	1
bleach* (1)	a small container
blocks, small (9)	4 or 5
cafeteria tray (or sheet cake pan) (9)	1
calculator (13)	1
calculator† (12)	1
cardboard, thin (7)	2 small pieces
cardboard* (11)	several pieces
cardboard boxes with lids (11)	2
carpet, strip about 3 cm × 15 cm* (9)	1
caulk* (11)	a small container
construction paper, colored* (11)	several sheets
containers for water, small enough to fit in cardboard box* (11)	1 or 2
cotton (6)	a handful
craft stick (8)	1
cups, 12 oz. plastic (5)	4
detergent* (5)	a small container
fertilizer* (1, 5)	a small package
(or other soil additives)* (2)	a small package
field guide to local flora and fauna (4)	1
to insects† (10)	1
filter paper (6)	4 pieces to fit petri dishes
flowerpots, small (2)	3
foam insulation* (11)	enough to line a cardboard box
foil* (11)	several sheets
food coloring, red (5)	a small bottle
fork (9)	1
glass jars, large (7)	2
globe (or map with contour lines) (4)	1
gloves (1, 6, 10)	1 pair per student
glucose tablets (5)	1 or 2
glucose test paper (5)	4 to 5 pieces
goggles (1, 5, 6)	1 pair per student
graduated cylinder (1, 5, 6)	1
graph paper (1, 2, 7, 13)	several sheets per student
graph paper† (4)	1 sheet
grass clippings (8, 9)	a few handfuls
grass seeds, rye (or fescue) and buffalo (2)	about 1 package of each
gravel (5)	a few handfuls
hand lens (2, 10)	1
ice cubes* (7)	about 10
index card (8)	1
knife (or single-edged razor blade) (1, 11)	1
lab apron (1, 5, 6)	1 per student
light source* (2)	1
limestone chips (or other rock fragments)* (6)	a handful
markers, colored (2, 3)	a few
metric ruler (1, 2, 4, 5, 8, 9, 11)	1

* Different materials are recommended for each student group; this item is one of several suggestions.
† This is an optional item for this activity.

Material (Chapter No.)	Quantity per Group
miscellaneous small classroom objects* (9)	a few
motor oil (or cooking oil)* (5)	a small container
newspapers, magazines, and other publications (14)	an assortment
notebook (all)	1 per student
onion bulbs, white, 1.5 to 2 cm in diameter (1)	about 9
paint* (11)	a small can
black and white latex* (7)	a small can of each
paper clip (8)	1
peanuts (8)	5 to 10
peas or beans, dried (9)	a handful
pen or pencil (all)	1 per student
pencil (8)	1 per student
pencils, colored (or markers) (13)	a few
petri dishes (6)	2
pH paper (6)	4 pieces
pie pans, aluminum (2)	3
plastic pan, 30 cm × 35 cm × 15 cm (or shoe box) (8)	1
plastic wrap, clear (7, 11)	4 to 6 squares, about 20 cm per side
poster board (3)	1 large sheet
rocks, small (8)	a handful
rocks, small* (11)	a handful
rubber bands (or string) (7)	2
sand (5)	a few handfuls
(8)	enough to fill a large plastic pan
scissors (14)	1 pair
scissors* (2)	1 pair
seeds, radish (or mustard) (6)	200
bean, corn, or other varieties* (6)	100
sod (or other plants)* (7)	enough to fill a jar
soil (5, 8)	a few handfuls
potting (2)	enough to fill 3 small flowerpots
sandy (9)	enough to fill a large cake pan
soil* (7, 11)	a few handfuls
spray bottle filled with water (9)	1
stakes (3, 10)	4
stirring rod (1, 5)	1
straightedge (or ruler) (7, 13)	1
string (or twine) (3)	about 50 m
(10)	about 10 m
tape (7, 11, 14)	1 roll
masking (2)	a few pieces about 10 cm in length
tape measure, metric (or meter stick) (3, 10)	1
test tubes, large (1)	about 9
test-tube rack (1)	1
thermometer (11)	1
outdoor (7)	2
thumbtack (5)	1
tiles, small* (11)	about a dozen
toothpicks (9)	about 100
twigs (8, 9)	a handful
vinegar* (5)	a small container
watch (or clock) (7, 8, 11)	1
water (5, 7, 9)	about 1 L
(6)	about 100 mL
distilled (1)	about 500 mL
water* (11)	about 1 L
wax pencil (1, 5, 6)	1

* Different materials are recommended for each student group; this item is one of several suggestions.

† This is an optional item for this activity.

C H A P T E R 1

Environmental Science: A Global Perspective

Planning Guide

CHAPTER COMPONENTS AND RESOURCES	½ Year Average	½ Year Advanced	1 Year Average	1 Year Advanced
Chapter Opener, p. 2				
EcoLog, p. 2	•	•	•	•
1.1 Understanding Our Environment, pp. 3–12	•	•	•	•
Case Study: *Lake Washington: An Environmental Success Story,* pp. 6–7	•	•	•	•
Making a Difference: *Chicken of the Trees,* pp. 30–31		•	•	•
Transparency 1: *The Biosphere*	•	•	•	•
Section Review, p. 12	•	•	•	•
1.2 Using Science to Solve Environmental Problems, pp. 13–19	•	•	•	•
Field Activity, p. 15			•	•
Section Review, p. 19	•	•	•	•
1.3 Making Environmental Decisions, pp. 20–24	•	•	•	•
Field Activity, p. 24			•	•
Section Review, p. 24	•	•	•	•
End of Chapter				
Chapter Highlights, p. 25	•	•	•	•
Chapter Review, pp. 26–27	•	•	•	•
Investigation: *An Onion Conundrum,* pp. 28–29	•	•	•	•
Videodiscovery STS Science Forums: Vol. 2—*Space Station*				•
Chapter Tests	•	•	•	•
PACING GUIDE	7–10 days	7–10 days	15–16 days	15–16 days

Materials Organizer

Activity	Materials	Advance Preparation
Investigation: *An Onion Conundrum,* pp. 28–29	goggles, gloves, lab apron, 9 white onion bulbs between 1.5 and 2.0 cm in diameter (unsprouted, mold-free, and of approximately the same size), graduated cylinder, stirring rod, beaker, wax pencil, 9 large test tubes, distilled water, fertilizer, bleach, test-tube rack, knife, notebook, pen or pencil, graph paper, metric ruler (per student group)	none

CHAPTER OVERVIEW

This chapter presents a broad overview of environmental problems and the tools used to solve them.

Section 1.1 identifies the main categories of environmental problems and gives these problems a global perspective by comparing developed and developing countries.

Section 1.2 describes the tools used by scientists to learn about the natural world and to solve practical problems such as how to protect the environment.

Section 1.3 presents an environmental decision-making model that students can use to analyze environmental issues.

RESOURCES

. . . for Teachers

Bonnet, Robert L., and G. Daniel Keen. *Environmental Science: 49 Science Fair Projects.* Blue Ridge Summit, PA: TAB Books. (This book lists projects about the environment that are designed for use in science-fair competitions.)

Sierra Club, Public Information, 730 Polk Street, San Francisco, CA 94109; 415-776-2211. (This organization has a wealth of free or inexpensive teacher resources.)

Stein, Edith Carol, in cooperation with the Environmental Data Research Institute. *The Environmental Sourcebook.* New York City, NY: Lyons and Burford, 1992. (This book is a useful reference on environmental issues, organizations, periodicals, books, foundation grants, and more.)

. . . for Students

"Dwarf Wedge Mussel Report," Keene High School, 43 Arch Street, Keene, New Hampshire 03431; 603-352-0640. (This report details the findings of the students who are described in this chapter of the student textbook.)

Silver, Cheryl Simon, with Ruth S. DeFries for the National Academy of Sciences. *One Earth, One Future: Our Changing Global Environment.* Washington, D.C.: National Academy Press, 1990. (This book clearly explains the complex web of processes that drives our planet's systems, and those forces that threaten it.)

Willers, Bill, ed. *Learning to Listen to the Land.* Washington, D.C.: Island Press, 1991. (This is a collection of essays by some of America's most provocative thinkers and writers. Together these essays provide a thought-provoking overview of pressing environmental issues.)

Films, Videos, Software, and Other Media

Eco-Rap: Voices from the Hood. Video. 38 minutes. The Video Project, 5332 College Avenue, Suite 101, Oakland, CA 94618; 800-4-PLANET. (This award-winning video follows a group of young men and women of various ethnicities as they learn about local environmental hazards and express their concerns using rap music.)

Interactive NOVA: Race to Save the Planet. CAV Videodisc, Level I, III. English/Spanish. Scholastic, P.O. Box 7502, Jefferson City, MO 65102; 314-636-5271. (This videodisc uses case studies and role-playing scenarios to help students learn complex concepts of global ecology.)

World Watch Database Diskette. Software. Macintosh or IBM. Worldwatch Institute, 1776 Massachusetts Avenue, NW, Washington, D.C. 20036-1904; 202-452-1999. (This is a set of 123 format spreadsheets that contain data on energy, agriculture, pollution, transportation, the economy, ecology, and resources.)

Getting Involved

A listing of environmental organizations, publications, and volunteer opportunities can be found in the **Laboratory and Field Guide.** Note especially those organizations with the goal of enhancing people's decision-making and problem-solving skills so that they may better solve environmental problems. Some examples include the Alliance for Environmental Education, the Sierra Club, and the Student Conservation Association.

USING THE QUOTATION

"All the flowers of all the tomorrows are in the seeds of today." Point out that proverbs are usually intended to teach a lesson. Ask: What lesson might this one be teaching? (*Sample answer: Our actions of today influence the future—positive actions lead to a positive future.*) Ask: How might this quotation relate to the environment? (*Sample answer: If we become educated about the world's problems today, we may be able to do something about them in the future.*)

USING THE ECOLOG

To assess students' knowledge about the subject of this chapter, have them answer the EcoLog questions on page 2. In answering these questions, students will reveal their prior knowledge, which is usually very basic and often flawed or even misconceived. Then, by revising their answers at the conclusion of the chapter, students will have the opportunity to eliminate any misconceptions and to see firsthand how their knowledge has grown.

UNDERSTANDING OUR ENVIRONMENT

FOCUS

Students follow the activities of a group of high school students who are helping to save an endangered species. This sets the stage for a discussion about environmental problems and what can be done to resolve them.

MOTIVATOR

Ask students to name some of the most important environmental problems that the Earth is experiencing today. List their answers on an overhead or on the chalkboard. *(Answers may include such worldwide issues as global warming or rain-forest destruction.)* Then ask students to name some environmental problems that are facing the local community, and list these problems also. Ask students to identify one of the local problems that is part of a larger global problem. For example, if smog is a problem in your area, students may point out that automobile emissions might also contribute to global warming. Then ask students to brainstorm ways that the local problem could be solved. *(Ideas might include carpooling or using mass transit whenever possible.)* Conclude by asking students to write an explanation of why environmentalists urge people to "think globally, act locally."

TEACHING STRATEGIES

WHAT ARE OUR MAIN ENVIRONMENTAL PROBLEMS? *page 5*

Real-World Connection

To help students become aware of real environmental problems in your community and the world, suggest that they start a notebook of clippings from magazines and newspapers. Their clippings should relate in some way to the environment. Bring in your old newspapers and magazines and, if necessary, occasionally ask other faculty volunteers to do the same so that all students have equal access to information. Suggest that students keep their notebook in their Portfolio and add to it throughout the course.

A GLOBAL PERSPECTIVE, *page 9*

Discussing the Issue

Ask students to discuss the connection between the activities of Americans and global environmental problems. Point out, for example, that the United States imports significant quantities of tropical hardwood from rain forests. To stimulate discussion, ask questions similar to the following: What could the average American do to help preserve rain forests? Have we Americans been using more than our fair share of nonrenewable resources like petroleum? If so, should we work on changing our wasteful lifestyles before criticizing other countries for their environmental problems? Or should we continue to develop technology to clean up the environment, and provide that technology free of charge to countries that experience the negative effects of our pollution? Suggest that students include their thoughts in their Portfolio. Then at different stages in the course, you could have students refer back to their thoughts and reflect on whether their opinions have changed based on new knowledge.

Using the Transparency

Figure 1-10 is available as a Teaching Transparency.

DEVELOPED AND DEVELOPING COUNTRIES, *page 10*

Drama Connection

Divide the class into small groups. Have half of the groups write and perform a skit about a typical day in the life of a teenager living in America. The skit should highlight the conveniences and ample resources available to citizens of a developed country. Have the other groups prepare a skit about what they imagine a day in the life of a teenager in a developing nation would be like. Tell these students to use their imaginations to envision what life would be like without the conveniences and resources that they now take for granted. After students have performed their skits, encourage class discussion about how each lifestyle affects the environment.

CLOSURE

Section Review

Answers to these questions are provided in the Answer Key on page T141.

Alternative Assessment

Provide students with a list of 20 environmental problems that include local, regional, and global environmental issues. Ask students to identify each problem as an example of resource depletion, pollution, extinction, or a combination of factors. Then have students pick 10 of these problems and describe specific human activities that contribute to them.

Reteaching

Help students realize their reliance on natural resources by first dividing them

into small groups, and then having group members discuss their use of natural resources. Direct student thinking by suggesting that they examine the connection between natural resources and their clothes, the contents of their backpacks, the method by which most students come to school, and common objects around the classroom. *(Cotton is a product in many clothes; petroleum is used to make the synthetic fibers in other types of clothing, plastic ballpoint pens, or parts of athletic shoes; coal is burned to produce the electricity used to operate classroom lights; wood is used as a building material for the school and desks, etc.)* Compile a class list of group findings on the chalkboard, and help students identify which of the listed resources are renewable and which are nonrenewable.

Extension

Suggest that students work in pairs to develop and conduct a survey to determine the attitudes of family members, classmates, neighbors, teachers, and friends toward specific environmental issues. Encourage students to include their survey results in their Portfolio.

SECTION 1.2

Using Science to Solve Environmental Problems

Focus

Students are introduced to the idea that environmental science is an applied science that uses information from many other sciences to solve practical problems. Then, real-world examples help students learn about scientific methods.

Motivator

Place a houseplant where everyone in the class can see it. Ask students to name several things that they have learned about plants in other science classes. *(Plants are green because they contain chlorophyll; plants use carbon dioxide and give off oxygen; etc.)* Point out that they learned these things through the study of science. Then ask students to describe everything that they know about this particular plant. As students describe the plant, listing its height, its color, and so on, point out that they are using one of the fundamental methods of science—observation. Then ask students questions such as how often this type of plant should be watered, how many hours of sunlight it must have, what kind of fertilizer is best for it, etc. When students respond that they have no way of knowing the answer, ask them how they could find out. Students will probably propose a simple experiment, such as "Get another plant just like it, and water one every day and the other once a week and see which one does best." Point out that experimenting is another essential scientific process and that when they observe and experiment, they are following a scientific method, just as scientists do.

Teaching Strategies

What Is Science? *page 14*

Meeting Individual Needs

Second-Language Learners In order for students to understand this section it is important that they understand the "language of science." Use a foreign-language dictionary to give students the words for such terms as *ecology, hypothesis, control group, graphs,* and *statistics.* Ask students if they have used these terms before in their native language. Tell students that the terms will be defined and explained in this part of their text.

Observing, *page 14*

Reinforcing Ideas

Ask students to bring paper and a pencil, then take the class outside to a natural area on the school grounds or to a nearby park. Tell students that they are to learn as much as possible about a living thing by simply observing it. Have students observe for at least 20 minutes and take careful notes or make sketches. Students may observe such things as the behavior of pigeons or other birds; the numbers of fish, snails, and other organisms in a stream or pond; or where foraging ants go when they leave the anthill.

When students return to the classroom, ask them to write a description of what they have learned by observation.

Also ask that they write a question that could be answered by more observation, such as: Does the cardinal I observed in that tree have predictable patterns of behavior—can it be regularly observed in this area at this time? Then ask students to write a plan describing how they could go about finding the answer. For instance, in the above example a student's plan might involve visiting the area at the same time every day to make further observations. Suggest that students include their observations and their plan in their Portfolio.

EXPERIMENTING, *page 16*

Meeting Individual Needs

Students Having Difficulty Help students understand that scientific methods like experimenting can be used for practical reasons, such as determining the best way to do something or the best product to buy. Ask students to name some inexpensive consumer products that they frequently buy. Then ask students to think of and carry out an experiment that could help them learn more about the product. For example, the experiment could be to test which brand of cola stays fizzy the longest or which brand of ballpoint pen will write for the longest period of time.

CLOSURE

Section Review

Answers to these questions are provided in the Answer Key on page T141.

Alternative Assessment

Ask students to imagine the following scenario: You are planning to start your own business offering environmental consulting services to builders in your area. Your goal is to help builders minimize negative environmental impacts at their project sites. One thing you need to know before beginning your work is the best way to prevent erosion on cleared land while a project is underway. Have students outline an experiment that could provide this information. Students' outlines should include how they would gather information in order to form a good hypothesis, the single variable that will be tested, and how they will set up a control. As an extension, suggest that students complete the experiment and describe the results. Encourage them to present their findings to a classmate. Then students could include a description of the experiment, a summary of their results, peer commentary, and response to that commentary in their Portfolio.

Reteaching

Remind students that science only seeks to answer questions about the natural world and how it works. There are many other types of questions that cannot be answered by using scientific methods. Ask students to give examples of a variety of questions that people might have and write the questions on the chalkboard. Then ask students which ones are questions that science could answer and which ones are not. *(Nonscientific questions may range from "Will I have a date next weekend?" to "What is the meaning of life?" Examples of scientific questions might be "What gasoline should I use in my car to get the best gas mileage and engine performance?" or "Does hot water really freeze faster than cold water?")*

Extension

Provide students with a copy of an article about an environmental issue from a magazine such as *New Scientist* or *Scientific American.* The student should highlight or underline parts of the article that demonstrate one of the scientific methods described in this section. For each highlighted section, the student should make a note in the margin indicating whether this shows observing, hypothesizing, experimenting, and so on.

SECTION 1.3

MAKING ENVIRONMENTAL DECISIONS

FOCUS

Students learn the importance of gathering information, considering values, and exploring consequences when making an environmental decision. They are encouraged to use the model to make their own decisions concerning environmental issues.

MOTIVATOR

Ask students to imagine what they would think if they heard that a major employer in their area was found to be dumping poisonous chemicals which in turn showed up in the local water supply. Ask: How would you recommend solving the problem? Then tell students that the company has stated that if it is forced to clean up its operations, it will simply close this plant and open one in another country with less strict environmental controls. Ask: Does this change your opinion about what should be done? Have students list a positive and a negative result of allowing the plant to continue operating the way it always has, and a positive and a negative result of closing the plant. Point out that knowing the best way to solve environmental problems is difficult because there are often both positive and negative consequences to all possible actions.

TEACHING STRATEGIES

AN ENVIRONMENTAL DECISION-MAKING MODEL, *page 21*

Using the Theme of *Interacting Systems*
Tell students that one of the basic concepts of environmental science is that "everything is connected to everything else." Have students brainstorm some examples of this concept. *Focus Question:* How is this basic concept related to the fact that it is often difficult for people to reach agreement about environmental issues? *(Taking action to improve the environment usually has repercussions—people may lose their jobs or land, or have to change their lifestyles. Likewise, if action is not taken, plants and animals may die; human health may be affected; and air, water, and land resources may be damaged or destroyed. Because people have different priorities, it is often difficult to reach an agreement.)*

A HYPOTHETICAL SITUATION, *page 22*

Meeting Individual Needs
Gifted Learners Encourage students to discuss a local environmental issue that is similar to the hypothetical one described in this section. (If your community is not currently involved in such a debate, you may wish to discuss a well-known national issue). Ask students to identify as many similarities as possible between the hypothetical situation and the real one. For example, ask: With regard to our local issue, are volunteer activists organizing to demand political action to solve the problem? Are opposing sides each predicting dire consequences if their plan of action is not adopted? Explain that debates surrounding environmental problems usually share similar characteristics, no matter what the specific controversy involves.

MAKE A DECISION, *page 24*

Using the Table:
Figure 1-29, page 23
Ask students to use the information provided in this section to make their own table like the one Michael Price made, shown in Figure 1-29. Students should be encouraged to use the values that they consider most important in making their table. Then they could state what their decision would be about how to vote in the referendum and write a justification for their decision. Students may include their table and their justification in their Portfolio.

CLOSURE

Section Review
Answers to these questions are provided in the Answer Key on page T141.

Alternative Assessment
Provide students with two articles from newspapers or magazines that present opposing points of view about a controversial environmental issue. *Garbage* magazine is an excellent source for such articles, since features written from different points of view are often presented in one issue. Ask students to write an evaluation of each article based on the following criteria: Which article presents more complete factual information that helps the reader decide the issue? What values seem to be most important to the writer of each article in deciding the issue? Which article gives the more reasonable predictions about outcomes if a particular plan of action is chosen or not chosen?

Reteaching
When the Salmon Runs Dry, a video from the Video Project, 5332 College Avenue, Suite 101, Oakland, CA, 94618; 800-4-PLANET, provides an example of the difficulty surrounding attempts to resolve conflict about an environmental issue. Showing this or a similar video may help students understand why environmental issues often stimulate so much controversy, and it may make them aware of methods of conflict resolution.

Extension
Cooperative Learning Activity Ask students to review the hypothetical situation in this section. Then ask for student volunteers to act out the following roles in an impromptu skit: a biologist studying the spotted warbler, an environmental activist demanding that all development cease in the warbler habitat area, a property owner upset about possible governmental regulation, several business owners, the town's mayor, and Michael Price—the citizen trying to make up his mind about how to vote in the referendum. Ask each player to state his or her position on the issue, and encourage other players to interrupt and argue whenever they feel their character would do so in real life. After the skit, ask for suggestions from the rest of the class about how each character could better encourage people with other points of view to be more cooperative and less confrontational.

Answers
Answers to *Chapter Highlights*, *Chapter Review*, and *Investigation* are provided in the Answer Key for Chapter 1, which begins on page T141.

TEACHING STRATEGIES FOR INVESTIGATION:

AN ONION CONUNDRUM, *pages 28–29*

Test the Water
Review the scientific method with students. Possible tests include varying amounts of water, varying amounts of fertilizer, and varying amounts of bleach (solutions from 0% to 10% should be adequate). Make sure all students set up a control.

CHAPTER 2

Living Things in Ecosystems

Planning Guide

CHAPTER COMPONENTS AND RESOURCES	$\frac{1}{2}$ Year Average	$\frac{1}{2}$ Year Advanced	1 Year Average	1 Year Advanced
Chapter Opener, p. 32				
EcoLog, p. 32	•	•	•	•
2.1 Ecosystems: Everything Is Connected, pp. 33–38	•	•	•	•
Transparency 2: *Levels of Organization in the Biosphere*	•	•	•	•
Making a Difference: *Butterfly Ecologist*, pp. 52–53			•	•
Section Review, p. 38	•	•	•	•
2.2 How Species Interact With Each Other, pp. 39–42	•	•	•	•
Field Activity, p. 40				•
EcoSkills: *Creating a Wildlife Garden*, pp. 396–397				•
Section Review, p. 42	•	•	•	•
2.3 Adapting to the Environment, pp. 43–46	•	•	•	•
Case Study: *Natural Selection and the Peppered Moth of England*, pp. 44–45	•	•	•	•
Transparency 3: *Natural Selection and Deer Populations*	•	•	•	•
Section Review, p. 46	•	•	•	•
End of Chapter				
Chapter Highlights, p. 47	•	•	•	•
Chapter Review, pp. 48–49	•	•	•	•
Investigation: *Showdown on the Prairie*, pp. 50–51			•	•
Chapter Tests	•	•	•	•
PACING GUIDE	7–9 days	6–7 days	15–16 days	15–16 days

Materials Organizer

Activity	Materials	Advance Preparation
Investigation: *Showdown on the Prairie*, pp. 50–51	3 prepared flowerpots (see note at right), 3 small flowerpots, 3 aluminum pie pans, potting soil, native grass seeds (buffalo is best), imported grass seeds (rye or fescue), tape and marker for labeling flowerpots, metric ruler, hand lens, notebook, pen or pencil, graph paper, one or more of the following (optional): light source, scissors, fertilizer, other soil additives (per student group)	Prepare 3 pots with soil and grass seeds as specified in step 2 on p. 50 of the student text. Note: Grass seeds typically sprout in a week to 10 days. (per student group)

CHAPTER OVERVIEW

In this chapter, students are introduced to the concept of ecosystems and to the idea that living things depend on each other and on their environment for survival.

Section 2.1 defines an ecosystem, highlighting the importance of biotic and abiotic factors. In addition, the concepts of habitat, niche, communities, and populations are introduced.

Section 2.2 describes the major types of species interactions, including predation, competition, and symbiotic relationships.

Section 2.3 explores the process of evolution by natural selection and explains the concepts of adaptation, coevolution, and extinction.

RESOURCES

. . . for Teachers

Devries, Philip J. "Singing Caterpillars, Ants and Symbiosis." *Scientific American,* Oct. 1992, pp. 76–82. (This article describes an unusual case of mutualism between caterpillars and ants.)

Facklam, Howard, and Margery Facklam. *Plants: Extinction or Survival?* Hillside, NJ: Enslow Publishers, 1990. (This book discusses how humans may be contributing to the extinction of many possibly useful plant species by depending on just a few plant species for food.)

Riss, Pam Helfers. "A Ratio Explanation for Evolution." *Science Scope,* Jan. 1993, pp. 36–44. (This article outlines a hands-on approach for studying the principles of human evolution.)

. . . for Students

Berger, Cynthia. "Donating Their Birdseed to Science." *National Wildlife,* Dec.–Jan. 1993, pp. 42–45. (This article describes how thousands of bird-watchers are doing scientific research right in their own backyards.)

Hoyt, Erich. *Environment Reference Series: Extinction A–Z.* Hillside, NJ: Enslow Publishers, 1991. (This book is a compilation of terms used in the study of extinction biology, with special emphasis on plant and animal extinction.)

Line, Les. "Curse of the Cowbird." *National Wildlife,* Dec.–Jan. 1994, pp. 40–45. (This article describes the brown-headed cowbird, which parasitizes the nests of other birds.)

Films, Videos, Software, and Other Media

In the Company of Whales. CAV Videodisc, Level I, III. Coronet/MTI Film and Video, P.O. Box 2649, Columbus, OH 43216; 800-777-8100.(This video explores the relationship between whales, humans, and marine ecosystems.)

Red Wolf. Video. 13 minutes. Churchill Media, 6901 Woodley Avenue, Van Nuys, CA 91406; 800-334-7830. (This video illustrates how red wolves have come close to extinction and analyzes how human attitudes toward predators in general have contributed to their losses.)

Relationships. CAV Videodisc, Level I. SYSCON Corporation, 2686 Dean Drive, Virginia Beach, VA 23452; 804-486-2656. (This videodisc investigates the myriad of relationships among organisms.)

Getting Involved

A listing of environmental organizations, publications, and volunteer opportunities can be found in the **Laboratory and Field Guide**. Note especially those organizations with the goal of maintaining natural ecosystems, such as the National Wildlife Federation and the Nature Conservancy.

USING THE QUOTATION

"When we try to pick out anything by itself, we find it hitched to everything else in the universe."

Ask a volunteer to read the quotation aloud. Then call on students to explain what they think Muir might have meant when he made the statement. Write the following list on the board in a circular pattern: sun, water, tree, carbon dioxide, oxygen, bird, fungi, soil, crickets, earthworm, owl, mouse, and snake. Ask students to brainstorm as many relationships between the listed items as they can, and have a volunteer draw lines between the related items. Ask students to explain how this "relationship map" illustrates what Muir was trying to say.

USING THE ECOLOG

To assess students' knowledge about the subject of this chapter, have them answer the EcoLog questions on page 32. In answering these questions, students will reveal their prior knowledge, which is usually very basic and often flawed or even misconceived. Then, by revising their answers at the conclusion of the chapter, students will have the opportunity to eliminate any misconceptions and to see firsthand how their knowledge has grown.

Ecosystems: Everything Is Connected

Focus

To introduce the concept that all parts of an ecosystem are in some way connected, the section opens with a discussion of how DDT caused unforeseen consequences in an island ecosystem. Students then learn about the different components that make up an ecosystem, including biotic and abiotic factors, and how ecosystems are structured into populations and communities. Finally, students investigate the difference between an organism's niche and its habitat.

Motivator

Draw a "connections web" on the chalkboard. Start with any common animal in nature, such as a blue jay, a raccoon, or a wasp. Write its name on the board and circle it. Then have students name interactions that this animal has with other plants or animals in its environment. For instance, a blue jay eats different insects, so you could draw a line connecting the bird to an insect. Be sure to extend the connections and include abiotic factors as well (the insect feeds on a plant, which in turn uses sunlight). Continue until the web becomes very complex. Point out that the web contains both living (biotic) and nonliving (abiotic) components and that every item in the web has the potential to be connected to many different things. Introduce the idea of an ecosystem, emphasizing that your web would represent only a fraction of the interactions happening in a natural ecosystem at any given time.

Teaching Strategies

Reinforcing Ideas

Use the following example to illustrate the complexity and interconnectedness of the components within an ecosystem. In San Diego, California, a marsh habitat that was home to two endangered bird species was destroyed to build a freeway. To get a permit to build the freeway, the city had to agree to "rebuild" the ecosystem for the birds.

After five years and $500,000, scientists and officials found that replacing an ecosystem is something that we do not know much about. For example, when the endangered birds were released into their re-created ecosystem, they would not nest because the marsh grass was not tall enough. The grass was stunted because a tiny beetle was not present in the new ecosystem. The beetle eats insects that feed on marsh grass; without the beetle to control the insect population, the marsh grass could not grow to its full height. This was only one of many problems that the city faced in its ecosystem re-creation project.

What Is an Ecosystem? *page 34*

Using the Illustration:
Figure 2-2, page 34

Ask students to make a list of likely interactions between the species shown in the soil ecosystem. Students should also explain the nature of the relationship. For example, a mole and an earthworm would interact because the earthworm is a primary food source for many types of moles.

Background: Soil in Old-Growth Forests

Scientists studying the soil in old-growth forests of the Pacific Northwest say that temperate forest soils have a biological diversity that rivals that found in tropical rain forests. In fact, researchers hypothesize that invertebrates in forest soils may be the most important factor in determining the long-term productivity of a forest. Soil arthropods, which are often about the size of the period at the end of this sentence, include beetles, centipedes, pseudoscorpions, springtails, and mites.

Using the Transparency

Figure 2-4 is available as a Teaching Transparency.

Using the Illustration:
Figure 2-4, page 36

Have students analyze the illustration, explaining the connection of each level to the preceding one. Then have students look at a series of pictures or slides, and ask them to classify the images using the terms *organism, population, community, ecosystem,* and *biosphere.* A series could consist of one individual of an animal species, such as a wildebeest *(organism);* a herd of wildebeests *(population);* a close-up of several species, such as wildebeests, lions, and giraffes *(community)*; scenery of the African savanna showing many animals *(ecosystem)*; and the Earth *(biosphere)*.

Cooperative Learning Activity
Divide the class into small groups, and have them construct a diorama of an ecosystem using modeling clay, paper, photographs from magazines, dried plants, and other materials. Encourage students to be creative. Each student in the group could be responsible for one type of organism, such as a decomposer, a plant, an herbivore, or a carnivore.

Meeting Individual Needs
Second-Language Learners Using magazine pictures or their own illustrations, have students create an ecosystem brochure that describes a chosen ecosystem. Their brochure could be in English or in their native language. Suggest that students begin the brochure with an exciting cover that clearly identifies the chosen ecosystem. Then, inside the brochure, students could show pictures of representative plants, animals, fungi, and bacteria. The habitat and niche should be noted for each organism. Students should also show pictures of the abiotic factors such as minerals, water, and rocks, and students should explain the role of those factors in the ecosystem. Suggest that students include their completed brochure in their Portfolio.

Meeting Individual Needs
Gifted Learners Ask students to imagine that they are on a committee that has the task of deciding whether a golf course can be built without creating havoc in the existing environment. Tell students to pick a specific location in their community for their proposed golf course and describe the ecosystem present at that location. Then students should analyze and describe all of the possible repercussions of the golf course on the ecosystem. Finally, students should determine if the golf course should be built. If not, they should explain why. If they decide that the golf course is a good idea, students should develop a plan showing how the course could be designed so that it has little impact on the environment. If a golf-course scenario isn't appropriate for your region, suggest a highway, parking garage, shopping center, or other development.

Using the Theme of *Energy*
Focus Question: Most ecosystems contain all of the following: plants, herbivores, carnivores, sunlight, and decomposers. Which supplies the most important source of energy? Justify your answer. *(Although a case could be made for any of the components listed, make sure that students understand that sunlight provides the initial energy on which almost all ecosystems depend.)*

Notable Quotes
"We need bats whether we like them or not . . ."
Merlin Tuttle

"The history of life on earth has been a history of interaction between living things and their surroundings."
Rachel Carson

POPULATIONS, *page 37*

Using the Photographs:
Figure 2-5, page 37
Ask students to examine the photograph of the penguins standing on an icy shoreline. Have them speculate about what other organisms might interact with the penguins. *(Students may be surprised to find that although penguins appear to live in barren surroundings, they are actually part of a complex community that includes microscopic algae, krill, baleen whales, squid, fish, toothed whales, leopard seals, and birds such as the albatross and the petrel.)* Ask interested students to do research on how these organisms interact. Students can present their findings in a poster diagram. Encourage students to include their poster in their Portfolio.

NICHE AND HABITAT, *page 38*

Multicultural Connection
To help students understand their uniqueness in the world, have a discussion of niches. Begin by describing your niche in life. Then have student volunteers share their niches by describing their place in their family, their home, their neighborhood, their age, their classes, and what they do each day. Students should give a reason why their niche is unique and important. This activity will help students understand and appreciate others of different backgrounds and cultures.

Cooperative Learning Activity
Divide the class into groups to do research on Biosphere II or Space Station Alpha (which is currently being planned by NASA). Once students have learned something about what it takes to run a self-sustaining colony, have them design their own colony. Students could form teams and design a colony for the moon, Mars, underwater, Antarctica, or a desert. Suggest that they create models; scale drawings or "blueprints"; reports on the recycling of air, water, and waste; and descriptions of the type of colony, including what populations will be represented and how those populations will be "managed." Encourage students to include information about their colony in their Portfolio.

CLOSURE

Section Review
Answers to these questions are provided in the Answer Key on page T142.

Alternative Assessment
Ask students to pick their favorite organism and write a "biography" for that organism that includes three chapters: one on habitat and niche, a second on the survival of that organism's population within

the community, and a third on abiotic characteristics that are crucial to the survival of the organism. Invite students to include their "biography" in their Portfolio.

Reteaching
Divide the class into groups of four to five students. Then have group members work together to devise a short skit, rap, song, or limerick that illustrates the terms used in this section.

Extension
Ask students to work in groups to make a photographic collage or video of several ecosystems near their homes. For example, students may choose a pond, a yard, a ravine, an empty lot, or a garden. The collage should include written captions, and the video should include student narration about populations, communities, and abiotic factors. Students may wish to include their work in their Portfolio.

SECTION 2.2

HOW SPECIES INTERACT WITH EACH OTHER

FOCUS
Five major types of species interactions will be explored in this section.

MOTIVATOR
Ask students to write a short paragraph describing a selected carnivore, such as a fox, through the eyes of its intended meal, such as a mouse.

TEACHING STRATEGIES

PREDATION, *page 39*

Discussing the Issue
There are several general principles of predation which state that (1) predators are usually beneficial organisms, (2) predators prey on "surplus" animals and do not cause a serious decline in the prey population (they will turn to other more accessible prey when the original prey is hard to capture), (3) predators do not usually cause species extinction, and (4) overpopulation of prey species will occur if not held in check by a predator species. Ask students to discuss how these patterns of predation can be affected by human activity.

Using the Photographs:
Figure 2-7, page 39
Have students look at the predators in Figure 2-7, and ask: What characteristics, or adaptations, does each predator shown use to capture its prey? *(The chameleon has a long, sticky tongue that can shoot out and capture prey; its coloration allows it to blend in with its environment; its eyes protrude from its head and can move independently to more easily spot prey; and the toes on its feet and its prehensile tail allow it to grasp branches so that it can remain immobile while waiting for prey. The golden eagle has large feet equipped with long, curved talons for grasping prey and a hooked bill for tearing prey. Its eyesight is very keen. Its long, broad wings enable it to carry off prey that weighs almost as much as it does. The starfish attaches itself to the clam with its many rows of suction-powered tube feet, after which the starfish's strong and flexible arms allow it to overpower and open the clam. The starfish then ejects its stomach to surround and digest the soft parts of the clam. The tawny coloration of the cougar allows it to stalk its prey undetected. It has sharp, retractable claws for grasping prey and powerful jaws equipped with daggerlike teeth for killing the prey. Its muscular body is ideal for running short distances quickly. The red fox has long, muscular legs that allow it to run for long distances. Although it is not detectable from the photograph, foxes have keen hearing, sense of smell, and eyesight. They also have sharp teeth for grasping prey.)*

COMPETITION, *page 40*

Using the Photographs:
Figure 2-8, page 40
Discuss with students the examples of competition in the pictures; then have them write two examples of competition

that they have observed in their daily lives. One should involve competition between two different species, and the other should involve competition between individuals of the same species. The competition described may include animals, plants, or humans. Ask students to clarify what resources the species are competing for and whether the resource is a necessity for life.

Discussing the Issue

Ask students if they have ever had any encounters with fire ants, and if so, ask volunteers to share their experiences. *(Students familiar with fire ants will probably mention the painful sting of the insect and the difficult nature of eliminating the pest.)* Explain to students that the imported fire ant is outcompeting native ants throughout the southern United States. Both native and nonnative ants inhabit the same niche. Ask students: Why might the imported fire ant be more successful in its new environment? *(The imported fire ants may be more successful because their natural predators, found in their native South American habitat, are not present.)*

Using the Theme of *Stability*

Focus Question: How might the introduction of a nonnative species, such as the kudzu vine, upset the stability of the relationships between native members of a community? *(An introduced species may have no natural predators in its new environment. Similarly, the organisms on which it feeds may have no natural defenses against it. Or, in the case of the kudzu vine, it outcompetes the other plant species. Thus, the introduced species may grow unchecked, altering the sizes of the populations with which it interacts.)*

PARASITISM, *page 41*

Reinforcing Ideas

Ask students: Would a giraffe browsing on an acacia tree be considered a parasite? Explain your answer. *(Students may suggest that the giraffe is indeed a parasite because it is living off the tissues of another organism without actually killing it. However, giraffes do not actually live in or on the tree and may obtain food from a variety of other sources. Therefore, giraffes are not considered parasites.)*

Language-Arts Connection

Have students write a humorous advertisement for a newspaper or a commercial for television about a parasite seeking a new host. The written advertisements can be combined in a newspaper format, and students can act out their television ads for the rest of the class. Encourage students to include their work in their Portfolio.

MUTUALISM, *page 41*

Background: Ants and Butterflies

Another interesting example of mutualism involves a certain species of ant and the larvae of the blue butterfly. The ants tend the larvae of the butterfly, providing them with protection from predators and parasites. In return, the ants receive a sugar solution, which is secreted from specialized glands in the larvae.

COMMENSALISM, *page 42*

Reinforcing Ideas

Explain to students that it is often difficult to distinguish between commensalism and mutualism because it is difficult to determine whether the second partner is benefiting from the relationship or not.

Meeting Individual Needs

Learners Having Difficulty Ask students to help you make a collage of photographs and words from magazines and newspapers that illustrate and describe interactions between organisms. Have available magazines, newspapers, and environmental newsletters, as well as poster board, scissors, and glue. When it is finished, display the collage in the classroom.

Art Connection

Ask interested students to create a portrait of a predator, a comic strip about a predator-prey relationship, or a watercolor depicting some form of mutualism. One example includes the sea anemone and the clown fish. Ask students to propose their own artistic endeavor if they wish, and encourage them to include their work in their Portfolio.

CLOSURE

Section Review

Answers to these questions are provided in the Answer Key on page T142.

Alternative Assessment
Ask students to devise a wildlife management plan for a nongame species found in the area. In preparing their plan, students could answer questions such as the following. What restrictions and regulations should be placed on the community? How should the nongame population be controlled? How should other competing species and predators be treated? How can a stable, natural population be achieved? *(Answers will vary. Students should demonstrate an understanding of the importance of preserving as many interrelationships as possible.)*

Reteaching
Write several short descriptions of animals that exhibit one of the five interactions between organisms. Read these descriptions to the class, and ask them to guess which type of interaction is represented, such as predator-prey interaction or competition between species for the same resources.

Extension
To examine the role of prey density (the amount of prey in a given area) on the success of a predator, try the following game. Blindfold a volunteer "predator." Loosely tape 25 poker chips to a large piece of cardboard. Have another student act as timer, and allow the "predator" one minute to pick up as many "prey" as possible. On the same piece of cardboard, increase the amount of prey by 15 each time for five more trials. Make note of the "predator's" results in a table, and then plot the data on a graph. Have students write a hypothesis explaining the results.

SECTION 2.3

ADAPTING TO THE ENVIRONMENT

FOCUS
This section begins by describing water-conservation techniques that allow the kangaroo rat to survive in the desert without ever actually drinking water. Then students are introduced to Darwin's theory of evolution by natural selection. This leads to a discussion of adaptation and how one species may influence the evolution or possible extinction of another species.

MOTIVATOR
Explain to students that within any population, such as the population of humans in your classroom, there are natural variations among characteristics of a species. Have students work with a partner to measure each other's shoe size in centimeters. In a table on the chalkboard, record the data for each student. Then construct a bar graph using the results. Explain to students that Darwin, in his theory of evolution by natural selection, used natural variation within a population to show how individuals with favorable characteristics have a better chance of survival and therefore are able to produce more surviving offspring.

TEACHING STRATEGIES

Biology Connection
After students have read the section introduction, explain that an important part of the kangaroo rat's survival strategy involves the use of metabolic water, or water produced as a result of cellular respiration. In order to obtain energy, the food eaten by the kangaroo rat is broken down into simple sugar molecules known as glucose. Glucose is carried through the bloodstream to individual cells, where it is burned for energy in a process known as respiration. Respiration, just like fire, requires oxygen to occur. Organisms, such as human beings and kangaroo rats, obtain oxygen through breathing. When oxygen combines with glucose, energy is given off, and carbon dioxide and water are formed. The kangaroo rat and other desert dwellers recycle this water and use it to help them survive. Ask students what they would see if they exhaled on a mirror or piece of glass. *(water vapor)* Explain that the water vapor that they see is one of the end products of cellular respiration.

Using the Case Study:
Natural Selection and the Peppered Moth of England, pages 44–45
Tell students that the trees changed color because light-colored lichens covering the trees were very sensitive to the air pollution that developed in industrial areas; they died, exposing the dark bark underneath. Students may wonder why the lichens didn't adapt to the new

conditions like the moths did. Explain that, even though dark moths were rare to begin with, the fact that there was some inheritable variation in color allowed natural selection to take place. In contrast, none of the lichens could survive the changing environment.

Using the Transparency

Figure 2-14 is available as a Teaching Transparency.

EXTINCTION, *page 46*

Using the Theme of *Patterns of Change*

Focus Question: What are some changes that could occur within an individual's lifetime that would allow an organism to adapt to changes in the environment? Can these changes be passed along to the organism's offspring? *(These changes could include occupying a slightly different habitat, migration, hibernation, obtaining food from a different source, and becoming active during different times of the day. Although these parents could pass these behaviors along to their offspring by example, the changes would not be passed along genetically because they do not involve changes to the parents' genetic material.)*

Debate

Set up debate teams, and allow students time to research and prepare for formal debate on this topic: "Should humans be allowed to affect the evolution of other organisms by selectively breeding them?" Subtopics might include the domestication of plants and animals, genetic engineering to produce disease-resistant plants, using bacteria for the recycling of wastes, and using bacteria for fighting disease.

CLOSURE

Section Review

Answers to these questions are provided in the Answer Key on page T143.

Alternative Assessment

Have students work in cooperative groups to prepare a newspaper that provides information about the extinction of species and incorporates the concept of natural selection. Topics could include extinction of dinosaurs, twentieth-century extinctions, human extinction written from the viewpoint of an insect, and prehistoric mass extinctions (of which there were several). Each student could produce one major news story in addition to helping with other features, such as headlines, cartoons, obituaries, advertisements, and advice columns. Print the papers for students to read.

Reteaching

After students have read and discussed the Case Study about the peppered moth (pages 44–45), have them experiment with paper moths to see if variation can help survival. Divide the class into small groups. Each group should make 50 moths from black paper and 50 from newspaper. Shake the "moths" in a container, and spread them out on newspaper (classified ads are best). One student will act as the predator. Another student should act as the timer. The predator picks up the moths one at a time for one minute. After each moth is picked up, the predator should look up and across the room. Repeat with the black paper as background. Record the frequency of each type of moth selected and the background. Ask: Which color of moth was most frequently selected and least selected on each background? Which species is most favored for survival and why? Students' findings should reinforce the findings outlined in the Case Study.

Extension

Read a selection or adopt main ideas from Darwin's *Origin of Species.* Ask students to write a review of the excerpt that was read. Some students may wish to read and review the entire book. Suggest that students include their review in their Portfolio.

Answers

Answers to *Chapter Highlights*, *Chapter Review*, and *Investigation* are provided in the Answer Key for Chapter 2, which begins on page T142.

TEACHING STRATEGIES FOR INVESTIGATION:

SHOWDOWN ON THE PRAIRIE, *pages 50–51*

Review the meaning of competition with students, emphasizing the different forms that competition can take. Stress both the scientific and economic (i.e., agricultural) importance of understanding the effects of competition on plant growth.

Helpful Hints

Grass seeds typically sprout in a week to 10 days. Check the package to assist you in judging how far in advance the seeds should be planted. Keep in mind that, depending on the types of seeds used, competition may affect germination rates.

Extensions

Suggest that interested students explore this topic further. For instance, they could investigate the effects of competition between grasses and other types of plants, such as garden vegetables or houseplants. They might also examine the effects of intraspecies competition by observing the growth of one species planted at various densities and noting differences in growth rates.

CHAPTER 3

How Ecosystems Work

Planning Guide

Chapter Components and Resources	½ Year Average	½ Year Advanced	1 Year Average	1 Year Advanced
Chapter Opener, p. 54				
EcoLog, p. 54	•	•	•	•
3.1 Energy Flow in Ecosystems, pp. 55–61	•	•	•	•
Points of View: *Where Should Wolves Roam?* pp. 76–77	•	•	•	•
Transparency 4: *A Food Chain*	•	•	•	•
Case Study: *DDT in an Aquatic Food Chain*, pp. 58–59		•	•	•
Transparency 5: *A Food Web*	•	•	•	•
Field Activity, p. 60			•	•
Section Review, p. 61	•	•	•	•
3.2 The Cycling of Materials, pp. 62–65	•	•	•	•
Transparency 6: *The Water Cycle*	•	•	•	•
Transparency 7: *The Carbon Cycle*	•	•	•	•
Transparency 8: *The Nitrogen Cycle*	•	•	•	•
Section Review, p. 65	•	•	•	•
3.3 How Ecosystems Change, pp. 66–70	•	•	•	•
Environmental Careers: *Research Wildlife Biologist*, pp. 374–375				•
Transparency 9: *Secondary Succession*	•	•	•	•
Case Study: *Fires in Yellowstone*, pp. 68–69		•	•	•
Field Activity, p. 70			•	•
Section Review, p. 70	•	•	•	•
End of Chapter				
Chapter Highlights, p. 71	•	•	•	•
Chapter Review, pp. 72–73	•	•	•	•
Investigation: *What's in an Ecosystem?* pp. 74–75	•	•	•	•
Chapter Tests	•	•	•	•
Pacing Guide	**9–10 days**	**8–10 days**	**16–18 days**	**16–18 days**

Materials Organizer

Activity	Materials	Advance Preparation
Investigation: *What's in an Ecosystem?* pp. 74–75	tape measure or meter stick, 4 stakes, string, poster board, markers or felt-tip pens of several different colors, notebook, pen or pencil (per student group)	Determine a location for conducting this investigation. You may wish to set up the plot ahead of time.

CHAPTER OVERVIEW

In this chapter, students learn about the flow of energy, the cycling of materials, and succession in ecosystems.

Section 3.1 describes the roles of sunlight, producers, consumers, and decomposers in food chains and food webs. Energy transfer among organisms at various trophic levels is also examined.

Section 3.2 focuses on the water, carbon, and nitrogen cycles. The effect of human activities on the carbon cycle and the importance of nitrogen-fixing bacteria are also discussed.

Section 3.3 describes primary and secondary succession, emphasizing the importance of pioneer species and explaining the relationship between primary succession and the formation of soil.

RESOURCES

. . . for Teachers

Art, Henry W., ed. *The Dictionary of Ecology and Environmental Science.* New York: Henry Holt, 1993. (This dictionary provides complete and authoritative explanations of more than 8,000 terms and concepts relating to all aspects of environmental science.)

Odum, Eugene P. *Ecology and Our Endangered Life-Support Systems.* 2nd edition. Sunderland, MA: Sinauer Associates, 1993. (This is a comprehensive, textbook-style book that is designed to be a citizen's guide to the principles of modern ecology as they relate to today's threats to our Earth.)

. . . for Students

Anderson, Margaret J. *Food Chains: The Unending Cycle.* Hillside, NJ: Enslow Publishers, 1991. (This book about food chains is part of a series on ecological communities and principles that is designed specifically for young people.)

Tudge, Colin. *Global Ecology.* New York City, NY: Oxford University Press, 1991. (This colorful and engaging book was written by an eminent scientist to accompany an ecology exhibit at London's Natural History Museum.)

Films, Videos, Software, and Other Media

Ecology Series: Food Chains/Nutrient Cycles. CAV Videodisc, Level 1. Coronet/MTI, 4350 Equity Drive, P.O. Box 2649, Columbus, OH 43216; 800-777-8100. (This videodisc helps students learn more about food chains, food webs, and nutrient cycles.)

Population Ecology. Software. Logal Software Inc., P.O. Box 1499, East Arlington, MA 02174-9850; 800-LOGAL-US. (With this software, students can simulate populations of organisms as they grow and interact in controlled ecoystems.)

Succession: From Sand Dune to Forest. Video. Encyclopedia Britannica, 310 South Michigan Avenue, Chicago, IL 60604-9839; 800-554-9862. (This video traces all of the stages in ecological succession, from beach to forests.)

Getting Involved

A listing of environmental organizations, publications, and volunteer opportunities can be found in the **Laboratory and Field Guide**. Note especially those organizations with the goal of maintaining healthy ecosystems, such as the Center for Holistic Resource Management, Conservation International, the Sierra Club, and the Society for Ecological Restoration.

USING THE QUOTATION

"You could cover the whole world with asphalt, but sooner or later green grass would break through."

Call on a student to read the quotation aloud. Then ask: What do you think Ehrenburg might have meant by this statement? *(Sample answer: He might have meant that human activities are only intrusions into the natural world. They are eventually replaced by the natural world that they pushed aside.)* Ask: Does Mr. Ehrenburg seem to think this ability represents something positive or something negative about the living world? *(Accept all reasonable answers.)* Then ask: What do you think this tendency says about the natural world and humans' relation to it? *(Sample answer: The Earth's natural forces are powerful and resilient. Thus, when humans alter nature in some way, those changes may not be permanent.)* Encourage students to express a variety of viewpoints. Make a list of their responses on a large sheet of paper, and bring it out again for discussion at the end of the chapter.

USING THE ECOLOG

To assess students' knowledge about the subject of this chapter, have them answer the EcoLog questions on page 54. In answering these questions, students will reveal their prior knowledge, which is usually very basic and often flawed or even misconceived. Then, by revising their answers at the conclusion of the chapter, students will have the opportunity to eliminate any misconceptions and to see firsthand how their knowledge has grown.

Energy Flow in Ecosystems

Focus

In this section a discussion of energy and energy transfers leads to the concept of food chains and the principle that sunlight is the energy source for nearly all living things. Students learn about the roles of producers, consumers, and decomposers in food chains and food webs. The section concludes with a discussion of energy pyramids.

Motivator

Ask students: What is made of the following ingredients?

- 1/2 bathtub full of oxygen
- 50 glasses of water
- 1/2 cup of sugar
- 1/2 cup of calcium
- 1/10 thimbleful of salt
- a small pinch of assorted elements such as phosphorus, potassium, nitrogen, sulfur, magnesium, and iron
- a mystery ingredient

(Students will probably not know the answer to this question.) Tell students: Believe it or not, these are the basic ingredients that *you* are made of. Continue by saying: Knowing that I was describing you, what do you think the mystery ingredient is? *(energy)* Then ask: Where does that energy come from? *(the food we eat)* Ask: Where does the energy in food come from? *(ultimately, the sun)* If students cannot answer these questions now, tell them not to worry. Explain that in this lesson they will learn the important role of the sun and how it influences the flow of energy in ecosystems.

Teaching Strategies

Life Depends on the Sun, *page 55*

Using the Photographs: *Figure 3-2, page 56*
Use the photographs shown in this figure to help students visualize how life depends on the sun. Ask: In this series, how does the wolf depend on the sun? *(The wolf depends on the energy that was passed from the sun to the grass because the deer ate some of that grass and derived energy from it, and the wolf will eat the deer and derive energy from the deer.)* Then ask: What would happen if the grass did not get enough sun, withered, and died? *(Both the deer and the wolf could starve to death.)*

Who Eats What, *page 57*

Meeting Individual Needs
Second-Language Learners To help students learn some of the new vocabulary in this section, direct their attention to Figure 3-4. Read the table out loud, discussing each portion. Then use magazine photos and other visual aids to provide more examples of each type of organism. Check student comprehension by holding up various photos and calling on students to explain whether the organism shown is a producer or a consumer and if it is a consumer, which type of consumer.

Energy Transfer: Food Chains, Food Webs, and Trophic Levels, *page 58*

Using the Case Study: *DDT in an Aquatic Food Chain, pages 58–59*
After students have read the Case Study, ask: How does the 10 percent rule relate to the concept of bioaccumulation? *(Each step in a food chain increases the concentration of a substance such as DDT tenfold.)* To help students visualize the process, copy the following food chain onto the chalkboard.

Algae and pond weeds
1 part in 1,000,000
↓
Invertebrates (zooplankton)
1 part in 100,000
↓
Minnows
1 part in 10,000
↓
Perch
1 part in 1,000
↓
Bass
1 part in 100

Next, use the food chain to demonstrate why the concentration of DDT is expected to be approximately 10,000 times as great in the bass as it was in the algae and pond weeds. Remind students that only chemicals that are not excreted by organisms or broken down by decomposers are subject to bioaccumulation.

Extending the Case Study: *DDT in an Aquatic Food Chain, pages 58–59*
Tell students that the pesticide DDT is just one example of a substance that is subject to bioaccumulation. Suggest that students do research to find out about at least three other substances that can bioaccumulate in food webs. *(PCBs, strontium-90, and mercury are possible answers.)* Encourage students to include their research findings in their Portfolios.

Food Chains and Food Webs, *page 59*

Using the Transparencies
Figures 3-6 and 3-7 are available as Teaching Transparencies.

Meeting Individual Needs
Learners Having Difficulty Direct students' attention to Figure 3-7. Use the diagram to review with students. Ask students to isolate a single food chain, other than

the one shown in Figure 3-6. Then help students identify the organisms that are producers or members of the first trophic level, herbivores or members of the second trophic level, carnivores or members of the third trophic level, and omnivores.

CLOSURE

Section Review

Answers to these questions are provided in the Answer Key on page T144.

Alternative Assessment

Divide the class into two groups: Group A and Group B. Tell students in Group A that their assignment is to determine the cost per ounce of rice, flour, dried beans, and rolled oats. Tell students in Group B that their assignment is to determine the cost per ounce of ground beef, pot roast, chicken, and pork chops. (You can provide students with this information or delegate one person from each group to call a grocery store.) Then ask each group to determine the number of calories per ounce for each type of food. (The packaging of some food items lists this information. In addition, reference books that list the calories of common food items are available at most libraries.) Next, students should find the cost of one calorie for each type of food by dividing the cost per ounce by the calories per ounce. Have groups record their findings in the form of a bar graph.

Finally, have students from Group A compare their results with those of Group B. *(Costs per calorie are lower for grains and other plant products than for meats.)* Ask: Why do you think this is the case? *(Because plants are lower on the food chain than animals. Since 90 percent of the energy of any trophic level is not transferred to the next, this means that more energy is needed to raise animals for meat than to raise an equivalent amount of grain or vegetables.)* Ask: What is the significance of your findings for humans? *(Humans could have more food for more people and spend less for it by eating lower on the food chain—being herbivores instead of carnivores.)* *Note:* Be sure that students understand the nutritional value of meat as a source of protein and other essential vitamins and minerals. Explain that vegetarians must be sure to eat the right combinations of fruits, nuts, vegetables, grains, etc., to get enough protein.

Reteaching

To help students draw connections between what they have been reading in this section and their own role (as a human) in a food web, perform the following demonstration. Gather a variety of human foods and cooking ingredients, including both plant and animal products. Then show the products one at a time, calling on a student to name the source of each product. List the product and its source on the chalkboard. When the list is complete, divide the class into small groups, and have them arrange the materials into a food-web diagram. Conclude by asking: In what ways is this human food web like the "natural" food webs discussed in this section? *(Sunlight is the source of energy; plants are the producers; humans are the "top carnivore" in the food chain; humans might be harmed by a substance like DDT that could become concentrated in the food chain, etc.)*

Extension

Cooperative Learning Activity Divide the class into five groups. Give each group one of the following locations, and tell them that their task is to determine what sort of food web exists in that location: a salt marsh in Florida; a coral reef in the Caribbean; the Sonoran Desert in Arizona; the Kodiak National Wildlife Refuge in Alaska; and the Ft. Pierre National Grassland in South Dakota. Encourage students to determine how humans have affected or been affected by the food web in their location. Suggest that students report their findings in the form of posters, oral presentations, dramatizations, or video documentaries. Encourage group members to include something representative of their project in their Portfolio.

SECTION 3.2

THE CYCLING OF MATERIALS

FOCUS

This section describes the water, carbon, and nitrogen cycles, including information about how human activities affect these cycles.

MOTIVATOR

A week or more before beginning this section, set up a "pond ecosystem" in a large, clean glass jar with a lid. To make your "pond," add the following to the jar: water plants, a snail or two, and pond water (or zooplankton and water). Then tightly seal the lid and place the jar in a sunny window.

When you are ready to begin teaching this lesson, show the class the jar ecosystem, and tell them it has been sealed for more than a week. No air can go in or out, and no food has been given to the organisms in the jar. Ask students: How have the organisms survived? Where do they get their food? their oxygen? their water? *(The animals get their food by nibbling on the plants. The plants get the carbon dioxide they need for photosynthesis from animal and plant cellular respiration. They use this to make glucose and starches. The animals get the oxygen they need to breathe from the byproducts of the plants' photosynthesis. Water is produced as the plants transpire—or give off water—through their leaves.)*

Tell students that our Earth is similar to the jar ecosystem in that all of the chemical elements that are needed to support life are contained in the Earth's soil, water, or atmosphere. The Earth uses the same elements over and over again in a series of cycles. Explain that this section is about three important cycles: the water cycle, the carbon cycle, and the nitrogen cycle.

TEACHING STRATEGIES

THE WATER CYCLE, *page 62*

Using the Transparencies

Figures 3-10, 3-11, and 3-13 are available as Teaching Transparencies.

Background: Biogeochemical Cycles

The cycling of materials is accomplished in huge cycles called biogeochemical cycles. Broken down into its component parts, *biogeochemical* means, literally, "life-earth-chemical." Different elements require different lengths of time to go from the living part of an ecosystem (an organism), to the nonliving part (soil, water, atmosphere, etc.), and back to the body of a living organism again. The amount of each element processed through a biogeochemical cycle is enormous. About 90 million metric tons (100 million tons) of nitrogen is cycled per year—90 percent of it by bacteria and 10 percent by human-produced fertilizers. Over 180 billion metric tons (200 billion tons) of oxygen are released each year by photosynthetic plants and algae, which in turn produce about 180 billion metric tons (200 billion tons) of organic matter. Although animals are usually thought of as returning CO_2 to the air, both plants and decomposers return more CO_2 than animals do (plants, 42 percent; decomposers, 46 percent; animals, 12 percent).

History Connection

Based on the following information, have each student write a short story describing who or what might have used the same carbon or oxygen atom or water molecule that he or she is using now. It takes a carbon atom about 300 years to complete one cycle; an oxygen atom, about 2,000 years; and a water molecule, about 2 million years.

Using the Theme of *Scale and Structure*

Focus Question: How does the structure of an ecosystem influence nutrient cycling? *(Structured interrelationships such as those that exist among soil, water, nutrients, producers, consumers, and decomposers allow for the transfer of nutrients from one thing to another in a chain of events. If one link in the chain is missing, the transfer cannot be completed. For example, in Chapter 12 students will learn that there are no natural decomposers for plastics because plastics are made not by nature, but by humans. As a result, plastics cannot be cycled because they cannot be decomposed.)*

HOW HUMANS ARE AFFECTING THE CARBON CYCLE, *page 63*

Notable Quotes

". . . we can never know how wide a circle of disturbance we produce in the harmonies of nature when we throw the smallest pebble into the ocean of organic life."
George Perkins Marsh

"In nature there are neither rewards nor punishments—there are consequences."
R. G. Ingersoll

THE NITROGEN CYCLE, *page 64*

Using the Diagram: *Figure 3-13, page 65*

Direct students' attention to Figure 3-13 and ask: If only bacteria can use nitrogen from the air, how do plants and animals take part in the nitrogen cycle? *(Plants absorb nitrogen from the soil and use it to make proteins. Proteins are passed to the next trophic level when herbivores eat the plants. Carnivores obtain their nitrogen from the herbivores they eat, and so on. The wastes of animals contain nitrogen that is returned to the soil or water for reuse by plants.)*

Reinforcing Ideas

Obtain samples of garden beans, soy beans, alfalfa, or clover with their root nodules intact. If microscopes are available, have students squash a nodule on a microscope slide and stain it with methylene blue. Then have students apply a cover slip and examine the nitrogen-fixing bacteria. Suggest that they draw what they see. Next, have students dry the plant samples and mount them on poster board along with their drawings of the nitrogen-fixing bacteria and samples or drawings of other members of the nitrogen cycle. They should title their poster The Nitrogen Cycle.

Using the Diagrams: *Figures 3-10, 3-11, and 3-13, pages 63–65*

Have students study the water, carbon, and nitrogen cycles shown on pages 63–65. Ask them to list the ways that the three cycles are similar and different. *(The cycles are similar in that all move their materials between different components of the biosphere, have no beginning or end, and conserve the materials in the cycle. Furthermore, carbon and nitrogen travel along similar pathways: from air to plants to animals and from plants and animals back to the soil and air by way of decomposers. The cycles differ in that the water cycle only slightly involves living organisms; in the nitrogen and carbon cycles, living things play key roles at almost every stage. Furthermore, water remains chemically intact throughout the cycle, in contrast to nitrogen and carbon, both of which become part of a number of different compounds during their respective cycles. Another difference is that plants obtain carbon dioxide from the air but they obtain their nitrogen from the soil.)*

Language-Arts Connection

Write the following poem on the board.

life
bountiful bright
growing cycling changing
sun plant animal nutrient
sinking decaying fading
scarce dark
death

Tell students that this is a poetic form called a diamante. The name reflects its diamondlike shape. Explain that it is a useful form for demonstrating shades of meaning. The poem on the board begins with the concept of life and ends with the concept of death, using the words in-between to connect the two. Ask: Is this poem an adequate summary for this section? Why or why not? *(Yes, because it shows the interconnectedness between living and once-living things.)* Suggest that interested students choose two opposite words that describe some concept they learned in this section and write a diamante of their own. Students should use the following form.

noun
adjective adjective
participle participle participle
noun noun noun noun
participle participle participle
adjective adjective
noun

Encourage students to include their diamante in their Portfolio.

CLOSURE

Section Review

Answers to these questions are provided in the Answer Key on page T144.

Alternative Assessment

A couple of days before you wish to conclude this section, obtain several small potted plants and water them. Then put each plant in a plastic bag large enough to cover the plant completely, and close the bag with a twist tie. Place the plants in a sunny window. When you are ready to conclude this section, bring the plants to class. Divide the class into groups, and give each group one plant. Then ask students to work together to explain how the plant might be seen as a miniature version of the water cycle. *(The plant takes in water from the soil. The water travels up to the leaves, where it is transpired, or "breathed out." The plastic bag prevents the water vapor from escaping—instead it adheres to the surface of the bag until it drips onto the plant or soil. When this happens, the water has made a complete cycle from the soil, to the plant, to the "atmosphere," and to the soil again, similar to the way water travels in the Earth's water cycle.)*

Reteaching

In a large beaker, prepare a 0.1 percent solution of bromothymol blue. (Bromothymol blue is a pH indicator; it is blue in a base and yellow in an acid.) Adjust the color with a small amount of baking soda until the solution turns a deep blue color. Divide the solution into three large test tubes or similar glass containers. Choose three students, and have each of them exhale through a straw into one of the test tubes until the solution in the tube turns yellowish green. Tell the class that the yellowish color indicates that carbon dioxide is present. **Safety Alert:** *Be sure that students do not inhale or ingest any of the water.*

Then: (1) seal the first tube with a test-tube stopper, and place it in a well-lit location; (2) put a sprig of elodea into the second tube; seal it and place it in the same location; (3) prepare the third tube in the same manner as the second, but cover the entire tube with foil, and place it in the same location. Review the carbon cycle with the class while students wait for results. The next day, the tube containing elodea and exposed to light (#2) will turn blue, while tubes #1 and #3 will stay yellowish. Help students use what they've learned about the carbon cycle to explain the results. *(Test tube #2 represents one portion of the carbon cycle. In that tube, the elodea plant took in the carbon dioxide that came from the students' breath and used it in the presence of sunlight as part of the photosynthesis process. If this cycle were complete, the carbon would be passed from the plant to an animal that consumed it. That animal would then release some of the carbon back to the atmosphere in the form of carbon dioxide, just as the student did.)*

Extension

Modern agriculture relies on the use of nitrogen-rich fertilizers. Ask students to find out how these fertilizers are produced and what impact they have on the environment. Suggest that students use their findings to answer the following question: Why would some environmentalists say that nitrogen fertilizers harm the environment twice? *(First, nitrogen fertilizers take energy to make. This energy usually comes from burning fossil fuels; in this way, nitrogen fertilizers affect the carbon cycle and contribute to the greenhouse effect. Second, nitrogen compounds run off fields and into bodies of water, where they can cause an overgrowth of algae and water plants. Those plants eventually die and decay. When this happens, bacteria use oxygen to decompose the plant remains. With an abundance of plant remains, so much oxygen is used by decomposers that there is not enough oxygen for other aquatic animals. Thus, nitrogen fertilizers also affect the oxygen cycle.)*

SECTION 3.3

HOW ECOSYSTEMS CHANGE

FOCUS

This section introduces the concept of succession. It explains the importance of pioneer species and distinguishes between secondary and primary succession. In the discussion of primary succession, students also learn about how soil is formed.

MOTIVATOR

Ask the class to imagine that on graduation day their school is bulldozed and a tall fence is placed around the bare school grounds. Then ask them to picture returning to the school grounds in 1 month, 1 year, 5 years, and 100 years. Call on individual students to describe what they think the school grounds would be like at these time intervals. *(Living things would return to the area, but the types of living things would change over the years. Weeds might be the first to appear, then brush and small trees. Finally, after 100 years, the area would be restored to its most natural state—for, example, in some areas it would be heavily wooded with full-grown trees.)* Explain that this process is called succession and that this section is all about what drives the forces of succession. (An alternative motivator would be to show a film or video about succession, such as *Succession: From Sand Dune to Forest,* which is listed in the Resources section of this Teacher's Edition, page T42.)

TEACHING STRATEGIES

SUCCESSION, *page 66*

Using the Theme of *Patterns of Change*

Focus Question: When clearing land for development, ranching, agriculture, or other uses, why is it important to preserve patches of the original habitat? *(By preserving patches of species, we are improving the chances for an ecosystem to return to its original condition. Those species could recolonize an area after a disturbance. But if patches are not preserved, those species, and their potential value to humans, could be lost forever.)* Tell students that they will learn more about this subject in Chapter 8: Land and Chapter 10: Biodiversity.

SECONDARY SUCCESSION, *page 67*

Using the Transparency

Figure 3-16 is available as a Teaching Transparency.

FIRE-MAINTAINED COMMUNITIES, *page 68*

Using the Case Study: *Fires in Yellowstone, pages 68–69*

Suggest that students divide into two groups and debate the pros and cons of the National Park Service's fire policy.

Extending the Case Study: *Fires in Yellowstone, pages 68–69*

Ask a forest-service firefighter to come to your class. Have students prepare a list of questions in advance. Their questions might include: Do any species benefit from forest fires? How are forest fires fought? What is done to prevent forest fires?

The Nature Conservancy has chapters that, in some states, perform controlled burning to preserve certain climax communities. If such a chapter exists in your area, contact the organization to see if your class could witness or help with a controlled-burn project and learn about succession ecology. If not, see if a representative from the organization could come to the class to speak about the group's work.

Multicultural Connection

Read to students the famous speech of Chief Seattle, a Native American chief of the Dwanish, Suquamish, and other tribes. One version of the speech follows.

You must teach your children that the ground beneath their feet is the ashes of our grandfathers. So that they will respect the land, tell your children that the Earth is rich with the lives of our kin.

Teach your children what we have taught our children, that the Earth is our Mother. Whatever befalls the Earth befalls the sons of the Earth. If men spit upon the ground, they spit upon themselves.

The Earth does not belong to man; man belongs to the Earth. All things are connected like the blood which unites one family.

Man did not weave the web of life; he is merely a strand in it. Whatever he does to the web, he does to himself!

Have students work together in small groups to relate the speech to the ecological principles outlined in this chapter. Students can present their findings in any form: written, visual, or oral. Encourage them to include their findings in their Portfolio.

Meeting Individual Needs

Gifted Learners Tell students to consider the following statement. "An ecosystem has historical aspects; the present is related to the past, and the future is related to the present." Ask: Is this statement true or false? *(The statement is true. The Earth's cycles and the predictable patterns of change such as those evident in succession connect an ecosystem's past to its present.)* Suggest that students think about the statement and then write a short paragraph outlining their thoughts. Encourage students to include their paragraph in their Portfolio.

CLOSURE

Section Review

Answers to these questions are provided in the Answer Key on page T144.

Alternative Assessment

Ask students: How does the complexity of a food web change during the succession process? Suggest that students answer by drawing food webs to reflect each stage in the succession sequence. Then encourage students to include their drawing in their Portfolio.

Reteaching

Place a piece of a rotting log in a terrarium. After giving everyone a chance to observe, touch, and smell the log, use the log as an example of (1) a food chain; (2) the cycling of materials; and (3) succession. *[(1) One possible example is tree → termite → spider → snake. (2) Carbon is released as CO_2 when the animals and decomposers consume the log and each other. Nitrogen and other nutrients are released by decomposers from the log and from animal wastes. (3) As the log rots, different communities of plants, animals, and microbes occupy it. They affect the physical and chemical composition of the log, which affects the types of organisms that can survive there.]*

Extension

Have students observe the process of succession during the school year. Early in the school year, get permission to mark off a section of the school grounds that is approximately 4 ft. × 4 ft. Post "Do Not Mow" signs in the marked-off area. Have students monitor the area by keeping a regular journal of the changes that they observe over time. Their journal can be written or recorded on videotape. Students may also wish to collect and preserve a few specimens of the plants and insects that occupy the area. Be sure that students carefully label their notes and collections with a complete date and a detailed description of their findings. The project could be continued from one school year to the next, with current students building on the documentation of previous students.

Answers

Answers to *Chapter Highlights*, *Chapter Review*, *Investigation*, and *Points of View* are provided in the Answer Key for Chapter 3, which begins on page T144.

TEACHING STRATEGIES FOR INVESTIGATION:

WHAT'S IN AN ECOSYSTEM?

pages 74–75

Advance Preparation

Pick a site that is fairly undisturbed. Tie 10 m sections of string to the stakes ahead of time (use twine if strong string is unavailable) because setting up the plot can be time-consuming. Use a hammer to secure the stakes into the ground. Before beginning the activity, discuss generally with students what organisms they might expect to find at their site and how those organisms might interact, as well as what abiotic factors might come into play.

Extension

Return to the site at a different time of day or during a different season. Often, marked shifts in biotic and abiotic components of the ecosystem can be observed.

CHAPTER 4

KINDS OF ECOSYSTEMS

PLANNING GUIDE

CHAPTER COMPONENTS AND RESOURCES	½ Year Average	½ Year Advanced	1 Year Average	1 Year Advanced
Chapter Opener, p. 78				
EcoLog, p. 78	•	•	•	•
4.1 Forests, pp. 79–89	•	•	•	•
Transparency 10: *Biomes of the World*	•	•	•	•
Transparency 11: *Layers of the Rain Forest*	•	•	•	•
Case Study: *Saving Costa Rican Biomes,* pp. 84–85		•	•	•
Section Review, p. 89	•	•	•	•
4.2 Grasslands, Deserts, and Tundra, pp. 90–99	•	•	•	•
Field Activity, p. 95			•	•
Points of View: *The Future of the Arctic National Wildlife Refuge,* pp. 118–119			•	•
Section Review, p. 99	•	•	•	•
4.3 Freshwater Ecosystems, pp. 100–104	•	•	•	•
Transparency 12: *Zones in a Lake*	•	•	•	•
Section Review, p. 104	•	•	•	•
4.4 Marine Ecosystems, pp. 105–112	•	•	•	•
Transparency 13: *Zones in the Ocean*	•	•	•	•
Section Review, p. 112	•	•	•	•
End of Chapter				
Chapter Highlights, p. 113	•	•	•	•
Chapter Review, pp. 114–115	•	•	•	•
Investigation: *Identify Your Local Biome,* pp. 116–117			•	•
Chapter Tests	•	•	•	•
PACING GUIDE	7–9 days	7–9 days	17–19 days	17–19 days

MATERIALS ORGANIZER

Activity	Materials	Advance Preparation
Investigation: *Identify Your Local Biome,* pp. 116–117	globe or map, field guide to local flora and fauna, binoculars (optional), notebook, pen or pencil, metric ruler, graph paper (optional) (per student group)	Determine a location for conducting this activity. For example, you could choose the schoolyard, a nearby park, or a residential yard. Get permission in advance if necessary.

CHAPTER OVERVIEW

In this chapter students are introduced to an array of land and water ecosystems found throughout the world.

Section 4.1 defines biomes and then discusses various forest ecosystems and the unique adaptations that plants and animals have developed to survive there.

Section 4.2 describes the factors that characterize grassland, desert, and tundra ecosystems as well as the adaptations that allow some living organisms to survive in each ecosystem.

Section 4.3 explores the characteristics of different freshwater ecosystems and compares adaptations of organisms within these ecosystems.

Section 4.4 discusses marine environments and looks at adaptations of some of the unusual organisms found within each.

RESOURCES

. . . for Teachers

American Forestry Association, Global ReLeaf Program, P.O. Box 2000, 1516 P Street, NW, Washington, D.C. 20013; 202-667-3300. (This organization can provide useful information about reforestation, wise forestry management, and tropical forest degradation.)

Cousteau Society, 930 West 21st Street, Norfolk, VA 23517; 804-627-1144. (This organization can provide additional information about marine organisms and ecosystems.)

Koch, Maryjo. *Pond, Lake, River, Sea.* San Francisco, CA: Collins Publishers, 1994. (This very unusual hand-lettered, hand-drawn text colorfully discusses water ecosystems and the creatures that inhabit them.)

. . . for Students

Findley, Rowe. "Will We Save Our Own?" *National Geographic*, Sept. 1990, pp. 106–136. (This informative article presents the past, present, and uncertain future of the old-growth forests of the Pacific Northwest.)

Matthews, Anne. *Where the Buffalo Roam.* New York City, NY: Grove Press, 1992. (This book chronicles a year's journey across the plains of America; it is rich in historical accounts and descriptive tales of today's grassland ecosystems.)

Mitchell, Andrew. *Vanishing Paradise —The Tropical Rainforest.* Woodstock, NY: Overlook Press, 1990. (This is a large book of eye-catching photographs that documents the rain-forest ecosystem.)

Films, Videos, Software, and Other Media

Land Above the Trees. Video. 20 minutes. Bullfrog Films, P.O. Box 149, Oley, PA 19547; 800-543-3764 or 215-779-8226. (This award-winning video provides a beautiful portrait of tundra ecosystems, highlighting the ecological connections and special adaptations of the tundra community.)

On Dry Land: The Desert Biome. CAV Videodisc, Level I, III. Coronet/MTI, P.O. Box 2649, Columbus, OH, 43216; 800-777-8100.(This video explores the major role played by deserts in the Earth's ecosystems.)

The Sea. Video. 27 minutes. Britannica Videos, 310 South Michigan Avenue, Chicago, IL 60604; 800-554-9862. (This video studies ocean-dwelling plants and animals, focusing on their interdependence and their influence on the conditions of the marine environment.)

Getting Involved

A listing of environmental organizations, publications, and volunteer opportunities can be found in the **Laboratory and Field Guide.** Note especially those organizations with the goal of maintaining natural ecosystems, such as the National Wildlife Federation and the Nature Conservancy.

USING THE QUOTATION

"In all things of nature there is something of the marvelous."

Call on one student to read the quotation aloud. Then ask students to write on a slip of paper an interesting fact about nature. *(Example: A horned lizard can squirt blood from its eyes.)* Next, ask students to exchange papers and describe in a few sentences the type of ecosystem that the animal or plant mentioned lives in. *(Sample answer: The horned lizard lives in a desertlike environment, with dry conditions and sparse vegetation.)*

USING THE ECOLOG

To assess students' knowledge about the subject of this chapter, have them answer the EcoLog questions on page 78. In answering these questions, students will reveal their prior knowledge, which is usually very basic and often flawed or even misconceived. Then, by revising their answers at the conclusion of the chapter, students will have the opportunity to eliminate any misconceptions and to see firsthand how their knowledge has grown.

SECTION 4.1

FORESTS

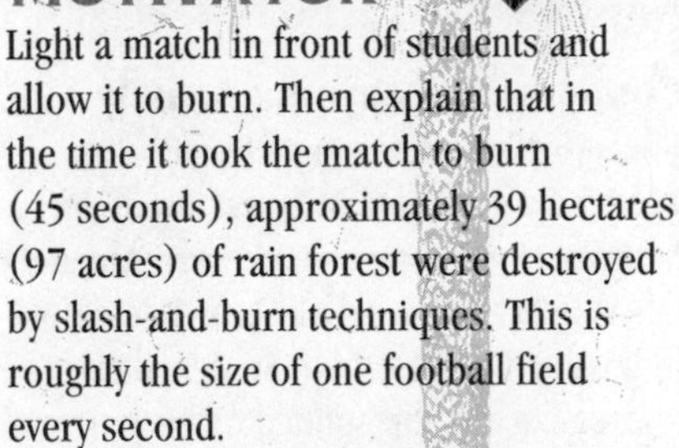

FOCUS
A descriptive tour through the tropical rain forest, temperate rain forest, temperate deciduous forest, and taiga introduces students to the characteristics of each type of ecosystem and to the concepts of biomes, climate, and plant and animal adaptations. Students also learn about the current threats facing tropical rain forests.

MOTIVATOR
Light a match in front of students and allow it to burn. Then explain that in the time it took the match to burn (45 seconds), approximately 39 hectares (97 acres) of rain forest were destroyed by slash-and-burn techniques. This is roughly the size of one football field every second.

TEACHING STRATEGIES

Using the Transparency
Figure 4-1 is available as a Teaching Transparency.

Reinforcing Ideas
Students may have difficulty distinguishing between the terms *biome* and *ecosystem*. This is not an easy distinction—in fact, the terms are not consistently employed even among scientists. Each term is defined by both abiotic and biotic factors, and each is characterized by the types of interactions among its component life-forms. To help students understand the difference, discuss the following hierarchy. The biosphere is divided into biomes, which are, in turn, divided into smaller ecosystems. Biomes, then, typically consist of separate but similar ecosystems.

TROPICAL RAIN FORESTS, *page 80*

Using the Transparency
Figure 4-4 is available as a Teaching Transparency

Background: The Limiting Factor Principle
Ecosystems operate on a principle known as the *limiting factor principle.* This principle states that too much or too little of an abiotic factor can limit or prevent the growth of a population of a species, even if all of the other abiotic factors are present in sufficient amounts. Limiting factors include temperature, water, light, and soil nutrients. After students have read about tropical rain forests, ask: Which of these factors is most likely to be a limiting factor in the tropical rain forest? *(Because the climate is very stable, with steady temperatures and rainfall, the limiting factor is most likely the lack of nutrients in the soil.)* As you proceed through the chapter, have students identify the limiting factor for each ecosystem encountered.

Field Activity

Forests throughout the world have many things in common, such as the diversity of their living inhabitants. Their soils, on the other hand, can be strikingly different. Each type of forest tends to have its own characteristic soil type. For example, the soil in tropical rain forests is acidic, with few nutrients. If possible, have students investigate the soil in two forested areas. Each region should feature a different dominant tree species. In each area have students dig a hole and measure a cross section of the soil. They should also draw a picture that illustrates the layers they see, and then use their drawings to compare the two different areas. Encourage sudents to note any small organisms that they uncover (without disturbing the organisms). As an extension, have students send their soil samples to a nursery to be analyzed for minerals and pH levels. Suggest that students include their field notes, drawings, and soil-test results in their Portfolio.

Using the Photograph:
Figure 4-5, page 82
Have students examine the photograph of the bat and the flower. Ask: What characteristics do these species have that allow them to thrive in the rain-forest environment? *(The bat and the flower are coadaped to each other—note the close fit between the shape of the flower and the shape of the bat's head. The bat gains a necessary resource from the mutualistic relationship, while the plant benefits by being pollinated.)*

Using the Theme of *Stability*
Focus Question: Scientists have long debated the reason for the great biodiversity that is found in tropical rain forests. One theory suggests that it is a combination of energy in the form of sunlight, available area, and stability of the climate. Why would stability of the climate promote the evolution of biodiversity? *(In temperate and polar regions, organisms are adapted to a wide range of seasonal variations in temperature. These organisms are then capable of occupying a larger geographical area. In contrast, the relatively stable climate in the tropics allows species to specialize in narrower pieces of the environment in order to outcompete their neighbors. It is this high degree of specialization that makes rain-forest inhabitants vulnerable to any changes.)*

Multicultural Connection
Point out to students that the destruction

of many ecosystems threatens the livelihood of groups of many indigenous people. Have students do research on some of these groups, such as the Yanomami of the rain forests in Brazil. Ask: How did these groups adapt to the ecosystem where they lived? What are the threats to their culture and way of life? Have students write a short report and share their findings with the rest of the class. Encourage students to include their report in their Portfolio.

Notable Quotes

"There is more information of a higher order and complexity stored in a few square kilometers of forest than there is in all the libraries of mankind."
Eugene Odum

"Humanity is cutting down its forests, apparently oblivious to the fact that we may not be able to live without them."
Isaac Asimov

Multicultural Connection

Many rain-forest fruits, such as kiwi and carambola ("star fruit"), are considered exotic throughout much of the United States. Most of these products, however, are quite common in other, more tropical, areas of the world. If possible, bring samples of rain-forest fruits (which are usually available at specialty grocery stores) or other rain-forest products to the classroom. Find out in advance where each product comes from and how it is used, and share this information with students. Some students, especially those whose families come from tropical areas, may regularly use a variety of rain-forest products. If so, suggest that they share their knowledge with the class. Invite students to include something from their presentation in their Portfolio.

Research Activity

Tell students that there are many everyday products that come from the rain forest. We take many of these products, such as coffee, bananas, and chocolate, for granted. Suggest that students research some of these products and share their findings with the class. Direct student thinking by providing the following categories: woods, houseplants, spices, fruits, vegetables, other foods, fibers, gums and resins, pharmaceuticals, and oils. Encourage students to show some rain-forest products when presenting their findings to the class. Invite students to include their findings in their Portfolio.

PLANT ADAPTATIONS, *page 82*

Biology Connection

As pointed out in the text, the soils of tropical rain forests are extremely poor and thin. As a result, plants must derive their nutrients from decaying matter on the forest floor. But how do the plants get these nutrients? The answer lies in a symbiotic relationship the plants have with mycorrhizal fungi. The fungus aids in the direct transfer of nutrients such as phosphorus, zinc, and copper. In return, the plant supplies organic carbon to the fungus. Although this type of relationship exists in other communities between plants and fungi, it is especially important in the nutrient-poor soils of the rain forest.

Cooperative Learning Activity

Divide students into four groups, and provide each group with a large sheet of poster board or butcher paper. Each sheet should be labeled with a type of forest biome discussed in this section. Ask students to illustrate in light colors the *vegetation* that inhabits the listed forest biome. Then have the groups switch posters and add, in darker colors, the appropriate *animals* for the biome. Have students switch illustrations once again, this time describing the *interrelationships* between the organisms shown. Finally, students could switch posters and label and describe some of the organisms' *adaptations*.

ANIMAL ADAPTATIONS, *page 82*

Meeting Individual Needs

Learners Having Difficulty To help students understand how animals are specially adapted to their environment, share the following example. Many primate species, such as monkeys, have a special adaptation for climbing trees, peeling fruit, and holding objects—their hands are equipped with opposable thumbs. Since humans are also primates, we share the characteristic of opposable thumbs. Conduct the following activity to show students what it is like not to have this adaptation. Tape the thumb of each student to the rest of his or her hand (making the thumbs immobile). You will, in effect, be creating "paws." Next, challenge students to crack open a peanut, peel a banana, or open a door. Have students discuss the difficulty primates would have living in the forest without their opposable thumbs.

TEMPERATE RAIN FORESTS, *page 84*

Cooperative Learning Activity

Recently the northern spotted owl has received a great deal of attention in the battle between loggers and environmentalists over the fate of old-growth forests in the Pacific Northwest. Students may not realize that the spotted owl is part of a complicated food web containing martens, flying squirrels, red-backed voles, Douglas firs, pseudoscorpions, termites, black and yellow mycorrhizal fungi, and spotted owls. Have students break into teams, with each team investigating the ecological niche of one of those organisms. Then have students work together to make a bulletin-board display. At the center of the display, place an illustration or photograph of a tree from an old-growth forest. Groups should add a picture of their organism to the display along with a written description of its niche and of its interaction with the tree.

Using the Case Study: *Saving Costa Rican Biomes, pages 84–85*

While the Costa Rican program has been a success and a model for other countries, point out that Costa Rica is a special case. It is a relatively prosperous and educated country, and it is largely free of the poverty and civil strife that typify many other Central and South American countries. For other countries, the setting aside of large amounts of land as ecological reserves may not be politically or economically realistic. Such an action might be a luxury that these countries could ill afford. Ask: Why would a poverty-stricken country be less likely to follow the Costa Rican example? *(Population pressure and poverty would force people to use whatever resources were available, including sensitive lands.)* Ask: What can we, as residents in developed nations, do to help less developed nations follow the Costa Rican example? *(Answers will vary. Some suggestions include offering grants, trading credits, economic assistance, or buying the land outright.)*

TEMPERATE DECIDUOUS FORESTS, *page 85*

History Connection

John Muir (1838–1914) was an American naturalist, explorer, and writer. His conservation efforts included helping persuade Congress to establish both Yosemite and Sequoia National Parks in 1890 and founding the Sierra Club in 1892. He traveled to many parts of the world, and is known for his work explaining Yosemite's glacial origins, as well as for his discovery of a glacier in Alaska that bears his name. After immigrating to the United States from Scotland at age 11, he developed his interests in botany and geology while growing up on a farm and later while attending the University of Wisconsin. Muir Woods, a redwood forest near San Francisco, was named after him in 1908. His books include *The Mountains of California* (1894), *Our National Parks* (1901), and *The Yosemite* (1912).

CLOSURE

Section Review

Answers to these questions are provided in the Answer Key on page T145.

Alternative Assessment

Divide the class into groups of five or six students. Give each group a detailed picture of a forest biome and ask them to describe the characteristics of the biome. *(Answers should include a description of where the biome is located, what types of animal and plant species are unique to it, and what special adaptations allow those organisms to survive there.)*

Reteaching

Some students may have trouble distinguishing between the different forest biomes and remembering where they are located. Ask volunteers to prepare a three to five minute special news report featuring each of the different forest biomes and the current threats or dangers facing it. Have students include the characteristics and location of each forest biome. Then have students present their news flashes to the class. Encourage all students to take notes and include them in their Portfolio.

Extension

Have students develop a travel brochure that advertises camping vacations in one of the four forest ecosystems. In addition to making a list of the supplies that campers will need, students should provide information about the climate (including any important seasonal variations such as a rainy season), a description of the plants and animals that campers might encounter (including any that might be dangerous or endangered), and recommendations on how campers can lessen the environmental impact of their activities. The brochures should include photos or illustrations, as well as a map showing the location of the forest described. Encourage students to include their brochure in their Portfolio.

S E C T I O N 4.2

GRASSLANDS, DESERTS, AND TUNDRA

FOCUS

Students are introduced to the factors that shape and characterize grassland, desert, and tundra ecosystems. They learn how the adaptations of plants and animals in each ecosystem help them to survive.

MOTIVATOR

About two weeks before beginning this section, plant grass in several aluminum cake pans. (A variety of grass seeds are available at most nurseries; pick a type that is native to your region.) Divide the class into groups and give each group a pan of the germinated grass. Tell students to carefully observe and record details about the grass's physical characteristics, characteristics about the soil, notes about some of the grass's adaptations, etc. Ask: How is the pan of grass similar to a real grassland ecosystem? *(The major flora in a grassland ecosystem is grass, and climatic conditions in the classroom may be similar to those in a grassland.)* How are they different? *(A real grassland is much more complex in terms of plant life, animal life, and biological interactions.)*

TEACHING STRATEGIES

TROPICAL SAVANNAS, *page 90*

Background: The Serengeti

Unlike tropical rain forests, where large mammals tend to be scarce, the tropical savanna is unparalleled in its population and diversity of large mammals. The major predators in the biome are lions, leopards, cheetahs, hyenas, and wild dogs. Each group of predators concentrates on a specific kind of prey. For example, lions hunt zebras, wildebeests, giraffes, and buffalo. Unfortunately, the tropical savannas are under assault from overgrazing and farming activities. Cattle, which are not native to the savannas, degrade the environment by making daily treks to watering holes. It is hoped that some African savannas will be preserved as game farms. Studies have shown that when herds of native antelopes are raised instead of cattle, ranchers receive a much higher meat yield. Also, antelopes need much less water and do not degrade the environment as much.

TEMPERATE GRASSLANDS: PRAIRIES, STEPPES, AND PAMPAS, *page 92*

Multicultural Connection

Students may be interested to know that the Native Americans who inhabited the plains of North America discovered an ingenious method for escaping prairie fires. They would start another small fire near their community. The small fire would burn out a safety zone in a short time. When the large fire bore down upon the community, it would pass by the already burned area. This method of escape was passed on to European settlers on the plains.

Using the Theme of *Interacting Systems*

When cattle or other nonnative species are introduced to grasslands, the result is often degradation of the environment. *Focus Question:* Why would grazers native to a grassland have less of an impact on the area than introduced grazers? *(When a community has remained stable for a long period of time, the animals and plants that comprise it are typically well adapted to each other's presence. A native grass, for instance, will have developed the ability to survive the amount of grazing that is normal for the region's native herbivores [otherwise it would have gone extinct in that region]. When a new species is introduced, different consumption patterns might be too severe for a native grass to withstand. If it is not able to adapt, the grass would probably die out. The loss of the grass species could then affect the entire ecosystem.)*

DESERTS, *page 94*

Background: Xerophytes and Their Special Adaptations to Dry Biomes

In all dry regions of the world, particularly deserts, there are plants that have evolved strategies that enable them to cope with water shortages. These plants are called *xerophytes*. They include cacti, euphorbs, and similar plants with thick, succulent stems that enable them to store water. Some xerophytes have evolved mechanisms that permit the curling or folding of leaves during times of water stress. This allows them to reduce the amount of leaf surface exposed to sunlight (and thus subject to water loss). Other xerophytes have reduced or eliminated their leaf surfaces altogether. Still others have developed extremely short reproductive cycles. They can take advantage of an infrequent rainfall by germinating, flowering, and producing seeds in about two weeks.

THREATS TO DESERTS, *page 97*

Reinforcing Ideas

Tell students that humankind does its share of damage to the desert ecosystem, both intentional and unintentional. Encourage discussion by asking: What effects could all-terrain vehicles have on a desert ecosystem? *(crushed vegetation, the destruction of bird eggs laid on the ground, pollution from leaking motor oil and discarded trash, etc.)* Next ask: Do you think that people should be allowed to seek out recreational opportunities wherever they want? *(Answers will vary—accept all reasonable responses.)*

Multicultural Connection

Just as plants and animals are adapted to desert life, people have also found a way to survive there. Have students do research on groups of people who live in the world's deserts, such as the San of the Kalahari, the Tuareg of the Sahara, or the Native Americans of the Southwest. Ask: How have these people adapted to survive in deserts? Suggest that students report their findings to the class and then include their report in their Portfolio.

TUNDRA, *page 97*

Art Connection

Review with students the traits that characterize the tundra as emphasized in this section. *(extreme cold, scant rainfall, sunless winter, permafrost, large numbers of swarming insects in the summer, and small plants with shallow roots)* Next, have students create a new living organism that could thrive in the Arctic tundra and then write an essay about their creature and its adaptations. *(Students should describe the creature's looks, size, color, fur, defense mechanisms, diet, and reproductive strategy.)* Encourage students to draw a sketch to accompany their essay. Share the new creatures with other class members, and then have students add their work to their Portfolio.

Using the Points of View:
The Future of the Arctic National Wildlife Refuge, pages 118–119
Have students work in groups to create an outline of the main points of view expressed in this feature. Once they have completed this task, ask them to analyze whether each point is adequately answered by the opponent's side. (For instance, the second position's response to the concern over the caribous is to note that the caribou population actually increased in the Prudhoe Bay area.) Suggest that students refer to Figure 1-24 to identify the type of value being invoked (environmental, economic, aesthetic, etc.). Then, for each point, have students decide which point of view is the most persuasive. *(Stress that there is often no right answer.)* Finally, based on their analysis, student groups should discuss the debate and decide which point of view is the most convincing. Then encourage each group member to write a short statement explaining which point of view is closest to his or her own belief about the situation.

CLOSURE

Section Review
Answers to these questions are provided in the Answer Key on page T146.

Alternative Assessment
Tell students to imagine that they are leading authorities on grassland, desert, and tundra ecosystems and that they have just been contacted by a major TV talk-show host to appear nationwide. The host will ask them five questions, which have been forwarded to the "authorities" in advance so that they may prepare their answers (see below). Give students two to three days, and then stage a mock interview where you are the TV host. Students may wish to create a videotape for their Portfolio.

Questions From TV Host
1. What can you tell our viewers about the extreme conditions of the tundra ecosystem? *(Answers will vary, but students should mention cold temperatures, scant rainfall, large numbers of insects in the summer, a permafrost layer on the ground, and low or small plants with shallow roots.)*
2. After visiting the tundra, spending time in the desert must have been a complete culture shock. Is this true, and why or why not? *(Answers will vary, but most students will say it's true, in that the temperatures are vastly opposite. Also, the plant and animal life are vastly different. Students may comment that both regions are similar in that neither experiences much rainfall.)*
3. Did you find different plant adaptations between the tundra foliage and the plants of the grasslands? *(Basically, tundra plants are low to the ground with shallow roots, while grasses are often tall with extended root systems.)*
4. Is there anything all three ecosystems have in common? *(yes, little rainfall)*
5. What was the most unusual plant or animal that you encountered in any of the three ecosystems, and how was it adapted to survive in its environment? *(Answers will vary.)*

Reteaching
Ask students to cut representative pictures from science or nature magazines and use them to create a colorful collage of a grassland, desert, or tundra ecosystem. Hang the collages in the classroom, and suggest that students include their collage in their Portfolio.

Extension 
Have students collect a variety of seeds from grasses native to their area and display the seeds on a poster board. Students could also write a short report to accompany their display. The report should indicate the major characteristics of each type of grass represented. Suggest that students include their display in their Portfolio.

SECTION 4.3

FRESHWATER ECOSYSTEMS

FOCUS

In this section, students are introduced to different freshwater ecosystems and their characteristic traits. Students also learn the survival adaptations of living organisms found in both moving and standing freshwater ecosystems.

MOTIVATOR

Two weeks ahead of time, fill two small containers with pond water that contains some algae. Add a dropperful of concentrated fertilizer to one, and let both containers sit on a windowsill or under a grow light for two weeks. Show the results to students and tell them that each sample contains pond water from the same source. Ask: What difference do you see between the two samples? *(The growth of algae in the sample with fertilizer should far exceed that in the sample without fertilizer.)* What could cause the differences that you see? *(Accept all reasonable responses, but explain to students that the difference is caused by fertilizer, which often gets into our water supply through runoff from lawns, gardens, and farms.)* Ask students: Do you think this is an unhealthy situation? *(Make sure students understand that fertilizer is one of the main pollutants of our freshwater systems. This is because it encourages rapid algal growth, and the algae then use up available nutrients. Algae overpopulation causes the algae to die in large numbers, and the decaying organisms use up all available oxygen. This could lead to the destruction of the lake or pond ecosystem, including all of its inhabitants.)*

TEACHING STRATEGIES

LAKES AND PONDS, *page 100*

Using the Transparency

Figure 4-38 is available as a Teaching Transparency.

Chemistry Connection

The molecular structure of water is such that it gives water some very unusual properties. For example, the solid form of water is less dense than the liquid. This is *not* true for almost all other substances. If water did not have this unique property, almost no life would exist in water outside the tropics. That is because in the winter, ice would sink, and lakes and ponds would freeze solid, killing all of the inhabitants.

Earth Science Connection

Massive sheets of ice once covered much of the Earth. As these glaciers retreated, they often gouged deep holes in the earth. Over time, many of these basins filled with water and became lakes. The Great Salt Lake in Utah is one example. It is the largest lake in the western United States, and it is also the saltiest. This is because the lake is the final destination for several streams and rivers, which deposit water and natural salts into the lake. Although some of that water evaporates, the salts are left behind. Thus, the lake becomes increasingly salty over time.

WETLANDS, *page 102*

Background: The Everglades

Nearly 90 years ago, an enthusiastic scientist with the United States Department of Forestry received a small packet of seeds in the mail from a colleague in Australia. The seeds were of the melaleuca tree. The scientist scattered the seeds across the Everglades in Florida, hoping that the water-loving trees would dry up the "mucky wasteland" to make way for agricultural and residential development. Since then, the melaleuca has run wild across thousands of acres of South Florida wetlands. The tree has no natural enemies there and has driven out many native plants and animals. Scientists are currently trying to eradicate the tree by experimenting with another import from Australia—a melaleuca-eating insect.

Research Activity

Florida is home to a diversity of wetlands, including mangrove swamps, salt marshes, wet prairies, freshwater marshes, cypress and tupelo swamps, and bottomland hardwood forests. The state contains more wetlands—over 11 million acres covering nearly 30 percent of the land—than any other state except Alaska. Ask students to research the unique mangrove swamps of Florida and the wildlife that live there. Encourage students to present their findings to the class in some creative way, and then include the findings in their Portfolio.

RIVERS, *page 103*

Using the Illustration: *Figure 4-45, page 104*

Tell students to observe the illustration of the three different arrowhead plants. Then ask: How does the plant adapt to the speed of the river's current that it grows in? *(In fast-flowing water, its leaves become long, narrow, and submerged, so racing water can flow over them; in a slower river, the arrowhead plant may produce rounded, floating leaves; in a calm river, the leaves of the plant are arrow shaped and remain above the water.)*

CLOSURE

Section Review

Answers to these questions are provided in the Answer Key on page T146.

Alternative Assessment

Have students imagine that they are goldfish journeying through three freshwater ecosystems: a pond, a lake, and a river. Ask students to describe what they see, including details about the characteristics of each body of water and the organisms they encounter along the way. *(Sample answer: The pond is shallow and calm. Light penetrates all the way to the bottom and plants are everywhere. I eat some insects for lunch, and in turn, I am almost eaten by a larger fish. Moments later, a hawk tries to grab me, but I just barely escape. The pond is not very big, and I've soon traveled all the way across it. I feel a patch of cooler water and follow it to its source; I have to swim harder to fight the current as I enter the river...)* Encourage students to include their description in their Portfolio.

Reteaching

Take students to a local pond, river, lake, or bog. Discuss the traits of the ecosystem and answer any questions that students may have.

Extension

Tell students to suppose that they are going fishing in a small freshwater pond located in their home state. Have them do research and then write a short paragraph about the types of fish they are likely to catch and the type of bait that they will need. Suggest that, if possible, students follow up their research with an actual fishing trip. They could then compare the results of the trip with research findings. Suggest that students include a description of their findings in their Portfolio.

Marine Ecosystems

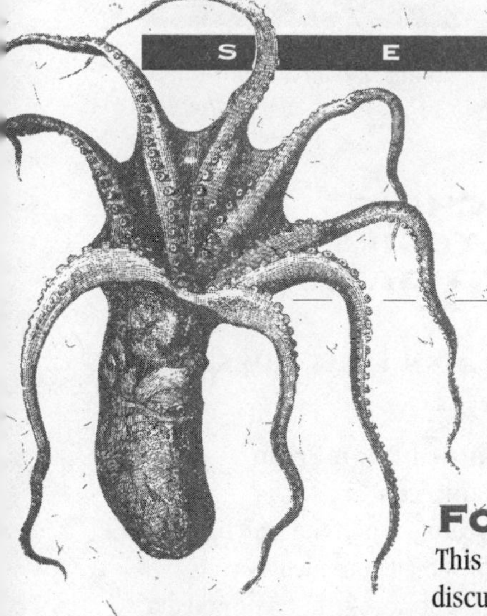

Focus

This section discusses the characteristics of marine ecosystems such as estuaries, coral reefs, and open oceans. Students also explore the adaptations of the organisms that inhabit each ecosystem and current threats to the ecosystems.

Motivator

At the beginning of class, have students close their eyes and sit quietly. Play a brief audiotape of a dolphin or whale song. Afterward, ask students: What is the largest animal on Earth? *(the blue whale)* Inform students that blue whales can weigh over 110 metric tons (over 242,000 lb.), which is about 15 times the weight of the largest African elephants. It is also larger than any dinosaur. Ask: Why can't animals of that size live on land? *(They would not be able to support their weight.)* Can you think of other organisms that might form part of the blue whale's marine community? *(Answers will vary. Blue whales eat small crustaceans called krill, the community's principal producers are phytoplankton, and fish feed on the phytoplankton and each other.)*

Teaching Strategies

Estuaries, *page 105*

Using the Diagram: *Figure 4-47, page 106*
Direct students' attention to the diagram. Ask: What unique challenges does an estuary ecosystem present to organisms? *(Sample answer: All organisms must be adapted to cope with varying levels of salinity.)*

Coral Reefs, *page 107*

Using the Map: *Figure 4-49, page 107*
Tell students to study the map showing the coral reefs of the world. Then ask: "Why are there no coral reefs located off the southern tips of South America, Africa, and Australia?" *(Those areas are too cold—corals like warm, well-lit waters.)*

Meeting Individual Needs
Second-Language Learners In this activity students will illustrate a coral reef for the classroom. To begin, tape a large sheet of butcher paper to one wall of the classroom. Label it "Coral Reef" and draw an outline of a coral-reef formation on the paper. Provide students with construction paper, scissors, markers, tape, glue, and pictures or drawings of coral reef inhabitants. Instruct students to pair up and select a living organism that is associated with the coral reefs. Have them illustrate or construct their animal, cut it out, and glue or tape it to the appropriate location on the "Coral Reef." Examples may include sharks, fish, eels, shrimp, sponges, rays, sea anemones, and urchins. When the project is finished, have students comment on the diversity of living things found in and around a coral reef. Suggest that students document their comments and include them in their Portfolio.

Real-World Connection
If possible, provide actual samples of coral (nonliving), shells, and preserved fish for students to observe. Encourage discussion about how each organism was adapted to survive in its ecosystem. You may also wish to arrange a field trip to a local aquarium or pet shop to study a miniature living marine biome.

The Ocean, *page 108*

Using the Transparency
Figure 4-53 is available as a Teaching Transparency.

Geography Connection
Throughout history, a disproportionate amount of human settlement has occurred in coastal areas. There are several reasons for this. First, having access to a body of water facilitates travel and trade with other areas. In addition, the fish and other organisms in oceans represent a valuable and readily available resource. Also, the climate in coastal areas is usually moderated by the nearby water body. It is interesting to note that, historically, the ocean and ocean-dwelling organisms have figured prominently in human culture. For instance, the ocean is a powerful symbol in many religions, and cowrie shells were perhaps the first form of established currency.

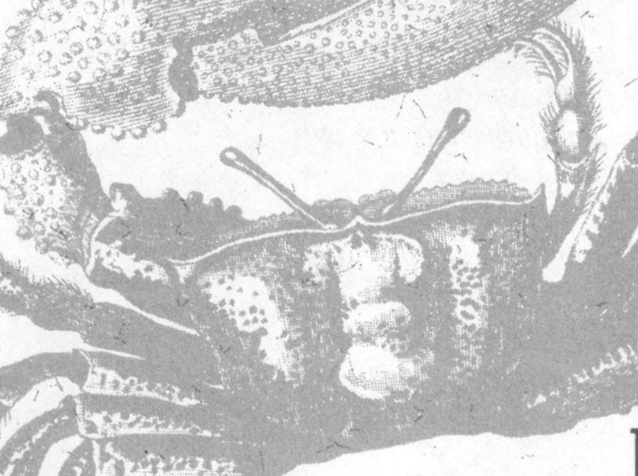

Research Activity
Have students do research and then write a comparative essay of the major oceans of the world. Ask them to include descriptive examples of living organisms unique to each ocean. Students could include their essay in their Portfolio.

Background: Vision in Deep-Sea Fish
The deep ocean is inhabited by many strange vertebrates such as the gulper eel, ceratoid angler fish, and chimaera shark. If not for bioluminescence, many deep-sea fish would be blind due to the absence of light and the impenetrable darkness of the benthic abyss. Perhaps the strangest of all deep-sea fish is the ipnops. Its eyes are found on the top of its head. But the shape of the eyes of this and many other deep-sea fish has undergone an important change; the eyes have become tubular, a configuration they share with a far different creature but one that also must capture as much light as possible, the nocturnal owl. But unlike the owl, deep-sea fish have developed an additional retina to cope with the dark ocean depths.

Notable Quotes
"Roll on, thou deep and dark
blue ocean—roll!
Ten thousand fleets sweep over
thee in vain;
Man marks the earth with
Ruin—his control
Stops the shore."
"George Gordon," Lord Byron

"The sea is tired. It's throwing back at us what we're throwing in there."
Senator Frank Lautenberg

CLOSURE

Section Review
Answers to these questions are provided in the Answer Key on page T146.

Alternative Assessment
Have students write and perform a skit in which eight to ten students operate a special submarine that takes a small group of guests (the rest of the class) on an underwater tour through the different depths of the ocean (from the surface to the benthic zone). Then have each student in the crew describe characteristics of the ocean at each level and facts about organisms they encounter. Have the guests devise questions for the crew members.

Reteaching
Show students a video about marine ecosystems. Two recommendations are *The Sea* (listed in the Resources section of this chapter, page T50) and *The Living Ocean.* The latter video highlights ocean inhabitants, discusses ocean formation, and examines the role of oceans in the biosphere. This video can be ordered from National Geographic Educational Services, P.O. Box 98019, Washington, D.C. 20090-8019; 800-368-2728.

Extension
Divide the class into three groups. Assign each group a portion of Section 4.4 to teach to the class. Group one would have the responsibility of teaching about the estuary biome. Group two would teach about coral reefs. Group three would teach about the ocean. Tell each group they must create a lesson plan for their topic that includes objectives, the actual lesson, a handout (a daily assignment), and a visual aid. Give students one or two days to prepare; then have students present their lesson to the class. Suggest that students include their lesson plan in their Portfolio.

Answers
Answers to *Chapter Highlights, Chapter Review, Investigation,* and *Points of View* are provided in the Answer Key for Chapter 4, which begins on page T145.

TEACHING STRATEGIES FOR INVESTIGATION:

IDENTIFY YOUR LOCAL BIOME, *pages 116–117*

Collect Information From Local Resources
Should the resolution of the map in the text prove insufficient to identify your local biome, check your library or media center for more detailed sources. Or ask a biology or environmental science professor at a nearby college. You may also use the deductive approach to determining your biome. Ask: What kind of climate do we have? What is our primary type of natural vegetation? Parks and preserves are good places to get an idea of the natural vegetation that was once widespread in your area.

A good source for climate data is the local weather service office. Any United States (not world) atlas will generally have comprehensive month-by-month climate data. If you live in a rural area for which information is not available, obtain data from three or four surrounding cities and interpolate from this data. To convert Fahrenheit to Celsius, subtract 32 and then multiply by $\frac{5}{9}$. To convert inches to centimeters, multiply by 2.54.

Perform Field Observations
In some cases, natural areas may be hard to find. In such cases, suggest that students visit an undeveloped preserve or park area. Remind students not to enter private property without permission from the owner.

CHAPTER 5

WATER

PLANNING GUIDE

CHAPTER COMPONENTS AND RESOURCES	½ Year Average	½ Year Advanced	1 Year Average	1 Year Advanced
Chapter Opener, p. 120				
EcoLog, p. 120	•	•	•	•
5.1 Our Water Resources, pp. 121–130	•	•	•	•
Environmental Careers: *Fish and Wildlife Trooper*, pp. 378–379			•	•
Transparency 14: *City Water-Treatment System*	•	•	•	•
Case Study: *The Ogallala Aquifer: An Underground Treasure*, pp. 126–127	•	•	•	•
Transparency 15: *Aquifer—An Underground Water Source*	•	•	•	•
EcoSkills: *Flushing Less Water*, p. 404			•	•
Section Review, p. 130	•	•	•	•
5.2 Freshwater Pollution, pp. 131–139	•	•	•	•
Field Activity, p. 133			•	•
Transparency 16: *Nonpoint Pollution at Home*	•	•	•	•
Case Study: *The Fight Over Pigeon River*, pp. 134–135	•	•	•	•
Making a Difference: *High School Chemist*, pp. 148–149			•	•
Transparency 17: *How Pollutants Get Into Groundwater*	•	•	•	•
Section Review, p. 139	•	•	•	•
5.3 Ocean Pollution, pp. 140–142				•
Field Activity, p. 141				•
Section Review, p. 142				•
End of Chapter				
Chapter Highlights, p. 143	•	•	•	•
Chapter Review, pp. 144–145	•	•	•	•
Investigation: *How Safe Is Our Groundwater?* pp. 146–147	•	•	•	•
Videodiscovery STS Science Forums: Vol. 1—*Water & Growth;* Vol. 2—*Chemicals in Drinking Water*			•	•
Chapter Tests	•	•	•	•
PACING GUIDE	7–9 days	6–8 days	14–16 days	14–16 days

MATERIALS ORGANIZER

Activity	Materials	Advance Preparation
Investigation: *How Safe Is Our Groundwater?* pp. 146–147	250 mL beakers (5); wax pencil; glucose solution; red food coloring; graduated cylinder; stirring rod; glucose test paper; 12 oz. plastic cups (4); thumbtack; gravel; sand; soil; metric ruler; tap water; notebook; pen or pencil; goggles; lab apron; optional substances: motor or cooking oil, vinegar, fertilizer, detergent, antifreeze (per student group)	Prepare the glucose solution by adding one glucose tablet to 200 mL water. Glucose tablets and test paper are available at most pharmacies.

CHAPTER OVERVIEW

In this chapter students are introduced to our water resources and the dangers that confront them.

Section 5.1 presents an overview of the various water resources.

Section 5.2 examines the problem of freshwater pollution.

Section 5.3 examines the problem of ocean pollution.

RESOURCES

. . . for Teachers

Olson, Erik D. *What's in the Water? A State-by-State Report on Groundwater Quality.* Washington, D.C.: Island Press, 1992. (This book, sponsored by the National Wildlife Federation, presents a state-by-state review of groundwater quality and pollution-protection programs.)

Save Our Streams Adoption Kit. Izaak Walton League of America, 707 Conservation Lane, Gaithersburg, MD 20878; 301-548-0150. (This information kit contains a wealth of fact sheets, brochures, pollution-prevention guides, references, and teaching strategies. It also provides data sheets and instructions for classroom water-quality testing projects and outlines how a class can raise money for various water projects.)

"Water: The Power, Promise, and Turmoil of North America's Fresh Water." *National Geographic Special Edition,* Nov. 1993. (The entire issue is devoted to water-related issues concerning North America.)

. . . for Students

la Riviére, J. W. Maurits. "Threats to the World's Water." *Scientific American,* Sept. 1989, pp. 80–94. (This article discusses how ignorance, overuse, and mismanagement have endangered many of the world's water resources. Unless proper steps are taken, severe shortages could soon occur.)

Postel, Sandra. *Last Oasis: Facing Water Scarcity.* New York City, NY: W. W. Norton & Company, 1992. (Part of the World-watch Environmental Alert Series, this book discusses the relationship between society and its water resources.)

Vesilind, Priit. "The Middle East Water—Critical Resource." *National Geographic,* May 1993, pp. 38–71. (This article discusses the tensions related to water use in the Middle East, and explains that maintaining adequate water supplies requires the nations in the Middle East to cooperate; other-wise, war could result.)

Films, Videos, Software, and Other Media

The Power of Water. Video. 59 minutes. National Geographic Educational Services, P.O. Box 98019, Washington, D.C. 20090-8019; 800-368-2728. (This former TV special shows some of the problems surrounding freshwater supplies and demonstrates what people are doing to solve those problems.)

When the Spill Hit Homer. Video. 27 minutes. The Video Project, 5332 College Avenue, Suite 101, Oakland, CA 94618; 800-4-PLANET. (This acclaimed video provides a firsthand account of how the *Exxon Valdez* oil spill affected the residents of Homer, Alaska.)

Getting Involved

- Use the following EcoSkills activity to enhance student interest in the subject of this chapter: *Flushing Less Water,* p. 404.
- A listing of environmental organizations, publications, and volunteer opportunities can be found in the **Laboratory and Field Guide.** Note especially those organizations with the goal of protecting water resources and marine life, such as American Rivers, American Water Resources Association, Clean Water Action, and the Cousteau Society.

USING THE QUOTATION

"All is born of water; all is sustained by water."

Encourage class discussion by asking what Goethe might have meant by this statement. *(Answers will vary. Sample answer: He might have meant that all life is dependent on water.)* Ask: How does the quotation stress the importance between our water resources and how we maintain them? *(Because all life depends on water, it is important that we do not overuse or pollute our water resources.)*

USING THE ECOLOG

To assess students' knowledge about the subject of this chapter, have them answer the EcoLog questions on page 120. In answering these questions, students will reveal their prior knowledge, which is usually very basic and often flawed or even misconceived. Then, by revising their answers at the conclusion of the chapter, students will have the opportunity to eliminate any misconceptions and to see firsthand how their knowledge has grown.

OUR WATER RESOURCES

FOCUS

Fresh water, crucial to the existence of human life, represents a tiny portion of the world's supply of available water. This fresh water has a number of sources, all of which are threatened to one degree or another by pollution or overuse.

MOTIVATOR

Use the following visual demonstration to emphasize how little fresh water is available in the world. Start with a gallon-sized jug of water. Explain to students that the jug of water represents all of the water on Earth. Then ask: Can we use all of this water as drinking water or as water for our crops? *(No; most of it is sea water.)* Into a transparent measuring cup, pour about 4 fluid oz. (half a cup) from the jug. Set the jug aside. Tell students that the water in the measuring cup represents the amount of fresh water available in the world. Explain: We can't use all of this water either, because most of it is locked in the world's icecaps. Then pour 2 mL of water from the cup into a small graduated cylinder. Explain: This is all of the fresh water that is left for us humans and other organisms.

TEACHING STRATEGIES

Meeting Individual Needs

Learners Having Difficulty Remind students about what they learned concerning the water cycle in Chapter 3. Encourage these students to examine the diagram of the water cycle on page 63 and relate its various phases to the main topics covered in this section.

Reinforcing Ideas

Explain that water is regarded as a *renewable* resource, meaning that when used, it is replenished by nature. On the other hand, if the water is poisoned or otherwise made unusable at a rate faster than nature can replenish it, then it becomes *nonrenewable.*

WATER, WATER EVERYWHERE, BUT . . . *page 122*

Language-Arts Connection

The title of this section is adapted from a famous passage in "The Rime of the Ancient Mariner" by Samuel Taylor Coleridge. An excerpt of the poem follows.

Water, water, everywhere,
And all the boards did shrink
Water, water, everywhere,
Nor any drop to drink . . .
And every tongue, through
utter drought,
Was withered at the root;
We could not speak, no more than if
We had been choked with soot.

Suggest that students write a short summary explaining what this passage has to say about water as a resource.

Cooperative Learning Activity

As a way of demonstrating why salt water is not suitable for drinking, have small groups of students carry out the following activity. Cut a potato into two equal pieces. Determine and record the mass of each piece. Then prepare two beakers as follows. In beaker 1, dissolve 10 g of salt into 250 mL of tap water. Tell students that this equals the approximate concentration of sea water. Into beaker 2 (the control), place 250 mL of tap water. Place one potato piece into beaker 1 and the other piece into beaker 2. Let the beakers stand overnight. The following day, have students observe the texture of each potato piece. Blot the pieces dry and determine the mass of each piece once again. Compare to the original masses. Tell students to discuss among their group what has happened, and why. *(The potato in the salt water shriveled up, while the other appeared unchanged. The salt has caused the potato's cells to lose water and shrink.)* Ask: Why can't people drink salt water? *(Their cells will also shrivel up.)*

Using the Theme of *Evolution*

Focus Question: What special adaptations enable certain plants and animals to tolerate salt water? Suggest that students do research to find several examples. *(Sample answer: Modern bony saltwater fish evolved from freshwater fish about 200 million years ago. To cope with the elevated salt content, they developed special cells in their gills that excrete salt back into the surrounding water. To counteract osmotic dehydration, they constantly gulp water. A class of plants called halophytes developed specialized structures for dealing with elevated salt concentrations. For example, saltbush has special cells that concentrate the salt and pump it into bladders. The bladders expand as the salt is pumped into them. Eventually they burst, and the salt is carried away.)*

Real-World Connection

Tell students that the state of California recently experienced a six-year drought—the most severe in that state's modern history. Explain that at the worst stage of the drought, serious water rationing was required. Residents in at least one community were allotted only 38 L (10 gal.) of water per day. Tell students that the average person uses more than 380 L (100 gal.) per day.

Ask: Could your household get by on 38 L per day? *(not without hardship)* Point out that upwards of 15 million Californians live in an area that is essentially desert. They must import most of their water, shuttling it through a series of pipelines and canals from sources up to 480 km (300 mi.) away. Unfortunately, research suggests that the last few decades in southern California (a period of explosive population growth) have been unusually wet. The "drought" may, in fact, have been a sign of the climate beginning to return to normal. If this is the case, then the current population is probably not sustainable, and future growth is out of the question.

Multicultural Connection
Much of the west coast of South America has a hyperarid climate. Years often pass without any rainfall. Yet in this hostile place a group of people called the Taukachi-Konkan flourished almost 4,000 years ago. This culture was able to exist because its people had mastered irrigation. An elaborate irrigation system captured the water of nearby mountain streams and distributed it among a network of terraced fields where crops were grown. The Taukachi-Konkan culture existed for several hundred years.

SURFACE WATER, *page 123*

Geography Connection
The Mississippi River system drains river water from about 30 states. In fact, its watershed covers about 40 percent of the land area of the continental United States. The Mississippi and its tributaries also supply fresh water to thousands of communities across the central United States. Examine Figure 5-5 and locate the major tributaries of the Mississippi. (You could also use maps from the Water and Soil Conservation Service.) Determine whether your community is in the Mississippi watershed. If so, what tributary is it near? Trace the path of water from your community to the Gulf of Mexico. Discuss how matter from your city can ultimately find its way to the Gulf Coast.

Using the Transparency
Figure 5-6 is available as a Teaching Transparency.

DAMS, *page 125*

Background: Dams vs. Fish
The construction of a series of hydroelectric dams on the Columbia River has caused the salmon population to decline dramatically. As salmon travel downstream they must pass through the hydroelectric turbines in each dam. Surprisingly, only about 15 percent of the fish die in passing through a given dam. However, since the fish must transit a series of dams, there is a major cumulative loss.

AQUIFERS ARE RUNNING LOW, *page 127*

Notable Quote
"When the well's dry, we know the worth of water."
Benjamin Franklin

Using the Case Study: ***The Ogallala Aquifer: An Underground Treasure, pages 126–127***
Point out that the decline of the Ogallala Aquifer threatens some towns and cities with the loss of their water supply. For example, within a few years, the water supply of Lubbock, Texas, could be depleted. If the city is unable to afford an alternative solution, it will face severe hardship. Ask: How does the expression "living beyond your means" apply to this situation? *(People who live beyond their means spend their money faster than they earn it and either deplete their savings or borrow money to make up the difference. In either case, bankruptcy awaits if the situation goes on long enough. People have "borrowed" water from the Ogallala at a rate faster than nature could "pay the bills," with the result that the area may soon be "bankrupt" of water.)*

Using the Transparency
Figure 5-10 is available as a Teaching Transparency.

Background: When Water Is Scarcer Than Oil
In the desert countries of North Africa and the Middle East, water is a scarce and valuable commodity. Throughout an area of more than 6.5 million sq. km (4 million sq. mi.), no permanent streams flow, with the exception of the Nile, which derives its water from the moist tropics far to the south. More than 1.6 million sq. km (1 million sq. mi.) of this region typically experiences no rainfall for years at a time. Oddly, though, much of this desert region is underlain by large aquifers. The water that these aquifers contain dates to much wetter times thousands of years ago. Occasionally the water table reaches the surface to form an *oasis.* Wells scattered thinly across the desert supply the rest of the water used. These wells are few and far between because in only a few places does the water table come close enough to the surface to be easily reached. Tell students that in some regions of Saudi Arabia and Kuwait, wells drilled in search of water more often strike oil, and it is considered to be quite inconvenient when this occurs.

DESALTING THE SEA, *page 128*

Meeting Individual Needs
Second-Language Learners Work individually or with small groups of second-language learners to recap the main points of each subsection. Ask students to describe, in their own words, these main points. Quiz students orally, as necessary, to ensure their understanding.

WATER CONSERVATION, *page 129*

Background: Xeriscape
A Xeriscape is a landscape that incorporates well-adapted plants and water-conserving measures in order to cut back on the amount of water used. Some desert cities with limited water supplies, such as Phoenix, Arizona, have enacted major restrictions on outdoor watering, forcing many people to abandon water-intensive landscaping in favor of Xeriscape.

CLOSURE

Section Review

Answers to these questions are provided in the Answer Key on page T147.

Alternative Assessment

Ask students to write a short poem, fable, song, or other creative work that focuses on water. Students' works might center on the indispensable nature of water, its changeability, its color, the way in which organisms respond when deprived of it, and so on. Encourage students to include their work in their Portfolio.

Reteaching

To emphasize that the limited amount of available fresh water is not very evenly distributed, use a detailed physical map of the world to highlight the areas where water is abundant and the areas where water is scarce. Point to Canada, and tell students that in this country hundreds of thousands of lakes scattered across thinly populated areas make up about 20 percent of the world's freshwater supply. Next, point to Lake Baikal in Russia (near the Mongolian border). Explain that this lake contains another 20 percent of the world's freshwater supply, but, again, the lake serves only a small population of people. Then point to the Amazon River and tell students that the Amazon Basin contains another 20 percent of the world's freshwater supply, but again, it supports only a small number of the world's people. Then point to China and India, countries with combined human populations of over 2 billion (and climbing). Explain that there are only a few rivers in these countries, and most are heavily polluted. Yet, the rivers supply water to huge populations of humans. Explain that southwestern Asia has a fast-growing population in a largely arid climate. Encourage discussion about this unequal water distribution and the problems it creates.

Extension

Have students record the amount of water they use when brushing their teeth, when showering or bathing, when washing dishes, and so on. Then ask students to devise a plan to reduce by one-third the amount of water they use to do those things. The plan should be reasonable and should not involve great sacrifice. After students have completed this activity and they realize that it did not involve a great sacrifice, ask: If everyone in the world followed your plan, how much water could be saved? *(This question is designed simply to stimulate class discussion; a concrete answer is not necessary.)* Suggest that students include their plan and notes in their Portfolio.

SECTION 5.2

FRESHWATER POLLUTION

FOCUS

The Earth's freshwater resources are being overloaded with the toxic byproducts of human activities. Recent remedial efforts have had some success in reducing the problem.

MOTIVATOR

Write the following categories in different columns on the chalkboard: the tap, a spring or well, a mountain stream, a nearby creek or lake, and a drainage ditch after a heavy rain. Point to each category and ask students whether they would consider drinking water from it. For each source, record the number of students who respond "yes." You will probably notice that students became progressively less willing to trust the quality of the water as you proceeded through the categories. Point out that they, like many people, instinctively recognize the likelihood that many of our freshwater sources are polluted. If any students answer "no" to the first category, tap water, ask them to explain. *(Many people avoid tap water because of the fear that it contains high concentrations of chemicals or pollutants.)* Tell students that in this section they will learn more about freshwater pollution, which has become a real problem. Tell students: The warning "don't drink the water," which is often given to Americans visiting developing countries, could someday apply here as well.

TEACHING STRATEGIES

POINT POLLUTION, *page 132*

Using the Case Study: *The Fight Over Pigeon River, pages 134–135*

The Pigeon River conflict is a classic case of clashing interests. On one side, the economic survival of a community is at stake. On the other side, the health and quality of life for another community are at stake. The conflict is complicated by (or perhaps would not have occurred, except for) the fact that the opposing parties are in different states. This situation is similar to conflicts that often occur across national borders. In such cases the options are limited. The aggrieved party can take legal action, which is time-consuming and very expensive, or the party can apply economic or political pressure, which isn't always possible. Point out to students that historically the paper industry has been a notorious polluter and that paper manufacturers have been the subject of many complaints. Explain that these manufacturers typically produce large volumes of noxious waste. The odor these wastes produce has been called "indescribably bad." The Pigeon River case shows that pollution doesn't always involve good guys versus bad guys. Sometimes the issues are just as cloudy as the waters of the Pigeon River.

NONPOINT POLLUTION, *page 133*

Real-World Connection

Invite students to look for sources of nonpoint pollution around their community as they go to and from school. Encourage them to think about what they see, and suggest that students include their thoughts in their notebook. Direct student thinking by asking questions such as the following: What is the nature of this pollution? Could it be better controlled? Is there anything you personally could do to reduce nonpoint pollution?

Point out that the most common pollutants are oil and other fluids that drip from automobiles in parking lots and on highways. Ask students: What kinds of waste might wash from the streets and yards of a typical neighborhood? *(fertilizers, pesticides, motor oil, gasoline, antifreeze, litter, animal waste)* Encourage students to include their thoughts and responses in their Portfolio.

Using the Transparency

Figure 5-20 is available as a Teaching Transparency.

WASTEWATER TREATMENT PLANTS, *page 134*

Background: Home Pollutants

Homes are a major source of water pollutants. Homeowners frequently dump chemicals such as paints, solvents, motor oil, and household cleaners down the drain or onto the ground—often in violation of state or local law. State and local agencies often provide guidelines for the disposal of common household chemicals that are hazardous. Invite students to find out what state and local laws govern the disposal of waste in your community.

PATHOGENS, *page 136*

Background: The Pathogens Among Us

Most people think of waterborne diseases as a problem reserved for developing countries, not one that Americans have to worry about. But, despite elaborate purification systems that virtually eliminate disease-causing organisms, occasionally problems occur. In 1993, thousands became ill when the water supply of the city of Milwaukee, Wisconsin, became contaminated by an outbreak of *Cryptosporidium,* a harmful microbe. Later that year a similar outbreak occurred in Round Rock, Texas. In 1993 and 1994, an epidemic of another, deadlier, water-borne illness—cholera—swept Latin America as far north as northern Mexico, raising fears that the United States could eventually be affected.

HOW WATER POLLUTION AFFECTS ECOSYSTEMS, *page 136*

Chemistry Connection

The degree to which an ecosystem is affected by a pollutant depends, in part, on the chemical characteristics of the pollutant. Pollutants that are fat soluble tend to accumulate in animal tissues, while pollutants that are water soluble will not. Pollutants that are chemically stable accumulate more readily than pollutants with a short life span. DDT, for example, decays very slowly, with a half-life of a few years. As students learned in Chapter 3, however, DDT can accumulate over time to reach toxic levels. Ecosystems also contain built-in mechanisms for detoxifying pollutants, up to a point. For instance, water hyacinths and certain anaerobic bacteria excel in removing toxins from an ecosystem. In fact, both have been employed in commercial waste-processing facilities.

THE SPECIAL PROBLEM OF GROUNDWATER POLLUTION, *page 139*

Background: Hidden Sources of Pollution

Underground storage tanks for gasoline, such as those used at almost every filling station, can present major groundwater-contamination problems. Especially problematic are older metal tanks and tanks that have been abandoned. Most modern storage tanks are made of fiberglass or some other durable, noncorroding synthetic material. In many cases, state or local statutes have required the replacement of metal tanks with noncorroding ones.

Using the Tranparency

Figure 5-26 is available as a Teaching Transparency.

Background: Bioremediation

In recent years, scientists have learned how to use living organisms to clean up toxic spills. Certain naturally occurring bacteria will feed on harmful organic chemicals, such as petroleum compounds or dioxin compounds. By selectively breeding these bacteria (which are harmless to humans), scientists have developed strains of bacteria that voraciously consume and render harmless a number of common pollutants. Such bacteria have been used successfully on oil and chemical spills at sea. But perhaps their most promising application is in the remediation of polluted groundwater.

Tainted aquifers are notoriously hard to "repair" because they are physically out of reach. Bioremediation gives environmental workers the tools they need to address this critical need. The aquifer to be remediated is "seeded" with a colony of the bacteria. The bacteria multiply exponentially as they encounter and consume the pollutant. When the pollutant is exhausted, the bacteria die.

Using the Theme of *Scale and Structure*

After sharing with students the preceding information about bioremediation, ask them the following questions. *Focus Questions:* Would bioremediation be as helpful in cleaning up heavy-metal pollutants as it is in cleaning up organic pollutants? Why or why not? *(No; while organic pollutants can be made harmless by breaking them down into simpler compounds, poisonous metals cannot. Being elements or mixtures of elements, they cannot be broken down further.)*

CLOSURE

Section Review

Answers to these questions are provided in the Answer Key on page T147.

Alternative Assessment

Tell students to imagine that they have been appointed to a blue-ribbon committee that is studying the problem of nonpoint pollution. Each member of the committee has been requested to come up with at least five recommendations for controlling the problem of nonpoint pollution. Invite students to place their recommendations in their Portfolio.

Reteaching

To ensure that students grasp the difference between point and nonpoint pollution, first review the definitions of each and then provide an example of each type of pollution. Quiz students by giving an example of each type of pollution and then selecting a student to identify it as either point or nonpoint pollution. The following are some examples you may wish to use: parking-lot runoff *(nonpoint),* untreated sewage *(point),* runoff from feedlots *(nonpoint),* a restaurant's drainage pipe emptying runoff into a river *(point),* the litter in storm runoff *(nonpoint).*

Extension

Randomly assign students one side or the other of the Pigeon River issue. Have each student draft a mock letter to the head of the Environmental Protection Agency explaining why his or her point of view should be adopted. Students should use logic and reason to bolster their case instead of emotional appeals. In addition, there should be some main goal to the letters. For example, a goal could be to reduce dioxin emissions by 75 percent, to return the river to its original state, or to guarantee that the plant be allowed to make reasonable changes in order to save jobs.

SECTION 5.3

OCEAN POLLUTION

FOCUS

Human pollution washed from the land into oceans, and pollution that is dumped directly from ships into the ocean, threaten the ocean's vitality, and ultimately, humans themselves.

MOTIVATOR

Provide each student with a plastic loop cut from a six-pack of drinks. (You should be able to get six loops per six-pack.) Have students slide the ring over their tightly closed fingers (excluding the thumb). The band should be just slightly below the first knuckles, where the fingers join the hand. Tell students that this simulates what happens when such rings get caught around the head of a bird, seal, or fish. Ask students to remove the loop without using the thumb or the other hand. Their struggle simulates the ordeal that marine animals face when they become entangled in our debris.

TEACHING STRATEGIES

HOW POLLUTANTS GET INTO OCEANS, *page 140*

Meeting Individual Needs

Gifted Learners Your school is most likely located near a stream or river. Tell students to imagine that someone has dumped a load of pollutants into a waterway near your school. Using a map or series of maps (starting with a large-scale local map and working up to smaller scales), have students trace the path that these pollutants would take on their way to the sea. Ask them to make a rough estimate of the time required for the pollutant to reach the sea. Imagine that the half-life of the pollutant is 15 days. How much of the pollutant will remain by the time it reaches the sea? *(Assuming that the gross distance via the stream is 1,600 km [1,000 mi.], adding about 25 percent for meanders, and using an average flow speed of around 3.2 km/h [2 mph], one possible answer is 625 hours, or a little over 26 days. It takes 1.7 half-lives for the pollutant to reach the sea, so the amount remaining is* $1/2^{(1.7)}$*, or 0.308, or about 31%.)*

Using the Theme of *Interacting Systems*
Students who live in Gulf Coast states may have heard about, or even participated in, "beach sweeps." These are periodic events during which the beaches are cleaned of litter and debris. *Focus Question:* The beaches of the Gulf of Mexico are easily fouled for a number of reasons; what might they be? Have students examine a map to answer this question. *(The Gulf is essentially an enclosed body of water with only two relatively narrow outlets; it is ringed by highly populated areas; many rivers drain into it; and it carries a very large volume of ships. Some students may also mention oil-drilling activity. Any floatable object that falls or is dropped into the Gulf stands a good chance of winding up on a beach.)* Remind students that recent agreements ban or greatly restrict the disposal of waste by ships at sea. The problem is that there is currently no way to effectively enforce these bans and restrictions.

Background: That Pesky Beach Tar
In the past, people who went to the beach often encountered blobs of sticky tar that seemed to be everywhere. Offshore drilling activity was often blamed for the tar. Actually, drilling activity contributed very little to the problem. Natural seepage from underwater oil deposits was often a source of beach tar, but the major source was oil tankers. Until recently, it was common for oil tankers to flush their holds using sea water. This practice, which dumped large amounts of oil into the ocean, was recently banned by international agreement. There has since been a substantial decline in the amount of tar floating at sea or covering beaches.

CLOSURE

Chapter Review
Answers to these questions are provided in the Answer Key on page T147.

Alternative Assessment
Tell students to imagine that the United States Postal Service has decided to issue a series of stamps that will encourage people to honor and respect our world's oceans. The postal service has commissioned your class to provide the designs. Students may design a single stamp or a whole series, and may communicate the design either in writing or with sketches. Suggest that students include their designs in their Portfolio.

Reteaching
On the chalkboard, list all of the major types of ocean pollution. Ask: Which pose a threat, directly or indirectly, to marine life? *(all)* Which pose a threat, directly or indirectly, to human life? *(All except garbage; it could be argued that most garbage is more of a nuisance than a threat.)*

Extension
Even though regulations forbid it, the dumping of waste by ships and offshore operations remains a serious problem. Ask students to suggest ways of increasing the rate of compliance with these regulations. *(Suggestions might include requiring ships to have certified waste-handling systems, an inventory system in which typical waste items must be accounted for, a system of unannounced spot inspections of ships and their waste-handling systems, a designated waste officer who is responsible only to regulatory authorities, severe penalties for violators, and so on.)*

Answers
Answers to *Chapter Highlights, Chapter Review,* and *Investigation* are provided in the Answer Key for Chapter 5, which begins on page T147.

TEACHING STRATEGIES FOR INVESTIGATION:

HOW SAFE IS OUR GROUNDWATER?
pages 146–147

Prepare
To prepare the glucose solution, add one glucose tablet to 200 mL of water. (Tablets and test paper are available at most pharmacies.) Large empty jars may be substituted for beakers.

On Your Own
To prepare test solutions, use 5 mL of motor oil, cooking oil, vinegar, liquid laundry detergent, or antifreeze per 200 mL of water. Ammonia can also be used in the same concentration. Use 3 tablespoons of powdered laundry detergent or 2 tablespoons of high-nitrogen-content fertilizer per 200 mL of water (mix the night before if possible). To detect the contaminants, use pH paper (turns red in vinegar and blue in fertilizer and ammonia). Detergent is detectable by covering the container and gently shaking it to produce suds. Motor oil, cooking oil, and antifreeze should be visible in the filtrate.

Contact your local Environmental Protection Agency, Department of Waste Management, or Health Department to find out how to properly dispose of motor oil and antifreeze. Do NOT allow students to pour these substances down the drain or into the garbage.

CHAPTER 

AIR

PLANNING GUIDE

CHAPTER COMPONENTS AND RESOURCES	½ Year Average	½ Year Advanced	1 Year Average	1 Year Advanced
Chapter Opener, p. 150				
EcoLog, p. 150	•	•	•	•
6.1 What Causes Air Pollution? pp. 151–156	•	•	•	•
Transparency 18: *Primary Air Pollutants*	•	•	•	•
Points of View: *Are Electric Cars the Cure for Air Pollution?* pp. 170–171	•	•	•	•
Transparency 19: *Thermal Inversion*	•	•	•	•
Transparency 20: *Formation of Smog*	•	•	•	•
Section Review, p. 156	•	•	•	•
6.2 Effects on Human Health, pp. 157–160		•		•
Field Activity, p. 158				•
Case Study: *Air Pollution in Poland*, pp. 158–159				•
Transparency 21: *Indoor Air Pollutants and Sources*		•		•
6.3 Acid Precipitation, pp. 161–164			•	•
Transparency 22: *Formation of Acid Precipitation*			•	•
Field Activity, p. 162				•
Transparency 23: *Acid Precipitation Worldwide*				•
Section Review, p. 164			•	•
End of Chapter				
Chapter Highlights, p. 165	•	•	•	•
Chapter Review, pp. 166–167	•	•	•	•
Investigation: *How Does Acid Precipitation Affect Plants?* pp. 168–169			•	•
Videodiscovery STS Science Forums: Vol. 2—*Mass Transit Pass*				•
Chapter Tests	•	•	•	•
PACING GUIDE	3–4 days	3–4 days	6–7 days	6–7 days

MATERIALS ORGANIZER

Activity	Materials	Advance Preparation
Investigation: *How Does Acid Precipitation Affect Plants?* pp. 168–169	goggles, gloves, lab apron, artificial acid precipitation, two small beakers, wax pencil, tap water, graduated cylinder, pH paper, radish or mustard seeds, notebook, pen or pencil, two petri dishes, thin layer of cotton, filter paper to fit petri dishes, optional: acid solutions, other seeds, limestone chips, and other rock fragments (per student group)	Prepare artificial acid precipitation by adding 5.5 mL of 0.01 M H_2S0_4 to 1 L of water. The pH of the resulting solution should be close to four.

CHAPTER OVERVIEW

In this chapter, students are introduced to the causes and effects of air pollution.

Section 6.1 defines air pollution and identifies its chief causes.

Section 6.2 examines the health-related effects of air pollution on humans. Indoor air pollution is also defined.

Section 6.3 discusses acid precipitation, its effects on ecosystems, and the international effort to solve this problem.

RESOURCES

. . . for Teachers

Acid Rain Abstracts Annual. New York City, NY: Bowker A & I, 1991. (This indexing journal provides a good reference for articles on acid rain in specific geographic regions.)

Lyons, T. J., and W. D. Scott. *Principles of Air Pollution Meteorology.* Boca Raton, FL: CRC Press, 1990. (The book includes information on air-quality modeling and provides a wealth of tables and figures, making it a useful teaching tool.)

Samet, Jonathan M., and John D. Spengler, eds. *Indoor Air Pollution: A Health Perspective.* Baltimore, MD: Johns Hopkins University Press, 1991. (This book is a compilation by various medical and scientific experts who discuss the results of current research on the quality of indoor air. Methods of measuring exposure to pollution, strategies for improving indoor air quality, and the health effects of specific pollutants are also discussed.)

. . . for Students

Cozic, Charles, ed. *Current Controversies: Pollution.* San Diego, CA: Greenhaven Press, 1994. For a catalog write: P.O. Box 289009, San Diego, CA 92198-9009; 800-231-5163. (This anthology of articles covers subjects such as: How serious is air pollution? How do corporations affect the environment? How effective is the Environmental Protection Agency?)

Gay, Katherine. *Air Pollution.* New York City, NY: Franklin Watts, 1991. (This book uses accessible language to discuss important issues such as air quality, smog, acid rain, global warming, ozone depletion, toxic emissions, nuclear waste, and indoor air pollution. A discussion of legislation and actions that individuals can take to improve the situation is also included.)

Films, Videos, Software, and Other Media

The Environment Series: Air. Software for IBM, Apple II, or Macintosh. Que, Inc., 338 Commerce Drive, Fairfield, CT 06432; 203-335-0906 or 800-232-2224. (This program simulates air-pollution problems and allows students to try different options to remedy them. The result is a lesson in balancing costs against improvements in air quality.)

Problems of Conservation: Acid Rain. Video. 18 minutes. Britannica Videos, 310 South Michigan Avenue, Chicago, IL 60604; 800-554-9862. (This video examines acid rain, how it has damaged trees and reduced populations of fish and other animals, and what steps can be taken to alleviate the problem.)

Problems of Conservation: Air. 2nd ed. Video. 18 minutes. Britannica Videos, 310 South Michigan Avenue, Chicago, IL 60604; 800-554-9862. (This video discusses the relationship between petroleum and pollution and offers some possible solutions to air-pollution problems.)

Getting Involved

A listing of environmental organizations, publications, and volunteer opportunities can be found in the **Laboratory and Field Guide.** Note especially those organizations that are concerned with air pollution, such as the Acid Rain Foundation, the Natural Resources Defense Council, the Sierra Club, and the World Resources Institute.

USING THE QUOTATION

"I thought I saw a blue jay this morning. But the smog was so bad that it turned out to be a cardinal holding its breath."

This quotation can be used as a motivating activity for Section 1. See *Motivator,* page T69.

USING THE ECOLOG

To assess students' knowledge about the subject of this chapter, have them answer the EcoLog questions on page 150. In answering these questions, students will reveal their prior knowledge, which is usually very basic and often flawed or even misconceived. Then, by revising their answers at the conclusion of the chapter, students will have the opportunity to eliminate any misconceptions and to see firsthand how their knowledge has grown.

What Causes Air Pollution?

Focus

Major air pollutants and their sources are defined in this section. Thermal inversions, which aggravate the pollution problem, are also explored.

Motivator

Call on one student to read aloud the quotation on page 150 of his or her text. Tell students that sometimes humor gets people's attention about a serious issue better than any other method. Divide the class into small groups, and tell them to use humor, as Cohen did, to complete the following sentence: You know the air is polluted when. . . . After a few minutes, have the class discuss each group's response. Then arrange the responses into a "Top Ten" list. Students may enjoy illustrating the list and using it to design a bulletin board that could remain up for the duration of this chapter.

Teaching Strategies

Using the Photograph: *Figure 6-1, page 151*

Point out that in Mexico City many factors combine to cause the city's notoriously poor air quality. First of all, Mexico City is the world's most populous metropolis, with at least 20 million inhabitants (the exact number is not known). Its roads are traveled by millions of vehicles every day, many of them older, poorly maintained models that spew pollutants into the air. Because of the city's high elevation (2,309 m or 7,575 ft.), even modern cars run less efficiently and pollute more than they would at sea level. The city is also the site of many heavy industries. These industries pump large amounts of pollutants into the air. Perhaps the most serious problem is that the city is located in a deep valley, surrounded by high mountains that effectively trap pollutants.

Ask any students who have been to Mexico City to describe the air quality. Encourage class discussion about what it would be like to live under such circumstances. *(Students who have been there will almost certainly have noted the poor quality of the air. Most students would find the air quality unbearable.)*

End your discussion on a positive note by telling students that city authorities have implemented many plans in recent years to improve Mexico City's air-pollution problems. Suggest that interested students consult recent periodicals to learn about some of those efforts, and then encourage those students to report their findings to the class.

Meeting Individual Needs

Second-Language Learners Work with these students individually or in a small group if possible. Ask them to summarize the main points conveyed by the photographs and diagrams in this chapter. In this way, you can make sure that the students understand many of the main points in the chapter, and you can build student confidence.

Using the Transparency

Figure 6-3 is available as a Teaching Transparency.

Controlling Air Pollution From Vehicles, *page 153*

Background: Unleaded Gas

To meet the tough clean air standards that took effect in 1975, manufacturers that sell cars in the United States began equipping cars with catalytic converters. Up to this time, all grades of gasoline contained a small amount of tetraethyl lead, $Pb(C_2H_5)_4$, a compound that suppressed engine knocking and lubricated engine valves. Because even a small amount of lead can render the metal compounds in the catalytic converters useless, unleaded gasolines were introduced. Then, in the 1980s, it became widely known that lead is a toxin. As a result, a phaseout of leaded gasolines was initiated. By 1990, they were no longer available in many areas of the United States, and they are slated to be phased out entirely by 1996.

Industrial Air Pollution, *page 154*

Using the Theme of *Interacting Systems*

Focus Question: Why might regions directly downwind from major producers of particulate pollutants often have measurably greater rainfall than the surrounding area? *(The particles act as condensation nuclei for water vapor. If enough condensation occurs, rain, sleet, or snow will fall.)*

Thermal Inversions, *page 155*

Using the Transparency

Figure 6-9 is available as a Teaching Transparency.

Meeting Individual Needs

Learners Having Difficulty The following is a simulation of a thermal inversion. Perform the demonstration without telling students what you are simulating. Fill the bottom 2 in. of an aquarium with ice. Note: Keep the aquarium away from any drafts during the demonstration.

Place a small plate in the center of the ice. Allow the ice to stand for about a minute in order to chill the surrounding air. Then set fire to one end of a piece of natural-fiber rope. After a few seconds, blow out the flame. The rope should smoke heavily. Gently place the piece of rope onto the plate. Ask students to observe what happens. *(The smoke pools in the aquarium and does not rise out of it.)* Ask: What might this demonstration simulate? *(thermal inversion)* Why? *(Warmer air above the smoking rope keeps the cooler air at the bottom of the aquarium from moving upward. Thus, the smoke is trapped below with the cooler air just as air pollutants are trapped in an urban environment.)* Point out that particular weather conditions contribute to the formation of inversions—cloud cover, which reduces warming at the Earth's surface, and lack of wind.

Smog, *page 156*

Using the Transparency

Figure 6-10 is available as a Teaching Transparency.

Multicultural Connection

Centuries ago, long before Europeans settled in North America, the area that became Los Angeles, California, was known for its smog. Native Americans knew the Los Angeles Basin as the "Valley of the Smokes" because the frequent thermal inversions trapped the smoke from their fires, casting a gray pall over the valley.

Language-Arts Connection

Divide the class into small groups, and have members of each group work together to discuss the following questions. What are the benefits of living in an industrialized country? How might those benefits contribute to air pollution? Do the benefits of having more goods and services available outweigh their negative effects? After the groups have had time to discuss these questions, tell students to return to their desks and complete the following writing assignment: Imagine that you are living 50 to 60 years from now and that you have grandchildren. How might your grandchildren answer the three questions your group discussed? Respond in the form of a letter from your grandchild to yourself, the grandparent. The letter should explain what life is like for the grandchild. Encourage students to include their letter in their Portfolio.

Using the Theme of *Energy*

Focus Question: How might conserving energy by taking advantage of the following activities reduce air pollution: carpooling, using electric trains for regular transportation, improving insulation in homes, using fluorescent bulbs for lighting, and recycling aluminum cans? *(Each of these is an energy-saving activity. Because some air pollution results from the production of energy in each case—whether in motor-vehicle exhausts or as the byproduct of industrial processes—less air pollution is created when less energy is expended.)*

Closure

Section Review

Answers to these questions are provided in the Answer Key on page T148.

Alternative Assessment

Ask students to prepare a script for a simulated TV news report on a fictitious smog or ozone alert in your area (or another area if that seems more appropriate). Students' scripts should be creative yet plausible and coherent. Ask students to describe the source of the pollutants and any aggravating factors. They should also include advice about how people can minimize their risk during the alert and about what individuals can do to help alleviate the problem. Suggest that students include their script in their Portfolio.

Reteaching

Write three categories on the blackboard: Industrial and Commercial, Transportation, and Noncommercial. Explain that these categories represent the different sources of pollutants. The first two categories are fairly self-explanatory. The third category might include such miscellaneous sources of pollution as backyard barbecues, lawn mowers, and wood-burning stoves. Encourage students to name as many causes of air pollution as possible, and help them place the pollutants in the appropriate categories. Finally, discuss with students the types of pollutants that each source usually emits and possible remedies for those pollutants. *(Industrial and commercial sources could emit particulates, carbon monoxide, sulfur dioxide, and hydrocarbons; transportation sources emit mostly carbon monoxide, carbon dioxide, nitrogen oxides, hydrocarbons, and particulates; and noncommercial sources usually emit carbon dioxide and particulates. Many pollutants can be reduced by using scrubbers, catalytic converters, and other pollution-control devices.)*

Extension

Many automakers have devised methods to reduce auto emissions. For example, in the 1970s Honda developed the CVCC (Controlled Vortex Combustion Chamber) engine. Aside from being exceptionally fuel efficient, this engine burned fuel so cleanly that it did not require a catalytic converter. Have students investigate other efforts to lower the emissions of cars and trucks. What improvements are being implemented now? How do they work? What improvements are planned to meet the tougher emission standards of the future? *(Students may mention electric cars, dual fuel systems that permit a car to use either natural gas or gasoline, and so on.)*

SECTION 6.2

EFFECTS ON HUMAN HEALTH

FOCUS

This section discusses the effects of air pollution on human health. Indoor air pollution is evaluated as a major source of health problems.

MOTIVATOR

Bring to class a portable oxygen tank (you can probably borrow one from a medical supply company, nursing home, or doctor's office). To begin the class, ask students how much they think pollution costs in terms of the dollars spent to combat the health problems caused by pollution. *(Most students will be uncertain.)* Hold up the oxygen tank. Mention that some patients with severe lung diseases like emphysema or chronic bronchitis use up to one tank of oxygen every day, at a cost of about $20 per tank. Explain that these diseases are not caused by air pollution, but pollutants certainly aggravate and worsen these and other respiratory illnesses. Continue by telling students that one study suggests that as many as 1 million Americans with lung diseases probably would not need to use oxygen from a tank were it not for air pollution. So, assuming that each person uses two oxygen tanks per week, pollution costs Americans a minimum of about $2 billion annually in oxygen costs alone! Explain that many of these oxygen-dependent people cannot afford to pay for the oxygen themselves. As a result, the general public assumes these costs through higher taxes and insurance rates.

TEACHING STRATEGIES

Notable Quotes

"Ill air slays sooner than the sword."
Ratis Raving, c. 1450

"Fresh air keeps the doctor poor."
Danish proverb

Background: Health Costs of Pollution

Researchers believe that about 15 percent of the 150,000 cases of lung cancer diagnosed each year in the United States are caused by air pollutants, including naturally occurring radon gas. The American Lung Association estimates that 120,000 Americans die annually of lung diseases caused by urban air pollution. In addition, millions of Americans suffer from asthma, emphysema, or bronchitis, which are aggravated by air pollution. Treatment of these diseases costs billions of dollars.

Using the Case Study:

Air Pollution in Poland, pages 158–159

Point out that the air pollution problems in Poland represent an extreme example of what can happen when environmental considerations are completely disregarded. Ask students to consider the statement made by one worker in the Case Study: "I'd rather have a sick, short life with a job than a longer, healthier life without work." Encourage students to discuss whether they agree or disagree with this worker's reasoning.

Remind students that these pollution problems are not restricted to Poland but occur throughout the world. Ask students to speculate about the reasons for this. *(Students may bring up some of the following points. Under communist rule, there was a strong push for industrialization at all costs. Because industry and the economy were centrally planned, local citizens had no voice in the matter. In addition, because all jobs were essentially controlled by the government, citizens had no choice—they had to take whatever jobs were available, and many of these were in heavy industry. With the absence of thorough regulatory agencies, the cheapest and dirtiest industrial methods were often used, and gross inefficiency and waste frequently went unpunished. Since all land and facilities were owned by the government, there was little incentive to keep them in good shape. In many countries, the cost of pollution is passed on to the general populace in the form of environmental and health problems, instead of being assumed by industries through investments in better technologies or procedures.)*

INDOOR AIR POLLUTION, *page 158*

Using the Transparency

Figure 6-13 is available as a Teaching Transparency.

Using the Diagram:

Figure 6-13, page 159

Have students work in small groups to name some of the things that might pollute the air in their school or home, using Figure 6-13 as a guide. Encourage students also to discuss methods for preventing or lessening the effects of such pollution. Suggest that students then use that information to make posters and display them in the classroom or hallway.

Real-World Connection

The Occupational Safety and Health Administration (OSHA) enforces regulations concerning a number of airborne substances in the workplace. If possible, invite an OSHA representative to speak to the class about such regulations. Or, you could ask a worker or an employer from the community to speak to the class about steps that must be taken to meet OSHA regulations and to control indoor air pollution in the workplace.

RADON GAS, *page 160*

Background: Radon

Radon, an odorless, invisible, radioactive gas, is a byproduct of the radioactive decay of radium, which is in turn a decay product of uranium and thorium. Uranium and thorium are present in trace quantities in soil and rock but are especially common in certain igneous rocks and the soils derived from them. There are three naturally occurring isotopes of radon, two of which are relatively common. The isotope formed from the decay of thorium, Rn-220, has a half-life of about 55 seconds. The isotope formed from the decay of uranium, Rn-222, has a half-life of about four days.

Real-World Connection

Radon gas detector kits are reasonably priced and readily available at hardware stores. Interested students could purchase one and test their home or the classroom, sharing their results with the class.

Cooperative Learning Activity

Divide the class into five groups, and assign each group one of the following topics. (1) Is radon a problem in your area? If so, where and why? (2) What are the natural sources of radon gas? Where do these sources come from? (3) How is radon detected? (4) What are the effects of radon on human health? (5) How can radon contamination be dealt with? When all groups have finished compiling their data, have a "radon-awareness day" during which each group shares its findings with the class. Suggest that students include their findings in their Portfolio.

Debate

The extent to which radon is harmful to our health is hotly debated. The Environmental Protection Agency (EPA) contends that exposure to indoor radon is responsible for thousands of lung-cancer deaths and that millions of homes have unnaturally high levels of the gas. Some critics disagree with the EPA's methods and question the motivation of EPA employees, however. Divide students into two groups. Have one group support the EPA's position that radon risks are high and should be taken seriously, while the other group takes the position that the radon risks are greatly exaggerated. Each group should then prepare a convincing argument for its point of view. Give students time to research the subject. Then each group should choose a spokesperson to articulate the group's point of view to the class. Finally, suggest that students record the main points of each group's argument, along with their own individual conclusions and supportive reasoning.

CLOSURE

Section Review

Answers to these questions are provided in the Answer Key on page T148.

Alternative Assessment

Divide students into groups, and encourage them to compose a rap song based on the theme of air pollution's effect on the lungs. Rhythm accompaniment could be in the form of coughing and wheezing sounds. An alternative would be for students to compose and read a funny poem or limerick or perform a short skit about the subject. Suggest that students include their song, poem, or script in their Portfolio.

Reteaching

For this demonstration, you will need an automobile and the key to start the engine. The car's engine should be completely cold when you begin. **Safety Alert:** *Exercise extreme caution when performing this demonstration. Do not let students have any role in the demonstration aside from observing. Try the demonstration with an adult partner before performing it for students.*

Have all students stand at least 6 m (20 ft.) from the tail end of the car. Using plenty of tape, attach a 13 gal. trash bag very securely to the car's tailpipe. Start the engine, and quickly join the students behind the car. The bag will fill quickly and then burst or tear. Turn off the car's engine and let the gas dissipate. **Safety Alert:** *Make sure that you and all of the students avoid breathing the fumes.*

Point out that the filled bag contained virtually no oxygen, but was instead filled mostly with harmful gases such as carbon dioxide. Tell students that if the bag filled in only a few seconds while the engine was just idling, imagine how much exhaust comes from a car as it travels down the highway. Explain that burning only 1 gal. of gas (3.8 L) produces 9 kg (almost 20 lb.) of carbon dioxide. Return to the classroom and review with students Figure 6-2, paying particular attention to the health effects of pollutants that come from cars.

Extension

Tell students to assume the role of investigative reporters. They should interview school personnel, such as board members, trustees, the principal, or a custodian, to determine if your school building has ever had asbestos-abatement work done on it or if the ventilation system has any means to control mold spores, bacteria, and other indoor air pollutants. If so, students should get the details and write a short article describing the situation. As an alternative, students could investigate two different office buildings in their community: a modern, sealed building and an older building that is ventilated primarily by windows. The buildings should be roughly the same size, have roughly the same number of occupants, and have no obviously biasing factors (such as one building being a storage place for chemicals). Students could interview occupants of each building about the incidence of allergies, contagious diseases, unexplained headaches, nausea, or other symptoms of sickness. By comparing the data from each building, students could conclude whether the building suffers from "sick-building syndrome." Students could sum up their findings in an "investigative report" for a student or imaginary newspaper. Suggest that students include their article in their Portfolio.

Acid Precipitation

Focus

This section explores the causes and effects of acid precipitation. The international efforts to control acid precipitation are also discussed.

Motivator

The night before you are to perform this demonstration, prepare 1 L of sulfuric acid solution with a pH of 1.5. (Do this by adding sulfuric acid a little at a time to water and checking it with pH paper until it has a pH of 1.5.) Pour equal parts of the solution into two beakers (beaker 1 and beaker 2). Place a few small pieces of limestone (or chalk) into beaker 2. Leave the solutions undisturbed overnight.

The next day, show students the beakers and tell them what is in each. Then explain that you are going to test the pH of each beaker, and ask if anyone can predict what you will find. *(The first beaker will be acidic and the second will not be acidic.)* Test the pH value of each beaker. Ask: How can you explain the difference between the two beakers? *(The limestone or chalk neutralized the acid in beaker 2.)* Tell students that beaker 1 represents acid rain that fell in West Virginia and that beaker 2 represents a remedy that was used to combat the effects of acid rain.

Ask students: What effect do you suppose acid rain, such as that simulated in beaker 1, could have on human structures and on the environment? *(Such acidity could destroy structures made of limestone and many other materials, such as a car's paint and finish. It could also directly damage vegetation and change lake chemistry, making lakes uninhabitable or degraded for many lake organisms.)* Make sure students understand that this is an extreme case and that most acid precipitation is not so severe. At the same time, reinforce the idea that acid precipitation is a serious problem, as students will learn in this section.

Using the Transparency

Figure 6-16 is available as a Teaching Transparency.

Background: The Source of the Trouble

Power plants are some of the largest contributors of pollutants that cause acid rain. Nitrogen oxides are formed when oxygen and nitrogen combine at high temperatures, such as during combustion. Sulfur dioxide, SO_2, is a noxious gas with a characteristic "rotten-egg" smell; it is toxic to both plants and animals. SO_2 is produced mainly by power plants that burn high-sulfur coal (coal that contains iron sulfides) to generate electricity. A large power plant may burn 9,070 metric tons (10,000 tons) of coal per day. If the coal contains 3 percent sulfur, about 544 metric tons (600 tons) of SO_2 are produced every day. Automobiles produce very little SO_2 but are the source of almost half of the nitrogen oxides released into the air.

Meteorology Connection

Ask students to think about why the Adirondack Mountains have suffered so much from the effects of acid rain. Display a map that shows the topography of the region. Point out that because of their topography, the Adirondacks consistently trigger rainfall and snowfall on a large scale. This is because air masses are forced upward by the mountains; as the air rises and cools, it retains less moisture. Furthermore, since prevailing winds blow from west to east, the Adirondacks are the first mountain range encountered by winds from the industrial belt of the upper Midwest (with their load of pollutants). Consequently, the Adirondacks receive the full, undiluted load of pollutants from the industrial belt in the form of acid precipitation.

Cooperative Learning Activity

Divide the class into groups of four to five students. Have each group collect samples of distilled water, tap water, rainwater (from multiple rainfalls, if possible), and surface waters (streams, lakes) in your community. When a sample is collected (students should not let samples sit), have students measure its pH with litmus paper or a pH meter and record their results. Students should notice some variation in the water samples. Ask them to account for this difference. *(Distilled and tap water have been treated—this process affects their pH; groundwater from wells has filtered through layers of rock and soil; rainwater pH is affected by weather patterns; and the pH of surface waters is affected by multiple chemical and biological processes. The most important feature of surface water that affects pH is its buffering capacity—its ability to neutralize acid. A water's buffering capacity depends largely on the geology of the area.)*

International Conflict and Cooperation, *page 163*

Meeting Individual Needs

Gifted Learners Have students research the effects of acid precipitation on historical structures such as statues, monuments, and buildings. Direct their research by mentioning that the famous Egyptian obelisk that now resides in New York City's Central Park has suffered more erosion in the last 100 years than in the previous 3,000. Encourage students to document their findings with pictures whenever possible. Also suggest that students find out about any actions that are being taken to protect these treasures. Students may also wish to research whether anything is being done by modern artists and architects to guard against damage by acid precipitation in the future. Suggest that students share their findings with the class in the form of a newspaper article or academic report, with supporting photographs or drawings. Encourage students to include their article or report in their Portfolio.

Using the Transparency
Figure 6-24 is available as a Teaching Transparency.

CLOSURE

Section Review
Answers to these questions are provided in the Answer Key on page T149.

Alternative Assessment
Tell students that acid precipitation has fatally damaged 50 percent of the trees in a hypothetical forest ecosystem. Ask students to predict the indirect effects of that precipitation on the ecosystem. Students can respond with a short paragraph or with a concept map. *(Sample paragraph: Any plants or animals that depend on the trees that are killed might die, persist in diminished condition or numbers, or be forced to migrate. Animals that preyed on these animals will be deprived of a food source, and may also die or migrate. Other plants and animals, freed from predation or competition, may thrive; certain insects might attack the remaining vegetation, selectively eating some species and leaving only unpreferred species. Because the soil will be exposed, erosion will take place at an accelerated rate, diminishing the ability of the land to support plant life and deteriorating water in lakes, rivers, and streams. Even if the soil is not eroded, the acid rainfall might release certain minerals from the soil, changing its fertility and damaging some plant species while aiding others.)*

Reteaching
Prepare index cards that list terms, places, objects, and concepts mentioned in this section. For example, you could list acid precipitation, fossil fuels, acid shock, fish, amphibians, Ruhr Valley, Helsinki Declaration, etc. Make one card for each student. Once each student has a card, call on a student to read his or her card aloud. Then let the other students who think their card is related to the initial one explain the relationship. Or you could call on a second student and have the class explain whether the two cards are related, and if so, how. Record student responses in the form of a concept map on the chalkboard.

Extension
Ask students to explain whether they think it would be better to prevent the release of acid-producing pollutants through expensive pollution-control equipment or to find ways of cleaning up the environmental effects of acid precipitation after it has occurred. *(Most students will probably choose the first option because it is more expensive to clean up pollution than to prevent it, and because cleaning up after the fact cannot bring back organisms that have died from the direct or indirect effects of acid precipitation.)* Suggest that students include their paragraph in their Portfolio.

Answers
Answers to *Chapter Highlights, Chapter Review, Investigation,* and *Points of View* are provided in the Answer Key for Chapter 6, which begins on page T148.

TEACHING STRATEGIES FOR INVESTIGATION:

HOW DOES ACID PRECIPITATION AFFECT PLANTS? *pages 168–169*

Prepare
As noted in the *Planning Guide* for this chapter (page T67), there is some advance preparation necessary for this investigation. Allow at least one week for the completion of this activity.

Other quick-germinating seeds can be used for this activity. Avoid lettuce, carrot, and other seeds which are so small that counting individual seeds and identifying obviously nonproductive seeds (empty hulls, etc.) is difficult.

Observe the Effects of Acid on Seeds
All leftover acid solution should be poured into a common container and neutralized to pH 7 by adding a dilute base, drop by drop, before pouring it down the drain.

Extension
Suggest that students design an experiment to test the effects of acid precipitation on car finishes. They could use an artificial acid precipitation solution and model-car paint finishes. They may also wish to do research to find out if, when, and where automobiles have been damaged by acid precipitation.

CHAPTER 7

ATMOSPHERE AND CLIMATE

PLANNING GUIDE

CHAPTER COMPONENTS AND RESOURCES	½ Year Average	½ Year Advanced	1 Year Average	1 Year Advanced
Chapter Opener, p. 172				
EcoLog, p. 172	•	•	•	•
7.1 The Atmosphere, pp. 173–175			•	•
Field Activity, p. 174				•
Transparency 24: *The Five Layers of the Atmosphere*			•	•
Section Review, p. 175			•	•
7.2 Climate, pp. 176–179			•	•
Environmental Careers: *Climate Researcher*, pp. 384–385				•
Section Review, p. 179			•	•
7.3 Greenhouse Earth, pp. 180–185	•	•	•	•
Transparency 25: *The Greenhouse Effect*	•	•	•	•
Case Study: *Computer Models and Earth's Future Climate*, pp. 182–183		•	•	•
Section Review, p. 185	•	•	•	•
7.4 The Ozone Shield, pp. 186–188	•	•	•	•
Transparency 26: *Chemical Reactions That Destroy Ozone*	•	•	•	•
Making a Difference: *Ozone Scientist*, pp. 194–195				•
Section Review, p. 188	•	•	•	•
End of Chapter				
Chapter Highlights, p. 189	•	•	•	•
Chapter Review, pp. 190–191	•	•	•	•
Investigation: *Global Warming in a Jar*, pp. 192–193		•	•	•
Videodiscovery STS Forums: Vol. 1—*Fossil Fuel & the Greenhouse Effect*			•	•
Chapter Tests	•	•	•	•
PACING GUIDE	**4–6 days**	**3–5 days**	**8–10 days**	**8–10 days**

MATERIALS ORGANIZER

Activity	Materials	Advance Preparation
Investigation: *Global Warming in a Jar*, pp. 192–193	2 large glass jars of the same size and shape, 2 outdoor thermometers, small pieces of thin cardboard, tape, clear plastic wrap, rubber bands or string, watch or clock, graph paper, straightedge or ruler, notebook, pen or pencil, and one or more of the following items: black and white latex paint, water, soil, sod or other plants, ice cubes (per student group)	none

CHAPTER OVERVIEW

In this chapter students learn about climate, weather, and the structure of the Earth's atmosphere. The protective role of the atmosphere is also discussed.

Section 7.1 explains how the Earth's atmosphere makes life possible, discusses how photosynthesis and respiration keep the amount of carbon dioxide in the air nearly constant, and describes the structure of the atmosphere.

Section 7.2 describes the causes and characteristics of seasons and climates.

Section 7.3 defines the greenhouse effect and explains that many scientists predict that human activities will cause the Earth's climate to grow warmer.

Section 7.4 shows how the ozone layer reduces the amount of damaging ultraviolet radiation that reaches the Earth's surface and demonstrates how artificial chemicals are probably damaging that protective layer.

RESOURCES

. . . for Teachers

Manabe, Sykuro, and Ronald J. Stouffer. "Century-Scale Effects of Increased Atmospheric CO_2 on the Ocean-Atmosphere System." *Nature,* July 15, 1993, pp. 215–218. (This article examines the possible climatic effects of enhanced levels of CO_2 in the atmosphere over the next few centuries.)

Zimmer, Carl. "The War Between Plants and Animals." *Discover,* July 1993, pp. 16–17. (This article examines the struggle between plants and animals and how this interaction has affected the climate.)

. . . for Students

Nye, Bill. "Big Energy in Thin Air." *Omni,* May 1994, pp. 4–7. (In this article, Bill Nye, television's "Science Guy," examines the powerful presence of the air all around us.)

Poore, Patricia, and Bill O'Donnell. "Ozone: Scam or Crisis?" *Garbage,* Sept./Oct. 1993, pp. 24–29. (In this article, both sides of the ozone debate are presented.)

Films, Videos, Software, and Other Media

Atmospheric Science (series). CAV Videodisc, Level 1. Coronet/MTI, P.O. Box 2649, Columbus, OH 43216; 800-777-8100. (This series explores all aspects of climate and weather, including global winds, the atmospheric energy balance, clouds and precipitation, and storms.)

Global Warming: Hot Times Ahead? Video, Film, or Level 1 Videodisc. Churchill Media, 6901 Woodley Avenue, Van Nuys, CA 91406-4844; 800-334-7830. (This award-winning program defines global warming and discusses why it may be happening and what possible effects could result from the phenomenon. Strategies for slowing the buildup of greenhouse gases are also discussed.)

SimEarth. Software for Macintosh or IBM. Maxis, 2 Theater Square, Orinda, CA 94563; 800-336-2947 or 510-254-9700. (This challenging and highly instructive program allows you to create a simulated world and alter it as you wish. The object is to create a world suitable for higher life-forms within a 10-billion-year span.)

Getting Involved

A listing of environmental organizations, publications, and volunteer opportunities can be found in the **Laboratory and Field Guide.** Note especially those organizations that strive to help protect the atmosphere, such as the American Forestry Association (Global ReLeaf), the Union of Concerned Scientists, and the World Resources Institute.

USING THE QUOTATION

"The atmosphere is the key symbol of global interdependence."

Ask students to define global interdependence. *(Sample answer: the condition in which every person or society is affected by every other)* Then ask students to explain the meaning of the quotation. *(Sample answer: Because the atmosphere circulates freely, it is not affected by borders. Anything dumped into the atmosphere in one country ends up affecting all countries. For the atmosphere to be maintained in "good working order" for all, everyone has to refrain from polluting it.)* Solicit other examples of global interdependence. *(Possible answers: pollution of rivers and oceans, pollution that can contribute to acid rain, etc.)*

USING THE ECOLOG

To assess students' knowledge about the subject of this chapter, have them answer the EcoLog questions on page 172. In answering these questions, students will reveal their prior knowledge, which is usually very basic and often flawed or even misconceived. Then, by revising their answers at the conclusion of the chapter, students will have the opportunity to eliminate any misconceptions and to see firsthand how their knowledge has grown.

THE ATMOSPHERE

FOCUS

The atmosphere, a thin envelope of gases surrounding the Earth, makes life on this planet possible.

MOTIVATOR

Our planet has sometimes been called Spaceship Earth. Ask your students to explore this concept. In what ways is the metaphor particularly apt? How does this concept provide a sense of global unity and responsibility? *(Like a spaceship, the Earth is totally self-contained, it is a fragile oasis of life in the hostile void of space, and it is traveling through space. Like passengers in a spaceship, the inhabitants of Earth are "stuck with each other," interdependent, and vulnerable.)*

TEACHING STRATEGIES

Meeting Individual Needs

Second-Language Learners Provide second-language learners with translations of the vocabulary terms of this section. For example, the Spanish equivalent of atmosphere is *atmósfera.*

Multicultural Connection

Virtually every culture has myths about the origin of the Earth. For example, according to Icelandic legends, the world was born of fire. Encourage students to compare selected myths with current scientific views about the Earth's origin. Suggest that students pick a myth and share it with the rest of the class in the form of a poster, a book of drawings, an animated story, or some other creative method. Suggest that students include something from their presentation in their Portfolio.

HOW PHOTOSYNTHESIS CHANGED THE ATMOSPHERE, *page 174*

Background: The Early Atmosphere

It is actually fortunate that the early atmosphere had little oxygen. If the early atmosphere had contained large amounts of oxygen, life might never have evolved. The oxygen would have destroyed the large organic molecules that had to form before life could evolve.

Outgassing, the release of gases from the Earth's interior by volcanic activity, is believed to have created the early atmosphere. Huge amounts of water vapor could have been pumped into the atmosphere by constant volcanic activity, causing violent torrential downpours.

At first the Earth's surface may have been so hot that the water turned to steam as soon as it struck the ground. But after millions of years, the surface cooled enough for water to begin to collect in low places. Many scientists believe that this is the way the oceans were formed.

Demonstration: What Percentage of Air Is Oxygen?

For this demonstration you will need the following materials: a pint-sized cylindrical glass jar, a low-voltage energy source (such as a 6-volt lantern battery), two 50-cm-long pieces of electrical wire—each with an alligator clip at one end, a steel wool pad, a shallow pan, and enough water to fill the pan 1–2 in. in depth.

Safety Alert: *Please exercise extreme caution while performing this activity.*

By combusting (combining with oxygen) a quantity of steel wool, this activity helps students realize how much oxygen is present in air. To begin, firmly connect the alligator clips to opposite sides of the steel wool pad. Stuff the pad into the bottom of the jar, taking care not to dislodge the alligator clips. Place the jar, mouth down, into a shallow pan that is partially filled with water. Be careful not to allow any water into the jar as you set it in the pan. Connect the free ends of the wires to your electrical source. Have students record their observations in their notebook.

Electrical resistance will cause the steel wool to heat up and glow red. The heat accelerates the oxidation process to the point that the steel wool actually burns. This is signified by occasional pinpoints of bright light. Leave the power source connected until you no longer see signs of combustion. Mark the level to which the water has risen. Ask students to explain what they have seen. *(The portion of the jar's volume that was occupied by oxygen is now occupied by water.)*

Have students compute the percentage of oxygen in air by volume. Assuming the jar is cylindrical, students can compute the volume by dividing the distance from the opening of the jar to the water line by the total inside depth of the jar. Ask: How does this activity demonstrate the oxygen content of the air in the jar? *(The oxygen is removed from the air as it combines with iron, forming iron oxide, or rust. Since no other gases in the air combine with the iron, other gases are not removed. Thus, by comparing the volume of the removed portion with the volume of air at the start, you can determine the percentage of oxygen present in air.)*

Using the Table:
Figure 7-3, page 175

Direct students' attention to the table. Many students will mistakenly believe that the atmosphere is composed mostly of oxygen; the high percentage of nitrogen may surprise them. Point out that nitrogen is released by many Earth processes and, because it is largely unreactive, accumulates continuously in the atmosphere. Remind students of what they learned in Chapter 3 about the role nitrogen plays in one of the Earth's most vital cycles; without it, life could not exist.

The Five Layers, *page 175*

Using the Transparency
Figure 7-4 is available as a Teaching Transparency.

Using the Diagram:
Figure 7-4, page 175

First direct students' attention to the photograph in Figure 7-1. Ask them if the layers of the Earth's atmosphere are visible in the photograph. *(not really)* Then ask students to look at the layers shown in Figure 7-4. Ask: How might the layering be determined if those layers aren't really visible? Do you think scientists have arbitrarily designated the layers? *(The layers are signified by differences in movement and composition. The designation is not arbitrary.)*

Background: The Troposphere
The troposphere is the layer in which almost all weather occurs. This is because the troposphere contains the most water vapor and is heated from below by contact with the ground and oceans. The tropopause, or boundary between the troposphere and the stratosphere, is highest at the equator (about 16 km, or 10 mi.) and lowest at the poles (about 10 km, or 6 mi.). Students have probably seen evidence of the tropopause without realizing it. The flattened "anvil" top seen in large thunderheads results when a cloud rises so high that it reaches the tropopause. At the tropopause, clouds encounter the fierce winds of the stratosphere, which shear off the top part of the clouds.

Using the Theme of *Scale and Structure*
Focus Question: The atmosphere is both highly chaotic and highly structured at the same time. How can anything have such apparently opposite traits simultaneously? *(Sample answer: It is a function of scale. Viewed as a whole, the atmosphere is highly structured. However, when observed close-up, it is chaotic and without apparent structure. The troposphere, for example, is a structural element. But to see it as such, you have to look at it in relation to the other atmospheric layers and the Earth's surface. If you look more closely at the troposphere, you see that it is composed of a highly dynamic mixture of gases that is undergoing constant change.)*

Art Connection
Suggest that students make paintings or drawings of the Earth as it might look if there were no atmosphere. Have students include their drawing in their Portfolio. *(Students' drawing should show a surface something like the moon with a black sky, no water or life, and occasional craters.)*

Closure

Section Review
Answers to these questions are provided in the Answer Key on page T150.

Alternative Assessment
Have students write a short story, poem, or script that describes an imaginary trip through the atmosphere from the ground to outer space. Students should describe the features, structure, and characteristics of the various layers of the atmosphere.

Reteaching
Construct a diagram of the different layers of the Earth's atmosphere. Elicit responses from students by pointing at the different layers and asking for information about the indicated layer. Prompt students if necessary.

Extension
Polar icecaps formed from snowfalls that were deposited over tens of thousands of years. Students may be interested to know that air trapped between the snowflakes formed bubbles as the snow was converted to ice. By sampling these air bubbles, scientists can determine what the composition of the atmosphere was at the time the snow fell. Ask students to speculate about how scientists might find such information useful. *(By correlating the atmospheric composition data with other information, scientists can learn about the climates and environments of the past. They may also determine the connection between atmospheric composition and global climate.)* Encourage interested students to find out more about sampling techniques that scientists use to learn about the atmosphere. Suggest that they include their findings in their Portfolio.

CLIMATE

FOCUS

A number of factors combine to cause the Earth's climates and seasons.

MOTIVATOR

Cooperative Learning Activity

Have students form small groups (up to five students) to design the perfect climate. Ask the groups to consider questions such as the following: Will your climate have seasons? If so, what kind of weather would typify each season? Students should describe their perfect climate by including information about its high and low temperatures, annual precipitation, wind speed, and so on. By working together, students will come to identify the factors that make climates—and by extension, Earth—habitable.

TEACHING STRATEGIES

Using the Theme of *Energy*

Focus Question: How is the weather an indirect form of solar energy? *(Sample answer: The sun drives the circulation of air, which causes weather.)*

LATITUDE, *page 177*

Background: Latitude and Climate

The closer to one of the poles a location is, the greater the variation between its winter and summer weather. For example, it is common for Fairbanks, Alaska, to experience an annual temperature range of about 83°C (150°F), with pleasant, mild summers but extremely cold winters. On the other hand, in Memphis, Tennessee, the temperature ranges only about 50°C (90°F) in an average year. Summers are warm to hot, and winters are moderate. Mexico City has an even narrower annual temperature range of about 25°C (45°F) and experiences relatively slight seasonal temperature differences.

Using the Diagram:

Figure 7-6, page 177

Ask your students if any of them have ever lived in a place at a very high or a very low latitude. Ask those students who respond to describe the climate of their former home. *(High-latitude climates typically change drastically from season to season. Low-latitude climates vary little throughout the year.)* Ask students who have lived in any very northerly areas to describe the pattern of light and darkness that comes with the changing seasons. *(Students should mention the extreme change in light and darkness between winter and summer.)*

ATMOSPHERIC CIRCULATION PATTERNS, *page 177*

Background: Air Masses and Climate

Many regions of the country have sayings such as, "If you don't like the weather, just wait five minutes and it'll change." This statement reflects the effect of *air masses* on the local climate. Air masses are large bodies of air that contain more or less the same temperature and humidity throughout. When an air mass moves across an area, it displaces the existing air mass, often causing a dramatic change of weather in the process. With hardly a moment's notice, mild, spring-like weather can give way to a frigid gale. This phenomenon is particularly pronounced in the middle latitudes, where air masses often clash. In certain mountainous areas, the movement of a Pacific air mass across the region can raise the temperature 25°C (45°F) in minutes.

Using the Diagram:

Figure 7-7, page 177

Have students examine the diagram. Ask: Based on what you see here, which way do the prevailing winds blow in the United States? *(from west to east)* The prevailing westerly winds of the middle latitudes cause weather systems and air masses to move generally from west to east. This affects the climate. For example, the climate of the West Coast of the United States is predominantly marine. The climate of the East Coast is decidedly not marine. Though influenced by the Atlantic Ocean, the climate is predominantly continental. Students should realize that the actual pattern of atmospheric circulation is much more complex than is shown in the diagram and that it changes with the seasons.

Background: Volcanoes and Weather

Natural events can also affect the weather. For example, the year 1816 (known as "the year without a summer") was characterized by unusually cold weather; the cause of this cold weather was believed to be a volcanic eruption. In 1815, Tambora, an Indonesian volcano, exploded violently, blasting an estimated 25 cu. km (6 cu. mi.) of ash and dust into the air. Dispersed by global winds, the ash cloud covered the Northern Hemisphere with a haze that significantly diminished the amount of sunlight reaching the Earth's surface. The result was the coldest year on record for many areas of Europe and North America.

OCEAN CIRCULATION PATTERNS, *page 178*

Geography Connection

Use the following example to illustrate the effects of ocean currents on climate. Tell students that San Francisco,

California, and Wichita, Kansas, are at about the same latitude, yet for the month of January, the average high temperature in Wichita is less than the average low temperature in San Francisco. On the other hand, for the month of July, the average low temperature in Wichita is about equal to the average high temperature in San Francisco. Ask students what might cause this peculiar effect. *(San Francisco's climate is moderated by the Pacific Ocean. Wichita, located in the middle of the continent, is far from any moderating influences.)*

Using the Diagram:

Figure 7-8, page 178

After students have examined the diagram, ask them how the Earth's climate would differ if the currents suddenly stopped flowing. *(The Earth might develop climatic zones paralleling the lines of latitude, with regular decreases in temperature from equator to poles. Weather and climate patterns would become less complex.)*

Seasonal Changes in Climate, *page 179*

Background: Latitude and Day Length

The tilt of the Earth's axis causes night and day to vary in length from season to season everywhere except the equator. The amount of change varies with distance from the equator. In far northern or southern regions the change in day length between summer and winter is quite dramatic. For example, at the summer solstice, Fairbanks, Alaska (latitude 64° N), experiences 24 hours of light, with only a brief twilight period. At the winter solstice, however, Fairbanks experiences almost 24 hours of darkness—the sun peeks over the horizon for only an hour or two around noon.

Meeting Individual Needs

Learners Having Difficulty Use a globe and a light source to graphically demonstrate how the Earth's tilt causes seasons. Place an upright reading lamp (with the shade removed) in the middle of the room. Turn off the lights. Circle the light source while holding the globe, taking care to keep the axis oriented in the same direction. Ask students to describe the way in which the pattern of light and dark changes as the "Earth" revolves. Then ask students to identify the "seasons," "equinoxes," and "solstices" as they occur. Turn the globe as you revolve, showing students how some areas can be in continuous darkness while others experience continuous sunlight.

Multicultural Connection

The Celts, an ancient group of people that once populated northern Europe, recognized two seasons: the "light" time and the "dark" time. Ask students to speculate about the basis for this classification. *(Because they lived in a northern area, the Celts experienced dramatic changes in day length from winter to summer. Summer days were extremely long, the "light" time, while winter days were extremely short, the "dark" time.)*

Closure

Section Review

Answers to these questions are provided in the Answer Key on page T150.

Alternative Assessment

Tell students to imagine that it is the distant future and they have been appointed to an advisory committee that is studying the feasibility of eliminating the Earth's seasonal differences. Currently under discussion is a plan to do this by moving the Earth closer to the sun. Tell students that they have just witnessed a two-hour presentation showing how scientists—whose technology and expertise is undeniable—propose to move the Earth.

Explain that the committee members must now decide whether moving the Earth closer to the sun will actually eliminate the seasons. Tell students to write a short, persuasive article explaining their viewpoint. The article should include an alternative plan if they do not agree that moving the Earth closer to the sun will change the seasons. *(Students should note that moving the Earth closer to the sun would not eliminate seasonal differences; it would merely make the Earth hotter all around. However, devising a way to eliminate the Earth's axial tilt would put an end to the seasons.)* Suggest that students include their article in their Portfolio.

Reteaching

So that students can better understand the effect of the Earth's circulation patterns on climate and weather, discuss how isolated events such as the eruption of Mount Pinatubo can cause changes to the weather all over the world. Emphasize that just as the circulating winds distribute heat and moisture, they also distribute the gases and ash ejected by volcanic eruptions. The ash particles reduce the amount of sunlight that reaches the Earth, causing cooling. The gases released by volcanoes cause other changes as well, such as enhancing the greenhouse effect (which students will study in the next section) and contributing to destruction of the ozone layer (which students will learn about in the last section of this chapter).

Extension

Tectonic activity will continue to rearrange the Earth's continents in the future. Europe and North America will continue to spread apart, allowing greater circulation between the Arctic and Atlantic Oceans. At the same time, Antarctica will move away from the South Pole. Ask students to imagine that they have been transported 50 million years into the future. How is the Earth different in terms of the events just described? Have students write a story, illustrated if they wish, about the Earth of the distant future. *(Students' answers will vary but should reflect the fact that the Earth will be much warmer and sea level higher because, unlike today, ocean currents will reach both polar regions, warming them. Additionally, Antarctica, having moved away from the South Pole, will no longer be an icebound continent. Students may also suggest that the continents will be rearranged and that some of today's prominent geographic features [like the Rocky Mountains] will be eroded.)* Suggest that students include their story in their Portfolio.

SECTION 7.3

GREENHOUSE EARTH

FOCUS

Students are introduced to the concept that certain trace gases in the Earth's atmosphere cause the atmosphere to trap heat like a greenhouse. Increasing levels of one of these trace gases (carbon dioxide) may cause a warming of the Earth's climate.

MOTIVATOR

Find a topographic map of a low-lying coastal region of the United States, such as Florida. Locate the 150 ft. contour, and make sure all students can see it. Explain that if the polar icecaps were to melt, everything below the 150 ft. contour would be underwater. Ask: What might make the polar icecaps melt? *(an increase in temperature at the poles)* Then ask if any students have heard of global warming—worldwide increases in average temperatures. *(Most students will say yes.)* Tell students that if global warming is occurring, a worst-case scenario suggests that water levels could rise to the point you showed on the map. Ask students to think about the effects of such an event on humans. You might close by telling students that this will happen someday, regardless of whether global warming is, in fact, occurring. Point out that Earth is still experiencing an ice age, and that this ice age will almost certainly come to an end sooner or later.

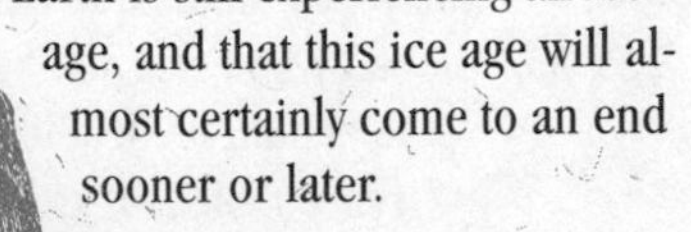

TEACHING STRATEGIES

THE GREENHOUSE EFFECT, *page 180*

Using the Photograph:
Figure 7-11, page 180
Point out that the greenhouse effect is normally a positive thing. It is estimated that without the greenhouse effect, the average temperature of the Earth would be about 33°C (60°F) lower. Ask students to consider how the Earth might be different in such a case. *(Sample answer: The planet would probably be too cold for life to exist.)*

Using the Transparency
Figure 7-12 is available as a Teaching Transparency.

GREENHOUSE GASES AND THE EARTH'S TEMPERATURE, *page 183*

Geography Connection
Different nations produce different amounts of greenhouse gases. Using an atlas or almanac, have students determine which nations use fossil fuels most heavily. *(developed nations such as the United States, Japan, and Germany)* Ask students to also consider how these figures have changed over the past few decades and why. *(Usage has increased markedly over the last few decades due to rapid industrialization and the greater use of the automobile.)* Then ask students to predict how the figures might change over the next few decades, and ask them to support their predictions. *(Usage will most likely decline in the decades to come as heavy industry is shifted to the developing nations and cars become more fuel efficient. Nations that now use relatively few fossil fuels, such as the developing nations, will most likely escalate their use of fossil fuels in the future as they become industrialized.)*

Meeting Individual Needs
Gifted Learners Ask students: What variables not listed in the text could possibly affect the global climate? What factors are probably not included in current climate models? *(Examples might include volcanic activity, asteroid impacts, massive die-offs of plants or animals due to disease, nuclear war, and so on.)* Students may have to do a bit of research to come up with a comprehensive list.

Reinforcing Ideas
If you have the computer program *SimEarth*, let students use it to create a simulated world. Have them manipulate the levels of atmospheric gases such as oxygen and carbon dioxide and observe how the simulation responds. Ask them to suggest other variables that could be manipulated and make predictions about the effects of doing so.

A WARMER EARTH, *page 184*

Discussing the Issue
Ask students: If global warming is such a hazard, why haven't we already switched from fossil fuels to fuels that produce no carbon dioxide when burned? *(Sample answer: Our society is heavily dependent on fossil fuels; switching to other fuels would be difficult and costly. Furthermore, with the exception of*

hydrogen, every fuel produces some carbon dioxide when burned.)

Using the Case Study: *Computer Models and Earth's Future Climate, pages 182–183*
Students may wonder why there is so much ambiguity among scientists about the subject of global warming. Point out that weather and climate are extremely complicated phenomena that are far from being fully understood and that even the most sophisticated climatic models are only crude approximations of the real thing. In part, models are limited by the amount of computational power available. Better models could be devised, but they would be so computation-intensive that even the fastest computers would take years, or even centuries, to run them.

Workable models must incorporate simplified representations of natural processes. Consequently, these models are never completely accurate. The only question is, how inaccurate are they? As a case in point, remind students that meteorologists have continually improved the models they use to forecast weather while employing ever more powerful computers to run these models. Yet these models are only modestly successful at accurately predicting weather more than two or three days in advance. And they often make major errors even in short-range forecasts. Without question, further refinements and faster computers will lead to a new generation of powerful, more accurate models.

Discussing the Issue

Tell students that there is no wrong answer to the question you are about to ask and that they should feel free to answer honestly. Then ask: Which of you are convinced that humans are causing the Earth's climate to undergo a long-term warming trend? Have all of the students who raise their hand form a group and work together to provide evidence to support their view. *(sample evidence: warmer-than-average temperatures over the last few years, declining snow fall, retreating glaciers, slightly rising sea levels)* Ask students who are not convinced to form another group. They should explain why they are not convinced and what kind of evidence they would like to have in order to confirm that the climate is indeed warming due to human activity. *(Students in this group may suggest that variations in global climate are natural and frequent and that our current warming phase did not just begin but had probably been underway for many decades before there was a significant contribution of CO_2 by humans. They may also mention the fact that the winter of 1993–1994 was one of the coldest on record for most of the Eastern United States. The evidence that these students request could include meteorological, oceanographic, and atmospheric data from the past and records of volcanic events.)*

Once students have discussed their opposing points of view, ask them to assume that global warming is actually happening—regardless of the cause. Then suggest that they work together as a class to develop a list of possible positive and negative effects associated with the phenomenon. Write their ideas on the chalkboard. *(Negatives: longer, hotter, dryer summers; diminished capacity to grow certain foodstuffs; more frequent, violent storms; disruption of ecosystems; higher cooling bills; increased problems with pest insects; flooded coastal areas due to higher sea levels. Positives: longer growing seasons; possible rainfall increase; opening of new agricultural lands; milder winters; lower heating bills.)*

CLOSURE

Section Review

Answers to these questions are provided in the Answer Key on page T150.

Alternative Assessment

Tell students to imagine that they are employed by the Environmental Protection Agency and have been assigned to design a public service advertising campaign to encourage the public to reduce carbon dioxide levels. Students' efforts should be persuasive and creative, yet factual and informative. Have students place their work in their Portfolio.

Reteaching

Have a "chalk talk" to clarify how the greenhouse analogy can be used to describe global warming. Using chalkboard drawings, explain that a greenhouse is able to capture heat because the light energy that enters the greenhouse through the glass is turned into heat energy. That energy remains in the greenhouse because the heat energy does not flow back through the glass as well as light energy does and because the glass prevents the warm air from blowing away. Similarly, in the vacuum of space there is no way for the Earth's heat to "blow away." And greenhouse gases in the atmosphere are, like the glass, largely impenetrable to the infrared radiation (heat) emitted by the Earth as a result of absorbing solar radiation. Thus, the radiation is prevented from escaping, and the heat energy is retained.

Extension

Have each student develop a plan of action that individuals or families could use to cut their carbon dioxide emissions by 50 percent. The plan should be detailed and reasonable. For example, instead of saying "cut car usage by 50 percent," students should suggest how a person could actually reduce his or her car usage. That is, students may suggest consolidating all errands into a single trip or making no unnecessary trips.

THE OZONE SHIELD

FOCUS

Students learn that the ozone layer serves as a protective shield against ultraviolet radiation. They are also presented with evidence that suggests humans may be causing damage to the ozone layer.

MOTIVATOR

Ask students how many of them have ever had a bad sunburn. Then ask: What caused the sunburn, and how long did it take to get it? *(Ultraviolet radiation in the sunlight causes the burn. Three or four hours in the summer sun would give most people a bad sunburn.)* Ask students to think about how life would be affected if it took only minutes, not hours, to get a bad sunburn. This could occur if the ozone layer, which screens out almost all of the sun's ultraviolet light, were seriously damaged. Explain that there is concern that the ozone layer is being significantly damaged by artificial chemicals and that this subject will be taken up in the section they are about to begin.

TEACHING STRATEGIES

Background: Ozone Formation

Ozone is constantly being created and destroyed by natural processes. In the stratosphere, some diatomic oxygen, O_2, is dissociated by high-energy solar radiation into two molecules of monatomic oxygen, O. O reacts with O_2 to form ozone, O_3. During the day, however, ozone has a short lifetime because it is rapidly degraded by solar radiation into O and O_2. But because new ozone is constantly being formed, the net level of ozone does not change much.

BREAKING APART CFCs, *page 186*

Using the Transparency

Figure 7-18 is available as a Teaching Transparency.

Real-World Connection

People who work with chlorofluorocarbons, such as air conditioner repair personnel, must be trained and licensed in the proper handling of CFCs. Strict guidelines must be followed. Invite someone who repairs air conditioners to your class to explain how the handling of CFCs has changed in recent years.

THE OZONE HOLE, *page 187*

Background: Ozone Distribution

The stratospheric ozone concentration varies with latitude and the time of year. Generally, the concentration is highest in the spring and lowest in the summer and fall. Ironically, ozone is most abundant over the poles (where fewer organisms exist) and least abundant over the equator (where more organisms exist). The peak equatorial ozone concentration is about 60 percent of the peak polar ozone concentration.

STOPPING THE OZONE EATERS, *page 188*

Background Information: Opposing Viewpoints

It is by no means certain that CFCs are causing significant damage to the ozone layer. Some experts cite the following points.

- The "ozone hole" over Antarctica is most likely a natural, seasonal phenomenon. For example, the lowest ozone level ever recorded over Antarctica occurred in 1958, years before CFCs became widely used.
- Natural sources of chlorine, the main ozone-damaging chemical in CFCs, exceed artificial sources. Volcanic eruptions, for example, can put hundreds of thousands of tons of chlorine into the atmosphere. Huge amounts of salt (sodium chloride) from sea water are also carried aloft by winds.
- Even the highest CFC levels in the stratosphere rarely exceed a few parts per trillion—an extremely minute concentration that is unlikely to have any significant effect.
- There is conflicting evidence about whether ultraviolet radiation levels have increased at the Earth's surface.

Debate

Randomly divide students into two groups. Tell one group that their assignment is to convince the other group that CFCs are a genuine and serious concern and that world governments are right to implement strategies for reducing and eliminating CFCs. Tell the other group that their task is to demonstrate that CFCs may not be as damaging as supposed and that governments should investigate the issue more fully before committing billions of dollars to switching to CFC substitutes. Allow time for both groups to gather as much evidence as possible to support their point of view. Each group should also develop a recommended plan of action for world governments. Then stage a debate so that both sides can be expressed. After the debate, encourage students to record their personal opinion (independent of any group or peer pressure) in a short paragraph. Students may want to include their opinions in their Portfolio. Reassure students that their point of view will not be judged.

Using the Theme of *Interacting Systems*

Focus Question: What effect, if any, might global warming have on ozone depletion, and why? *(It would probably accelerate it. A warmer climate would increase the incidence of thunderstorms and hurricanes, whose powerful updrafts can pump sea-level CFCs into the stratosphere.)*

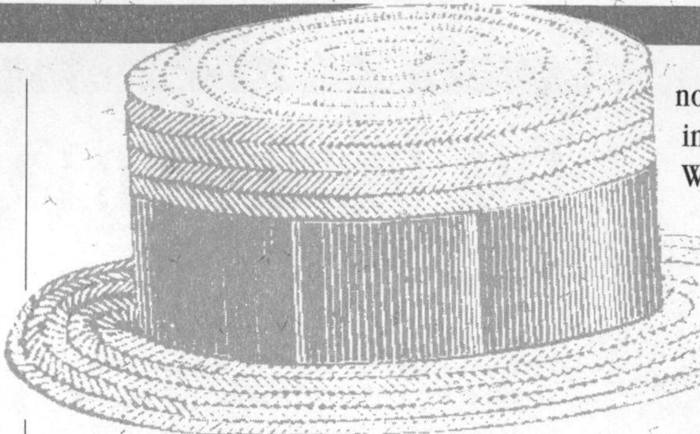

CLOSURE

Section Review

Answers to these questions are provided in the Answer Key on page T150.

Alternative Assessment

Tell students to imagine that they live in a world in which the ozone layer has been destroyed. Have them write a diary entry describing a typical day in their life. *(Answers should demonstrate that students understand the protective role of the atmosphere.)*

Reteaching

Some students may have trouble understanding why, if CFC use is decreasing, the ill effects of CFCs on the ozone could actually be increasing. Explain that it can take many years for CFC molecules to reach the stratosphere. The CFCs causing trouble today may have entered the atmosphere 15 or more years ago, during a time when CFC use was increasing. Ask students to predict how this phenomenon will affect the future level of CFCs in the stratosphere. *(There will be a lag time between the actual reduction of CFCs at ground level and the reduction in the stratosphere. CFC levels may continue to increase for a few more years, level off, and then start to decrease.)*

Extension

Encourage students to do research to find out what non-CFC alternatives are available for existing air conditioning and refrigeration systems. Suggest that they try to determine how much it would cost governments to require an immediate, mandatory switch to such products. Also encourage students to find out how the non-CFC systems compare to existing systems. In other words, ask: What are the advantages and disadvantages of such products? Encourage students to share their findings with the class in a creative way. This may be an appropriate item for students to include in their Portfolio.

Answers

Answers to *Chapter Highlights, Chapter Review,* and *Investigation* are provided in the Answer Key for Chapter 7, which begins on page T150.

TEACHING STRATEGIES FOR INVESTIGATION:

GLOBAL WARMING IN A JAR,
pages 192–193

Demonstrate the Greenhouse Effect

This investigation works well as a cooperative learning activity. Divide the class into small groups of three to four students.

Choose a Variable to Test

Be sure that each group has written a clearly stated hypothesis before designing the experiment. When they begin designing, remind students of the need to establish a valid control. Approve these on an individual basis, making sure that each student clearly understands the purpose of the project.

Design an Experiment

Suggest that students follow these steps when designing their experiment:

- design a hypothesis and state it clearly
- develop a mental image of what is happening in the bottle
- make and record observations
- process data
- share or communicate results
- compare the results with original hypothesis
- refine the hypothesis

CHAPTER 

LAND

PLANNING GUIDE

CHAPTER COMPONENTS AND RESOURCES	½ Year Average	½ Year Advanced	1 Year Average	1 Year Advanced
Chapter Opener, p. 196				
EcoLog, p. 196			•	•
8.1 The City, pp. 197–203			•	•
Transparency 27: *Land Use in the United States*			•	•
Section Review, p. 203			•	•
8.2 How We Use Land, pp. 204–212			•	•
Transparency 28: *Clear-Cutting Versus Selective Cutting*			•	•
EcoSkills: *Planting a Tree*, pp. 398–399			•	•
Section Review, p. 212			•	•
8.3 Public Land in the United States, pp. 213–218				•
Environmental Careers: *Environmental Lawyer*, pp. 380–381			•	•
Transparency 29: *Public Land Holdings in the United States*				•
Field Activity, p. 215				•
Case Study: *Public Lands for the Asking*, pp. 216–217				•
Section Review, p. 218				•
End of Chapter				
Chapter Highlights, p. 219			•	•
Chapter Review, pp. 220–221			•	•
Investigation: *Mining for Peanuts*, pp. 222–223			•	•
Making a Difference: *Big-City Farmer*, pp. 224–225			•	•
Chapter Tests			•	•
PACING GUIDE	0 days	0 days	5–6 days	5–6 days

MATERIALS ORGANIZER

Activity	Materials	Advance Preparation
Investigation: *Mining for Peanuts*, pp. 222–223	plastic pan, 30 cm × 35 cm × 15 cm (12" × 14" × 6"), or shoe box; metric ruler; clean sand; soil; 5 to 10 peanuts; grass clippings, twigs, small rocks, etc.; index card; paper clip; craft stick; watch or clock; notebook; pencil (per student group)	none

CHAPTER OVERVIEW

This chapter discusses how land is used and overused, and emphasizes the value of land-use planning.

Section 8.1 describes urbanization and emphasizes how land-use planning can help relieve the problems associated with the urban crisis.

Section 8.2 discusses various ways that land is used to support human needs.

Section 8.3 explores the different uses of public land in the United States and discusses how different priorities for land use can cause conflicts.

RESOURCES

. . . for Teachers

Naar, Jon, and Alex J. Naar. *This Land Is Your Land.* New York City, NY: HarperPerennial, 1993. (This book uses a case-history approach to document the condition of endangered ecosystems. It details some actions that humans can take to repair these ecosystems.)

Sierra Club Public Affairs, 730 Polk Street, San Francisco, CA 94109; 415-776-2211. (This organization has a series of booklets that provide detailed information about each type of public land in the United States.)

United States Department of the Interior, Bureau of Land Management, 1849 C Street, NW, Washington, D.C. 20240; 202-208-5717. (This governmental office can send free information about BLM lands.)

. . . for Students

Conniff, Richard. "Federal Lands." *National Geographic,* February 1994, pp. 2–39. (This article discusses the effects of heavy pressure from many different usages of public lands in the western United States.)

Dolan, Edward F. *The American Wilderness and Its Future: Conservation Versus Use.* New York City, NY: Franklin Watts, 1992. (This book gives a captivating, engaging, and informative discussion of land use in the United States.)

Mitchell, John G. "Our National Parks: Legacy at Risk." *National Geographic,* October 1994, pp. 2–55. (This article examines the status of U.S. national parks.)

Films, Videos, Software, and Other Media

The Forest Through the Trees. Video with study guide. 58 minutes. The Video Project, 5332 College Avenue, Suite 101, Oakland, CA 94618; 800-4-PLANET. (Narrated by Sydney Pollack, this award-winning video examines the controversy surrounding old-growth forests in the United States.)

Ghosts Along the Freeway. Video. 10 minutes. The Video Project, 5332 College Avenue, Suite 101, Oakland, CA 94618; 800-4-PLANET. (This video focuses on two urban neighborhoods affected by nearby superhighways.)

Replanting the Tree of Life. Film or video. 20 minutes. English and French versions. Bullfrog Films, P.O. Box 149, Oley, PA 19547; 800-543-3764. (This award-winning production examines the value of trees and forests, and emphasizes the need for caring for these vital resources.)

Getting Involved

- Use the following EcoSkills activity to enhance student interest in the subject of this chapter: *Planting a Tree,* pp. 398–399.
- A listing of environmental organizations, publications, and volunteer opportunities can be found in the **Laboratory and Field Guide.** Direct students toward organizations that help protect land resources. Some examples include the American Forestry Association, the National Arbor Day Foundation, the Natural Resources Defense Council, and the Student Conservation Association.

USING THE QUOTATION

"The size of the parcel of land matters less than the relationship of the people to it."

Ask students to describe, in their own words, the meaning of the quotation. *(Sample answer: The worth or productivity of a piece of land is measured by how effectively it can be used, not simply by its size, because a large plot of land used improperly could yield less than could a small plot used properly.)* Continue by asking: How will this have more and more relevance in the coming decades? *(In the future, people will have to get more out of less land as the human population grows and the amount of usable land shrinks.)*

USING THE ECOLOG

To assess students' knowledge about the subject of this chapter, have them answer the EcoLog questions on page 196. In answering these questions, students will reveal their prior knowledge, which is usually very basic and often flawed or even misconceived. Then, by revising their answers at the conclusion of the chapter, students will have the opportunity to eliminate any misconceptions and to see firsthand how their knowledge has grown.

S E C T I O N 8.1

The City

Focus

This lesson focuses on how rapid urbanization has strained land resources. Methods for dealing with the resulting problems are also presented.

Motivator

Write the following questions on the chalkboard: What is a city? Where do all of the goods and services that city dwellers use come from? What percentage of the world's population lives in cities? What are some of the problems that city dwellers face? Then divide the class into small groups, and have them work together to answer the questions. After 15 minutes or so, encourage class discussion. Without providing any answers, tell students that they will learn more about the city and the resources required to sustain it in this section. After teaching this section, you may wish to have students divide into groups once again and revisit the above questions.

Teaching Strategies

The Urban-Rural Connection, *page 198*

Using the Transparency

Figure 8-2 is available as a Teaching Transparency.

The Urban Crisis, *page 199*

Background: The Largest City on Earth

The largest city on Earth is probably Mexico City, which is the capital of Mexico. Although the Tokyo-Yokohama megalopolis is officially larger (28.5 million people versus 24 million people), most experts agree that population data for Mexico City do not adequately reflect the millions of squatters that live at the city's fringes. In fact, no one knows the exact population of Mexico City. Some estimates range as high as 40 million people.

Using the Theme of *Scale and Structure*

Focus Question: Is the largest city necessarily the most densely populated? *(no)* The following examples should help students realize why. With a population of about 14.6 million people, New York City is the largest city in the United States. It has a population density of only 11,500 persons per sq. mi. In contrast, the Tokyo area, with 28 million people, has a population density of about 27,500 people per sq. mi. And Hong Kong, the most crowded city on Earth, crams 5.6 million people into an area of only 23 sq. mi., giving it the amazing population density of almost 250,000 people per sq. mi. If everyone were spread out evenly, each person would have an area equivalent to a square of slightly more than 10 ft. per side. Invite students to find out the population density of your community.

Meeting Individual Needs

Learners Having Difficulty Divide the class into small groups, and give each group a few current newspapers or magazines. (If you do not live in an urban region, do not use local newspapers, but instead use newspapers from a large city.) Have students search through these resources for articles that describe the urban crisis in some way. Typical articles that fall into this category include stories about increasing crime, homelessness, soaring taxes, decaying neighborhoods, gangs, graffiti, traffic congestion, and so on. After students have had sufficient time to look through the magazines and newspapers, have a class discussion about what the urban crisis is and what is being done to alleviate it. Students may wish to work together to make a bulletin-board display or a poster that reflects the grim and the hopeful sides of the urban crisis.

Suburban Sprawl, *page 200*

Meeting Individual Needs

Gifted Learners As a method of controlling the growth of cities, some states or counties have imposed urban limit lines around swelling population centers. Urban limit lines are simply boundaries beyond which a municipality is not permitted to grow. These limit lines effectively contain urban sprawl and force the full development and redevelopment of the urban areas they contain. A major disadvantage of this approach is that it forces the cost of housing to rise by increasing the price of available land and restricting the available supply of houses. Suggest that students find out more about urban limit lines and incorporate their findings in the form of a written or videotaped news report about the subject. If you live in an urban or suburban area, suggest that students contact city planners to find out if urban limit lines have ever been considered for the area. Encourage students to include their news report in their Portfolio.

LAND-USE PLANNING, *page 201*

Real-World Connection

Invite a city planner to your classroom. Ask that person to discuss the problems and challenges of his or her job. Encourage students to prepare questions in advance for the guest speaker. Sample questions are: How does the planning department envision the city will look in 5, 10, or 20 years and beyond? How large will the city be? What changes will have to be made to the infrastructure? How will urban problems be prevented?

MASS TRANSPORTATION, *page 202*

Background: Why Cities Need Mass Transportation

The downtown areas of many large cities, and the roads leading to them, are frequently congested to the point of gridlock. A major cause of this problem is heavy traffic into and out of the city. For example, more than a million cars head into Manhattan (a region covering only about 62 sq. km, or 24 sq. mi.) every weekday morning and leave every weekday night. Traffic is so bad that it is often faster to walk a given distance than to drive. Fortunately, Manhattan has a subway system, which many people rely on rather than private transportation. There is also an extensive system of shuttle trains that take many commuters to and from their homes in the suburbs. Were it not for mass-transportation systems, traffic in Manhattan, bad as it is now, would be much worse.

Background: Why Some Cities Don't Have Mass Transportation

Proposed mass-transportation systems must generally be authorized by referendums, and their high construction and operating costs often result in their defeat at the polls. Most mass-transit systems in America and abroad are subsidized. Americans are less willing to support heavy subsidies than their European counterparts, so the cost of the mass-transit service is much higher for riders in America than for those in Europe.

INNER-CITY RENOVATION, *page 202*

Cooperative Learning Activity

Tell students to imagine that they have been assigned to a planning commission in charge of renovating a severely deteriorated inner-city area. Explain that their planning options are unlimited but should include the use of incentives to encourage new investment and new residents. Ask students to form groups of three or four to consider the problem and develop a plan that involves at least five renovation ideas. **Note:** To add interest, assign each group a specific area of your town or city.

OPEN SPACES, *page 203*

Math Connection

How much open space does a city need? Suppose that an independent consulting firm has recommended that, in order to maintain a desirable urban setting, a city should set aside at least 1 acre of parkland for every 200 citizens. Imagine a city that has a population of 100,000, is growing at a rate of 3 percent per year, and currently has 400 acres of parkland. To the nearest acre, how much parkland will the city need to acquire over the next year? *(100,000 people × 1.03 = 103,000; 103,000 people ÷ 200 persons/acre of parkland = 515 acres. The city now has 400 acres of parkland, so it should add 115 acres over the next year.)*

CLOSURE

Section Review

Answers to these questions are provided in the Answer Key on page T151.

Alternative Assessment

Ask students to design a model community that is to be home for 1,000 people. Students may select a specific geographic area in which to locate the community, or they may invent one. *(Student designs should anticipate future population growth in the region. In addition, they should demonstrate some concern for alleviating the problems that often occur in urban areas. For example, the plan could include open areas and bike paths. Students should also mention the need for community services such as electricity, water, trash pick-up, police and fire protection, etc.)* Suggest that students include their design in their Portfolio.

Reteaching

Write the following terms on the board: urbanization, urban crisis, loss of forests and farmland, suburban sprawl, and infrastructure. Work with your class to make a concept map using these terms.

Extension

Tell students to imagine that they are candidates for the office of city manager for their community (or a community of their choice). Have each student compose a campaign speech in which he or she promises to address certain critical needs of the community. Students' speeches should cite at least three specific real-world problems of some significance and provide solutions to each. To gather information, students may want to look through recent editions of the local paper from their community. Suggest that students include their speech in their Portfolio.

How We Use Land

Focus

People use the land for many purposes. A major challenge of our time is to move from a lifestyle based on nonsustainable land-use practices to one based on sustainable land-use practices.

Motivator

Ask your class: How many trees have we used up lately? Explain that you are going to conduct an activity to figure out how much paper your school uses each month and use that data to estimate how many trees were felled to make that paper. Ask students how much paper they typically use in a month. Answers will vary. Suggest a figure of 100 pages of typical letter-sized, 20# bond per student per month. Multiply this figure by the number of students in your school. Assuming 1,000 students, then 100,000 pages were used. A 5 ft. section of a typical tree yields about 10,000 sheets, and the typical tree has about 30 ft. of usable trunk. So, each tree yields about 60,000 sheets. 100,000 sheets ÷ 60,000 sheets/tree = 1.67 trees. Ask: How many trees did we use last year? Assuming a nine-month school year, students at the school used 9 × 1.67, or slightly over 15, trees.

Teaching Strategies

Notable Quotes

"A continent ages quickly once we come."

Ernest Hemingway

"The horizons are still mine. The ragged peaks, the cactus, the brush, the hard brittle plants, these are mine and yours. We must be humble with them."

Simon Ortiz

Harvesting Trees, *page 205*

Background: Harvesting Old-Growth Forests

Old-growth forests contain stands of trees that are hundreds or even thousands of years old. In the United States, the last remaining old-growth forests are found in the humid regions of the Pacific Northwest. Old-growth trees are magnificent specimens that may be 90 m (300 ft.) tall and 15 m (50 ft.) in circumference. Continual logging has reduced these forests to a fraction of their former size and threatens their existence altogether. Because of their importance, old-growth forests have become a backdrop for one of the fiercest environmental battles raging today. Environmentalists contend that the loss of these forests would represent more than an aesthetic loss. The entire regional ecosystem could be damaged. Salmon fishers also oppose the logging of old-growth forests because without the trees to hold soil in place, salmon spawning grounds in rivers and streams are often covered by mud and debris. Thus, with the loss of the old-growth forests, the salmon-fishing industry could be destroyed in some areas. On the other hand, loggers contend that the logging ban would eliminate thousands of jobs and cause economic hardship in many areas. Students will learn more about this controversy in the Points of View: *Owls Versus Loggers,* on pages 276–277 of their textbook.

Deforestation, *page 205*

Using the Transparency

Figure 8-11 is available as a Teaching Transparency.

Protecting Forests, *page 207*

Multicultural Connection

Wangari Maathai (pictured in Figure 8-12) is a Kenyan environmentalist who has won international recognition for her conservation and humanitarian efforts. As founder of the Green Belt Movement, she not only has been instrumental in coordinating a massive network of community reforestation projects in Kenya, but also has been involved in starting similar movements in a dozen other countries, including the United States. Maathai's work has helped slow environmental degradation as well as help provide income and firewood for many poor rural dwellers.

At times a political dissident, Maathai is also a strong believer in the power of the individual: "When any of us feels she has an idea or an opportunity, she should go ahead and do it. . . . One person *can* make a difference."

Multicultural Connection
The National Center for the Production of Native Images has created a videotape of an innovative program called the Earth Bridge Project. The project involves blending multicultural instruction with lessons on ecology. This is done by uniting students from a variety of cultures and allowing them to spend several weeks in the woods, where they learn survival tactics and take part in workshops on various environmental and cultural subjects. To order the *Earth Bridge* video, contact The Video Project, 5332 College Avenue, Suite 101, Oakland, CA 94618; 800-4-PLANET.

RANCHING, *page 208*

Using the Theme of *Energy*
Overgrazing has largely destroyed the native grass cover of much of the American West and has allowed nuisance plants such as mesquite, juniper, and cactus to invade. How might this change the flow of energy through the ecosystem? *(There is less energy flow from plants to animals because fewer animals are able to feed on the nuisance species than on grass. There may be a smaller population of ground-dwelling animals as well, resulting in a diminished flow of energy between first-order consumers and second-order consumers.)*

Background: Property Rights
In the United States, landowners are traditionally given very strong property rights. Essentially, they may do with their land as they wish with few restrictions. The United States is unusual among nations in granting such exclusivity and decision-making power to landowners. For example, European countries typically require landowners to observe strict land-use guidelines and to permit at least limited access by the public where appropriate. Some people would like to see similar guidelines in the United States. But, because landowners usually have strong influence in Congress and in state legislatures, legislators have been unwilling or unable to redefine property rights. Supporters of strong property rights argue that private ownership promotes sound land-management practices because people generally take better care of what belongs to them than what belongs to someone else. Critics of strong property rights argue that the profit motive drives landowners to take shortcuts and seek maximum possible short-term yield from the land at the expense of its long-term health.

Real-World Connection
Have students arrange an interview with a landowner or developer in your area. Students should inquire about the size and type of the land, the land's use today and in the past, the extent to which the land is available for public use, and the regulations that govern the land's use. Students may find it helpful to practice interviewing techniques with each other beforehand. Have students report their findings to the class.

CLOSURE

Section Review
Answers to these questions are provided in the Answer Key on page T151.

Alternative Assessment
Tell students to imagine that they have been transported back in time. They are now settlers of the American wilderness. They must build a home and supply themselves with food by hunting and raising animals and growing and gathering plants. Tell students to write a letter to an imaginary friend or relative on the East Coast. The letter should describe the student's impact on the local environment. It should also be specific and realistic. Examples of descriptions that students might include in their letters are how much land must be cleared, what those cut trees will be used for, what animals might be displaced, where building materials and furnishings will come from, where fuel will come from, what will happen to waste, where water will come from, how he or she will maintain an adequate food supply, etc. Invite students to include their letter in their Portfolio.

Reteaching
Work with students to construct a concept map to represent the ideas presented under the heading "Harvesting Trees," pages 205–207. At a minimum, the map should use all of the vocabulary terms under the heading. A sample concept map is shown on page 219. Have students summarize the concept map in writing.

Extension
Tell students that a large tract of pristine desert has just been sold to a development company, and that this company plans to build a residential community. The community will include a golf course, a tennis complex, swimming pools, and a track for off-road bicycles. In addition, each home will feature lush landscaping and large lawns. Ask students to (a) draw on the knowledge they gained in Chapter 4 about desert ecosystems to outline the probable effects of this development on the plants and animals living in the area and (b) develop a list of suggestions, in the form of a letter, that will guide the development company in making the community more environmentally sound. Suggest that students include their letter in their Portfolio.

Public Land in the United States

Focus

Large tracts of this country's land are owned and managed by the United States government for use by the American people. This section discusses the various ways in which this public land is used and the problems arising from its usage.

Motivator

Tell students to imagine that as a result of recent legislation, all national parks and wilderness areas have just become available for private development. Pass out a slip of paper to each student with the following message written on it: *Dear Friend, It has recently come to our attention here at Wilderness Properties, Inc. that large amounts of prime real estate property are now available in our national parks and wilderness areas. The land is being offered for a mere $100 per acre. This property will easily triple in value in the next year! If you are interested in an investment at this time, please check the box below, and you will be added to our list of preferred clients.*

Encourage students to respond to the request by checking the box or by writing "no, thank you." Then collect the slips of paper, and count the number of students who responded positively. Express this as a percentage of the class. Discuss with students the likely outcome of this scenario. Ask: Would the national interest be better served by such a policy? *(Opinions will vary. Sample answers: No, because public lands would become private. Yes, because taxes may decrease due to reduced government expenses.)* Would you be better served? *(Opinions will vary.)*

Teaching Strategies

This Land Belongs to You and Me,
page 214

History Connection

In the late 1970s and the early 1980s a movement that came to be known as the Sagebrush rebellion sought to reduce federal land ownership in West. Sagebrush "rebels" argued that public management by distant bureaucracies was invariably mismanagement. They contended that only a private landowner with a vested interest could be expected to properly manage the land. Critics charged that the rebellion's prime motivation was greed: It hoped to force the sale of vast federally owned lands to private interests at incredibly low prices. The only people that stood to gain, they said, were the "rebels" themselves. The Sagebrush Rebellion ultimately failed, but many of its ideas were adopted by the Wise-Use Movement, a private-property-rights organization that is currently promoting its agenda in about 40 states.

Meeting Individual Needs

Second-Language Learners Ask second-language learners to summarize, in their own words, the main points of this section. Suggest that, as an aid, they also construct a concept map using the section's vocabulary terms. Have students whose English skills are better developed assist students whose English skills are not as well developed. Encourage students to include their concept map in their Portfolio.

Managing Our Public Land,
page 214

Using the Transparency

Figure 8-21 is available as a Teaching Transparency.

Using the Case Study: *Public Lands for the Asking, pages 216–217*

Most students will be surprised to find that the situation in the Case Study exists, and some may wonder why something hasn't been done about it. Point out that the mining interests that would be affected by any changes are very influential. Western senators, whose states would be affected, generally vote as a bloc on this issue and use their influence to induce other senators to vote the same way. As a result, no changes are made. Critics of reform point out that paying royalties would be an economic hardship and would cause many operations to fail. Ask: Is this a valid argument? *(Not really; mineral companies, such as oil companies, that operate on nonpublic land pay royalties yet survive and make a profit.)* The mining reform issue was recently revived when a Canadian company acquired a large tract of western land that likely contains billions of dollars worth of gold and other minerals.

Being a foreign-owned mine, the extracted wealth would flow out of the United States. Ask: Is this a "fair deal" for the United States? *(Opinions will vary. Most students will probably say no.)*

Background: The Tragedy of the Commons

Again and again throughout history, resources shared by all and managed by none have been abused, often to the point of their destruction. This has been called the "tragedy of the commons." The phenomenon is partly the result of greed and partly the result of the sense of urgency that develops when people think they are in competition with others for a resource. For example, a person might think: If I don't use this, then someone else will. So they end up using more of the resource than they would have otherwise. When many people use this reasoning, a resource can be depleted quickly. Another aspect of the tragedy is the collective damage that is done to a resource when many people individually inflict a small amount of damage. For example, the atmosphere is owned by none and used by all. Each person adds some burden of pollution to it and thinks that his or her tiny addition won't matter. But when millions of tiny events are combined, the effect is significant. Ask: What is the solution to the tragedy of the commons? *(Sample answer: individual awareness of the problem coupled with the obligation to act properly; a management system that provides fair access but controls exploitation)*

WILDERNESS, *page 215*

Discussing the Issue

In order to encourage students to think seriously about the philosophy behind land-management practices, pose the following questions: Why do we need to preserve and protect the wilderness, anyway? Don't the needs of people take precedence over trees and animals? Students should realize that these are legitimate questions that some people ask in all earnestness. Encourage students to discuss all possible responses to these questions.

CLOSURE

Section Review

Answers to these questions are provided in the Answer Key on page T152.

Alternative Assessment

Ask students to write a "want ad" for a national park, national forest, or wilderness area. Students should describe the qualities that each area should have. *(Sample answer: National Park Wanted: Region should have substantial historical, natural, ecological, scenic, or geological value. Must be willing to remain largely undeveloped. Must be able to handle crowds without faltering.)* Suggest that students include their ad in their Portfolio.

Reteaching

Suggest that students read John G. Mitchell's article: "Our National Parks: Legacy at Risk," *National Geographic,* October 1994, pp. 2–55. This is an article about the woes of the United States national park system.

Extension

Tell students to picture the following scenario. A group of cave explorers has just discovered a huge, previously unknown cavern with a unique ecosystem

containing a variety of organisms. The landowner, Wilma Smith, has a dilemma. She can either sell rights to the cave to Underground Resorts, Inc. (which has made a generous offer) or sell the rights to the United States government, which has offered less money but will turn the site into a national park. Suggest that students write a letter to Wilma, expressing their opinion about what she should do. Encourage students to include their letter in their Portfolio.

Answers

Answers to *Chapter Highlights, Chapter Review,* and *Investigation* are provided in the Answer Key for Chapter 8, which begins on page T151.

TEACHING STRATEGIES FOR INVESTIGATION:

MINING FOR PEANUTS,
pages 222–223

Create a Model Site

Students should make very accurate, scale-model drawings of their sites (or take instant photos) to show the locations of the peanuts and to document the condition of the site prior to mining.

Mine for Ore

Students may be tempted to treat the mining portion of this activity less seriously than they should. Make sure that they approach the matter scientifically and proceed systematically. Ask: How do the time limits imposed affect the outcome? *(Less care is taken in extracting the ore and properly disposing of the waste.)* How is this like real life? *(Companies typically work as quickly as they can to keep expenses low and sometimes "cut corners" in the process.)*

C H A P T E R 

FOOD

PLANNING GUIDE

CHAPTER COMPONENTS AND RESOURCES	½ Year Average	½ Year Advanced	1 Year Average	1 Year Advanced
Chapter Opener, p. 226				
EcoLog, p. 226		•	•	•
9.1 Feeding the People of the World, pp. 227–230		•	•	•
Transparency 30: *World Grain Production*		•	•	•
Section Review, p. 230		•	•	•
9.2 Agriculture and Soil, pp. 231–239		•	•	•
Transparency 31: *Arable Land Loss Through the Year 2000*		•	•	•
Field Activity, p. 232				•
Transparency 32: *Structure of Fertile Soil*		•	•	•
Transparency 33: *Worldwide Soil Erosion*		•	•	•
Case Study: *Aquaculture: A Different Kind of Farming,* pp. 236–237		•	•	•
Section Review, p. 239		•	•	•
9.3 Pest Control, pp. 240–244		•	•	•
EcoSkills: *Eliminating Pests Naturally,* pp. 390–391			•	•
Section Review, p. 244		•	•	•
End of Chapter				
Chapter Highlights, p. 245		•	•	•
Chapter Review, pp. 246–247		•	•	•
Investigation: *The Case of the Failing Farm,* pp. 248–249			•	•
Points of View: *Genetically Engineered Foods,* pp. 250–251			•	•
Videodiscovery STS Science Forums: Vol. 1—*Food Production & the Environment;* Vol. 2—*Transgenic Organisms, Eliminate Mosquitos*			•	•
Chapter Tests		•	•	•
PACING GUIDE	0 days	6–7 days	8–10 days	8–10 days

MATERIALS ORGANIZER

Activity	Materials	Advance Preparation
Investigation: *The Case of the Failing Farm,* pp. 248–249	cafeteria tray or sheet cake pan; sandy soil; metric ruler; small blocks; twigs; toothpicks; fresh grass clippings; dried peas or beans; spray bottle filled with water; fork; narrow strip of carpet; miscellaneous materials, such as classroom objects; notebook; pen or pencil; paper towels or rags (per student group)	none

Chapter Overview

This chapter should give students a sense of how difficult it is to feed the world's growing human population.

Section 9.1 analyzes why people go hungry, defines the green revolution, and explains how new technology has helped feed the world's people.

Section 9.2 discusses the kind of land, soil, and farming methods needed to productively grow crops. The problems of soil erosion, salinization, and desertification, as well as possible solutions to these problems, are also addressed.

Section 9.3 focuses on pest control. Students learn the value of pesticides, the problems associated with pest control, and possible alternatives to pesticides.

Resources

. . . for Teachers

Soil and Water Conservation Society, 7515 Northeast Ankeny Road, Ankeny, IA 50021-9764; 800-843-7645 (This agency can provide a variety of educational information free of charge.)

United States Department of Agriculture (USDA) Soil Conservation Service, P.O. Box 2890, Washington, D.C. 20503; 202-205-0026. (This government agency can provide free publications about the land and soil in your region.)

. . . for Students

Billings, Charlene W. *Pesticides: Necessary Risk.* Hillside, NJ: Enslow Publishers, 1993. (This book gives an overview of pesticides and their effects, presents opposing views about the use of pesticides, and finally emphasizes the wise use of pesticides.)

Faeth, Paul. "Building the Case for Sustainable Agriculture." *Environment,* Jan.–Feb. 1994, pp. 16–20, 34–39. (This article details the result of scientific research conducted in parts of India, Chile, and the Philippines to determine whether current farming practices or alternative, organic ones would be more profitable and more sustainable.)

Films, Videos, Software, and Other Media

The American Experience: Rachel Carson. Video. 60 minutes. PBS, 1320 Braddock Place, Alexandria, VA 22313; 800-424-7963. (This documentary tells the story of Rachel Carson, whose book about the unregulated use of pesticides in this country caused a revolution in the government's environmental policy and the public's thinking about ecology.)

Save the Earth: Feed the World. Video. 60 minutes. Annenberg CPB Series. Scholastic Software, 740 Broadway, New York City, NY 10003; 800-LEARNER. (This video tells how Australia, Indonesia, West Africa, and the American Midwest rediscovered traditional farming practices.)

USDA IVEN Sharedisc II. CAV Videodisc, Level I. Image Premastering, 724 North 1st Street, Minneapolis, MN 55401; 800-966-2932 or 612-341-3431. (This videodisc provides instruction on soil conservation, agricultural management, food science, nutrition, pest management, and other topics.)

Getting Involved

- Use the following EcoSkills activity to enhance student interest in the subject of this chapter: *Eliminating Pests Naturally,* pp. 390–391.
- A listing of environmental organizations, publications, and volunteer opportunities can be found in the **Laboratory and Field Guide.** Note especially those organizations that have sustainable food production as their goal. Some examples include the Center for Holistic Resource Management, Conservation International, the National Coalition Against the Misuse of Pesticides, and the World Resources Institute.

Using the Quotation

"In simplest terms, agriculture is an effort by man to move beyond the limits set by nature."

Ask one student to read the quote aloud. Ask: What limits do you think the author is referring to? *(sample answer: limited availability of natural resources)* How does agriculture help farmers overcome these limits? *(Agriculture raises the productivity of an area for a given crop.)* Even with the improvements of agriculture, what elements of nature must farmers still contend with? *(weather, pests, soil type and quality)*

Using the EcoLog

To assess students' knowledge about the subject of this chapter, have them answer the EcoLog questions on page 226. In answering these questions, students will reveal their prior knowledge, which is usually very basic and often flawed or even misconceived. Then, by revising their answers at the conclusion of the chapter, students will have the opportunity to eliminate any misconceptions and to see firsthand how their knowledge has grown.

SECTION 9.1

FEEDING THE PEOPLE OF THE WORLD

FOCUS

In this section, students learn about the nutritional needs of humans, about some of the reasons why providing food for all of the world's people is so difficult, and about some of the methods used to combat world hunger.

MOTIVATOR

Prepare for this class by writing the numbers one, two, or three on pieces of paper so that one-third of the class will get the number one, one-third will get the number two, and one-third will get the number three. As students enter the classroom, have them pick a number. Then divide the class according to those numbers. Distribute two beans to each student in group three, three beans to each student in group two, and four beans to each student in group one. Then announce that students in group one represent wealthy people in a developed country, students in group two represent average people in a developed country, and students in group three represent the average person in a developing country. The beans signify the daily food supply per person available in their section of the world. Then inform students that a food supply represented by three beans is the minimum number of calories needed per day for good health. Ask: What do you think about the distribution of "food"? Encourage discussion.

Note: If possible, instead of using beans actually serve students the following meals. Students in group three will be served one scoop of rice with a vegetable sauce and a glass of water. Students in group two will receive a scoop of rice with a vegetable sauce, a piece of fruit, a small dessert item, and a glass of milk or water. Students in group one will receive a large piece of meat (such as chicken), two vegetable items (such as broccoli and mashed potatoes), a dinner roll, a dessert item, and a glass of tea. By the time the meal is over, some students will be conspicuously hungry, while others will probably be overfed. Be prepared for some students to be angry about the injustice and unfairness of this activity. Explain that the activity is a fair representation of what happens in the real world. Every day there are many people in the world who go hungry, while others receive more than their share of the food supply.

TEACHING STRATEGIES

Notable Quote

"Growing crops is easier than building castles, yet there are many castles in our world and not enough crops."
Petak Saeoung

WHAT PEOPLE EAT, *page 228*

Multicultural Connection

In many countries around the world meals consist of one staple food that is eaten with seasoned vegetables and a small amount of meat. The most important of these staples by far is rice. Rice is eaten by over half of the world's population and is especially important in Asia. In parts of China and India, however, wheat in the form of noodles, steamed buns, and flat griddle breads is the preferred staple. In Central America, corn (made into tortillas) is more common. Potatoes are an important staple in European countries, along with breads made from wheat, rye, and other grains. Ethiopians rely on several staple foods, including millet (a cereal grass) and a unique bread called *injera.* Injera is similar to a huge, spongy pancake. Ethiopians put small amounts of spicy stew and vegetables on the injera to create a complete meal.

Using the Photograph: *Figure 9-3, page 228*

Students may benefit from additional information about this subject. Explain that poor nutrition inevitably causes health problems and possibly death if not corrected. Tell students that kwashiorkor is one disease caused by a severe protein-deficiency. Another disease, called marasmus, is caused by a diet low in both calories and protein. Both diseases are found among populations where poor nutrition abounds. Most victims of these diseases are babies who have never been breast-fed or who no longer receive sufficient nutrition from breast milk because their mothers are malnourished. The common characteristics of a child with kwashiorkor are a bloated abdomen, diarrhea, loss of hair, and liver damage in addition to wide eyes and a thin body. These children are also very lethargic. If not treated, the condition can lead to mental retardation and stunted growth—both of which can be reversed if treated in time with a balanced diet.

WHY PEOPLE GO HUNGRY, *page 228*

Helping Students Make a Difference

Students who are interested in the subject matter of this chapter and who would like more ideas about how to get involved might like to hear some of these "Earth-friendly" tips.

- Eat more vegetables and grains and less meat.
- Avoid eating overly processed food,

such as "junk food."
- Encourage cafeteria personnel to serve healthy foods.
- Prepare a meatless meal at least once a week.
- Incorporate more fresh food and less convenience food into your meals.
- Grow your own organic vegetable garden.
- Buy produce and other items from local growers and manufacturers. This reduces the amount of energy used to deliver a product.
- Give some of your home-grown produce to a local soup kitchen or another organization that can use it.
- Volunteer at local food banks and other charities that help hungry and homeless people.

History Connection

Students may be interested to know about the great potato famine in Ireland. Point to Ireland on the world map, and then explain that Ireland's problem began in the early 1800s when the region's human population started growing rapidly. Much of Ireland's population was poor, and potatoes were the primary means of sustenance. When a disease wiped out virtually all of Ireland's potato crops between 1845 and 1847, the results were disastrous. About 750,000 people died of starvation or because their bodies were too weak to fight disease. Hundreds of thousands of people fled the area. Stimulate class discussion by asking: What could Irish officials have done to prevent this disaster? *(Answers will vary. Sample answer: They could have encouraged a more diverse crop base; then, when the potato crops failed, people still would have had something to eat.)*

THE GREEN REVOLUTION, *page 229*

Notable Quotes

"If you eat, you're involved in agriculture."
Bumper sticker

"If the Green Revolution has created food for millions, it is because science has tamed nature."
New York Times editorial, August 29, 1986

Using the Transparency

Figure 9-4 is available as a Teaching Transparency.

Background: Norman Borlaug

In 1970, Aase Lionæs of the Nobel Prize committee said of Norman Ernest Borlaug, "More than any other single person of this age, he has helped to provide bread for a hungry world." Borlaug is an American agricultural scientist who received the Nobel Peace Prize in 1970 for developing high-yield strains of wheat that reduced or eliminated the threat of famine in many developing countries. The scientist was also awarded this highly esteemed award for his leadership of the green revolution.

In his acceptance speech Borlaug stated that the green revolution has given humans only temporary success in the war against hunger. He admitted that the revolution did not solve all of the problems of food production and distribution, but, he said, "it is far better for mankind to be struggling with new problems caused by abundance rather than with the old problem of famine."

CLOSURE

Section Review

Answers to these questions are provided in the Answer Key on page T152.

Alternative Assessment

Tell students to imagine that they are part of a small, self-sufficient community that lives on an isolated island. Explain that in recent years, a fungus killed all of the island's rice crop. Rice was the primary crop and the major food source for everyone on the island, and without adequate rice many people died. The community's leaders finally solved the problem by importing rice from other nations. On this island, importing goods from other countries is traditionally unacceptable except in extreme emergencies, so community leaders have called on you (the students) to devise a plan to keep the problem from ever recurring. Divide the class into small groups to develop a plan. *(Plans will vary. The obvious solution is that in addition to rice, islanders could plant other crops that are high in nutritional value. Then, when one crop fails, people will still have other nutritious foods to eat. Encourage students to think beyond this answer to come up with innovative solutions. For example, students may suggest sending a team of scientists to find out what caused the fungus to spread so rapidly. Then they could take measures to prevent such an event from happening again.)*

Reteaching

Invite a county agricultural extension agent, high school or college agriculture teacher, farmer, or other agricultural specialist to speak to the class about farming methods currently used in your area. You may wish to invite several people with different perspectives on agricultural methods. Suggest that students prepare in advance a list of questions about the green revolution, sustainable agriculture, food and politics, and other issues.

Extension

The government has recently approved a hormone that can increase the milk production of cows. This approval has generated much controversy. Have students research the nature of the controversy and prepare a report that highlights several viewpoints. Students could conclude their report with a short paragraph describing whether they think the government should allow the use of hormones in dairy cows. Suggest that students include their report in their Portfolio.

AGRICULTURE AND SOIL

FOCUS

This section focuses on soil and its role in agriculture. Students learn how soil forms and how it can be eroded, salinized, or turned into desert. Students also explore some methods for solving these serious problems.

MOTIVATOR

Ask students: What kind of landscape do you think of when you hear of Iraq, Syria, Palestine, Lebanon, or other countries in that part of the world? You may wish to point out the area on a world map. *(Students will probably mention deserts or something similar.)* Continue by saying: Would you believe that these countries were once part of an area called the Fertile Crescent? The area was also called the "cradle of civilization" because it was so suitable for growing lush gardens and greenery that many people wished to live there. Explain that the area is still relatively fertile, but, as they all know, the countries mentioned above are certainly no longer known for lush vegetation. Ask students: What might have caused this dramatic change in landscape? Hint: It relates to how humans managed the soil. *(This was one of the first areas of the world to discover irrigation. Over the years, the area was overirrigated, causing soil problems such as salinization.)* Tell students that in this section they will learn more about how humans overuse agricultural land and about what can be done to prevent such overuse.

TEACHING STRATEGIES

Using the Transparency

Figure 9-7 is available as a Teaching Transparency.

FARMING, *page 231*

Meeting Individual Needs

Second-Language Learners Ask students from other countries to report on the dominant type(s) of farming in their native land. They should include information about the crops planted, the farming equipment used, popular farming techniques, etc. Students may wish to use visuals (photographs, drawings, etc.) to supplement their discussion. Suggest that students include their report in their Portfolio.

FERTILE SOIL, *page 232*

Meeting Individual Needs

Learners Having Difficulty Inexpensive soil testing kits, which measure pH and the levels of several key nutrients, are available at most nurseries. Help students work in groups to test samples of fertile topsoil, infertile topsoil, and subsoil. They should take note of the soil's color, texture, moisture content, organic content (a hand lens will be helpful), and consistency. Have students record the findings of their tests. Ask: Which soil sample was the most fertile? Which was the least fertile? Which was in-between?

Using the Transparencies

Figures 9-11 and 9-13 are available as Teaching Transparencies.

TOPSOIL EROSION: A GLOBAL PROBLEM, *page 234*

Notable Quote

"The nation that destroys its soil destroys itself."

Franklin D. Roosevelt

Language-Arts Connection

John Steinbeck's *The Grapes of Wrath* describes some of the conditions of the Dust Bowl, a name given to the midwestern United States during the 1930s. The name describes the condition that resulted when poor agricultural practices combined with severe drought set the stage for severe wind erosion of the region's topsoil. John Steinbeck's *The Grapes of Wrath* describes this time period with poignancy and powerful language. Readings from the novel may help students understand the terrible impact of the Dust Bowl on the people who lived in the region at the time.

SOIL CONSERVATION, *page 236*

Multicultural Connection

George Washington Carver was an internationally known African American scientist who lived from 1856 to 1943. Carver researched and promoted methods of soil conservation and techniques for improving crop production. Carver sought to teach more productive agricultural practices to Southern—particularly African American—farmers. He did so by writing and distributing literature about applied agriculture, setting up exhibits and demonstrations at conferences, and giving public lectures.

As a slave in Diamond Grove, Missouri, Carver first learned about agriculture on a farm. He went on to obtain a master's degree in agricultural science, teach at two universities, head a university's agriculture department, and direct a state agricultural station. As a member of the Commission on Inter-Racial Cooperation and the Young Men's Christian Association (YMCA), Carver also spent much time working to improve race relations. Carver received many awards in his lifetime. For instance, in 1916, he was named a fellow of London's Royal Society of Arts. In 1923, he received recognition from the National Association for the Advancement of Colored People (NAACP) for distinguished service in agricultural chemistry. In 1939, Carver received the Theodore Roosevelt Medal for his valuable contributions to science.

SUSTAINABLE AGRICULTURE, *page 238*

Using the Theme of *Interacting Systems*

Focus Question: How might a river, such as the Mississippi River, increase the fertility of a region? *(Every time a river floods its banks, it contributes rich organic matter and soil to an area called the flood plain. The water in the flood plain eventually drains away, but much of the soil and organic matter that the river was carrying remains behind and increases the fertility of the region's soil.)* Ask: How might flood-control measures decrease the fertility of a region? *(When a river is leveed to prevent devastating floods, the river's "load" can no longer be deposited, and the region's soil will not receive periodic enrichment.)*

Extending the Case Study: *Aquaculture: A Different Kind of Farming, pages 236–237*

Find out if there are any aquaculture operations in your area. If so, arrange for a tour of the facility or invite a representative from the facility to speak to the class about the benefits and drawbacks associated with aquaculture.

PREVENTING SALINIZATION, *page 238*

History Connection

Tell students: It is said that after the Romans defeated the Carthaginians, they salted the soil of Carthage. Ask: Why might the Romans have done this? *(Sample answer: By salting the soil, the Romans made the land unsuitable for growing crops. This action was an attempt to decrease the self-sufficiency of the Carthaginians.)*

OLD AND NEW FOODS, *page 239*

Background: A New Green Revolution?

Some people believe that we are now on the verge of another green revolution—this one due, not to an increase in crop yields, but to an increase in the variety of crops. Today, scientists are discovering (or rediscovering) the value of wild plant species. An estimated 75,000 species of plants are said to be edible, and many of them are quite nutritious, such as the beans of the Kalahari Desert and the lake algae found in Chad and Mexico. This discovery is remarkable, considering that over half of the food eaten by humans today comes from just three plants—wheat, rice, and maize. Most of the rest of our food comes from fewer than 20 other species. The wild plant species have other benefits also. An estimated 600,000 wild plant species are said to offer either nutritional, medicinal, or material benefits to humans.

CLOSURE

Section Review

Answers to these questions are provided in the Answer Key on page T153.

Alternative Assessment

Tell students that they have just been called in as consultants to a major agricultural company; the firm is about to begin operations on a new tract of farm land and is concerned about maintaining the soil's quality. The students should draft a letter of recommendations to the company, explaining some procedures the company could implement to help prevent soil erosion, desertification, and salinization. The letter should also address the company's intention to maintain the fertility of the soil, as well as inform the company of ways to minimize the environmental impact of their operations. Encourage students to include their letter in their Portfolio.

Reteaching

Ask students what other words they think of when they hear the word "sustain." *(sample answers: maintain, support, prolong, endure, nourish)* Then ask students to think about what is meant by the expression "sustainable agriculture." *(Sample answers: Sustainable agriculture means agriculture that will endure, farming practices that can be supported year after year, farming practices that can be maintained within natural ecosystems, etc.)*

Extension

The USDA Soil Conservation Service provides soil surveys for each state, county, and district. See the Resources section of this chapter, page T94, for the agency's address and phone number. These surveys contain maps, tables of information on the properties and capabilities of each soil type, aerial photographs of the land, etc. Suggest that students study this information and then report on the soil conditions in your area. Interested students can include their findings in their Portfolio.

SECTION 9.3

Pest Control

Focus

In this section some of the problems associated with the use of pesticides are discussed. Alternative pest control methods and pesticide regulations are also explored.

Motivator

Show students the PBS special *The American Experience: Rachel Carson.* Information on obtaining a copy of this video is found in the Resources section of this textbook, page T94.

Teaching Strategies

Notable Quote

"One for the maggot, one for the crow, one for the cutworm, and one to grow."

Farmer's Almanac

Discussing the Issue

Inform students that in 1987, the Environmental Protection Agency ranked pesticides in foods as the third most serious threat to this nation's environment. The National Academy of Sciences suggests that 4,000 to 20,000 cases of cancer each year may be linked to pesticide use in the United States. Ask students: Do you think the dangers of pesticide use outweigh their benefits to agriculture? Suggest that students record their answer to this question in their Portfolio.

Pollution and Persistence, *page 242*

Discussing the Issue

Tell students: Pesticides that have not been tested or are known to cause cancer may be made and exported to other countries by pesticide companies. It is possible that crops treated with these pesticides are then sold to consumers in this country. Encourage class discussion.

Resistance, *page 243*

Using the Theme of *Evolution*

Focus Question: Why is the continual use of the same insecticides not a solution to the pest-control problem? Suggest that students use their knowledge of evolutionary processes, insects, and pesticides to answer this question. *(Often, a few insects will carry a gene that makes them resistant to a pesticide. These insects pass on that gene to their young, while many insects without the gene die from the pesticide. Because insects reproduce quickly, there will soon be many insects that are resistant to the pesticide and few that are not.)*

Cooperative Learning Activity

Divide the class into groups. Ask: Are all insects bad? Tell students to think of three insects or other arthropods that are beneficial to humans and to justify their choices. *(Spiders, praying mantises, and ladybugs all prey on pest species. Many arthropods are pollinators, and others help decompose dead organisms or aerate soil.)* Encourage students to draw a picture of their beneficial arthropods in action. Tell students that many pesticides are dangerous to beneficial as well as harmful species, and instruct students to write a few sentences about what consequences could follow the extinction of their chosen species.

Debate

Divide the class into two groups, and tell one group to take the pro position and the other the con position for the following topic: The benefits of pesticides outweigh their harmful effects. Give students sufficient time to prepare for the debate; then moderate while students debate the issue.

Meeting Individual Needs

Gifted Learners Tell students that studies by the National Academy of Sciences show that the Federal Insecticide, Fungicide, and Rodenticide Act (FIFRA) of 1972 is the most poorly enforced environmental law in our legal system. This law allows the Environmental Protection Agency to license the use of pesticides that have not been fully tested. It also allows pesticides that may not be safe to remain on the market. Encourage students to look into this situation by getting to know more about the law. Suggest that they find out if there are any organizations or individuals trying to change the law. Students could also find out if there are any organizations or individuals trying to keep the law as it is. Recommend that interested students take a stand and support one side or the other by writing letters, volunteering their time, etc.

Biological Pest Control, *page 243*

Notable Quote

"Bugs are not going to inherit the Earth. They own it now. So we might as well make peace with the landlord."

Thomas Eisner

CLOSURE

Section Review

Answers to these questions are provided in the Answer Key on page T153.

Alternative Assessment

Tell students that your next-door neighbors, the Browns, have a dog that spends part of its time indoors and a rabbit that lives in a cage outside. The Browns have noticed that at certain times of the year flies, fleas, and mosquitoes are very troublesome. They don't want to use pesticides to kill the pests because the last time they did that, both animals got sick. The Browns have asked you for advice, and you would like help from your students. Divide the class into small groups, and have each group recommend a plan of action for the Browns. *(Possible plan: Keep garbage cans tightly closed; don't leave pet food out for very long; clean up dog and rabbit droppings frequently; empty and refill the animals' water dishes daily; don't allow standing water to remain in outside containers; bathe both animals regularly; vacuum indoors regularly and change the vacuum-cleaner bag each time.)* Suggest that interested students consult pages 390–391 in their textbooks for more "environmentally friendly" pest-control ideas.

Reteaching

Divide the class into five equal groups. Choose one person from each group to represent a mole. The other students will represent insects. Have a student pass out a small card with "DDT" written on it to each "insect." Inform students that each insect has consumed a molecule of DDT while eating plants. Tell students that the "moles" will now consume the insects. Have the "insects" give the DDT cards to the student representing the mole in their group. Now tell students that you are a peregrine falcon and that you are about to consume the moles. Collect all of the DDT cards. Point out how many DDT "molecules" you have gained. Tell students that bioaccumulation works in the same way. The pesticide is not biodegradable, so it stays in the body of the animal that eats it.

Extension

Have students conduct an experiment to answer the following question: Will the aroma of peppermint repel ants? To begin, divide the class into groups of four, and provide each group with the following materials: one leaf of a peppermint plant, a mortar and pestle, two saltine crackers, a shoe box, a 16 in. piece of clear plastic wrap, tape, and three ants in a petri dish with a lid on it. Tell students not to open the petri dish until they are told to do so. Be sure to choose ants that are easy to control or that do not sting. **Safety Alert:** *Tell students not to handle the ants. If a student is allergic to ant stings, he or she should be exempted from this activity.*

Tell students to tape the crackers to opposite sides of the short end of the shoe box near the bottom. Then they should crush the peppermint leaf using the mortar and pestle and put the leaf pieces in a semicircle on the bottom around one of the crackers. Next, have students carefully release the ants into the shoe box. Tell students to quickly put the plastic wrap over the top of the box and tightly tape it in place. Have students record their observations and conclusions.

Answers

Answers to *Chapter Highlights, Chapter Review, Investigation,* and *Points of View* are provided in the Answer Key for Chapter 9, which begins on page T152.

TEACHING STRATEGIES FOR INVESTIGATION:

THE CASE OF THE FAILING FARM,
pages 248–249

Background

Encourage discussion about reasons for the increased rate of erosion on Mr. Katawa's farm. *(Dry weather tends to dehydrate soil, and when crops grow poorly, the roots are less likely to hold the soil in place. The animals on the farm probably added to the problem by churning the top layer of soil when they walked on it. Together, these factors left the soil very dry and loose; the runoff from the wet weather and snowmelt that followed easily washed away this soil.)*

Erosion in Action

Be sure that students have the hills placed properly and that they water the soil just enough to start the process of erosion (otherwise they may have to rebuild the model for other experiments). It is a good idea to have paper towels or rags handy for cleanup.

Model Experiments

Materials that could be added to the soil to change the rate of erosion include sand, composted organic materials, and bits of clay. Encourage creativity, and don't give too many hints, but be sure the students have reasons for their proposed modifications. Other possible variations include terracing the hills, strip cropping (planting alternating rows of different crops), adding additional fence rows, creating channels for water runoff, and reducing the amount of disturbance from the animals. Once the students have evaluated their changes to the farm, ask them to write a letter back to Mr. Katawa stating their findings. Also, discuss with students how their findings relate to problems of erosion in other settings, such as urban landscaping and golf course maintenance.

CHAPTER 10

BIODIVERSITY

PLANNING GUIDE

CHAPTER COMPONENTS AND RESOURCES	½ Year Average	½ Year Advanced	1 Year Average	1 Year Advanced
Chapter Opener, p. 252				
EcoLog, p. 252	•	•	•	•
10.1 Biodiversity at Risk, pp. 253–260	•	•	•	•
Transparency 34: *History of Mass Extinctions*	•	•	•	•
Field Activity, p. 254				•
Transparency 35: *Numbers of Known Species*	•	•	•	•
Case Study: *Whooping Cranes: Regaining Lost Ground?* pp. 256–257	•	•	•	•
Section Review, p. 260	•	•	•	•
10.2 Public Policy, pp. 261–266	•	•	•	•
Environmental Careers: *Environmental Filmmaker*, pp. 372–373			•	•
Case Study: *The Snail Darters and the Dam*, pp. 264–265	•	•	•	•
Section Review, p. 266	•	•	•	•
10.3 The Future of Biodiversity, pp. 267–270	•	•	•	•
Transparency 36: *Species Hot Spots Around the World*	•	•	•	•
Points of View: *Owls vs. Loggers*, pp. 276–277	•	•	•	•
Section Review, p. 270	•	•	•	•
End of Chapter				
Chapter Highlights, p. 271	•	•	•	•
Chapter Review, pp. 272–273	•	•	•	•
Investigation: *Backyard Diversity*, pp. 274–275			•	•
EcoSkills: *Building a Bat House*, pp. 400–403				•
Chapter Tests	•	•	•	•
PACING GUIDE	9–11 days	8–10 days	16–17 days	16–17 days

MATERIALS ORGANIZER

Activity	Materials	Advance Preparation
Investigation: *Backyard Diversity*, pp. 274–275	gloves, tape measure or meter stick, 4 stakes, string (about 10 m), hand lens, notebook, pen or pencil, field guide to insects (optional) (per student group)	Determine two locations for conducting this investigation. For example, you could choose a Xeriscaped park and the schoolyard. Get permission in advance if necessary.

CHAPTER OVERVIEW

In this chapter students are introduced to the concept of biodiversity, current threats to biodiversity, and various efforts to preserve it.

Section 10.1 shows how humans are contributing to the loss of biodiversity. The section also explains why it is important that we preserve biodiversity.

Section 10.2 describes public policies designed to protect biodiversity and discusses conflicts that often arise when legislation is enacted.

Section 10.3 explores nonlegislative programs that can help preserve biodiversity and analyzes the effects of human needs on biodiversity.

RESOURCES

. . . for Teachers

Barker, Rocky. *Saving All the Parts.* Washington, D.C.: Island Press, 1993. (This book examines the relationship between endangered species protection and the economic well-being of resource-dependent communities.)

U.S. Fish and Wildlife Service, Division of Endangered Species, Arlington Square Building, 4401 Fairfax Drive, Mail Stop 452 ARLSQ, Arlington, VA 22203; 703-358-1711.(This government agency can provide you with information on endangered species and refer you to the regional office closest to you for information on endangered species in your area.)

. . . for Students

DiSilvestro, Roger L. *Audubon Perspectives: Fight for Survival.* New York City, NY: John Wiley & Sons, 1990. (This book is filled with beautiful photographs of diverse life-forms around the world, and it includes an interesting narrative describing the plight of several of the world's endangered species.)

Wilson, Edward O. *The Diversity of Life.* Cambridge, MA: The Belknap Press, 1992. (This informative and engaging book provides detailed scientific information about biodiversity and the biodiversity crisis.)

Films, Videos, Software, and Other Media

Biosphere Reserves in Tropical America. Video. 26 minutes. Conservation International, 1015 18th Street, NW, Suite 1000, Washington, D.C. 20036; 202-429-5660. (This documentary takes viewers on a tour of five Latin American biosphere reserves, showing that innovative thinking can help preserve biodiversity.)

The Last Show on Earth. Video. 102 minutes. Bullfrog Films, P.O. Box 149, Oley, PA 19547; 800-543-3764. (With modern music and well-known actors, this video celebrates the efforts of individuals who are working to save endangered species; it also illustrates the causes of extinction.)

Saving Endangered Species. Video. 30 minutes. National Audubon Society, 950 Third Avenue, New York City, NY 10022; 212-979-3000. (This video outlines the Endangered Species Act and discusses how students can make simple lifestyle changes to reduce their impact on endangered wildlife.)

Getting Involved

• Use the following EcoSkills activity to enhance student interest in the subject of this chapter: *Building a Bat House,* pp. 400–403. The bat-house activity would make an excellent class project.

• A listing of environmental organizations, publications, and volunteer opportunities can be found in the **Laboratory and Field Guide**. Note especially those organizations with the goal of protecting biodiversity and natural resources, such as Defenders of Wildlife, National Audubon Society, Rainforest Action Network, Sierra Club, and Student Conservation Association.

USING THE QUOTATION

". . . in wildness is the preservation of the world."

Call on a student to read the quotation aloud. Then ask the class: What do you think Thoreau meant by "wildness"? *(Sample answer: Thoreau might have meant those things that are not controlled by human beings.)* Ask: How does Thoreau relate the preservation of the world to "wildness"? *(Sample answer: Perhaps Thoreau meant the secret to maintaining life on this planet lies in the proper functioning of natural systems.)*

USING THE ECOLOG

To assess students' knowledge about the subject of this chapter, have them answer the EcoLog questions on page 252. In answering these questions, students will reveal their prior knowledge, which is usually very basic and is often flawed or even misconceived. Then, by revising their answers at the conclusion of the chapter, students will have the opportunity to eliminate any misconceptions and to see firsthand how their knowledge has grown.

SECTION 10.1

Biodiversity at Risk

Focus

A discussion of the dinosaurs' demise introduces students to the terms *mass extinction* and *biodiversity*. Then students learn that habitat destruction, hunting, and the introduction of exotic species may be contributing to another mass extinction. Finally, students are asked to consider some of the reasons why many people are concerned about the loss of species on Earth.

Motivator

Divide the class into groups of three or four students. Then have group members work together to answer the following question: If you could get rid of one species of plant or animal on Earth, what would it be? *(Students will probably name species that can cause harm or discomfort to humans, such as mosquitoes, bees, or poison ivy.)* Once group members have settled on one species, ask them to share that answer with another group, and ask for that group's answer in exchange. Tell students that now their task is to list some reasons that the given species should be saved instead of destroyed. *(Sample answer: Bees are important for many reasons, including their ability to pollinate flowers [some of which could not reproduce without the bees] and to produce honey [which some animals depend on as a food source], etc.)* Note: Some students may maintain that there is no good reason to save certain species. Accept this answer, and tell the class that many people share a similar belief. Then lead into a class discussion about when we should dedicate money, time, and resources to protecting a species, when we should not, and who should make those decisions. Encourage a variety of opinions, and tell students that they will learn more about the subject in this section.

Teaching Strategies

Meeting Individual Needs

Second-Language Learners For students whose first language is not English, find out what the word for "extinction" is in their language. For example, the word in Spanish is *extinción*, in French it is *extinction,* and in German it is *das Aussterben.* Provide foreign-language dictionaries so that students can look up the meaning of the word if necessary.

Using the Transparency

Figure 10-1 is available as a Teaching Transparency.

Using the Graph:

Figure 10-1, page 253

Use Figure 10-1 to reinforce the fact that extinction is a natural process that occurs over time. Call on a student to read the caption aloud. Then tell students that the increasing number of extinctions today has caused some scientists to assert that we are in the middle of another mass extinction. Ask: Do you think a change in global climate could be causing a mass extinction today? Why or why not? *(Most students will probably say that they do not think a change in climate could be causing the current loss of species because they have not noticed a change in the world's climate. Some students may say that global warming and the destruction of the ozone layer could be causing the world's extinctions.)* Ask students to consider what role humans might play in the current loss of species. *(Encourage class discussion, and assure students that as they progress through the chapter they will be better equipped to answer these questions.)*

A World Rich in Biodiversity, *page 254*

Using the Transparency

Figure 10-2 is available as a Teaching Transparency.

Using the Chart:

Figure 10-2, page 254

Students may be interested to know that almost all animals with feathers or fur have been named and classified by scientists. In fact, researchers classify only about three to five new species of birds each year and about 20 species of animals with fur per year. Approximately half of those species are not newly named at all, but simply reclassified based on updated scientific findings. Numerous insect species, on the other hand, are named for the first time each year. Ask students to speculate about why this discrepancy might exist. *(Generally, birds and other animals are more visible to scientists than are insects; thus, they were studied and classified before the smaller insect species. Insects and arthropods often require microscopic examination before identification and classification can be made. In addition, birds and other large land-dwelling animals are traditionally more appealing to humans than are insects. Most importantly, there are many more insect species than there are other animal species.)* You may also wish to point out that marine organisms are not as well studied as land organisms because their habitat is less accessible.

Background: How Many Species Are There on Earth?

Determining the total number of plant and animal species on Earth is neither a simple nor a precise task. As mentioned on page 254, estimates of the total number of species vary from 10 million to 100 million. Students may wonder how such a disparity could exist. Explain that there is no foolproof method for counting all of the species on Earth. Instead, scientists develop systems for making educated guesses. The following are two such systems.

- By thoroughly sampling organisms living in a relatively unstudied region, scientists determine how many of those species are unclassified. That number is then used to make estimates of the proportion of unclassified organisms in all areas.
- By working with the premise that far more large species have been identified than small ones, scientists find a ratio between the number of large and small organisms in a well-studied area to determine the total number of species.

Notable Quotes

"Biodiversity is our most valuable but least appreciated resource."
Edward Osborne Wilson

"One planet, one experiment."
Edward Osborne Wilson

Debate

Divide students into two groups. Tell students that you are going to read a position statement about whether extinction is really a serious problem. Explain that you want one group to support the position and the other group to oppose it in a formal debate. Then read the following position statement.

The problem of extinction is overrated because the species that are dying out are typically very small, both in size and in numbers, and live only in a very limited area. The rain forests, where most of the dying species live, are so diverse that every single acre contains species found nowhere else. In such a place, any human act is going to make some species extinct. It cannot be avoided. Similarly, in the temperate zones there are many species of small animals such as beetles or minnows that are found only in very specialized environments within a small area. The loss of some of these species—naturally or with human assistance—is unavoidable and probably does not greatly affect the ecosystem. When species with major roles in the ecosystem start to die off, then and only then should we start to worry.

The closer we look, the more we realize that there are far more species than was once thought, and so we simply are more aware of a problem that has been occurring all along. There is no way to avoid eliminating some species. We probably could stand to be a little more careful, and we definitely should not go out of our way to eliminate species, but neither should we regress to the Stone Age to save a few worms and guppies.

Research Activity

Tell students that the science of systematically naming and recording species is known as *taxonomy*. The system we now use was officially begun in 1758 by a Swedish naturalist named Carolus Linnaeus. Have interested students research the system established by Linnaeus. Suggest that students locate a key for native plants in your area and demonstrate to the class how it can be used to identify a familiar plant. (Keys for identifying organisms are available in many libraries and from the biology or botany departments of many high schools, colleges, and universities.) Tell students to read the instructions carefully before attempting to use the key and to refer to the glossary for any unfamiliar terms.

How Are Humans Causing Extinctions? *page 254*

Extending the Photograph: *Figure 10-3, page 255*

Several resources are available for students who would like to learn more about the near extinction of the Florida panther. One interesting account can be found on pages 249–273 of the book *Meant to Be Wild* by Jan DeBlieu, Fulcrum Publishing, Golden, Colorado, 1991.

Extending the Case Study: *Whooping Cranes: Regaining Lost Ground? pages 256–257*

Encourage students to use the decision-making model from Chapter 1 to analyze the following situation.

A large, undeveloped tract of land in your area (such as a former ranch or park) was recently divided and sold to 50 people. Most of those people bought the land with the intention of having a house or business built there, but newly released research indicates that the land is vital habitat for the endangered whooping crane and other wildlife. To protect the wildlife, a local environmental group is pressuring officials to rezone the area so that the new landowners cannot build on their property. Ask students: Would you support this action? *(Stress that there are no right or wrong answers, only individual opinions.)*

Using the Theme of *Patterns of Change*

Focus Question: Do you think modern technology has influenced the acceleration of species extinction? Explain. *(Students will probably suggest that technology has allowed more rapid human expansion and development, which are the primary causes of species extinction.)*

EXOTIC SPECIES, *page 258*

Background: Exotic Species

Exotic species is one term among many used to describe a species living in—but not native to—an area. Some other common terms are *introduced species, immigrant species,* and *alien species.*

THE VALUE OF BIODIVERSITY, *page 258*

Cooperative Learning Activity

Divide the class into three groups, and assign each group one of the following topics: (1) saving species preserves ecosystems; (2) practical uses of species; and (3) aesthetic reasons to preserve species. Tell students that in three (or more) days there will be a "Why Should We Care?" fair. Their assignment is to develop a creative means to convince the class that their topic is a powerful reason for preserving biodiversity. For example, students could pass out flyers, show a video, perform a skit, make a mural or a series of posters, or even create a T-shirt design. Each group should have a slogan or logo, and their presentation should be supported with facts and data. Suggest that students begin their project by reading the section in their textbook that has the same heading as their topic (pages 258–260). After students are finished, suggest that they include something from their project in their Portfolio.

SAVING SPECIES PRESERVES ECOSYSTEMS, *page 258*

Meeting Individual Needs

Learners Having Difficulty Students may be interested to know that scientists picked up the word *keystone* from architectural use. In masonry, a keystone is the central stone of an arch. Until the keystone is in place, a true arch will not support itself. Thus, the architectural formation is entirely dependent on the keystone, just as an ecosystem is dependent on its keystone species. The origin of this term, provided as an analogy, may help students better visualize the importance of keystone species to an ecosystem. If possible, provide students with a drawing or photograph of an arch with an obvious keystone to enhance their mental image. See below.

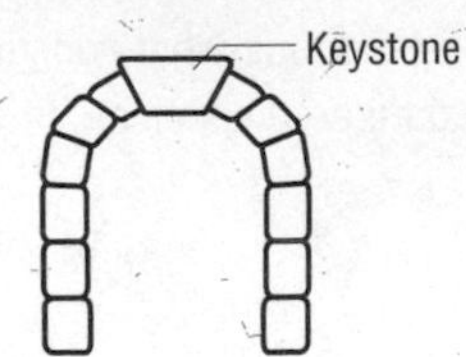

Reinforcing Ideas

Write the following sentence on the board: *Sometimes one species is so valuable to an ecosystem that the ecosystem cannot function properly if that species (the keystone species) dies out.*

Have students illustrate the sentence or write a short story or poem describing it. Encourage creativity, and suggest that students include their work in their Portfolio.

Using the Theme of *Energy*

Focus Question: What happens to the energy within an ecosystem when a keystone species is driven to extinction? *(The flow of energy is disrupted, and the natural equilibrium of the ecosystem is disturbed.)*

CLOSURE

Section Review

Answers to these questions are provided in the Answer Key on page T154.

Alternative Assessment

Tell students to pretend that a teacher in a nearby school asked each of them to talk to a fifth-grade class about biodiversity. Their assignment is to plan how they intend to teach the material. *(Answers will vary. An acceptable plan would define and explain the value of biodiversity, mention some possible reasons for the current decline in biodiversity, and include some information about efforts to preserve biodiversity.)*

Reteaching

Divide the class into groups of three or four students, and give each group the name of an exotic organism that has negatively affected the natural environment of your city, state, or region. Also provide students with a brief summary of the damage that the organism has caused since its arrival in the area. Have students use the information to create a TV news program, commercial, or rap song to show how the exotic species upset the natural equilibrium of an ecosystem.

Extension

Ask students to do research on a plant or animal that has become extinct. Suggest that they analyze the factors that contributed to the species' demise and create a documentary to present their findings. Encourage students to include any ideas they have about what could have been done to improve the organism's chances of survival. Students could also present their findings to the class in the form of a report, skit, or poster.

SECTION **10.2**

Public Policy

Focus

Students are introduced to national and international policies that are designed to protect biodiversity. They learn how the Endangered Species Act defines an endangered or threatened species and why establishing public policy to protect biodiversity is not always an easy task.

Motivator

Survey the class to identify students who have (or have had) tropical fish, parrots, boa constrictors, pythons, orchids, or other exotic animals or plants. Being careful not to embarrass anyone, ask where the exotic organisms came from. If they say "the pet shop" or "the nursery," ask them where the pet shop or nursery got the organisms. *(Most students will not know the answer.)* Tell students that many of the exotic species are imported directly from tropical rain forests and other habitats in foreign countries to the United States and other developed nations. This activity has driven some of the species to near extinction. Although laws now protect many of the heavily traded plants and animals, some are still illegally captured and sold. Therefore, anyone who purchases an exotic species might be contributing to the demise of that species in the wild. To avoid contributing to the problem, students should verify where an exotic species came from before purchasing it and purchase only plants or animals that have been bred and raised in the United States. Tell students that in this section they will learn about how laws and international cooperation can help slow or halt this problem.

Teaching Strategies

The Endangered Species Act, *page 261*

Meeting Individual Needs

Gifted Learners In March 1994, a federal appeals court ruled that the Fish and Wildlife Service had overstepped its bounds in defining what constitutes harm to an endangered species. The court said that an endangered species, not the ecosystem of which it is a part, must be kept from harm. Suggest that students do research to find out any recent changes to or controversy surrounding the Endangered Species Act. Students could write a critique of current policies and include the critique in their Portfolio.

Using the Case Study:
The Snail Darters and the Dam, pages 264–265

Call on one or more students to read the case study aloud. Then ask: Do you think protecting the habitat of the snail darter was worth the time and expense? *(At least some students will probably answer no.)* If students answer "no," ask: What if the organism was a larger, better-known species, such as a dolphin? *(More students will probably answer yes.)* Then ask: Where should we draw the line, and who should determine whether a species is worth protection or not? Encourage class discussion.

Worldwide Efforts to Prevent Extinctions, *page 264*

Using the Photograph:
Figure 10-12, page 265

Divide the class into small groups of three or four students to answer the question in the caption for this photograph. Encourage students to use the decision-making model from Chapter 1 to formulate their answers.

Real-World Connection

Many environmental organizations based in the United States cooperate with other countries. Invite a representative from such an organization to speak to the class about the international efforts of that organization to protect biodiversity. Or contact one of the environmental organizations listed in the **Laboratory and Field Guide** to see if they have

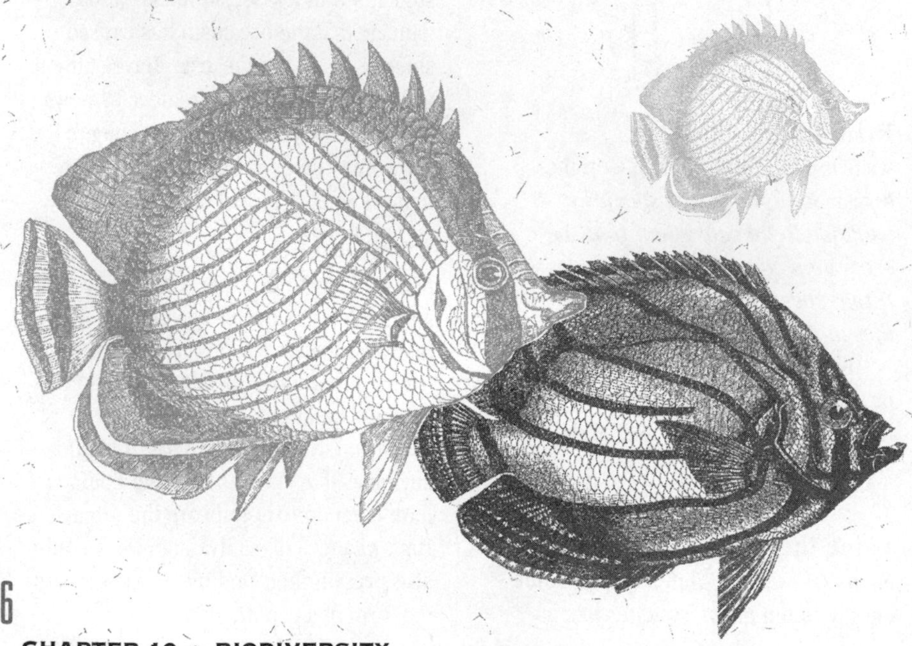

a film or video about their efforts. (One of the many videos available from Conservation International is listed in the Resources section of this chapter, page T102.)

Prevention of Poaching, *page 265*

Using the Graph:

Figure 10-13, page 266

Use Figure 10-13 to help illustrate the positive result of the 1989 ivory ban.

The Biodiversity Treaty, *page 266*

Social-Studies Connection

When President Clinton signed the Biodiversity Treaty in 1993, it was still uncertain how the international community would raise the funds and allocate the money for protecting the biodiversity of developing nations. However, according to Madeleine Albright, United States ambassador to the United Nations, the Clinton administration felt that the best course of action was to sign the treaty and play a leadership role in its enactment. Suggest that students do research to find out what leadership steps have been taken by the United States since Clinton signed the treaty. Encourage them to include their findings in a "news report" or a critical review, and suggest that they add their report or review to their Portfolio.

Closure

Section Review

Answers to these questions are provided in the Answer Key on page T154.

Alternative Assessment

Ask students to imagine that they work for the World Wildlife Fund, IUCN, Conservation International, or another organization that helps preserve the world's biodiversity. Have students write a short drama, make an illustration, or design a comic strip that describes their efforts to save a plant or animal species.

Reteaching

Some students may have difficulty understanding why habitat loss is a problem for many animal species. They may ask: Why don't the animals just move across the street or to some nearby area that has not been developed? To illustrate the problem, offer students the following scenario, which is based on actual studies. A group of scientists surveyed the number of deer in a natural habitat slated for development. They counted 20 deer. Then they counted the number of deer in the area immediately surrounding the development site. They counted 50 deer. Immediately after the land was bulldozed, the scientists surveyed the remaining natural habitat. They determined that there were now 70 deer. Two months later they surveyed the site once more and found that there were 50 deer again.

Explain that if organisms move across the street or to nearby land, they will find other organisms there. Competition will result in the death of many organisms.

Extension

Tell students to suppose that they are public affairs agents for the U.S. Fish and Wildlife Service. Their assignment is to develop a campaign to educate the general public about the laws surrounding the capture, trade, or use of threatened species and endangered species. Encourage students to present their public service announcement in the form of a flyer, poster, TV or radio commercial, or newspaper advertisement. Suggest that students include their announcement in their Portfolio.

S E C T I O N 10.3

The Future of Biodiversity

Focus

The section begins with a discussion of nonlegislative programs designed to help save individual species. The new emphasis on protecting entire ecosystems rather than individual species is then discussed. Finally, students are encouraged to realize that the needs of humans cannot be ignored when attempting to slow the loss of biodiversity on Earth.

Motivator

Tell students that the class will begin with an impromptu skit. The skit involves a corporation called Fun Park USA that plans to improve the future of your community by building a huge amusement park on several thousand hectares of land in a rustic and beautiful area close to your town.

Ask for four volunteers to role-play the scenario, and then assign each volunteer one of the following roles: (1) a struggling public relations executive whose business will fail if his or her number-one account, Fun Park USA, is unable to go through with its project; (2) a lawyer who is fighting for Fun

Park's right to use the land as it pleases; (3) an environmentalist who says that the development of Fun Park will have a tragic impact on the biodiversity of the entire state; and (4) a city representative who has tried in vain for several years to get a local species classified as endangered. Tell the actors to present the scenario as a short skit in which they demonstrate the conflict and then resolve it in some way. Give them several minutes to plan how they will proceed. When the actors have finished their skit, ask the other students in the class if they are pleased with the results. Discuss their answers, and ask students how a different conclusion to the skit might have changed their opinions.

TEACHING STRATEGIES

CAPTIVE-BREEDING PROGRAMS, *page 267*

Background: Zoos and Wildlife Parks

Once viewed solely as places of entertainment, zoos and wildlife parks today are taking on a much larger role. They allow people the opportunity to see animals that they may never be able to see in the wild. They can also make the plight of an endangered species more relevant to a person who has never seen that species except in a newspaper or magazine picture. Zoos can also help revitalize animal populations by establishing captive-breeding programs. Captive-breeding programs are not limited to zoos and wildlife parks, however; they are also conducted at preserves and special breeding centers.

Unfortunately, inducing wild animals to breed in captivity is a long-term, expensive undertaking even when successful. And it is difficult to release animals into the wild in a manner that gives them a reasonable chance of survival.

Nevertheless, there are a number of species that are either already extinct in the wild or definitely headed for extinction, and they can be saved only if attempts to breed them in captivity succeed. The California condor, the whooping crane, and the black-footed ferret are all examples of animals helped by captive-breeding programs.

Reinforcing Ideas

Students may appreciate another example of a successful captive-breeding program. One little-known but interesting example involves Przewalski's horse. Przewalski's horse is the only undomesticated horse species to survive from the Ice Age to the twentieth century. It is a short, stocky animal with a golden brown coat, a black Mohawk-style mane, and a long, black tail. In 1879, Russian scholar Nikolai Przewalski discovered the horse in the Gobi Desert. This rare animal is thought to be the true ancestor of all domestic horses. Przewalski's horse is now extinct in the wild, but between 1897 and 1902, western collectors and zoos bred 50 of the adult horses in captivity. As a result, there are more than 500 of these unique horses alive today, and plans are underway to restore some of them to wilderness areas in their Mongolian homeland.

Using the Theme of *Evolution*

One alternative some scientists have considered for preserving subspecies threatened with extinction is to crossbreed them with closely related subspecies (not separate species, because they cannot interbreed successfully). Although this introduces "impurities" into a subspecies' gene pool, it may keep the population from becoming extinct. *Focus Question:* How might the mixing of populations be a natural part of the evolutionary process? *(If two populations that represented separate subspecies merged into one, the original subspecies might then be considered extinct, and the new population could constitute a new subspecies.)* You may also wish to discuss the ethics of this situation with students, emphasizing that there are not right or wrong answers, only individual opinions.

GERM-PLASM BANKS, *page 268*

Background: Germ-Plasm Banks

Genetic storage facilities, such as those described in the student text, go by many different names—germ-plasm banks, gene banks, seed banks, and genetic libraries are just a few. Gathering, collecting, and maintaining the never-ending supply of samples for these facilities is an enormous project, and there is rarely enough space or staff to meet the demand. In addition, since refrigeration is usually necessary for proper storage of these materials, one power failure could permanently damage huge supplies of genetically priceless materials.

THE ECOSYSTEM APPROACH, *page 268*

Using the Transparency

Figure 10-17 is available as a Teaching Transparency.

Geography Connection

Have students do research and then use magazine photos or personal illustrations to create a large bulletin-board display, mural, or poster portraying the physical aspects of one or more of the biodiversity "hot spots" shown in Figure 10-18 on page 269. Suggest that students include their work in their Portfolio.

Using the Theme of *Interacting Systems*

Focus Question: How are growing human populations affecting the wildlife in national parks and wildlife preserves? *(Increasing human use of these facilities for pleasure or economic benefit is disrupting the habitat of wildlife within the parks and preserves.)* Suggest that interested students do research to learn more about this subject. They may want to begin by reading about Yosemite National Park and the decision of its managers to limit the total number of visitors allowed into the park on any one day. Suggest that students summarize their findings in a brief paragraph, poem, or drawing, and include it in their Portfolio.

Multicultural Connection

Cultural views have enabled many indigenous societies to preserve the biodiversity of their region. For example, the Kuna tribe of Panama has strict rules involving the harvest of lobsters, turtles, and game animals. These rules tend to discourage overkill and waste. The Wet'suwet'en and Gitksan peoples of western Canada protect their staple food, salmon, with religious customs. They believe that salmon spirits offer their bodies to humans for food but will punish humans who waste or misuse this gift or disrupt salmon habitats.

Ask students to discuss how cultural views about wildlife have affected the survival or demise of plants or animals in your state, the nation, or other countries. *(Students may refer to organisms that are revered and protected, such as the bald eagle or a state flower or bird. In addition, they may mention the extinction or near extinction of a species considered undesirable by a culture, such as a cougar or shark.)*

Using the Points of View:

Owls vs. Loggers, pages 276–277

Cooperative Learning Activity Divide the class into two teams. Encourage each team to elect a leader, a recorder, presenters, and researchers. Tell students that each team will debate one side of the owls vs. loggers debate. Then assign the "People and Jobs Should Come First" position to one team and the "Owls and Forests Should Come First" position to the other team. Give teams one class period to research their position. Emphasize to students that they are to use specific and logical justifications that are based on economic and scientific data, as well as emotional, historical, and social justifications.

CLOSURE

Section Review

Answers to these questions are provided in the Answer Key on page T154.

Alternative Assessment

Tell students: You are a wildlife biologist who has studied a rare species of alligator for five years. You submitted paperwork showing that there is a dire need to classify the alligator as an endangered species, but you just learned that the U.S. Fish and Wildlife Service has determined that protecting this species is not as high a priority as protecting other species. Instead of giving up, you decide to think of another program to help the alligator. Record your ideas and a recommended plan of action. Encourage students to include their plan in their Portfolio.

Reteaching

If there is a botanical garden, zoo, captive-breeding center, germ-plasm bank, or wildlife preserve nearby, arrange to take students on a tour of the facility. Many of these organizations have a tour guide or other representative who could talk to students about any work the facility might have done for threatened or endangered species. If you do not have one of these facilities nearby, a representative from a university agriculture or botany department or from a nursery may be able to come speak to the class about breeding threatened or endangered species and releasing them into the wild.

Extension 

Suggest that students research the National Seed Storage Lab (1111 S. Mason Street, Fort Collins, CO 80521-4500; 303-495-3200) or another such facility. Encourage students to then make a small seed bank with a shoe box and some seeds. Students could include the seed bank in their Portfolio.

Answers

Answers to *Chapter Highlights, Chapter Review, Investigation,* and *Points of View* are provided in the Answer Key for Chapter 10, which begins on page T154.

Teaching Strategies for Investigation:

Backyard Diversity, *pages 274–275*
This investigation can be performed as a cooperative learning activity or as a take-home assignment for each student.

Background

Before students perform the investigation, discuss with them the importance of detailed field notes. Explain that every scientific investigation requires thorough, descriptive field notes of the study sites. Quick, general field notes often lead to false conclusions.

Observe, Record, and Think

Xeriscaped lawns generally have no pesticide residues. For this reason, these areas are especially interesting for this investigation. If the schoolyard is Xeriscaped, encourage its use. Insect populations in these areas are often extremely diverse. If time permits, encourage students to use a field guide to identify the insects they see. The following field guides are available in most bookstores and libraries.

Leaky, Christopher. *Peterson Field Guide to Insects.* Boston, MA: Houghton-Mifflin, 1987.

Milne, Lorus J. *Audubon Society Field Guides to North American Insects and Spiders.* New York City, NY: Facts on File, Inc., 1986.

O'Toole, Christopher. *The Encyclopedia of Insects.* New York City, NY: Facts on File, Inc., 1986.

State a Hypothesis

If students are having trouble writing a hypothesis, you may want to offer additional suggestions, such as the following.

- Foot traffic across an area decreases insect diversity. (Or students might even propose that foot traffic increases insect diversity.)
- Proximity to a building decreases (or increases) insect diversity.
- Organic garbage increases insect diversity.

If students have access to the necessary information, they may wish to test hypotheses about pesticides, watering, etc.

Extensions

Have students mark off three sites (with the same dimensions as in the investigation) that are at varying degrees of proximity to an area that has been disturbed in some way by humans. For example, one plot would be right next to the disturbed area, one could be 20 m away, and one could be 40 m away. Ideally, the three sites would be fairly similar in terms of vegetation and terrain. Students should record data on these sites as in the investigation and then answer the following questions: Did you find differences in either the number of insects or the number of insect types observed? *(The answer will most likely be yes.)* Why do you think this pattern existed? *(The degree of human disturbance decreases farther away from the highway, for instance.)*

Students could also carry out the investigation with attention to the diversity in plant species instead of insect species. Ask: Are the patterns observed in insect diversity the same as those observed in plant diversity? Why or why not? *(Human disturbance affects plants as well as animals, but plant diversity may not change as quickly as insect diversity.)*

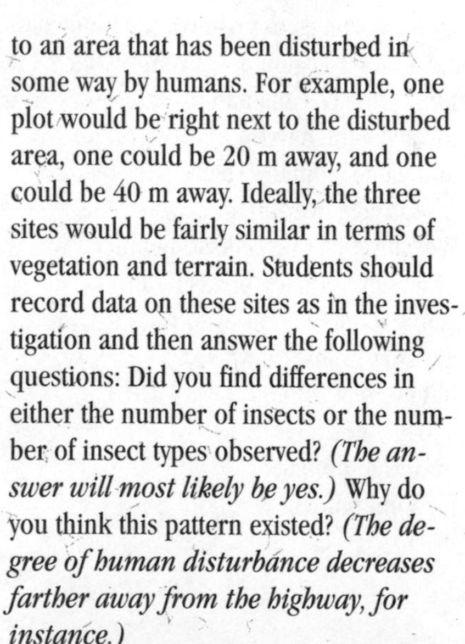

CHAPTER 11

ENERGY

PLANNING GUIDE

CHAPTER COMPONENTS AND RESOURCES	½ Year Average	½ Year Advanced	1 Year Average	1 Year Advanced
Chapter Opener, p. 278				
EcoLog, p. 278	•	•	•	•
11.1 Fossil Fuels to Electricity, pp. 279–282	•	•	•	•
Transparency 37: *Energy Used in the United States and the World*	•	•	•	•
Transparency 38: *Coal-Fired Power Plant*	•	•	•	•
Transparency 39: *Formation of Oil and Gas Deposits*	•	•	•	•
Section Review, p. 282	•	•	•	•
11.2 Nuclear Energy, pp. 283–286	•	•	•	•
Transparency 40: *Nuclear Fission*	•	•	•	•
Transparency 41: *Nuclear Power Plant*	•	•	•	•
Making a Difference: *Environmental Activist*, pp. 300–301			•	•
Section Review, p. 286	•	•	•	•
11.3 A Sustainable Energy Future, pp. 287–294	•	•	•	•
EcoSkills: *Boosting Your Home's Energy Efficiency*, p. 389			•	•
Case Study: *A Super-Efficient Structure*, pp. 288–289	•	•	•	•
Transparency 42: *Solar Cell*	•	•	•	•
Field Activity, p. 291			•	•
Transparency 43: *Hydroelectric Dam*	•	•	•	•
Section Review, p. 294	•	•	•	•
End of Chapter				
Chapter Highlights, p. 295	•	•	•	•
Chapter Review, pp. 296–297	•	•	•	•
Investigation: *Solar Design*, pp. 298–299	•	•	•	•
Videodiscovery STS Science Forums: Vol. 1—*Home Power Generation*				•
Chapter Tests	•	•	•	•
PACING GUIDE	9–11 days	8–10 days	17–18 days	17–18 days

MATERIALS ORGANIZER

Activity	Materials	Advance Preparation
Investigation: *Solar Design*, pp. 298–299	2 identical cardboard boxes with lids; knife or single-edged razor blade; clear plastic wrap; thermometer; tape; metric ruler; watch or clock; notebook; pen or pencil; one or more of the following: cardboard, foam insulation, caulk, paint, foil, tiles, water containers, colored construction paper, rocks, soil, etc. (per student group)	none

CHAPTER OVERVIEW

In this chapter students are introduced to the fundamentals of energy use in our society. Fossil fuels, nuclear energy, and alternative sources of energy are explored.

Section 11.1 explains the process of generating electricity. Fossil fuels are defined, and the concepts of renewable and nonrenewable sources of energy are introduced.

Section 11.2 explores the process of nuclear fission and assesses its benefits and drawbacks as a source of energy. Nuclear fusion is also discussed.

Section 11.3 introduces the concept of energy conservation and explores various alternative energy sources for the future.

RESOURCES

. . . for Teachers

Brower, Michael. *Cool Energy: Renewable Solutions to Environmental Problems.* Revised edition. Cambridge, MA: MIT Press, 1992. (This book reviews progress in the field of renewable energy technologies, including solar, wind, biomass, hydroelectric, and geothermal technologies.)

Burmeister, George, and Frank Kreith. *Energy Management and Conservation.* Washington, D.C.: National Conference of State Legislatures, 1993. (This book highlights the energy-conservation programs of the six most energy-efficient states. It includes clear examples of innovative energy-saving tactics and in-depth analyses of the United States energy consumption, alternative fuel issues, etc.)

. . . for Students

Cozic, Charles, ed. *Current Controversies: Energy Alternatives.* San Diego, CA: Greenhaven Press, 1994. For a catalog write P.O. Box 289009, San Diego, CA 92198-9009; 800-231-5163. (This is an anthology that includes several points of view on topics such as the following: Should the United States decrease its use of fossil fuels? Is nuclear power a viable energy alternative? What are the alternatives to gasoline?)

Inhaber, Herbert, and Harry Saunders. "Road to Nowhere," *The Sciences,* Nov.–Dec. 1994, pp. 20–25. (This enlightening article discusses how attempts to conserve energy often backfire and lead to increased energy consumption.)

Films, Videos, Software, and Other Media

Biogas From the Sea. Video. 29 minutes. Bullfrog Films, P.O. Box 149, Oley, PA 19547; 215-779-8226 or 800-543-3764. (This video spotlights German and Italian scientists who are converting excess marine algae from the Venice Lagoon into biogas, a versatile, clean-burning fuel.)

Nuclear Energy in View Series. Video. (Videos range from 11 to 18 minutes in length.) Britannica Videos, 310 South Michigan Avenue, Chicago, IL 60604; 800-554-9862. (Videos from this series can be bought individually, or the whole series can be purchased at a discounted price. The videos explore the positive and negative effects of nuclear energy and its associated technologies.)

Power Struggle. Video. 58 minutes. Bullfrog Films, P.O. Box 149, Oley, PA 19547; 215-779-8226 or 800-543-3764. (Narrated by Meryl Streep, this award-winning video has been specially edited for schools and divided into two parts: *Energy Supply* [34 minutes] and *Energy Efficiency* [23 minutes].)

Getting Involved

- Use the following EcoSkills activity to enhance student interest in the subject of this chapter: *Boosting Your Home's Energy Efficiency,* page 389.
- A listing of environmental organizations, publications, and volunteer opportunities can be found in the **Laboratory and Field Guide.** Note especially those organizations with the goal of promoting the efficient use of resources, such as Greenpeace USA, the Nuclear Information and Resource Service, the Union of Concerned Scientists, and the World Resources Institute.

USING THE QUOTATION

"The law of conservation of energy tells us that we can't get something for nothing, but we refuse to believe it."

Explain that the law of conservation of energy states that when physical and chemical changes take place in a closed system, no amount of energy is created or destroyed; it simply changes form. Thus, the energy input is always equal to the energy output—you can't get something for nothing. Encourage discussion about what Asimov might have meant by his statement, using "perpetual-motion" machines as an example.

USING THE ECOLOG

To assess students' knowledge about the subject of this chapter, have them answer the EcoLog questions on page 278. In answering these questions, students will reveal their prior knowledge, which is usually very basic and often flawed or even misconceived. Then, by revising their answers at the conclusion of the chapter, students will have the opportunity to eliminate any misconceptions and to see firsthand how their knowledge has grown.

SECTION 11.1

FOSSIL FUELS TO ELECTRICITY

FOCUS

Students analyze the important role of fossil fuels in our society. They also learn about renewable and nonrenewable resources.

MOTIVATOR

Ask students to take a moment to imagine what life would be like without fossil fuels. Explain that fossil fuels are materials such as coal, oil, and natural gas that are burned to supply heat or generate electricity. Suggest that students run through their daily routine from the moment they wake up in the morning to the time they go to sleep and think about how fossil fuels affect that routine. Ask: How many of the things that you do each day depend on fossil fuels in some way? Solicit examples from students. *(There are countless examples, such as blow-drying one's hair, making toast, and driving to school.)* Then encourage discussion about how life would be different if fossil fuels did not exist.

TEACHING STRATEGIES

Meeting Individual Needs

Second-Language Learners Have these students create their own translation glossary of the vocabulary terms in this chapter. The glossary should consist of two sections, as in a real translation dictionary. In one section the vocabulary term in English is accompanied by a definition in the students' native language. In the other section the vocabulary term in the students' native language is accompanied by a definition in English.

ELECTRICITY: ENERGY ON DEMAND, *page 280*

Using the Transparency

Figure 11-2 is available as a Teaching Transparency.

Meeting Individual Needs

Learners Having Difficulty After students have read page 280 of their text, demonstrate how electricity is generated by constructing a simple generator. To do so, wrap approximately 100 or so turns of thin, insulated copper wire around a paper-towel tube. Connect the free ends of the wire to a galvanometer or a multimeter (set to the lowest voltage scale). Place a strong bar magnet into the tube, and move it back and forth along the axis of the tube. Direct students' attention to the voltage output. Students should notice that the voltage output is "positive" when the magnet is moved one way and "negative" when the magnet is moved the other way. This indicates the direction in which the electrons are moving through the wire. It is also an example of alternating current, the type of electric current that we use in our homes.

FOSSIL FUELS, *page 281*

Using the Transparencies

Figures 11-4 and 11-5 are available as Teaching Transparencies.

Using the Diagram: *Figure 11-4, page 281*

Most of the electric power used in this country is generated at plants such as the one shown in the figure. Tell students that the typical power plant converts into electric energy only about 25 to 30 percent of the energy contained in the fossil fuels it uses. Ask students to examine the diagram to see if it contains any clues as to where the "lost" energy goes. *(The lost energy is dissipated as waste heat in the following ways. (1) Not all of the heat from the burning fuel is transferred to the water; much escapes through the walls of the combustion chamber. (2) Heat is lost through the walls of the boiler and the pipes carrying steam to the turbine. (3) The turbine is not 100 percent efficient at capturing the kinetic energy of the steam. (4) The heat energy remaining in the steam after it passes through the turbine is of no use, so it is deliberately dissipated to the outside environment; this represents wasted energy. (5) Friction in the turbine and generator also diverts some energy. (6) Not all of the kinetic energy of the spinning generator is converted to electricity. (7) Finally, electrical resistance in the generator and associated circuitry dissipates additional energy.)*

Using the Theme of *Patterns of Change*

Focus Question: Why might burning fossil fuels actually be a way of recycling them? *(Burning fossil fuels results in the release of carbon dioxide. Organisms then take in carbon dioxide. Eventually, those organisms die and decay, contributing carbon compounds to the soil. If left alone, those carbon compounds will, in time, be incorporated into fossil-fuel deposits again. This is an example of the carbon cycle that students learned about in Chapter 3.)* Point out to students that burning fossil fuels accelerates the carbon cycle by putting more carbon dioxide into the air than "normally" exists.

Multicultural Connection

Long before the automobile came along, people used petroleum for various purposes. Native Americans obtained petroleum from natural seeps and shallow wells and used it as a fuel and medicine. Thousands of years ago people in the Middle East used pitch, a solid form of petroleum, as a fuel, caulking for boats, and paving material. Ancient Egyptians used pitch as a preservative in the mummification process. American colonists used petroleum as a lubricant and liniment.

DWINDLING SUPPLIES OF FOSSIL FUELS, *page 282*

Using the Graph: *Figure 11-7, page 282*

Make sure students realize that projections of the life span of fossil-fuel reserves are based largely on deposits not yet discovered. It is theoretically possible (but highly unlikely) that little or no additional reserves will be found, greatly shortening the life span of the resource. On the other hand, it is not at all unlikely that recoverable fossil-fuel deposits actually exceed the estimated amounts by as much as a factor of two or three. Ask students: Why might it be dangerous for people to assume that the number and size of recoverable fossil-fuel deposits have been underestimated? *(People would be lulled into a false sense of security. There would be even less incentive to conserve fossil fuels and to search for alternative energy sources. Fossil-fuel use would continue to increase, accelerating the damage being done to the environment.)* Point out that even if the estimates are low, they would probably be off by only a few years because fossil-fuel use continues to increase by about 4 percent every year.

Background: The Arab Oil Embargo

The Arab oil embargo of 1973 showed how vulnerable the United States is to disruptions of its oil supply. In retaliation for American support of Israel in the 1973 Yom Kippur war, a number of Arab nations suspended shipments of crude oil to the United States. This action, which reduced the oil supply in the United States by less than 10 percent, was extremely disruptive. Gasoline shortages were common. Many service stations closed, and long lines formed at those that remained open. Gasoline prices skyrocketed, and the federal government moved to ration gasoline. Military action to restore the flow of oil was discussed. The repercussions of the embargo were long lasting. It galvanized a struggling conservation movement and led to remarkable improvements in American energy efficiency. More recently, the Persian Gulf War slightly disrupted the supply of Middle Eastern oil and sent prices sharply upward.

Language-Arts Connection

Ask your students to complete the following sentence: "Basing an economy on oil is like . . ." Students' sayings should use humor to point out the absurdity of basing an entire world economy on a finite and rapidly vanishing resource. Display the most creative sentences.

CLOSURE

Section Review

Answers to these questions are provided in the Answer Key on page T155.

Alternative Assessment

Ask students to write a story about a fictional event, set in the future, in which oil supplies suddenly dwindle to nothing. A sample title could be "The Day the Oil Ran Out." Well-crafted stories will accurately portray the chaos and economic meltdown that would result.

Reteaching

To check your students' grasp of renewable and nonrenewable resources, ask them whether the following are renewable or nonrenewable resources and why: wood *(renewable—trees eventually grow back)*; water *(renewable—water is endlessly recycled in nature)*; oxygen *(renewable—it is recycled by life processes)*; iron *(nonrenewable—iron is not replenished by nature at a usable rate)*; coal *(nonrenewable—not replenished by nature at a usable rate)*. Of the resources that are renewable, are they renewable under every circumstance? *(no—if they are used up faster than they are replenished, they are nonrenewable)* If time were not a limiting factor, how would the definitions of renewable and nonrenewable resources differ? *(All materials would ultimately be renewable because everything is recycled by nature—some things are just recycled more slowly than others.)*

Extension

Pose the following problem to students. It is estimated that about 1,000 years are required for nature to replace the oil used by humans in just one day. Knowing this and assuming a constant rate of usage, determine (a) the time required to replenish the oil used by humans in one hour; (b) the time required to replenish the oil used by humans in an average lifetime of 75 years; (c) the rate at which humans could use oil to match the rate of natural replenishment (assuming that about 25 billion barrels are used per year); (d) the United States yearly share of this (it uses about 28 percent of the world total). *(**a:** about 42 years; **b:** 75 yrs. × 365 day/yr. × 1,000 yrs. per day of consumption = 27,375,000 yrs.; **c:** rate of use = 365,000 × rate of replenishment, so 25,000,000,000 barrels/yr. ÷ 365,000 = 68,493 barrels/yr.; **d:** 68,493 barrels/yr. × 0.28 = 19,178 barrels/yr.)*

NUCLEAR ENERGY

FOCUS

The advantages and disadvantages of nuclear fission as a means of producing electricity are examined in this chapter. Nuclear fusion is also considered as a possible source of electricity for the future.

MOTIVATOR

Copy the following messages onto slips of paper so that half of the class will receive message A and the other half will receive message B. Then randomly distribute the slips of paper, and tell students not to read aloud or discuss their message.

(A) *Greetings, friend. As mayor, I am pleased to announce that a high-tech industrial facility has been proposed for our community. The facility will employ many people, is clean and non-polluting, is part of an industry that has an outstanding safety record, and produces a commodity that will be indispensable to hundreds of thousands of people. Can I count on your support?*

(B) *Greetings, friend. As mayor, I am pleased to announce that a nuclear power plant has been proposed for our community. Can I count on your support?*

Tell students to decide whether to support the proposal described on the paper. Then ask: Who received message A? Half of the class should raise their hands. Then ask: How many of you who received message A support the proposal? Record the number on the chalkboard. Ask: How many do not support the proposal? Again, record the number on the chalkboard. Repeat the procedure for message B. When finished, it should be clear that fewer students support proposal B than proposal A. Read both messages aloud, and tell students that both messages are about exactly the same project. Ask: Why, then, was there such a difference in support between proposals A and B? Encourage discussion about the perception of nuclear power. Close by stating that in this section students will learn more about nuclear power.

TEACHING STRATEGIES

HOW NUCLEAR ENERGY WORKS, *page 283*

Using the Transparencies

Figures 11-8 and 11-9 are available as Teaching Transparencies.

Using the Diagram:
Figure 11-8, page 283

Ask students to compare the nuclear power plant in this figure with the fossil-fuel power plant shown in Figure 11-4. Students should see that the overall process is very similar. In both systems, reactions—chemical in one case, nuclear in the other—are simply a source of heat. This heat is used to boil water to make steam, which drives a turbine.

SO WHY AREN'T WE USING MORE NUCLEAR ENERGY? *page 285*

Background: Breeder Reactors

One of the criticisms against nuclear power is the relative scarcity of fuel. Depending on whose estimate you accept, the world has anywhere from a 20-year to 200-year supply of uranium-235 at current rates of consumption. A special type of reactor called a breeder reactor can help overcome this problem. Breeder reactors convert uranium-238, which is relatively common but not fissionable (usable as nuclear fuel), into plutonium-239, which is fissionable. They do this while generating usable power. Well-designed breeder reactors can actually produce more usable nuclear fuel than they consume. One major concern about breeder reactors is that the plutonium they produce is a key ingredient of atomic bombs. This connection leads to fears that breeder reactors could be the target of terrorist attacks.

Using the Theme of *Energy*

After sharing the above background information with students, pose the following *Focus Question:* Given that breeder reactors create more useful energy (stored in the form of plutonium) than they use (stored in the form of uranium), do breeder reactors violate the law of conservation of energy, which states that energy can neither be created nor destroyed? Why or why not? *(No, they do not. A breeder reactor does not create energy. It simply uses energy to take some of the energy that is stored, but not useful, in uranium, and store it in plutonium, a form that is easier to use.)*

Debate

Tell the class to imagine that a nuclear

power plant has been proposed to replace an existing, aging power plant in their community. Then randomly divide the class in half. Tell half of the class that they are to defend the proposal, while the other half will oppose it. Allow time for students to build a case for their position. To make it interesting, you might suggest that groups have each member assume the role of a person who might actually be involved in such a debate. For example, members could be utility company representatives, design engineers, unemployed workers, concerned citizens and parents, property owners, and business owners. Another option is for students to hold a simulated town meeting in which the issue is discussed and ultimately voted on.

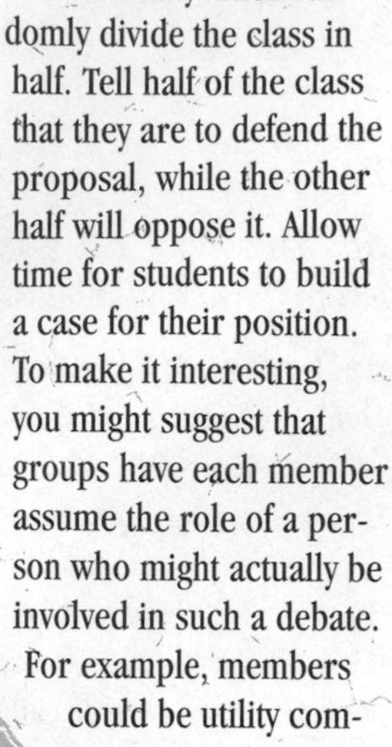

NUCLEAR FUSION, *page 286*

Background: Fusion Technologies

After 40 years of experimentation, scientists have succeeded in carrying out fusion on a very limited scale. So far, however, usable amounts of energy have not been produced. There are two basic fusion-reactor designs: the magnetic-confinement method and the pellet-fusion method. The most successful magnetic-confinement system, called a tokamak (a Russian acronym for "toroidal magnetic chamber"), uses intense magnetic fields and a strong electric current to confine deuterium plasma in a toroidal (doughnut-shaped) field while heating it to ignition temperatures of 100 million °C (180 million °F). The pellet-fusion method uses a spherical array of powerful, simultaneously firing laser beams to implode a small deuterium-enriched pellet with such force that its atoms fuse. Huge amounts of energy are needed to do this. The combined power of the lasers exceeds, for an instant, the power output of the entire United States. Both methods are believed to be ultimately workable, but unforeseen problems could still arise that prove unsolvable. In any event, because of the numerous major problems yet to be solved, commercial nuclear fusion, if it occurs, is many years in the future.

Meeting Individual Needs

Gifted Learners Suggest that students research nuclear fusion and produce a poster, report, or short presentation describing the main methods of nuclear fusion and what a fusion power plant might look like. Students may want to focus on the incredibly complicated technical problems that are posed by fusion and to discuss how those problems are being solved. Suggest that students include their findings in their Portfolio.

CLOSURE

Section Review

Answers to these questions are provided in the Answer Key on page T155.

Alternative Assessment

Have students prepare a brochure or advertisement for either a pro-nuclear or anti-nuclear organization. The goal of the document should be to recruit more members to the organization, so the item should be persuasive. Tell students that their document should be supported with facts and sound reasoning, not emotional appeals. Encourage students to include their persuasive document in their Portfolio.

Reteaching

To demonstrate how rapidly a fission chain reaction can get out of control, carry out the following demonstration. Cut a hole slightly larger than a Ping-Pong ball in the side of a cardboard box. Then line the bottom of the box with mouse traps, set the traps, and carefully balance a Ping-Pong ball on each trigger. **Safety Alert:** *Do not allow students to perform this demonstration. Use extreme care to avoid triggering the mousetraps prematurely.* Tape a transparent lid (such as a piece of Plexiglass) on top of the box. Next, toss a Ping-Pong ball through the hole, and watch what happens. Explain that each mousetrap represents an atom of uranium-235 and each Ping-Pong ball represents a neutron released during fission. The mousetrap going off and sending the Ping-Pong ball flying represents an atom fissioning and releasing a neutron, which goes on to trigger the fissioning of other atoms.

Extension

Tell students that they have been appointed to a national committee whose task is to select a method of dealing with nuclear wastes. The committee has several options: (a) do nothing; (b) send the waste into space; (c) dump the waste in the deepest, most isolated part of the ocean; (d) dump the waste in Antarctica; (e) pay a developing country to dispose of it; (f) bury it in a stable geologic formation in an isolated area; (g) study the issue further. Students should select an option and justify their decision with sound reasoning. Without being told, students should come to realize that there are two main issues involved: safety and economy. The best solution takes both issues into account.

A SUSTAINABLE ENERGY FUTURE

FOCUS

Alternatives to fossil fuels are described in this section, along with methods of conserving energy. Finding ways to guarantee the sustainability of resources is also emphasized.

MOTIVATOR

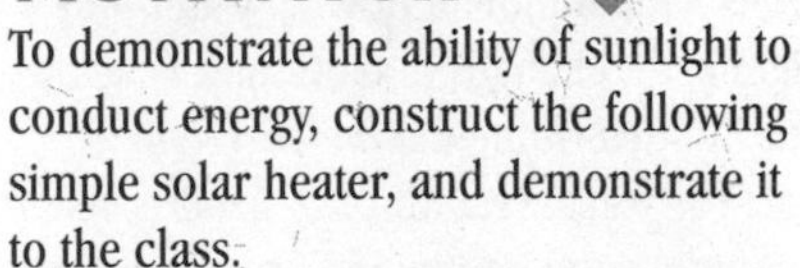

To demonstrate the ability of sunlight to conduct energy, construct the following simple solar heater, and demonstrate it to the class.

Materials: a shallow cardboard box or box lid at least .27 sq. m (3 sq. ft.) in area, clear plastic wrap, transparent tape, clear plastic tubing (about 8 m, or 26 ft.), scissors, heavy black paper, glue, 2 pails, motor- or hand-driven pump (suitable for pumping water) **Note:** Clear plastic tubing and hand-driven pumps are available at most hardware or auto parts stores.

Your finished solar heater will look like that shown at right. Begin by lining the cardboard box with the black paper and gluing it firmly in place. Poke holes in opposite sides of the box to accommodate the tubing. Arrange the tubing in S-shaped curves inside the box so that it covers as much of the available surface as possible but does not overlap. Leave about 1.5 m (5 ft.) of tubing protruding from each end. Securely tape the tubing to the inside of the box so that it is flat against the black paper lining. Connect the pump to one end of the tubing. (**Note:** In lieu of using a pump, you may use the siphon principle to operate the solar heater. Simply make sure that the source end is at least 2 ft. higher than the destination end. You will have to start the flow with suction.) Securely cover the top of the box with clear plastic wrap. Place your solar collector in a sunny place, positioned so that it faces the sun as directly as possible. Drop the end of the tubing attached to the pump into a bucket of cool water. Place the other end of the tubing in an empty bucket. Turn the pump on. Invite students to compare, by touch, the temperature of the water in the source bucket to the water in the destination bucket. Then measure the temperature of each with a thermometer, and encourage students to discuss the findings.

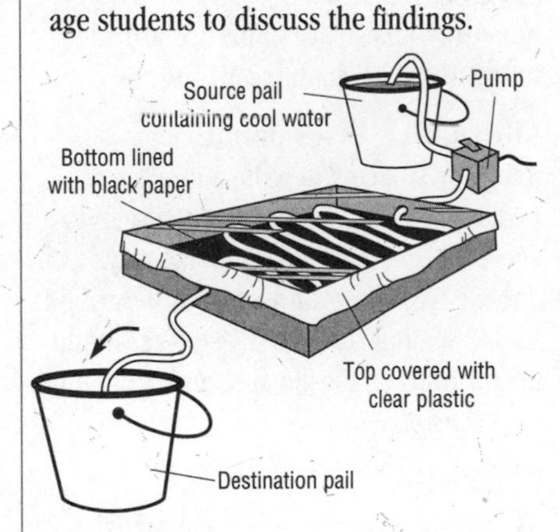

TEACHING STRATEGIES

ENERGY CONSERVATION, *page 287*

Real-World Connection
Many utility companies offer energy audits, in which trained technicians survey a home or business to determine its energy efficiency and suggest improvements. If such a service is available in your community, invite an energy auditor to come to your class and describe an energy audit. Invite him or her to perform an energy audit of your classroom.

Using the Case Study: *A Super-Efficient Structure, pages 288–289*
Call students' attention to the energy saved by each of the devices mentioned in the article. For example, the fluorescent light bulbs save 57 watts of electricity, and the refrigerator uses 92 percent less electricity than a standard refrigerator. Remind students that no sacrifice in living standards is incurred by implementing these energy-saving ideas. Explain that figures such as these are averages of the amount of energy that could be saved using current technology. Ask: Which of the institute's systems are applicable everywhere? Which are not? Why? *(The solar-heating and electrical-generating systems and the solar clothes dryer would not be applicable everywhere because many places do not get as much wintertime sunshine as do the Rocky Mountains. Every other system could be applied anywhere.)* Ask: What is the main point of the Case Study? *(that large amounts of energy could be saved with a little additional effort and expense and no loss of comfort)* Amory Lovins coined the term "negawatts" to refer to the energy saved through increased efficiency. Ask: How are negawatts like new sources of energy? How will these negawatts help us meet the higher energy demands of the future? *(Negawatts are the functional equivalent of additional energy that can be used somewhere else. The more negawatts that are "generated," the greater the demand that can be met without having to find new sources of energy.)*

Solar Energy, *page 288*

Using the Diagrams: *Figure 11-15, page 289, and Figure 11-16, page 290*

Have students compare Figures 11-15 and 11-16. Ask students: Which solar heating system is more efficient and why? *(The passive system; it requires no additional input of energy to operate. The active system, however, uses some energy to achieve essentially the same effect.)* Which system could be more easily retrofitted to an existing house? Why? *(The active system; no major structural work would be required. The passive system, on the other hand, would work only in a house that was designed from the ground up to include it.)*

Math Connection

Have students obtain old utility bills for their household and use them to estimate how much energy and money could be saved if their house was half as energy efficient as the structure in the Case Study. *(Electricity bills would probably be reduced by 80 percent and water bills would be reduced by 25 percent in the average home. Actual figures will vary, depending on specific circumstances.)*

Using the Transparency

Figure 11-17 is available as a Teaching Transparency.

Wind Energy, *page 291*

Cooperative Learning Activity

Have students form groups of four or five to construct a miniature windmill using common materials that they have at home or can find at school. (Students may use any materials for the blades and axle as long as they are not pre-assembled components.) The windmill will be made to lift a load as it turns. This will be accomplished by winding a length of string, with test weights attached, around a spool that forms part of the windmill's axle. Use steel washers of uniform size as your test weights. Students should begin by establishing a plan and drawing a basic diagram of their windmill. Windmills should measure no more than 60 cm (24 in.) in height. Use a large box fan to simulate wind. Have students test their windmills to determine the maximum load that they can lift. You might conduct two or three trials, allowing students to make adjustments and modifications between trials. The design that lifts the heaviest load (and is therefore the most efficient) is the winner.

Hydroelectricity, *page 292*

Using the Transparency

Figure 11-19 is available as a Teaching Transparency.

Closure

Section Review

Answers to these questions are provided in the Answer Key on page T156.

Alternative Assessment

Ask your students to write a mock proposal for an alternative energy system to serve their community. Students may choose from any of the systems described in this section. The energy source should be appropriate for the area and economically feasible.

Reteaching

As a class, construct a concept map that will identify the fundamental differences between the energy sources described in this section and those described in the first two sections of the chapter. *(The map should show that the energy sources in this section are renewable, while those in the previous two sections are nonrenewable.)* Then lead a discussion about a sustainable energy future. Explain that a sustainable energy future is one in which a given energy usage pattern could be maintained indefinitely. This is not possible with nonrenewable energy resources—they will run out someday—but is possible only with renewable energy resources.

Extension

Ask students to determine which of the alternative energy sources mentioned in this section are suitable for your area or are already in use. Tell students to assume the following when making their decision: 200 days of sunshine are required for solar energy systems, and an average wind speed of over 14 km/h (9 mph) is required for wind energy systems. Geothermal energy is generally an option only in mountainous areas. Hydroelectricity is an option only in regions with abundant flowing water and significant changes in elevation. (Use an atlas to determine the meteorological data for your area. Maps of geothermal potential can be obtained from encyclopedias or most comprehensive energy-related texts.)

Answers

Answers to *Chapter Highlights, Chapter Review,* and *Investigation* are provided in the Answer Key for Chapter 11, which begins on page T155.

Teaching Strategies for Investigation:

Solar Design, *pages 298–299*

Make a Model of a Passive Solar Home

Have students record the air temperature outside their homes and compare it to the inside temperature to determine the amount of heating that takes place.

How Can You Improve Your Home's Performance?

Suggest that students also measure how efficiently their models retain heat after the sun "goes down" by measuring the temperature of the model at intervals after the sunlight has been blocked. Students should find that large amounts of tile or stone retain heat well.

CHAPTER 12

WASTE

PLANNING GUIDE

CHAPTER COMPONENTS AND RESOURCES	½ Year Average	½ Year Advanced	1 Year Average	1 Year Advanced
Chapter Opener, p. 302				
EcoLog, p. 302	•	•	•	•
12.1 Solid Waste: The Throwaway Society, pp. 303–309	•	•	•	•
Transparency 44: *Sanitary Landfill*	•	•	•	•
Environmental Careers: *Landfill Manager*, pp. 376–377			•	•
Transparency 45: *Solid-Waste Incinerator*	•	•	•	•
Section Review, p. 309	•	•	•	•
12.2 Solid Waste: Options for the Future, pp. 310–315	•	•	•	•
Case Study: *Paper or Plastic?* pp. 312–313	•	•	•	•
EcoSkills: *Making Your Own Compost Heap*, p. 394–395			•	•
Section Review, p. 315	•	•	•	•
12.3 Hazardous Waste, pp. 316–322	•	•	•	•
Case Study: *Love Canal: A Toxic Nightmare*, pp. 318–319		•	•	•
Transparency 46: *Hazardous-Waste Deep-Well Injection*	•	•	•	•
Points of View: *Should Nuclear Waste Be Stored at Yucca Mountain?* p. 328–329			•	•
Field Activity, p. 322				•
Section Review, p. 322	•	•	•	•
End of Chapter				
Chapter Highlights, p. 323	•	•	•	•
Chapter Review, pp. 324–325	•	•	•	•
Investigation: *Does Fast Food Have to Be Wasteful?* pp. 326–327		•	•	•
Videodiscovery STS Science Forums: Vol. 1—*Computers & the Paperless Classroom;* Vol. 2—*Nuclear Waste Disposal*			•	•
Chapter Tests	•	•	•	•
PACING GUIDE	7–9 days	7–9 days	14–16 days	14–16 days

MATERIALS ORGANIZER

Activity	Materials	Advance Preparation
Investigation: *Does Fast Food Have to Be Wasteful?* pp. 326–327	notebook, pen or pencil, calculator (optional) (per student group)	If this activity is to be done as a class, locate at least one fast-food restaurant, and then obtain necessary permissions and arrange for transportation to and from the restaurant(s).

Chapter Overview

In this chapter students are introduced to our society's waste problems.

Section 12.1 explores the different kinds of solid wastes and discusses problems related to the disposal of these wastes.

Section 12.2 explains how producing less waste, recycling, buying recycled products, composting, and changing the types of materials we use can help alleviate the waste problem.

Section 12.3 discusses how hazardous waste is produced and disposed of, and considers ways to dispose of such material more safely.

Resources

. . . for Teachers

"A Guilt-Free Guide to Garbage." *Consumer Reports.* February 1994, pp. 91–113. (This article provides an individual with practical knowledge for dealing with garbage.)

Bender, David, ed. *Opposing Viewpoints: The Environmental Crisis.* San Diego, CA: Greenhaven Press, 1991. For a catalog contact: Greenhaven Press, P.O. Box 289009, San Diego, CA 92198-9009; 800-231-5163. (This set of 16 educational pamphlets helps students explore the various sides of environmental issues. Several of the pamphlets are especially relevant to waste-related issues.)

Gourlay, K. A. *World of Waste: Dilemmas of Industrial Development.* Atlantic Highlands, NJ: Zed Books, 1992. (This book examines the environmental costs of our increasing production, consumption, and disposal of waste.)

. . . for Students

Carless, Jennifer. *Taking Out the Trash: A No-Nonsense Guide to Recycling.* Washington, D.C.: Island Press, 1992. (This book is full of helpful hints and facts about recycling.)

EarthWorks Group. *The Recycler's Handbook.* Berkeley, CA: EarthWorks Press, 1990. (This friendly book includes page after page of useful recyling tips.)

Films, Videos, Software, and Other Media

Garbage: The Movie—An Environmental Crisis. Film, Video, or Level I CAV Videodisc. 24.5 minutes. Churchill Media, 6901 Woodley Avenue, Van Nuys, CA 91406-4844; 800-334-7830. (This award-winning production is hosted by students for students. It examines the problems of the environment and solid wastes, and looks at some promising solutions for those problems.)

Toxic Waste. Film, Video, or CLV Videodisc (all come with study guide). 60 minutes. Bullfrog Films, P.O. Box 149, Oley, PA 19547; 800-543-3764. (This acclaimed production uses scientific principles to explain real-world problems related to toxic waste.)

Getting Involved

• Use the following EcoSkills activity to enhance student interest in the subject of this chapter: *Making Your Own Compost Heap,* pages 394–395.

• A listing of environmental organizations, publications, and volunteer opportunities can be found in the **Laboratory and Field Guide**. Note especially those organizations with the goal of promoting waste-reduction techniques and understanding hazardous wastes, such as the Citizens Clearinghouse for Hazardous Wastes and the National Recycling Coalition.

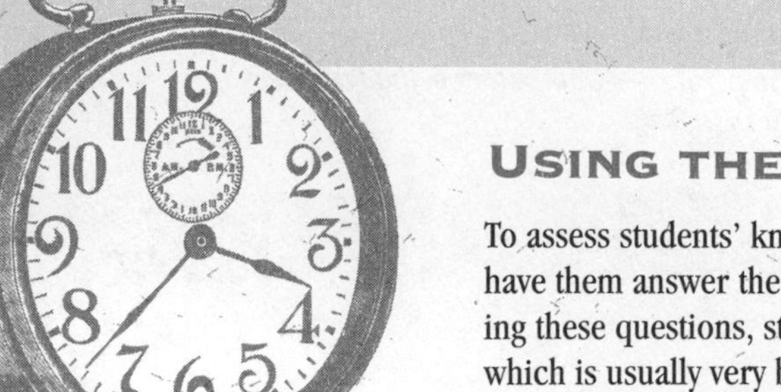

Using the Quotation

"Nothing can be forgotten, only left behind."

Call on a student to read the quotation aloud. Then ask the class what Joy Harjo might have meant by the statement. Encourage class discussion. Accept all reasonable answers, and tell students to keep the quotation in mind as they progress through the chapter.

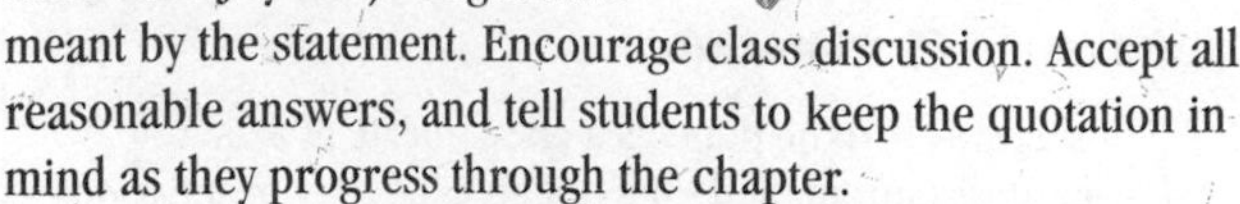

Using the EcoLog

To assess students' knowledge about the subject of this chapter, have them answer the EcoLog questions on page 302. In answering these questions, students will reveal their prior knowledge, which is usually very basic and often flawed or even misconceived. Then, by revising their answers at the conclusion of the chapter, students will have the opportunity to eliminate any misconceptions and to see firsthand how their knowledge has grown.

SOLID WASTE: THE THROWAWAY SOCIETY

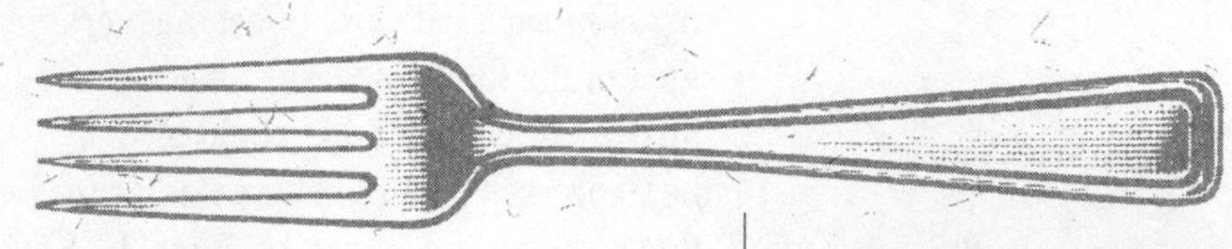

FOCUS

This section encourages students to think about where trash comes from and what happens to trash after it is disposed of. Students also learn how the disposal of certain kinds of human-produced materials such as plastics affects the environment.

MOTIVATOR

Cooperative Learning Activity Divide the class into several groups. Give each group two packaged products, one that is packaged in many layers (such as a packet of gum) and one that is not (such as a large box of detergent). For each product, have students calculate the percentage of mass for the consumable product itself and for the packaging. Then ask them to compare the two products. Finally, encourage students to discuss the benefits and drawbacks of each type of packaging. *(Sample answer: A product placed in bulk in one cardboard box decreases the total amount of packaging used and saves landfill space. In addition, the cardboard could come from recycled materials. However, the product may spoil faster, it may take up more cabinet space, or it may be inappropriate for sharing with others.)*

TEACHING STRATEGIES

Using the Photograph:

Figure 12-1, page 303

Encourage students to do research to find out specifically which landfill or incinerator their trash goes to.

Notable Quotes

"We live in a world of things, and our only connection with them is that we know how to manipulate or consume them."
Erich Fromm

"A society in which consumption has to be artificially stimulated in order to keep production going is a society founded on trash and waste . . ."
Dorothy L. Sayers

"Waste not, want not."
Benjamin Franklin

Using the Photograph:

Figure 12-3, page 304

Students may be interested to know about the NIMBY phenomenon. Explain that NIMBY stands for "Not In My Backyard," and it represents the attitude of people who do not want a landfill close to their home. Ask students why a landfill is considered undesirable by some. *(fear of traffic, noise, and dust; aesthetic concerns; declining property values; groundwater contamination; and other pollution)* Then ask: What would happen if everyone took the NIMBY point of view? *(Landfills would have to be located far from people's homes and other places that humans frequent. As a result, garbage-disposal trucks would have to transport the waste farther from its source, causing additional environmental impact because of the energy required to do so.)*

ALL WASTES ARE NOT EQUAL,

page 305

Background: Plastics

Plastics are petroleum-based polymers that consist of long chains of molecules. Most of these human-produced molecular chains are very difficult to break. In fact, most plastics used today are not readily biodegradable—they take 200 to 400 years to degrade. In the United States in 1990, 25 million barrels of oil were used to produce plastic packaging alone. Plus, the plastics industry is one of the leading producers of hazardous wastes. When deposited in a landfill, toxic cadmium and lead compounds that are sometimes added to plastics as binders or for other purposes can leach out. If the landfill is unlined or if the liner is damaged, these compounds can ooze into groundwater and surface water and cause contamination.

In addition to plastics, polymers are also used to make cloth (such as polyester), carpets, and a myriad of other common consumer products. These materials endure in landfills. Alternatively, the petrochemicals from which they are made can cause a separate environmental problem—they release toxins into the air when incinerated.

Using the Graph:

Figure 12-4, page 305

Use Figure 12-4 to help students visualize the tremendous increase in municipal solid waste between 1960 and 1990. Have students copy the graph onto a sheet of paper and then assume a similar rate of solid-waste increase until the year 2000. Ask: How much trash did the average person dispose of daily in 1980? *(about 1,500 g)* How much will that person dispose of in the year 2000? *(about 2,500 g)* Invite students to talk about the implications of such an increase in the volume of trash. Ask students: Where will we put all this trash? *(In a growing number of landfills and incinerators; some students may mention the possibility that we could run out of landfill space.)* Encourage students to brainstorm ways we can reduce the volume of trash we produce. *(reduce, reuse, recycle)*

WHERE OUR TRASH GOES, *page 306*

Notable Quote

"Everybody used to believe in the Trash Fairy—put your trash on the curb and it's gone."

Natalie Roy

LANDFILLS, *page 306*

Using the Transparencies

Figures 12-10 and 12-13 are available as Teaching Transparencies.

Multicultural Connection

Materials from construction sites often take up an enormous amount of landfill space. Frequently, these materials are items that, with a little creative thinking, can be used again. For example, a group of Japanese engineers knew that the construction of a large tunnel in an urban area would result in massive quantities of "muck" (the earth and rocks removed to form the tunnel cavity). Ordinarily, this material would be transported a long distance to a rapidly filling landfill. Instead of using this approach, however, the engineers used the muck to reclaim a portion of land that was previously unusable by the city. In doing so, the engineers provided two valuable services: they created urban land space that could be used as parkland or for development, and they saved precious landfill space.

Meeting Individual Needs

Learners Having Difficulty Have each student write down the last three things that he or she threw away. Compile a list on the board of all the items, divided by category: paper products, metals, glass, plastics, organic wastes, and other waste. (You may wish to have students do this in groups.) Compute the percentage of items in each category to get a general idea of the kinds of wastes that are thrown away most (this could also be expressed as a pie chart similar to those shown in Figures 12-7 and 12-8). Discuss with students what will happen once all of these items reach a landfill or an incinerator: which categories will burn the easiest in an incinerator *(paper products)*, which categories will decompose the fastest in a landfill *(organic wastes)*, etc.

INCINERATORS, *page 309*

Social-Studies Connection

Invite students to create a class presentation in which one student portrays a radio talk-show host whose program is focused on a new solid-waste incinerator to be built near a residential part of town. The other students should prepare arguments for or against the incinerator and then "call in" their opinions to the radio station. Encourage students to take a variety of points of view.

CLOSURE

Section Review

Answers to these questions are provided in the Answer Key on page T156.

Alternative Assessment

Have students create an illustration, a poem, or a short story to contrast natural and synthetic materials. *(Answers should reflect the following. Natural materials are biodegradable, which means that microorganisms can readily decompose them into simpler chemicals that other living things can consume. Synthetic materials are not found in nature, but instead are made by humans in a laboratory. As a result, microorganisms have not*

developed ways to break them down.) Suggest that students include their work in their Portfolio.

Reteaching

Have students help you collect trash discarded in the classroom and set up a display. Label the items as natural or synthetic, biodegradable or nonbiodegradable, recyclable or nonrecyclable.

Extension

Invite an official from your area's solid-waste department to come and talk to the class. Prior to the visit, students should use the knowledge they gained from this section to create a list of questions to ask the visitor. For example, students may have questions about landfills, incinerators, groundwater and surface-water contamination by leachate, etc. as they apply to your area.

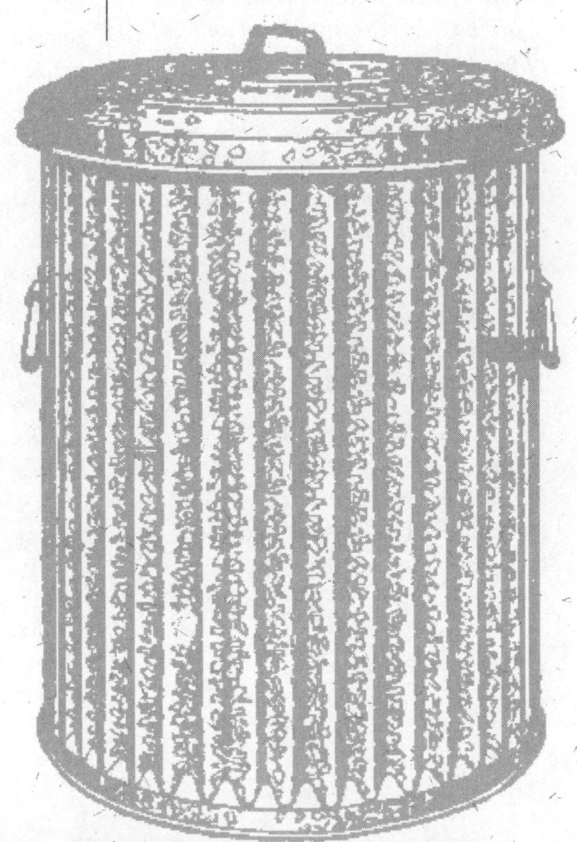

SECTION 12.2

SOLID WASTE: OPTIONS FOR THE FUTURE

FOCUS

In this section students discuss ways to reduce the amount of waste sent to landfills and incinerators. Some options include: producing less waste, recycling, buying recycled products, composting, and changing the types of materials we use.

MOTIVATOR

Bring the following items to class: a sheet of white office paper, a plastic soft-drink bottle, a glass container, a tire, and a bowl containing fruit and vegetable peelings and fresh grass clippings. Hold up the sheet of paper and ask students: What is this? *(paper)* Reply by saying: It may look like paper now, but tomorrow it could very well be part of a home's insulation, bedding for farm animals, or toilet paper. Hold up the plastic bottle and say: This bottle could become carpet, insulation in your jacket, stuffing in your pillow, part of a car's interior, or even part of a window. This glass could be added to asphalt to make new streets, and this tire could help fuel a power plant or become part of the soles on your shoes. Holding up the organic matter, tell students: Grass and fruit and vegetable wastes like these help a steam-powered generator supply electricity to over 30,000 homes in Florida. Have students help you make a concept map or a flowchart reflecting what the products you showed them, as well as others, can become when recycled.

TEACHING STRATEGIES

RECYCLING, *page 311*

Notable Quotes

"Solid wastes are only raw materials we're too stupid to use."
Arthur C. Clarke

"Recycling is where the environment and the economy meet."
Phil Bailey

History Connection

Tell students that in 1994, there were about 12,000 community recycling programs in existence in the United States. Compare that with more than 21,000 salvage organizations that existed during World War II, when the population was about half of what it is now. Explain that cooking fat was saved and used for making explosives and aluminum chewing-gum wrappers were saved for airplane fuselages. Ask: Why was recycling so important during World War II, and why are people once again beginning to recycle? *(During WWII people were faced with shortages of certain products. They reacted by reusing the useful materials in discarded products. Today, largely because of increasing human populations, many countries are worrying about national and worldwide shortages of raw materials and thus are enacting recycling programs.)* Suggest that interested students research salvage operations of WWII and prepare a written, visual, or audiovisual report to reflect their findings. Recommend that students include their report in their Portfolio.

Background: Curbside Recycling

Curbside recycling began in the city of Baltimore, Maryland, in 1974. By 1992, there were about 4,000 curbside recycling programs in the United States. Many programs arose because municipalities began to run out of landfill space. One of the greatest problems faced by recycling programs today is finding buyers for the products made from the recycled materials, which are sometimes more expensive.

Until 1994 when President Clinton mandated that the federal government (the largest purchaser of paper in the nation) use recycled paper, our government did little to encourage manufacturers to use recycled materials. Today, the most efficiently recycled material is aluminum. Recycling aluminum saves 95 percent of the energy needed to process virgin aluminum. About 1,500 aluminum cans are recycled every second in the United States. Despite this seeming success, enough aluminum is thrown out each day to rebuild the entire United States commercial airfleet every 3 months.

Meeting Individual Needs

Gifted Learners Have students interview your school's custodian or administrative staff to find out if the school has a recycling program. If there is no program, suggest that students write a proposal to get one started. The proposal should include the types of materials that could be recycled from the school's waste, a list of the companies available within the area that have pick-up or drop-off recycling programs, and a description of how the program could be initiated and maintained. Students should also use persuasive-writing skills to convince school administrators of the validity of such a project. If a recycling system is already in place, students can verify the efficiency of the system and make recommendations for its improvement, if necessary. Or students could help a nearby school or business set up a recycling program.

Using the Theme of *Scale and Structure*

Ask students: Do you or your family have a daily newspaper delivered to your home? If so, how long does it take for those papers to pile up in a wastebasket or recycling stack? *(Most students will have noticed that the newspapers pile up quickly.)* Tell students to imagine newspapers piling up in the homes of the millions of other people across the country who receive a daily newspaper. Ask: How might newspapers contribute to the nation's waste-disposal problems? Students should recognize that a newspaper may seem to be an insignificant item, but when multiplied by the entire population of newspaper readers, its environmental impact becomes significant. For instance, newspapers take up more than 40 percent of landfill space. On the other hand, less than one metric ton of recycled newspaper saves 17 trees and 2.4 cu. m (3 cu. yd.) of landfill space. And if each person recycled his or her daily paper, four trees would be saved per person each year.

Students may be interested to know that there has been some talk in recent years of distributing newspapers via computer. Encourage interested students to research this subject and write a short paragraph describing his or her opinion of such a venture. Suggest that students include their paragraph in their Portfolio.

Research Activity

Suggest that students find out how paper is made and then make some. Detailed instructions can be found in many activity books on recycling. In general, the process involves shredding paper, mixing it with water in a blender, draining the mixture, and spreading it over a mesh screen that is placed on a cloth towel. When dry, the mixture can be lifted from the screen in a sheet, and voilà—recycled paper! Encourage students to include their recycled paper in their Portfolio.

Using the Flowchart: *Figure 12-16, page 311*

Call on a student to describe the "flow" of a material as it is recycled by following the arrows shown in the illustration. Ask students to discuss what effect they think an increase in recycling will have on jobs. *(Many jobs will be created in the recycling industry, including drivers and collectors for curbside pickup and workers at recycling facilities. However, jobs will also be lost, including those involved in managing landfills or in obtaining raw materials such as wood and mined metals. Plans to implement recycling programs often involve trying to predict complex economic variables.)*

COMPOSTING, *page 312*

Background: Composting

Each year in the United States, people bag 10.8 million kg (24 million lb.) of leaves and grass clippings and send them to landfills. At the same time, farmers and homeowners buy huge amounts of commercial synthetic fertilizers to add to their crops and lawns. Many of these products are made from petrochemicals and can be toxic to the local environment. If yard wastes were composted, the byproducts could add valuable nutrients to topsoil and serve as a soil enricher, thereby reducing the amount of synthetic fertilizer necessary. Plus, those useful items would no longer be taking up dwindling landfill space.

CHANGING THE MATERIALS WE USE, *page 313*

Using the Case Study: *Paper or Plastic? pages 312–313*

Call on several students to read the Case Study. Then encourage class discussion. Ask: Do you prefer paper or plastic bags? Why? Write *paper* on one side of the chalkboard and *plastic* on the other side, and then list student responses in the appropriate categories. Make sure students include information about ways they can reuse both types of bags. Then lead students in a discussion of why "scientific" research done by interested parties often cannot be trusted. Encourage students, when examining research data, to find out if the researcher had any special interest in the outcome. Students should also find out as much as possible about the researcher's methods and conclusions.

CLOSURE

Section Review

Answers to these questions are provided in the Answer Key on page T157.

Alternative Assessment

Ask students to draw a concept map showing some solid-waste disposal concerns and options for reducing those concerns in the future. *(Some concerns include: running out of landfill space, using synthetic products that cannot biodegrade, and the contamination of water supplies by landfill leachate. Some options for reducing those concerns include: producing less waste, recycling, buying recycled products, composting, and changing the materials we use.)*

Reteaching

Arrange a class visit and tour of a local recycling center or an establishment that makes new products out of recycled materials. Ask the tour guide to discuss the cycle of a product from its original state to a recycled product and the importance of consumer demand for recycled products. Encourage students to ask questions as you proceed through the tour.

Extension

Cooperative Learning Activity Working together in small groups, students could develop a campaign to educate the rest of the school about the importance of recycling. One group could develop a plan to establish or expand a school-wide recycling program, another could make educational posters, another an informative pamphlet, etc. Encourage students to think of innovative ways to reduce the number of resources they need to complete their project. For example, a pamphlet could be copied onto recycled paper, and then, rather than distributing one to each student, a "take-one" display including limited numbers of the pamphlet could be installed at various locations in the school. Suggest that students include something from their project in their Portfolio.

HAZARDOUS WASTE

FOCUS

In this section students learn what constitutes hazardous waste. Through a discussion of the Superfund Act, they discover how difficult it is to dispose of hazardous waste. Finally, students learn how these dangerous materials are best managed.

MOTIVATOR 

Bring in the following (or similar) common household products: a can of paint and bottles of furniture oil, nail-polish remover, motor oil, and weedkiller. Being careful not to embarrass anyone, ask students if they have ever used any of the products you are displaying. If so, ask: How did you dispose of the product when you were finished with it? Tell students that most people simply toss the empty or partially empty container into the trash. If some of the product remains, he or she sometimes pours that extra substance down the drain. A few people may take those substances to a hazardous-waste collection facility.

Ask students: Why would anyone take these common materials to a hazardous-waste collection facility? Instead of answering, encourage students to look at the labels of the products, most of which probably have warning labels and hazardous-product symbols. Ask: What possible effects could those chemicals, when discarded, have on the environment? *(One example is: They could contaminate local water sources, possibly harming marine life, birds, pets, and other organisms and microorganisms.)* Then tell students that by taking their hazardous household wastes to a special collection facility they could eliminate the risk of such dangers.

TEACHING STRATEGIES

Language-Arts Connection

Suggest that students read "Pandora's Poison," by Eric F. Coppolino, *Sierra,* Sept.–Oct. 1994, pp. 41–45, 74–77. This article gives a startling history of PCBs in this country. Have interested students assume the role of investigative reporters. Their assignment is to determine whether the article is written fairly and accurately and then write another article asserting their conclusion. Students' research should include a trip to the library for other articles and books on the subject. Students may also wish to contact the Environmental Protection Agency, the companies listed in the article, and any other businesses or agencies that may be involved in the production or cleanup of PCBs for "their side of the story." Recommend that students include their article in their Portfolio.

Background: Household Items Made With Hazardous Materials

Students may be surprised to know that the following hazardous substances, most of which are thought to be carcinogenic or mutagenic, are found in common household products.

Substance	Products
Benzene	antiperspirants, deodorants, detergents, oven cleaners, shampoo
Phenols	glue, antibiotic cream, flavorings, pet shampoos, drain cleaners, shaving cream, disinfectants
Toluene	contact cement, detergent, perfume, degreasers, dandruff shampoo
Tricholoroethylene (TCE)	upholstery cleaners, degreasers, waxes
PCBs (polychlorinated biphenyls)	electrical appliances
Arsenic	weedkiller, wood preservative
Cadmium	paints, lawn treatments, textiles
Mercury	insecticides, batteries, antiseptics

Meeting Individual Needs

Second-Language Learners

Technical chemical terms for some hazardous wastes may be difficult for students whose native language is not English. Simplify some of those terms by referring to them generically as poisonous chemicals, harmful metals, materials that are dangerous to health and the environment, etc.

HOW CAN WE MANAGE HAZARDOUS WASTES? *page 318*

Using the Theme of *Energy*

Focus Question: Can the potential energy in discarded hazardous wastes be harnessed? Explain. *(Yes. Chemicals, even toxic ones, are capable of interacting with other chemicals to produce energy through chemical reactions. In addition, when a toxic waste is burned [by incineration, for example], the burning process produces energy. Both kinds of energy can potentially be harnessed to create new chemical compounds or to generate electricity.)* Invite students to find out what new technologies are needed to enable us to safely use the potential energy of hazardous wastes. Suggest that they interpret their findings in some creative way and then include the result in their Portfolio.

Real-World Connection

Direct students' attention to Figure 12-21 on page 317. Ask students if there are any Superfund sites nearby. If so, suggest that students form teams to investigate the situation. They could read regional newspapers and other periodicals, and contact the nearest office of the Environmental Protection Agency for more information about the site. Students could present their findings in the form of a news report, video, poster, etc. Suggest that students include something from the project in their Portfolio. Note: If there is not a Superfund site nearby, have students pick one of the map sites to research.

LAND DISPOSAL, *page 320*

Using the Transparency

Figure 12-22 is available as a Teaching Transparency.

Using the Points of View: *Should Nuclear Waste Be Stored at Yucca Mountain? pages 328–329*

Have students debate whether nuclear wastes should be stored at Yucca Mountain. Encourage each side to prepare a host of scientific, economic, and personal reasons for their position.

Art Connection

The United States Department of Energy is experimenting with designs that will warn people living 10,000 years from now that the material buried at Yucca Mountain is extremely dangerous. To stimulate students' thoughts, ask: What was human society like 10,000 years ago? If you saw hieroglyphs or cave paintings from that long ago, would you be able to interpret them correctly? Considering the rate of technological change, what do you think society will be like 10,000 years from now? Can you be sure any sign, symbol, construction, or other warning will be understood by humans in the far future? Why or why not? These concerns aside, invite students to use their imaginations to draw or build models of a warning sign or symbol that could warn humans living 10,000 years from now to avoid the dangerous materials buried beneath Yucca Mountain. Suggest that students write an explanation to accompany their Yucca Mountain warning designs and include both the design and explanation in their Portfolio.

CLOSURE

Section Review

Answers to these questions are provided in the Answer Key on page T157.

Alternative Assessment

Bring a book such as *Clean and Green: The Complete Guide to Nontoxic and Environmentally Safe Housekeeping* (by Annie Berthold-Bond, Woodstock, NY: Ceres Press, 1990) and a hazardous household-cleaning substance to class. First have students examine the list of chemicals in the product and determine how hazardous they are. Also ask students to determine how the product should eventually be disposed of. (Be sure you know these specifics about the product in advance.) Then have students design an experiment to test the effectiveness of the principal product against alternatives given in the book. (Have students wear gloves and safety goggles.) Encourage students to carefully document their results and, when finished, summarize their thoughts. Direct student thinking by asking: Were your organic alternatives as effective as the principal cleaner? If not, does the hazardous nature of the principal cleaner discount any of its advantages as a cleaning agent? Why or why not? What is your priority: safety or effective cleaning?

Reteaching

Contact your city's or state's environmental agency to find out if they have any films or videos about hazardous waste in your area. If not, obtain a video from the list provided in the Resources section of these teacher's notes, page T120, or from a local library or university.

Extension

Have students write letters to their local or state environmental agency to find out how much of the household waste found in landfills is considered toxic. Suggest that students also ask if any environmental problems can be attributed directly to household hazardous wastes. If so, what are those problems, and what is being done to alleviate them? Students should also ask if there is a toxic-waste collection facility in the area and, if not, write letters requesting such a facility.

Answers

Answers to *Chapter Highlights*, *Chapter Review*, *Investigation*, and *Points of View* are provided in the Answer Key for Chapter 12, which begins on page T156.

TEACHING STRATEGIES FOR INVESTIGATION:

DOES FAST FOOD HAVE TO BE WASTEFUL? *pages 326–327*

Collect Data

Remind students to take careful and accurate notes. Under no circumstances should they rely on their memories.

To achieve enhanced accuracy, students may want to visit the restaurant at two different times (the same amount of time each visit): during a very slow time and during a very busy time. To determine the "average" hourly business of the restaurant, students should average the results of each visit.

Remind students of the need to account for the waste generated by take-out business. To do this, note the number of eat-in *and* take-out customers, adding them together to find the total. Express this as a ratio (total customers to eat-in customers), and multiply all solid-waste totals by this ratio (assuming no difference in wastes between eat-in and take-out orders). For instance, if the students count four eat-in customers and three take-out customers, then the eat-in solid-waste totals should be multiplied by seven-fourths, or 1.75.

CHAPTER 13

POPULATION GROWTH

PLANNING GUIDE

CHAPTER COMPONENTS AND RESOURCES	½ Year Average	½ Year Advanced	1 Year Average	1 Year Advanced
Chapter Opener, p. 330				
EcoLog, p. 330			•	•
13.1 How Populations Change in Size, pp. 331–335			•	•
Case Study: *Overshooting the Carrying Capacity*, pp. 332–333				•
Field Activity, p. 334				•
Section Review, p. 335			•	•
13.2 A Growing Human Population, pp. 336–341			•	•
Transparency 47: *Increase in Human Population*			•	•
Case Study: *The Stories Behind the Statistics*, pp. 338–339				•
Field Activity, p. 339				•
Transparency 48: *Three Stages of Demographic Transition*			•	•
Transparency 49: *Population Patterns in Developing Countries*			•	•
Transparency 50: *Population Histograms*			•	•
Section Review, p. 341			•	•
13.3 Problems of Overpopulation, pp. 342–346				•
Field Activity, p. 343				•
Section Review, p. 346				•
End of Chapter				
Chapter Highlights, p. 347			•	•
Chapter Review, pp. 348–349			•	•
Investigation: *What Causes a Population Explosion?* pp. 350–351			•	•
Points of View: *Should Population Growth Be Limited?* pp. 352–353				•
Chapter Tests			•	•
PACING GUIDE	0 days	0 days	5–6 days	5–6 days

MATERIALS ORGANIZER

Activity	Materials	Advance Preparation
Investigation: *What Causes a Population Explosion?* pp. 350–351	notebook, graph paper, pen or pencil, straightedge or ruler, colored pencils or markers, calculator (per student group)	You may wish to perform the calculations for this investigation before assigning it to students so that you can monitor their progress.

CHAPTER OVERVIEW

In this chapter, students are introduced to general concepts of population growth and are urged to relate those concepts to human population growth.

Section 13.1 describes the factors that affect any population's size and explains why populations grow and what factors limit that growth.

Section 13.2 describes how human population growth has changed over time, discusses the stages of population growth, and identifies the factors that shape different growth rates.

Section 13.3 describes some of the problems that stem from the explosive growth of the human population and presents some possible solutions to those problems.

RESOURCES

. . . for Teachers

Ehrlich, Paul R., and Anne H. Ehrlich. *The Population Explosion.* New York City, NY: Simon and Schuster, 1990. (This book presents an overview of some of the ecological and social problems that stem from human overpopulation.)

McCuen, Gary E. *Population and Human Survival.* Hudson, WI: Gary E. McCuen Publications, 1993. (This book is part of a series called *Ideas in Conflict,* which presents counterpoints, debates, opinions, commentary, and analysis for use in high school classrooms.)

The Population Reference Bureau, 1875 Connecticut Avenue, NW, Suite 520, Washington, D.C. 20009; 202-483-1100. (This organization publishes many curriculum and resource materials for teachers; it also conducts workshops for teachers.)

Zero Population Growth, 1400 16th Street, NW, Suite 320, Washington, D.C. 20036; 202-332-2200. (This organization develops population education programs for schools and offers many resource materials for teachers; it also conducts workshops for teachers.)

. . . for Students

Gallant, Roy A. *The Peopling of Planet Earth: Human Population Growth Through the Ages.* New York City, NY: Macmillan, 1990. (This book offers an easy-to-read overview of the strict scientific perspective of human population growth. Numerous photographs and diagrams make the text particularly approachable.)

Newton, David E. *Population: Too Many People?* Hillside, NJ: Enslow Publishers, 1993. (This book, which was written for high school students, presents many viewpoints on the issue of whether there is a population problem. The author encourages the reader to learn the facts about population growth and then make his or her own informed decision.)

Films, Videos, Software, and Other Media

Paul Ehrlich's Earth Watch. Video. 17 minutes. Anson Schloat, 175 Tompkins Avenue, Pleasantville, NY 10570; 800-833-2004. (This video presents an ecological perspective of human population growth, emphasizing the environmental destruction and species extinction that could result if our population is not controlled.)

World Population. Video. 6 minutes. Zero Population Growth, 1400 16th Street, NW, Suite 320, Washington, D.C. 20036; 202-332-2200. (This video presents a dramatic, graphic simulation of human population growth from the year A.D. 1 to the present and beyond, with projections to the year 2050.)

Getting Involved

A listing of environmental organizations, publications, and volunteer opportunities can be found in the **Laboratory and Field Guide.** Note especially those organizations that are concerned with the human population explosion, such as Population-Environment Balance, the Population Reference Bureau, Worldwatch Institute, and Zero Population Growth.

USING THE QUOTATION

"People are everywhere. Some people say there are too many of us, but no one wants to leave."

Ask students if they can relate to the quotation. *(Most students will probably answer yes.)* Continue by asking what images the quotation brings to mind. *(Students will probably mention shopping malls, concerts, airports, etc.)* Then ask students to apply the quotation to the Earth. Ask: Do you think human overpopulation is a problem? *(Encourage a variety of opinions.)*

USING THE ECOLOG

To assess students' knowledge about the subject of this chapter, have them answer the EcoLog questions on page 330. In answering these questions, students will reveal their prior knowledge, which is usually very basic and often flawed or even misconceived. Then, by revising their answers at the conclusion of the chapter, students will have the opportunity to eliminate any misconceptions and to see firsthand how their knowledge has grown.

How Populations Change in Size

Focus

This section introduces students to the general characteristics of populations. Students learn that there are natural limits to population growth. These limiting resources are described in relation to the concept of carrying capacity as the ultimate determinant of population size.

Motivator

Tell students: Today we are going to discuss something very important; in fact, what we will cover is so important that I have been asked to invite other classes to join the discussion. Then write on the chalkboard the names of several other teachers in your school and the number of students in those teachers' classes so that it is obvious that there will be far too many people to fit in your classroom. For example, if your classroom seats 30 students and you have 29 students, you may write: Mr. Jones, 28 students; Ms. Smith, 31 students; Mrs. Blackburn, 25 students; and Mr. Marcus, 23 students. Tell your students that you need their help in preparing the room to accommodate the visiting classes. Explain that each person will need someplace to sit and something to write on, and it is important that every student is able to see, hear, and concentrate on the important subject matter. Then have students discuss what preparations can be made.

If students suggest that what you really need is more space, tell them that you have checked, but all of the school's larger rooms are in use. Pretty soon students should begin to realize that there simply are not enough resources to satisfy the needs of all of the visiting students. At this time, explain that there really won't be any visitors and that you were simply trying to make a point. Ask: How might you compare our situation to what sometimes happens in the natural world? *(The classroom visitors would have outnumbered the resources available in the classroom just as a natural population of organisms can grow to exceed the life-supporting resources provided by their environment. A limited amount of natural resources is what keeps the size of a population under control.)*

Teaching Strategies

Math and Geography Connections

Invite students to substitute actual demographic data into the word equations given on pages 331–332 of their texts. Students can find information on the number of births, deaths, immigrants, and emigrants in an almanac. Suggest that students obtain data for as many different countries as possible and then use their findings to compare the population situation in each nation. They may wish to show their findings on a copy of a simplified world map, which they could then include in their Portfolio.

How Fast Can a Population Grow? *page 332*

Meeting Individual Needs

Learners Having Difficulty To help students understand the difference between arithmetic growth and exponential growth, write the following numbers on the chalkboard: 1, 2, 3, 4, 5, 6. Explain that this series of numbers is increasing at a slow and steady pace, by one numeral each time. Ask students what number would follow *(7)*, and have them give you a few more numbers to add to the series. Then write: 1, 2, 4, 8, 16, 32. Ask students what number would follow *(64)*, and have them give you a few more numbers to add to the series. Then tell students that in the same number of steps, each series of numbers has grown at a very different rate—when there were only 6 in the first example, there were already 32 in the second example. Thus, although the second series is constantly increasing just as the first did, it is increasing at a progressively faster rate. This is because each number in this series is double the previous number. Tell students that the second example illustrates exponential growth, which characterizes human population growth today.

Using the Case Study:

Overshooting the Carrying Capacity, pages 332–333

Call on one or more students to read the Case Study aloud. Then call on another student to draw a graph on the chalkboard that reflects the reindeer's population curve over time. Next, call on another student to draw a line that indicates the island's carrying capacity for the reindeer. *(250 reindeer)* Invite students to discuss why the reindeer population exploded on the island when the caribou population was stable. Write their responses on the chalkboard. *(The list may include factors such as differences in food sources, differences in disease vulnerability, or parasitism between two species.)*

Notable Quote

"As more individuals are produced than can possibly survive, there must . . . be a struggle for existence, either one individual with another of the same species, or with the individuals of distinct species, or with the physical conditions of life."
Charles Darwin

Using the Graph: *Figure 13-3, page 333*

Use Figure 13-3 to illustrate and reinforce the dramatic increases seen in the exponential growth of populations. Invite students to describe the curve and explain why the growth it shows is exponential. Ask: What would the curve look like if it showed arithmetic growth? *(It would not be as steep; it would rise more gradually.)* You may want to actually draw such a curve on the chalkboard. Ask students if they think the graph could show the actual population curve for a ladybug population. *(no)* Why? *(Limits on resources that ladybugs need, as well as predation, help control the ladybug population.)* Ask: What might their actual curve look like? *(It would be much flatter.)*

WHAT LIMITS POPULATION GROWTH? *page 334*

Using the Graph: *Figure 13-5, page 335*

Have students study the graph and explain what caused the initial population crash of the Australian rabbit population. *(There were more rabbits than there were resources to support them. In other words, the population exceeded its carrying capacity.)* Have students analyze the subsequent fluctuations in the rabbit population. Ask: What do you observe about these fluctuations? *(The fluctuations become less extreme.)* Then ask: If the graph could be extended to include more time, how do you predict the population line would change? *(The population line should eventually straighten out and more or less conform to the line that indicates its carrying capacity.)*

Using the Theme of *Stability*

Focus Question: How might an introduced species disrupt the stability of a normally functioning ecosystem? Suggest that students use the graph in Figure 13-5 as a basis for their answer. *(Students should recognize that introduced species, such as the Australian rabbits, disrupt the interactions among organisms in an ecosystem, thus destabilizing it. With the absence of native predators or other controls, introduced species are likely to overexploit resources, leading inevitably to a decrease in the size of the population. Once the species population begins to stabilize, it may be incorporated into the ecosystem, living within the habitat's ability to support it.)*

CLOSURE

Section Review

Answers to these questions are provided in the Answer Key on page T158.

Alternative Assessment

Present the following scenario to students: You have just been offered a temporary part-time job. The job will last one month, and your boss has given you two salary options. You can either receive $10 a week with a $5 per week raise every week until the end of the month, or you can receive one penny for your first day on the job, and double the previous day's pay for each of the remaining 30 days of the month. Ask students to determine which salary option they would prefer. Provide students with calculators to calculate their weekly and daily salaries. Note: It is not necessary for students to perform all of the calculations in order to make their decision. It should quickly become obvious that the "double penny" option yields a much higher salary. *(The weekly payment option would yield the student $70 for the month. The "double penny" option would yield the student $21,474,836.47. Thus, most students will prefer the second option of payment.)*

Next, have students explain how the payment options can be compared to arithmetic and exponential growth. *(The weekly payment option is an example of arithmetic growth. The "double penny" option is an example of exponential growth.)* Finally, ask students to explain which payment option is a better reflection of human population growth, and why. *(Since human populations are increasing exponentially, the "double penny" option is the better reflection of human population growth.)*

Reteaching

Invite students to think about the resources (both vital and nonvital) that they have and use at home. As a class, have students suggest a list of these resources, and write them on the chalkboard. *(Some examples include space and water [vital] and TV and stereo [nonvital].)* Being careful not to embarrass anyone, encourage students to talk about any competition that may occur in their home for these resources. For example, do people fight over or have to wait to use the bathroom in the morning? Have students describe what might happen if relatives moved into their house. Ask: How would adding more people affect your home's resources? Your own quality of life? What would you say is your home's carrying capacity for people? Then explain how this example can be compared to the use of resources and the quality of life experienced by people throughout the world as the human population increases.

Extension

Have students design a comic strip to show what might happen when a species exceeds its carrying capacity. Suggest that they include their comic strip in their Portfolio.

SECTION 13.2

A Growing Human Population

Focus

Patterns of human population growth and decline are analyzed in this section. Students also learn about the stages of population growth.

Motivator

Hold up an apple, and tell students that the apple represents the Earth. Then slice the apple into quarters, and hold up three of the quarters. Ask: What do these three quarters represent? *(the world's oceans)* Ask: What does the remaining quarter represent? *(the world's land)* Set aside the three "oceans," and slice the remaining quarter in half. Hold up one of the pieces, and tell students that this piece represents land that cannot be used by people; it includes the world's polar areas, swamps, deserts, and very high or very rocky mountains. Set that piece down, and hold up the other half. Ask: What fraction of the Earth do I have left? *(one-eighth)* Slice that piece into four sections, and hold up three of them. Explain that these pieces represent the land on which are built the cities and suburbs, highways, and all other places where people live and work but do not usually grow food. Hold up the remaining piece. Ask: Now what fraction of the Earth remains? *(one-thirty-second)* Carefully peel the skin from that piece and then hold up the skin. Tell students that this tiny bit of peel represents the portion of the Earth's surface that can be effectively used to produce food for the world's people. Explain that this area is a fixed resource—it probably won't get much bigger, but it could get smaller. At the same time, we have an ever-increasing number of humans to feed. Thus, it is essential to protect land and other fixed resources, such as our air and water.

Teaching Strategies

From Hunting and Gathering to Agriculture, *page 336*

Background: Malthus

A clergyman and economist named Thomas Malthus first warned of the dire effects of overpopulation in 1798. Malthus pointed out that while the human population tends to grow exponentially, agricultural production tends to increase arithmetically. The result, he said, was that every population tends to grow faster than its food supply until reduced by starvation, disease, and war. In practice, food production increased faster than human population for part of the twentieth century. But in principle, Malthus was right. Despite modern advances in agricultural production, such as the green revolution of the 1960s and 1970s, agricultural production has leveled off or actually declined in some parts of the world. This downward trend is due largely to farmland overuse that has resulted in nutrient depletion, erosion, and contamination with pesticides and other chemicals.

Multicultural Connection

Anthropologists believe that the world's first humans probably had only one name, a first name. But as populations grew, last names helped distinguish people who shared the same first name. Some historians believe that when China formally began the use of surnames about 2,000 years ago, people were asked to choose a last name from the words of a short poem. Wing, which means "warm," and Lee, which means "pear tree," are two examples of Chinese surnames. Japanese surnames, which were widely adopted in the 1800s, reflect the Japanese reverence for nature. For example, Hana, which means "flower," and Togukawa, which means "virtuous river," are typical last names. Suggest that interested students research their own last name and include in their Portfolio a drawing or short paragraph reflective of their findings.

Using the Transparency

Figure 13-7 is available as a Teaching Transparency.

Using the Graph: *Figure 13-7, page 337*

Zero Population Growth's short video, *World Population* (referenced in the Resources section of this textbook, page T128), would complement this graph nicely. The video presents a dramatic, graphic simulation of how human populations have grown from the year A.D. 1 to the present and beyond, with projections to the year 2050.

After viewing the graph and video, encourage students to discuss what might have caused the human population explosion and what trends may be causing human populations to continue to expand. Rather than confirming the students' hypotheses, assure students that they will read more about the exploding human population and its possible causes in the next few pages of their textbook.

The Population Continues to Grow, *page 338*

Background: The Industrial Revolution

During the Industrial Revolution, people flocked to the cities for jobs in factories. Sometimes so many people arrived in cities so quickly that overcrowding led to epidemics of disease. Research into the causes of these epidemics led to improvements in sanitation and nutrition and helped lower infant-mortality rates and increase life-expectancy rates. Furthermore, it was about this time that scientists discovered that bacteria were the cause of many diseases. This discovery led to the prevention, treatment, and cure of many illnesses. The Industrial Revolution also led to improvements in transportation, which made medicines and other goods that improved the odds of a person's survival available to more people. Thus, in the last 300 years, advances in health care, transportation, and agriculture (discussed earlier in this section) have fueled the human population explosion.

THE DEMOGRAPHIC TRANSITION, *page 338*

Social-Studies Connection

Divide your class into groups, and suggest that students interview an elderly friend, neighbor, or relative. Encourage them to prepare a questionnaire about how the size of that person's extended family has changed over time. For example, students could ask: How many children did your grandparents have? How many children did your parents have? How many children did you have? How many children do your children have? How many children do your brothers and sisters have? etc. In order to build confidence, students may wish to prepare themselves for this exercise by forming pairs and rehearsing. When students are finished conducting their interviews, ask them to show how their findings relate to Notestein's three stages of demographic transition. Suggest that they include their findings in their Portfolio.

Using the Case Study: *The Stories Behind the Statistics, pages 338–339*

Have students create a time line showing the events mentioned in the Case Study that have affected birth rates in the United States. Students should also indicate the average number of children born per family during each time. Interested students may wish to add additional events to the time line.

Extending the Case Study: *The Stories Behind the Statistics, pages 338–339*

Add the 1990s and the year 2000 to the time line, and have students hypothesize about what is and what will be influencing American birth rates in the future. Students may wish to do research before making these additions to their time line. Suggest that students include their time line in their Portfolio.

Meeting Individual Needs

Second-Language Learners Instead of having these students make a time line to illustrate the events mentioned in the Case Study, have them use magazine photographs and other illustrations to make a poster or bulletin-board display that graphically illustrates how the human population has changed through the years. Suggest that students include their poster or something from their display in their Portfolio.

Using the Transparencies

Figures 13-10, 13-11, and 13-13 are available as Teaching Transparencies.

CLOSURE

Section Review

Answers to these questions are provided in the Answer Key on page T158.

Alternative Assessment

Cooperative Learning Activity Have students work in groups to design concept maps that show Notestein's three stages of population growth and two examples of each stage.

Reteaching

Perform this activity to help students understand the effect of family size on the total human population. Draw a line down your chalkboard to divide it in half. Make sure you have enough room to draw a seven-generation family tree on each side of the line. Then draw a small circle at the top of each half. Tell students that the circle on the left side of the line represents a person who had two children and that the circle on the right represents a person who had three children. Show this by drawing two circles beneath the one on the left and three circles beneath the one on the right. Now tell students that each of the people on the left had two children, while each of the people on the right had three children. Continue forming the family trees in this same manner until you have seven generations. Now students should begin to see the difference that one additional child per generation can make. Note: Students may feel uncomfortable about the size of their family or the number of children that they may wish to have one day. Explain that the purpose of this activity is simply to build awareness, not to create feelings of guilt or to cause people to cast judgments on others.

Extension

Review with the class the effects of the agricultural revolution and the Industrial Revolution on human population growth. Tell students that we are now in the midst of what many people call the Technological Revolution or the Information Revolution. Invite students to brainstorm a list of characteristics of this modern revolution and the effects they think each characteristic will have on population growth, society, and the environment. Interested students can do research to find out more about this modern phenomenon and present their findings in some visual form. Suggest that students include their visual in their Portfolio.

SECTION 13.3

PROBLEMS OF OVERPOPULATION

FOCUS

Students encounter some of the problems associated with human overpopulation.

MOTIVATOR

Tell students that recent surveys indicate that most Americans believe that overpopulation is only a problem in developing countries. They argue that our population is growing by only about 1.1 percent per year, compared to 3.1 percent in Kenya. Thus, our attention should be focused on slowing population growth in developing countries without worrying about the growth in this country. Ask students: Based on what you learned about exponential growth in the first section of this chapter, do you think there is an overpopulation problem in this country? *(Yes, even the smallest fraction of steady growth leads eventually to doubling and redoubling as long as can be imagined.)*

Write "150 million" on the chalkboard. Tell students that this was the population of the United States in 1950. Then write "258 million." Explain: This is the number of people who lived in the United States in 1993. Then tell students that our population is now increasing by nearly 3 million people per year. Write "520 million" on the board. Say: If current rates of growth continue, the population of this country will equal 520 million people in just 64 years. Explain that these figures make the United States one of the fastest growing countries in the industrialized world. Plus, this country leads the entire world in consumption of goods and services. Ask: What problems might these patterns cause in the future? *(shortages of resources, disease, crowding, social unrest, harm to other organisms and to the stability of ecosystems, etc.)* Explain that in this section students will learn about some of these problems and what can be done to solve them.

TEACHING STRATEGIES

A SHORTAGE OF FUELWOOD, *page 342*

Multicultural Connection

A man named Bahuguna, who lives in India, realized that people who live in the rural areas of his country depend on fuelwood for survival. He also realized that where trees are rapidly cut down for fuelwood and not replaced, the environment is quickly degraded and unable to support the future needs of the people. As a result, the Indians suffer tremendous poverty. To rectify this problem, Bahuguna started a movement called *Chipko,* which means "to hug" (as in "to hug a tree") in Hindi. The name is especially appropriate because supporters are often seen joining hands to surround, or collectively "hug," a tree in order to protect it from being cut down by loggers. But the purpose of the Chipko movement is not so much to keep loggers from felling trees as it is to teach everyone the value of trees and the necessity of replanting them. Bahuguna visits thousands of rural areas each year to teach the people that for every tree cut down, at least one must be planted to replace it. Countless trees have been planted through this program, and some deforested areas are being reclaimed. The trees planted are rapidly growing trees that can be used for fuelwood. As these are used, forests that would have otherwise been felled for fuelwood can be spared. The efforts of Bahuguna and other people who are involved in the Chipko movement give rural Indians hope for a better future.

SOCIAL UNREST, *page 344*

Notable Quote

"The bird of war is not the eagle but the stork."

Charles F. Potter

ENVIRONMENTAL REFUGEES, *page 345*

Meeting Individual Needs

Gifted Learners Encourage students to learn about how the influx of vast numbers of environmental refugees has caused social disruptions in developed countries. Students may do research in the library, clip articles from newspapers and magazines, or watch television news programs and documentaries to collect information about how some residents of developed countries react to refugees in their area (e.g., "skinheads" in Germany and their relationship with environmental refugees from surrounding countries) or about the effects that huge numbers of refugees have on the economies and social service expenditures of developed nations. Students may create written reports or produce their own TV documentary to present the complexities of these issues to the class.

SOLVING THE PROBLEMS OF OVERPOPULATION, *page 346*

Discussing the Issue

Tell students that population-control measures often cause ethical and religious controversies. For example, do couples have the right to have as many children as they want, or should they be encouraged (or forced) to have fewer children to ensure that the human population doesn't overtax the carrying capacity of the Earth? Divide the class into two teams, and have them argue their points of view while you (or a student) moderate. Record major points of the debate for later evaluation.

CLOSURE

Section Review

Answers to these questions are provided in the Answer Key on page T158.

Alternative Assessment

Tell students that they are public affairs agents for the government. Encourage students to use what they learned in the first section of this chapter about exponential growth and what they learned in this section about the problems that a country can suffer due to overpopulation to convince the public that controls on population growth should be taken seriously in this country. They should then make a persuasive poster, which can be included in their Portfolio.

Reteaching

Read students the following quotation from biologist Paul Ehrlich: "A baby born in the United States will damage the planet 20 to 100 times more in a lifetime than a baby born into a poor family in an LDC [less developed country]. Each rich person in the United States does 1,000 times more damage than a poor person in an LDC." To help students understand why this is true, ask them to help you make a bulletin board to illustrate the following facts: the United States represents only 4.3 percent of the world's population, but produces about 21 percent of the world's goods and services, uses about one-third of the world's processed mineral resources and about one-fourth of the world's nonrenewable energy, and produces at least one-third of the world's pollution and trash.

Extension

It is well known that as standards of living increase, human population growth slows. People in developed countries use about 83 percent of the world's resources, have few children, and enjoy a relatively high standard of living. People in developing countries are allotted a paltry 1.4 percent of the world's resources, have large families, and have a poor standard of living. Invite students to debate the following: Should people in developed countries conserve resources so that a greater proportion can be used to raise the standard of living (and lower the birth rate) in developing countries? Remind students that this sharing may result in a lower standard of living for people in developed countries. Have one student act as moderator and list the arguments for each point of view.

Answers

Answers to *Chapter Highlights, Chapter Review, Investigation,* and *Points of View* are provided in the Answer Key for Chapter 13, which begins on page T158.

TEACHING STRATEGIES FOR INVESTIGATION:

WHAT CAUSES A POPULATION EXPLOSION? *pages 350–351*

Background

Before beginning this investigation, make sure that students thoroughly understand how a histogram is structured. Emphasize the division between left and right for gender, and explain that each histogram represents only one year. Also review with students the assumptions that are listed under "Background" on pages 350–351. Students will not correctly complete this investigation unless they accept the given assumptions. Test this understanding by asking questions similar to the following. If there are 2,000 five- to nine-year-olds in the year 2010, how many ten- to fourteen-year-olds will there be in the year 2015? *(2,000)* If there are 5,000 twenty- to twenty-four-year-olds in the year 2010 and each woman in that age group has three children, how many zero- to four-year-olds will there be in the year 2015? *(7,500)*

Predict Population Changes

The method of "moving up" age groups to complete the tables is critical; make sure that students understand this process. Also, if time is limited for this investigation, step 8 may be assigned as homework, provided that each student has a copy of the data to take home.

Extensions

Suggest that students extend the investigation to the end of the twenty-first century; this will be especially interesting for Group C's population because that population stops increasing in size during that time. Discuss with students the significance of a stable population size.

CHAPTER 14

TOWARD A SUSTAINABLE FUTURE

PLANNING GUIDE

CHAPTER COMPONENTS AND RESOURCES	½ Year Average	½ Year Advanced	1 Year Average	1 Year Advanced
Chapter Opener, p. 354				
EcoLog, p. 354			•	•
14.1 International Cooperation, pp. 355–359			•	•
Case Study: *A Whale of an International Controversy,* pp. 356–357			•	•
Section Review, p. 359			•	•
14.2 Environmental Policies in the United States, pp. 360–364			•	•
Making a Difference: *Building Toward the Future,* pp. 370–371	•	•	•	•
Field Activity, p. 362				•
EcoSkills: *Environmental Shopping,* pp. 392–393				•
Section Review, p. 364			•	•
End of Chapter				
Chapter Highlights, p. 365			•	•
Chapter Review, pp. 366–367			•	•
Investigation: *Be an Environmental Journalist,* pp. 368–369			•	•
Chapter Tests			•	•
PACING GUIDE	1–2 days	1 day	4–5 days	4–5 days

MATERIALS ORGANIZER

Activity	Materials	Advance Preparation
Investigation: *Be an Environmental Journalist,* pp. 368–369	newspapers, magazines, and other publications; tape; scissors; notebook; pen or pencil (per student group)	Make sure the assortment of publications contains sufficient articles on environmental issues.

CHAPTER OVERVIEW

In this chapter, students are introduced to the concept of sustainable living, and they learn how public policy can make a difference in the health of our natural systems.

Section 14.1 discusses the need for international cooperation in resolving environmental problems.

Section 14.2 describes opportunities to influence public decisions at the local, state, and national levels.

RESOURCES

. . . for Teachers

Chiras, Daniel D. *Lessons from Nature: Learning to Live Sustainably on the Earth.* Washington, D.C.: Island Press, 1992. (This book defines sustainability in a variety of contexts and presents numerous examples of positive actions taken by individuals, corporations, and government.)

Orloff, Neil. *The Environmental Impact Statement Process.* Washington, D.C.: Information Resources Press, 1978. (This book provides a very detailed, step-by-step explanation of the EIS process, including "how-to" guidelines for the public.)

. . . for Students

Berry, Joyce K., and John C. Gordon, eds. *Environmental Leadership: Developing Effective Skills and Styles.* Washington, D.C.: Island Press, 1993. (This book features personal accounts from a diverse group of successful environmental leaders. These leaders discuss their own career paths as well as skills and personal characteristics that are useful for leadership positions in the conservation field.)

The Earthworks Group. *The Student Environmental Action Guide.* Berkeley, CA: Earthworks Press, 1989. (This popular book lists activities anyone could do to protect the Earth.)

National Commission of the Environment. *Choosing a Sustainable Future.* Washington, D.C.: Island Press, 1993. (This report, prepared by a private-sector initiative convened by the World Wildlife Fund, outlines viable, long-term economic and environmental strategies for the United States.)

Films, Videos, Software, and Other Media

50 Simple Things Kids Can Do to Save the Earth. Video and activity book. Churchill Media, 6901 Woodley Avenue, Van Nuys, CA 91406-4844; 800-334-7830. (This popular book and award-winning video, which is hosted by Sara Gilbert, Brian Green, and other young celebrities, is an excellent tool for motivating students.)

One Second Before Sunrise: A Search for Solutions. Video. 60 minutes. Bullfrog Films, P.O. Box 149, Oley, PA 19547; 215-779-8226 or 800-543-3764. (Hosted by Lynn Redgrave, this award-winning series of five separate programs is available with study guides for schools. The series details local, national, and global solutions to environmental problems.)

Get It Together. Video. 28 minutes. Study guide included. The Video Project, 5332 College Avenue, Suite 101, Oakland, CA 94618; 800-4-PLANET. (This empowering student-produced video highlights the contributions of diverse young people who are getting involved in their communities to protect the environment.)

Getting Involved

- Use the following EcoSkills activity to enhance student interest in the subject of this chapter: *Environmental Shopping,* pp. 392–393.
- A listing of environmental organizations, publications, and volunteer opportunities can be found in the **Laboratory and Field Guide.** Note especially those organizations with the goal of promoting sustainability and fostering a healthy relationship between humans and nature, such as the Worldwatch Institute, the National Center for Appropriate Technology, and the Union of Concerned Scientists.

USING THE QUOTATION

"Act, act, act. You can't just watch."

Call on a student to read the quotation aloud. Ask: How would you describe the tone of the statement? *(urgent)* How does Serrano's statement apply to the environment? *(Sample answer: The urgent tone conveys the message that we must act quickly to solve environmental problems before they get worse.)* What does his word *act* mean to you? *(Answers will vary, but* act *could mean campaigning for political candidates, designing a recycling program, or volunteering time to a local environmental group.)*

USING THE ECOLOG

To assess students' prior knowledge about the subject of this chapter, have them answer the EcoLog questions on page 354. In answering these questions, students will reveal their prior knowledge, which is usually very basic and often flawed or even misconceived. Then, by revising their answers at the conclusion of the chapter, students will have the opportunity to eliminate any misconceptions and to see firsthand how their knowledge has grown.

INTERNATIONAL COOPERATION

FOCUS

A discussion of the Earth Summit is used to help students understand how nations can work together to preserve and protect natural resources. The section also highlights the difficulties surrounding attempts at international cooperation.

MOTIVATOR

Many environmental problems are international in scope and require international cooperation to address. Air pollution, emission of greenhouse gases, loss of topsoil, disposal of nuclear waste, ocean pollution, and depletion of forests are just a few problems that have international implications. The Chernobyl disaster that spewed radioactive particles into the air and contaminated food crops and milk supplies in Germany and in Scandinavian countries is a dramatic example. Have students call out other examples to help you create a table or concept map on the chalkboard. The graphic should illustrate how the actions of one nation can affect the environmental health of many others. After a while, students should realize that all countries contribute to environmental problems and all, in turn, are affected. It is for this reason that international cooperation is so important.

TEACHING STRATEGIES

Multicultural Connection

Organize a research project or a means of exchanging ideas and information with students from another part of the country or from another nation. One organization that connects classrooms around the world so that they may work together to analyze water quality, for instance, is the Global Rivers Environmental Education Network (GREEN), 721 East Huron Street, Ann Arbor, MI 48104; 313-761-8142.

Notable Quotes

"It is in the knowledge of his follies that man shows his superior wisdom."
French proverb

"Putting ecological politics into action means approaching life with imagination and intelligence, knowledge and emotion, responsibility and culture. It fights against bureaucracy and ideology, uniformity, authoritarianism, and any attempt to eliminate diversity and autonomy. And to everyone it offers a new friend: the Earth."
Mario Signorino

Social-Studies Connection

Cooperative Learning Activity In his famous essay "The Tragedy of the Commons," biologist Garrett Hardin describes a time in medieval Europe when the inhabitants of a village shared pastureland known as the "commons." Farmers quickly realized that the more cattle they grazed on the commons, the more they stood to gain. Consequently, farmers added more and more cattle to the area until it was destroyed by overgrazing. Hardin suggested that because the commons were public, no one took responsibility for keeping the area in good condition. Many people argue that this "tragedy" is common to public lands all over the world.

Have students divide into small groups and brainstorm answers to the following questions: What could the king have done to make sure that the commons would be there for future generations of farmers? *(Accept all reasonable responses. Students may suggest that the land could have been broken up into individual plots because private ownership would have made the villagers responsible for their own land.)* In some areas, resources may move around (such as terrestrial wildlife or resources from aquatic ecosystems). In these cases, what solutions can you suggest? *(Answers will vary. Managing these resources may require a consensus among all involved parties.)*

Social-Psychology Connection

Divide students into groups, and ask each group to analyze one common resource in the school environment (library, cafeteria, classroom space, etc.). Tell students to observe and record student behavior in one of these common areas for 15 minutes. Direct student thinking by asking questions similar to the following: What are some of the ways that the students you observed used this resource? Were there instances of individuals using the resource for individual gain but to the detriment of others? When this occurs, what happens to the resource? How could the resource be protected for everyone's common good? Each group should report its findings to the class. Suggest that students include their findings, as well as their commentary, in their Portfolio.

The Earth Summit: Promise and Problems, *page 355*

Extending the Case Study: *A Whale of an International Controversy, pages 356–357*

Read a description of the whale in Herman Melville's novel *Moby Dick*. Have students describe how the description affects their attitude toward the whale in the story. Ask: Do you think people's attitudes toward whales have changed since *Moby Dick* was written? *(possibly)* To what do you attribute this change? *(Answers will vary. Use this question to generate discussion about the power of various media to influence social attitudes.)*

Using the Theme of *Patterns of Change*

Focus Question: How has the Earth Summit changed global environmental policy? *(Answers will vary, but students should recognize that although international cooperation existed before this conference, environmental policy-making on a global scale was limited. Thus, this conference has paved the way for positive patterns of change.)*

Wealthy and Poor Nations, *page 359*

Discussing the Issue

Ask students to list the five most important factors in their lives, in order of importance. Material goods are usually listed low on the list, below friends, family, and a good education. Tell students that a recent study on happiness led to this conclusion: "Above the poverty level, the relationship between income and happiness is remarkably small." Oxford psychologist Michael Argyle found that the main factors that influence happiness have nothing at all to do with income and consumption. Family members, work, leisure time, and friendships are much more important. Encourage discussion.

Closure

Section Review

Answers to these questions are provided in the Answer Key on page T160.

Alternative Assessment

Have students create a television news report and commentary on the Earth Summit. It should summarize the major accomplishments and limitations of the event, and should include the students' opinions on the different needs of wealthy and poor nations when faced with environmental issues. Students may

wish to transcribe or videotape their presentation for inclusion in their Portfolio.

Reteaching

To provide more material for a discussion about the whale as a symbol of how nations must work together to solve environmental problems, show students the video *Orca: Killer Whale or Gentle Giant?* This 26-minute video can be ordered from The Video Project, 5332 College Avenue, Suite 101, Oakland, CA 94618; 800-4-PLANET.

Extension

Have students make a list of all of the appliances that they have in their homes. Lists could include television, radio, blow dryer, microwave, refrigerator, freezer, and lamps. Students should then categorize these items as either "essential" or "luxury." Then have students interview a senior citizen to find out which of the appliances the older person had access to when he or she was growing up. If the senior citizen did not have a certain appliance, students could find out what that person used instead. After the interview, suggest that students reevaluate the classifications they made in their original list. Ask: Did your thinking change, and if so, why? *(Answers will vary.)* Finally, challenge students to give up one of these items for a week.

Environmental Policies in the United States

Focus

In this section, students are introduced to government laws and policies. They also learn how people can implement change at the national, state, and local levels. Finally, some of the shortcomings at each level of government are discussed.

Motivator

Ask students to think about the environmental issues discussed in this course. Stress that there are no right or wrong answers to the following questions, only individual opinions. Ask: Which of the environmental issues that you learned about during this course did you find the most interesting? the most important? the most relevant to your life? Then tell students to choose one of those issues. Be prepared to give several examples, should students need prodding.

Once each student has identified an issue, tell students to identify which level of government—local, state, or federal—is likely to be the most involved in making decisions about that issue. Then have students write a simple, short letter addressed to a government official at that level. Have students include a brief summary of the issue, a statement explaining his or her position on the issue, and a request for action from the official. Encourage students to use an emotional tone in the letter. Suggest that students include their letter in their Portfolio.

Teaching Strategies

Background Information: Environmental Legislation

Environmental legislation is not a static phenomenon. As new scientific information becomes available, the public may see apparent threats to our environment and demand protection. Also, as weaknesses and loopholes in the law become apparent, amendments and revisions are necessary. Many of our environmental laws, such as the Clean Air Act, the Endangered Species Act, and the Clean Water Act, require periodic review and revision.

Using the Table:

Figure 14-6, page 361
Ask students to bring old issues of periodicals and newspapers from home. (**Note:** If students began a notebook of environmental articles earlier in the course [as was suggested in the Teaching Strategies for Chapter 1], they simply need to bring their notebooks to class.) Divide students into small groups and have them flip through the reading materials, noting any articles that relate to an environmental problem. Then, have students choose at least two articles, and suggest that they use Figure 14-6 to determine the federal agency or law that would address the reported issue or problem. Encourage group discussion by asking questions such as the following: Is the issue or problem a result of someone breaking the law? Should the law be changed to deal more effectively with the problem or issue? Is the resource being used up or destroyed, so as to prevent use by future generations?

Environmental Impact Statements, *page 361*

Meeting Individual Needs

Gifted Learners Have students research some of the principal steps involved in producing an environmental impact statement. (One possible source is listed on page T136, in the Resources for Teachers section of this textbook.) Then, if possible, have students use what they've learned to conduct their own investigation of a proposed development in their community. With data gathered from a tour of the proposed site, news reports, and interviews with the builder, business owners, and other concerned parties, students could then write their own environmental impact statement. Suggest that students include their impact statement in their Portfolio.

Using the Photograph:

Figure 14-7, page 362
Cooperative Learning Activity Divide students into equal-sized groups (five groups of five or six, for instance) to develop brief position statements for different interested parties in the Platte River controversy. Groups could be assigned the position of an Audubon Society member, an unemployed dam construction worker, a member of the Chamber of Commerce, an owner of a motorboat company, a rancher whose land would be flooded, or a regional wildlife commissioner. Then reorganize the groups so that each new group has at least one student representing one of the former groups, and encourage discussion within the new groups. Finally, discuss the activity and the controversy as a class.

Influencing National Policy, *page 362*

Notable Quotes

"Be great in act, as you have been in thought."
William Shakespeare

"Never doubt that a small group of thoughtful, committed citizens can change the world, indeed it is the only thing that ever has."
Margaret Mead

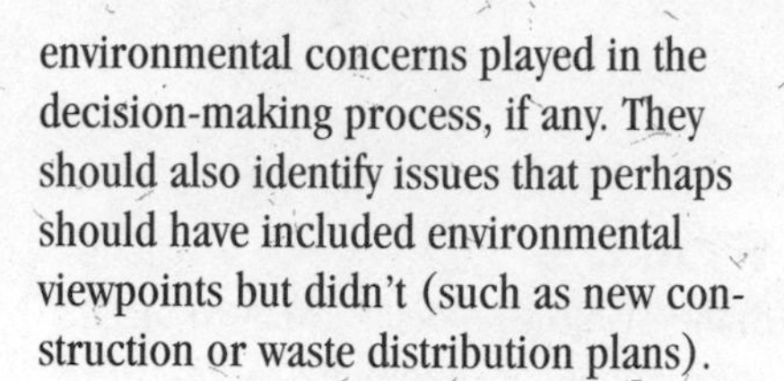

Meeting Individual Needs

Second-Language Learners Have interested students create a bumper sticker, an advertising slogan, or a brochure that relates to an environmental problem of interest to them. Their projects can be in English, their native language, or both. Suggest that students include their project in their Portfolio.

Meeting Individual Needs

Learners Having Difficulty Have students read *The Lorax* by Dr. Seuss (New York City, NY: Random House, 1971). Then suggest that students work in groups to create a children's book that has the environment as its underlying theme. Encourage students to share their books with younger siblings. You may also wish to make these books available to nearby elementary schools. Suggest that students include their children's book in their Portfolio.

LOCAL COMMUNITIES, *page 363*

Research Activity

Have students investigate the lives of individuals who have made a difference by improving the environment. The subjects can be well known, such as John Muir or Rachel Carson, or they can be people known primarily in their local community. Students should include in their report a description of how the persons became involved in environmental issues and what steps they took to reach their goals. Suggest that students include their report in their Portfolio.

Real-World Connection

Invite a city-council representative or community member to speak to your class about citizen participation in local issues. Have the speaker explain to students what they can do as individuals, and as a group, to affect environmental ordinances and policies in their community.

INTO THE FUTURE, *page 364*

Discussing the Issue

Tell students that this activity will focus on whether a higher tax should be placed on gasoline. Divide the class into four groups. Give each group member a card with the name of something that will be affected by a higher tax, such as public transportation, air pollution, low-income people, high-income people, shopping malls, parking lots, carpooling, suburban sprawl, and foreign-oil dependence. With each group member acting as the person, object, or agency listed on his or her card, each group must reach a consensus about whether a gasoline tax should be implemented.

CLOSURE

Section Review

Answers to these questions are provided in the Answer Key on page T160.

Alternative Assessment

Inform students that they will be interviewed as experts on environmental policies and policy making. Give them a few days to prepare, and then "interview" them, asking questions drawn from the text of this section, such as: What are environmental impact statements? *(outlines of the environmental effects of proposed projects or legislation)* What is a major obstacle that local environmental movements face? *(the lack of community-level coordination)* Allow time for classmates to pose questions to each other as well.

Reteaching

Invite a social-studies or civics teacher to your classroom to discuss how a bill becomes a law. When the guest has finished his or her discussion, work with the class to create on the chalkboard a flowchart of the steps taken to turn a bill into a law.

Extension

Have motivated students attend a city-council meeting. Often, decisions that involve environmental issues will be discussed, although the environmental point of view will not be presented explicitly. Have students write a short report on the meeting, including what was discussed, what was decided, and the role that environmental concerns played in the decision-making process, if any. They should also identify issues that perhaps should have included environmental viewpoints but didn't (such as new construction or waste distribution plans).

Answers

Answers to *Chapter Highlights, Chapter Review,* and *Investigation* are provided in the Answer Key for Chapter 14, which begins on page T159.

TEACHING STRATEGIES FOR INVESTIGATION:

BE AN ENVIRONMENTAL JOURNALIST, *pages 368–369*

Be an Investigative Reporter

Help students understand the goal of this investigation by choosing a sample article to discuss with the class. Highlight for students the ways in which environmental issues are presented and the opinions that are expressed. Note any biases you observe, and emphasize the roles of tone and style in communicating ideas to the reader.

If students are having trouble selecting an issue to concentrate on, the National Wildlife Federation's Environmental Hotline (202-797-6655) describes recent and upcoming action on federal environmental legislation. Or, you could allow students to choose one of the pieces of environmental legislation that is mentioned in this chapter.

Let the World Know

If students have access to a desktop publishing program, suggest that they use the program to make their article graphically appealing. Then, interested students could submit their article to the school newspaper for publication.

ANSWER KEY

CONTENTS

CHAPTER 1

SECTION 1.1

CASE STUDY, *page 7*

1. Sample answer: Scientists working in Dr. Edmondson's lab observed, analyzed, and determined the cause of the lake pollution. Edmondson used his experience to assess the seriousness of the situation. Then he wrote to James Ellis, describing the problem and outlining a solution. Ellis successfully campaigned for the passage of a bill that would allow the city to apply Edmondson's solution. All along, news reporters educated the public about the situation, and public forums were held to let Seattle residents voice their opinions on the plan. Once the bill was passed, scientists and engineers devised methods to reroute the sewage.

2. Both groups made observations and collected data to document the existence of an environmental problem. They also reported their findings to officials who could do something to solve the problem.

SECTION 1.1 REVIEW, *page 12*

1. Resource depletion, pollution, and extinction. The rapid consumption of fossil fuels is an example of resource depletion; contamination of waterways by pesticides is an example of pollution; and the death of all passenger pigeons is an example of extinction.

2. The population crisis is the result of the human population growing larger than the Earth can support. The consumption crisis is the result of people using, wasting, or polluting natural resources faster than they can be renewed, replaced, or cleaned up. Developing countries, population crisis; developed nations, consumption crisis

3. Answers will vary but may include preserving habitats, converting garbage into harmless substances, using nonrenewable resources sparingly, and using renewable resources no faster than they can be replaced.

4. Answers will vary. Iron is one possible answer because if humans use it too rapidly, it cannot be replaced at a usable rate.

SECTION 1.2

SECTION 1.2 REVIEW, *page 19*

1. Pure science involves answering questions about how the natural world works, while applied science uses the information provided by pure science to solve problems. Applied science

2. Ecology is the study of the interrelationships among living things. Environmental science draws heavily on the information gained by ecology. Yet ecology differs from environmental science in that environmental science is an applied science that focuses on solving problems.

3. Answers will vary but should include three of the following: observing, hypothesizing, predicting, experimenting, organizing and interpreting data, using graphics, and sharing information.

4. Testing a single variable and using a control. It is important that a single variable be tested to establish a one-to-one correspondence between possible causes and observed effects. The control is important because it accounts for all factors except the variable being tested.

5. Sample answer: Is it ethical to keep animals in zoos? Scientific information that a person could use to formulate an answer might include data about the health, life span, reproductive habits, and general behavior of animals kept in captivity and of animals living in the wild.

SECTION 1.3

SECTION 1.3 REVIEW, *page 24*

1. A sample answer could include economic, social, and recreational values. Economic values relate to jobs, commerce, and living standards. Social values concern attitudes, traditions, and common knowledge. Recreational values relate to how a person wants to spend his or her free time.

2. Answers will vary. One such situation would be when the economic interests of a housing developer conflict with the desire of conservationists to preserve wildlife habitat. Another example would be when the recreational values of all-terrain-vehicle users conflict with the desire of naturalists to protect ecosystems.

3. Answers will vary. Students should demonstrate the ability to think independently of the example given in the textbook. Charts should be clear and concise.

ECOLOG *page 25*

1. Sample answer: Examples of major environmental problems are water, soil, and air pollution; the extinction of plant and animal species; the depletion of the Earth's natural resources; and overpopulation.

2. Sample answer: Science can help people identify environmental problems, hypothesize about the causes and effects of those problems, test hypotheses, and analyze the resulting data. The findings can help solve environmental problems.

CHAPTER REVIEW *page 26*

1. Sample answers:

a. A natural resource is any natural substance that living things use; a nonrenewable resource is a natural resource that cannot be replaced at a usable rate.

b. The population crisis involves the excessive growth of the human population, while the consumption crisis involves the excessive use or destruction of natural resources.

c. Pure science investigates the natural world, while applied science applies scientific knowledge and method to solve practical problems.

d. Ecology is the broad study of how living things are related to each other, while environmental science is the specific

application of ecology and other sciences to problems that arise from human interactions with the environment.
2.–12. Answers can be found on pages 26–27.
13. Sample answer: Environmental concerns might require that I sacrifice some of the goods, services, and activities that I am used to. For example, I would like to buy a car to replace my bicycle, but the bicycle is better for the environment.
14. Unintentional pollution involves pollutants that are accidental byproducts of human activities. Intentional pollution is the deliberate creation of toxic substances that poison the air, water, or soil.
15. Answers will vary. Some students may say that pure science is more important because it provides a solid base from which problems can be solved and technology and innovation can arise. Other students may assert that applied science is more important because it is only by applying science that problems are solved.
16. Geology, the science of physical nature and history of the Earth, provides environmental scientists with information about rocks, minerals, fossil fuels, land and water formations, etc. Oceanography supplies information about the physical, chemical, and biological aspects of the oceans.
17. Creating a table is a systematic method for defining and analyzing complicated issues to determine which aspects are the most important.
18. Travel by private vehicles decreases and travel by mass transit and walking or cycling increases as the city grows more crowded. Mass transit
19. Renewable and nonrenewable resources are defined by how long they will be usable to human beings. A renewable resource is one that will remain usable for many human generations. Because the sun will not burn itself out for many millions of years, sunlight can be considered a renewable resource.
20. The structure of the biosphere allows life to exist on Earth. For example, the atmosphere protects us from harmful radiation and effectively traps oxygen and other chemicals that are necessary for life to exist. Within the biosphere, living things live in a delicate balance with one another and nonliving things. Even slight changes in the biosphere's structure could jeopardize life on Earth.
21. If a resource cannot be replenished at a usable rate, the resource is considered nonrenewable. For example, coal takes millions of years to form. So, if humans use up all coal supplies tomorrow, no more coal will be available for several million years. On the other hand, a resource that can be replenished rapidly, such as a tree, is considered renewable. However, if trees are cut down faster than they can be replanted and grown to a useful size, they would become a nonrenewable resource.
22. Answers will vary. Students should demonstrate the ability to interpret demographic information and to compare and contrast data.

INVESTIGATION *pages 28–29*

1. The scientific problem is why Omar's onions did not take root.
2. Possible variables include the amount of water, the amount of organic fertilizer, and the amount of bleach in the groundwater underneath the farm. Sample hypothesis: The onions failed to take root because the new watering system did not provide the onions with adequate moisture.
4. Sample size may vary but should consist of at least three onions for each experimental condition. A control is necessary because it provides a standard against which change can be measured.
6. Tables will vary depending on the chosen variable and experimental conditions.
7. Student answers will vary but could reflect the following findings. For maximum growth, onions must be kept wet at all times; growth in the presence of fertilizer will probably be maximum at zero to half the recommended concentration but will depend on the fertilizer used; growth in bleach will typically be less at higher concentrations. Ensure that student reports are clear, concise, and complete.
8. Sample answer for the hypothesis listed in item 2 above: The amount of water is the variable that changed, and all other variables remained constant. Changing only one variable is necessary to identify clearly the factor that is inhibiting the growth of the roots.
9. To test the effects of acid rain or salt accumulation on the onion bulbs, students could use the procedure outlined in this investigation and incorporate solutions of varying levels of acidity or salinity.

Caption, *page 29:* The team labeled the test tubes to identify the different solutions clearly.

CHAPTER 2

SECTION 2.1

SECTION 2.1 REVIEW, *page 38*

1. Biotic components include earthworms, insects, snakes, moles, bacteria, and a variety of plants and fungi. Abiotic components include sunlight, temperature, soil, minerals, rocks, and moisture.
2. Sample answer: Fleas living on a dog
3. Sample answer: A roadrunner's niche is foraging on the ground for insects and a variety of small vertebrates. Roadrunners build shallow nests in trees or cactus clumps. The female lays two to six eggs in the spring, and both parents care for the young. The habitat of the roadrunner is the deserts of the southwestern United States and Mexico. The roadrunner is usually found on the ground.
4. Abiotic factors. A community includes all of the organisms, or biotic factors, that interact with one another in an area, but it does not include the abiotic factors.

SECTION 2.2

SECTION 2.2 REVIEW, *page 42*

1. Answers will vary. One example is a fox eating a rabbit.
2. Mutualism occurs when both species involved in the relationship benefit. Parasitism occurs when one species benefits from, and the other is harmed by, the relationship.
3. If the parasite kills its host, it will lose not only its habitat, but also its food source.
4. One method would be to attempt growing members of the pair both separately and as a pair. Some variable, such as growth or reproduction, could be measured to determine whether each member performed better (or the same) as a pair or as an individual.

SECTION 2.3

CASE STUDY, *pages 44–45*
Caption, *page 44:* The moths that are seen most easily
1. Yes; having made observations, he developed a hypothesis, performed an experiment to test his hypothesis, and found that the results supported his predictions.
2. As the trees became lighter in color, the better-camouflaged pale moths would become more prevalent than dark moths.

SECTION 2.3 REVIEW, *page 46*
1. Factors in the environment, such as temperature, availability of food or water, and the number and type of predators, place demands on the members of a species. Because there is natural variation among individuals of a species, some individuals may have a hereditary characteristic that enables them to better overcome the constraints imposed by the environment. This individual will be more likely to survive and reproduce, thus causing the number of individuals with the beneficial hereditary characteristic to increase with each generation.
2. Evolution, in this context, is simply change in the genetic characteristics of a population from one generation to the next. Natural selection is one process that results in evolution.
3. No; coevolution occurs when two or more species evolve in response to one another.
4. There may be some individual insects with a hereditary characteristic that enables them to survive exposure to the insecticide. Those insects will be able to survive and reproduce, and their offspring born with this characteristic will also survive and reproduce. Those not born with the characteristic will die out. Eventually, the population will be largely immune to the effects of the insecticide.

ECOLOG *page 47*

Sample answer: If all lions were removed, the population of prey animals would probably grow. Eventually, these prey animals might become so numerous that they would deplete resources that they and other organisms need to survive. The proportion of sick and feeble prey would rise because these animals are the favorite targets of the lions. Alternatively, competing predators, such as the hyenas, would become more numerous.

CHAPTER REVIEW *page 48*

1. Sample answers:
a. Biotic factors are the living components of an ecosystem; abiotic factors are the nonliving components.
b. A population is composed of all of the members of a given species living in an area. A community is a group of interacting populations of different species.
c. A niche is an organism's way of life, including all of its interactions with its environment; a habitat is where an organism lives.
d. Parasitism is a relationship in which one organism gains at the expense of another. Mutualism occurs when each organism in a relationship benefits from the other.
e. An adaptation is a hereditary change in a species that helps it to survive in its environment. Extinction is the death of a population or an entire species, often due to the failure of that species to adapt to changes in the environment.
2.–12. Answers can be found on pages 48–49.
13. No; it is an example of parasitism. The tapeworm does not attack and eat the cow, which would be an example of predation. Instead, it draws sustenance from the cow at the cow's expense.
14. An ecosystem is similar to a mechanical system in that all of the parts of the system interact. Removing key parts of either system would lead to a "malfunction."
15. Possible answers include predation: killing and eating other animals; parasitism: being attacked by mosquitoes; mutualism: maintaining *E. coli* in our intestines; commensalism: being home to many microorganisms that live on our bodies; competition: using resources that might be used by other organisms.
16. Species can compete without meeting by utilizing the same resource at different times, such as one species of insect feeding on a plant during the day and another species of insect feeding on the same plant at night. Or, two species of plants that flower at the same time may be in competition for pollinators.
17. The populations of species A and B fluctuate together, which could reflect mutualism; however, it might also be interpreted as a highly specialized type of parasitism or predation. A and C are opposite, so they could be in competition, in which a different species monopolizes the resources depending on conditions. Or, species A might prey on C. As the number of predators (A) falls, the number of prey (C) would rise. Then as A begins to feed on increased numbers of C, the population of A would increase and the population of C would fall.
18. Commensalism; if the organisms were competing for resources, then the extinction of one would make more resources available for the other; thus, it should thrive. If species A derives benefit from species B and species B becomes extinct, then species A might also become extinct.
19. Humans might be influencing the course of evolution by causing many species to become extinct. At the same time, humans are causing other species to flourish that might otherwise have lingered in minor roles or become extinct.
20. Evolution by natural selection occurs over many successive generations. Each generation represents a certain amount of time. The greater the amount of time that passes, the greater the number of generations that are represented, and the greater the possibility for evolution to occur.
21. Because they influence the ability of an organism to survive. For example, humans would not exist if they did not have air to breathe or water to drink.
22. Eight

INVESTIGATION *page 50*

2. These pots provide information on the growth that can be expected in the absence of competition from the other species; they are also useful for identifying the species in the mixed pot.
3. Sample hypothesis: Growth will be impaired by the presence of another species.
5.–7. Answers will vary; usually, one of the two species will grow less in the presence of a competitor than it would by itself.
8. Lab conditions will not mimic exactly conditions in the wild. For example, temperature may vary a great deal in nature but remain constant in the lab. One species might be better able to survive changes in temperature. Also, predators or other factors may affect growth rates in the wild.
9.–15. Answers will vary. As experimental conditions begin to resemble the ideal growing conditions for one of the species, that species should be more successful as a competitor.
16. In a laboratory experiment a limited number of variables are examined, while

in nature there are usually many more variables that may affect competition among species. More complex experiments could be performed to test the reactions of several species to several variables in different combinations; also, a field study that examines native soils might be informative.

CHAPTER 3

SECTION 3.1

CASE STUDY, *page 59*

1. If DDT were water soluble, it would be "washed out" of the animal's body by its excretory system instead of accumulating in its fatty tissues. Therefore, DDT would not accumulate in food chains.
2. The FDA could ban the importation of fruits and vegetables treated with DDT and of meats that contain DDT. The United States could also negotiate with the governments of foreign countries that use DDT and more strictly enforce its laws concerning the use of DDT and similar chemicals.

FIGURE 3-8 CAPTION, *page 61*
Because energy is lost at each trophic level, organisms at higher trophic levels need more food to support themselves than do organisms at lower tropic levels. Thus, each trophic level can support fewer organisms than the level below it.

SECTION 3.1 REVIEW, *page 61*
1. Sunlight is the source of energy for nearly all producers in nearly all ecosystems on Earth. Because it provides the energy used by producers to make food, the sun is also the indirect source of energy for almost all consumers.
2. There could be no sustainable life on Earth without producers. If all producers were suddenly removed from the Earth, consumers and decomposers could continue to live for a short time by feeding on each other. However, all life soon would cease as the available food energy ran out.
3. Herbivores are organisms that eat only plants. Omnivores eat both plants and animals.
4. The crabeater seal is both predator and prey in the food web. It eats krill and is eaten by leopard seals and killer whales.
5. Yes. Because about 90 percent of the energy is lost at each step in a food chain and plants are the first step in a food chain, about 10 times as many humans could be supported by 20 acres if they ate only plants.

SECTION 3.2

FIGURE 3-11 CAPTION, *page 64*
Plants take in carbon from the atmosphere during photosynthesis; organisms then eat the plants and release the carbon to the atmosphere via cellular respiration.

FIGURE 3-13 CAPTION, *page 65*
When animals die or deposit urine or dung, the nitrogen they obtained from eating plants or other animals is converted by bacteria into ammonia. Some ammonia is absorbed by the soil and some is released into the atmosphere as nitrogen gas.

SECTION 3.2 REVIEW, *page 65*
1. In the atmosphere and in the model water cycle, evaporation releases water vapor into the air. The vapor condenses into liquid water in the cooler parts of the jar, just as it does in the cooler parts of the atmosphere.
2. Producers take in carbon dioxide from the air and use it to make simple sugars during a process called photosynthesis. Many sugar molecules can then be linked together to form starches, which the plant uses to store energy. The energy and carbon in these molecules are used by animals that eat the plants.
3. Humans are adding carbon to the atmosphere by burning large quantities of fossil fuels. The result is an increase in CO_2 levels in the atmosphere, which may contribute to global warming.
4. When animals exhale, they release carbon back into the atmosphere in the form of carbon dioxide.
5. Answers will vary. Concept maps should be logical and should include the required terms.

SECTION 3.3

FIGURE 3-16 CAPTION, *page 67*
Once mature pine trees shade the forest floor, oak tree seedlings outcompete pine seedlings because the oaks require less sunlight to grow.

CASE STUDY, *page 69*
1. The National Park Service might allow a natural fire to burn in Yellowstone if the fire did not jeopardize the lives, homes, or businesses of humans. Doing so would permit the natural process of secondary succession to occur.
2. Answers will vary. Students should demonstrate the ability to use logical reasoning to give a clear and concise viewpoint.

SECTION 3.3 REVIEW, *page 70*
1. Primary succession occurs on bare rock where nothing has ever grown before, while secondary succession takes place on soil where other communities have lived before. Primary succession involves a sequence of stages that is similar to secondary succession and that ends in a climax community.
2. Forest fires are a necessary part of secondary succession in some areas because certain organisms reproduce only after a forest fire has occurred. For example, the jack pine can release its seeds only after being exposed to the intense heat of a fire, Also, periodic, smaller fires might prevent major fires.
3. Lichens are pioneer species that can grow on bare rock. They gradually break down the rock and help start the process of soil formation.
4. Secondary succession; primary succession occurs only where no ecosystem has existed before.

ECOLOG *page 71*
1. Sample answer: The animals eaten by a hawk may be carnivores, herbivores, or omnivores. These organisms obtain energy either by eating plants directly or by eating animals that eat plants. Plants are producers; they obtain their energy directly from sunlight. Therefore, a hawk and all of the consumers in a food chain containing the hawk obtain their energy indirectly from sunlight.
2. Sample answer: A nitrogen atom that was once part of a dinosaur may reappear in a sandwich through the nitrogen cycle. The dinosaur got the nitrogen atom by eating a plant or by eating an animal that had eaten a plant. When the dinosaur died, the nitrogen in its body was converted into ammonia by decomposers. That ammonia could have been released into the air or acted upon by nitrogen-fixing bacteria on the roots of certain plants. Those bacteria would then render the ammonia usable to any plant, such as wheat. Thus, eventually, a wheat plant could have taken up the

nitrogen atom that once existed in the dinosaur and ended up in a sandwich made with wheat bread.

CHAPTER REVIEW *page 72*

1. Sample answers:

a. An organism that uses energy from the sun in order to make food is a producer, while an organism that gets its energy by eating other organisms is a consumer.

b. A carnivore is a consumer that eats only other consumers, while an omnivore is a consumer that eats both consumers and producers.

c. The carbon cycle is the means by which carbon atoms are recycled in the environment, while the nitrogen cycle is the means by which nitrogen atoms are recycled.

d. A food web shows all of the feeding relationships in an ecosystem. A food chain is a single strand in a food web.

2.–13. Answers can be found on pages 72–73.

14. The products of cellular respiration are the starting materials for photosynthesis, and vice versa.

15. As energy is transferred through food chains in an ecosystem, the amount of energy available decreases at each successive trophic level. After several trophic levels, there is no more energy available.

16. They break down dead organisms, which returns the nutrients contained in the organisms to the soil and water.

17. There is no starting or ending point. Water molecules are not created or destroyed on a large scale in nature. Instead, they evaporate, condense, and evaporate again continually.

18. Carbon dioxide is taken from the atmosphere by producers during photosynthesis and returned to the atmosphere by consumers during cellular respiration.

19. The graph shows that the location of the Earth's water varied during the time indicated. When the sea level was low, the atmosphere may have contained a high concentration of water vapor or there may have been a great deal of water in the form of ice.

20. Producers make food that they and other organisms use to live. Decomposers, on the other hand, obtain their food by breaking down dead organisms.

21. The reason could be that nitrogen-fixing bacteria live within the roots of mesquite trees.

22. They are similar in that both involve change over time. They are different in that evolution involves changes in the inherited characteristics of a species, whereas succession involves changes in the species present.

23. Because organisms at higher trophic levels depend either directly or indirectly on primary consumers, carnivores might not survive if the primary consumers were eliminated unless there were omnivores upon which to prey. The producers would most likely remain intact.

24. One might first estimate the volume of the lake. Then one could actually take a measured sample of the lake, physically count the number of organisms in that sample, and then use proportions to calculate the number of organisms in the entire lake. One would also have to determine whether the density of organisms varied in different parts of the lake.

INVESTIGATION *pages 74–75*

6. Answers will vary depending on choice of symbols, but students should indicate that plants are producers, animals are consumers, and fungi and most soil organisms are decomposers.

7. Answers will vary depending on the particular circumstances, but students should demonstrate the ability to make careful and thorough observations.

10. Sample answer:
grass→grasshopper→bird;
rosebush→aphid→ladybug

11. Producers. Sample explanation: Some consumers may draw resources from outside the site, or producers may be small and therefore difficult to observe and count accurately.

12. Sample answer: Mushrooms and other fungi; bacteria and fungus spores that are invisible to the naked eye and earthworms and other soil organisms that are probably present underground

13. Decomposers break down rotting organic material into basic, reusable substances; dead organic matter would eventually pile up, and living producers would use up the soil nutrients, leaving the site unable to sustain any life.

14. Sample answer: Certain organisms were observed where the sunlight was at its brightest, the soil was dry, and few rocks were present, whereas others were observed in shady, moist, rocky areas. By growing, moving, and releasing wastes, living things modify the soil and air, create shade, and affect erosion rates.

15. Answers will vary. Students might move rocks to a different location to observe the effects on organisms that live under the rocks. Or students could create an area of dense shade and investigate the effects on the organisms present.

Caption, *page 75:* The symbols for grass, low brush, trees, wildflowers, and lichen represent producers. The symbols for ants, other insects, and birds represent consumers.

POINTS OF VIEW *page 77*

1.–2. Answers will vary. Students should demonstrate the ability to use logical reasoning to give a clear and concise viewpoint.

CHAPTER 4

SECTION 4.1

CASE STUDY, *page 85*

1. Sample answer: Serious conservation measures are often costly in the short term. They also frequently generate tremendous opposition from people who will no longer derive economic benefit from the protected lands.

2. If people are proud of their country's natural treasures, they are more likely to support government conservation efforts. Officials have also encouraged Costa Rican residents to appreciate the value of tourist dollars.

SECTION 4.1 REVIEW, *page 89*

1. Temperature and precipitation

2. The location of temperate rain forests near the ocean creates a moderate and moist climate that allows for the growth of an abundance of forest plants. Taiga is located further north, where it experiences extremely cold temperatures, low precipitation, and a short frost-free growing season.

3. Temperate deciduous forest, because heavy rainfall leaches most of the nutrients from the soil of a tropical rain forest

4. Answers will vary. Concept maps should be clear and logical.

Section 4.2

Section 4.2 Review, *page 99*

1. Both tundras and deserts are products of severe climates, contain fragile ecosystems that are slow to heal, and have low species diversity.
2. Sample answer: The needs of the city, particularly for water, are greater than the desert environment can support. In addition, the fragile desert ecosystem is easily damaged by development.
3. Grasslands have deep, rich topsoil because the grass roots effectively hold soil in place, preventing soil erosion and nutrient depletion. In addition, because trees are scarce in grassland regions and the land is relatively flat, it is not difficult to convert the land to farmland.
4. Fire kills many sprouting trees and shrubs and the above-ground growth of grass plants. The charred remains return nutrients to the soil. However, the roots of grass plants are specially adapted to survive fire undamaged, so after a fire the roots use the new supply of nutrients to rapidly send up new shoots.

Section 4.3

Section 4.3 Review, *page 104*

1. The areas with the most sunlight; sunlight drives photosynthesis in plants, which form the base of all freshwater-ecosystem food chains.
2. Answers will vary but students should recognize that wetlands along river bottoms have several important environmental functions, such as providing habitat for wildlife, trapping carbon that would otherwise be released into the atmosphere as carbon dioxide, absorbing floodwaters, filtering out pollutants, and providing many commercially important products.
3. Many of these lakes were formed by glaciation that took place during past ice ages; glaciation did not affect the lower altitudes or latitudes.
4. Runoff might wash some of the pesticides and fertilizers into nearby streams and rivers. If these chemicals accumulate in high enough concentrations, they can affect the plants and animals living there.

Section 4.4

Section 4.4 Review, *page 112*

1. Some of the factors are water temperatures that are too high or too low, water drainage from freshwater sources, agricultural runoff, oil spills, sewage, ozone depletion, overfishing, dredging, careless divers, shipwrecks, and souvenir collecting.
2. They are vulnerable because dense human populations often develop and pollute them. Understanding this is important because humans depend on the rich diversity of estuaries for many reasons, including food.
3. They would die because the algae in corals need sunlight to photosynthesize and thus live only in shallow waters that receive plenty of sunlight.
4. Because most nutrients come from land, these waters are often nutrient poor. In addition, sunlight penetrates only the surface layers—the depths of the ocean are perpetually dark. Plant life depends on sunlight and nutrients to grow, so plants exist only in small numbers in the surface layer. Because the food supply in the depths consists mainly of dead organisms that fall from the surface, only limited populations of organisms can be supported below.

EcoLog *page 113*

1. Sample answer: The ecosystem in which I live is characterized by grasslands; heavy clay soil; mesquite trees; wildflowers; flat plains; cattle; coyotes; various types of birds such as flycatchers, meadowlarks, and hawks; and mild temperatures. Yes, I think there are ecosystems similar to mine in other parts of the world.
2. Sample answer: Cottonwood trees grow in my area. They tend to grow on the banks of streams because there they are guaranteed plenty of water. Cottonwoods need a lot of water because their huge leaves lose large amounts of water through transpiration, allowing the tree to grow very fast. If temperatures become very hot or water becomes scarce, cottonwoods respond by dropping their leaves. This allows them to retain more water. Few plants can grow in the deep shade cast by the cottonwood, reducing the number of plants that could possibly crowd out the cottonwood.

Chapter Review *page 114*

1. Sample answers:

a. Savannas are grasslands in tropical regions near the equator; steppes are grasslands that are found in colder regions, such as Russia and Ukraine.

b. Biomes are regions with distinctive climates and organisms. They contain many ecosystems, each with similar vegetation.

c. Sunlight can penetrate the waters of the littoral zone, which exists around the edges of a body of water. The benthic zone consists of very deep waters that are beyond the reach of sunlight.

d. Taiga has very long, snowy, and cold winters, but it is still warm enough there for trees to grow. Because of the extremely cold, dry, and windy climate, tundra contains only tiny, ground-hugging plants (no trees).

2.–13. Answers can be found on pages 114–115.
14. Most of the light is screened out by the dense foliage overhead, leaving insufficient light for plant growth on the forest floor.
15. The extensive root systems can hold soil in place, effectively preventing soil erosion.
16. During both, an animal retreats to its den and enters a dormant state in order to survive unfavorable conditions.
17. Tundra is underlain by permafrost. Decomposition occurs very slowly in these frozen conditions.
18. An estuary receives nutrients from both freshwater and saltwater sources.
19. Grasslands occur in a temperate climate characterized by average temperatures and low moisture; tropical rain forests exist in tropical climates that are characterized by high temperatures and moisture; the tundra exists in an arctic climate that is characterized by low temperatures and moisture. Deserts are found in both tropical and temperate climates; this is because if an area is dry enough, it will be a desert regardless of the temperature.
20. Rivers that have rapid drops in elevation, such as those in mountains, are associated with high speeds. These rivers usually have rocky beds and many rapids and drop-offs. As a result, the water is well aerated. When rivers emerge from the mountains and slow down to flow over flatter land, they usually warm up and lose oxygen because warm water holds less oxygen than does cold water.
21. The algae provide the corals with oxygen and carbohydrates, and the corals provide protection and carbon dioxide for the algae. If the reef sank, the algae would no longer be able to perform photosynthesis, and the corals would be deprived of oxygen and carbohydrates.
22. Answers will vary but should accurately represent the location of the ecosystems.

INVESTIGATION *page 116*

1.–2. Answers will vary depending on your particular circumstances.
3. Climatograms should be set up in a fashion similar to that shown in the sample on page 116.
4.–10. Answers to items 4–10 will vary depending on your particular circumstances. Students should be accurate and comprehensive in their responses and should be able to cite evidence to support their conclusions. Students should be careful to distinguish native plants and animals from those that have been introduced. For item 10, it is unlikely that the local climate will exactly match those shown in the chapter.
Caption, *page 117:* Temperate deciduous forest

POINTS OF VIEW *page 119*

1. Answers will vary. Students should offer a logical explanation for their opinion.
2. The following factors could be responsible for the increase in the caribou population: a decrease in the number of predators, an extended period of milder-than-normal weather, a decrease in the incidence of disease, a change in the migratory patterns of the caribous, or the dependence of the caribous on human handouts.

CHAPTER 5

SECTION 5.1

CASE STUDY, *page 127*

1. There is no logical source for watering the recharge zone. Even if a lake could be used for the task, actually getting the water from the lake to the recharge zone would be an extremely slow and labor-intensive project.
2. Sample answers: Yes, because residents of the Great Plains are the ones who are draining the aquifer, and thus they should be the ones who conserve it. OR: No, much of the water from the Ogallala is used to grow crops that people all over the country, if not the world, use. Thus, it should be the responsibility of everyone who benefits from the Ogallala to find ways that Ogallala water can be conserved.

SECTION 5.1 REVIEW, *page 130*

1. Only a tiny fraction of the Earth's freshwater supply is available for our use, and it is unevenly distributed.
2. Groundwater is water that is stored underground. It can take millions of years to replenish a groundwater source, so once a source is used up, it may not be replenished within our lifetimes.
3. Answers will vary but may include installing low-flow faucets, watering lawns at night, gardening with native plants, and putting a bottle in a toilet tank.
4. All countries that use a river's water can claim some right to it as a resource; therefore, nations should strive to reach compromises that reflect the interests and needs of all nations in question.
5. Sample answer: Life on Earth would change drastically for species that depend on fresh water, and many organisms would die. There would most likely be widespread panic and extreme competition for remaining water supplies. (This might include warfare among humans.) Humans would probably attempt to desalinate ocean water and to seed clouds in order to produce rain.

SECTION 5.2

CASE STUDY, *page 135*

1. Even though the paper mill had been identified as the source of pollution, socioeconomic considerations such as the possible loss of jobs and the health of the local economy hindered measures that would have changed the factory's operations.
2. Answers will vary. Students should demonstrate the ability to use logical reasoning to give a clear and concise viewpoint.

SECTION 5.2 REVIEW, *page 139*

1. The origin of point pollution can be identified and measures can be taken to treat the polluted water. Nonpoint pollution comes from many different sources that cannot always be identified.
2. Pathogens are disease-causing organisms that pollute water sources via untreated sewage or animal feces. They cause diseases such as cholera, hepatitis, and typhoid.
3. Pesticides, chemical fertilizers, and other agricultural chemicals. It is difficult to clean up because it is difficult to reach underground contaminants that are hidden among sand grains and in tiny rock fissures.
4. Sewage and fertilizer runoff, which contain phosphorus and nitrogen, can run into rivers or lakes, where they can promote the rapid growth of algae. Oxygen from the water is consumed when the algae die and decompose, suffocating organisms that rely on the water's oxygen.
5. Answers will vary; the diagrams should demonstrate how small amounts of a pollutant at lower trophic levels accumulate into greater and greater concentrations with each successive trophic level.

SECTION 5.3

SECTION 5.3 REVIEW, *page 142*

1. Most ocean pollution comes from the land. It is mainly garbage (especially plastic), sewage sludge, and oil.
2. The three major sources of oil pollution are small spills from tankers, leaks from offshore rigs, and oil spilled while loading and unloading oil tankers.
3. Because it is lethal to many animals. It comes from land and boats. It can be stopped by passing new legislation, better enforcement of existing legislation, improving international cooperation, and increasing public awareness about the problem.
4. Answers will vary but may include poisoning of marine plants and animals by various toxins, disruption of marine ecosystems, and the decline and possible extinction of marine organisms.
5. Sample answer: I could educate my friends about the problems associated with ocean pollution, cut back on the amount of plastics and other wastes that I discard, and participate in beach cleanup programs.

ECOLOG *page 143*

1. Answers will vary depending on the sources that serve your community. Communities typically use either groundwater or surface water.
2. Sample answer: The water we drink at home is reasonably pure because it comes from a municipal supply. Regulations strictly control the amount of impurities in public water supplies. However, many municipal water supplies still have a certain degree of contaminants, as students will learn in the Making a Difference feature on pages 148–149.

Chapter Review *page 144*

1. Sample answers:

a. Surface water includes all above-ground sources such as lakes, ponds, rivers, and streams. Groundwater is water in underground sources such as aquifers.

b. The entire region that is drained by a river is its watershed. A recharge zone is the region that water seeps through before reaching a groundwater supply.

c. Point pollution comes from a single, identifiable source, while nonpoint pollution comes from many sources.

d. Water pollution includes any factors that degrade the quality of a water source. Thermal pollution occurs when excessive amounts of heat are introduced into a body of water.

2.–11. Answers can be found on pages 144–145.

12. Sample answer: Accessibility problems, high costs associated with transporting the ice long distances and against adverse weather conditions, ownership disputes, and storage problems

13. Nonpoint pollution, because its sources are not as easily identified and controlled

14. Distillation and reverse osmosis. In distillation, heat is used to evaporate the water, leaving the salts behind. In reverse osmosis, pressure forces water through a semipermeable membrane that allows the passage of water but not salts.

15. Bioaccumulation causes certain toxins to become more concentrated with every trophic level. Thus, slight concentrations of mercury that came from factory emissions could have been taken up by organisms at a low trophic level. Those organisms could then have been eaten by a fish at a higher trophic level, which would then have a higher concentration of the mercury. That fish could have been eaten by another fish, again increasing the concentration of mercury. Finally, a human could have eaten that fish and ended up with such a high level of mercury that he or she experienced mercury poisoning.

16. Because most of our drinking water comes from freshwater sources, the pollution of freshwater sources presents a more immediate danger to us.

17. August; May; July and October; by storing water from months with a surplus of water for use in times of deficit

18. Sample answer: Rain washes the oil off the driveway and into the street; the oil flows with water down into street drains and is eventually washed into a stream; the stream joins a river and the oil is washed out to the ocean with the river water.

19. Natural eutrophication is a slow, stable process that allows for gradual changes in the types of plants and animals that live there. Because artificial eutrophication is an accelerated process, there is not enough time for the water body and its inhabitants to adapt to the new conditions. Thus, artificial eutrophication is more disturbing to the environmental stability of the ecosystem.

20. 2,500,000,000 L; 660,501,981.5 gal.

Investigation *page 146*

3. Observations should be clear and concise.

5. Sample prediction: The substances that are dissolved in water (glucose and food coloring) will not be successfully filtered.

7. Observations should be clear and concise.

8. Glucose is present. It cannot be seen, but it is detectable with test paper.

9. The soil was removed; it is not visible. Food coloring was not removed; it is still visible.

10. Students should note that substances dissolved in the water (food coloring and glucose) pass through the filter. They may conclude from this observation that any hazardous chemicals that dissolve in water pose a threat to our groundwater sources.

11. Answers will vary depending on the substance chosen. In most cases, if a substance is a liquid or dissolves in water, it will still be present in the water after filtration.

12. Observations and interpretations should be clear and concise.

13. Answers will vary. Sample answer: Water is an excellent solvent, so harmful substances such as fertilizers, insecticides, and hazardous wastes can potentially be carried by water through the ground. Precautions for keeping groundwater clean include recycling motor oil instead of dumping it, minimizing the use of fertilizers and pesticides, and protecting groundwater recharge zones from pollution and hazardous materials.

Chapter 6

Section 6.1

Section 6.1 Review, *page 156*

1. Secondary pollutant. The primary pollutant is SO_2. The SO_2 reacts with O_2 in the air to form SO_3.

2. By giving the Environmental Protection Agency the authority to regulate auto emissions; by mandating the installation of scrubbers, electrostatic precipitators, and other pollution-control devices

3. Dense traffic combined with a sunny climate and the city's location in a coastal valley

4. The pollution came from the large amounts of coal that were burned in homes and by industries in London at the time.

5. Sample answer: I think the best way to reduce air pollution would be for the government to fund further research and development of nonpolluting fuels. I also think that the government should offer incentives to encourage people to reduce their use of fossil fuels and to urge industries to develop more energy-efficient operations. Answers will vary. Student models should be clear and logical, and their decision should be clearly stated.

Section 6.2

Case Study, *page 159*

1. The more sulfur the coal contains, the more sulfur dioxide it releases when burned. Sulfur dioxide irritates the respiratory system of humans, so increased sulfur dioxide levels will make respiratory problems even more serious.

2. Sample answer: I think a free-market economy would solve some of Poland's pollution problems because competition would probably force manufacturers to make their manufacturing processes more efficient. I also think that people would have greater control over the condition of their environment because purchasing power would allow them to take business away from companies that were damaging the environment.

Section 6.2 Review, *page 160*

1. It may be difficult because there are often many possible causes for a disease, some of which are difficult to isolate and

test. Scientific conclusions cannot be made about untested factors.
2. Asbestos fibers can cut and scar lung tissue. Because this lung tissue is additionally weakened by smoking, a smoker exposed to asbestos is even more likely to develop lung cancer.
3. Sample answer: It is not likely that my neighbor's house has sick-building syndrome because that syndrome usually occurs in modern, tightly sealed buildings, but his home is older and well ventilated.

SECTION 6.3

SECTION 6.3 REVIEW, *page 164*

1. Any human activity that involves the use of fossil fuels, such as using an internal combustion engine or burning coal, could contribute to acid precipitation.
2. Sometimes the country that is causing the acid precipitation is not affected by it, but a neighboring country suffers from its negative effects. Therefore, if a country is to solve the problems caused by acid precipitation, it must work with the country that is the source.
3. Normal rainwater contains dissolved CO_2, which is mildly acidic.
4. Lime must have a pH above 7 because a pH of greater than 7 is required to counteract the acid in an acidic lake, which would have a pH of less than 7.

ECOLOG *page 165*

1. Sample answers: There is an air-quality problem in our area. I have noticed smog on some days, heard of "ozone alerts," and even had difficulty breathing once in a while. OR: The air quality in our area is excellent. Because of its location on an open plain, we are not subject to thermal inversions. In addition, light automobile traffic and few industries mean that pollutants do not accumulate in this region's air.
2. Sample answer: The air pollution in our area is due in part to our community's location in a valley. Pollutants are sometimes trapped over our city by thermal inversions. In addition, the many vehicles and industrial facilities in our area contribute to smog and other air-pollution problems.

CHAPTER REVIEW *page 166*

1. Sample answers:
a. Air pollution, which is the degradation of air quality, is caused largely by automobile exhaust.
b. Acid shock occurs when acidic snow that has accumulated all winter melts and rushes into lakes and other bodies of water.
c. Thermal inversion occurs when a layer of cool air near the ground is covered by a layer of warmer air above.
2.–13. Answers can be found on pages 166–167.
14. Carpooling reduces the number of cars on the road and, therefore, reduces the amount of pollutants released into the air.
15. Disagree; although most pollution is caused by burning materials, pollutants such as volatile organic compounds do not result from burning materials.
16. Radon is not generated by human activities but is found naturally in the environment.
17. In Los Angeles, pollutants cannot escape sideways because the region is surrounded by mountains on three sides and by the ocean on the fourth side. Prevailing winds from the west often prevent pollutants from escaping over the Pacific Ocean.
18. Because lime is basic, it neutralizes the acidic pH of the water.
19. Near major metropolitan areas; there must be major industry located near Birmingham.
20. No; while reductions in the total amount of pollutants produced may continue to decline, pollutants cannot be expected to decline to zero because internal-combustion engines will always produce some pollutants.
21. Fossil fuels are a major contributor to air pollution, and air pollution has been known to aggravate many common illnesses and diseases. Thus, if alternative forms of energy reduce the amount of air pollution, the health of many people could improve. They could then live longer, and the average life expectancy could rise.
22. "Killer fogs" were primarily caused by the heavy use of coal. Under certain conditions, the dense smoke from burning coal would join with dense fog to cause stifling, poisonous conditions. London is an example of a city that experienced killer fogs. In recent decades, the use of coal has tapered off greatly. As a result, "killer fog" conditions no longer occur.

INVESTIGATION *page 168*

6. Students should notice that soaking the seeds softens the seed coat.
9. Students will probably notice that many of the seeds in water have germinated, while few, if any, of the seeds in the artificial acid precipitation have germinated.
10. It should be apparent that the "acid precipitation" prevented the seeds from germinating. Based on student results, the farmer would probably lose all 10,000 plants.
11. Sample hypothesis: Less damage would occur to a plant if its water supply were exposed to limestone.
12. Sample procedure: Prepare beakers as before, but add an additional beaker with artificial acid precipitation and 1.2 g of limestone chips.
13. Sample prediction: Germination will occur in the beaker with water only and in the beaker with the artificial precipitation and limestone chips.
14. Answers will vary. Be sure that students clearly document their results.
15. Answers will vary. Student answers should provide a clear and concise interpretation of their results.
16. Sample answer: The results of this experiment show that acid precipitation affects seed germination—this is known to be true in the real world. The experiment is not realistic; in nature, seeds would not soak for 24 hours in a pure "acid precipitation" solution. Instead, seeds would probably be in soil, and that soil would probably contain some compounds that could at least partially neutralize the acid. To make the experiment more realistic, I could grow the seeds in soil.

POINTS OF VIEW *page 171*

1. Answers will vary. Students should demonstrate the ability to use logical reasoning to give a clear and concise viewpoint. Alternative solutions for reducing air pollution might include educating people about the seriousness of air-pollution problems and providing incentives for people to use mass transportation. Also, cities could set up "no drive" days. This would involve asking or requiring city dwellers to leave their cars at home one day a week and to use mass transportation or carpool during high-pollution months.
2. Answers will vary. Students should demonstrate the ability to use logical reasoning to give a clear and concise viewpoint.

CHAPTER 7

SECTION 7.1

SECTION 7.1 REVIEW, *page 175*

1. It contains gases that are essential to living things and also screens out much of the harmful radiation from the sun.
2. The evolution of early plants
3. The troposphere
4. The process of cellular respiration

SECTION 7.2

FIGURE 7-9 CAPTION, *page 179*

Westerly winds pick up moisture over the Pacific Ocean. As the air rises to cross the Sierras, it cools, causing most of the moisture to precipitate out as rain or snow. Upon descending the leeward side of the mountain, the moisture-depleted air warms up and absorbs moisture from the land it passes over.

SECTION 7.2 REVIEW, *page 179*

1. Because of its shape and tilt, the Earth is more or less perpendicular to the sun's rays at the equator, whereas at the poles the Earth is nearly parallel to the sun's rays. As a result, the rays do not spread out when they strike the Earth at the equator, but they do spread out greatly at the poles.
2. Dublin experiences much milder temperatures and more rainfall than does Moscow.
3. The equator would probably have essentially the same climate as it does now, but the temperate regions would differ in that they would no longer have any seasonal climatic variations.
4. Answers will vary. Concept maps should be clear and logical.

SECTION 7.3

CASE STUDY, *page 183*

1. Students may mention such factors as ash and gases from volcanoes, changes in solar energy due to the sunspot cycle or other natural fluctuations, absorption of CO_2 by the oceans, changes in ocean currents, changes in the amount of woodlands or developed areas, and increased snowfall at higher latitudes due to warmer temperatures.
2. Sample answer: The fact that there is considerable disagreement among scientists is a sign that some skepticism might be justified. It would probably not be sound to ignore those scientists who predict global warming, but neither should we blindly accept their predictions. We should realize that predictions will improve as the information they are based on improves, so we should continue to be informed about the subject.

SECTION 7.3 REVIEW, *page 185*

1. Radiation from the sun passes through the glass, is absorbed by the materials inside the greenhouse, and is converted into heat. Since heat energy doesn't pass through glass as easily as light energy does, much of the heat stays in the greenhouse, warming it.
2. The large-scale burning of fossil fuels and the clearing and burning of tropical forests
3. The strong correlation between the average global temperature and the amount of carbon dioxide in the atmosphere. Since the amount of CO_2 is increasing, the temperature must also increase if the correlation is valid.
4. A disruption in weather patterns

SECTION 7.4

SECTION 7.4 REVIEW, *page 188*

1. A form of oxygen with molecules made of three oxygen atoms. By shielding the Earth's surface from most of the sun's harmful ultraviolet radiation
2. Plastic foams, aerosols, refrigerants, and chemical solvents
3. Because it reacts with ozone, causing it to break apart
4. Answers will vary, but students may mention the need to wear protective clothing, wear sunglasses, reduce unprotected exposure to the sun, use more sunscreen, and modify their diet.

ECOLOG *page 189*

1. Sample answer: This phenomenon is called the greenhouse effect because it is similar to the way in which a greenhouse traps heat.
2. Sample answer: Ozone in the upper atmosphere serves an extremely useful purpose: screening out much of the sun's deadly ultraviolet radiation.

CHAPTER REVIEW *page 190*

1. Sample answers:

a. Weather is the state of the atmosphere at any one moment. Climate is the average weather over a long period of time.

b. The atmosphere is a thin layer of gases that surrounds the Earth. Ozone is an important component in one atmospheric layer, the stratosphere.

c. The greenhouse effect is the trapping of heat energy by certain gases, causing the Earth's atmosphere to warm up. Global warming is the result of an intensified greenhouse effect caused by artificially increased levels of certain gases.

d. The troposphere is the layer of the atmosphere closest to the Earth's surface. Life exists in the troposphere, and weather occurs there. The stratosphere is the atmospheric layer above the troposphere. Essentially no weather occurs there, and conditions are too harsh to support life.

2.–12. Answers can be found on pages 190–191.
13. Increased warmth could cause ice and snow to melt and flow into rivers, streams, lakes, and oceans, which might then flood. Major weather patterns could change, causing more precipitation in some areas and droughts in others.
14. Because the site was in the middle of the Pacific Ocean, far away from forests and cities (which could skew the readings), Keeling reasoned that average carbon dioxide levels for the entire Earth could be measured at Mauna Loa.
15. A warmer Earth could cause the polar icecaps to melt. The water thus released would flow into the oceans, increasing their volume and therefore raising sea levels.
16. NASA scientists reviewed data that had been sent back to Earth by the *Nimbus 7* weather satellite since its launch in 1978. They found evidence that supported the reports of ozone thinning.
17. The location indicated by the green line. The cities differ in climate because one (green line) is probably located in a coastal region where temperatures remain fairly constant throughout the year, and the other (red line) is probably located inland, where it experiences more dramatic changes in the weather.
18. Sample answer: Since carbon dioxide is a trace gas, the effect, if any, on most organisms will probably be negligible. Organisms that use carbon dioxide (such

as lime-secreting organisms and plants) may experience enhanced growth, given the greater amount of carbon dioxide available. The effects of global warming will probably be more pronounced. Assuming that global warming takes place over a long span of time, some highly specialized organisms would probably become extinct (such as ice-dwelling organisms), but most would probably be able to adapt—behaviorally and through natural selection. The adaptations, in some cases, might be so extensive that the species would become, in effect, new species. In that respect, global warming would also cause "extinctions."

19. Clearing large regions of rain forest increases the amount of carbon dioxide in the atmosphere because there are fewer trees to remove carbon dioxide from the air by photosynthesis. Burning large regions of rain forest increases the amount of carbon dioxide in the atmosphere because burning plants release carbon dioxide.

20. Because global warming would change weather patterns, possibly increasing the number of natural disasters. This would change the amount of risk faced by homeowners and therefore affect the liability of the insurance company.

INVESTIGATION *page 192*

Caption, *page 192:* The uncovered jar is the control. It enables one to determine the normal temperature of a jar placed in the sun, which is then compared to the "greenhouse" temperature.

7. Sample hypothesis: The effect of global warming will be greater in dry regions than in humid regions.

8. Sample answer: I would take a soil sample, divide it into two equal parts, place one part in each jar, and label one jar A and the other B. (A will be the control.) Then I could place a thermometer in each soil sample and make sure both samples are the same temperature. Next, I would put the two jars in the sun for exactly 1 hour, recording the temperature at 5-minute intervals. Keeping the jars in the sun, I would then add a small (measured) amount of water to the soil in B until it felt damp to the touch. I would then record their temperatures at 5-minute intervals for 1 hour. Next, I would repeat the same experiment, this time increasing the amount of water in sample B by an amount equal to that added in trial #2. Finally, I would record the temperature in each jar at 5-minute intervals for 1 hour.

9. Sample prediction: I predict that the temperature of dry soil in a covered jar will increase more rapidly than the temperature of wet soil in a covered jar when both jars are placed in sunlight.

13. Sample conclusion: The dry soil became warmer than the wet soil. Therefore, I can conclude that my hypothesis is valid.

14. Students will probably find that moisture, darkness or lightness of surfaces, and thermal inertia (in the form of ice) all significantly affect the rate of greenhouse warming.

15. Sample answer: My experiment reflects the likelihood that arid regions would be more affected by global warming than would humid regions. The experimental model is very simplistic and does not account for the interaction between arid and non-arid regions. Furthermore, few regions will have soil as moist as that in the model.

Caption, *page 193:* All of the materials could be useful, depending on the hypothesis.

CHAPTER 8

SECTION 8.1

SECTION 8.1 REVIEW, *page 203*

1. Although cities take up only a small portion of the world's total land space, the people who live in cities require large amounts of food, water, fuel, raw materials, and finished goods from non-urban lands.

2. New technology has enabled farms to be operated by fewer people, and improved transportation systems have eliminated the need for consumer items to be produced locally. Therefore, there are fewer jobs available in non-urban areas than in the past, and people have moved to the cities, where jobs and opportunities are concentrated.

3. Urban crisis results when more people live in a city than that city's infrastructure can support. Improved mass-transit systems, urban renewal and renovation projects, and long-term land-use planning are all examples of what can be done to relieve the urban crisis.

4. Sample answer: Yes, laws should limit suburban sprawl because if cities continue to expand at present rates, the ability of non-urban lands to supply the world with food and other vital services could be at risk.

SECTION 8.2

SECTION 8.2 REVIEW, *page 212*

1. Reforestation is when trees are replanted, either naturally or by humans, to replace trees that have died or been cut down. It is important because it renews a resource and helps restore natural areas to functioning, productive ecosystems.

2. By limiting the size of grazing herds, fencing areas and digging multiple watering holes to prevent overgrazing, eliminating rangeland weeds, and planting new vegetation where necessary, these management techniques can protect rangeland vegetation and soil from permanent destruction.

3. Sample answer: Recycle products made from minerals (thereby reducing the need for more mining) and enact mining legislation that requires practices which protect the environment, such as reclamation.

4. Because it is so difficult to assess the economic value of a natural area, politicians, environmentalists, and other concerned parties often have trouble convincing the general public that protecting natural areas is worthwhile. Thus, when making decisions about how land will be used, government officials sometimes support plans that will satisfy the easily definable economic needs of the community and give little attention to the long-term protection of the land and land resources upon which the community depends.

SECTION 8.3

CASE STUDY, *page 217*

1. Sample answers: Yes, this is a good solution because it seems to solve several problems at once. The reclamation would improve the health of the affected ecosystem and invigorate local economies by creating new jobs. Plus, the higher cost of mining may encourage more responsible mining practices. OR: No, this would not be a good solution because the higher cost of mining would lead to a serious reduction in the number of mining jobs. It is unlikely that reclamation and restoration projects would create enough jobs to replace those that would be lost.

2. Answers will vary. Students should use logical reasoning to give a clear and concise viewpoint.

Section 8.3 Review, *page 218*

1. Answers will vary. One benefit of the multiple-use strategy is that it allows a large number of people to use public land. One disadvantage is that managing the land to be "all things to all people" can result in its overuse and destruction.

2. Sample answer: I can use some public lands, such as national parks, for recreation. Some of the products I use, like electronic equipment, have mineral components that may have been mined from public land. Wilderness areas and historical monuments provide me with opportunities to learn more about the natural heritage and history of an area.

3. When Europeans first arrived in North America, they settled in the East. There they rapidly cleared natural forest land and used that land for private enterprises such as farming, ranching, and mining. Before the same thing could happen in the West, however, the federal government was able to claim a great deal of the land for natural resource protection and for multiple use.

4. Answers will vary. Students should use logical reasoning to give a clear and concise viewpoint.

EcoLog *page 219*

1. Sample answer: Yes; in order for me to get the food I eat, farmers must cultivate the land and ranchers must use it to raise animals. Plus, many of the products that I use are made of minerals that are mined from the land. Also, recreational areas must be set aside for people like me to enjoy. Solid waste and pollution also result from many of the things that I do, such as throwing away trash that must be hauled to a landfill, riding in a car, and using electricity.

2. Sample answer: By planning how to best use the land, humans may be able to prevent serious problems like the urban crisis or environmental degradation. On the other hand, not planning almost certainly guarantees that cities will experience problems associated with the urban crisis and that the environment will suffer.

Chapter Review *page 220*

1. Sample answers:

a. The process of restoring an area to the condition it was in before it was mined is called reclamation.

b. Land-use planning helps us determine how to manage the land to satisfy as many needs as possible.

c. Suburban sprawl has increased many urban areas to several times the size they were 50 years ago.

d. Hospitals, roads, and power lines are all part of a city's infrastructure.

2.–12. Answers can be found on pages 220–221.

13. Suburban sprawl involves the spread of homes and businesses across land that could have been used for agricultural purposes. Also, row after row of homes and businesses take up more space than a multiple-use structure such as a skyscraper does. In addition, more roads and infrastructure are needed to support a sprawling city.

14. Urbanization results when technology enables non-urban industries to be operated by fewer people and when transportation systems eliminate the need for consumer items to be produced locally. When this happens, there are fewer jobs available in non-urban areas, and people move to cities in search of work.

15. Cities sometimes grow so rapidly that their infrastructure is unable to keep pace. This causes problems such as overcrowding in schools and hospitals, traffic congestion, deteriorating roads, and increasing numbers of homeless people whom the city cannot support.

16. Many environmentalists object to clear-cutting because it removes a large number of trees at one time. These trees cannot regenerate rapidly, and some tree species may never grow back. This causes the displacement and sometimes the demise of wildlife and may have lasting effects on the forest ecosystem. It also destroys the beauty of many natural areas. Some environmentalists prefer selective cutting because fewer trees are removed at one time, and thus the impact to the forest is less dramatic.

17. By managing the land, people can better protect its long-term usefulness.

18. About 48 percent, including urban, cropland, and rangeland areas. Probably close to 100 percent of the land serves the needs of city dwellers in one way or another. None

19. Answers will vary. Students should use logical reasoning to give a clear and concise viewpoint.

20. Sample answer: Over the last 200 years, increasing human populations have led to the overuse of all types of land. For instance, the need to feed a growing human population has led to the misuse of some agricultural land; the infrastructure of many urban areas has been overloaded, causing urban crises all over the world; and "protected" areas such as national parks and wildlife refuges are being scarred by mining and ranching activities as well as by increasing numbers of tourists.

21. A recycling program that recycles newspaper and other paper products could lead to a reduction in the number of trees that are cut down in a forest, which might allow the area to undergo natural reforestation.

22. Answers will vary depending on student selections, but students should demonstrate creative and logical thinking skills.

Investigation *page 223*

13. Generally, students will dig individual holes to extract nuggets of ore. In real practice, however, a company either would use strip-mining techniques or would bore shafts into the mineral deposit.

14. The cost of reclamation could conceivably surpass the cost of the mining operation itself.

15. Probably not. The "miners" from some companies were probably faster, more skilled, or less concerned about damage to the land than others, or their sites may have contained more "ore."

16. Answers will vary. Students may suggest that the miners carefully document the state of the land before they mine, work slowly and carefully, and plan every step of the operation in advance.

Chapter 9

Section 9.1

Section 9.1 Review, *page 230*

1. They may be too poor to buy food, or available food may not be transported to

people who need it.
2. Benefit: more crops have been produced, and thus more people can be fed. Drawback: the new crops require more fertilizer and pesticides than the previously grown varieties, and these products can cause environmental problems.
3. Human population growth was increasing faster than grain production.

SECTION 9.2

CASE STUDY, *page 237*

1. Because large numbers of fish are confined in small spaces, they are more easily caught or "harvested" than fish in open waters. Also, feeding, reproduction, and other life processes of fish can be monitored in aquaculture environments so that fish are made a more reliable "crop."
2. No; not all areas have access to wetlands or appropriate water sources.

SECTION 9.2 REVIEW, *page 239*

1. Living organisms such as fungi, bacteria, and other microorganisms contribute by decomposing organic matter. Earthworms, insects, and other small animals help to break up the soil and create air spaces as they move through it. Animal waste contributes nutrients to the soil, and nitrogen-fixing bacteria convert nitrogen from the atmosphere into a form that plants can use.
2. Because irrigation water is taken from rivers or the groundwater, which are saltier than rainwater. When this water evaporates, it leaves salts behind. Also, irrigation causes the groundwater level to rise. Then groundwater is drawn up to the surface by capillary action. When it evaporates, the groundwater leaves behind the salts it contains.
3. Because overgrazing, the use of trees and shrubs for fuel, and farming methods that involve shorter fallow periods are all contributing to the loss of topsoil. Without sufficient topsoil, land can easily deteriorate to the point that it becomes a desert.
4. Answers will vary. Concept maps should be logical and should include the required terms.
5. One possible reason is that large areas of rain forest are being cut down in Central America. Intact forests hold topsoil in place, especially in regions such as Central America that experience high amounts of rainfall.

SECTION 9.3

SECTION 9.3 REVIEW, *page 244*

1. Because cultivated plants are usually planted as a large group of one species, while wild plants usually occur in mixed communities with many other plant species. Consequently, pests that feed on cultivated species do not have to spend much time looking for a host plant, but can instead stay in the cultivated area, where it is easy for them to feed and multiply.
2. Because they do not break down immediately into harmless chemicals. Instead they accumulate in the water and soil, as well as in animal tissues.
3. If an insect population contains some individuals that possess a gene protecting them from a pesticide, those individuals do not die when the pesticide is applied. Instead, they survive and reproduce offspring that may possess the same protective gene. Eventually, only those individuals possessing the protective gene are left, and the entire insect population is resistant to the original pesticide.
4. A pesticide, by definition, is a substance that kills pests. Pheromones do not kill pests. Therefore, pheromones can be used for pest control, but they are not pesticides.
5. Answers will vary. Some common pesticides are diazinon, malathion, dursban, acephate, and sevin. To reduce the amount of pesticides used at home, a person could keep pet bedding clean, keep pets bathed and outside, keep windows and doors closed, keep garbage cans closed tightly, and empty standing water so that mosquitoes have less of an opportunity to breed.

ECOLOG *page 245*

1. Sample answer: Natural disasters can wipe out food crops, causing serious food shortages. Poverty can prevent people from obtaining enough food as well. Also, some people are far from farming regions or live in areas with inadequate transportation systems. Any breakdown in transportation leads to food shortages. Other people live in regions that are affected by war or civil strife, which can disrupt the distribution of food.
2. Sample answer: Farmers can use plant extracts that repel or kill pests, organisms that feed on or cause disease in pests, or products that confuse pests or disrupt pest breeding cycles. Farmers can also alternate their crops.

CHAPTER REVIEW *page 246*

1. Sample answers:
a. Malnutrition is the chronic deficiency of certain essential nutrients.
b. Erosion is the wearing away of topsoil due to the action of wind and water.
c. Desertification is the degradation of land to the point that it becomes a desert.
d. Resistance involves the ability to tolerate something, such as insects tolerating a particular pesticide.
2.–13. Answers can be found on pages 246–247.
14. If enough of the soil erodes, farming becomes impossible.
15. Excessive pesticide use can cause insects to develop resistance, pollute the soil and water, and harm nonpest organisms that are exposed to sufficient amounts.
16. The persistence caused it to accumulate in the environment and in the tissues of organisms that either directly or indirectly ingested it. Some organisms that were high on the food chain, such as penguins, pelicans, and eagles, ended up with high levels of DDT. As a result, they experienced reproduction problems that led to their near extinction in some areas.
17. Pathogens can be used to infect pests with diseases that can kill them.
18. Irrigation can lead to the salinization of soil; it may also deplete water sources, divert the normal flow of streams and rivers, or disrupt ecosystems.
19. Each time a new pesticide was applied, the insects (having no resistance) were controlled, so crop yields rose. By the second year of pesticide use, the insects had acquired resistance to the pesticide, so the crop yield dropped. Increasing the dosage had only a slight effect on the insect population and crop yields.
20. Insects reproduce rapidly and in very large numbers (millions or even billions of offspring). So even if just a few insects develop natural resistance to a pesticide, that resistance will be passed on so quickly that a sizable resistant population will exist before the species population becomes endangered.
21. Each time energy is converted from one form into another, some energy is lost as heat. For example, when one organism is eaten by another, some energy is lost. Therefore, each trophic level has less available energy, and more energy is available at lower trophic levels than at higher trophic levels. In contrast, although an organism in a low trophic level might contain

only a tiny amount of DDT, organisms higher up the food chain will consume many organisms containing DDT. Because of this and because DDT is not easily broken down, more and more of it will end up in higher trophic levels.
22. Pesticide usage is causing insects to evolve at an accelerated rate because it strongly selects for a certain trait (resistance) by eliminating the vast majority of nonresistant individuals. Out of a population of billions, only a handful of nonresistant individuals may survive. Resistance, along with any other unique traits of the survivors, will be passed on to offspring, greatly magnifying the traits represented in the population.
23. Humans as primary consumers, because when food supplies are limited, a population of primary consumers is better off than a population of secondary consumers
24. There are many historical examples of environmental degradation. You might suggest such examples as Easter Island, the dust bowl of the 1930s, the deforestation of Europe, or the killer fogs of London.

Investigation *page 249*

5. The soil washes down the slopes.
6. The bottom of the stream fills with soil, and the displaced water spreads over the banks to form a wider and shallower stream.
7. The fork loosens the soil, causing it to erode faster.
8. Erosion increases because the alfalfa no longer holds the soil in place.
9. Answers will vary, but students should notice that all modifications could help lessen erosion if properly implemented.
10. Answers will vary. For suggestions on experimental modifications, see Teaching Strategies for the Investigation on page T100.
11. Avoid overgrazing, rotate the crops, plant wind breaks, terrace the crops, plant crops in rows that follow the contours of the land, and add green belts.

Points of View *page 251*

1. Identifying the source of the foreign genes would remove some of the mystery from the process and may increase public confidence. At the same time, medical concerns, such as the possibility of allergic reactions, would be addressed. Consumers could make informed decisions about whether or not to buy the genetically engineered produce.
2. Answers will vary. Students should demonstrate the ability to use logical reasoning to give a clear and concise viewpoint.

Chapter 10

Section 10.1

Case Study, *page 257*

1. A single storm, drought, or disease could easily wipe out 250 whooping cranes. Thus, the bird's survival is in no way guaranteed.
2. Answers will vary. Students should demonstrate the ability to use logical reasoning to give a clear and concise viewpoint.

Section 10.1 Review, *page 260*

1. By eliminating the particular environmental conditions that the species requires for survival
2. Each species has a role to play in its ecosystem. Even though that role may not always be obvious to us, the disappearance of a species causes its natural ecosystem to change in some way. Therefore, by protecting individual species, we may be preserving the entire ecosystem.
3. Answers will vary. Some practical disadvantages mentioned in the section include the loss of species for food, medicine, and other products.
4. Answers will vary. Students should demonstrate the ability to use logical reasoning to give a clear and concise viewpoint.

Section 10.2

Case Study, *page 265*

1.–2. Answers will vary. Students should demonstrate the ability to use logical reasoning to give a clear and concise viewpoint.

Figure 10-13 Caption, *page 266*

The price of ivory rose steadily before 1989 because ivory was becoming more rare. With the 1989 ivory ban, the sale of all ivory was made illegal, so ivory could be sold only to people who were willing to break the law. Since few people were willing to do so, the price of ivory fell sharply (indicated by the downward plunge on the graph), and the poaching of elephants decreased noticeably.

Section 10.2 Review, *page 266*

1. An endangered species is one with a population so small that it is likely to become extinct in the near future unless action is taken immediately. A threatened species is one that may be common in some areas but is likely to become endangered if protective action is not taken.
2. Because they require a large range to obtain adequate food, find a suitable mate, and rear their young. Or a species may be wiped out on a small plot of land because of unusually harsh weather or disease.
3. Answers will vary. Students should demonstrate the ability to use logical reasoning to give a clear and concise viewpoint.
4. With a limited ivory ban, poachers could easily sell ivory on the world market simply by attaching forged documents to it. With a total ban on ivory, however, the only ivory that could be sold was to people who were willing to break the law. Since few people were willing to do so, the price of ivory fell sharply, and the poaching of elephants decreased noticeably.

Section 10.3

Section 10.3 Review, *page 270*

1. Germ-plasm banks
2. By protecting entire ecosystems, many species are protected, not just those that have been thoroughly studied and listed. Also, because only a tiny fraction of the species in danger of extinction can be listed officially, protecting an entire ecosystem is a much more practical method of preserving species. Finally, protecting entire ecosystems contributes to the total health of the Earth's environment.
3. Sample answer: Many are located fairly close to the equator, in regions with a tropical climate; many are located on islands; and South America appears to contain more hot spots than any other continent.

EcoLog *page 271*

1. Sample answer: Today most species become extinct as a result of human activities, such as hunting and the destruction of habitat.
2. Sample answer: Yes, because by saving a species, we help preserve the entire ecosystem in which the organism lives. In addition, a species may have commercial or medical importance, so losing species could mean losing resources.

Chapter Review *page 272*

1. Sample answers:

a. Hunting refers to the legal killing of animals for food, sport, or profit. Poaching refers to the illegal hunting of animals that are protected by law.

b. A native species is a species that lives in the area in which it evolved and that is adapted to the local ecosystem. An exotic species is a species that is not native to an area.

c. An endangered species is one with a population so small that it is likely to become extinct in the near future unless action is taken immediately. A threatened species is one that may be common in some areas but is likely to become endangered if protective action is not taken.

2.–11. Answers can be found on pages 272–273.

12. Legal hunting is regulated. Government agencies such as fish and wildlife commissions work with hunters to control the percentage of game animals that are killed each year.

13. Exotic species are species that are introduced to areas other than their natural habitat.

14. Sample answer: Protecting land for an endangered species may anger or disappoint residents who would prefer to see that land used in another way—such as for jobs, housing, or other development.

15. Sample answer: Setting aside wilderness areas may cause nearby communities to lose jobs and tax revenue, protecting rain forests may mean that farmers' livelihoods are threatened, and prohibiting the sale of elephant tusks may affect a poor family's struggle to survive.

16. Charts should show the following percentages (rounded): mammals, 23%; birds, 35.6%; reptiles, 5%; amphibians, 3.75%; fish, 32.5%.

17. Due largely to the activities of humans, the present rate of extinctions is probably higher than it has been at any time in the last 65 million years. Extinctions in the past, on the other hand, were probably due to natural causes, such as changes in climate.

18. Natural selection is a slow and gradual process that occurs because of incremental responses to changes in the environment. The effect of natural selection is to allow only "properly equipped" individuals and species to survive. By definition, the process is selective. Unlike natural selection, the forces that are causing the present mass extinction kill vast numbers of species and individuals in a very short period of time and are largely nonselective.

19. During photosynthesis, plants release water into their surroundings (cooling and raising the humidity of the surrounding air) and take up carbon dioxide (a greenhouse gas). A large forested area can transpire enough water to cause rainfall and take up enough carbon dioxide to affect the global balance of this greenhouse gas. Removing the forest negates these effects: locally the climate becomes drier and hotter, and globally the climate becomes warmer due to greater amounts of CO_2 in the atmosphere.

20. 53 percent

21. Answers will vary depending on the particular circumstances of your area.

Investigation *page 275*

8. Sample hypothesis: Watering an area increases the number of insects found there.

9. Sample description: I could find two sites of equal size, similar terrain, similar vegetation, similar sunlight exposure, etc., keeping all these variables constant. Then I could sample the two sites for insect number and diversity, using methods similar to those used in the investigation. Assuming that the two sites were comparable in insect number and diversity, I could carry out the experiment by watering one site on a regular basis and not watering the other. After a specified period of time, say three months, I could sample the two sites again. If the watered site had a significantly greater number and diversity of insects than the unwatered site, I could conclude that my hypothesis was supported by the data. If there was no significant difference, or if the unwatered site had a greater number and diversity of insects, then I could conclude that my hypothesis was not supported by the data.

Points of View

Owls vs. Loggers, *page 277*

1. Answers will vary but students' recommendations should take into account the conflicting needs and desires of both sides of the issue.

2. Answers will vary. Some possible responses include: How much of the old-growth ecosystem does the owl really need to survive? What other effects on the environment (such as air and water quality) would cutting more of the forest have? Since scientists cannot destroy forests to test their hypotheses, answers to these questions must be theoretical or based on experimental models.

Chapter 11

Section 11.1

Section 11.1 Review, *page 282*

1. Fossil fuels are burned. This releases energy in the form of heat, which is used to turn water into steam. The steam is then directed against the blades of a turbine, causing it to spin. The turbine is connected to a generator. The motion of the generator's coils of wire through a magnetic field causes electricity to be generated.

2. Renewable resources are replenished naturally at a rate greater than they are used. Nonrenewable resources are those that are used up faster than they are replenished.

3. The dramatic increase after 1950 was in response to an increased demand for oil. The demand resulted from the rapid industrialization and from the increase in the number of automobiles that took place at this time.

Section 11.2

Section 11.2 Review, *page 286*

1. They are similar in that both use some sort of reaction to heat water into steam that can be used to drive a turbine.

2. Advantages: They produce a lot of energy with a small quantity of fuel and they do not produce gases such as carbon dioxide. Disadvantages: They produce hazardous waste, their fuel supply is limited, they are expensive to construct and maintain, and they have at least a small potential of failing catastrophically and causing tremendous damage.

3. Fission breaks atomic nuclei apart. Fusion joins small atomic nuclei to create heavier nuclei.

4. Student models should be clear and logical, and their decision should be clearly stated.

SECTION 11.3

CASE STUDY, *page 289*

1. At higher latitudes, the sun is much lower in the sky in the winter, and the days are quite short. Thus, there would not be enough solar energy available to provide adequate heating.

2. Sample answer: Incentives (such as tax breaks or subsidies) could be offered for building or retrofitting homes to be energy efficient. Or disincentives, in the form of taxes or penalties, could be levied against non-efficient homes. Or it could simply become law to require all new homes to be energy efficient. An all-out campaign to popularize energy efficiency might be useful as well.

SECTION 11.3 REVIEW, *page 294*

1. Sample answer: I could conserve energy by walking, using public transportation, or riding a bike rather than driving. I could also adjust the thermostat to run the furnace or air conditioner less.

2. Advantages: It is inexhaustible, free, nonpolluting, and safe. Disadvantages: It requires sunny weather and a large sunlit surface to work. Areas where it could work, such as deserts, are usually far from population centers.

3. They are similar in that both are essentially inexhaustible, nonpolluting, and available to some degree everywhere. Some differences include: solar energy is radiant energy, while wind energy is kinetic energy (the energy of something in motion); wind energy is an indirect form of solar energy and could not exist without it; wind can be used directly to do work, while solar energy must be used indirectly.

4. Answers will vary. Concept maps should be logical and should include the required terms.

5. Answers will vary. Students should select sources appropriate for their regions.

ECOLOG *page 295*

1. Sample answer: Most electricity comes from power plants with generators that are powered by fossil fuels. Hydroelectric dams and nuclear-fission plants are also used to power electric generators.

2. Sample answer: Renewable energy sources, such as solar, wind, and geothermal sources, will be most important in the future. Scientists may have learned how to make nuclear fusion a viable energy source by that time as well.

CHAPTER REVIEW *page 296*

1. Sample answers:

a. Renewable resources are those that are naturally replenished at a useable rate. Nonrenewable resources are not replenished by nature at a useable rate.

b. Nuclear fission is the process of breaking the nucleus of an atom apart. Nuclear fusion is the process of combining small atomic nuclei to form a heavier nucleus.

c. Active solar heating systems require collection devices, fans, and pumps to convert and distribute solar enery. Passive solar systems require no additional devices to function.

2.–12. Answers can be found on pages 296–297.

13. In many places geothermal energy is so far from the Earth's surface that it is effectively out of reach.

14. Yes; because the sun provides the energy that drives the wind and the water cycle and because neither wind nor hydro energy could exist without the sun

15. Countries with other energy supplies would use those first before using nuclear power, which is very expensive. In addition, if countries with few energy resources relied too heavily on fossil fuels, they could become vulnerable to the demands of other nations.

16. Answers will vary. Consistently cloudy or very northerly locations cannot depend heavily on solar energy.

17. Water must fall some distance to provide useful energy. This only happens in hilly or mountainous areas.

18. Higher oil prices stimulated more oil exploration. Thus, more oil was found. A worldwide tax might reduce exploration by reducing demand for the oil and therefore the amount of profits that could be earned. If demand is reduced but exploration continues at more or less the same rate, the number of known reserves might grow.

19. Globally, the Earth's heat that is lost to space or used by humans is continually replenished. However, locally, geothermal energy can be extracted much faster than it can be replenished.

20. An energy source that has a negative net energy yield would be useless because it would require more energy to process than it would yield.

21. Fossil fuels are formed from the remains of ancient organisms, which got their energy from the sun, either directly (in the case of photosynthesizers) or indirectly (in the case of everything else). The energy is "stored" in the form of molecular bonds in the fossil fuels.

22. Humans have used increasingly more energy throughout history. The greater our usage of energy, the higher our level of technology and our standard of living.

23. No. An energy source with a net energy yield of greater than 100 percent would have to somehow yield more energy in using it than it contained, a violation of the law of conservation of energy.

INVESTIGATION *page 299*

3.–4. Answers will vary.

5. Sample design, using sample hypothesis given in item 4 of the student text: Construct two houses that are identical except that one has smaller windows on the side of the home facing the sun. Expose each house to exactly the same conditions, and measure the temperature of the interior of each house after a fixed period of time. If the temperature is greater in the house with larger windows, conclude that the data supported the hypothesis. If the temperature does not vary or is lower in the house with larger windows, conclude that the data did not support the hypothesis.

6.–8. Answers will vary.

9. 10 years

CHAPTER 12

SECTION 12.1

SECTION 12.1 REVIEW, *page 309*

1. Because there are no natural microorganisms capable of decomposing nonbiodegradable wastes, the wastes do not break down easily in the environment. As a result, they persist as waste, whereas materials made from natural resources can be decomposed.

2. Solid waste is any discarded material that is not a liquid or gas; municipal solid waste is solid waste that is produced by households and businesses.

3. A modern landfill is lined with clay and synthetic materials to prevent leachate from leaking into the surrounding ground. Garbage is dumped and compacted on top

of this liner, and a system is installed to collect any leachate that forms from the waste. Vent pipes are also installed to carry methane out of the landfill. Once removed, the methane is either released into the air or burned to produce energy.
4. Advantages: It reduces the amount of solid waste sent to landfills, and it can be used to generate electricity. Disadvantages: Once the materials are burned, the ashes (which are more toxic than ordinary waste) must be disposed of in a landfill; hazardous materials can be released into the air; and communities that build incinerators may not be as motivated to recycle or reduce their wastes.
5. Sample answer: One nonbiodegradable product that I use is a plastic shopping bag. A biodegradable substitute for that product is a reusable canvas shopping bag.

Section 12.2

Case Study, *page 313*
1. By choosing recycled and recyclable products from one producer, environmentally conscious shoppers force other producers to either develop similar products or lose a share of the market.
2. A person should care because that product will eventually be discarded as waste, and some products, once discarded, have a greater impact on the environment than do other discarded products.

Section 12.2 Review, *page 315*
1. Sample answer: Use cloth towels instead of paper towels, give my newspaper to someone else to read after I have read it, recycle, and compost
2. The main drawback is that they never completely decompose; instead, they simply break down into smaller bits of plastic. One potential benefit, however, is that the bits of plastic become so small that they cannot harm an animal that might otherwise have choked on the material.
3. Composting decomposes organic wastes into a material that can be used for other purposes, such as garden fertilizer. In this way food and yard wastes are recycled.

Section 12.3

Case Study, *page 319*
1. Answers will vary. Students should demonstrate the ability to use logical reasoning to give a clear and concise viewpoint.
2. For more than 20 years after chemical wastes were first buried in Love Canal, the wastes resurfaced as thick, black sludge and contaminated air and water supplies. Today the area is clean only because of costly efforts to remove the wastes. But even cleaning up Love Canal didn't completely get rid of the wastes. They were simply moved to another location.

Section 12.3 Review, *page 322*
1. Hazardous waste is waste that is toxic, corrosive, or prone to explosion.
2. Land disposal (deep-well injection or surface impoundment). One danger of this method is that disposal sites may develop leaks and contaminate the surrounding environment. Another danger is that the material never goes away—it is simply stored. Thus, unforseen changes in the environment—such as an earthquake—could cause serious contamination.
3. Manufacturers could redesign factories to produce less hazardous waste or none at all; companies could cooperate to reuse discarded hazardous wastes; and some chemicals could be treated with other chemicals to make them less hazardous.
4. Answers will vary. Students should demonstrate the ability to use logical reasoning to give a clear and concise viewpoint.
5. Sample answer: Yes; one household hazardous waste that I have disposed of is house paint. Next time, I could take unused paint to a hazardous-waste collection facility rather than pouring it down the drain or dumping the can into the trash.

EcoLog *page 323*
1. Paper
2. Answers will vary. Some waste-reducing activities include recycling, using products more efficiently (such as writing on both sides of paper), composting, and choosing more "Earth-friendly" products when shopping.

Chapter Review *page 324*
1. Sample answers:
   a. Biodegradable materials are natural materials that can be broken down by living organisms, while synthetic materials are artificially made and cannot be broken down by living organisms.
   b. Solid waste is any discarded material that is not a liquid or a gas. Municipal solid waste is the solid waste produced by homes and businesses.
   c. Leachate results when hazardous waste materials are deposited in a landfill and water filters through the landfill, picking up contaminants from the waste. Hazardous waste is any waste material that is toxic, highly corrosive, or explosive.
2.–13. Answers can be found on pages 324–325.
14. The agricultural industry
15. They contain hazardous seepage from the landfill and keep it from contaminating the environment.
16. Compost consists of organic waste that has been allowed to decompose. It can be used to nourish the soil, and it is material that will not have to be disposed of in a landfill.
17. It allows the EPA to sue owners of hazardous-waste sites for the cost of cleaning up the site. The act also creates a fund to pay for hazardous-waste cleanup.
18. Motor oil that is dumped onto the ground can seep into groundwater that may be part of a city's or individual's water supply. Oil that is dumped down a sink or gutter will probably end up in a stream or river that may contribute to a city's supply of drinking water.
19. Between 1970 and 1975. During a recession, people have less money to spend. In such a situation, people are more likely to conserve or reuse resources, and fewer items end up as waste. After the recession was over, people had more spending money and less incentive to conserve or reuse.
20. At least 18 percent
21. Recycled products may look the same, be just as durable, and maintain the same level of quality as products made with new materials. Yet, because they require less energy and few, if any, new materials to make, they are actually better than products made entirely with new materials.
22. The complete ban of plastic would have a devastating effect on both the environment and society because plastics are widely used for many important purposes. For example, plastic liners help prevent landfill leachate from seeping into our groundwater supplies and causing contamination. Plastics also have many medical applications because they are not harmful to the body and can be formed into any shape.
23. About 15 years
24. Some products take 100 years or more to degrade (if they can ever degrade). If a person purchases many of

these long-lasting products, his or her shopping habits will have a long-term negative effect on the environment. These products will endure as waste in landfills or wherever they are ultimately deposited. If a person avoids purchasing such products, however, his or her shopping habits may have a long-term positive effect on the environment. This is because he or she may have contributed less waste that will still be around in 100 years.
25. In 1994 the states with the most hazardous-waste sites included New Jersey, Pennsylvania, California, New York, and Michigan. The states with the fewest hazardous-waste sites included Nevada, Hawaii, Mississippi, North Dakota, and Wyoming. Based on this information, it is likely that the number of hazardous-waste sites in a state is related to the number of people in a state and the amount of industry present there.

INVESTIGATION *page 327*

3.–4. Answers will vary, but student observations should be clear and complete.
5. Answers will vary, but useful data will depend on careful and thorough observations. Ensure that students use the given formula to arrive at their final estimate.
6.–7. Answers will vary, but useful data will depend on careful and thorough observations.
8. Answers will vary depending on specific circumstances. Letters should be clear, concise, and complete.

POINTS OF VIEW *page 329*

1. Answers will vary. Student models should be clear and logical, and their decision should be clearly stated.
2. Answers will vary depending on your particular circumstances. If there is a power plant in your community, students should demonstrate the ability to use logical reasoning to give a clear and concise viewpoint.

CHAPTER 13

SECTION 13.1

CASE STUDY, *page 333*

1. They are similar in that they both show exponential growth in the size of a population. The graph of the reindeer population, however, shows what happens when a population exceeds its carrying capacity—a sharp decline. Figure 13-3 shows no decline.
2. The predators probably would have kept the reindeer population from growing too large, so their population would have stabilized instead of crashing.

SECTION 13.1 REVIEW, *page 335*

1. Limiting resources for animals include food, water, shelter, and nesting sites. Limiting resources for plants include water, sunlight, and soil nutrients.
2. Because as the number of individuals increases, competition for resources increases while the availability of resources usually stays the same. The larger the population, the more individuals there are that will die from lack of resources. When this happens, birth rates fall, and the population decreases and finally stabilizes.
3. If resources become more abundant, then the carrying capacity of the environment will probably increase. Conversely, if resources become degraded or destroyed, then the carrying capacity of the environment will decrease.
4. Answers will vary. Concept map should be logical and should include the required terms.
5. The availability of resources such as food, water, shelter, and medicine

SECTION 13.2

FIELD ACTIVITY, *page 339*

Answers will vary. Most urban areas will experience an increased rate of growth in coming years. This will affect all natural resources, including land, water and energy supplies, and the plants and animals native to the area.

CASE STUDY, *page 339*

1. Sample answer: Immigrants directly influence the population by adding to the total number of people living in the United States. Indirect effects include the increased use of available resources and infrastructure.
2. Sample answer: The main factors will probably be the availability of jobs and of space.

FIGURE 13-13 CAPTION, *page 341*

Those young people will likely cause a huge increase in the populations of developing countries because many of them will eventually have children.

SECTION 13.2 REVIEW, *page 341*

1. Children could provide additional labor for the farm, increasing the family's productivity.
2. Improvements in agriculture, health care, sanitation and hygiene, and nutrition
3. The average included the large numbers of children who died, thus skewing the numbers.
4. Sample answer: Educated women find that they have alternative opportunities and can contribute to their family's economic prosperity by holding a job and by having fewer children; men and women discover that they need not have a large number of children to ensure that a few survive; and they also learn about methods of contraception.

SECTION 13.3

SECTION 13.3 REVIEW, *page 346*

1. In many countries most of the population is young and of childbearing age. Even if these young adults choose to have only one or two children, there are so many of them that the population will continue to increase throughout their reproductive years and for the foreseeable future.
2. Sample answer: As the human population increases, wilderness is converted to farmland, water is diverted for irrigation or polluted by industries (destroying aquatic ecosystems), and natural areas are used by people who need a place to live (thus displacing wildlife).
3. Friction between people competing for the same resources, which sometimes results in increased crime and violence
4. Answers will vary. Concept maps should be logical and should include the required terms.
5. Criteria may include availability of resources, degree of environmental degradation, global standards of living, and number of conflicts over natural resources.

EcoLog *page 347*

1. Sample answer: No one knows for sure what the Earth's carrying capacity is, but if the human population exceeds the number of people who could be supported by the Earth's resources, there would be too many people.
2. Sample answer: As a region's population outgrows its resources, it comes to depend on the resources of other areas. Thus, any area may be greatly affected by population growth somewhere else. Also, overpopulation might affect distant areas if the environment is severely damaged and both widespread pollution, carried by air and water currents, and environmental refugees result.

Chapter Review *page 348*

1. Sample answers:
a. Carrying capacity is the maximum population that an environment can support indefinitely. The biotic potential of a species is the rate of population growth that would occur if all offspring survived and reproduced at their theoretical maximum rate.
b. Environmental refugees are people who move from one country to another because of environmental problems, such as pollution, desertification, and overcrowding, in their home country. Hunter-gatherers are nomadic people who move from one location to another in search of food and other resources.
c. Demographic transition refers to changes in population growth patterns that occur at different stages of a country's technological development. The agricultural revolution refers to changes in the amount of food produced as the result of new agricultural technologies.
2.–10. Answers can be found on pages 348–349.
11. Increased competition for resources could push the habitat beyond its carrying capacity, damaging all species; the new species could outcompete the existing species, causing the population of the existing species to fall; or both species might coexist in smaller numbers.
12. There may not be enough resources in the habitat to support the entire population, so many individuals may die. In addition, the habitat may become degraded by the overuse of its resources. This would lead to a lower carrying capacity for the area.
13. Farming increased the carrying capacity of the environment and allowed more people to survive to adulthood, causing the population to grow.
14. People choose to limit the size of their family because they feel more confident that their children will survive to adulthood; because more people live in urban areas, where children are not needed as labor as much as they might be in agricultural settings; because they have greater access to birth control; and because they usually marry and have children at a later age.
15. Educational and employment opportunities in cities; shortages of fertile land in rural areas due to their overuse
16. Population 1 will grow faster because it has a larger portion of members who are of prime reproductive age.
17. Population and economic growth will slow down and, eventually, stabilize.
18. Under ideal conditions, more offspring survive than are needed to maintain a stable population, causing each succeeding generation to be larger than the previous generation.
19. When someone in an overpopulated region coughs, it is more likely that others will be nearby to be infected; this causes diseases to spread more quickly. Increasing the availability of general health care would slow the spread of disease in these areas, as would simpler methods such as covering the mouth when coughing.

Investigation *page 351*

2. Student histograms should be similar to the general shape of the histogram for Kenya (page 350), with the number of 0–4 year old males and females both equal to 5,000. A correct histogram can be found in the **Laboratory and Field Guide** on page 149.
4. Group A: 12,500; Group B: 7,500; Group C: 5,000; Group D: 3,000
5.–6. If done correctly, student tables will reflect the following data: 0–4 year olds, A.D. 2050—Group A: 125,000; Group B: 50,000; Group C: 8,000; Group D: 8,000. Total, A.D. 2050—Group A: 596,312; Group B: 290,375; Group C: 112,000; Group D: 96,500. Completed tables can be found in the **Laboratory and Field Guide** on pages 150 and 159.
7. Answers can be found in the **Laboratory and Field Guide** on page 160.
8. Completed histograms can be found in the **Laboratory and Field Guide** on page 160.
9. The number of children per woman had a larger effect on population growth.
10. No; the population doubled in less than 50 years even for the population with the lowest growth rate. This is because even if people choose to have only one or two children, there are still so many people having children that the population will continue to increase. However, the growth of populations with only two children per family will taper off as the number of women of child-bearing age stabilizes.

Points of View *page 352*

1. Answers will vary. Student models should be clear and logical, and their decision should be clearly stated.
2. Answers will vary. Students who advocate education may argue that citizens are likely to cooperate more fully with noncoercive policies. Students who feel that penalties and incentives are more effective may state that people generally have such strong desires to produce many offspring that population cannot be realistically controlled if left up to individual choice.

Chapter 14

Section 14.1

Case Study, *page 357*

1. Sample answer: They could make the public in individual countries aware of issues and then let those citizens encourage the policymakers to uphold the regulations. They could also hold meetings in which representatives from each concerned nation could express their concerns about proposed regulations. Those nations could then work together to both draft and enforce international regulations that would be acceptable to all concerned parties.
2. Sample answer: The leaders of some nations believe that by complying with the regulations of the IWC, they will harm the livelihood of those people who depend on harvesting whales for a living. But more importantly, many of them fear that by

complying, they will lose some of their sovereignty, or power to meet the needs of their nation's people.

Section 14.1 Review, *page 359*

1. Sample answer: The fact that many nations were talking for the first time about environmental issues. This is important because it is the first step in cooperating to solve global problems.

2. Answers will vary. Issues might include: the greater use of renewable energy, because burning fossil fuels contributes to widespread pollution and possibly to global warming; the pollution of water, land, and air, because these are serious problems that often cannot be resolved without international cooperation; population control, because the Earth has a limited carrying capacity; and soil conservation, because increasing development and poor agricultural practices are depleting the soils on which humanity depends.

3. Answers will vary. One possibility is that wealthy countries could share their resources (such as money and technology) to help poorer countries develop industries and other livelihoods that are environmentally sound.

Section 14.2

Section 14.2 Review, *page 364*

1. It can specify the loss of habitat, changes in air and water quality, and other environmental changes that may directly and indirectly affect people. Citizens can comment on the statement and indicate how they think the project affects them.

2. Local level. Many decisions, such as those about preservation and development, are made most frequently at the local level.

3. Answers will vary. Concept maps should be logical and should include the required terms.

4. Answers will vary. Students should demonstrate the ability to use logical reasoning to give a clear and concise viewpoint.

EcoLog *page 365*

1. Sample answers: I think that there will be a much larger human population on Earth, and as a result, existing environmental problems will be worse. On the other hand, I think that international cooperation, technological improvements, and a better understanding among humans about the environment will increase our ability and willingness to solve environmental problems. By considering the environment when I make decisions about the type of lifestyle I would like to lead and by educating my friends and family about environmental issues, I think that I can contribute to a healthier future for Earth.

2. Sample answer: The local level, because local governments are often more responsive to individuals than are state governments or the federal government

Chapter Review *page 366*

1. An environmental impact statement (EIS) is a report that must be filed by a federal agency to accompany any proposed legislation or project that would significantly affect the quality of the environment. An EIS states the need for a project, describes its environmental impact, and explains how the impact can be minimized. After an EIS is drafted, it must be available to the public for at least 90 days before the project begins. The public then has at least 45 days to comment on the project. Comments made by the public must be taken into consideration when the final EIS is written.

2.–12. Answers can be found on pages 366–367.

13. They bring together international policymakers who can cooperate to resolve important environmental issues.

14. The federal government establishes minimal guidelines for environmental quality in order to allow state and local governments control over issues that affect the environment on a smaller scale.

15. Environmental issues may have a higher priority in wealthy nations because basic political and economic issues may be less pressing there than in poor nations, where governments' most urgent concerns may be feeding their people or resolving armed conflict.

16. One way to exert influence on the national level is to call or write U.S. representatives and senators. Another way is to join organizations that lobby Congress. At the state level, citizens can attend public hearings on proposed projects that affect the environment. Citizens can also contact their state representative or appropriate state department. On the local level, citizens can organize among themselves, forming coalitions to influence local government.

17. Sample explanation: One reason is that top-level politicians may be more concerned about preventing a foreign power from trying to conquer the United States than are citizens. This may be because if the United States is confronted by another nation, high-level politicians will be held responsible for guaranteeing the welfare of the United States. Citizens may not consider preparation for a potential war a priority when faced with daily realities of poor health, poverty, and environmental degradation. Another motivation for politicians to invest federal money into the military is that the military is an important component of the U.S. economy.

18. Sample answer: One could stress the long-term perspective that without a healthy environment they will not have a healthy nation.

19. Local action can have a global impact in many different ways. For example, local activities, if publicized, can inspire other communities to take similar action. Also, local movements begun at the grass-roots level may spread to become larger national or even international movements. Also, if a community has a resource or other feature that the larger world depends on or is affected by, the actions of the community may have direct effects on the world at large.

20. Students should find that this claim is a valid one. If the United States had followed current patterns of environmental responsibility 100 years ago, its environmental record today would probably be much improved. However, students should suggest that by learning from the mistakes of developed nations, developing nations can better protect the world as a whole.

Investigation *page 368*

Answers will vary depending on specific issues analyzed. Ensure that students have briefly summarized the article, stated what the issue is about, explained who is involved in the controversy, and listed the article's source. They should also demonstrate an understanding of how environmental issues can be best discussed in writing. For more suggestions, refer to Teaching Strategies for the Investigation on page T140.

HOLT

ENVIRONMENTAL SCIENCE

Karen Arms

This book was printed with soy-based ink on acid-free recycled content paper, containing MORE THAN 10% POSTCONSUMER WASTE.

HOLT, RINEHART AND WINSTON
Harcourt Brace & Company
Austin • New York • Orlando • Atlanta • San Francisco • Boston • Dallas • Toronto • London

STAFF CREDITS

Requests for permission to make copies of any part of the work should be mailed to:

Permissions Department
Holt, Rinehart and Winston, Inc.
6277 Sea Harbor Drive
Orlando, Florida 32887-6777

For permission to reprint copyrighted material, grateful acknowledgment is made to the following source:

The Cousteau Society, Inc.: From "Coastal Areas Swamped by Development" by Paul Clancy from *Calypso Log,* vol. 21, no. 2, April 1994. Copyright © 1994 by The Cousteau Society, Inc.

Printed in the United States of America

ISBN 0-03-003133-8

2 3 4 5 6 032 98 97 96 95

ACKNOWLEDGMENTS

Teacher Reviewers

Pedro Alaniz
South San Antonio High School
San Antonio, Texas

Lyn Bayer
Horticulture Department
West County Technical School
Chesterfield, Missouri

Kimberly Berg
Salado High School
Salado, Texas

Claudia Fowler
Science Coordinator/
Educational Services
Louisiana Public Broadcasting
Baton Rouge, Louisiana

William Glover
Austin High School
Austin, Texas

James Kraft
Green Bay East High School
Green Bay, Wisconsin

Janis Lariviere
Westlake High School
Alternative Learning Center
Austin, Texas

Clifford Lerner
Keene High School
Keene, New Hampshire

Sheila Lightbourne
Choctawhatchee High School
Ft. Walton Beach, Florida

Michael W. Lubich
Mapletown High School
Greensboro, Pennsylvania

Elizabeth A. Moore
Oak Ridge High School
Orlando, Florida

Sandra Seim Tauer
Derby Middle School
Derby, Kansas

John Michael Trimble
Corona Del Sol High School
Tempe, Arizona

Celia Wesenberg
Ponderosa High School
Shingle Springs, California

University and Government Reviewers

Hugh C. Allen
Miami-Dade Community College
Miami, Florida

David M. Armstrong, Ph.D.
University of Colorado
Boulder, Colorado

Judith Banister, Ph.D.
Center for International
Research
U.S. Bureau of the Census
Washington, D.C.

Bruce Briegleb
National Center for
Atmospheric Research
Boulder, Colorado

Larry Canter, Ph.D.
Environmental and
Groundwater Institute
University of Oklahoma
Norman, Oklahoma

Peter Connell, Ph.D.
Lawrence Livermore
National Laboratories
Livermore, California

Thomas J. Givnish, Ph.D.
Environmental Studies
University of
Wisconsin–Madison
Madison, Wisconsin

Robert Goodland, Ph.D.
Department of Environment
The World Bank
Washington, D.C.

Thomas H. Gorey
Bureau of Land Management
Office of Public Affairs
Washington, D.C.

David B. Green, Ph.D.
Natural Science Division
Pepperdine University
Malibu, California

John Haaga, Ph.D.
Committee on Population
National Research Council
National Academy of Sciences
Washington, D.C.

Robert J. Heinsohn, Ph.D.
Department of
Mechanical Engineering
Penn State University
University Park, Pennsylvania

Scott E. Hygnstrom, Ph.D.
Department of Forestry,
Fisheries and Wildlife
University of Nebraska
Lincoln, Nebraska

Hugh Iltis, Ph.D.
Department of Botany
University of
Wisconsin–Madison
Madison, Wisconsin

Harvey M. Jacobs, Ph.D.
Department of Urban
and Regional Planning and
Institute for Environmental
Studies
University of
Wisconsin–Madison
Madison, Wisconsin

John L. Kermond, Ph.D.
NOAA–Office of
Global Programs
Silver Spring, Maryland

Mark Kirkpatrick, Ph.D.
Department of Zoology
University of Texas at Austin
Austin, Texas

Karen O. Levy
U.S. EPA–Office of Policy
Analysis Review
Washington, D.C.

David Lombard, Ph.D.
U.S. Department of Energy
Washington, D.C.

Douglas MacCleery
U.S. Department of Agriculture
Timber Management
Washington, D.C.

Joe R. McBride, Ph.D.
Department of Forestry
University of
California at Berkeley
Berkeley, California

Stephen F. Marshall
Department of
Atmospheric Science
University of Washington
Seattle, Washington

Andrew Mason, Ph.D.
Program on Population
East-West Center
Honolulu, Hawaii

Leonard R. Massie, Ph.D.
Department of Agricultural
Engineering
University of
Wisconsin–Madison
Madison, Wisconsin

Molly Harriss Olsen
President's Council on
Sustainable Development
Washington, D.C.

Nestor R. Ortiz, Ph.D.
Nuclear Energy Technology Center
Sandia National Laboratories
Albuquerque, New Mexico

Georgia Parham
U.S. Fish and Wildlife Service
Washington, D.C.

Wayne Pferdehirt
Solid and Hazardous
Waste Education Center
University of
Wisconsin–Madison
Madison, Wisconsin

Kenneth Potter, Ph.D.
Department of Civil and
Environmental Engineering
University of
Wisconsin–Madison
Madison, Wisconsin

Bobby E. Price, Ph.D., P.E.
Department of Civil Engineering
Louisiana Tech University
Ruston, Louisiana

G. Allen Rasmussen, Ph.D.
Department of Range Science
Utah State University
Logan, Utah

Jimmy Richardson, Ph.D.
Department of Soil Science
North Dakota State University
Fargo, North Dakota

Armin Rosencranz, Ph.D.
Pacific Environment and
Resources Center
Sausalito, California

Norman Rostocker, Ph.D.
Department of Physics
University of California, Irvine
Irvine, California

Daniel Sivek, Ph.D.
Wisconsin Center of
Environmental Education
Learning Resources Center
University of
Wisconsin–Stevens Point
Stevens Point, Wisconsin

Wayne B. Solly
Chief: Branch of
Water-Use Information
U.S. Department
of the Interior
Geological Survey
Atlanta, Georgia

William Thwaites, Ph.D.
Department of Biology
College of Sciences
San Diego State University
San Diego, California

Kurt Usowski
U.S. Department of Housing
and Urban Development
Office of Policy and
Development Research
Washington, D.C.

William Vencill, Ph.D.
Crop and Soil Sciences
University of Georgia
Athens, Georgia

Dennis Yockers, Ph.D.
Wisconsin Center of
Environmental Education
Learning Resources Center
University of
Wisconsin–Stevens Point
Stevens Point, Wisconsin

Ali Azimi-Zonooz, Ph.D.
Department of Civil and
Materials Engineering
University of Illinois at Chicago
Chicago, Illinois

Curriculum Consultants

John F. Disinger, Ph.D.
School of Natural Resources
Ohio State University
Columbus, Ohio

Harold R. Hungerford, Ph.D.
Department of Curriculum
and Instruction
Southern Illinois University
Carbondale, Illinois

John Padalino
National Science Teacher's
Association
Task Force on Environmental
Education
Pocono Environmental Center
Dingmans Ferry, Pennsylvania

Field-Test Teachers

Karolyn Adams
San Marcos High School
San Marcos, Texas

Jim Cramer
Brookfield East High School
Brookfield, Wisconsin

Cheryl Frazier
Madison High School
San Antonio, Texas

Donna Kerlin
Altoona Area School District
Altoona, Pennsylvania

Sheila Lightbourne
Choctawhatchee High School
Ft. Walton Beach, Florida

Betty Neitzke
Minnetonka High School
Minnetonka, Minnesota

George Newberry
Sparrows Point High School
Baltimore, Maryland

Tracy Patsch
Nyack High School
Nyack, New York

Marie E. Rediess
Algonac High School
Algonac, Michigan

Barbara Rothstein, Ph.D.
North Miami Beach High School
Miami, Florida

Diane Savage
Nashua High School
Nashua, New Hampshire

Shirley Schoenberger
Santa Rita High School
Tucson, Arizona

Anne Tweed
Eaglecrest High School
Aurora, Colorado

Carol Wagner
Pflugerville High School
Pflugerville, Texas

Douglas Young
Coronado High School
Lubbock, Texas

Contributing Writers

Letitia Blalock
Marshall Frech
Natalie Goldstein
Ann Hoffman Harris
Jacquelyn Jarzem, Ph.D.
Mitchell Leslie

HOLT ENVIRONMENTAL SCIENCE

CONTENTS

▲ To learn how mussels have been useful in studying river pollutants, turn to page 3.

CHAPTER 1

CHAPTER 2

▲ When species interact, even lions have something to fear. Learn more on page 38.

▲ To find out why jack pines depend on forest fires for survival, see page 68.

▲ This biome supports more animal and plant species than any other in the world. Read more on page 80.

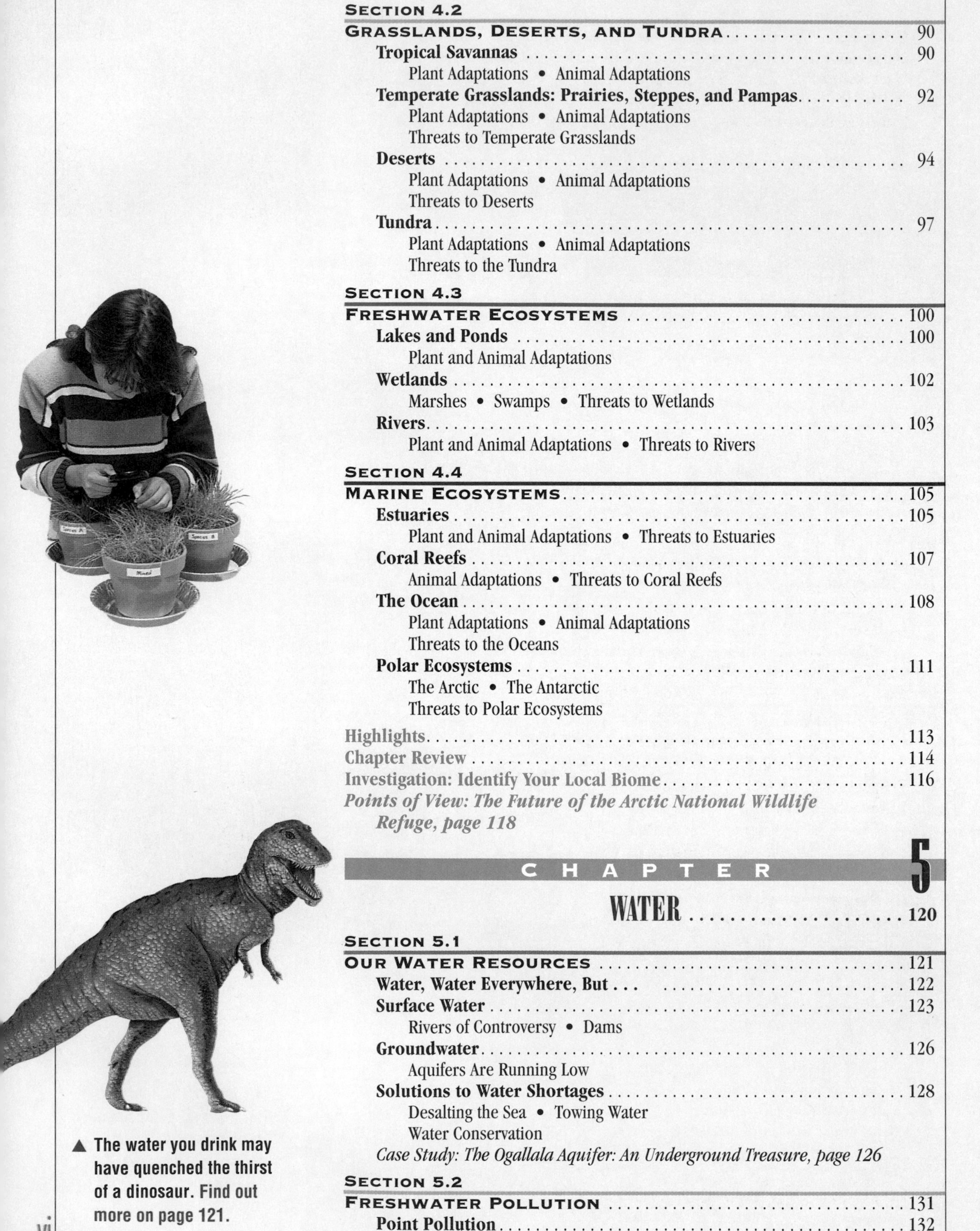

▲ The water you drink may have quenched the thirst of a dinosaur. Find out more on page 121.

▲ **Auto exhaust may have helped kill these trees.** Learn how on page 161.

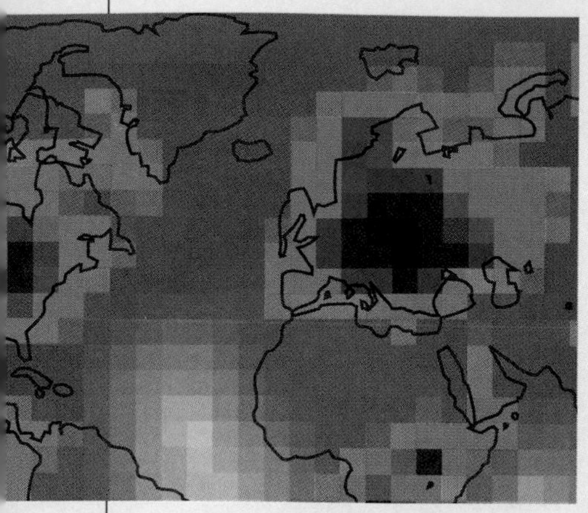

▲ Turn to page 182 to find out how computers are being used to study global warming.

▲ Did you know that American suburbs hold more people than the cities and rural areas combined? Find out more on page 200.

▲ What can be done to stop these rampaging beasts? Turn to page 240 for details.

▼ Chances are, an endangered species lives near you. Learn more on page 262.

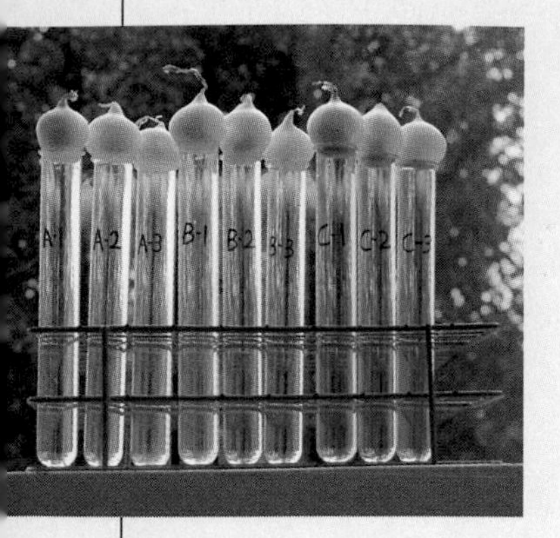

CHAPTER 11

ENERGY 278

▲ Farms that harvest the wind? See page 291.

CHAPTER 12

WASTE 302

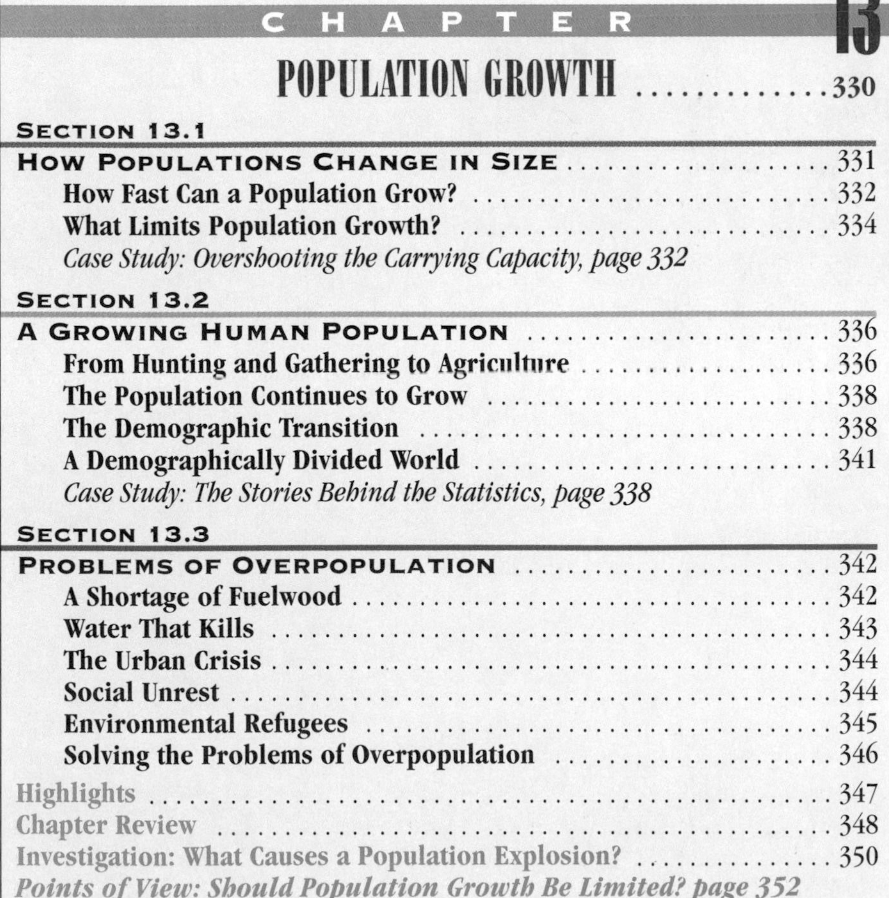

▼ How long do you think this ear of corn has been in a landfill? You'd be surprised. See page 308.

▲ How have Australian rabbits helped us learn how many people can fit in our world? Find out on page 335.

▲ A house built of old tires and soda cans? See page 370.

▲ Play in the dirt and get paid for it! James Bailey tells you how on page 376.

▲ Why build a house for bats in your yard? Find out on page 400.

CHAPTER 14

TOWARD A SUSTAINABLE FUTURE........354

ENVIRONMENTAL CAREERS

ECOSKILLS

REFERENCE SECTION

INVESTIGATIONS

▲ Take a shot at revolutionizing the fast-food industry on page 326.

FEATURES

CASE STUDIES

▲ This fish almost cost the taxpayers 50 million dollars. Find out how on page 264.

POINTS OF VIEW

MAKING A DIFFERENCE

▲ Wolves set free in your backyard? You decide on page 76.

To the Student

Like all other sciences, environmental science is a process of satisfying curiosities about why things are the way they are and about how things happen the way they do. **For example, in studying environmental science, you may discover the answers to the following questions.**

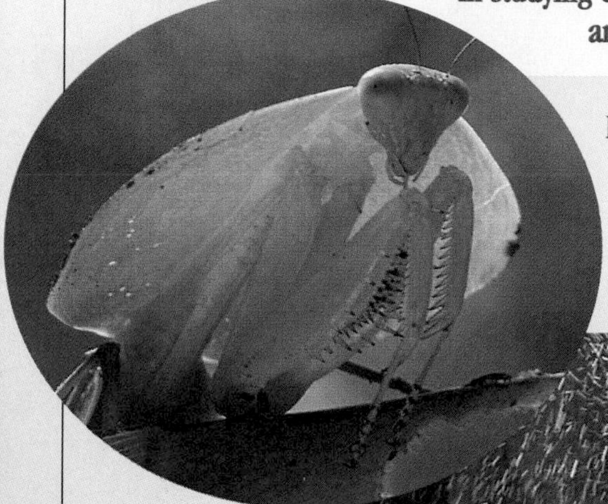

How could the demise of this seemingly unimportant insect cause severe damage to the rain forest in which it lives?

How could the watering of this lawn affect the water quality of a nearby stream?

How could a population of iguanas help save a rain forest from destruction?

How could recycling an aluminum can help save fossil fuels and reduce both air and water pollution?

How could the exhaust from these cars in New York contribute to the decline of salmon in Canada?

How could the birth of these otters indicate a healthy ecosystem?

You may not have expected to have the questions above answered in your study of environmental science. Nevertheless, the answers to these questions and many more like them help define this unusual and exciting area of study. In learning about the various aspects of our environment, you will quickly come to understand how interdependent life on Earth is.

In many cases, we know so little about environmental interactions that we can't even begin to predict long-term effects. For example, it took nearly 15 years of study before we understood the relationship between the pesticide DDT and the declining populations of bald eagles. It may take even longer to fully understand the relationship between CFCs and ozone. And what usually happens is that the answer to one question leads to a string of new questions.

Perhaps the most important question to ask at the beginning of this course is: **What do you hope to get out of this environmental science text?**

You may be interested in science and want to know more about the inner workings of our environment. Or you may be interested in learning more about human impact on the environment and what we can do to reduce the negative consequences. You may even want to know more about the environment firsthand so that you can decipher environmental issues for yourself, rather than simply accepting someone else's point of view.

Challenge

Regardless of your reasons for taking this course, my challenge to you is to think for yourself. In reading this textbook, you not only will learn a lot about science, you will learn about the complex issues facing our environment. You will explore different points of view and be exposed to a variety of differing opinions. Don't feel that you have to accept any particular opinion as your own. As your knowledge and skills in environmental science grow, so will your ability to draw your own conclusions.

Karen Arms

Safety Symbols

The following safety symbols will appear in this text when you are asked to perform a procedure requiring extra precautions. Once you have familiarized yourself with these safety symbols, turn to pages 409–412 for safety guidelines to use in all of your environmental science laboratory work.

WEAR APPROVED CHEMICAL SAFETY GOGGLES. Wear goggles when working with a chemical or solution, when heating substances, or when using any mechanical device.

WEAR A LABORATORY APRON OR LABORATORY COAT. Wear a laboratory apron or coat to prevent chemicals or chemical solutions from contacting skin or street clothes.

WEAR GLOVES. Wear gloves when working with chemicals, stains, or wild (unknown) plants or animals.

SHARP/POINTED OBJECT. Use extreme care with all sharp instruments such as scalpels, sharp probes, and knives. Do not cut objects while holding them in your hand; always place them on a suitable work surface. Never use double-edged razors in the laboratory.

ELECTRICAL HAZARD. To avoid electric shock, never use equipment with frayed cords. Tape electrical cords to work surfaces to ensure that equipment cannot fall from a table. Also, never use electrical equipment around water or with wet hands or clothing. When disconnecting an electrical cord from an outlet, grasp the plug rather than the cord.

DANGEROUS CHEMICAL/POISON. Always wear appropriate protective equipment, including eye goggles, gloves, and a laboratory apron, when working with hazardous chemicals. Never taste, touch, or smell any substance, and never bring it close to your eyes unless specifically instructed to do so by your teacher. Never return unused chemicals to their original containers. Do not mix any chemicals unless your teacher tells you to do so. Also, never pour water into a strong acid or base because this may produce heat and spattering. Instead, add the acid or base slowly to water. If you get any acid or base on your skin, flush the area with water, and contact your teacher right away. Finally, report any chemical spill to your teacher immediately.

FLAME/HEAT. Whenever possible, use a hot plate for heating rather than a laboratory burner. Use test-tube holders, tongs, or heavy gloves to handle hot items. Do not put your hands or face over any boiling liquid. When heating chemicals, be sure the containers are made of heat-proof glass. Also, never point a heated test tube or any other container at anyone. Be sure to turn off a heat source when you are finished with it.

GLASSWARE. Inspect glassware before use; never use chipped or cracked glassware. Do not attempt to insert glass tubing into a rubber stopper without specific instructions from your teacher. Clean up broken glass with tongs and a brush and dustpan. Discard the pieces in a labeled "sharps" container.

PLANTS. Do not ingest any plant part used in the laboratory, especially seeds. Do not rub any sap or plant juice on your skin, eyes, or mucous membranes. Wear disposable polyethylene gloves when handling any wild plant. Wash hands thoroughly after handling any plant part. Avoid the smoke of burning plants. Finally, do not pick wildflowers or other plants unless instructed to do so by your teacher.

LIVE ANIMALS. Do not touch or approach any animal in the wild. Always obtain your teacher's permission before bringing any animal into the school building. Handle animals only as your teacher directs. Always treat animals carefully and with respect.

BIOHAZARD. Wear appropriate personal protection, including disposable neoprene gloves and other gear provided by your teacher. Clean your work area with disinfectant before you begin and after you complete the investigation. Do not touch your face or rub your skin, eyes, or mucous membranes. Wash your hands thoroughly after use. Dispose of materials as instructed by your teacher.

Also read the General Guidelines for Laboratory Safety on pages 409–412.

CONCEPT MAPPING

A WAY TO CONNECT IDEAS

This book will introduce you to new ideas and information about environmental science. A concept map can help you understand these ideas by showing you how they are connected to each other. Concept mapping is a visual method of establishing relationships between concepts, so it helps you to see the "big picture." In a concept map, ideas are expressed as words or phrases that are connected by lines explaining their relationships. By understanding these relationships, you will also be better able to remember the concepts.

Practice is the key to good concept mapping—the more you do it, the better you'll become at relating concepts and ideas to each other. Your concept map may look different from those drawn by your classmates, even if you're mapping the same concepts. That's okay—different people may see different relationships between concepts.

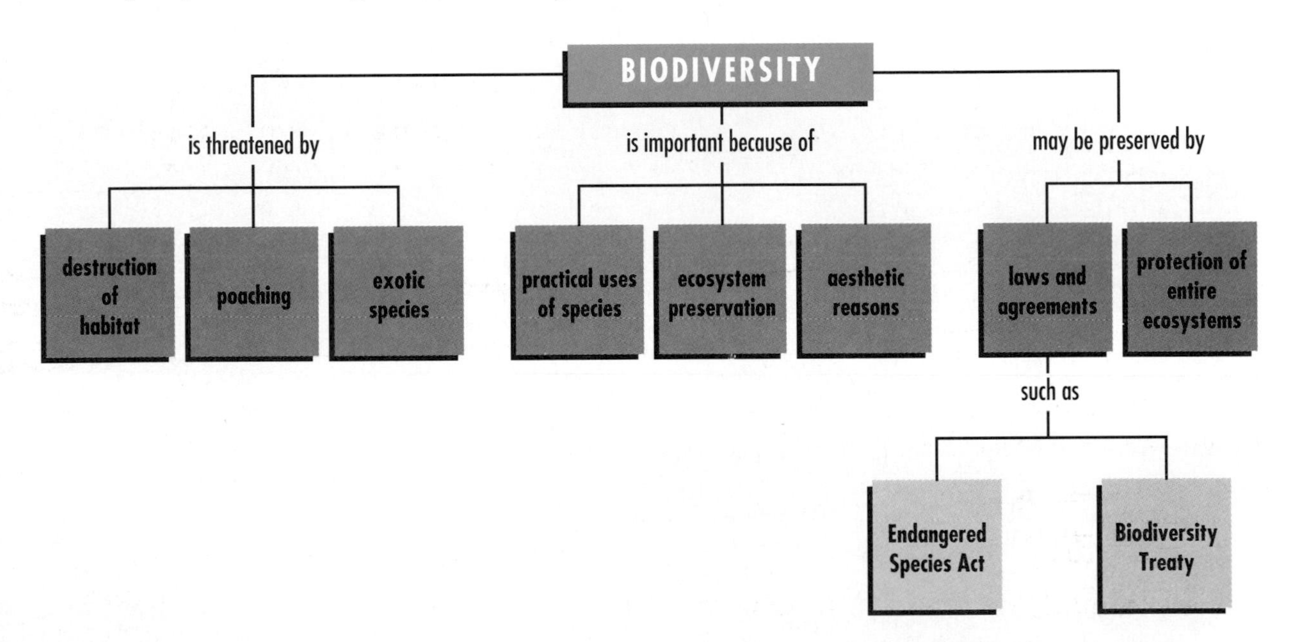

HOW TO MAKE A CONCEPT MAP

1. DETERMINE THE MAIN IDEAS OR CONCEPTS.

Make a list of the main concepts or topics in the material for which you are making a concept map. For your first few maps, you might find it helpful to write each concept on a separate piece of paper. This way, you can arrange the concepts in as many ways as you like in order to find the correct relationships. After you've done a few concept maps this way, you can go directly from writing your list to actually making the map.

2. PLACE THE CONCEPTS IN ORDER FROM THE MOST GENERAL TO THE MOST SPECIFIC.

Put the most general concept at the top of your concept map. Then look at the other concepts and determine how they relate to the most general one. On your map, arrange the concepts in order from general to specific, keeping in mind that some concepts may be equally specific. When you have mapped all of the concepts, draw a circle or box around each one.

3. DETERMINE THE RELATIONSHIPS BETWEEN THE CONCEPTS.

On your map, connect all of the related concepts with lines. On each connecting line, write a linking word or short phrase that explains the relationship.

TRY A CONCEPT MAP!

Look again at the example on this page. Then draw a concept map using the following terms: fish, all living things, plants, insects, trees, animals, flowers, Earth, birds. Find as many links between the concepts as possible.

CHAPTER 1

ENVIRONMENTAL SCIENCE: A GLOBAL PERSPECTIVE

"All the flowers of all the tomorrows are in the seeds of today."

CHINESE PROVERB

SECTION 1.1

UNDERSTANDING OUR ENVIRONMENT

SECTION 1.2

USING SCIENCE TO SOLVE ENVIRONMENTAL PROBLEMS

SECTION 1.3

MAKING ENVIRONMENTAL DECISIONS

EcoLog

Before you read this chapter, take a few minutes to answer the following questions in your EcoLog.

1. What do you consider to be our major environmental problems?
2. How can science help solve environmental problems?

SECTION 1.1

UNDERSTANDING OUR ENVIRONMENT

AFTER READING THIS SECTION YOU SHOULD BE ABLE TO

❶ describe the three categories into which most environmental problems fall.

❷ explain how the population crisis and the consumption crisis contribute to environmental problems.

❸ distinguish between renewable and nonrenewable resources.

The students you see in Figure 1-1 are searching the Ashuelot River in New Hampshire for dwarf wedge mussels, a kind of mollusk. The mussels, once abundant in the river, are now in danger of dying off completely—and the students from Keene High School want to know why.

To find out what's causing the decline in the number of mussels, the students test water samples from different parts of the river. Could the problem be that sewage is flowing into the river and contaminating the water? Or perhaps fertilizer from a nearby golf course is causing algae in the river to grow rapidly and use up much of the oxygen that the mussels need to survive? Another possible explanation is that a small dam on the Ashuelot River is disrupting the reproduction of the mussels.

Figure 1-1 **These students from Keene High School in Keene, New Hampshire, are counting the number of dwarf wedge mussels in this part of the Ashuelot River. They hope that the data they collect will help save this endangered species.**

Each year the Keene High School students present their findings to the U.S. Fish and Wildlife Service and local conservation officials. Their work has been highly praised and widely recognized. Yet the students hope for a more meaningful reward—the recovery of an endangered species, thanks in part to their efforts.

The students' work is part of a relatively new field—**environmental science,** the study of how humans interact with the environment. A major focus of environmental science is solving environmental problems. The term **environment** refers to everything that surrounds us. Our environment includes the natural world as well as things produced by humans.

Solving Environmental Problems

Forty years ago, the average citizen did not think very much about "the environment." During the economic boom years after World War II, many people were happy to move out of the crowded cities into the suburbs that were covering the old pastures and hillsides. Rapidly expanding industries made jobs plentiful, and bigger cars and household electrical appliances made life more comfortable.

In the last few decades, however, many people began to wonder if we might have to pay too high a price for our comfortable lifestyle. They began to notice that more and more rivers and lakes were becoming too polluted to swim in. A brown cloud of air pollution like the one in Figure 1-2 hung over many cities. News reports sounded the alarm of health risks from chemicals in the soil, in the water, and even in the food at the grocery store.

Figure 1-2 Chicago is just one of the many cities in the United States with air pollution.

These days, every informed citizen thinks about the changes in our environment. No one wants to drink poisoned water or breathe polluted air. Many communities are having a hard time figuring out what to do with their garbage. Some people, like the students at Keene High School, have begun to notice that animals they once took for granted are hardly ever seen anymore.

Environmental problems like these can be complicated. Preventing pollution can be expensive, and cleaning it up once it has happened can be even more expensive. Protecting an endangered species may threaten jobs and economic growth in a region. Some problems that affect the entire planet—such as destruction of our forests—seem so complicated that individuals may think there is nothing they can do about them.

But environmental problems can be solved. For example, Seattle's Lake Washington is cleaner and healthier now than it was 25 years ago, as the Case Study on pages 6–7 describes. The bald eagle, shown in Figure 1-3, is now making a comeback from the danger of extinction. Thanks to emissions controls and catalytic converters, new automobiles discharge only a fraction of the air pollutants given off by older models.

Nevertheless, our environmental problems are huge, and they require careful attention and action. The twenty-first century will be a crucial time in human history, a time when we must find solutions that allow people on all parts of our planet to live in a clean, healthy environment and have the resources they need for a good life.

Figure 1-3 The bald eagle was once on the brink of extinction. Today it is making a comeback, thanks to efforts to preserve its habitat and reduce pollution from the pesticide DDT. (The use of DDT has been banned in the United States since 1972.)

What Are Our Main Environmental Problems?

Your community may be facing several different environmental problems right now. Perhaps there is a debate about whether to ban septic tanks and hook everyone up to sewer lines. Or perhaps there is discussion about where the town is going to build a new landfill. The local news may mention from time to time that your city is under an ozone advisory and ask citizens to limit driving. Property owners may be arguing with environmentalists about the importance of a rare bird or insect. Even though there seems to be an unlimited number of environmental problems, almost all of them fall into one of three categories: resource depletion, pollution, and extinction.

Resource Depletion

Any natural substance that living things use can be considered a **natural resource.** Natural resources include such things as sunlight, air, water, soil, minerals, plants, animals, forests, and fossil fuels. A resource is depleted when a large part of it has been used up.

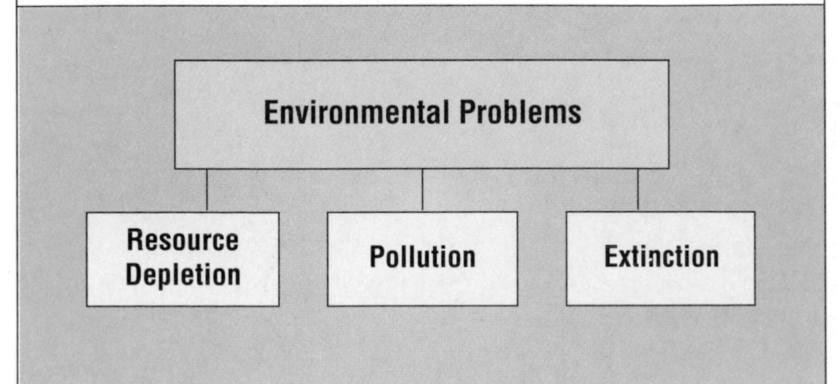

Figure 1-4 Environmental problems tend to fall into the three categories shown at left.

Some resources cannot be replaced. These are called **nonrenewable resources.** No more copper will be formed deep in the Earth after all of the copper ore has been removed from mines like the one shown in Figure 1-5. Similarly, fossil fuels like oil and natural gas were formed over millions of years. If we use them all up in a few hundred years, there will be no more for our descendants to use. The more wasteful and careless we are about how we use these nonrenewable resources, the sooner we will use them up.

Other resources are continually being replaced, even as they are being used. These are called **renewable resources.** No matter how much solar energy we use, sunlight will continue to stream toward the Earth as long as the sun shines. Trees that are cut down to be used for furniture or firewood can be replaced by new trees

LAKE WASHINGTON:
An Environmental Success Story

Seattle is located on a narrow strip of land between two large bodies of water. To the west is Puget Sound, which is part of the Pacific Ocean, and to the east is Lake Washington, a deep, 27-mile-long freshwater lake. During the 1940s and early 1950s, cities on the east side of Lake Washington completed 11 sewer systems that emptied into Lake Washington. Unlike raw sewage, this sewage was "clean" water that did not present a threat to human health. For this reason, both citizens and civic leaders were surprised by research in 1955 showing that the treated sewage was threatening their lake. Scientists working in Dr. W. T. Edmondson's lab at the University of Washington found a bacterium, *Oscillatoria rubescens,* that had never been seen in the lake before.

This was not good news. The sewage was releasing large amounts of phosphates from human wastes and from detergents into the lake. Phosphate acts as "fertilizer" for bacteria and algae. These organisms can grow so rapidly when they are fertilized this way that they cloud the water and form dense mats of green scum. Bacteria that decompose algae when they die use up so much of the oxygen in the water that fish begin to die from suffocation.

Dr. Edmondson knew that in several lakes in Europe, pollution by sewage had been followed by the appearance of *Oscillatoria* and then severe deterioration of the lakes, which became cloudy, smelly, and unable to support fish. When they detected *Oscillatoria* in the lake, the scientists realized that they were seeing the beginning of this process.

About this same time, Seattle set up the Metropolitan Problems Advisory Committee, chaired by James Ellis. Dr. Edmondson wrote Ellis a letter explaining what was happening to the lake and what could be expected in the future if action was not taken. The best solution to the problem seemed to be to quit dumping the sewage into Lake Washington. Instead, the sewage could be collected and carried around the lake to be emptied deep into Puget Sound. Although this may seem like saving one body of water by polluting another one, it was actually a good choice. The sewage had to go somewhere, and diluting it

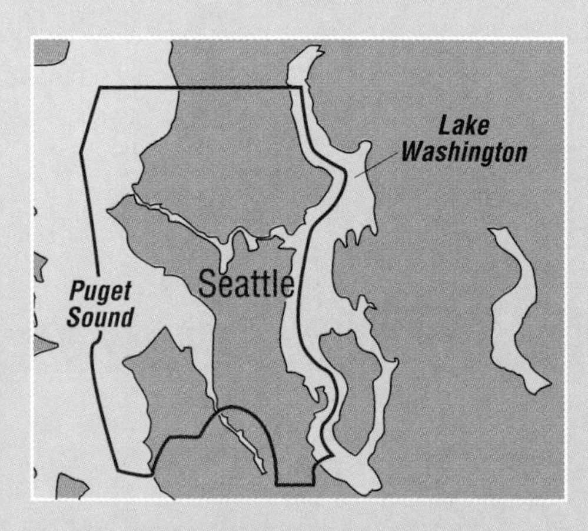

Today, treated sewage from the east side of Seattle is pumped across to Puget Sound, where phosphates and other chemicals become safely diluted in the marine waters.

that grow in their place. A corn crop like the one shown on the next page can be used as food for people or animals and can even be used to make ethanol to power cars. New corn crops can be planted each year.

However, even renewable resources can be depleted. For example, if trees are cut down faster than new ones can grow, we will run out of trees. The damage may be even worse than simply having to wait several decades for new trees. If all of the trees on a mountain are cut down, wind and rain may carry away all of the fertile topsoil. If this happens, new trees may not grow in this place for generations to come. We must be careful about how we use renewable resources, and learn to use them at a rate that allows them to renew themselves.

Figure 1-5 After all the copper has been removed from this mine, no more copper will be formed. To avoid running out of nonrenewable resources like copper, we must use them wisely and recycle whenever possible.

in the Pacific Ocean works much better than letting it build up in an enclosed lake.

All of the smaller cities around the lake had to work together to connect their sewage plants to large lines that would carry the treated sewage to Puget Sound. Since there was no legal way for cities to do this at the time, Ellis successfully worked for passage of a bill in the state legislature to set up boards to handle projects of this kind. Planners estimated that each household would have to pay about $2 per month extra on their wastewater bill to pay for the project. (In the end, the amount turned out to be slightly less.) Newspaper articles and letters to the editor addressed the issue. Public forums and discussion groups were held.

The first sewage plant was hooked up in 1963. Today, the lake is clearer than it has been since scientists began their studies of the lake in the 1930s.

The story of Lake Washington is a fine example of how environmental science and public action work together to solve environmental problems. The science was essential to understand a healthy lake ecosystem, to document changes that were beginning to cause problems, and to make predictions about what would happen if changes were made or if nothing was done. Engineers could offer practical solutions to the problem of moving the sewage. Legislators and civic leaders addressed the legal problems. Volunteers, local media, and local activists provided public education and pressed to get the problem solved quickly. The clear blue waters of Lake Washington stand as a monument to citizens' desires to live in a clean, healthy environment and their ability to work together to make it happen.

Scientists measure the clarity of the lake water with a device called a Secchi disk.

Sailing is just one of the popular activities that Seattle residents can enjoy on Lake Washington.

THINKING CRITICALLY

1. *Analyzing Processes* Explain how each person and group played a crucial role in the cleanup of Lake Washington.
2. *Analyzing Relationships* How was the scientists' work similar to the work of the Keene High School students you read about in this section?

Answers to Thinking Critically can be found on page T141.

Figure 1-6 **This corn crop is a renewable resource. If we use renewable resources no faster than nature replaces them, there will be enough of these resources in the future.**

Pollution Pollution is the poisoning of our air, water, or soil. Some pollutants are the unintentional byproducts of human activities, such as the nitrous oxides and carbon monoxide from automobile exhaust. Other pollutants, such as pesticides, are toxic substances that humans have intentionally created. Unfortunately, many can harm the environment as they build up in the water and soil.

Many pollutants are dangerous to human health. For example, radiation released from a nuclear power plant accident may cause the rate of cancers to increase, and mercury dumped into the oceans can cause birth defects and nerve damage.

Chicken of the Trees

How can giant lizards help save a rain forest? **Turn to pages 30–31 for the answer.**

Extinction Scientists estimate that thousands of species, such as the birds shown in Figure 1-7, are becoming extinct every year—many without ever having been named or studied. Extinction means that the last individual member of a species has died and the species is gone forever. Although species have become extinct throughout the history of the Earth, they are probably disappearing faster today than at any other time in history.

The rapid disappearance of so many species from the Earth is considered to be one of the most significant environmental challenges we face today. Most species that are becoming extinct are dying because their natural homes are being destroyed or polluted.

Figure 1-7 **These birds will never be seen on Earth again, because they are now extinct.**

a. passenger pigeon
b. dodo
c. great auk
d. Hawaiian o-o-a-a
e. dusky seaside sparrow
f. Atitlan giant pied-billed grebe

Figure 1-8 The amount of carbon dioxide in our planet's atmosphere is increasing, which may cause the climate of the Earth to change. The burning rain forests in Brazil (left) and the cars in Atlanta, Georgia (below), both contribute to the problem.

A Global Perspective

Many environmental problems are global problems. The sulfur dioxide released by coal-fired electric generators in the American Midwest comes back to Earth as acid rain falling on Canada. The millions of cars driven by people in America and Europe are increasing the amount of carbon dioxide in the atmosphere, which may cause the climate of the entire Earth to change. The destruction of tropical rain forests in South America and elsewhere is also contributing to the increase in carbon dioxide. Chemicals called chlorofluorocarbons, produced in many countries all over the world, are harming the ozone shield in the atmosphere, which helps to protect all of Earth's inhabitants from the sun's harmful rays.

When thinking about global environmental problems, it may help to visualize the **biosphere,** the thin layer of life around the Earth. The biosphere extends from the surface of the Earth to about

Figure 1-9 Trees and other plants remove carbon dioxide from the air.

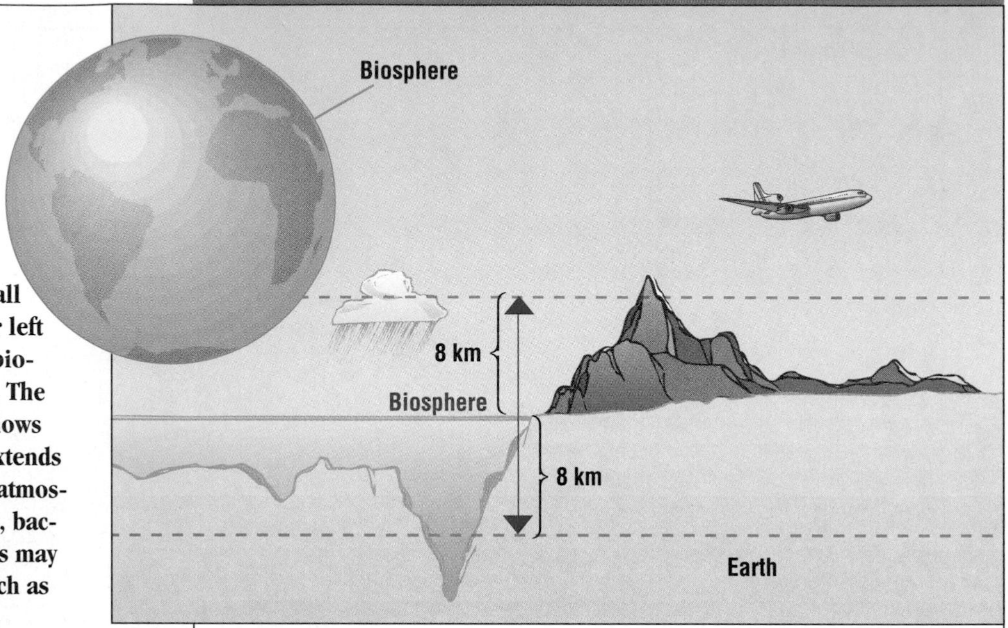

Figure 1-10 The small drawing at the upper left shows how thin the biosphere (in green) is. The larger illustration shows that the biosphere extends about 8 km into the atmosphere, where insects, bacteria, and plant seeds may be found, and as much as 8 km into the ocean.

8 km (about 5 mi.) above the surface and to about 8 km into the deepest part of the ocean. Although this may seem like a large area, Figure 1-10 shows just how thin this layer of life is. Within the biosphere, all living things—including humans—exist in a delicate relationship with each other and with the nonliving things necessary for their survival.

Developed and Developing Countries Although all humans live in the biosphere, people in different countries have different immediate needs and priorities. Most of the world's nations can be categorized into one of two groups: developed countries and

Figure 1-11 People in developing countries such as Guatemala (above) have different priorities from people in developed countries such as Japan (right). All people, however, will share the consequences of global environmental problems.

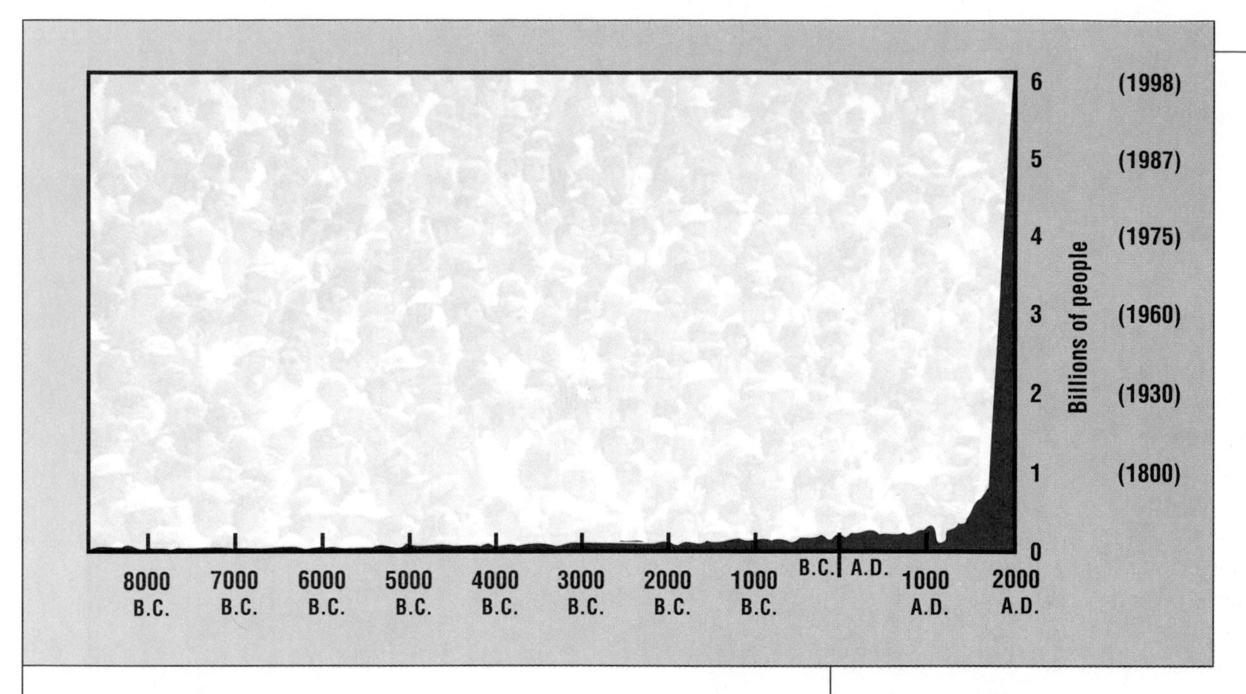

Figure 1-12 The human population is increasing exponentially.

developing countries. The highly industrialized countries, whose citizens have high average incomes, are known as **developed countries.** The major developed countries are the United States, Canada, Japan, Australia, New Zealand, and the countries of western Europe.

The **developing countries** are less industrialized, and their citizens have a much lower average income. For example, the average farmer in India makes only $130 a year, and the typical city dweller in Kenya earns only $40 a year. However, some developing countries are not as desperately poor as others. The economies of countries such as Malaysia, Mexico, and South Korea are growing rapidly.

The Root of Environmental Problems

Almost all environmental problems trace their origins back to two root causes. First, the number of people is growing too quickly for the Earth to support. (See Figure 1-12.) This is the **population crisis.** Second, people are using up, wasting, or polluting natural resources faster than they can be renewed, replaced, or cleaned up. This is the **consumption crisis.**

When there are too many people, there are not enough natural resources for everyone to live a healthy, productive life. As people struggle for survival in severely overpopulated regions, forests are stripped bare, topsoil is exhausted, and animals are driven to extinction. Malnutrition, starvation, and disease are constant threats. The population crisis is most severe in the developing countries. Even though there are not enough resources for everyone now, the human population continues to grow most rapidly in the poorest countries. (See bottom graph in Figure 1-13.)

The consumption crisis is most severe in the developed nations. The population has stabilized or is growing slowly in

Figure 1-13 Developing nations account for most of the population growth, and developed nations account for most of the consumption of resources.

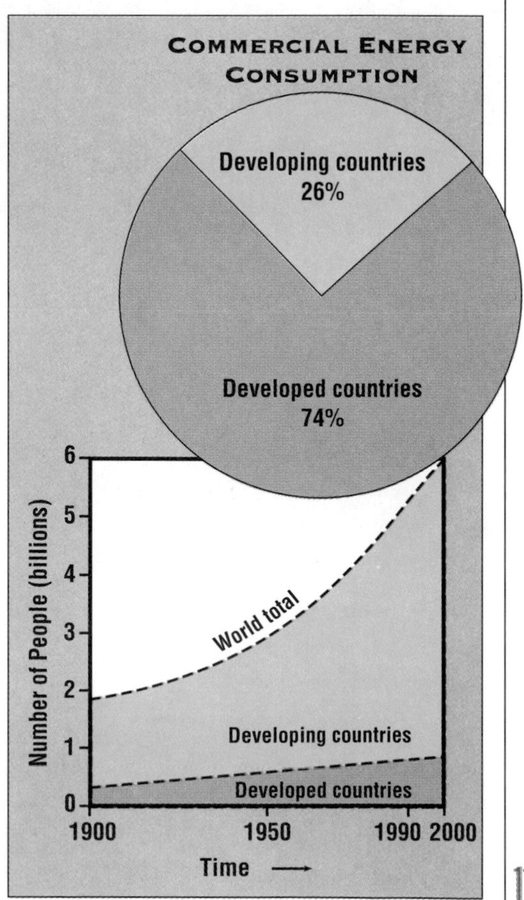

Answers to Section Review can be found on page T141.

Figure 1-14 **These photographs show a few of the ways we can work toward a sustainable future. Setting aside wildlife refuges helps preserve threatened species. Recycling means we won't run out of nonrenewable resources as quickly. And finding replacements for polluting technology, such as a vehicle that runs on electricity instead of gasoline, will help keep our air cleaner.**

Section Review

1. Name the three categories into which most environmental problems fall. Give an example of an environmental problem that illustrates each of these categories.
2. Explain the difference between the population crisis and the consumption crisis. Which countries are most affected by each of these crises?
3. List three types of actions that can contribute to a sustainable world.

Thinking Critically

4. *Applying Ideas* Give one example of a nonrenewable resource not mentioned in this chapter, and explain why it is a nonrenewable resource.

these countries, but the citizens are using far more than their share of the Earth's resources. Developed nations use up about 75 percent of the resources used every year, even though they make up only about 20 percent of the world's population.

A Sustainable World

The goal of environmental problem solving is to achieve a **sustainable world**, a world in which human populations can continue to exist indefinitely with a high standard of living and health. In a sustainable world, habitats would be preserved and garbage would be turned into harmless substances. Nonrenewable resources would be used sparingly and efficiently. And renewable resources would be used no faster than they could be replaced, so there would always be enough for generations that follow.

SECTION 1.2

USING SCIENCE TO SOLVE ENVIRONMENTAL PROBLEMS

AFTER READING THIS SECTION YOU SHOULD BE ABLE TO

❶ distinguish between pure and applied science.

❷ describe scientific methods.

❸ explain the uses of tables, line graphs, bar graphs, and pie charts.

In order to solve environmental problems, we must first understand our environment. The most effective tool that humans have developed for accurately understanding the natural world is science. There are two basic types of science: pure science and applied science.

Pure science seeks to answer questions about how the natural world works. Physics and biology are examples of pure sciences. These sciences try to answer questions like, "How does the sun produce light?" or "Why do insects and birds have different kinds of wings?" **Applied science** uses the information

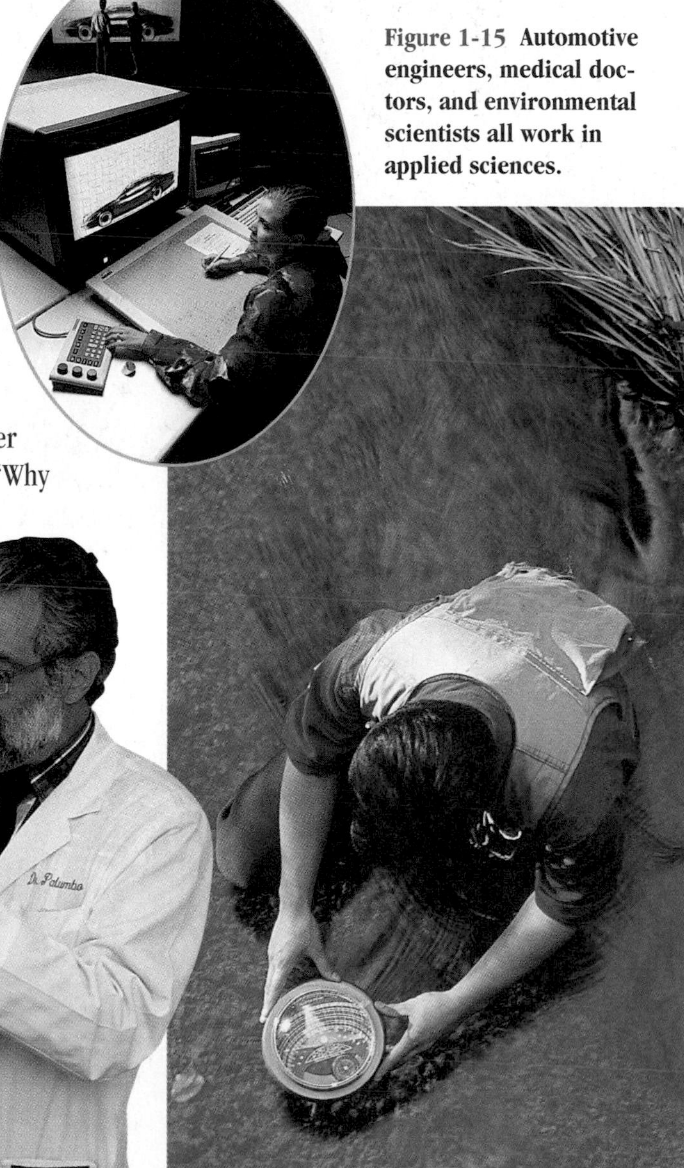

Figure 1-15 Automotive engineers, medical doctors, and environmental scientists all work in applied sciences.

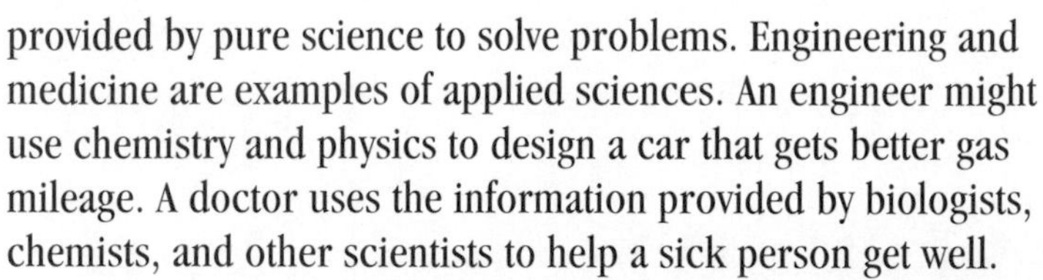

provided by pure science to solve problems. Engineering and medicine are examples of applied sciences. An engineer might use chemistry and physics to design a car that gets better gas mileage. A doctor uses the information provided by biologists, chemists, and other scientists to help a sick person get well.

Environmental science is an applied science. Many different sciences contribute to environmental science and are needed to solve environmental problems. Chemistry is used to understand the nature of pollutants. Botany and zoology provide information needed to protect species from extinction. Meteorology explains air and atmosphere. An environmental scientist may use information provided by almost all of the pure sciences at one time or another. As you read this textbook, you will be studying the results of the many different sciences that contribute to solving environmental problems.

One of the most important foundations of environmental science is **ecology.** Ecologists study how living things are related to each other—how they interact and how they depend on each other. An ecologist might study, for example, the relationship between bees and the plants they pollinate, or how bacteria break down the bodies of dead animals and return the nutrients to the soil.

Figure 1-16 **Science is not just something you *know*. These working scientists are *doing* science to find out the answers to questions about how the natural world works.**

What Is Science?

Science is really two things. On the one hand, it is something you *know*. It is all of the scientific information gathered by scientists throughout human history. This vast body of knowledge is passed down from generation to generation.

Science is also something you *do*. Figure 1-16 shows scientists at work "doing science." It is a way of getting the answers to questions about the natural world around us. In order to find these answers, scientists use methods—called *scientific methods*—that have been found to provide accurate, reliable answers to their questions.

Observing All science begins with observation. When we observe the natural world, we use our senses to discover what it is like. The sky is blue. Lemons have a sour taste. Rocks lying in the sunshine feel warm. Mockingbirds imitate the songs of other birds. Sulfur springs smell like rotten eggs. Scientists also use equipment that extends or enhances their senses. For example, a biologist might use underwater microphones to hear whale songs. Doctors routinely use X rays to view bones inside living bodies.

Science always begins with and rests on a solid foundation of observation. Good observations include thorough and accurate measurements. How tall is the tree? How long did the deer wait by the pond? How many birds return to the same site each year? For information like this to be scientifically valuable, it must be carefully recorded so that it can be shared with others. Many scientists have answered important and fundamental scientific questions by simply observing an organism or a natural process very carefully for many years.

Remember the students at Keene High School? As part of their research, the students count the number of mussels at different sites along the Ashuelot River. This is observation. It is also observation when they take various measurements of water quality—when they measure the number of bacteria, the amount of oxygen, and so on.

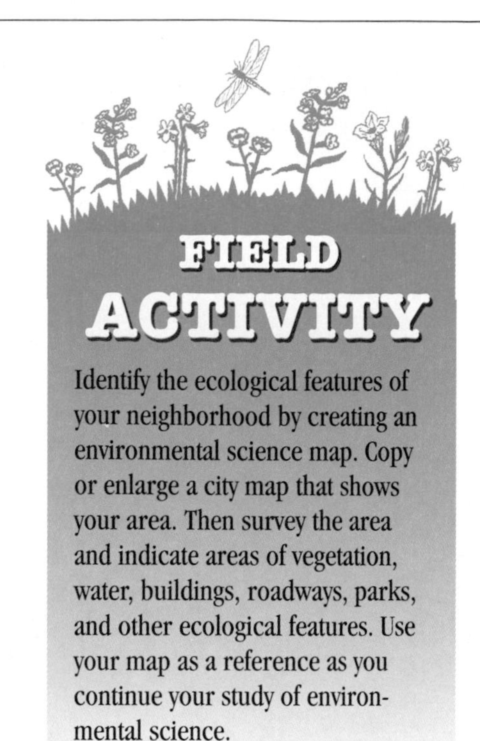

FIELD ACTIVITY

Identify the ecological features of your neighborhood by creating an environmental science map. Copy or enlarge a city map that shows your area. Then survey the area and indicate areas of vegetation, water, buildings, roadways, parks, and other ecological features. Use your map as a reference as you continue your study of environmental science.

Hypothesizing and Predicting Observations give us answers to questions about the natural world, but they almost always give rise to still more questions. When a scientist wants to know the answer to a very specific question, forming a hypothesis that can be tested is usually the best way to find the answer. A **hypothesis** is the scientist's prediction of what the correct answer will be, based on what he or she has already learned.

Take the dwarf wedge mussel situation on the Ashuelot River as an example. A biologist familiar with various mussel species might study the data collected by the students and recall that declines in mussel populations are often the result of some kind of reproductive failure. The biologist might state the following hypothesis: *Reproductive failure is what is causing the reduction in the number of dwarf wedge mussels in the Ashuelot River.*

To test this hypothesis, the biologist might make the following prediction: *Over time, the proportion of older mussels will increase and the proportion of eggs, larvae, and young mussels will decrease.* This would indicate that the mussels are not reproducing successfully.

If the biologist finds fewer young mussels over time, the hypothesis has been supported. If, on the other hand, the biologist observes no changes in the age distribution of the mussels and finds instead the normal number of eggs and young, the hypothesis has not been supported.

Figure 1-17 These students from Keene High School are making scientific observations when they count the number of mussels at each site on the Ashuelot River.

Experimenting Many times, the questions that arise from observations cannot be answered conclusively from general observations. In these cases, scientists may do experiments to find answers. In an **experiment,** a hypothesis is tested under controlled conditions. There are two essential ingredients to a good experiment: a single variable is tested, and a control is used.

In an experiment, two groups or situations are studied. These two groups must be identical in every respect except one. For example, a biologist might wish to know if the level of cadmium (a potentially harmful heavy metal) detected by the Keene High School students in the river water is sufficient to harm mussels. Hypothesizing that the level of cadmium is in fact toxic to mussels, the biologist predicts that a significant number of mussels exposed to this level of cadmium will die. (This experiment could not be done with an endangered species!)

The biologist would then collect a number of common mussels and place equal numbers in two aquariums, like the one shown in Figure 1-18. First, the biologist would ensure that the conditions in the two aquariums were identical in all respects—identical water temperature, food, plants, hours of light, natural substances present in the water, and so on.

Then just enough cadmium would be added to one aquarium to equal the concentration observed by the students in the river water. The amount of cadmium in the water is the variable. The

Figure 1-18 In an experiment, a scientist tests a hypothesis under controlled conditions. Studying mussels in an aquarium rather than in the river allows a scientist to examine the effect of only one variable at a time.

group of mussels to which cadmium has been added is the experimental group, and the group of mussels to which cadmium has not been added is the control group. If the "no-cadmium" mussels thrive while most of the "cadmium" mussels die, then the biologist's hypothesis is supported by observations.

The key to the success of this experiment and all other experiments is changing only one variable and having a control group. What if the aquarium in which most of the mussels died had cadmium in the water *and* also happened to be 5°F warmer? The biologist would have no way of knowing if the cadmium or the higher temperature caused the mussels to die. What if the biologist had not used a healthy control group for comparison? The mussels in the cadmium tank could be dying from some other factor, such as being fed the wrong food.

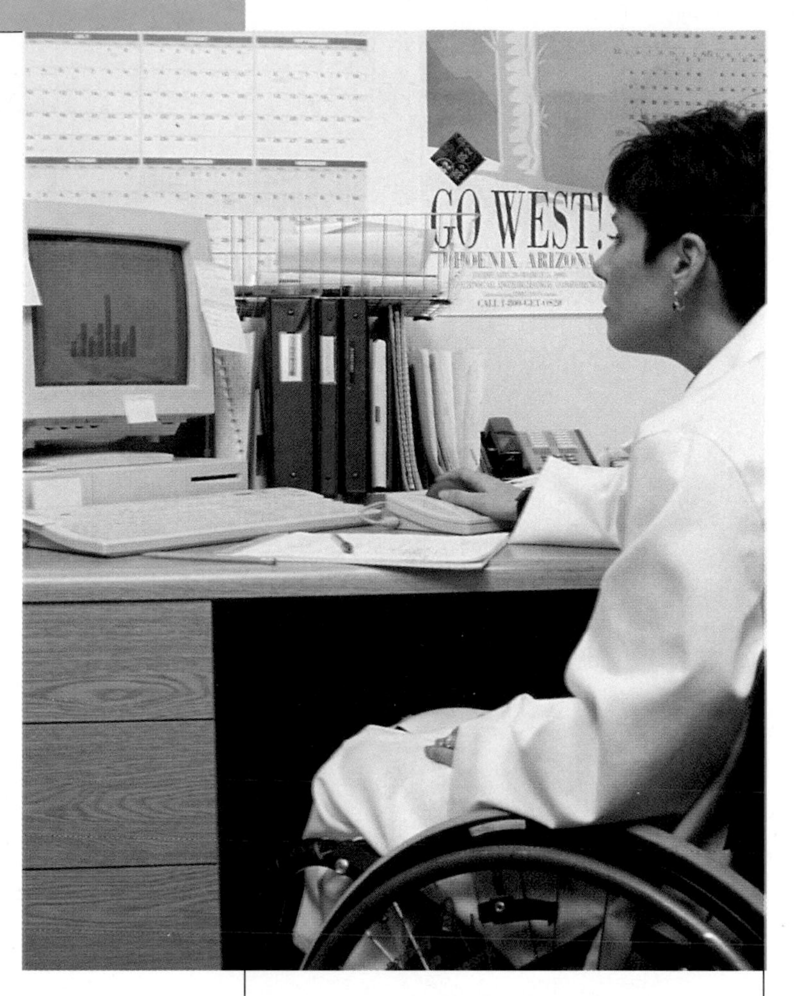

Figure 1-19 Scientists often use mathematics to determine whether the results they get are meaningful or just the result of chance.

Organizing and Interpreting Data

One of the most important parts of doing science is sharing what has been learned with others. This begins with keeping very careful and accurate records. Then the scientist must evaluate the data and decide if the hypothesis is supported. For example, what if the biologist observed that three mussels died in the control tank, but five died in the tank containing cadmium. Would that be a meaningful difference? What if two mussels died in the control tank and 10 in the cadmium tank? One factor that would affect the answer to this question would be, "How many mussels were in each tank to start with?" In other words, what percentage of mussels died?

Scientists often use mathematics to determine whether their observations or experimental results are meaningful or just the result of chance or coincidence. The mathematical discipline of statistics gives the scientist tools in the form of mathematical formulas to determine if the difference between the results in the control group and the experimental group is significant.

Using Graphics and Sharing Information

Organizing data into graphic illustrations helps scientists analyze the data and explain it clearly to others. Tables are a good way to summarize data. Figure 1-20 summarizes fecal coliform bacteria counts by the Keene High School students over a four-year period. The number of fecal coliform bacteria may indicate the amount of sewage present in water.

Figure 1-20 Tables organize and summarize information.

Four-Year Fecal Coliform Averages (bacteria colonies per 100 mL of water)

	1990	1991	1992	1993
Week 1				
Site 1	15	13	12	7
Site 2	39	16	5	78
Site 3	88	7	6	45
Week 2				
Site 1	5	17	23	11
Site 2	37	17	63	26
Site 3	29	13	27	6
Week 3				
Site 1	7	174	20	43
Site 2	81	54	10	23
Site 3	19	17	15	75

Site 1	11
Site 2a	28
Site 2b	8
Site 3	2
Site 4	9
Site 5	6
Site 6	6
Site 7	3
Site 8	1
Site 9	2
Site 10	1
Site 11	0
Site 12	0

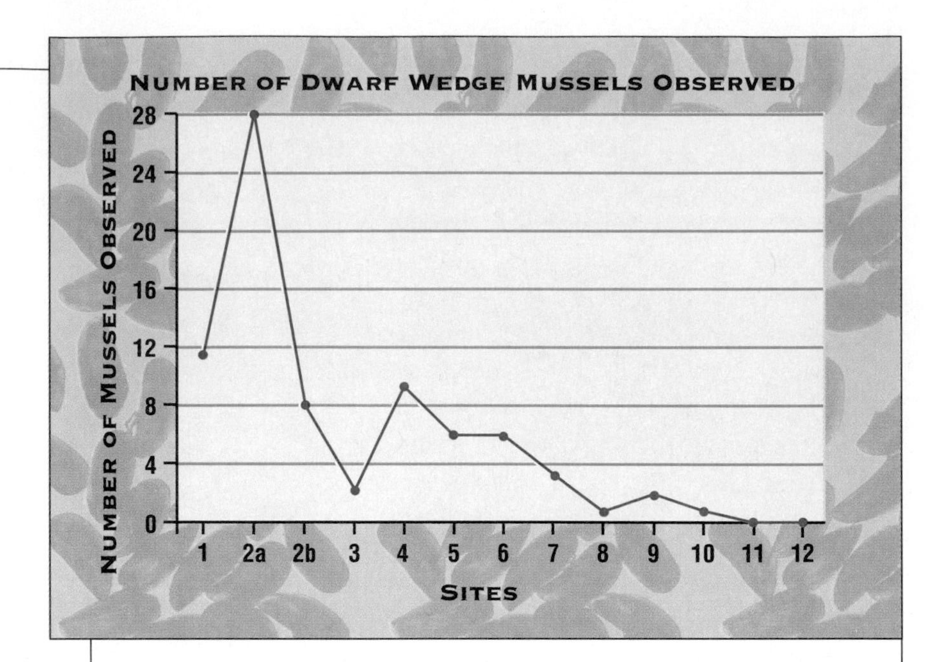

Figure 1-21 This shows how the same data might look in a table (left) and in a line graph (right). Both show how the number of mussels decreased as the students moved downstream.

Graphs are often used by scientists to convey comparisons or trends that they have discovered. For this reason, graphs are especially useful when the scientist is discussing the conclusions that can be drawn from the research. The right side of Figure 1-21 is a line graph that shows the change in the number of mussels observed by the students as they moved downstream. Bar graphs using single bars are often used in the same way as line graphs, but bar graphs are especially effective at showing several comparisons at once. For example, Figure 1-22 allows you to see how the phosphate levels in the Ashuelot River varied at three different sites and on three different dates when samples were collected.

Pie charts show percentages, with the entire circle representing 100 percent. A single pie chart can show relative percentages of different factors, but two pie charts can convey even more

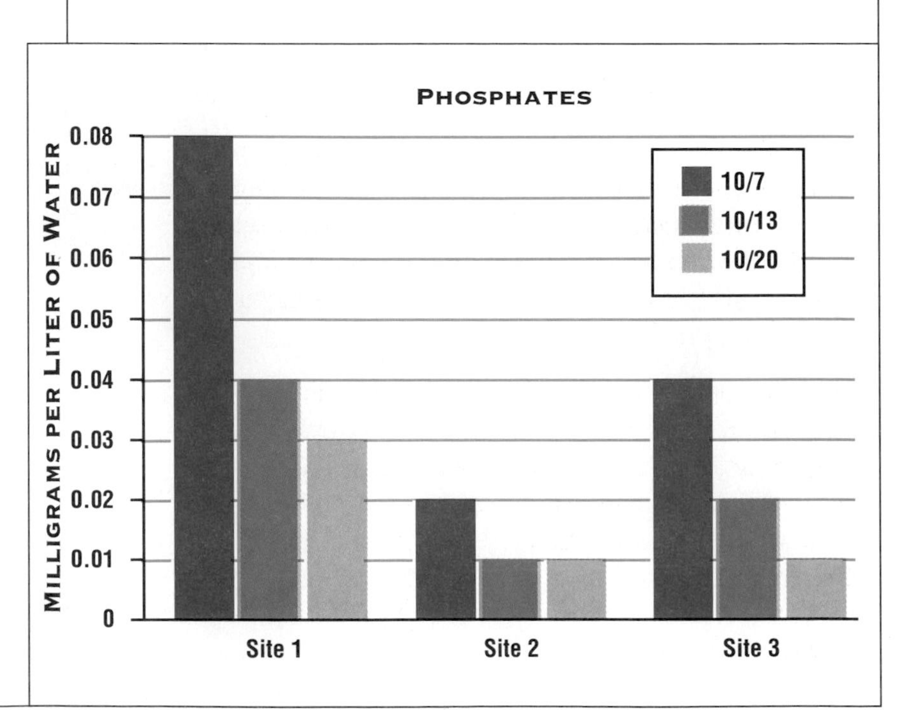

Figure 1-22 This bar graph allows you to compare the amount of phosphates measured by the Keene students at three different sites and on three different dates.

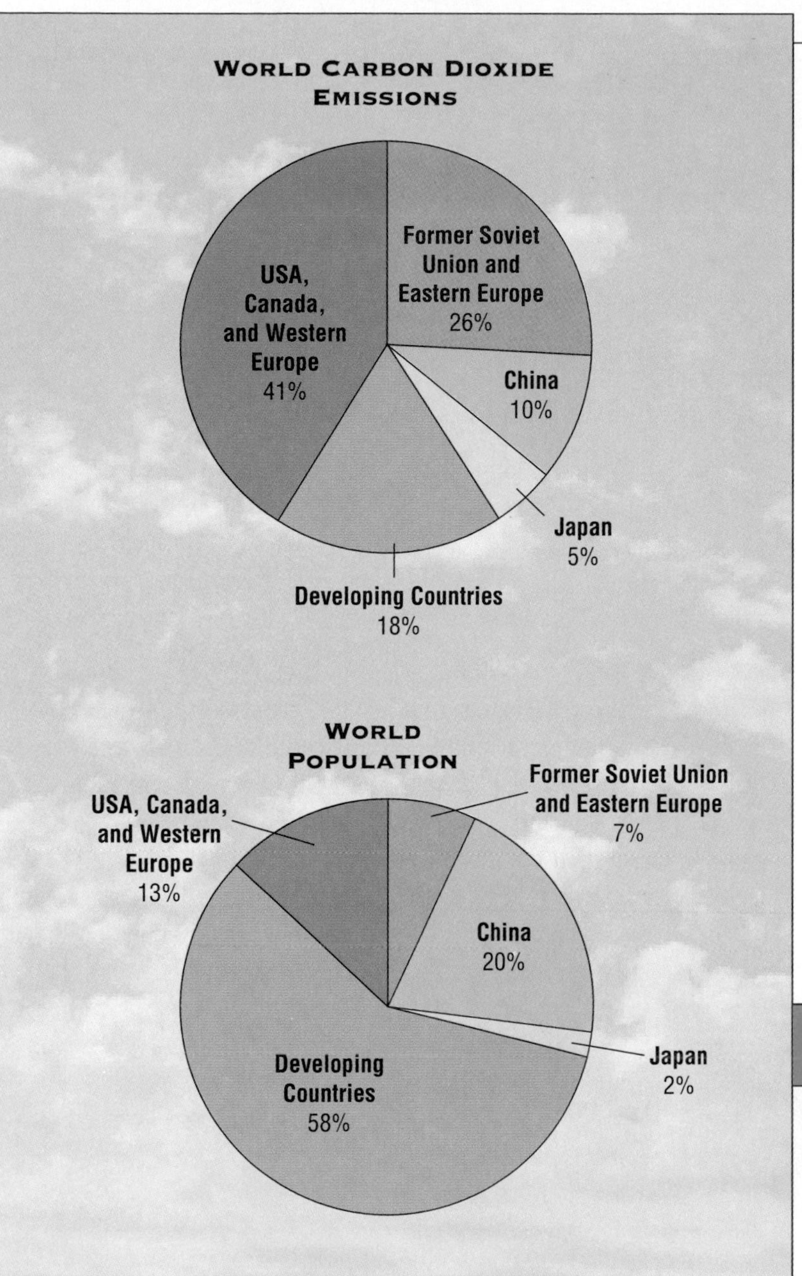

Figure 1-23 Pie charts show percentages, with the entire circle representing 100 percent.

information. For example, the pie charts in Figure 1-23 show the relative human populations in different parts of the world compared with the relative amounts of CO_2 produced in these regions. By studying the charts, the reader can see that the developed nations produce much more CO_2 per person than the developing nations.

Communicating Results

After the scientist has analyzed his or her data and has determined that what has been learned from the work is important enough to be of interest to other scientists, the results must be published. A scientific article must tell the reader what the question to be answered is, why the question is important or relevant, background information, a precise description of how the work was done, the data that were collected, and the scientist's evaluation of what the data mean.

Answers to Section Review can be found on page T141.

SECTION REVIEW

1. What is the difference between pure and applied science? What type of science is environmental science?
2. Explain the relationship between ecology and environmental science. How are they different?
3. Name three ways scientists answer questions about the natural world.
4. What are the two essential components of an experiment? Explain why each is important.

THINKING CRITICALLY

5. *Analyzing Processes* Science is only one kind of human knowledge. Can you think of a nonscientific question that might use scientific information as part of the answer?

SECTION 1.3

MAKING ENVIRONMENTAL DECISIONS

AFTER READING THIS SECTION YOU SHOULD BE ABLE TO

1. use a decision-making model to make a decision about an environmental issue.
2. name values that are important in making decisions about the environment.

Scientific research is an essential first step in solving environmental problems. A sound scientific basis is necessary before any action should be taken. However, many other factors must also be considered. How will the proposed solution affect people's lives? How much will it cost? Is it ethically right? Questions like these require an examination of values—what we consider important. What values should be considered when making decisions that affect the environment? Figure 1-24 lists some values that often affect environmental decisions. You can probably think of others as well.

Figure 1-24 Can you think of values not shown in this table?

Answer to caption:
Answers will vary. Samples include: religious, political

Values That Affect Environmental Decision Making

Value	Definition
Aesthetic	What is beautiful or pleasing
Economic	Gain or loss of money or jobs
Environmental	Protection of natural resources
Educational	Accumulation and use of knowledge
Ethical/Moral	What is right or wrong
Health	Maintenance of human health and prevention of sickness or disability
Recreational	Providing for human leisure activities
Scientific	Knowledge gained by scientific research
Social/Cultural	Maintaining human communities and respecting their values and traditions

Figure 1-25 Solutions to environmental problems begin with science, but other factors must also be considered. The value we give to the beauty of this view, for example, influences whether we want to keep our air free of smog that might obscure it.

An Environmental Decision-Making Model

Making decisions about environmental issues is usually difficult and may even seem overwhelming. It helps to have a systematic way of analyzing the issues and figuring out what is important to you. Figure 1-26 shows a decision-making model that could prove helpful to you.

The first step in making an environmental decision is to gather all the available information. In addition to watching television reports and reading newspapers, magazines, and books about environmental issues, it helps to listen carefully to well-informed people on all sides of an issue. Then consider which values apply to the issue. Explore the consequences of each option. Finally, evaluate everything and make a decision.

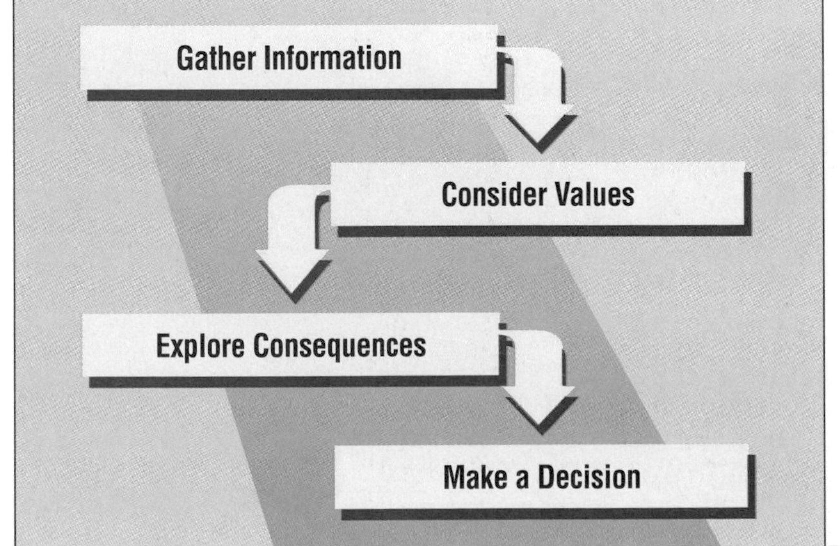

Figure 1-26 A decision-making model

A Hypothetical Situation

Consider the following hypothetical example. In the town of Pleasanton, in Valley County, biologists from the local college have been studying the spotted warbler (a bird). The warblers have already disappeared from several areas around the state, and their numbers are declining in Valley County. The biologists warn government officials in the county that if something is not done, the bird may be listed as an endangered species by the U.S. Fish and Wildlife Service.

Figure 1-27 **Rapid growth often spills out of city limits and can destroy the habitats of rare species.**

Pleasanton is growing rapidly, and much of the new development, like that shown in Figure 1-27, is occurring outside the city limits. Valley County already has strict environmental controls on building, but these controls seem to be inadequate to protect the birds. Several environmental groups band together and propose that the county buy several hundred acres of land known to be critical habitat for the spotted warbler and place strict controls on development on several hundred acres immediately around the preserve. To encourage growth in less sensitive areas, they propose that environmental restrictions on building be made less stringent than they are currently in those areas.

The environmental groups obtain enough signatures to put the issue to a vote. Not surprisingly, many property owners in the area of the proposed preserve are opposed to the plan. They say that they will suffer serious financial losses if they have to sell their land to the county rather than developing it. On the other hand, local leaders warn that if the warbler population falls so low that

Figure 1-28 **The map shows the imaginary plan to set aside protected habitat for the spotted warbler. The photograph shows a protected habitat in Austin, Texas.**

it must be listed as an endangered species, the recovery plan required by law may impose even stricter controls on development, threatening the economic health of the county. Many citizens of Pleasanton look forward to hiking and camping in the proposed nature preserve. Other residents are bothered by the idea of any government regulation of the use of private property.

How to Use the Decision-Making Model

The hypothetical situation in Pleasanton can be used to illustrate how to use the decision-making model. Michael Price is a voter in Valley County. This is how he decided which way to vote on the referendum to make a nature preserve for the spotted warbler.

Figure 1-29 **This is the table Michael made to help him decide how to vote on the referendum.**

Should Valley County Set Aside a Nature Preserve?

	Environmental	Economic	Recreational
Positive short-term consequences	Habitat destruction in the nature preserve area is slowed or stopped.	Landowners whose property was bought by the county receive a payment for their land. Property outside the preserve area can be developed with fewer restrictions.	Parts of the preserve are made available immediately for hiking and picnicking.
Negative short-term consequences	Environmental controls are made less stringent in parts of the county outside the preserve area.	Property owners inside the preserve area do not make as much money as if they had developed the land. Taxpayers must pay increased taxes to buy land for the preserve.	None
Positive long-term consequences	The population of spotted warblers increases, and the bird does not become endangered. Other species of plants, animals, fungi, and bacteria are also protected, even though it has not been possible to study them all. An entire ecosystem is preserved, rather than just a single species.	Property near the preserve increases in value because it is near the preserve. Businesses are attracted to Valley County because of its natural beauty and recreational opportunities, resulting in job growth. The spotted warbler is not listed as an endangered species, avoiding stricter federal controls on land use.	Large areas of the preserve are available for hiking and picnicking. Landowners near the preserve may develop campgrounds with swimming, fishing, and bike trails available on land adjacent to the preserve.
Negative long-term consequences	Other habitat outside the preserve may be damaged by overdevelopment.	Taxpayers must continue to pay for the maintenance and upkeep of the preserve.	None

FIELD ACTIVITY

Go to the local library or a book store. Spend some time looking through magazines, newspapers, and books that focus on environmental issues. Make a list of possible resources you may wish to use as you complete this course. It is useful to include the author's name along with the title of a book or article.

Answers to Section Review can be found on page T141.

SECTION REVIEW

❶ List three types of values, besides environmental values, that might be considered when making an environmental decision. Explain each one.

❷ Give two examples of situations in which environmental values could come into conflict with other values.

THINKING CRITICALLY

❸ *Making Decisions* Make a table like the one Michael made, showing the positive and negative consequences of setting aside the nature preserve. Do you think your chart would have resulted in Michael making a different decision? Why or why not?

Gather Information Michael studied the warbler issue carefully—watching local news reports, reading the newspaper, and attending forums where the issue was discussed. Several of the arguments on both sides of the issue make sense to him.

Consider Values To help make up his mind about how he will vote on the referendum, he makes the table shown in Figure 1-29 on page 23. There are many different values that he could have considered when making his table. Environmental values are key to the debate of course, and economic values play an important role in these types of decisions. Since he loves the outdoors and likes the idea of having the nature preserve available for hiking and camping, he decides to add recreational value to his table.

Explore Consequences As he studied his table, Michael decided that in the short term, positives and negatives seem almost equally balanced. A few people would suffer economically, but others would gain financially from the plan. Taxpayers would have to pay for the plan, but all the citizens of the county would have access to land that was previously off-limits as private property. Finally, some parts of the county would have more environmental protection and some would have less.

It was looking at the long-term consequences, however, that enabled Michael to make up his mind about how he would vote. Michael decided that the long-term benefits of the plan outweighed the long-term negative consequences. The idea of a bird becoming extinct saddened him. And he thought it would be smarter and less costly to protect it now rather than later, when protection might be much more difficult. He also liked the idea of working toward a future for his community where natural beauty and recreational opportunities would be preserved. By ensuring that Valley County remained a pleasant place to live and work, he also thought that in the long run the plan would bring economic gains for all of the county's citizens.

Make a Decision Michael chose to vote for the plan to set aside a nature preserve. Other people, looking at the same table listing pros and cons of the plan, might have voted differently. If you were a voter in Valley County, how would you have voted?

Throughout this book, you will be asked to use this decision-making model to make decisions about environmental issues. As you construct your tables to evaluate the issues, remember to include the values that you think apply to the issue, and to consider both the short-term and long-term consequences of each option.

CHAPTER 1

HIGHLIGHTS

SUMMARY

- Concerns about the environment have increased in the last few decades as problems of resource depletion, pollution, and extinction have become more common.
- Natural resources can be classified as renewable or nonrenewable. Care must be taken so that renewable resources are not depleted faster than they can be replaced.
- Environmental problems have global scope. The activities of one area may affect other locations or the Earth as a whole.
- The two major root causes of environmental problems are the population and consumption crises.
- Environmental science is an applied science that incorporates many of the pure sciences, including ecology, to help understand and solve environmental problems.
- Science involves observing, hypothesizing, and experimenting. Results are organized, presented, and communicated to the scientific community at large.
- Making environmental decisions involves an examination of values. All sides of an issue must be carefully weighed.

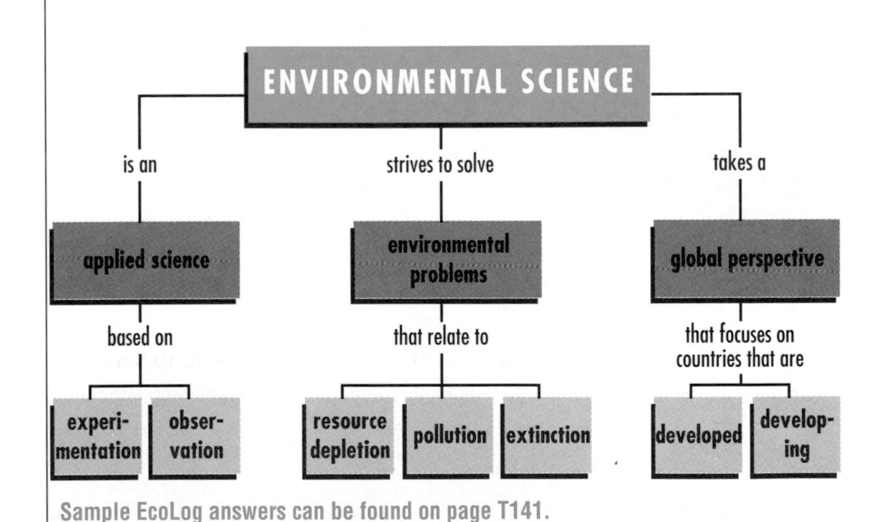

Sample EcoLog answers can be found on page T141.

EcoLog

Now that you've studied this chapter, revise your answers to the questions you answered at the beginning of the chapter, based on what you have learned.

❶ What do you consider to be our major environmental problems?

❷ How can science help solve environmental probems?

Vocabulary Terms

applied science (p. 13)
biosphere (p. 9)
consumption crisis (p. 11)
developed countries (p. 11)
developing countries (p. 11)
ecology (p. 14)
environment (p. 4)
environmental science (p. 4)
experiment (p. 16)
hypothesis (p. 15)
natural resource (p. 5)
nonrenewable resource (p. 6)
population crisis (p. 11)
pure science (p. 13)
renewable resource (p. 6)
sustainable world (p. 12)

CHAPTER 1

REVIEW

UNDERSTANDING VOCABULARY

1. For each pair of terms, explain the difference in their meanings.
   a. natural resource
      nonrenewable resource
   b. population crisis
      consumption crisis
   c. pure science
      applied science
   d. ecology
      environmental science

RELATING CONCEPTS

Answers not indicated can be found on pages T141–T142.

2. Copy the unfinished concept map below onto a sheet of paper. Then complete the concept map by writing the correct word or phrase in each box containing a question mark.

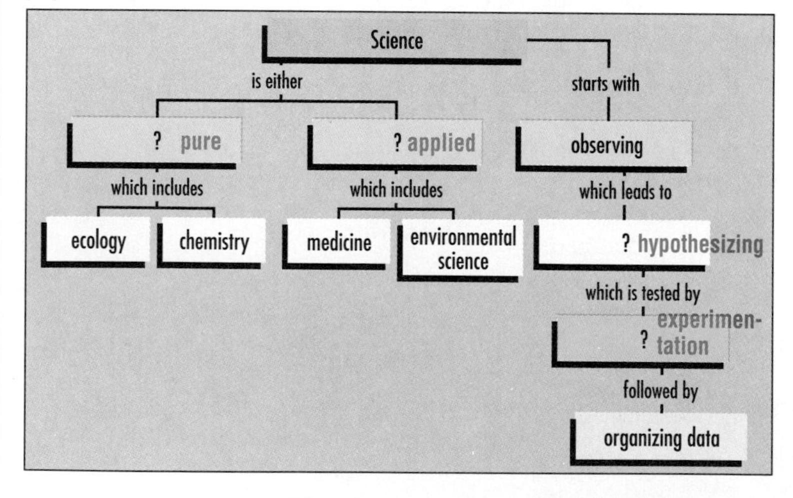

UNDERSTANDING CONCEPTS

Multiple Choice

3. When we talk about the environment, we are usually referring to
   a. people, plants, and animals.
   b. land, air, and water.
   c. streets, bridges, and buildings.
   (d.) all of the above

4. The environmental problems created late in the twentieth century are
   a. impossible to solve.
   (b.) partly a result of increased human population.
   c. not as bad as those created early in the twentieth century.
   d. both **a** and **b** above

5. All of the following would be considered resource depletion EXCEPT
   a. Texas petroleum running out.
   b. a rain forest in Brazil being burned.
   c. plants not growing in overused fields.
   (d.) solar energy not reaching the Earth's surface because of haze.

6. Though extinction is a natural process, it has become a problem because
   a. environmentalists have brought it to our attention.
   b. unknown organisms are now becoming extinct.
   (c.) the rate of extinction has increased drastically.
   d. the types of organisms becoming extinct are popular.

7. The consumption crisis is more severe in
   a. developing countries.
   (b.) developed countries.
   c. rural areas.
   d. complex ecosystems.

8. If a scientist's experiment did not support the hypothesis, the scientist should
   (a.) publish the results of the experiment anyway.
   b. consider the results abnormal and continue working.
   c. find a way to rationalize the data.
   d. try another scientific method.

9. The mathematical discipline frequently used by scientists to analyze data is called
   a. statistics.
   b. calculus.
   c. trigonometry.
   d. algebra.
10. The use of bar graphs is especially effective for
   a. summarizing data.
   b. conveying comparisons or trends.
   c. showing several comparisons at once.
   d. representing percentages.
11. To fully understand a complex environmental issue, you may need to consider
   a. economics.
   b. values.
   c. politics.
   d. all of the above
12. When you investigate an environmental problem, the final step is to
   a. list the pros and cons.
   b. evaluate the issues.
   c. make a decision.
   d. consider the consequences.

Short Answer

13. How could environmental concerns conflict with your desire to improve your standard of living?
14. Pollution can be intentional or unintentional. Explain.
15. Which is more important, pure science or applied science? Explain your answer.
16. How might environmental science involve the pure sciences of geology and oceanography?
17. How does making a table help clarify the values and concerns you have when making a decision?

Interpreting Graphics

18. **Examine the graph below.** It illustrates how the mode of transportation used varies with the working population of a city. What general trends does the graph show? Which mode of transportation increases the fastest as population density increases?

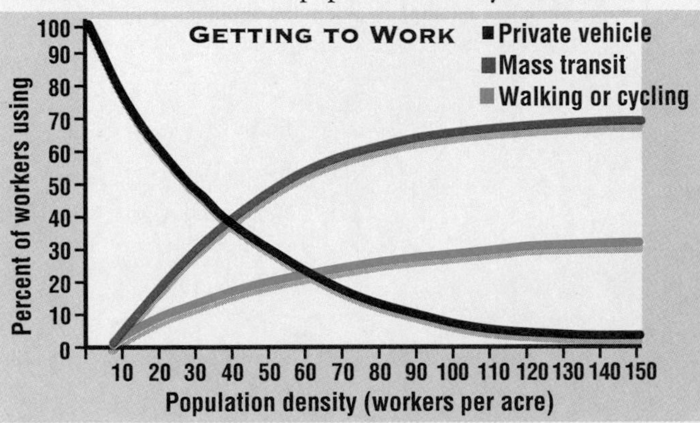

Thinking Critically

19. ***Analyzing Ideas*** Once the sun exhausts its fuel and burns itself out, it cannot be replaced. How then can sunlight be considered a renewable resource?

Themes in Science

20. ***Scale and Structure*** The biosphere is a thin layer of life surrounding the Earth. How does the size of the biosphere relate to the fragile nature of life on Earth?
21. ***Interacting Systems*** How does human consumption relate to whether resources are renewable or nonrenewable?

Cross-Discipline Connection

22. ***Demographics*** Obtain copies of the latest census report for your community. Prepare a table that shows the percentages of various categories of people in this population. The categories should be based on one demographic characteristic, such as ethnic group, political or religious affiliation, income, etc. How does the composition of your community compare to that of the United States as a whole?

Discovery Through Reading

For an interesting look at how global resources are affected by rapidly rising human populations, read "Putting the Bite on Planet Earth," by Don Hinrichsen, *International Wildlife,* Sept.–Oct. 1994, pp. 36–45.

PORTFOLIO ACTIVITY

Do a special project on the pure and applied sciences. Prepare a poster to illustrate the kinds of activities done by scientists in each area, and show the connections between and among different specialties that cooperate when focusing on environmental science.

CHAPTER 1

INVESTIGATION

AN ONION CONUNDRUM

You recently obtained a job working for an environmental consulting firm, the Environmental Brain Trust. Your firm, famous for "brainstorming," has recently received an anxious plea for help from Omar Frizbee, owner of Omar's Onions. The distraught Omar described how his onions failed to take root. He had tried everything, including the latest watering system and the finest organic fertilizer. Omar suspected that runoff from a nearby bleach company had contaminated the groundwater underneath his farm.

Answers can be found on page T142.

MATERIALS

- white onion bulbs between 1.5 and 2.0 cm in diameter (unsprouted, mold-free, and of approximately the same size)
- graduated cylinder
- stirring rod
- beaker
- wax pencil
- large test tubes
- distilled water
- fertilizer
- bleach
- test-tube rack
- knife
- notebook
- pen or pencil
- graph paper

OBSERVATIONS OF ONIONS

1. Omar has done the first step for you: he has observed that his onions have failed to take root. What will you do next? Get together with your team, and discuss the mystery of the failed crop. Decide what scientific problem you need to solve, and write it in your notebook.

MAKE A HYPOTHESIS

2. It's up to you and your co-workers to develop a hypothesis to explain the mystery of the failed crop. First, list all of the variables that might have affected Omar's onions, based on his description. Working with your team, pick one variable that you think is a likely cause of the problem, and develop a hypothesis. Write your hypothesis in your notebook. Remember: A hypothesis is a prediction of what the correct answer will be, based on what has already been learned.
3. Share your hypothesis with the class. Your teacher will write it on the board. There should be several hypotheses for each variable.

TEST THE WATER

4. The next step is to conduct a test to see if your hypothesis is correct. You will investigate the growth of onions by growing bulbs in test tubes filled with different growing solutions. The components of the solutions depend on which variable your team is testing. With your team, decide on the amounts (concentrations) of the substance in your test solutions.

   Here are some other questions you need to answer before you begin: What will your sample size be? Is it necessary to set up a control? Why or why not? Make a complete list of your procedures.
5. Fill the test tubes with the solutions that you are testing. Label each test tube, and set the test tubes in a rack. Remove the onions' outer layers, and cut off a 2 mm slice from the bottom of each onion. Insert the onion bulbs into the test tubes so that the bottom of the bulb is immersed in the solution. Set the test-tube rack in a sunny location.

COLLECT SOME DATA

6. Over a five-day period, make note each day of the root growth of the onions. Measure the growth using a metric ruler, and record your data in a table like the one shown below. (Include more columns if you use more than three test tubes.) After you have collected all of your data, make a line graph comparing root growth over time in the solutions you tested.

	Length of Roots (in centimeters)		
Day	Solution A	Solution B	Solution C
1			
2			
3			
4			
5			

WHAT HAPPENED?

7. Now you need to analyze your findings. Was your hypothesis correct? Write a report that includes your hypothesis, data, and analysis to send to your client, Omar.
8. In your experiment, what variable did you change, and what others did you hold constant? Why is it important to make sure that only one variable is changed for your test?

FIND OUT MORE

9. You may want to find out more about what effect other environmental pollutants can have on plant growth. What about acid rain or salt accumulation? How could you test the effects of these variables using onion bulbs? For the best results, follow the method you used above: *observe, hypothesize, test,* and *analyze.* Consult with your teacher before you try any experiment.

Here is one team's experimental setup. Why did they label their test tubes?

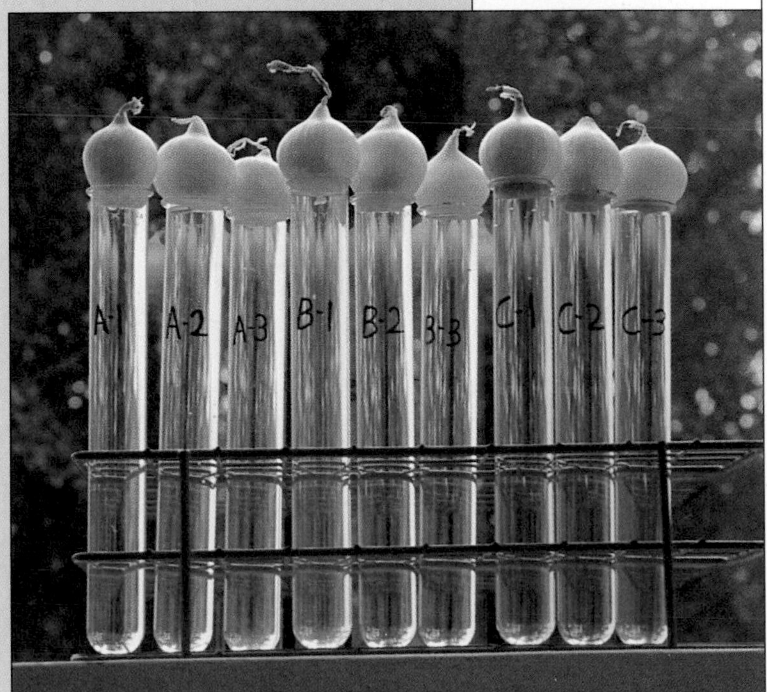

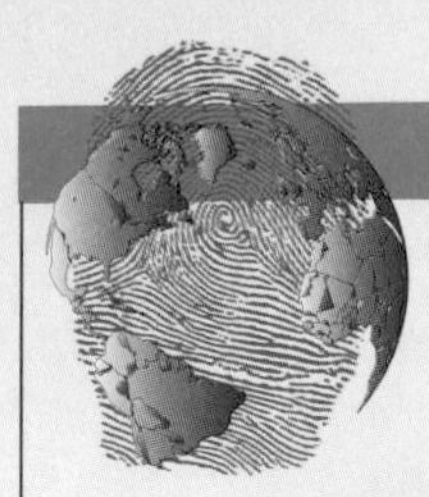

Making a DIFFERENCE

Chicken of the Trees

In the stillness of predawn, the air warms over the Carara Biological Preserve in Costa Rica. Several thousand eggs in sun-heated incubators just below the surface of the earth stir in response. Within these eggs are tiny iguanas—creatures that will eventually emerge and grow to a length of 1.5 to 2.0 m (5 to 6.5 ft.) and weigh up to 6 kg (13 lb.).

What's going on here? Well, these giant lizards are being raised so that they can be released into the rain forest. It's part of a project headed by German-born scientist Dagmar Werner. Her goal is to help restore an iguana population that has become nearly extinct in the past several decades. The lizard has suffered from the effects of hunting, pollution, and habitat destruction by people who clear the rain forest for farming. Since the 1980s, Dagmar has improved the iguanas' chances of survival by breeding them and releasing thousands of young iguanas into the wild. She also trains other people to do the same.

This lizard has been a favored source of meat among native rain-forest inhabitants for thousands of years.

Because she knew there was no time to lose before the iguana became extinct, Dagmar took an immediate and drastic approach to solving the lizard's problems. She combined her captive-breeding program at the preserve with an education program that shows farmers that there is more than one way to make a profit from the rain forest. Instead of raising cattle (and cutting down rain forest to do so), she encourages local farmers to raise iguanas, which can be released into the wild and sold for food. Known as "chicken of the trees," this lizard has been a favored source of meat among native rain-forest inhabitants for thousands of years.

Dagmar Werner has established an innovative method to help raise the number of iguanas living in the wild.

Dagmar and an associate discuss how the iguana can be "farmed."

But convincing farmers to use her methods hasn't been easy. According to Dagmar, "Many locals have never thought of wild animals as creatures that must be protected in order to survive. That's why so many go extinct." To get her message across, Dagmar has established the Fundación Pro Iguana Verde (the Green Iguana Foundation). This organization sponsors festivals and education seminars in local communities. These activities promote the traditional appeal of the iguana, increase civic pride in the animal, and heighten awareness about the iguana's economic importance.

By demonstrating that the needs of all concerned parties can be met when attempting to save an endangered species, Dagmar Werner has revolutionized the concepts of species preservation and economic development. This hard-working, unassuming scientist has hit upon a solution that may encourage farmers throughout Central America to "have their lizards and eat them too."

Dagmar's "iguana ranch" preserve has artificial nests where females can lay their eggs in a predator-free environment. After they hatch, the young lizards are placed in a temperature- and humidity-controlled incubator and given a special diet. As a result, the iguanas grow faster and stronger and are better protected from predators than their noncaptive counterparts. Ordinarily, less than 2 percent of all iguanas survive to adulthood in the wild, but Dagmar's iguanas have an 80 percent survival rate. Dagmar knows this because after she releases the iguanas into the rain forest, the lizards are tracked and monitored to determine whether they have successfully adapted to their less-controlled environment.

With Dagmar's methods, farmers can release many iguanas into the wild and earn a good living.

Iguanas at Carara Biological Preserve in Costa Rica

C H A P T E R 2

LIVING THINGS IN ECOSYSTEMS

"When we try to pick out anything by itself, we find it hitched to everything else in the universe."

JOHN MUIR
AMERICAN EXPLORER AND NATURALIST

EcoLog

Before you read this chapter, take a few minutes to answer the following question in your EcoLog.

- If all the lions on an African savanna were killed or removed, how do you think their absence might affect the other living things on the savanna?

ECOSYSTEMS: EVERYTHING IS CONNECTED

AFTER READING THIS SECTION YOU SHOULD BE ABLE TO

1. distinguish between the biotic and abiotic factors in an ecosystem.
2. explain the terms *population* and *community.*
3. distinguish between habitat and niche.

The environment is so complex and interconnected that scientists don't yet completely understand how it works. This becomes clear to us when human actions have unexpected effects on the environment, as they did on the Southeast Asian island of Borneo. The events on Borneo are shown in Figure 2-1. In 1955 the World Health Organization used the pesticide DDT to kill the mosquitoes that carry the disease malaria. The DDT killed the mosquitoes and relieved the malaria

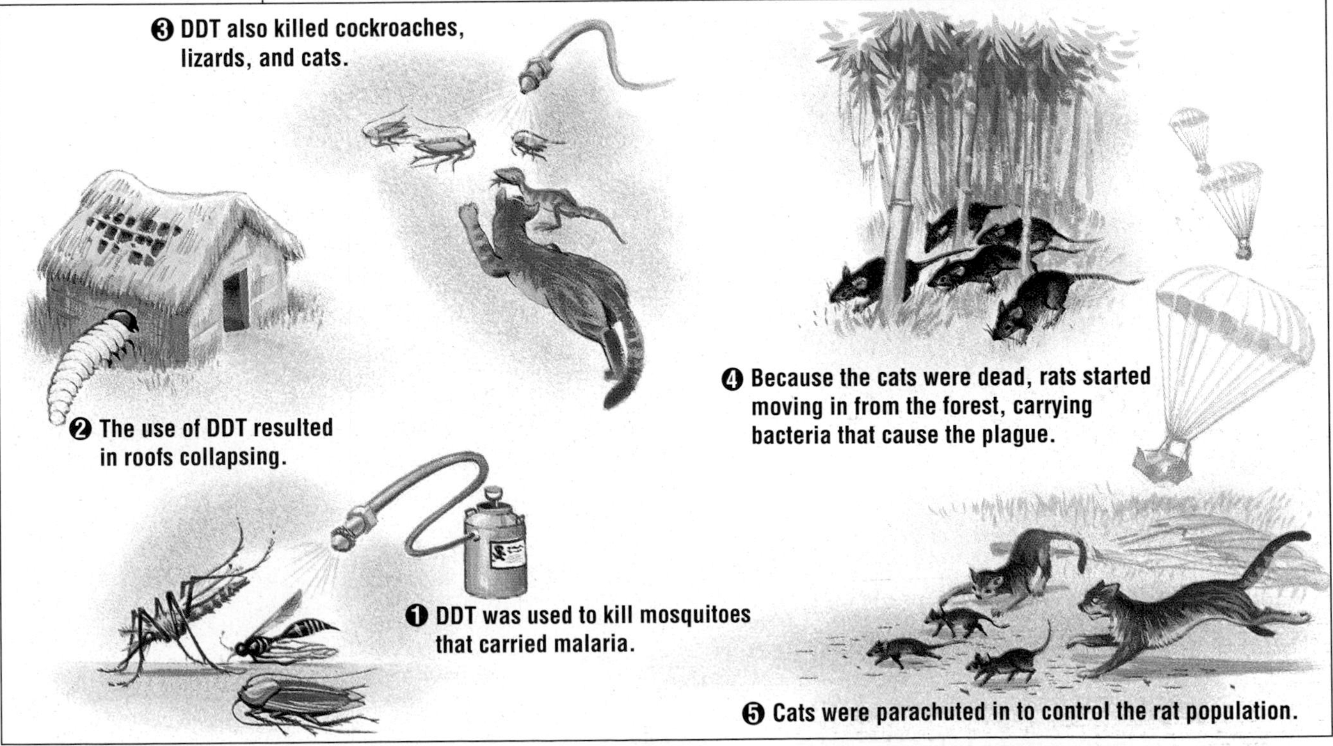

Figure 2-1 The use of DDT to control malaria in Borneo started a chain reaction that led to roofs collapsing and the decimation of the island's house-cat population.

problem on Borneo, but it also caused an undesirable chain reaction on the island.

First, the thatch roofs on the houses of Borneo started collapsing. What could this have to do with DDT? The DDT had killed the wasps that ate thatch-eating caterpillars. Without the wasps around, the caterpillars multiplied and devoured the thatch roofs.

Meanwhile, the DDT also landed on Borneo's cockroaches. The cockroaches were eaten by geckos (a kind of lizard). The geckos suffered nerve damage from the pesticide, causing their reflexes to become slower. Because the nerve-damaged geckos moved so slowly, most of them were caught and eaten by house cats. After the cats ate the geckos, they also suffered from the DDT and died in great numbers. Without the cats around, rats started moving in from Borneo's forests. On the rats came fleas, which carried the bacteria that cause the plague. Finally, officials resorted to parachuting healthy cats into Borneo to control the rat population!

The unforeseen chain of events on Borneo occurred because the living things on the island were connected to each other in an ecological network called an ecosystem.

What Is an Ecosystem?

An **ecosystem** includes all the different organisms living in a certain area, along with their physical environment. Figure 2-2 shows some of the inhabitants of a soil ecosystem. Other examples of

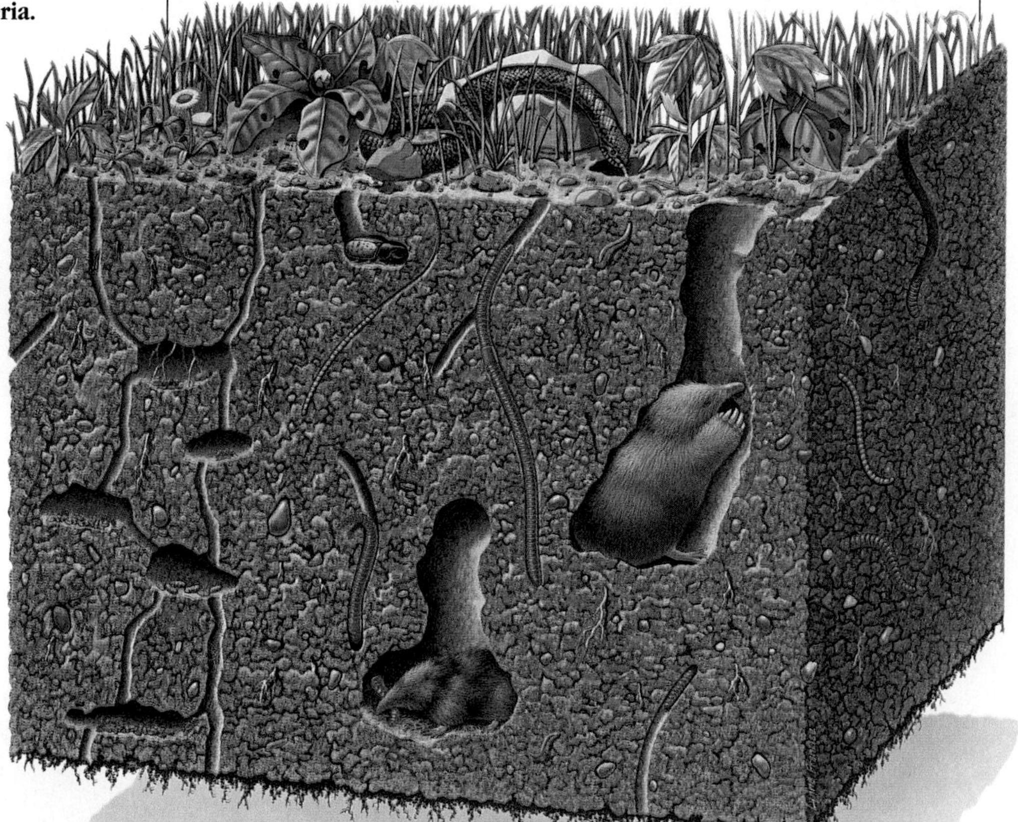

Figure 2-2 This soil ecosystem includes a variety of organisms—earthworms, snakes, moles, insects, plants, fungi, and bacteria.

ecosystems are shown in Figure 2-3. For convenience, ecologists often regard an ecosystem as an isolated unit, but ecosystems usually do not have clear boundaries. Things move from one ecosystem to another. Soil and leaves from a forest might wash into a lake, for instance, and birds might migrate from their winter homes in one ecosystem to their summer homes in another ecosystem.

An ecosystem is composed of both biotic and abiotic factors. **Biotic factors** are the living parts of an ecosystem—the animals, plants, and microorganisms. These biotic factors interact with each other in complex ways, and they also interact with the nonliving parts of the ecosystem—the **abiotic factors.** Some abiotic factors

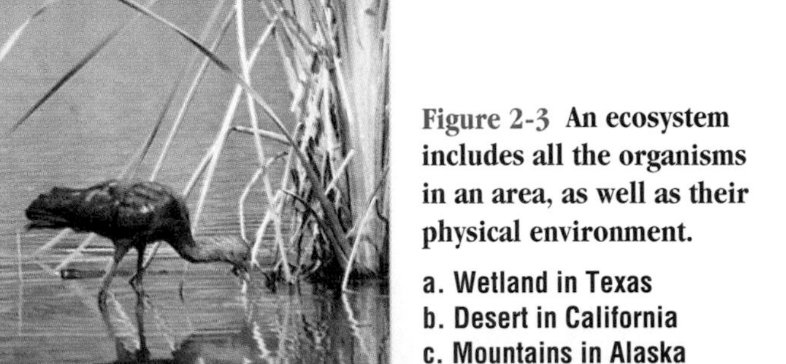

Figure 2-3 An ecosystem includes all the organisms in an area, as well as their physical environment.

a. Wetland in Texas
b. Desert in California
c. Mountains in Alaska
d. Polar region in Canada
e. Coral reef in Fiji

that influence living things are temperature, sunlight, humidity, water supply, soil type, and mineral nutrients such as nitrogen, phosphorus, and sulfur.

To appreciate how all the things in an ecosystem are connected, think about how an automobile works. The engine alone is made up of hundreds of individual parts, all working together to make the automobile run. If even one part breaks, the car might not run. Similarly, if one part of an ecosystem is destroyed—as it was in Borneo—the entire ecosystem can be affected. When thinking about ecosystems, remember that everything is connected.

If you look at Figure 2-4, you will see how an ecosystem fits into the organization of living things. The illustration shows the

Figure 2-4 **An individual organism is part of a population, a community, an ecosystem, and the biosphere.**

Figure 2-5 Each of these photographs shows a population of organisms.

a. Colonies of bacteria in a petri dish
b. Penguins on an island in the South Atlantic Ocean
c. Impala on a plain in Kenya
d. Bluebonnets in a Texas field

different levels of organization, from the individual organism to the entire biosphere.

Organisms and Species An **organism** is one individual living thing. You are an organism. An ant crawling across the floor is an organism, an ivy plant on the window sill is an organism, a gorilla in the rain forest is an organism, and a bacterium living in your intestines is an organism.

A **species** is a group of organisms that are able to reproduce together and that share common genes and therefore resemble each other. All humans, for instance, are members of the species *Homo sapiens*. All domestic dogs belong to the species *Canis familiaris,* and all gorillas belong to the species *Gorilla gorilla.*

Populations A **population** is a group of individuals of the same species living in a particular place. Some examples of populations include the bullfrog population of a pond, the lion population of a savanna, the bluebonnet population of a field, and even the bacterium population of a petri dish. Examples of populations are shown in Figure 2-5. The factors that influence the growth of populations are discussed in Chapter 13.

Communities Just as an individual organism doesn't exist in isolation, neither does a population. Every population is part of a **community**, a group of interacting populations of different species. All of the living inhabitants of an ecosystem make up a community. A pond community, for example, includes the populations of all the different plants, fish, insects, amphibians, and microorganisms that live in and around the pond.

Butterfly Ecologist

Some monarch butterflies travel 3,200 km (2,000 mi.) to reach their wintering grounds in Mexico, only to find their habitat destroyed. Turn to pages 52–53 to see how one young man is helping to save the monarchs' habitat.

Figure 2-6 **A lion's niche includes all of its relationships with its environment.**

a. Tick on a lion's face
b. Lion fighting with hyenas
c. Lions feeding on kill
d. Lions drinking at water hole

Answers to Section Review can be found on page T142.

Section Review

1. What are the biotic components of the soil ecosystem in Figure 2-2? What are the abiotic components?
2. Give an example of a population that was not mentioned in this section.
3. Choose an animal that was not mentioned in this section, and discuss the difference between that animal's habitat and its niche.

Thinking Critically

4. *Analyzing Relationships* What part of an ecosystem is not part of a community? After you figure that out, write your own definition of the term *community,* using the terms *biotic factors* and *abiotic factors.* Your definition should be different from the one given in this textbook.

Niche and Habitat

Consider a lion living on the savanna in eastern Africa. How does the lion fit into its ecosystem? Lions survive by killing and eating other animals, such as gazelles, zebras, and wildebeests. After the lions have killed an animal and eaten their fill, scavengers such as jackals, vultures, and hyenas devour what is left of the carcass. Bacteria, fungi, and insects in the soil also feed on the carcass, causing it to rot.

The lion itself is food for other organisms. Small animals like ticks, fleas, biting flies, and mosquitoes drink the lion's blood. The lion's dung serves as food for organisms that live in the soil. When the lion dies, it may be eaten by the same scavengers that once fed on the leftovers of its kills.

All of the lion's relationships with its environment—both the living and nonliving parts—make up what is called its niche. An organism's **niche** is its way of life. The lion's niche includes the relationships already described, as well as many others, such as when and how often it reproduces, how many offspring it has, what time of the day it is most active, and where it finds shelter. You can think of an organism's niche as its "profession," or how it "makes a living." Some parts of a lion's niche are shown in Figure 2-6.

The actual place an organism lives is called its **habitat.** The lion's habitat, for example, is a savanna. A howler monkey's habitat is a rain forest, a cactus's habitat is a desert, and a water lily's habitat is a pond. An organism's habitat may be thought of as its "address."

HOW SPECIES INTERACT WITH EACH OTHER

AFTER READING THIS SECTION YOU SHOULD BE ABLE TO

- explain the five major types of species interactions and give examples of each.

In Section 2.1 you saw a few of the ways that species within an ecosystem can affect each other, both positively and negatively. In this section you will look more closely at five major types of species interactions: predation, competition, parasitism, mutualism, and commensalism.

PREDATION

In **predation**, one organism kills and eats another organism. The organism that is eaten is called the **prey**, and the one that does the eating is called the **predator.** Familiar examples of predation include lions feeding on zebras, cougars eating deer, snakes consuming mice, and birds eating insects. The blue whale, the largest animal on Earth, is also a predator, because it feeds on tiny krill (shrimplike marine animals). Figure 2-7 shows various predators in action.

Figure 2-7 **When one organism kills and eats another, it is called predation.**

a. Chameleon catching a grasshopper
b. Starfish opening a clam
c. Golden eagle carrying away a prairie dog
d. Cougar attacking a mule deer
e. Red fox with a ground squirrel

FIELD ACTIVITY

You can study competition between bird species at home or at school. Build a bird feeder using a plastic milk jug, a metal pie pan, or some other inexpensive material. Fill the feeder with bread crumbs, sunflower seeds, or commercial birdseed. Observe the birds that visit the feeder. Sit quietly in the same spot, and make observations at the same time each day for several days in a row. Keep a record that includes data about the kinds of birds that use the feeder, the kinds of seeds that the birds prefer, factors that affect how much the birds eat, and which birds are better competitors for the birdseed.

Predators sometimes limit the population size of their prey. Often, though, predators have little effect on the numbers of their prey because they tend to feed on whatever prey species is abundant at the time. Predators also tend to feed on old and weak individuals who are more likely to die soon anyway.

COMPETITION

When hyenas fight with lions over the same animal carcass, they are in competition for food. Such contests can turn into fierce battles in which hyenas or, more rarely, lions can be killed. The relationship between the lions and the hyenas is called competition. **Competition** is a relationship between species in which they attempt to use the same limited resource. Another example of competition is the relationship between two species of plants competing for the limited amount of sunlight that reaches the floor of a forest. Some examples of competition are shown in Figure 2-8.

Species can compete even if they never come into contact with each other. Suppose one insect species feeds on a certain plant during the day, and another feeds on the same plant during the night. Because they use the same food source, the two species are competitors, even though they never come into direct contact with each other. Similarly, two plant species that flower at the same time and depend on the same pollinators are in competition for pollinators, even if they don't compete in any other way.

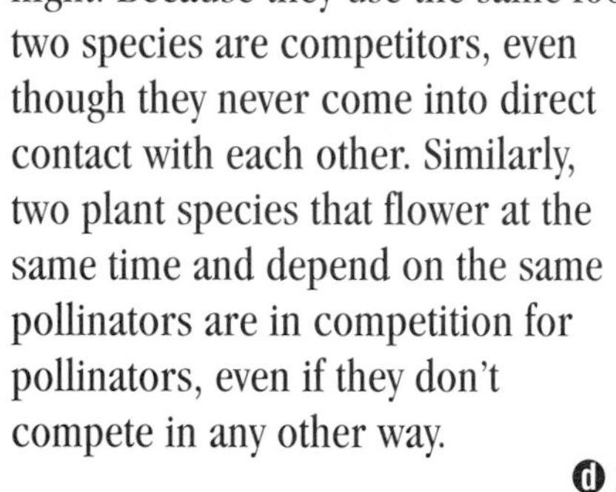

Figure 2-8 Competing species attempt to use the same limited resource.

a. Imported fire ants compete with native ants for territory.
b. Two young plants on a forest floor compete for sunlight.
c. Imported Kudzu vines cover native plants.
d. Pandas compete with humans for bamboo.

Figure 2-9 These organisms are parasites. The mistletoe is a parasite because it grows into the tissues of the tree and uses the tree as a source of nutrients.

Parasitism

What do ticks, fleas, tapeworms, heartworms, blood-sucking leeches, and mistletoe have in common? They are all **parasites**—organisms that live in or on another organism and feed on it without immediately killing it. Figure 2-9 shows some examples of parasites. The organism the parasite takes its nourishment from is known as the **host.** The relationship between the parasite and its host is called **parasitism.**

How is parasitism different from predation? The main difference is that parasites, unlike predators, usually do not immediately kill their hosts. Another difference is that a parasite lives in or on the host for part of its life. Most animal populations are negatively affected by parasites. Animals may be weakened by parasites, making them more vulnerable to predators.

Mutualism

Although you are probably not aware of their existence, billions of bacteria live in your intestines. They do not make you sick. Instead, they are your partners in digesting food. They break down food that you would otherwise be unable to digest. They also produce necessary substances your body cannot make. For instance, they supply you with vitamin K, which is essential for proper blood clotting. And you give them something in return—a warm, dark, food-rich environment in which to live.

The relationship between you and your intestinal bacteria is known as mutualism. **Mutualism** is a cooperative partnership between two species. Another example of mutualism can be found in Central America, where acacia trees are covered with ants, as you can see in Figure 2-10 on the next page. The acacia trees provide food (nectar and protein bodies on the leaves) and nesting

ECO-FACT

One South American mite species lives entirely on the blood it sucks from the hind feet of soldier army ants.

sites (hollow thorns) for the ants. In exchange, the ants defend the acacia trees from herbivores such as grasshoppers and beetles. The ants will even attack mammals that interfere with the trees. In one study, ecologist Daniel Janzen removed the ants from some acacias by spraying them with insecticides. He found that the trees grew more slowly and were heavily eaten by herbivores. Janzen had demonstrated that the ants provide an important service for the acacia trees.

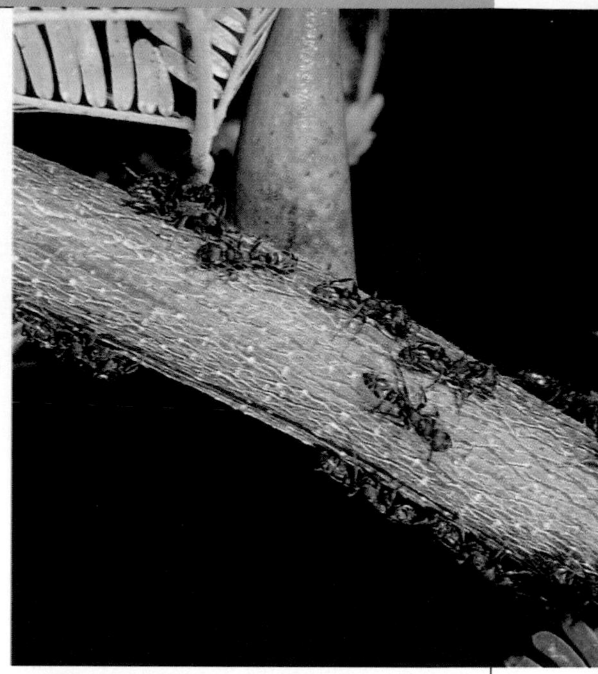

Figure 2-10 These ants defend the acacia tree against herbivores. In return, the acacia trees provide food and nesting sites for the ants. This is an example of the relationship called mutualism.

Wildlife Garden

If you think you might enjoy watching species interact with each other, see pages 396–397 to find out how to create your own wildlife garden.

Commensalism

Commensalism is perhaps the rarest and strangest type of species interaction. **Commensalism** is a relationship in which one species benefits and the other is neither harmed nor helped, such as the interaction between remoras and sharks, shown in Figure 2-11. Remoras are fish that attach themselves to sharks and feed on scraps of food left over from the shark's meals. Although the sharks do not seem to benefit by this relationship, they aren't harmed either.

Figure 2-11 This remora attached itself to the shark and benefits by eating food left over from the shark's meals. The shark is neither helped nor harmed by the remora. Commensalism is the name of this kind of relationship.

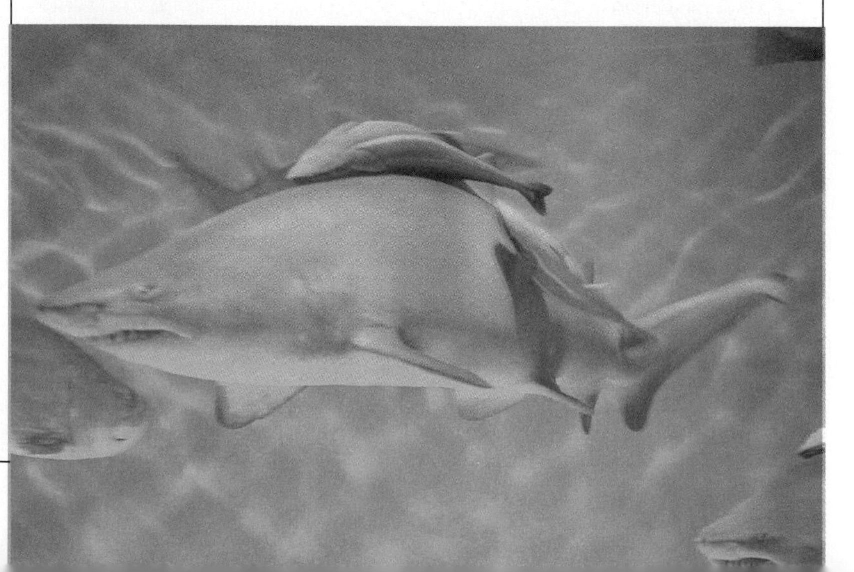

Answers to Section Review can be found on page T142.

Section Review

1. Give one example of a predator-prey relationship that is not mentioned in this chapter.
2. What is the difference between mutualism and parasitism?

Thinking Critically

3. *Inferring Relationships* Why is it advantageous for a parasite to leave its host alive?
4. *Analyzing Relationships* How could you show that a suspected case of mutualism was not a case of commensalism?

SECTION 2.3

ADAPTING TO THE ENVIRONMENT

AFTER READING THIS SECTION YOU SHOULD BE ABLE TO

1. explain the process of evolution by natural selection.
2. explain the concept of adaptation.
3. explain the concept of coevolution.
4. define the term *extinction.*

Organisms tend to be well suited to their environments. Kangaroo rats, for instance, are well suited to the deserts of the southwestern United States and northern Mexico, where there is little water. A kangaroo rat is shown in Figure 2-12. One of the ways kangaroo rats cope with this dry environment is by conserving the water in their bodies. They eliminate very little water in their urine and feces. They do not sweat, so they don't lose water in that way. And they are active at night, when it is cooler. In fact, kangaroo rats are so effective at saving water that they never need to drink.

Figure 2-12 **Kangaroo rats are adapted to the deserts of the southwestern United States and northern Mexico. They are so good at conserving water in their bodies that they never need to drink.**

EVOLUTION BY NATURAL SELECTION

How does this close match between organisms and their environments come about? In 1859 an English naturalist named Charles Darwin proposed an answer to this question. Darwin observed that members of a population differ from each other in form, physiology, and behavior. Some of these differences are hereditary—that is, passed from parent to offspring. Darwin proposed that the environment exerts a strong influence over which individuals have offspring. Some individuals, because of certain traits, are more likely to survive and have offspring than other individuals. Darwin used the term **natural selection** to describe the unequal survival and reproduction that results from the presence or absence of particular traits.

Darwin proposed that over many generations natural selection causes the characteristics of populations to change. A change in the genetic characteristics of populations from one generation to the next is known as

Evolution by Natural Selection
1. All organisms have the ability to produce more offspring than can possibly survive. For instance, a female cod (a type of fish) lays millions of eggs. If all these eggs hatched and the young grew to adulthood, the world would be flooded with cod. Of course, most of them don't survive.
2. The environment contains things that kill organisms. The environment is often hostile: hot or cold, dry or flooded. Predators are common, and the resources needed to survive and reproduce—food, water, living space, light—are often in short supply. Therefore, individuals compete for these limited resources. As Darwin put it, there is a "struggle for existence" among individuals.
3. Individuals vary, or differ, in their traits. They may differ in size, coloration, running speed, resistance to disease, and many other traits. This variation must be inherited for it to influence natural selection.
4. Some inherited traits give individuals an advantage in coping with environmental challenges, allowing them to survive longer and produce more offspring. Organisms with these traits are "naturally selected for."
5. Because individuals with advantageous traits have more offspring, each new generation contains a greater proportion of the offspring with these traits than did the previous generation. Since these traits are inherited, offspring of individuals with the advantageous traits become more numerous in the next generation. Individuals without the trait become less numerous. Gradually, over generations, the population contains more and more individuals with the advantageous trait. This process is evolution by natural selection.

Figure 2-13 These are the major points of Darwin's theory of evolution by natural selection.

evolution. According to Darwin's theory, the process of natural selection is responsible for evolution. Figure 2-13 summarizes the main points of Darwin's theory of evolution by natural selection.

Imagine a group of deer living in a lowland area where the temperatures are warm most of the year. What if some of these deer

NATURAL SELECTION AND THE PEPPERED MOTH OF ENGLAND

In the 1950s, H.B.D. Kettlewell of England's Oxford University developed and tested a hypothesis concerning a species of moth called the peppered moth. His results led to a very convincing demonstration of natural selection in action.

Kettlewell knew several things about the peppered moth. He knew it was nocturnal (active at night), that it rested on tree trunks during the day, and that some species of birds ate the moths.

In addition, Kettlewell learned a few things about the moth's history. Until the mid-nineteenth century, only pale, speckled peppered moths had ever been reported. Then in 1845, a dark gray specimen was found. Over the years, dark gray moths were seen more frequently, and by the 1950s nearly all peppered moths in some areas were dark gray. Kettlewell found this piece of information intriguing. He wondered why the dark moths had become more common in some parts of the country.

As it turned out, the increase in dark moths paralleled an increase in the number of polluting factories in England. In many English forests, air

Which moths do you think would be most likely to be eaten by birds?

Figure 2-14 In this imaginary situation, the thicker fur of the mountain deer is an adaptation to a cold climate. An adaptation increases an organism's chance of survival and reproduction in a particular environment.

became separated from the others high in the mountains, where the temperatures are cold most of the year? (See Figure 2-14.) Many of the deer might die in the cold. However, a few deer might have thicker fur that kept them warmer. Those deer would be more likely to survive and have offspring. The offspring with thicker fur

Answers to Thinking Critically and to Case Study caption can be found on page T143.

pollution from these factories had turned the trunks of trees from white to black. Kettlewell hypothesized that the dark moths had become more common because they were camouflaged against the blackened trunks and therefore less likely to be seen by predators. Pale moths, on the other hand, were obvious against the dark background and were more likely to be seen by predators.

Kettlewell chose two forests in which to test his hypothesis. The first, near the industrial city of Birmingham in central England, was heavily polluted; both the tree trunks and moths there were dark. The second forest, in southern England's Dorset County, had not been affected by pollution; there both the tree trunks and moths were pale. Into each forest, Kettlewell released equal numbers of pale and dark moths (marked underneath with dots of paint for later recognition). He set up a series of traps to recapture the surviving moths. In unpolluted Dorset, twice as many pale moths as dark moths were recaptured. Near Birmingham, twice as many dark moths were caught.

Although Kettlewell had demonstrated that moths matching the color of the local tree trunks were more likely to survive, he had not yet shown that this was due to predators. So he placed moths on tree trunks and with hidden cameras filmed the reactions of birds. In both forests, birds more frequently seized and ate the moths that didn't match the color of the tree trunks.

Kettlewell concluded that birds were the agents of natural selection that drove the evolution of dark coloration in these moths. In polluted areas, birds found and killed more of the moths with the gene for pale coloration. A higher proportion of the moths with the gene for dark color survived. So when this group of moths reproduced, the gene for dark color was inherited by many of their offspring. The new generation contained more dark moths than the parent population, which demonstrated evolution—a genetic change in a population from one generation to the next.

THINKING CRITICALLY

❶ *Analyzing Methods* Did Kettlewell use scientific methods to test his hypothesis? Explain.

❷ *Predicting Outcomes* What do you think would happen to the moths if the pollution disappeared?

Figure 2-15 **Cabbage plants evolved a chemical defense, and cabbage butterfly caterpillars evolved a way to break down the defense. This is an example of coevolution.**

would be more likely to survive to produce offspring that also had thicker fur. With each generation, a greater proportion of the deer would have thicker fur. The deer's thicker fur would be an **adaptation,** an inherited trait that increases an organism's chance of survival and reproduction in a certain environment. Another example of an adaptation is described in the Case Study on pages 44–45.

Coevolution

An organism's environment includes not only physical aspects such as climate, but also other organisms, which can be strong forces in natural selection. When two or more species evolve in response to each other, it is called **coevolution.** Examples of coevolution may be found between predators and their prey. Natural selection favors predators that are efficient at capturing prey. As a result of this selective pressure, predators have evolved numerous adaptations for finding, capturing, and subduing their prey. Similarly, prey have evolved various ways of avoiding, escaping, and fighting off predators. An example of predator-prey coevolution is found in the relationship between crabs (predators) and marine snails (their prey). Indo-Pacific crabs have stronger claws than Caribbean crabs, and Indo-Pacific snails have thicker shells than Caribbean snails. Presumably, the Indo-Pacific crabs and snails have evolved these characteristics in response to each other. Scientists think that a coevolutionary "arms race" between the crabs and snails has been going on for a long time.

Coevolution also occurs between plants and herbivores. Since plants cannot move, they must defend themselves from herbivores in other ways. Some plants produce poisonous chemicals that discourage or even kill herbivores. In response, some herbivores evolved the ability to detoxify these chemical defenses. Cabbage butterfly caterpillars, for example, have the ability to break down the chemical defenses of plants that defend themselves with mustard oils—chemicals that are harmless to humans but poisonous to many insect species. Because the caterpillars can break down mustard oils into nontoxic compounds, they can feed on plants that other insects cannot. (See Figure 2-15.) In this example, the plants evolved a chemical defense against plant-eating insects, and in response, the insects evolved a method of dealing with the chemical defense.

Extinction

The irreversible disappearance of a population or a species is called **extinction.** A species is extinct when the last individual member dies. Though extinction is a natural biological process, human population growth is causing species to become extinct at an alarmingly rapid rate. This problem is discussed in Chapter 10.

Answers to Section Review can be found on page T143.

Section Review

1. What role does the environment play in Darwin's theory of evolution by natural selection?
2. Many people confuse the terms *evolution* and *natural selection.* What is the difference between the two terms?

Thinking Critically

3. *Analyzing Processes* A population of rabbits evolves thicker fur in response to a colder climate. Is this an example of coevolution? Explain your answer.
4. *Inferring Relationships* In places where a certain insecticide (substance that kills insects) has been used for 10 or more years, the insecticide loses its effectiveness. Offer an explanation of this based on what you learned about Darwin's theory.

CHAPTER 2

HIGHLIGHTS

SUMMARY

- Ecosystems are ecological units that include all the different organisms living in a certain area, along with their physical environment.
- Living things are organized from the smallest to the largest unit in the following ecological levels: organism, population, community, ecosystem, and biosphere.
- Each organism has a niche, or "profession," and a habitat, or "address."
- Species interact with each other in five major ways—predation, competition, parasitism, mutualism, and commensalism.
- Charles Darwin proposed that organisms become well suited to their environments through evolution by natural selection.
- Coevolution is a type of evolution by natural selection in which two or more species evolve in response to each other.

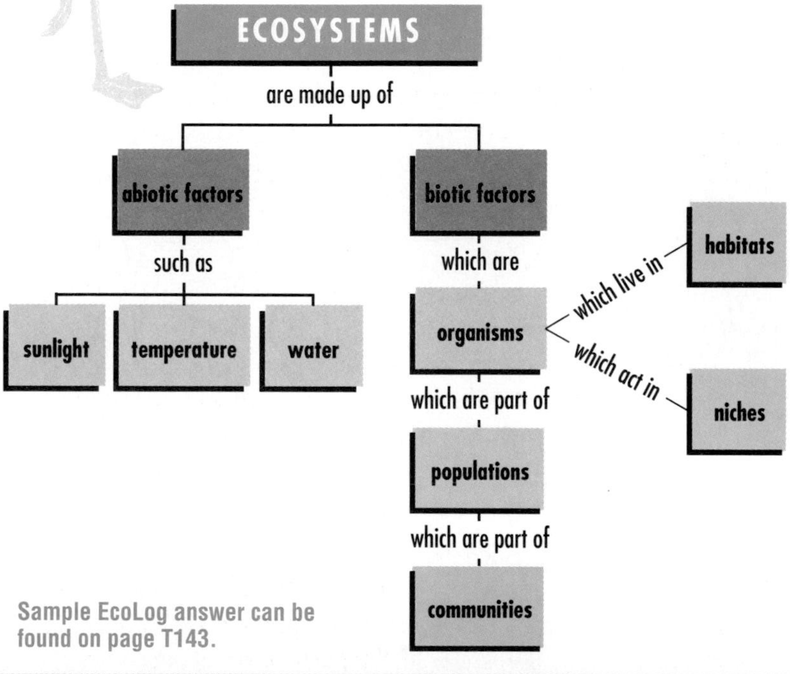

ECOSYSTEMS
are made up of
abiotic factors
such as
sunlight
temperature
water
biotic factors
which are
organisms
which live in
habitats
which act in
niches
which are part of
populations
which are part of
communities

Vocabulary Terms

abiotic factors (p. 35)
adaptation (p. 46)
biotic factors (p. 35)
coevolution (p. 46)
commensalism (p. 42)
community (p. 37)
competition (p. 40)
ecosystem (p. 34)
evolution (p. 44)
extinction (p. 46)
habitat (p. 38)
host (p. 41)
mutualism (p. 41)
natural selection (p. 43)
niche (p. 38)
organism (p. 37)
parasites (p. 41)
parasitism (p. 41)
population (p. 37)
predation (p. 39)
predator (p. 39)
prey (p. 39)
species (p. 37)

Sample EcoLog answer can be found on page T143.

EcoLog

Now that you've studied this chapter, revise your answer to the question you answered at the beginning of the chapter, based on what you have learned.

- If all the lions on an African savanna were killed or removed, how might their absence affect the other living things on the savanna?

REVIEW

UNDERSTANDING VOCABULARY

1. For each pair of terms, explain the difference in their meanings.
   a. biotic factors
      abiotic factors
   b. population
      community
   c. niche
      habitat
   d. parasitism
      mutualism
   e. adaptation
      extinction

RELATING CONCEPTS

Answers not indicated can be found on page T143.

2. Copy the unfinished concept map below onto a sheet of paper. Then complete the concept map by writing the correct word or phrase in each box containing a question mark.

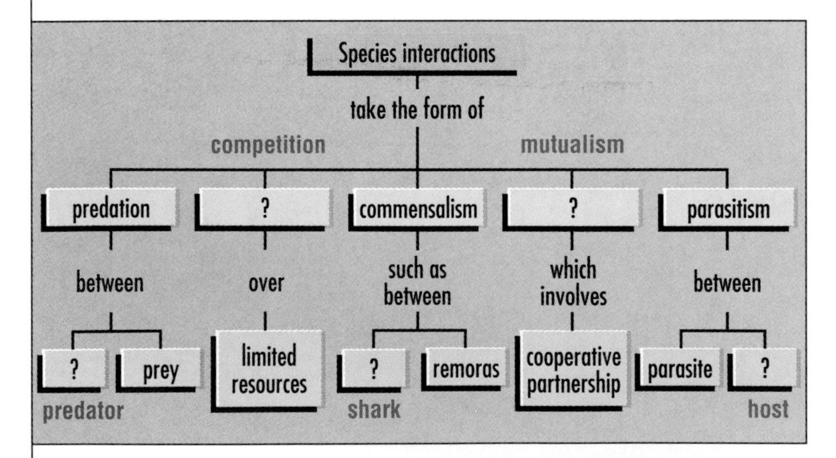

UNDERSTANDING CONCEPTS

Multiple Choice

3. Which is NOT an abiotic factor in an ecosystem?
   a. temperature
   b. air pressure
   c. sunlight
   (d.) bacteria

4. Which of these sayings most closely describes how an ecosystem operates?
   a. A stitch in time saves nine.
   (b.) A chain is only as strong as its weakest link.
   c. It's always darkest before the dawn.
   d. A penny saved is a penny earned.

5. Select the correct order from the choices below.
   a. ecosystem, population, community, organism
   b. population, organism, ecosystem, community
   (c.) organism, population, community, ecosystem
   d. community, population, organism, ecosystem

6. Which of these organisms could belong to the same population?
   (a.) two bluebonnet plants
   b. a gorilla and an orangutan
   c. Joey and his cat
   d. a zebra and a wildebeest

7. Which is NOT one of the major points in the theory of evolution by natural selection?
   a. Organisms have the ability to produce more offspring than can survive.
   b. Limited resources produce a struggle for existence.
   (c.) A community includes populations of different species.
   d. Organisms with advantageous traits produce more offspring.

8. Which is an example of coevolution?
   (a.) flowers that can be pollinated by only one insect species
   b. deer that live in a cold region and have thick fur
   c. dark gray moths that are found near Birmingham, England
   d. desert rats that do not sweat

9. Which of these are LEAST likely to belong to the same species?
   a. all mountain gorillas
   (b.) all predators
   c. all African elephants
   d. all *Escherichia coli* bacteria
10. Which is an example of competition?
   (a.) two species of insects that feed on the same rare plant
   b. a lake near a forest in the northeastern United States
   c. an African lioness feeding her cubs
   d. a tick living on a dog
11. Ants and acacia trees have a mutualistic relationship because
   (a.) they benefit each other.
   b. they are part of the same ecosystem.
   c. they are both adapted to a humid climate.
   d. the ants eat parts of the acacia tree.
12. Which of these statements is true of parasitism?
   a. The presence of a parasite does not affect the host.
   b. Parasitism is identical to predation.
   (c.) The presence of a parasite may make an animal more susceptible to predation.
   d. Both the parasite and the host benefit from parasitism.

Short Answer

13. A tapeworm lives inside the intestines of a cow and feeds by absorbing food that the cow has already digested. Is this an example of predation? Explain your reasoning.
14. How is an ecosystem similar to a mechanical system such as an automobile engine?
15. As a human, what types of interactions do you participate in with other species? Give two examples.
16. Explain how two species can compete for the same resource even if they never come into contact with each other.

Interpreting Graphics

17. Examine the graph below. Each line represents a different species. What type of interaction could be taking place between species A and B? between A and C? Explain the reasoning behind each of your answers.

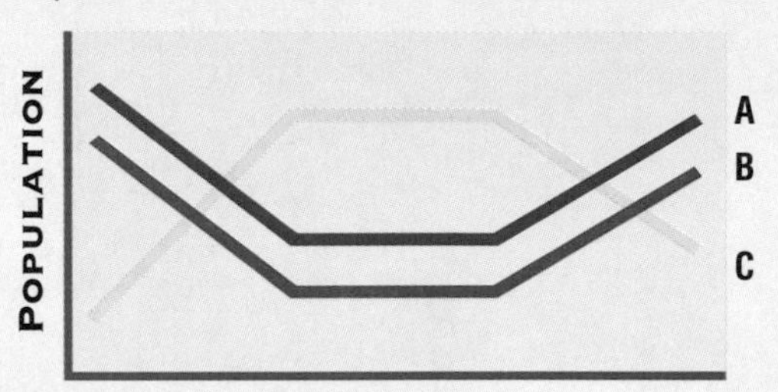

Thinking Critically

18. ***Analyzing Relationships*** Imagine that one species becomes extinct as a direct result of the extinction of another species. Which relationship did the two species more likely have, competition or commensalism? Explain your reasoning.
19. ***Inferring Relationships*** How might humans be affecting the course of evolution?

Themes in Science

20. ***Evolution*** Why is time an important factor in the process of evolution by natural selection?
21. ***Interacting Systems*** Why are abiotic factors just as important to an ecosystem as biotic factors?

Cross-Discipline Connection

22. ***Math*** Suppose that scientists introduced a breeding pair of one species into a certain habitat. This species doubles its population size each year. Another species, native to the habitat, keeps its numbers at a stable 500 individuals from one year to the next. In how many years will the introduced species have more members than the native species?

Discovery Through Reading

Read about how some animals protect themselves from predators by congregating in large herds. See "Survival by the Numbers" in *National Wildlife*, Oct.–Nov. 1993, pp. 4–13.

PORTFOLIO ACTIVITY

Do a special project about the ecosystem in which you live. Try to identify all of the biotic and abiotic factors in your ecosystem. List some of the interactions that take place among the biotic factors. Then, using your data, develop a series of storyboards to show what might happen if one factor were suddenly removed from this ecosystem.

CHAPTER 2 INVESTIGATION

SHOWDOWN ON THE PRAIRIE

In a quiet rural area in the Old West, a struggle for survival is going on. In the prairie, exotic or imported grass species are gradually replacing the native grasses. When the plant community changes, it affects the entire ecosystem, possibly threatening the survival of many types of native wildlife.

You are a plant ecologist studying ways to restore native grasses to the prairie. You want to learn why one type of grass might outcompete another, and whether different species succeed better under different conditions. In order to answer these questions, you and a team of other biologists will first do a simple plant competition experiment. You will then devise your own experiment to test the factors that affect competition among grasses.

Answers to this Investigation can be found on page T143.

MATERIALS

- flowerpots prepared by teacher
- 3 small flowerpots
- 3 aluminum pie pans
- potting soil
- native grass seeds (buffalo is best)
- imported grass seeds (rye or fescue)
- tape and marker for labeling flowerpots
- metric ruler
- hand lens
- graph paper
- one or more of the following (optional): light source, scissors, fertilizer, other soil additives
- notebook
- pen or pencil

CHECK OUT THE COMPETITION

1. In your notebook, make a table like the one shown here to record data. Add additional days if desired.

	Species A Alone		Species B Alone		Plants in Mixed Pot			
					Species A		Species B	
	Max. Height	No. of Leaves	Max. Height	No. of Leaves	Max. Height	No. of Leaves	Max. Height	No. of Leaves
Day 1								
Day 2								
Day 3								
Day 4								
Day 5								
Day 6								
Day 7								

2. Each team of biologists will be given three flowerpots filled with potting soil. One pot has been planted with grass seeds of native species A, another contains grass seeds of exotic species B, and the third contains equal numbers of seeds of species A and B mixed together. The pots with only one species of grass are the controls in this experiment. What is the purpose of having these two pots? Write your answer in your notebook.
3. How do you think the growth of each species in the mixed pot will compare to its growth when grown alone? Write your hypothesis in your notebook.
4. Check the pots each day. When the soil seems dry, add the same amount of water to all three pots. All pots should be watered at the same time.

5. As the grasses begin to sprout, count the number of leaves and measure the height of the tallest leaf in millimeters. In order to distinguish the two species in the mixed pot, compare them with the grasses in the two individual pots using a hand lens. Most grass species will differ in the amount and kind of hair, hooks, or spines covering them and in the presence or absence of glands. Color and size are not as reliable for identifying species. Record your observations each day.
6. After a period of 7 to 10 days, use your data table to make two line graphs. One graph should compare the height of plants in the three pots over time, and the other should compare the number of leaves.
7. How did the growth of each species in the absence of competition differ from its growth in the presence of competition? Was your hypothesis supported?

WEED OUT THE VARIABLES

8. Why is it difficult to simulate competition in a laboratory? In other words, why would the grass that survived best in your competition experiment not necessarily survive best in the wild?
9. Many factors might give one species a competitive edge over another. Select one of the factors listed below to investigate, or come up with your own.
    - amount and frequency of watering
    - amount of light
    - grazing (can be simulated by clipping the grasses at timed intervals)
    - level of nitrates in the soil (can be increased by adding fertilizer)
10. Write down a hypothesis regarding the effect of the factor you chose. For example, you might expect grass species A to outcompete grass species B during drought conditions.
11. In your notebook, describe an experiment that will test your hypothesis. Base the design on the preliminary competition experiment you did. Keep in mind that whatever modification you make must be applied to all three pots.
12. Based on your hypothesis, predict what you expect to happen. For example, you might expect grass species A to produce more and longer leaves than species B when equal numbers of seeds of the two species are grown together in a pot that is watered infrequently. Write your prediction in your notebook.

PUT THE COMPETITORS TO THE TEST

13. Carry out your experiment. Record your results in a table similar to the one used for the preliminary experiment.
14. Illustrate the results of your experiment using graphs of the data collected.
15. Compare your results with those obtained by other teams. Does one species outcompete the other under most conditions, or does each one do better under its own preferred conditions? Using the information from all experiments done by the class, propose some explanations as to why one species might outcompete another under various conditions.
16. Why is it difficult to draw conclusions concerning plant competition in a natural setting based on simple laboratory experiments? What additional laboratory and field studies could be done to help develop a plan for restoring native grasses to an area that has been taken over by exotics?

Examine the details of the grass leaves to identify the two species in the mixed pot.

Native grasses have been restored to this Kansas prairie that had been invaded by exotic species.

Making a DIFFERENCE

Butterfly Ecologist

Imagine millions of butterflies swirling through the air like autumn leaves, clinging in tightly packed masses to tree trunks and branches, and covering low-lying forest vegetation like a luxurious, moving carpet. According to Alfonso Alonso-Mejía, this is quite a sight to see.

Every winter Alfonso climbs up to the few remote sites in central Mexico where about 200 million monarch butterflies spend the winter. He is researching the monarchs because he wants to help preserve their habitat and the butterflies themselves. His work is also helping him earn a Ph.D. in ecology from the University of Florida.

Alfonso's work is helping Mexican conservationists better understand and protect monarch butterflies.

Monarchs are famous for their long-distance migration. Those that eventually find their way to Mexico come from as far away as the northeastern United States and southern Canada, some of them having traveled an amazing 3,200 km (2,000 mi.) before reaching central Mexico.

Unfortunately, the habitat that the monarchs travel such distances to reach is increasingly threatened by logging and other human activities. Only nine of the monarchs' wintering sites remain. Five of those are set aside as sanctuaries for the butterflies, but even these sanctuaries are endangered by people who cut down fir trees for firewood or for commercial purposes.

Alfonso's research has led to important discoveries about the monarch butterfly.

Alfonso's work is helping Mexican conservationists better understand and protect monarch butterflies. Especially important is Alfonso's discovery that monarchs depend on bushlike vegetation, called understory vegetation, that grows beneath the fir trees.

Alfonso's research showed that when the temperature dips below freezing (as it often does at the high altitude sites where the monarchs winter), understory vegetation can mean the difference between life and death for some monarchs. This is because low temperatures (–1 to 4°C, or 30 to 40°F) limit the monarchs' movement. In fact, the butterflies are not even able to fly. At extremely cold temperatures (–7 to –1°C, or 20 to 30°F), monarchs resting on the forest floor are in danger of freezing to death. But as long as there is understory vegetation, the monarchs can slowly climb the vegetation until they are at least 10 cm (4 in.) above the ground. This tiny difference in elevation can provide a microclimate that is warm enough to ensure the monarchs' survival.

The importance of understory vegetation was not known before Alfonso did his research. Now, thanks to Alfonso's work, Mexican conservationists will better protect the understory vegetation.

Alfonso is undoubtedly on his way to becoming a world-class monarch butterfly ecologist. When he completes his Ph.D., Alfonso Alonso-Mejía will devote himself to preserving the Mexican habitat of the monarch butterfly.

For More Information . . .

If you are interested in a nationwide tagging program to help scientists learn more about the monarch's migration route, write to Texas Parks and Wildlife, Nongame Program, 4200 Smith School Road, Austin, TX 78744.

The monarch butterflies that Alfonso studies migrate up to 3,200 km (2,000 mi.) before reaching the remote mountainous locations in central Mexico. ▼

CHAPTER 3

How Ecosystems Work

SECTION 3.1

Energy Flow in Ecosystems

SECTION 3.2

The Cycling of Materials

SECTION 3.3

How Ecosystems Change

"You could cover the whole world with asphalt, but sooner or later green grass would break through."

Ilya Ehrenburg, Russian Writer

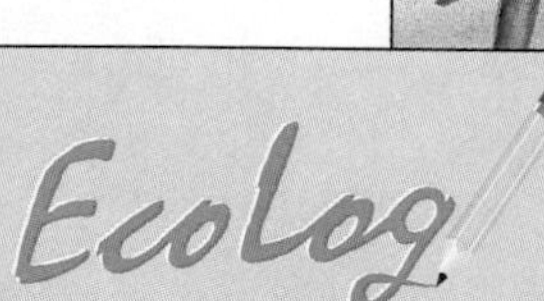

Before you read this chapter, take a few minutes to answer the following questions in your EcoLog.

❶ What does sunlight have to do with the amount of food available to a hawk, which eats only meat?

❷ How is it possible that an atom of nitrogen in your sandwich was once part of a dinosaur's body?

ENERGY FLOW IN ECOSYSTEMS

AFTER READING THIS SECTION YOU SHOULD BE ABLE TO

❶ describe the roles of producers and consumers.

❷ trace the transfer of energy from the sun to producers and from producers to consumers.

❸ differentiate among different types of consumers.

❹ draw a food chain and a food web.

❺ explain why an energy pyramid is a good representation of trophic levels.

Just as a car cannot run without fuel, an organism cannot survive without a supply of energy. How organisms meet this need for energy greatly affects the structure of ecosystems. Where does an organism's energy come from? The answer to that question depends on the organism, but the ultimate source of energy for almost all organisms is the sun.

LIFE DEPENDS ON THE SUN

Plants, algae, and some kinds of bacteria can capture solar energy and store it as food. On a sunny spring day, for instance, a clover plant is producing food in its leaves through the process of photosynthesis. The leaves absorb sunlight, which drives a series of chemical reactions that require water and carbon dioxide. The result is the production of starch, an energy-rich food.

Figure 3-1 These plants are using sunlight to produce food in their leaves.

Figure 3-2 Almost all organisms depend on the sun for energy. Most producers, like the grass shown here, get energy directly from the sun.

Figure 3-3 The tube worms shown below are part of an ecosystem that is dependent on bacteria.

When a rabbit eats the clover plant, it gets energy from the food the clover made during photosynthesis. Then, if a coyote eats the rabbit, some of the energy is transferred from the rabbit to the coyote.

In this example, the clover is the **producer.** A producer is an organism that makes its own food. Producers are also called autotrophs (self-feeders). Both the rabbit and the coyote are **consumers,** organisms that get their energy by eating other organisms. Consumers are also called heterotrophs (other-feeders).

The important thing to remember is that almost all organisms get their energy from the sun. Producers get energy *directly* from the sun. Consumers get energy *indirectly* from the sun by eating producers or other organisms that eat producers.

An Exception to the Rule: Deep-Ocean Ecosystems The previous paragraph said that *almost* all organisms get their energy directly or indirectly from the sun. That's because in 1977 scientists discovered areas on the bottom of the ocean off the coast of Ecuador that were teeming with life, even though sunlight didn't reach down that far. They found rich communities of fish, worms, clams, crabs, mussels, and barnacles living around cracks in the ocean floor through which hot water escaped. (See Figure 3-3.) These deep-ocean ecosystems exist in total darkness, where photosynthesis cannot occur. So where are the organisms getting their energy? As it turns out, there are bacteria that make food from hydrogen sulfide, which is present in the hot water that escapes from the cracks in the ocean floor. These bacteria are producers that have evolved a way to make food without sunlight. They are eaten by animals and thus support a rich ecosystem.

Who Eats What

Figure 3-4 shows who eats what in an ecosystem. Consumers that eat only producers are called **herbivores** (plant-eaters). Rabbits are herbivores, as are cows, sheep, deer, grasshoppers, and many other animals. Consumers that eat only other consumers, such as lions and hawks, are **carnivores** (flesh-eaters). You already know that humans are consumers, but what kind of consumers are we? After all, most humans eat *both* plants and animals. We are considered to be **omnivores** (eaters of all). Bears and pigs are other examples of omnivores. Consumers that get their food by breaking down dead organisms, causing them to rot, are **decomposers**. Bacteria and fungi are examples of decomposers. The decomposers make it possible for the nutrients contained in the rotting material to return to the soil or water.

The CALL of the WILD

To some, the howl of a wolf is a thrilling sound. To others, it signals death and destruction. Turn to pages 76–77 to read about the controversy surrounding gray wolves, carnivores of the American Northwest.

Respiration: Burning the Fuel

So far, you have learned how organisms get energy. Now, how do they use the energy they get? Take yourself as an example to understand the process. Suppose you have just eaten a large meal. That food contains a lot of energy. Your body gets the energy out of the food by using the oxygen you breathe to break down the food molecules. In this way your body "liberates" the energy stored in the food.

The process of breaking down food to yield energy is called **cellular respiration.** It is similar to combustion in an automobile

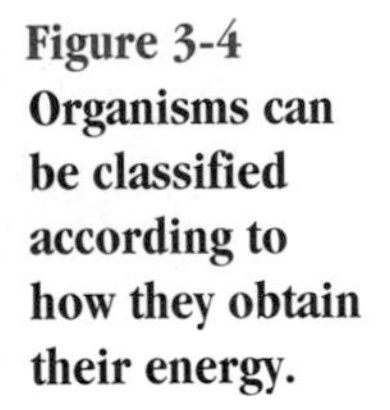

Who Eats What in an Ecosystem

Name	Energy Source	Examples
Producer	Makes its own food	Grasses, ferns, trees, algae, some bacteria
Consumer	Gets energy by eating other organisms	
Herbivore	Eats only producers	Cows, sheep, deer, grasshoppers
Carnivore	Eats only other consumers	Lions, hawks, wolves, pike
Omnivore	Eats both producers and consumers	Bears, pigs, humans
Decomposer	Breaks down dead organisms in an ecosystem, returning nutrients to the soil or water	Fungi, bacteria

Figure 3-4 Organisms can be classified according to how they obtain their energy.

$$6CO_2 + 6H_2O + \text{light} \longrightarrow C_6H_{12}O_6 + 6O_2$$

PHOTOSYNTHESIS

$$C_6H_{12}O_6 + 6O_2 \longrightarrow 6CO_2 + 6H_2O + \text{energy}$$

CELLULAR RESPIRATION

Figure 3-5 Notice that cellular respiration is essentially photosynthesis in reverse.

engine, in which gasoline is burned to obtain the energy to run the car. As you can see in Figure 3-5, cellular respiration is essentially the reverse of photosynthesis. (Don't confuse cellular respiration, which occurs within the cells of the body, with respiration, another name for breathing.) During cellular respiration, sugar and oxygen combine to yield carbon dioxide, water, and, most important, energy.

You use a portion of the energy obtained through cellular respiration to carry out your daily activities, such as walking, running, reading this book, thinking—and going to the refrigerator for more food. The energy is also used to make more body tissue, so that you grow, and some of it may be stored as fat or sugar.

All living things use cellular respiration to get energy from food molecules. Even organisms that make their own food through photosynthesis use cellular respiration to obtain energy from the food.

ENERGY TRANSFER: FOOD CHAINS, FOOD WEBS, AND TROPHIC LEVELS

Each time one organism eats another organism, a transfer of energy occurs. We can trace the paths that energy follows as it travels through an ecosystem by studying food chains, food webs, and trophic levels.

DDT IN AN AQUATIC FOOD CHAIN

Something strange was happening in the estuaries near Long Island Sound in the 1950s and 1960s. Carnivorous birds of prey such as ospreys and eagles that fed on fish in the estuaries had high concentrations of the pesticide DDT in their bodies. But when the water in the estuary was tested, it had low concentrations of DDT.

What accounted for the high levels of DDT in the birds? Poisons such as DDT that dissolve in fat can become more concentrated as they move up an aquatic food chain in a process called *bioaccumulation.* When the pesticide enters the water, small aquatic organisms such as algae and bacteria take in the poison. When fish eat the algae and bacteria, the poison dissolves into the fat of the fish rather than diffusing back into the water. Each time a bird feeds on a fish, the bird accumulates more DDT in its fatty tissues. At Long Island Sound, DDT concentrations in fatty tissue were magnified almost 10 million times from the bottom to the top of the food chain in some estuaries.

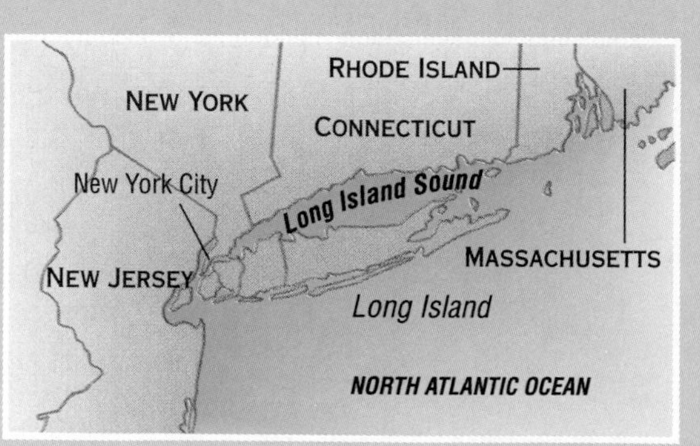

Long Island Sound lies between Long Island and Connecticut.

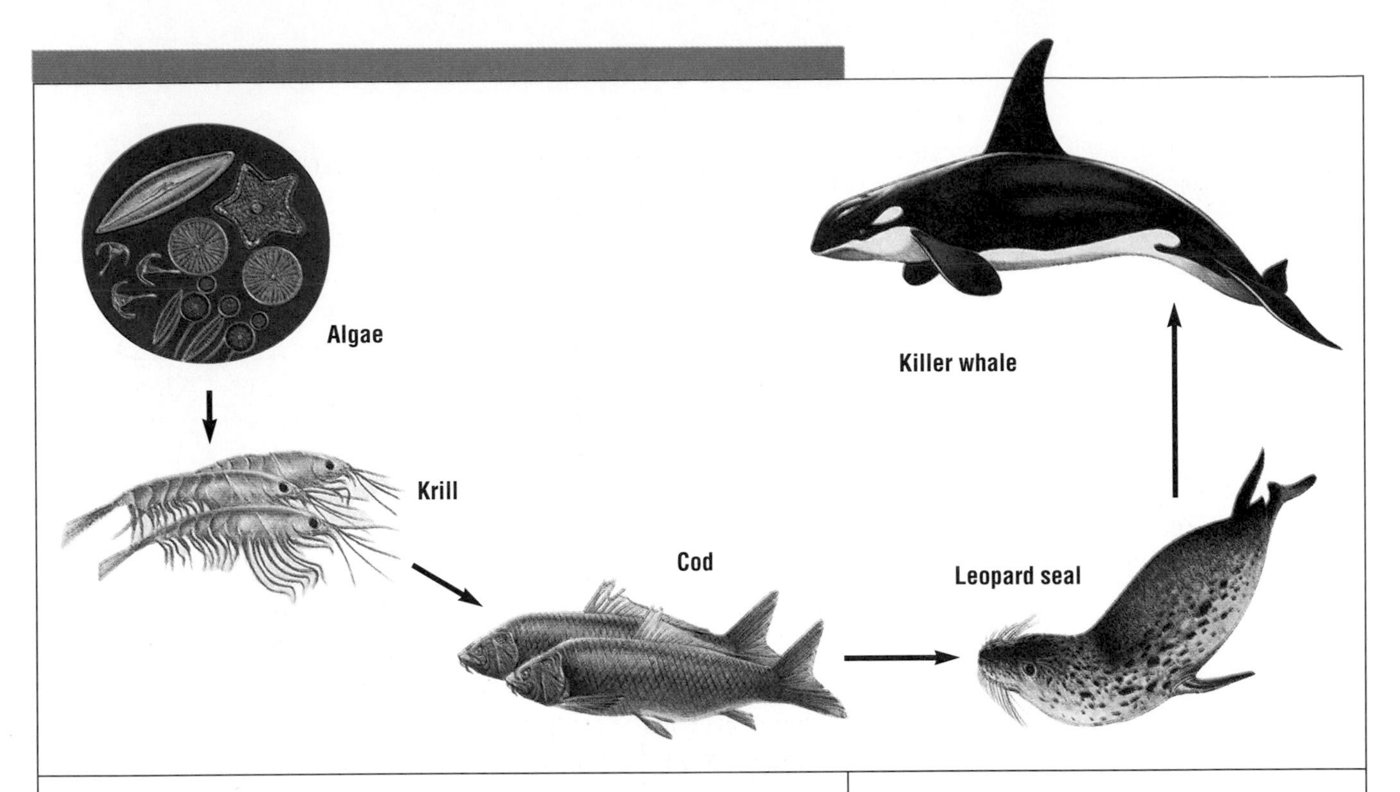

Food Chains and Food Webs A **food chain** is a sequence in which energy is transferred from one organism to the next as each organism eats another. Figure 3-6 shows a typical food chain in an ocean ecosystem. Algae are eaten by krill, which are eaten by cod, which are eaten by leopard seals, which are eaten by killer whales. (Poisons can also be passed on in a food chain, as discussed in the Case Study about DDT below.)

Figure 3-6 Energy is transferred from one organism to another in a food chain. Algae are the producers in this ocean food chain.

Answers to Thinking Critically can be found on page T144.

This bioaccumulation of toxins damages the carnivore at the top of the food chain. It may kill the carnivore, weaken its immune system, or impair its ability to reproduce successfully. High concentrations of DDT weaken the shells of bird eggs. When the eggs break, the chick embryos die; this causes a tremendous drop in the population of the bird species.

The United States government recognized bioaccumulation as a major side effect of the use of DDT and in 1972 banned its sale except for emergency use. The aquatic food chains immediately started to recover. Unfortunately, the food chains are still not totally free of DDT.

The pesticide breaks down very slowly in the environment. Also, DDT is still legal in some countries, where it is used in large quantities. As a result, migratory birds may be exposed to DDT while wintering in Latin America or other locations outside the United States.

An osprey catches a perch

1. *Analyzing Processes* DDT does not dissolve readily in water. If it did, how would the bioaccumulation of the pesticide in organisms be affected?
2. *Inferring Relationships* Suggest some specific measures that the United States could take to stop the DDT pollution that is still occurring.

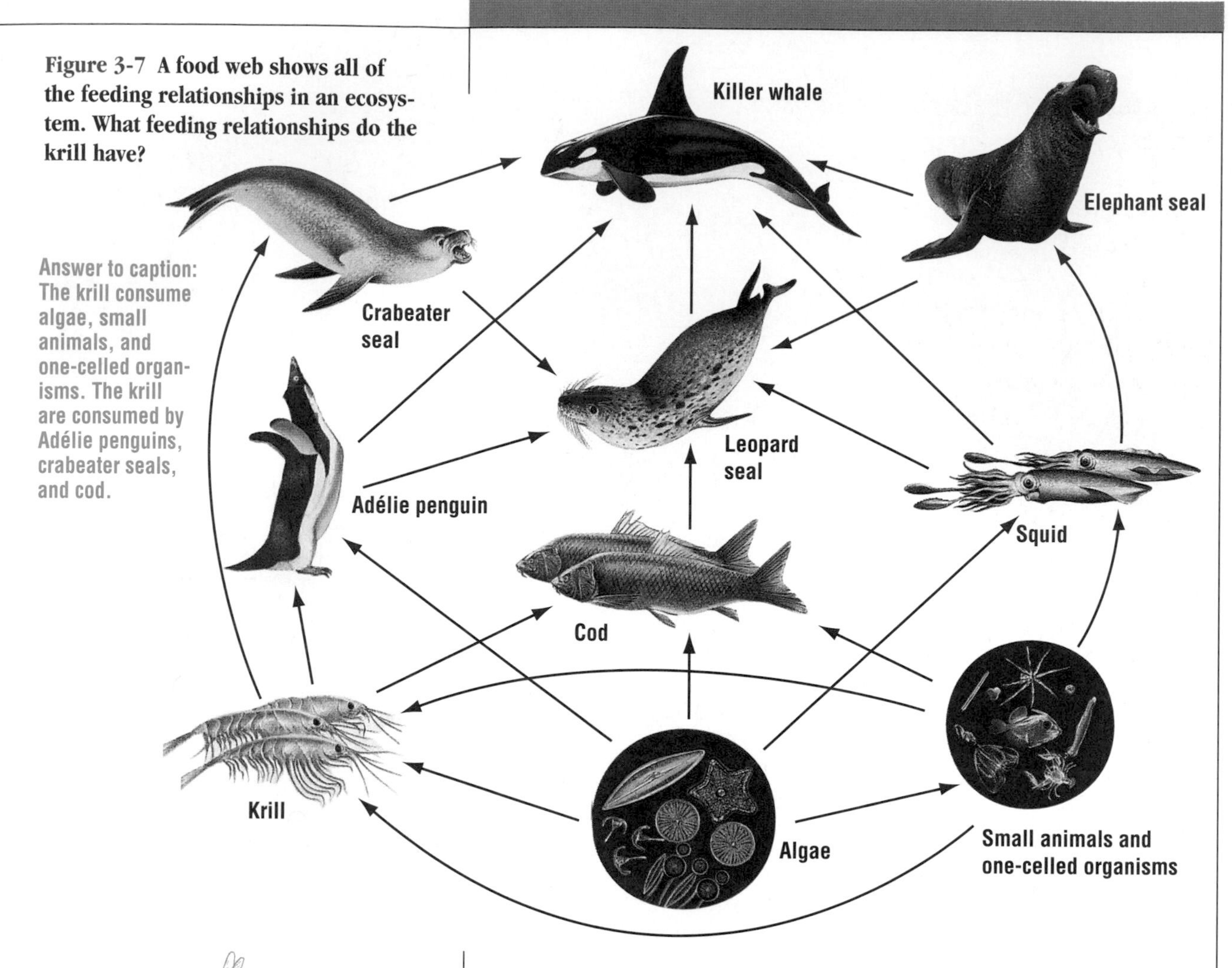

Figure 3-7 A food web shows all of the feeding relationships in an ecosystem. What feeding relationships do the krill have?

Answer to caption: The krill consume algae, small animals, and one-celled organisms. The krill are consumed by Adélie penguins, crabeater seals, and cod.

Entire ecosystems are much more complicated than a simple food chain. For one thing, ecosystems almost always contain many more species than those present in a single food chain. In addition, most organisms, yourself included, feed on more than one kind of food (life would be pretty dull if you didn't). So a more accurate illustration of what organisms eat in an ecosystem is a **food web**, such as the one shown in Figure 3-7. A food web shows all of the feeding relationships in an ecosystem. Notice that the food chain in Figure 3-6 is just one strand in the entire food web.

FIELD ACTIVITY

In a park, yard, or lot, measure out a 1-square-meter area. You may use sticks and string to mark off the area. Using glass jars or sealable plastic bags, carefully collect 5 plant species and 3 invertebrate species. Use a field guide to identify the different species, and then return the invertebrates to their original location. Draw a possible food chain or food web that includes the species you have collected.

Trophic Levels Each step in the transfer of energy through an ecosystem is known as a **trophic level**. In the ocean example, the algae are in the bottom trophic level, the krill are in the next level, and so on. Each time energy is transferred, less of it is available to organisms at the next trophic level.

What accounts for the decreased amount of energy at each trophic level? Organisms use much of their energy to carry out the functions of living—moving around, finding food, and so on. This energy is converted into heat and is lost from the food chain. About 90 percent of the energy at each trophic level is used up in this way. The remaining 10 percent becomes part of the organism's body, stored in its molecules. This 10 percent is all that is available to the next trophic level when one organism consumes another.

One way to visualize the reduced amount of energy from one trophic level to the next is to draw an energy pyramid like the one shown in Figure 3-8. Each layer in the energy pyramid represents one trophic level. Producers form the base of the pyramid, the lowest trophic level. Herbivores make up the second level. Carnivores that feed on herbivores form the next level, and carnivores that feed on other carnivores make up the top level. A pyramid is a good way to illustrate trophic levels because a pyramid gets smaller toward the top. At each higher level, there is less energy available.

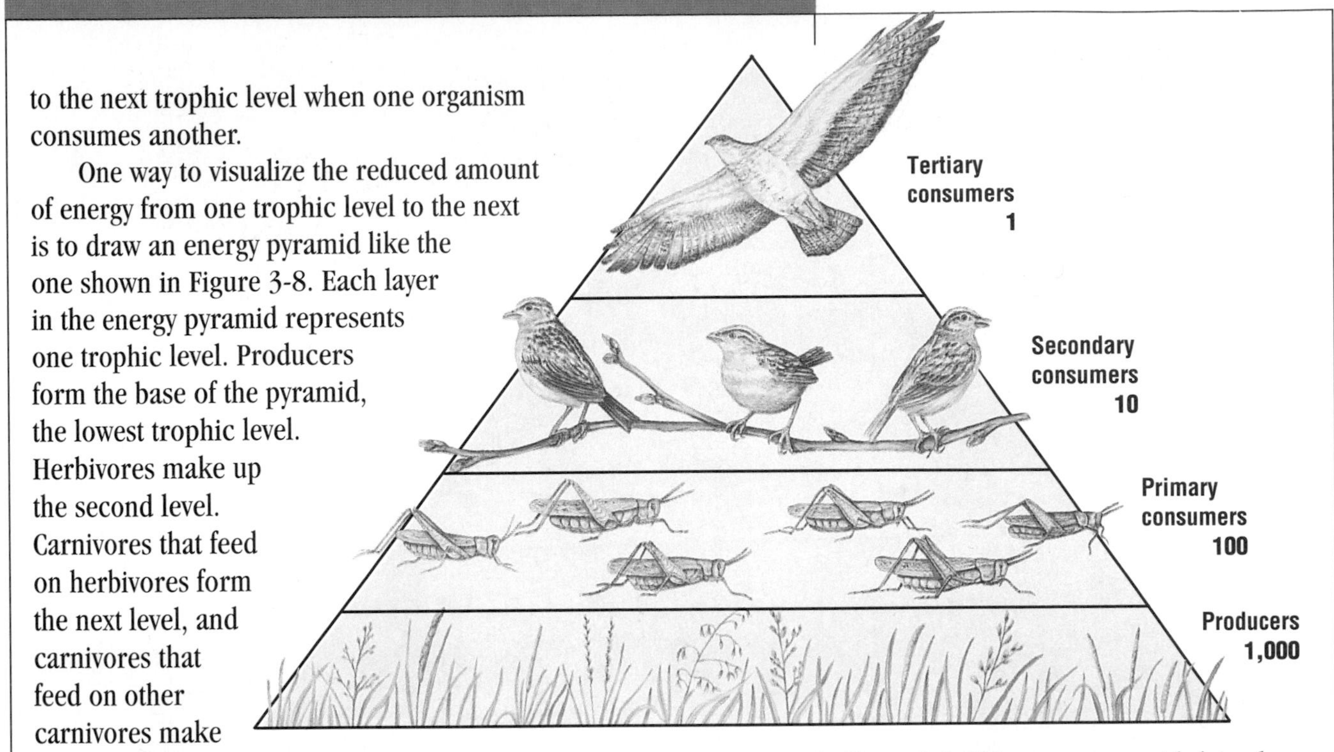

Figure 3-8 This energy pyramid shows how energy is lost from one trophic level to the next. The grass at the first level stores 1,000 times more energy than the hawk at the top level. Why are there fewer organisms at the top of the pyramid?

Another way to understand the reduced amount of energy at higher trophic levels is to think of how it could be similar to the flow of money. Let's say you pay $15 for a CD by your favorite band. How much of that $15 do you think is actually paid to the band members? Very little, because your money passes through several intermediates, each of whom takes a cut, before the money reaches the band. First, the store where you bought the CD keeps a large percentage of the money to cover its costs and make a profit. The record company then gets a share of the money, as does the band's agent. In the end, only a few dollars of your $15 reach the band. Similarly, as energy passes through an ecosystem, each trophic level takes some of the energy and uses it up.

How Energy Loss Affects an Ecosystem

How does the decreased amount of energy at each trophic level affect the organization of an ecosystem? First, the energy loss usually results in fewer organisms at the higher trophic levels. For example, zebras and other herbivores outnumber lions on the African savanna by about 1,000 to 1. There simply aren't enough herbivores to supply food for more lions.

Second, the loss of energy from trophic level to trophic level may place a limit on the number of trophic levels in an ecosystem. Ecosystems seldom have more than four or five trophic levels. Because of the reduced energy from one trophic level to the next, there is not enough energy left to support higher trophic levels.

Section Review and caption answers on page T144.

Section Review

1. What role does sunlight play in an ecosystem?
2. Could there be life on Earth without producers? Explain.
3. Contrast herbivores and omnivores.

Thinking Critically

4. *Interpreting Graphics* Look at Figure 3-7. What feeding relationships does the crabeater seal have?
5. *Analyzing Processes* Use what you learned about trophic levels to answer the following question. Could more people be supported by 20 acres of land if they ate only plants instead of eating both meat and plants? Explain your reasoning.

THE CYCLING OF MATERIALS

AFTER READING THIS SECTION YOU SHOULD BE ABLE TO

❶ describe the water, carbon, and nitrogen cycles.

❷ explain how humans are affecting the carbon cycle.

❸ explain the importance of nitrogen-fixing bacteria.

What will happen to the next ballpoint pen you buy? You'll probably use it until its ink supply runs out and then throw it away. The pen will end up in a landfill, where it may remain for centuries. The plastic and steel it contains will never be reused. By contrast, materials in ecosystems are used again and again; otherwise, they would soon be gone, and life could no longer exist. In this section, you will read about three different cycles that allow materials to be reused—the water cycle, the carbon cycle, and the nitrogen cycle.

THE WATER CYCLE

Water is the substance most essential to life. Fortunately, water is not usually destroyed; it just moves from place to place in the process shown in Figure 3-10. In this process, called the **water cycle**, water moves between the atmosphere and the Earth. If you would like to see how the water cycle works, make a miniature version of the cycle by following the instructions in Figure 3-9.

The sun provides the energy that drives the water cycle. Heat from the sun evaporates water from the oceans, from lakes and rivers, from moist soil surfaces, from the leaves of plants, and from the bodies of other organisms.

As water vapor cools in the atmosphere, it condenses and forms tiny droplets in clouds. When the clouds meet cold air, the water returns to the Earth again in

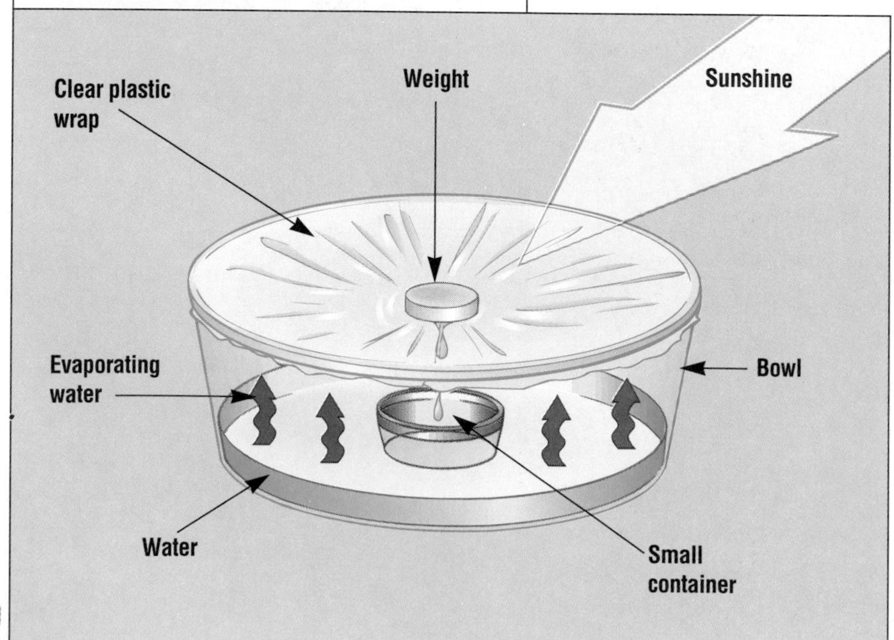

Figure 3-9 **Make a model of the water cycle as shown in the diagram. Leave it in the sun so that the heat will evaporate the water. The water vapor will condense on the cool plastic wrap and fall into the small container.**

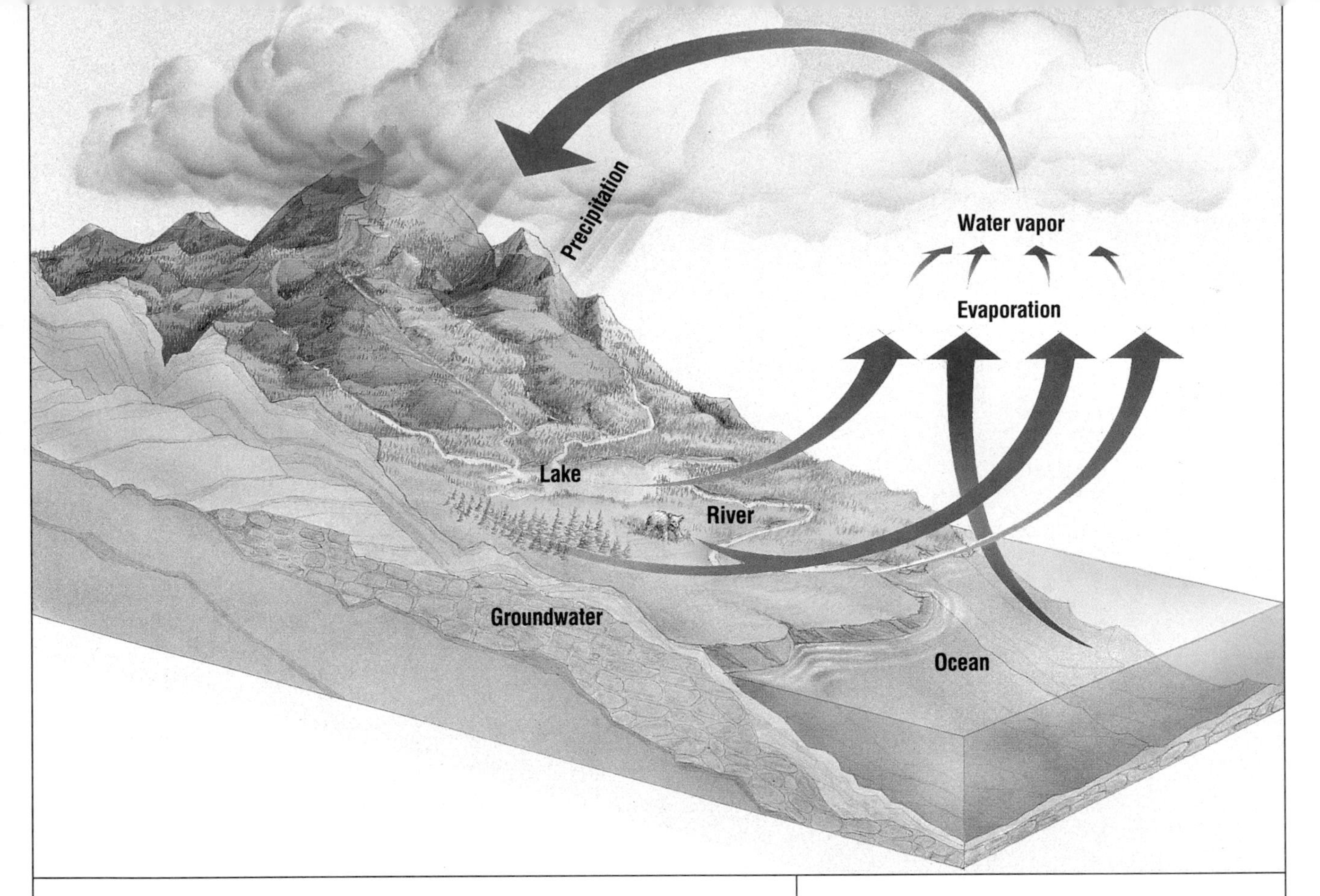

the form of **precipitation**—rain, sleet, or snow. Since oceans cover most of the planet, most precipitation falls on the oceans.

The precipitation that falls on land may just evaporate again into the atmosphere. Or it may collect in streams and rivers that flow into the oceans. Or the precipitation may soak into the soil. Water that soaks into the soil may be used immediately by plants, or it may seep down through the soil and rocks until it reaches a layer of rock or clay where it can go no farther. This layer of underground water is called *groundwater.* Groundwater may flow great distances underground and rise to the surface again, or it may be pumped up far from where it first entered the soil.

Figure 3-10 **The sun drives the water cycle, which moves water between the Earth and the atmosphere.**

On a typical day in the United States, about 16 trillion liters (4.2 trillion gallons) of rain or snow falls.

The Carbon Cycle

Organisms are composed mainly of complex molecules containing carbon. Carbon enters an ecosystem when producers take in carbon dioxide from the atmosphere during photosynthesis. When consumers eat the producers, they get carbon. As the consumers break down the food molecules during cellular respiration, the carbon is released back into the atmosphere as carbon dioxide. Photosynthetic organisms also release carbon dioxide during cellular respiration. Figure 3-11 on the next page shows a simplified version of the carbon cycle.

How Humans Are Affecting the Carbon Cycle Fossil fuels such as coal, oil, and natural gas are essentially stored carbon, left over from the bodies of plants and animals that died millions of years ago and were trapped underground. When we burn

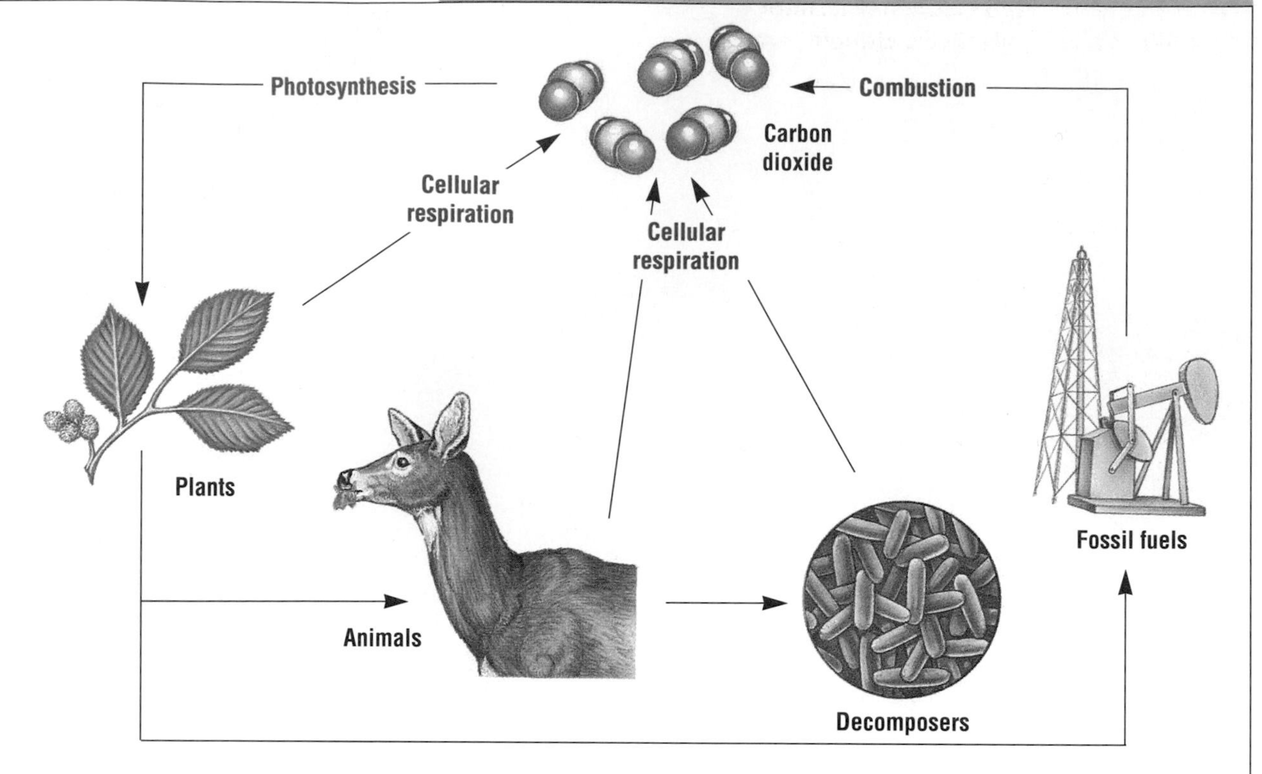

Figure 3-11 This diagram illustrates the carbon cycle. What role do plants play in the carbon cycle?

Answer to above caption on page T144.

Figure 3-12 These swellings on the roots of a soybean plant contain nitrogen-fixing bacteria.

fossil fuels, we release this carbon into the atmosphere. It returns to the atmosphere as carbon dioxide. The problem is that we are burning such large quantities of fossil fuels that the concentration of carbon dioxide in the atmosphere is increasing. In Chapter 7, you will read about the relationship between the increased levels of carbon dioxide and global warming.

The Nitrogen Cycle

All organisms need nitrogen to build proteins. Nitrogen is around us in vast quantities; nitrogen gas makes up 78 percent of the atmosphere. However, the only organisms that can use nitrogen gas directly from the atmosphere are a few species of bacteria known as **nitrogen-fixing bacteria**. All other organisms are dependent upon these bacteria for their nitrogen. Nitrogen-fixing bacteria are a crucial part of the nitrogen cycle, which is illustrated in Figure 3-13. These bacteria take nitrogen gas from the air and transform, or "fix," it into a form that plants can use.

Nitrogen-fixing bacteria live within the roots of a few plants such as beans, peas, clover, and alder trees. The bacteria use sugars provided by the plants and, in exchange, produce ammonia, a form of nitrogen that plants can use. The bacteria and plants have one of the relationships

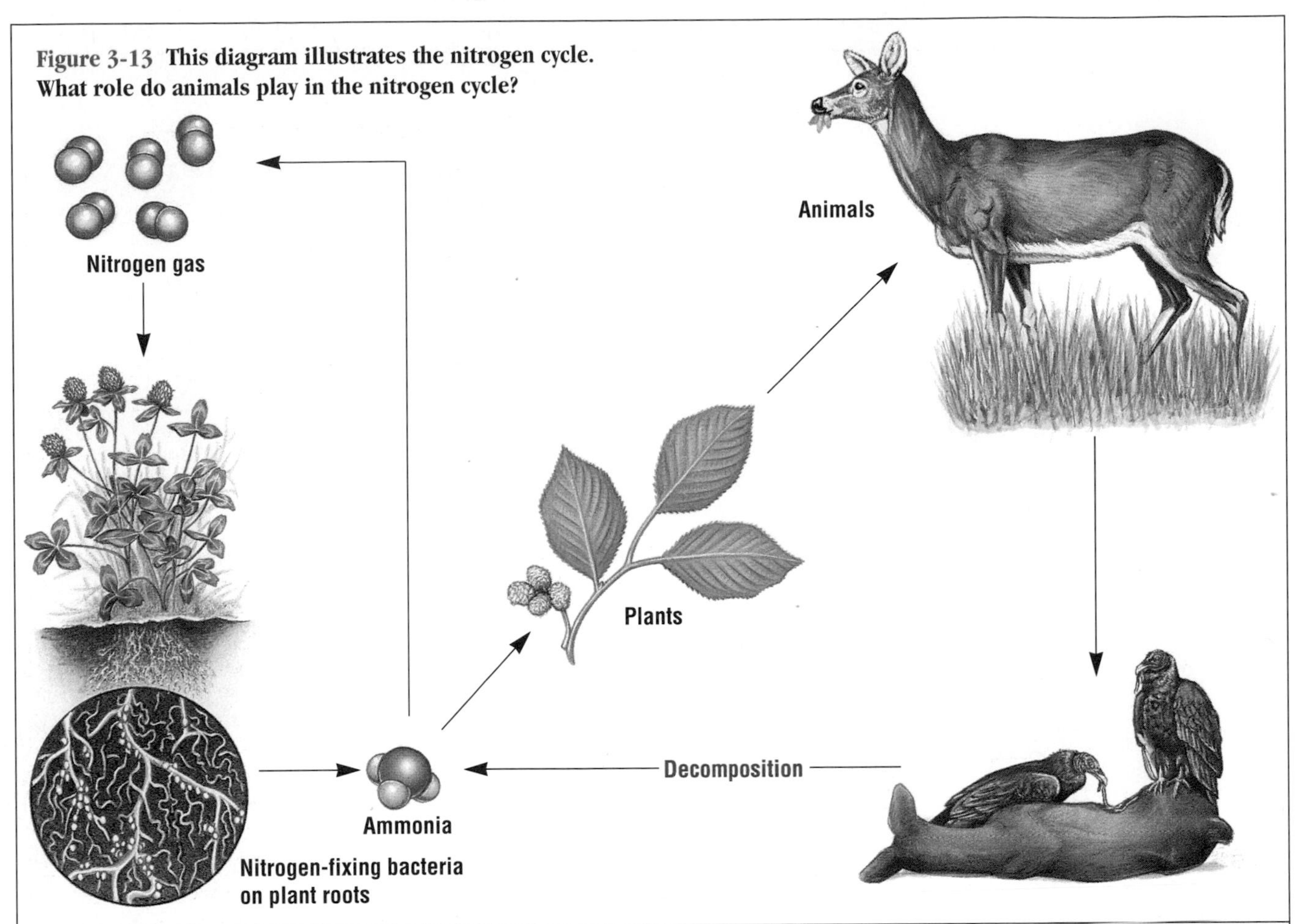

Figure 3-13 This diagram illustrates the nitrogen cycle. What role do animals play in the nitrogen cycle?

you learned about in Chapter 2—a mutualistic relationship in which both organisms benefit from the association. The excess nitrogen fixed by the bacteria is released into the soil. In addition, some nitrogen-fixing bacteria live in the soil rather than within roots, and these bacteria also add nitrogen to the soil.

Plants that don't have nitrogen-fixing bacteria in their roots get nitrogen from the soil. Animals get nitrogen by eating plants or other animals, both of which are sources of usable nitrogen.

Closing the Nitrogen Cycle In the nitrogen cycle, nitrogen moves back and forth between the atmosphere and living things. You've learned how nitrogen gets from the atmosphere to living things—through nitrogen-fixing bacteria. But how does the nitrogen return to the atmosphere?

Once again, it is bacteria, along with fungi, that are essential to the nitrogen cycle. These decomposers break down wastes (urine, dung, leaves, and other plant parts) and dead organisms, returning the nitrogen they contain to the soil in the form of ammonia. If it weren't for decomposers, much of the nitrogen in ecosystems would be locked away in wastes, corpses, and castoff parts such as tree branches. After decomposers have returned the nitrogen to the soil, other bacteria transform it into nitrogen gas, which returns it to the atmosphere. Then the cycle begins again.

Answers to caption and Section Review can be found on page T144.

Section Review

1. Explain how the model water cycle shown in Figure 3-9 is like the real water cycle shown in Figure 3-10.
2. Describe the role of producers in the carbon cycle.
3. How are humans altering the carbon cycle?

Thinking Critically

4. *Interpreting Graphics* Examine Figure 3-11. What role do animals play in the carbon cycle?
5. *Relating Concepts* Create a concept map that describes the water cycle. Be sure to include the following terms: water, groundwater, atmosphere, lakes, rivers, precipitation, plants, animals, and humans.

HOW ECOSYSTEMS CHANGE

After reading this section you should be able to

❶ describe secondary and primary succession.

❷ explain the importance of pioneer species.

❸ explain how soil is formed.

At some time in your life you may have walked through a forest that was hundreds of years old. If your great-grandmother and great-grandfather had walked in the same forest, they probably would have seen the same kinds of plants and animals that you saw. However, the area was not always a forest. A thousand years ago it might have been a meadow or even a shallow lake. The process responsible for the change is known as succession.

Succession

Succession is a regular pattern of changes over time in the types of species in a community. The process of succession may take hundreds or thousands of years.

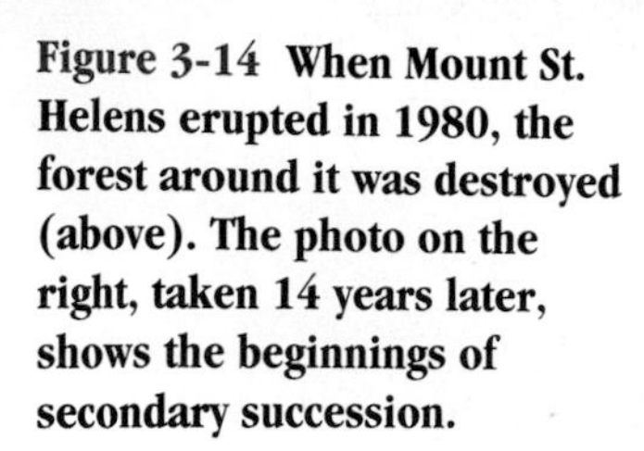

Figure 3-14 When Mount St. Helens erupted in 1980, the forest around it was destroyed (above). The photo on the right, taken 14 years later, shows the beginnings of secondary succession.

What causes succession? Each new community that arises makes it more difficult for the previous one to survive. In a pine forest, for example, the tall pine trees shade the ground and make it impossible for new pine seedlings to grow. But the seedlings of other trees, like oaks and maples, can grow with less light, so they begin to replace the pines.

The final, stable community that forms if the land is left undisturbed is called the **climax community**. Even though a climax community continues to change in small ways, the type of community remains similar through time. For example, a maple forest will remain a maple forest as long as no serious disturbances occur.

Secondary Succession In 1980 a volcano called Mount St. Helens erupted in Washington State. Over 18,000 hectares (about 44,460 acres) of forest were burned and flattened. If you visited Mount St. Helens today, you would find that the forest has already begun to regenerate through succession. (See Figure 3-14.) When succession occurs where an ecosystem has previously existed, such as in the burned area around Mount St. Helens, it is called **secondary succession**.

Another example of secondary succession is old-field succession, which occurs when farmland is abandoned. Figure 3-16 shows old-field succession on an abandoned farm in Tennessee. When a farmer stops cultivating a field, grasses and weeds quickly move in. These plants are called **pioneers**, the first organisms to colonize any newly available area and start the process of succession. The pioneer grasses and weeds grow rapidly and produce many seeds.

What happens when a hurricane ravages an ecosystem? This kind of question could keep a biologist very busy. **To find out more, see pages 374–375.**

Figure 3-15 These pine trees shade the ground and make it impossible for young pines to grow there. The young trees you see are the seedlings of maple trees, which thrive in the shade.

Answer to caption on page T144.

Figure 3-16 Why do young oak trees begin to appear around Year 20 in this example of secondary succession?

Then taller plants, such as perennial grasses, move in. These plants shade the ground, which keeps light from the shorter pioneer plants. The long roots of the taller plants also take up much of the water in the soil, depriving the pioneer plants of adequate water. The pioneer plants soon die from lack of sunlight and water.

Later, as succession continues, the taller plants themselves are deprived of light and water. This occurs as small shrubs or trees sprout and mature.

Finally, slower-growing trees such as oak and hickory or beech and maple move in and take over, blocking out the sunlight to the smaller trees. After a century, perhaps, the land returns to the climax community that was there before the farmers cleared it.

ECO-FACT

About 90 percent of all forest fires in the United States are caused by human beings.

Fire-Maintained Communities

Natural fires caused by lightning are a necessary part of secondary succession in some communities. Some species of trees, such as the jack pine, can release their seeds only after they have been exposed to the intense heat of a fire. Minor forest fires remove accumulations of brush and deadwood that would otherwise contribute to major fires that burn out of control. And some animal species depend on occasional fires because they feed on the

FIRES IN YELLOWSTONE

What happens when a fire breaks out in a national park? You might expect all available firefighters to rush to the scene. However, since 1972 the policy of the National Park Service has been to manage the national parks as naturally as possible. Since fire is a natural force in park ecosystems, this policy includes allowing most fires caused by lightning to burn. The only lightning-caused fires that are put out are those that threaten lives, property, uniquely scenic areas, or endangered species. All human-caused fires are put out.

Between 1972 and 1987, over 200 naturally caused fires burned in Yellowstone National Park. None of the fires grew excessively large, and all of them went out by themselves. The fire policy of the National Park Service was not really put to the test until the summer of 1988.

Conditions in Yellowstone during the summer of 1988 were unusual, because very little rain had fallen during the normally wet summer months. In fact, that summer turned out to be the driest on record, and several of the previous winters had also been drier than normal. All that was needed to set the dry forest ablaze was a few sparks, which were provided that summer by lightning.

By mid-July, Yellowstone had several fires that had been blazing for nearly a month and 11 others that had come and gone. At this time, the National Park Service began fighting the fires. Several days later in another part of the park, careless humans complicated the control efforts by starting what turned out

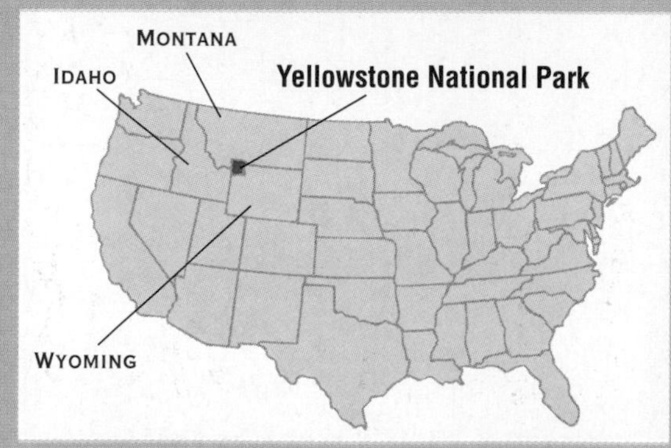

Yellowstone National Park, which spreads across three states, is the oldest national park in the world.

vegetation that sprouts after a fire has cleared the land. Therefore, foresters sometimes allow natural fires to burn unless they threaten human life or property. (See the Case Study below about fires in Yellowstone National Park.)

Primary Succession When succession occurs in areas where no ecosystem existed before, it is called **primary succession**. Primary succession occurs on new islands created by volcanic eruptions, for example, and in areas exposed when a glacier retreats.

Primary succession is much slower than secondary succession because it begins where there is no soil. It takes several hundred to several thousand years to produce fertile soil naturally.

Imagine that a glacier melts and exposes an area of bare rock. The first pioneer species to colonize the bare rock will probably be bacteria and lichens, which can live without soil. Lichens, as shown in Figure 3-17, are important early pioneers in primary succession. They are the rough, flaky patches that you see on the sides of trees

Figure 3-17 Lichens are able to live without soil, which makes them important contributors to primary succession.

Answers to Thinking Critically can be found on page T144.

A fire rages in Yellowstone National Park

to be the largest fire of all. Even more fires sprang up around the park in the passing weeks. Some of them persisted through the fall months, and a few were still burning in the winter.

About 25,000 firefighters controlled the fires and managed to protect major tourist sites from significant damage, but the price tag on the firefighting efforts came to $120 million. Final surveys showed that about 35 percent of the park's total land area had been blackened by the fires.

Some politicians and journalists criticized the National Park Service for allowing such widespread destruction. They claimed that the service's policy on fires was irresponsible and dangerous.

Two commissions of experts scrutinized the fire policy. They concluded that the policy was sound, but they recommended stricter limits on when and how long such fires should be allowed to burn. These recommendations were incorporated into Yellowstone's revised fire management plan and approved in April of 1992.

THINKING CRITICALLY

1. *Understanding Processes* Use what you have learned in this chapter about succession and fire-maintained communities to explain why the National Park Service might allow a naturally occurring fire to burn in Yellowstone.
2. *Expressing Viewpoints* Based on what you have read in this Case Study, do you think that the National Park Service's policy for controlling fires adequately protects humans and their property? Explain.

FIELD ACTIVITY

Explore two or three blocks in your neighborhood, and find evidence of succession. Make notes about the location and the evidence of succession that you observe. Pay attention to sidewalks, curbs, streets, vacant lots, and buildings, as well as parks, gardens, fields, and other open areas. Create a map from your data that identifies where succession is taking place in your neighborhood.

Answers to Section Review can be found on page T144.

SECTION REVIEW

❶ How is primary succession different from secondary succession? How is it similar?

❷ Explain why it may be damaging in the long run to put out some forest fires.

❸ What important role do lichens play in primary succession?

THINKING CRITICALLY

❹ *Analyzing Processes* Over a period of 1,000 years, a lake becomes a maple forest. Is this primary or secondary succession? Explain your answer.

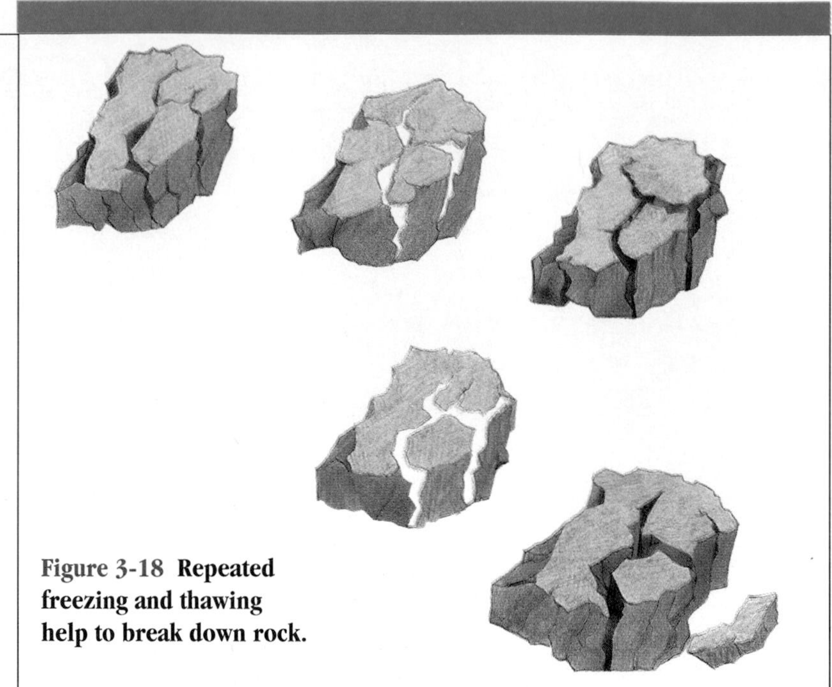

Figure 3-18 Repeated freezing and thawing help to break down rock.

and rocks. A lichen is actually two organisms, a fungus and an alga, that have a mutualistic relationship with each other. The alga photosynthesizes, while the fungus absorbs nutrients from rocks. Their action begins to break down the rock.

Water may freeze and thaw in cracks, breaking up the rock still further. (See Figure 3-18.) Soil slowly accumulates as dust particles in the air are trapped in cracks in the rock and as the dead remains of lichens and bacteria accumulate. Mosses may later gain a hold, breaking up the rock even more. When the mosses die, they decay and are added to the growing pile of soil. Thus, fertile soil is formed from the broken rock, decayed organisms, water, and air.

After some soil is formed, the seeds of small plants are able to germinate and grow. From this point on, the process is similar to secondary succession.

Primary succession can also be seen in any city street. Mosses, lichens, and weeds establish themselves in cracks in a sidewalk. Fungi and mosses invade a roof that needs repair. Even New York City would eventually turn into a rock-filled woodland if it were not constantly cleaned and maintained.

Figure 3-19 Primary succession on a city sidewalk

CHAPTER 3

HIGHLIGHTS

SUMMARY

- For the majority of the Earth's living things, the ultimate source of energy is the sun. Producer organisms harness the sun's energy directly. Consumer organisms use the energy indirectly, by eating other organisms.
- Only about 10 percent of the energy that an organism consumes is stored and therefore transferred when that organism is eaten. The rest is used up in carrying out life functions.
- Materials in ecosystems are endlessly recycled by natural processes.
- Carbon, water, and nitrogen are three materials essential for life, and each follows a recognizable cycle.
- The organisms in a community will follow a regular pattern of change over time. This process is called succession.
- Eventually, a final, stable community takes hold and persists more or less permanently until disturbed once again.

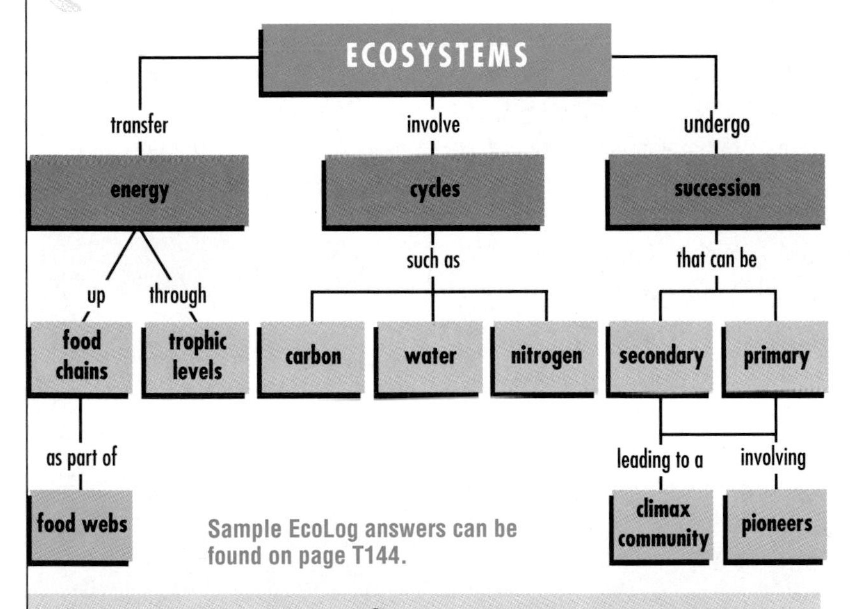

Sample EcoLog answers can be found on page T144.

EcoLog

Now that you've studied this chapter, revise your answers to the questions you answered at the beginning of the chapter, based on what you have learned.

1. What does sunlight have to do with the amount of food available to a hawk, which eats only meat?
2. How is it possible that an atom of nitrogen in your sandwich was once part of a dinosaur's body?

Vocabulary Terms

carnivore (p. 57)
cellular respiration (p. 57)
climax community (p. 67)
consumer (p. 56)
decomposer (p. 57)
food chain (p. 59)
food web (p. 60)
herbivore (p. 57)
nitrogen-fixing bacteria (p. 64)
omnivore (p. 57)
pioneers (p. 67)
precipitation (p. 63)
primary succession (p. 69)
producer (p. 56)
secondary succession (p. 67)
succession (p. 66)
trophic level (p. 60)
water cycle (p. 62)

CHAPTER 3

REVIEW

UNDERSTANDING VOCABULARY

1. For each pair of terms, explain the difference in their meanings.
   a. producer
      consumer
   b. carnivore
      omnivore
   c. carbon cycle
      nitrogen cycle
   d. food web
      food chain

RELATING CONCEPTS

2. Copy the unfinished concept map below onto a sheet of paper. Then complete the concept map by writing the correct word or phrase in each box containing a question mark.

Answers not indicated can be found on page T145.

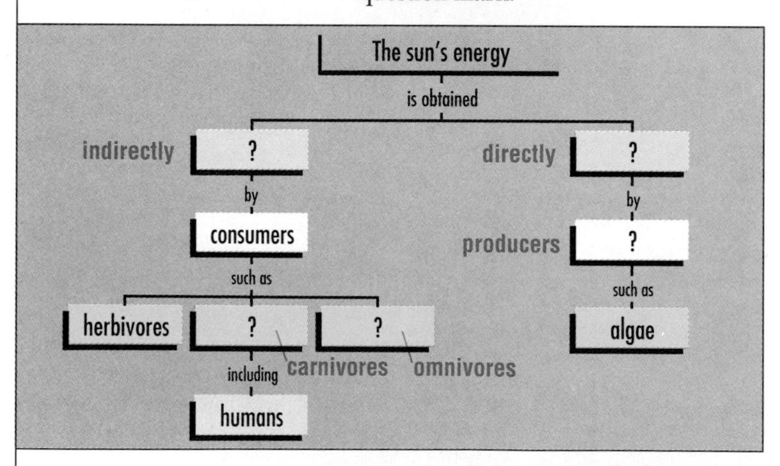

UNDERSTANDING CONCEPTS

Multiple Choice

3. Which is NOT true of consumers?
   a. They get energy indirectly from the sun.
   b. They are also called heterotrophs.
   (c.) They make their own food.
   d. They sometimes eat other consumers.
4. Which is correctly arranged from lowest trophic level to highest?
   a. bacteria, frog, eagle, mushroom
   b. algae, deer, wolf, hawk
   (c.) grass, mouse, snake, eagle
   d. grass, bass, minnow, snake
5. Communities of bacteria have been found living hundreds of feet underground. Which is a proper conclusion to draw from this?
   a. Somehow, they are conducting photosynthesis.
   b. They are living on borrowed time.
   c. They were somehow introduced by human activities.
   (d.) They use some energy source other than sunlight.
6. Which of these pairs of organisms probably belong to the same trophic level?
   (a.) humans, bears
   b. bears, deer
   c. humans, mushrooms
   d. both a and c
7. The energy lost between trophic levels
   a. can be captured only by parasitic organisms.
   b. cools the surrounding environment.
   (c.) is used in the course of normal living.
   d. disappears forever.
8. From producer to *secondary* consumer, about how much energy is lost?
   a. 10 percent
   b. 90 percent
   (c.) 99 percent
   d. 100 percent
9. Which is NOT true of the nitrogen cycle?
   a. Animals get nitrogen by eating plants or other animals.
   (b.) Plants generate nitrogen in their roots.
   c. Nitrogen moves back and forth between the atmosphere and living things.
   d. Decomposers break down waste to yield ammonia.

10. The water cycle
    a. was instituted to save scarce water.
    b. is the process by which water is used and reused in nature.
    c. is the cycle by which water is broken down chemically and turned into other compounds.
    d. is the cycle in which water evaporates and is lost forever.
11. Vegetation slowly returns after a woodland is destroyed by fire. This is an example of
    a. primary succession.
    b. secondary succession.
    c. a climax community.
    d. the carbon cycle.
12. Simple organisms colonize the barren lava produced by a volcanic eruption. This is an example of
    a. primary succession.
    b. secondary succession.
    c. the rock cycle.
    d. a climax community.
13. Which of these are MOST likely to be the pioneer organisms on an area of bare rock?
    a. saplings
    b. shrubs
    c. lichens
    d. perennial grasses

Short Answer

14. Explain the relationship of cellular respiration to photosynthesis.
15. Why can't there be an unlimited number of trophic levels?
16. Why are decomposers an essential part of an ecosystem?
17. Is there a starting point to the water cycle? an ending point? Explain.
18. Describe the role of carbon dioxide in the carbon cycle.

Interpreting Graphics

19. **Examine the graph below.** It shows the sea level over a period of thousands of years, some time in the past. Since water is not usually destroyed but instead recycled through the water cycle, what does this graph tell you about the location of the Earth's water during the time indicated? Where might the water have been?

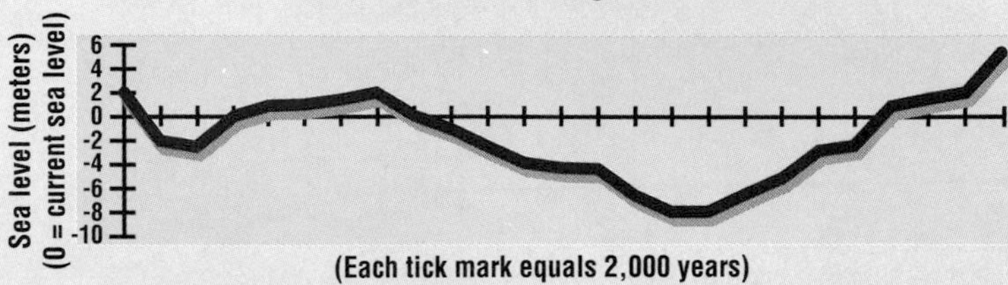

Thinking Critically

20. ***Comparing Processes*** How are producers and decomposers in some ways opposites of each other?
21. ***Inferring Relationships*** Abandoned fields in the southwestern part of the United States are often taken over by mesquite trees, which can grow in nutrient-poor soil. If the land is later cleared of mesquite, the soil is often found to be enriched with nitrogen compounds, becoming once more suitable for crops. What might be the reason for this phenomenon?

Themes in Science

22. ***Evolution*** Compare the process of succession with the process of evolution. How are the two processes similar? How are they different?
23. ***Interacting Systems*** Suppose that a plague eliminates all the primary consumers in an ecosystem. What will most likely happen to the other trophic levels?

Cross-Discipline Connection

24. ***Math*** Suppose you wanted to know the number of microorganisms that are present in a lake. Describe a procedure and the mathematical method you might use to get a good estimate.

Discovery Through Reading

Find out more about the role fire plays in maintaining ecosystems in "Fires of Life" by Doug Stewart, *National Wildlife,* Aug.–Sept. 1994, pp. 30–36.

PORTFOLIO ACTIVITY

Do a special project on succession. Find areas in your community that have been cleared of vegetation and left unattended at different times in the past. Ideally, you should find several areas that were cleared at different times, from recently to decades ago. Photograph each area, and arrange the pictures to show how succession takes place in your geographic region.

CHAPTER 3

INVESTIGATION

WHAT'S IN AN ECOSYSTEM?

How well do you know the environment around your home or school? You may walk through it every day without noticing most of the living things it contains or thinking about how they survive. Ecologists, on the other hand, spend their careers studying such details. Ecology is the study of the relationships between living things and their environments. In this Investigation, you will play the role of an ecologist by observing an environmental site in detail. You will collect data about both its physical features, or *abiotic* factors, and the organisms that live there, or *biotic* factors. You will then examine the interrelationships among these factors.

Answers to this Investigation can be found on page T145.

MATERIALS

- tape measure or meter stick
- 4 stakes
- string (about 50 m)
- poster board
- markers or felt-tip pens of several different colors
- notebook
- pen or pencil

MAP A SITE

1. CAUTION: Before beginning your field study, review the safety guidelines on pages 409–412. Remember to approach all plants and animals with caution.
2. With a tape measure or meter stick, measure a 10 m × 10 m square site to be studied. Place one stake at each corner of the site. Loop the string around each stake, running the string from one to the next, to form boundaries for the site.
3. Mark off as large a square as possible on the poster board to serve as a map of your site. Draw the physical features of the site on the map. For example, show the location of streams, sidewalks, trails, or large rocks, and indicate the direction of any noticeable slope.
4. Think of a set of symbols to represent the organisms at your site. For example, you might use green triangles to represent trees, blue circles to represent insects, brown squares to represent animal burrows or nests, and so on. At the bottom or side of the poster board, make a key to your symbols.

Use stakes and string to mark off a site that you will observe in detail.

5. Draw your symbols on the map to show the location and relative abundance of organisms. If there is not enough space on your map to indicate the specific kinds of plants and animals you observed, record them in your notebook.
6. Which symbols represent producers, which represent consumers, and which represent decomposers? Write your answers in your notebook.
7. In your notebook, record any observations of organisms interacting with each other or with their environment. (For example, note if you see insects feeding on plants or seeking shelter under rocks.)
8. After the maps and observations have been completed, rewind the string and remove the stakes. Do not destroy or remove anything from the site.

ANALYZE RELATIONSHIPS

9. Return to the classroom and display your site map. Use your site map and those of your classmates, together with your notes, to answer the following questions. Write your answers in your notebook.
10. List two food chains found on the maps, each containing a producer and two consumers.
11. Remember that as energy is transferred through food chains in an ecosystem, the amount of energy available decreases at successively higher trophic levels. Given this fact, would you expect there to be more producers or more consumers in the sites you mapped? Do your observations confirm this expectation? Consider both the number of organisms and their biomass (volume and weight). If your observation does not match your expectation, suggest an explanation for the situation you observed.

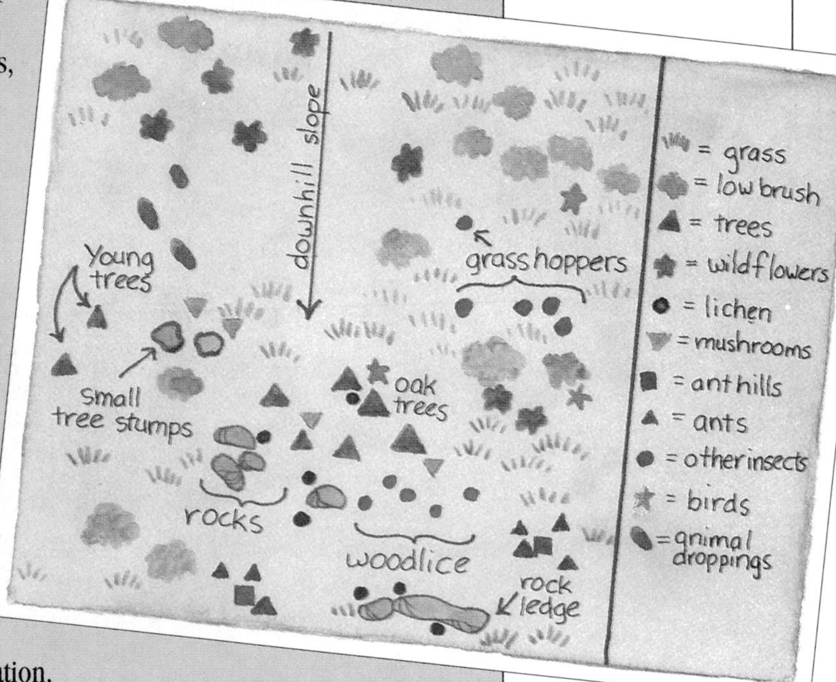

In this site map, which symbols represent producers, and which represent consumers?

12. What decomposers did you observe? What additional decomposers might be present, though not visible?
13. What role do decomposers play in the environment? What would happen to the producers and consumers if there were no decomposers?
14. How are the biotic factors (organisms) affected by the abiotic factors (nonliving things)? How are the abiotic factors affected by the biotic factors?

EXPLORE FURTHER

15. Based on what you have learned, think of a question that explores how the components of the ecosystem you observed interact with each other. For example, you might want to consider the influence of humans on the site, study a particular food chain in more detail, or explore the effect of physical features such as water or sunlight on the growth or behavior of organisms. Describe how you would investigate this question.

POINTS OF VIEW

Where Should the Wolves Roam?

In the 1920s the gray wolf was exterminated from much of the northwestern United States. Ranchers and federal agents killed the animal to protect livestock, which sometimes fell prey to the wolf. Today the gray wolf is an endangered species. The U.S. Fish and Wildlife Service has devised a plan to reintroduce the gray wolf to Yellowstone National Park, central Idaho, and northwestern Montana. The goal is to establish a population of at least 100 wolves at each location by 2002. However, some ranchers and hunters are uneasy about the plan, and some environmentalists and wolf enthusiasts think that the plan doesn't go far enough to protect wolves. Read the following points of view, and then analyze the issue for yourself.

WOLVES SHOULD BE REINTRODUCED

Many environmentalists and scientists believe that the reintroduction plan could bring Yellowstone National Park and two other areas into ecological balance for the first time in 60 years. They believe that the wolves will eliminate old and weak animals in elk, moose, and deer populations and help keep the populations of these animals from growing beyond the resources of their environment.

In response to ranchers' concerns that wolves will kill livestock, some biologists offer evidence that most wolves living near areas with adequate populations of deer, elk, moose, and other prey do not attack livestock. The scientists assert that wolf attacks on livestock are neither as widespread nor as serious as generally believed.

In response to fears that the wolves pose a threat to humans, proponents of the plan say that there have been no substantiated or documented attacks on humans by healthy wolves in North America. Wolves are shy animals that prefer to keep their distance from people, supporters of the plan say.

Most wolf enthusiasts readily admit that there are places where wolves belong and places where wolves do not belong. They believe that the selected reintroduction zones offer environments where wolves can carry out a vital role without becoming a serious liability to humans.

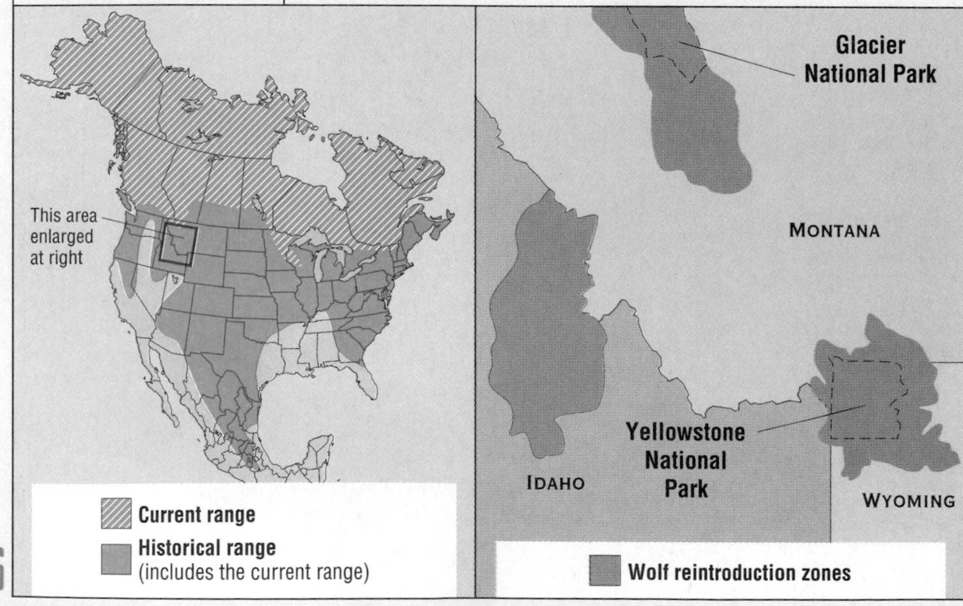

The gray wolf has lost most of its breeding range in the contiguous United States (left). Under the current plan, wolves will be reintroduced to parts of Montana, Wyoming, and Idaho (right).

A naturalist at the International Wolf Center in Montana describes wolf behavior to visitors.

Wolves Should Not Be Reintroduced

Some opponents of the reintroduction plan argue that wolves should not be classified as endangered at all. According to data from biologists, there are 1,500 to 2,000 wolves in Minnesota, 6,000 to 10,000 in Alaska, and 40,000 to 50,000 in Canada. With such numbers, many people feel that the animal should not receive the special treatment given to endangered species.

Hunters oppose the plan because they worry that wolves will kill a large share of the total number of animals that can be hunted. They cite studies that say large game animal populations cannot withstand hunting by both humans and wolves. This is a problem, they say, not only because it limits their hunting success, but also because hunting has a significant positive impact on the economies of the Western states. Furthermore, hunting is a valuable tool that park rangers use to control wildlife populations and to obtain funds for programs that are designed to help wildlife, such as habitat improvement and wildlife studies.

Ranchers are concerned that wolves will kill livestock. These losses could result in a tremendous financial burden to ranchers. A compensation program established by an environmental group to help reimburse ranchers will be in effect only as long as the wolf is classified as an endangered species. Ranchers point out that when wolves are plentiful enough to warrant removal from the endangered species list, the problem of wolf predation will be the most critical and yet compensation will no longer be available.

Gray wolf in Montana

Analyze the Issue

1. ***Expressing Viewpoints*** Supporters of the reintroduction plan suggest that much of the controversy surrounding the reintroduction of the gray wolf stems from a fear of wolves created by myths and legends about "the big, bad wolf." Do you think it is possible that society has given the gray wolf a negative reputation that it simply does not deserve? Support your opinion with examples.

2. ***Analyzing Viewpoints*** Opponents of the reintroduction plan often say that defenders of the gray wolf have had little, if any, direct experience with the animal and its ferocious instinct to kill. Do you think a person needs to have direct experience with a wolf to have an informed opinion about this issue? Explain.

Answers to Analyze the Issue can be found on page T145.

CHAPTER 4

KINDS OF ECOSYSTEMS

SECTION 4.1
FORESTS

SECTION 4.2
GRASSLANDS, DESERTS, AND TUNDRA

SECTION 4.3
FRESHWATER ECOSYSTEMS

SECTION 4.4
MARINE ECOSYSTEMS

"In all things of nature there is something of the marvelous."

ARISTOTLE, GREEK PHILOSOPHER

EcoLog

Before you read this chapter, take a few minutes to answer the following questions in your EcoLog.

1. What are the general characteristics of the ecosystem in which you live? Are there ecosystems similar to yours in other parts of the world?
2. Describe a plant or animal that lives in your area, and describe its survival strategy.

FORESTS

AFTER READING THIS SECTION YOU SHOULD BE ABLE TO

❶ define *biome.*

❷ compare and contrast the world's forest biomes.

❸ describe plant and animal adaptations in each kind of forest.

Earth is covered by hundreds of types of ecosystems. For convenience, ecologists divide these into a few biomes. **Biomes** are areas that have distinctive climates and organisms. Each biome contains many individual ecosystems. The locations of the major biomes are shown in Figure 4-1. Biomes are named according to their plant life because the plants that can grow in an area determine what other organisms can live there. But what determines which plants can grow in a certain area? The main determinant is climate. *Climate* refers to weather conditions in an area—temperature, precipitation, humidity, and winds—over a long period of time. Temperature and precipitation (rain, sleet, and snow) are the two most important factors in a region's climate. These and other aspects of climate are discussed in greater detail in Chapter 7.

In this chapter, you'll take a tour through the major biomes of the world—from lush rain forests to scorching deserts, and from the depths of the ocean to the icy polar regions. When reading about each biome, notice the adaptations of the organisms to their very different environments.

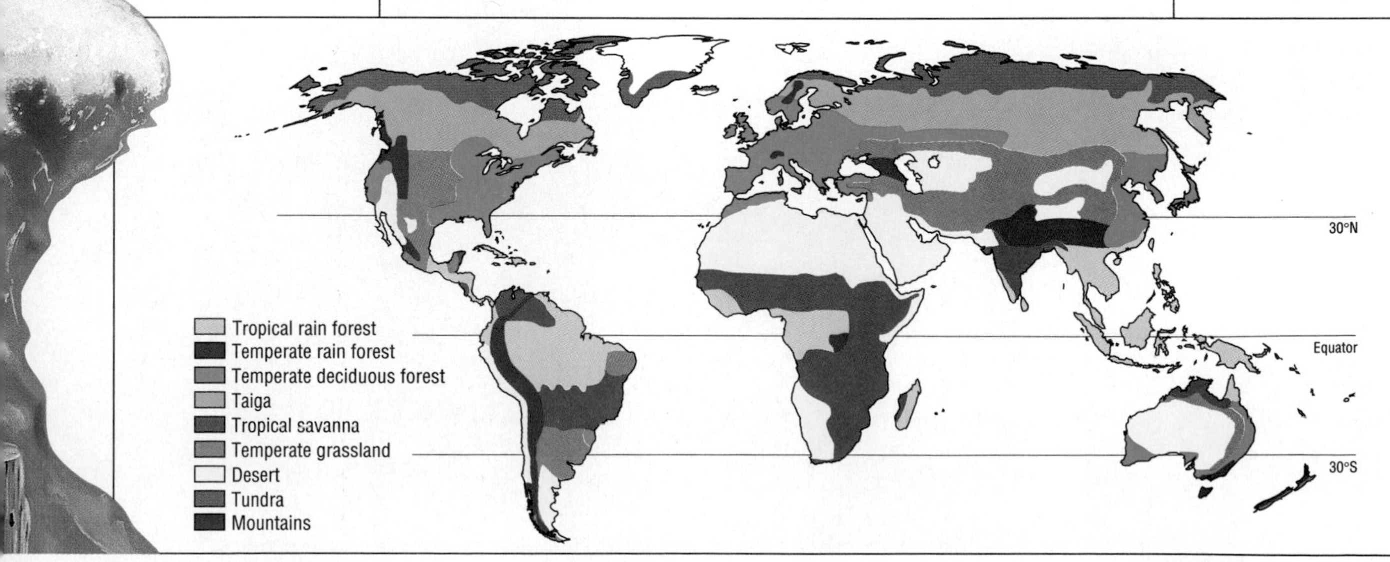

Figure 4-1 The world's ecosystems can be grouped into regions known as biomes, based on the types of plants and animals present.

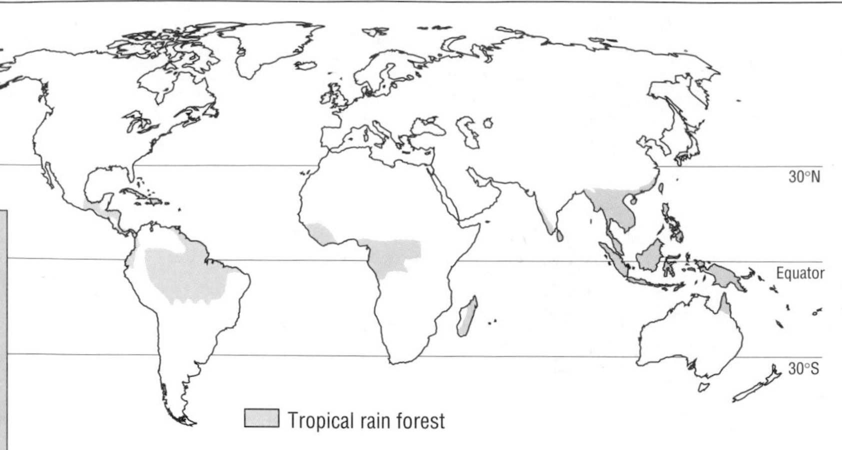

Figure 4-2 The world's tropical rain forests are characterized by heavy rainfall and fairly constant warm temperatures year-round.

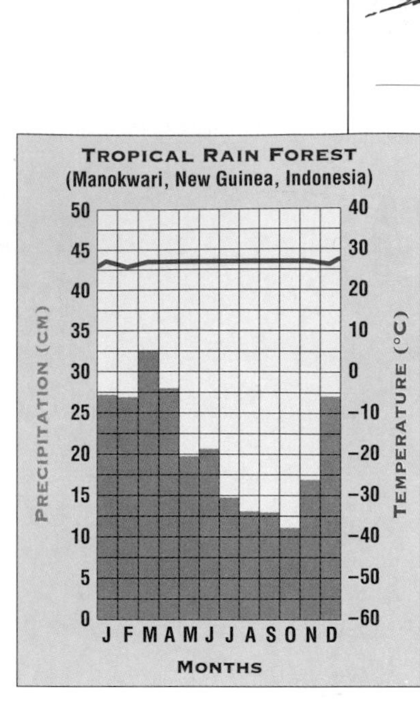

Figure 4-3 Tropical rain forests have the greatest biological diversity of any biome.

Tropical Rain Forests

The air is hot and heavy with humidity. You walk through the shade of the rain forest, stepping carefully over tangles of roots and vines, and brushing past enormous leaves. Life is all around you, but you see little on the surprisingly bare forest floor. Birds call, and monkeys chatter high above.

Tropical rain forests occur in a belt around the Earth near the equator, as shown in Figure 4-2. They are always humid and

warm and get about 250 cm (100 in.) of rain a year. Because they are near the equator, tropical rain forests get strong sunlight year-round, maintaining a climate with little seasonal variation.

The climate is ideal for growing plants. In fact, there are more different species of plants growing in tropical forests than in any other biome on Earth. While 1 hectare of temperate forest contains about 10 species of trees, the same area of a tropical rain forest may contain 100 species.

You would think that this profusion of plants grows on rich soil, but that is not so. Rapid decay of plants and animals returns nutrients to the soil, but these nutrients are just as quickly picked up by the plants. What nutrients remain are washed away by rainfall, so the soil is usually thin and poor. Many of the trees form aboveground roots that grow sideways from the trees, providing extra support for the trees in the thin soil.

The world's largest flower, *Rafflesia keithii*, Borneo

Monkey ladder vine, Trinidad

Golden lion marmoset, Brazil

Young scarlet macaws, Peru

Figure 4-4 The vegetation of the rain forest forms distinct layers. The plants of each layer are adapted to a particular level of light.

Plant Adaptations In tropical rain forests, plants grow in layers, as shown in Figure 4-4. Trees more than 30 m (100 ft.) tall form a dense **canopy** that absorbs at least 95 percent of the sunlight. Above the canopy, the tallest trees emerge into direct sunlight. Below the canopy, little light reaches the understory, and only smaller trees and shrubs adapted to shade can grow there. Herbs have large flat leaves that capture the small amount of sunlight that penetrates to the forest floor.

When fallen trees create an opening in the canopy, tree seedlings are adapted to grow very quickly to outcompete other trees for sunlight. Many plants, such as orchids and monkey ladder vines, use the tall tree trunks for support high in the canopy where there is light for photosynthesis.

Animal Adaptations The incredible diversity of rain-forest vegetation has led to the evolution of the greatest diversity of animals anywhere on Earth. Almost all rain-forest animals are specialists: each is adapted to exploit a specific resource in a particular way to avoid competition. For example, there are several species of birds called antwrens that eat insects, but each species catches insects in a different layer of forest vegetation. Another example is flowering plants that can be pollinated by only one species of insect, bat, or bird that can reach the flower's nectar.

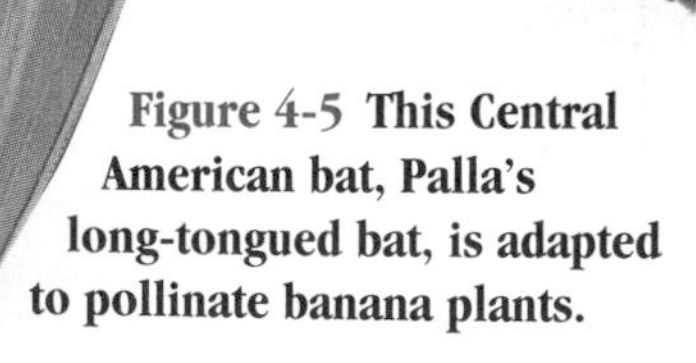

Figure 4-5 This Central American bat, Palla's long-tongued bat, is adapted to pollinate banana plants.

Figure 4-6 This Brazilian leafwing butterfly is well hidden from predators.

Some rain-forest animals have evolved elaborate methods for escaping predators, and others have evolved equally sophisticated methods for snaring prey. Camouflage is common. Insects, such as the butterfly shown in Figure 4-6, may be shaped like leaves or twigs. Some frogs blend in perfectly with plants; others have poisons in their skin that are advertised with bright colors to warn predators.

Figure 4-7 **The Costa Rican hillside shown above was cleared for cattle ranching.**

Threats to Rain Forests

Tropical rain forests used to cover about 20 percent of the Earth's surface. Today they cover about 7 percent. Every year an area of tropical rain forest the size of North and South Carolina combined is destroyed by logging or cleared for ranches and farms. (See Figure 4-7.) With the disappearance of these and other habitats, plants and animals become extinct. And native peoples, such as the Malaysian nomads shown in Figure 4-8, are often displaced, their culture and traditions lost.

You can help save rain forests by not buying rain-forest woods such as teak and mahogany. Instead, consider buying products that promote sustainable use of rain forests, such as nuts, fruits, and rubber. You can also support organizations that help preserve tropical forests. To learn more about a country that is saving some of its rain forests, read the Case Study about Costa Rica on pages 84–85.

Figure 4-8 **The nomadic people shown at left are attempting to block the logging of the Malaysian rain forests where they have lived for generations.**

Figure 4-9 A temperate rain forest in Olympic National Park, Washington

Temperate Rain Forests

Temperate rain forests occur in North and South America, Australia, and New Zealand. The Pacific Northwest is home to North America's only temperate rain forest, where tree branches are draped with mosses and tree trunks are clothed in lichens. The forest floor is blanketed by delicate ferns. Towering, 300-ft.-tall evergreen trees such as Sitka spruce and Douglas fir dominate the

Saving Costa Rican Biomes

Today human actions are altering virtually every biome on Earth. Many countries are attempting to preserve particularly interesting ecosystems as national parks and wildlife reserves. This concern is clearly evident in Costa Rica, a small country in Central America. Once almost entirely covered by magnificent forests, in the past 10 years Costa Rica has suffered a higher rate of deforestation than any other country. But Costa Rica still has many unique and beautiful ecosystems, and Costa Rican leaders hope to protect them.

About 12 percent of Costa Rica is set aside in national parks and preserves. Officials hope to boost that percentage to 25 percent by the end of the century. The protected areas include every major biome type in tropical Central America. Sections of rain forests, deciduous forests, coral reefs, volcanic mountain regions, and páramo (high open plains) are all protected in the nation's program.

One goal is to link many of the parks and preserves with forest corridors. The corridors will allow species a larger area from which to choose their breeding partners. This is an important step in helping species remain healthy and fertile because it allows for a constant exchange of genetic material.

Another goal of the program is to encourage practices that do not cause permanent damage to the country's land and wildlife, such as sustainable agriculture. An increasing number of Costa Rican wildlife specialists are helping to further this effort.

Other Costa Rican residents are taking a scientific interest in the diversity of plants and animals

Costa Rica is in Central America.

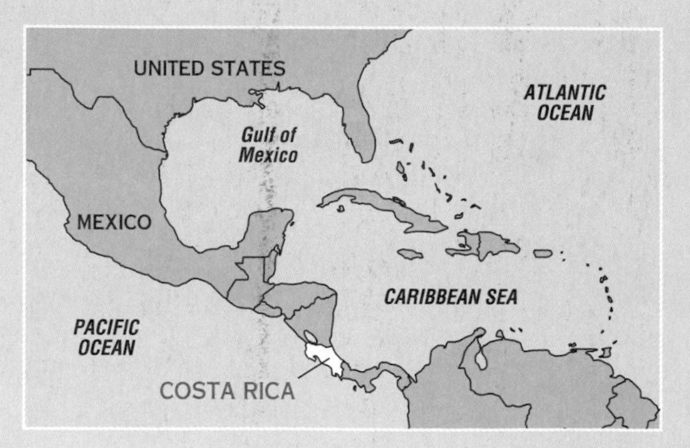

forest. Moisture pervades everything in this cool, humid forest.

Although the forest is located at about 50° to 60° north latitude, it never freezes because nearby Pacific Ocean waters moderate the temperature. In this area, winds travel from west to east. Ocean winds pick up moisture that is dropped on the coastal forests because the Olympic Mountains block their passage eastward. The high rainfall and moderate temperatures have created an ideal ecosystem for the abundant growth of forest plants.

This area of the United States has become extremely controversial in recent years because of conflicts over logging. You can read about the logging controversy in Chapters 8 and 10.

TEMPERATE DECIDUOUS FORESTS

To walk through a North American deciduous forest in the fall is to immerse yourself in color. Leaves in every shade of orange, red, and yellow crackle beneath your feet. The forest is quieter now than it was in the spring. Most birds have flown south. You see only chipmunks and squirrels gathering and storing the food they'll need during the long, cold winter.

Answers to Thinking Critically can be found on page T145.

These tourists are enjoying the beauty of Manuel Antonio National Park in Costa Rica.

that live in the country's rain forests. For example, a group of residents is helping an American pharmaceutical company study rain-forest plants and animals in the hopes of finding ingredients for medicines. A portion of the American company's licensing fees and possible future profits will be used to help preserve the forest's resources.

Organizations around the world also recognize the value of Costa Rica's resources. Environmental groups help Costa Rica obtain funds for protecting its natural treasures. Students help by establishing international children's rain forests. "Adopt-an-Acre" programs, which enable a person to adopt one or more acres of rain forest, successfully draw hundreds of thousands of dollars each year.

Costa Rica is also encouraging tourism. Tourist dollars provide a strong incentive for Costa Ricans to conserve the natural resources of their country. Tourist interest and appreciation also lead residents to take pride in those treasures.

These are just a few of the many aspects of Costa Rica's conservation program. The fact that the program is ambitious and yet successful has captured the world's attention. Many other countries are now taking similar steps to ensure the conservation of their own unique natural resources.

THINKING CRITICALLY

1. *Making Inferences* Why might a country's policy makers hesitate before enacting serious conservation measures?
2. *Understanding Relationships* Why would encouraging national pride in a country's resources be part of a conservation program? What else have Costa Rican officials done to change the attitudes of Costa Rican residents?

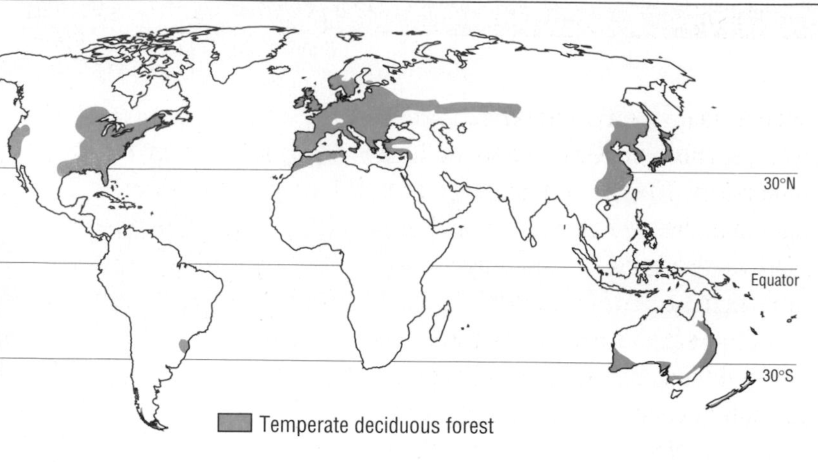

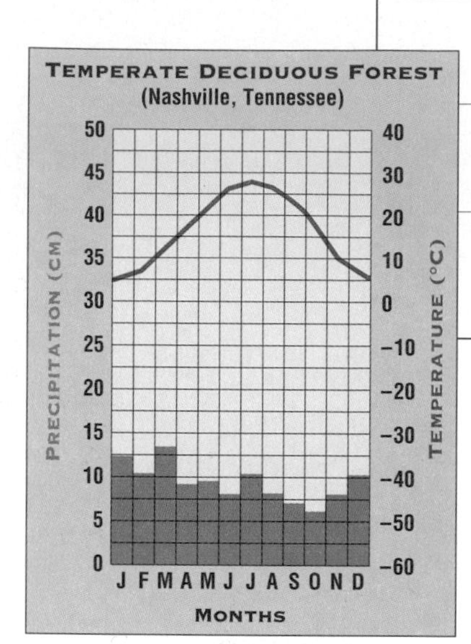

Figure 4-10 Temperate deciduous forests experience extreme changes in temperature from summer to winter.

In **temperate deciduous forests,** trees drop their broad, flat leaves each fall. These forests once dominated vast regions of the Earth, including parts of North America, Europe, and Asia. (See Figure 4-13.) Deciduous forests generally occur between 30° and 50° north latitude, so seasonal variations can be extreme, and the growing season lasts only four to six months. Summer temperatures can soar to 35°C (95°F). Winter temperatures often plummet well below freezing.

Deciduous forests are moist, receiving 75–250 cm (30–100 in.) of precipitation annually. The rain and snow aid in the decomposition of dead organic matter, such as fallen leaves, and make the soil in deciduous forests rich and deep.

Plant Adaptations Like rain forests, the plants of deciduous forests grow in layers. The forest canopy is dominated by tall trees

Figure 4-11 A temperate deciduous forest in Woodstock, New York, in summer, fall, and winter.

such as maple, oak, and birch. Small trees, shrubs, and bushes abound in the understory. The forest floor gets more light than that of a rain forest, and more plants—ferns, herbs, and mosses—grow there as a consequence.

Temperate-forest plants are adapted to survive seasonal changes. In winter, when ice locks up moisture in the soil, deciduous trees shed their leaves. Herb seeds and rhizomes (underground stems) become dormant in the ground, somewhat insulated by the cover of leaf litter and snow. In spring, as sunlight increases and temperatures rise, trees put out new leaves, seeds germinate, and rhizomes and roots put forth new shoots.

Animal Adaptations There are numerous habitats for animals in a deciduous forest, generally organized by layers of vegetation. Squirrels are adapted to exploit the nuts, seeds, and fruits in the treetops. Bears feast on the leaves and sweet berries of forest plants. Deer and other browsers nibble leaves from trees and shrubs.

Many birds nest in the relative safety of the canopy. Most of these birds are migratory. Because they cannot survive the harsh winters, each fall they fly south to warmer weather and available food. Each spring they return north to nest in temperate forests and feed on the abundance of food. Animals that do not migrate use various strategies for surviving the winter. Mammals such as bears and squirrels become inactive. Insects enter a state of very low metabolic activity.

Figure 4-12 **Deer, squirrels, and bears are among the numerous animals of the temperate deciduous forest.**

Figure 4-13 **The original forests of the United States have dwindled since Europeans first settled North America.**

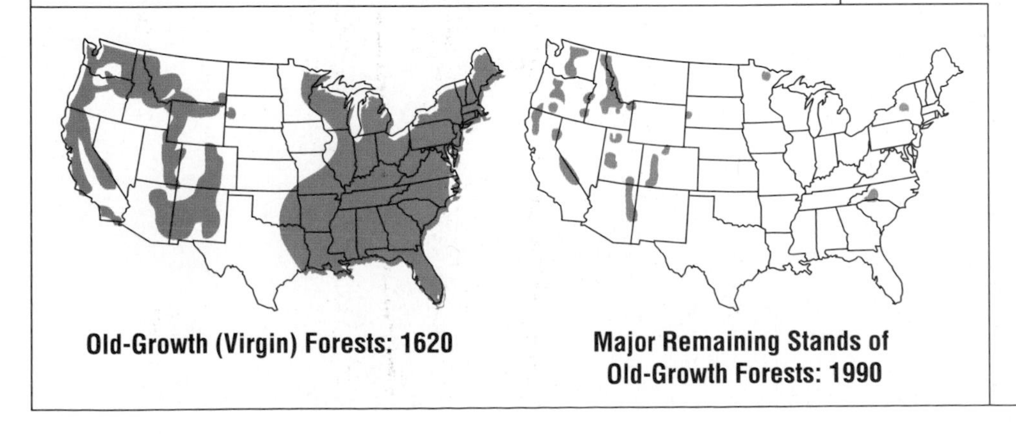

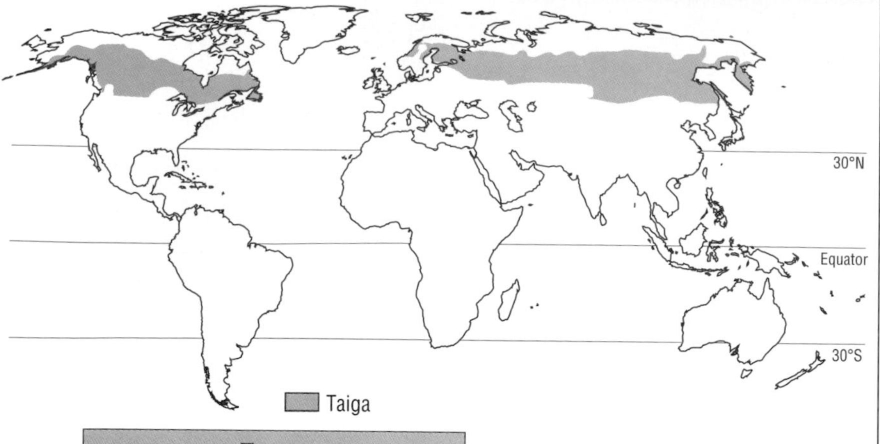

TAIGA

In taiga, the terrain is rough and the forest floor is sparsely vegetated. From the ground, many trees seem like straight, dead shafts of bark and wood—until you look up and see the tops of the trees.

Taiga is the northern coniferous forest, which stretches in a broad band across the northern hemisphere just below the Arctic Circle. As shown in Figure 4-14, winters are long (6 to 10 months) and extremely cold, with average subfreezing temperatures that often plummet to –20°C (–4°F). The frost-free growing season may be as short as 50 days, depending on latitude. Plant growth, however, is enhanced by constant daylight during the summer. Most of the precipitation falls as snow.

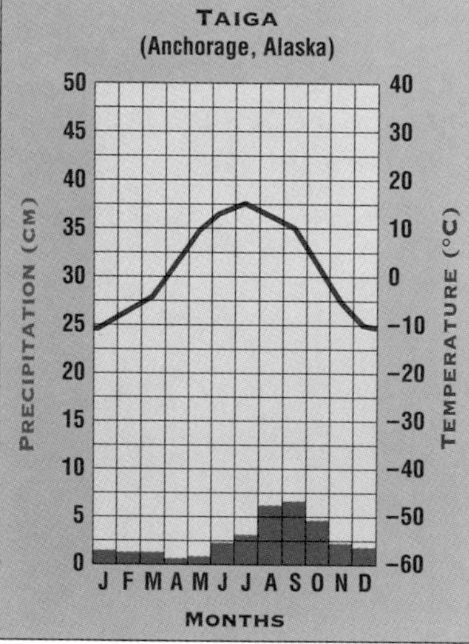

Figure 4-14 The taiga has long, cold winters and low precipitation.

Plant Adaptations A conifer is a tree whose seeds grow in cones. Most conifers do not shed their needle-shaped leaves, which help them survive harsh winters. The leaves' narrow shape and waxy coating retain water for the tree when the moisture in the ground is frozen. A conifer's typical cone shape helps it shed snow whose weight would otherwise crush it. These and other adaptations are shown in Figure 4-16. Dominant tree species include pine, hemlock, fir, and spruce.

Conifer needles contain acidic substances, and when they die and fall they acidify the soil. Most plants can't grow in acidic soil,

Figure 4-15 The taiga is characterized by conifers, trees whose seeds grow in cones.

which is one reason the forest floor is bare of most plants except blueberries, a few ferns, and mosses. Soil forms slowly because the climate and acidity of the leaf litter hinder decomposition.

Animal Adaptations The taiga is dotted with lakes and swamps that in summer attract birds that feed on insects, fish, or other wetland organisms. Many birds migrate south to avoid winter in the taiga. Some year-round residents, such as shrews and voles, may burrow underground during the winter, insulated by the deep snow cover. Moose and arctic hares eat what vegetation they can find. As shown in Figure 4-18, hares have adapted to avoid predation by lynx, wolves, and foxes by shedding their brown summer fur and growing white fur that camouflages them against the winter snow.

Figure 4-16 **Coniferous trees have adaptations that help them survive the snow and extreme cold of winter.**

Answers to Section Review can be found on page T145.

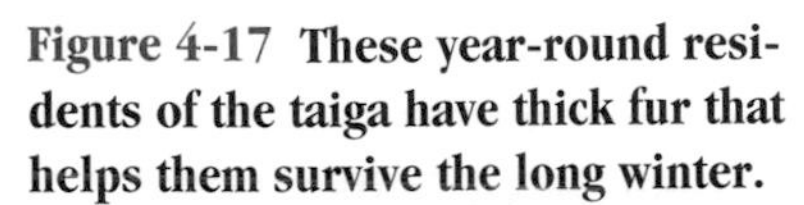

Figure 4-17 **These year-round residents of the taiga have thick fur that helps them survive the long winter.**

Figure 4-18 **The hare's fur changes color according to the season to help camouflage it from predators.**

SECTION REVIEW

1. Which aspects of climate are most important to forest ecosystems?
2. How does a temperate rain forest differ from the taiga?
3. Which would be better suited for agriculture, the soil of a tropical rain forest, or that of a temperate deciduous forest? Explain.

THINKING CRITICALLY

4. *Relating Concepts* Draw a concept map about the four types of forest ecosystems. Include at least one plant or animal from each forest type.

SECTION 4.2

GRASSLANDS, DESERTS, AND TUNDRA

AFTER READING THIS SECTION YOU SHOULD BE ABLE TO

❶ describe the factors that shape and characterize grassland, desert, and tundra ecosystems.

❷ explain how the adaptations of plants and animals in each ecosystem help them to survive.

In climates where there is less rainfall, forests give way to savannas and grasslands, which in turn give way to deserts. As precipitation decreases, so too does the diversity of species present. But while the number of different species may be small, the number of individuals of each species present may be staggering. Far to the north, another type of "desert" occurs—the tundra. Like the desert, little precipitation occurs in the tundra, but unlike the desert, temperatures stay very cold year-round.

Figure 4-19 **Tropical savannas have periods of heavy rainfall followed by periods of drought.**

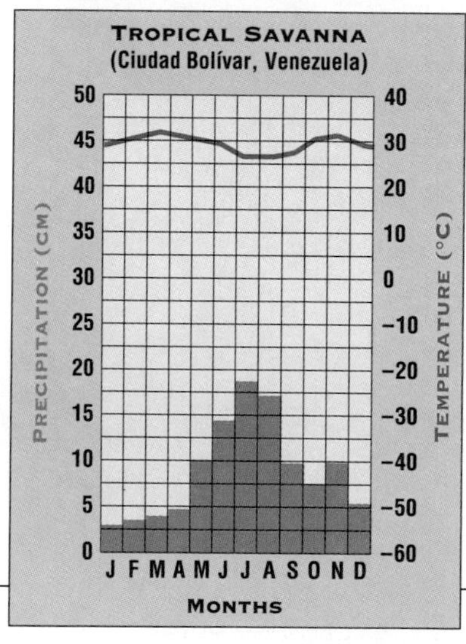

TROPICAL SAVANNAS

The West African plains, called **tropical savannas,** contain the greatest collection of grazing animals on Earth—along with the magnificent predators that hunt them. Tropical savannas are found in the tropics, near the equator. Yet because they occur inland, they get too little rain for many trees to grow. As Figure 4-19 shows, rain falls mainly at certain times of the year. Grass fires may sweep across the savanna during the dry season.

Plant Adaptations

Savanna trees and grasses have large underground root systems that survive fire, so plants regrow quickly after a fire. The root systems also help the plants survive during the parched dry season. The coarse savanna grasses have vertical leaves that further help them conserve water. Trees and shrubs often have thorns or razor-sharp leaves that deter hungry herbivores.

Figure 4-20 Tropical savanna, Tarangire National Park, Tanzania

Animal Adaptations Large, grazing herbivores have adopted a migratory way of life; they follow the rains to areas of newly sprouted grass. Some predators follow this mobile food source. Many savanna animals give birth only during the rainy season, when food is most abundant and the young are more likely to survive. Herbivores avoid competition for food by eating vegetation at different heights: small gazelles graze grasses, black rhinos browse on shrubs, and giraffes feed on tree leaves.

Figure 4-21 Savanna herbivores reduce competition for food by feeding on vegetation at different heights. Giraffes feed on tree leaves, while impala graze on grasses.

Figure 4-22 To find freshly sprouted grasses, wildebeest migrate in huge herds over long distances. The wildebeest shown below are crossing the Mara River in Kenya.

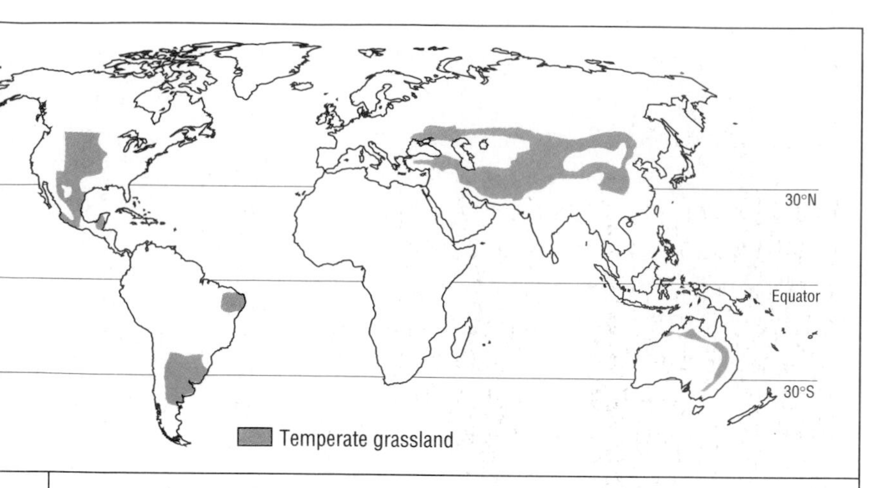

Figure 4-23 **Temperate grasslands are characterized by low rainfall, periodic droughts, and high temperatures.**

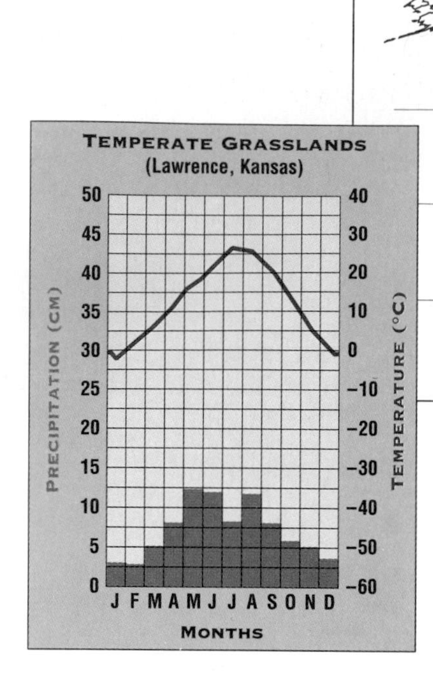

Figure 4-24 **Grasslands occur on many continents and are known by several different names.**

TEMPERATE GRASSLANDS: PRAIRIES, STEPPES, AND PAMPAS

In eighteenth-century North America, Conestoga wagons were swallowed up by the 12-ft.-tall grasses of the tallgrass prairies. Oxen sometimes died of thirst as wagon trains pushed west over the vast, dry shortgrass prairies. Although they seemed formidable, grasslands have the most fertile soil of any biome. It is no surprise, then, that most grasslands have been replaced with crops of corn, soybean, and wheat. The world's grasslands once covered about 42 percent of the total land surface of the Earth. Today, temperate grasslands occupy only about 12 percent of the Earth's surface. (See Figure 4-23.)

Temperate grasslands are found in the interiors of continents where there is too little rainfall for trees to grow. In addition to the prairies, there are the steppes of Russia and Ukraine and the pampas of South America. Mountains may play a crucial role in maintaining grasslands. For example, in North America, rain clouds from the west

Steppe, Mongolia

Pampa, Argentina

are blocked by the Rocky Mountains, so the shortgrass prairie east of the mountains gets only about 25 cm (10 in.) of rain a year. Rainfall increases as you move eastward, so taller grasses and some shrubs can grow. Heavy precipitation is rare; sizzling temperatures in summer make the grasslands a tinderbox. Thus, fire is common in the grassland biome.

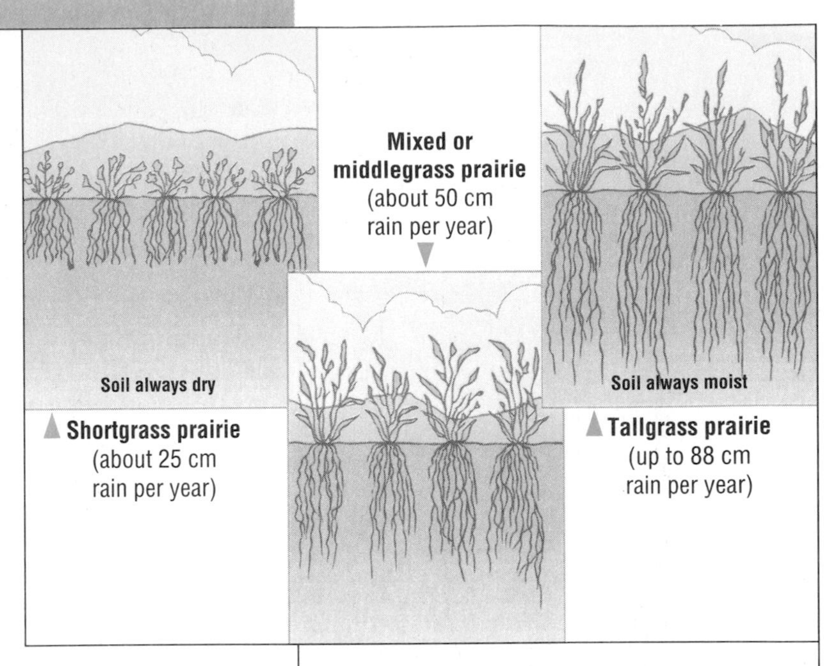

Figure 4-25 **The height of prairie plants as well as the depth of roots depends on the amount of rainfall.**

Plant Adaptations Prairie grasses are perennials, surviving from year to year. Their root systems form dense mats that survive drought and fire and hold the soil in place. The amount of rainfall determines what types of grasses will grow in an area. Figure 4-25 shows how root depth and grass height vary with rainfall. Few trees survive on the prairie because of drought, fire, and the constant battering of winds that roar over the land.

Animal Adaptations Young grasses are nutritious food for grazers such as pronghorns and the 60 million American buffalo (also called bison) that once roamed the American prairie. The buffalos' thick coats helped them survive the severe winters on the plains. They used their large heads to plow through snow and find plants to eat. Other plains animals, such as badgers, prairie dogs, and even owls live protected in underground burrows. The burrows shield them from fire and the elements and protect them from predators on the open grasslands.

Threats to Temperate Grasslands Cultivation and overgrazing have changed the grasslands. The grain crops that have replaced native grasses cannot hold the soil in place as well because their roots are shallow, and soil erosion results. Overgrazed grasses are constantly chewed down and cannot regenerate or hold the soil, causing further erosion.

Figure 4-26 **The prairie dogs shown below live in the Black Hills of South Dakota.**

Figure 4-27 **Because the soil is so fertile, much of the world's grasslands have been plowed under for crops. At left is a wheat field near Guymon, Oklahoma.**

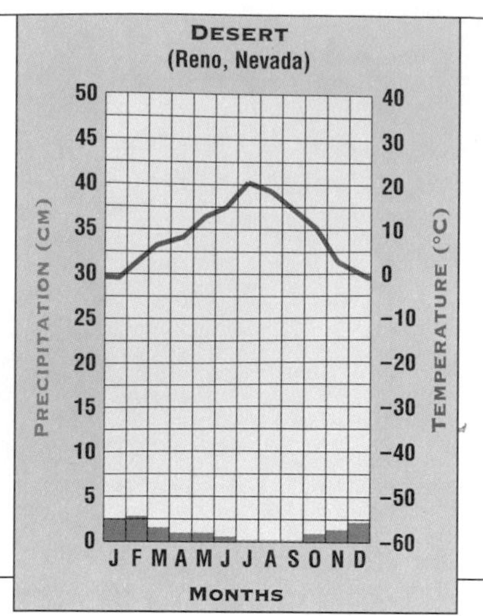

DESERTS

When some people think of deserts, they think of "Lawrence of Arabia" riding a camel over towering sand dunes. Other people might picture the Sonoran desert with its mighty saguaro cacti, the California desert graced with Joshua trees, or the magnificent rock formations of Monument Valley in Arizona and Utah. There are many different kinds of deserts, but the one thing they have in common is that they are the driest places on Earth.

Deserts are defined as areas that receive less than 25 cm (10 in.) of precipitation a year. Hot deserts occur between 15° and 35° north and south latitude around the Tropic of Cancer and the Tropic of Capricorn. Winds in these latitudes are very dry. Some deserts, as in the southwestern United States, were formed in the rain shadow of mountains that block the passage of rain-filled clouds. (See Figure 4-30.) Because deserts get so little rainfall, few nutrients are washed out of the soil, but the dryness also hinders decay of dead organic matter. Thus the soil is rich in minerals, but very poor in organic matter.

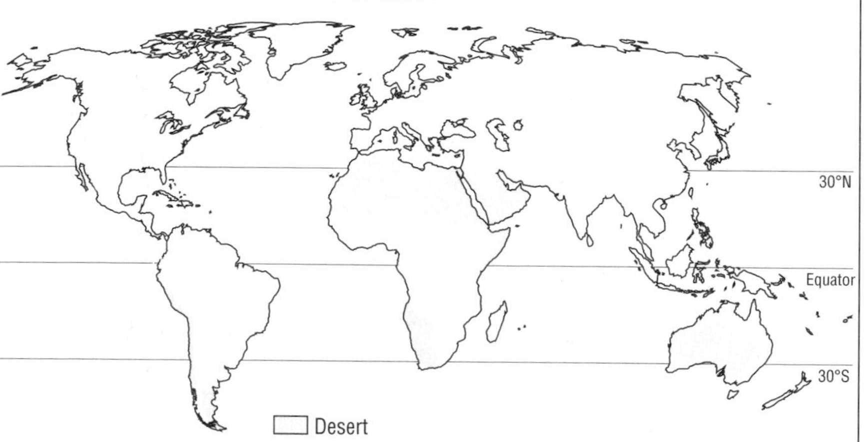

Figure 4-28 Deserts are the driest places on Earth.

Figure 4-29 Desert organisms are adapted to survive in a place with little water.

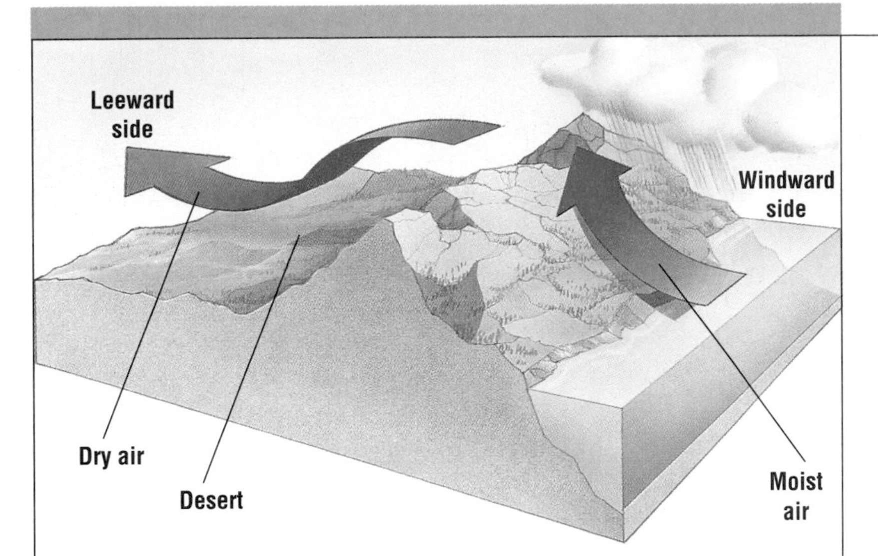

Figure 4-30 **Mountains sometimes block the passage of rain, creating a "rain shadow" where deserts form.**

Plant Adaptations Adaptations for getting and conserving water characterize all desert plants. Succulents and cacti have thick, fleshy stems and leaves that store water and that have a waxy coating to prevent water loss. Cactus spines are adaptations that deter thirsty animals from devouring the plant's juicy flesh. However, some birds build nests in cacti, where the thorns help protect them. (See Figure 4-31 on the next page.) Other plants store water in underground bulbs. Rainfall rarely penetrates deeply into the soil, so many plants' roots spread out widely just under the surface to catch as much rain as possible. The roots of some of these plants contain toxins that prevent other plants from growing nearby and competing for water. These plants

FIELD ACTIVITY

Create a miniature desert by growing a small cactus garden. Purchase two or three small cactus plants, or take several cuttings from a large cactus. To take cuttings, carefully break off the shoots growing at the base of the parent cactus. Place the plants in rocky or sandy soil similar to the soil in a desert. Keep the cacti in bright sunlight, and water them infrequently.

Scorpion with young, Arizona

Flapnecked chameleon, Botswana

Figure 4-31 **Cactus spines protect cacti from animals, but these curve-billed thrashers have overcome the defense.**

use a survival strategy called **drought-resistance:** they can live through the worst desert conditions.

Instead of resisting dry conditions, some desert plant species are adapted to escape drought. When it gets too dry, the plants die, dropping seeds that lie dormant in the soil until the next rainfall. Then new plants germinate, grow, and bloom very rapidly, before conditions worsen again.

Animal Adaptations Reptiles, such as lizards, snakes, and Gila monsters, have dry, scaly skin that prevents water loss. Amphibians, such as the spadefoot toad shown in Figure 4-32, survive scorching desert summers by **estivating**—burying themselves in the ground and sleeping through the dry season. Desert insects and spiders are covered with thick body armor that helps them retain water. In addition, most desert animals are at least partially nocturnal (active at night).

Figure 4-32 **Spadefoot toads survive dry periods by burying themselves and entering "summer sleep," or estivation.**

Figure 4-33 **The sidewinder has a unique way of moving so that only small portions of its body are in contact with the hot sands at any one time. This photograph was taken in the Namib Desert of Namibia.**

Threats to Deserts Residential development continues to encroach upon desert areas in the American West. Fierce political battles over water rights often ensue. Off-road and all-terrain vehicles kill desert vegetation, destroying the habitat of endangered animals such as the desert tortoise.

TUNDRA

Tundra is a biome without trees, where grasses and tough shrubs grow in the frozen soil that circles the Earth north of the Arctic Circle. As shown in Figure 4-34, summers are short in much of the tundra so that only the top few inches of soil thaw. Underneath lies the **permafrost,** permanently frozen soil. Because permafrost is impermeable, the tundra becomes dotted with bogs and swamps

TUNDRA
(Barrow, Alaska)
PRECIPITATION (CM): 0 5 10 15 20 25 30 35 40 45 50
TEMPERATURE (°C): −60 −50 −40 −30 −20 −10 0 10 20 30 40
J F M A M J J A S O N D
MONTHS

Figure 4-34 **Next to deserts, the tundra is the driest place on Earth.**

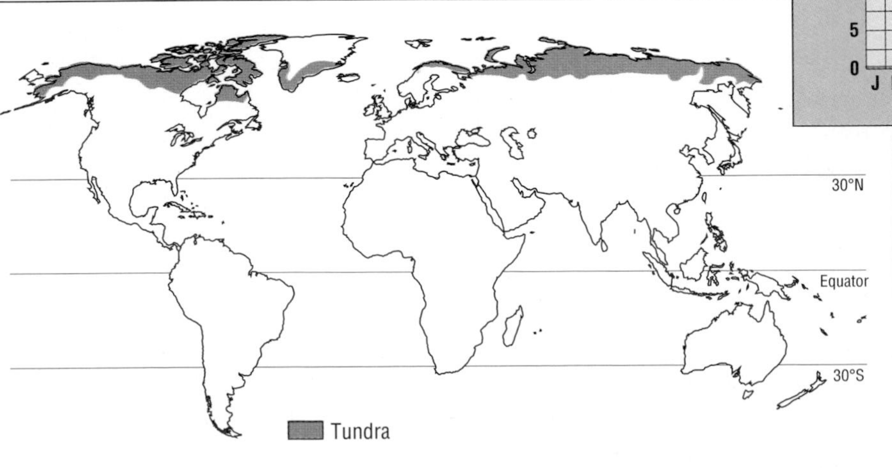

Figure 4-35 During its brief summer, the Alaskan tundra comes alive with many plants and lichens.

Figure 4-36 By growing close to the ground, dwarf shrubs and trees, such as these dwarf willows in Banks Islands, Canada, are protected from the battering wind.

when the top layer of soil thaws. In summer, these are ideal breeding grounds for huge numbers of swarming insects such as mosquitoes and blackflies and for the many birds that feed on them.

Plant Adaptations Mosses and lichens, which can grow without soil, cover acres of rocks in the tundra. Where soil exists, it is thin, and plants have wide, shallow roots that also anchor them against the arctic winds. Most flowering plants of the tundra, such as moss campion and gentian, are tiny. Hugging the ground keeps them out of the wind and helps them absorb heat from the sunlit soil. Woody plants and perennials such as willow and juniper have adapted dwarf forms, growing flat or trailing along the ground. Tundra plants make optimal use of the brief spell of summer sunshine, growing and flowering very quickly. (See Figure 4-35.)

Figure 4-37 **Many animals, such as caribou (right) and geese (below), return to the tundra each year.**

Animal Adaptations Millions of migratory birds breed in the tundra in summer. Food is abundant in the form of plants, mollusks, worms, and especially insects. Other tundra migrants are the caribou in North America and the reindeer in northern Europe. Being excellent hunters, wolves prey on caribou, deer, and moose as well as smaller animals such as lemmings, mice, and rabbits. Rodents burrow underground during the winter, though they are still active. Many year-round residents, such as arctic foxes, don white fur or feathers as winter camouflage. Animals that stay the winter are extremely well insulated. The musk ox's dense, shaggy coat makes the lumbering herbivore impervious to the cold.

Threats to the Tundra The tundra is one of the most fragile biomes on the planet. The food chains are relatively simple and thus easily disrupted. Because conditions are so extreme, the land is easily damaged and slow to recover. Until recently, these areas were undisturbed by humans. But oil has been found in some tundra regions, as at Prudhoe Bay in northern Alaska. Many fear the impact that oil extraction and transport will have on the land.

OIL EXPLORATION

Do oil and ice mix? Will oil exploration do irreparable damage to the tundra? **See pages 118–119.**

Answers to Section Review can be found on page T146.

SECTION REVIEW

1. How are tundras and deserts similar?
2. Some people have questioned the wisdom of building large cities in desert regions. Do you agree? Explain your reasoning.

THINKING CRITICALLY

3. *Identifying Relationships* Former grasslands are among the most productive farming regions. What do you think are some reasons for this?
4. *Analyzing Relationships* How can fire, a destructive event, actually be beneficial to the grassland ecosystem?

FRESHWATER ECOSYSTEMS

AFTER READING THIS SECTION YOU SHOULD BE ABLE TO

❶ describe the characteristics of the different freshwater ecosystems.

❷ compare survival adaptations of organisms in moving and standing freshwater ecosystems.

Freshwater ecosystems include the standing waters of lakes and ponds, the moving waters of rivers and streams, and the areas where land and water come together, known as wetlands. Fresh water is water that contains relatively little dissolved salt. The plant and animal life found in a freshwater ecosystem depends on the depth of the water, how fast the water moves, and the amount of mineral nutrients, sunlight, and oxygen.

LAKES AND PONDS

In the shallow areas close to the shores of lakes and ponds, aquatic life is diverse and abundant. This nutrient-rich area is known as the **littoral zone.** Farther out from the shore, the open water that gets enough sunlight for photosynthesis is dominated by tiny plants and animals known as phytoplankton and zooplankton. As shown in Figure 4-38, the types of organisms present depend on the amount of sunlight available.

Some bodies of fresh water have areas so deep that there is too little light for photosynthesis to occur. Dead plants and animals that drift down from above are decomposed by bacteria, and a few fish adapted for cooler, darker water also live here. Eventually, dead and decaying organisms reach the **benthic zone,** the bottom of a body of water, which is inhabited by decomposers, insect larvae, and clams.

A lake with a large amount of

Figure 4-38 A pond or lake ecosystem is structured according to how much light is available.

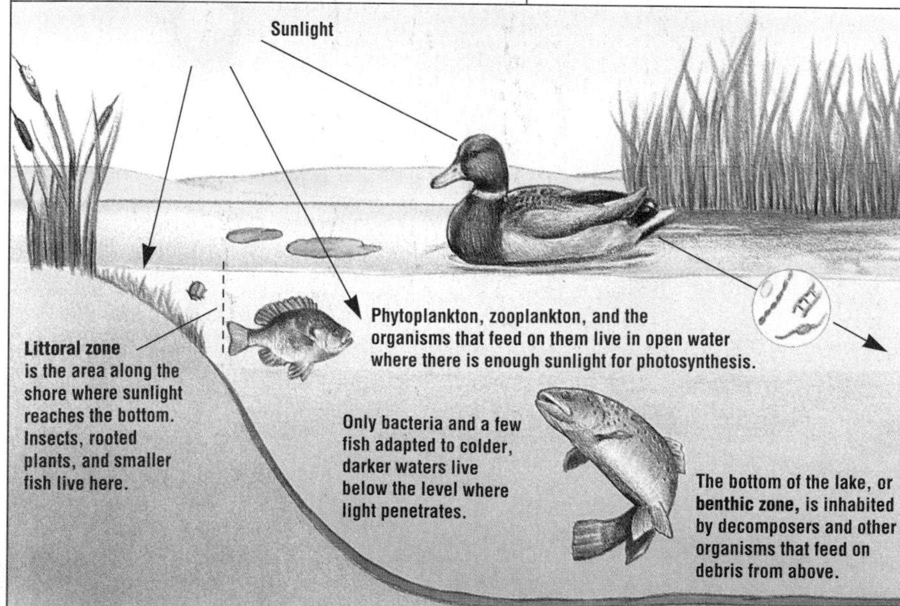

Figure 4-39 Lakes, such as this one in Maine, are standing waters. Rivers are moving waters.

plant nutrients is known as a eutrophic lake. As the amount of plants and algae grows, the number of bacteria feeding on the decaying organisms also grows. These bacteria use up the oxygen dissolved in the lake's waters. Eventually, the diversity of species declines. Lakes naturally become eutrophic over a long period of time. However, the process of eutrophication can be accelerated by runoff containing sewage, fertilizers, and animal wastes.

Plant and Animal Adaptations Along the shore, plants such as cattails and reeds are rooted in the bottom mud, their upper leaves and stems emerging above the water. In deeper water, floating plants such as pond lilies are found. Animals such as water beetles trap air in the hairs under their bodies because they cannot get air directly from the water. Whiskers help catfish sense food as they swim over dark lake bottoms. Fish are adapted to certain temperature

Figure 4-40 In addition to fish, organisms such as plants, insects, and amphibians are also adapted to living in or near the water in lakes and ponds.

Figure 4-41 **This map shows the locations of wetlands in the United States.**

ranges: lake trout need cold water, while bass prefer warmer waters. In regions where lakes partially freeze in winter, amphibians burrow into the littoral mud.

WETLANDS

Wetlands are areas of land that are covered with water for at least part of the year. The two types of freshwater wetlands are marshes, which contain non-woody plants, and swamps, which contain woody plants or shrubs. (Estuaries, which are a type of saltwater wetland, are discussed in the next section.) Wetlands perform several important environmental functions. Many of the freshwater game fish caught in the United States each year use the wetlands for feeding and spawning. In addition, these areas provide a home for native and migratory wildlife, including endangered and threatened species. Wetland vegetation traps carbon that would otherwise be released as carbon dioxide, which may be linked to rising atmospheric temperatures. Finally, wetlands remove pollutants from the water, control flooding by absorbing extra water when rivers overflow, and produce many commercially important products such as cranberries, blueberries, and peat moss.

Figure 4-42 **A marsh is a type of wetland that contains non-woody plants. This is a marsh in New Jersey.**

Marshes In the shallow water of marshes, plants such as reeds, rushes, and cattails are rooted in the rich bottom sediments, but their leaves are above water. The benthic zone is rich, containing plants and numerous decomposers and scavengers. Waterfowl such as grebes and ducks have beaks adapted for eating marsh vegetation. Water birds such as herons have spearlike beaks that they use to grasp small fish and frogs. Marshes attract many nesting birds such as blackbirds.

There are several kinds of marshes, each characterized by its salinity. Brackish marshes have slightly saline water; tidal marshes contain saltier water. In each marsh type, organisms are adapted to live within the ecosystem's range of salinity (salt content). The Everglades in Florida is the largest freshwater marsh in the United States.

Figure 4-43 Plants such as these bladderworts in a Louisiana swamp have underwater, "bladder-shaped" leaves that they use to trap insects.

Swamps Swamps occur on flat, poorly drained land, often near streams. Swamps are dominated by shrubs or water-tolerant trees such as red maple, cedar, oak, or cypress, depending on the latitude and climate in which the swamps occur. Mangrove swamps occur in warm climates near the ocean, so their water is saline. Swamps are the ideal habitat for many amphibians such as green frogs and salamanders, and they attract birds such as wood ducks that nest in hollow swamp trees near or over water.

Threats to Wetlands Wetlands were previously considered to be wasteland—noxious breeding grounds for pesky insects. Thus they've been relentlessly "improved"—drained and cleared for farms or residential or commercial development. The importance of wetlands as purifiers of wastewater and absorbers of otherwise hazardous flood waters is now recognized. Wetlands are also vitally important as habitats for wildlife. Herons, storks, and other birds depend on wetlands for nesting. Wetlands are also home to many amphibians and reptiles, such as alligators and crocodiles. The federal government and most states now prohibit destruction of some wetlands.

RIVERS

Many rivers originate from snowmelt in mountains. At its headwaters, a river is usually very cold and highly oxygenated, running swiftly through a shallow riverbed. As it tumbles down the mountain, a river may broaden, become warmer, lose oxygen, and flow more slowly. A river's characteristics may vary with changes in the land and climate through which it flows. Runoff, for example, may wash nutrients and sediment from the surrounding land into a river. These materials can affect the growth and health of the organisms in the river.

Figure 4-44 **The water flow of a river slows and the habitat changes as turbulent headwaters broaden into wide, slow-moving water downstream.**

There are more species of fish in the Amazon Basin than in any other river system on Earth—even more species than in the Atlantic Ocean.

Answers to Section Review can be found on page T146.

SECTION REVIEW

1. Which of the life zones of a lake, the area with sunlight or the area without, is likely to be more biologically diverse? Why?

2. Imagine that a city wishes to clear and drain the wetlands along a river bottom to control flooding and improve the river's water quality. Is this a good idea? Why or why not?

THINKING CRITICALLY

3. *Identifying Relationships* Most of the naturally occurring lakes in North America are found either high in the mountains or at latitudes above about 45°N. How might you explain this?

4. *Analyzing Processes* Explain how fertilizing your yard and applying pesticides can affect the health of a river ecosystem.

Plant and Animal Adaptations Near the churning headwaters, mosses anchor themselves to rocks by rootlike structures called rhizoids. Mayfly nymphs use hooks on their legs to cling to any stable surface. Frogs use the suction cups on their toes to maintain stability in these turbulent waters. Trout and minnows are adapted to thrive in the cold, oxygen-rich headwaters. Trout have streamlined bodies that present less resistance to the strong current, and they're powerful swimmers. Further downstream, catfish and carp prefer the warmer, calmer waters. Carp are adapted to gliding over the river bottom but not to swimming against the current. Here, freshwater aquatic plants such as water crowfoot set roots down into the rich sediment. Figure 4-45 shows how the arrowhead plant's leaf shape varies according to the strength of a river's current.

Threats to Rivers Industries use river water in manufacturing processes and as a receptacle for waste. For ages, people have used rivers to "take away" their sewage and garbage. These practices have polluted rivers with toxins, killing river organisms and making river fish inedible. Today, runoff from the land puts pesticides and other poisons into rivers and coats riverbeds with toxic sediments. Dam building alters river flow and may destroy fish habitat.

Figure 4-45 **The arrowhead plant's leaves differ according to the speed of the water it grows in.**

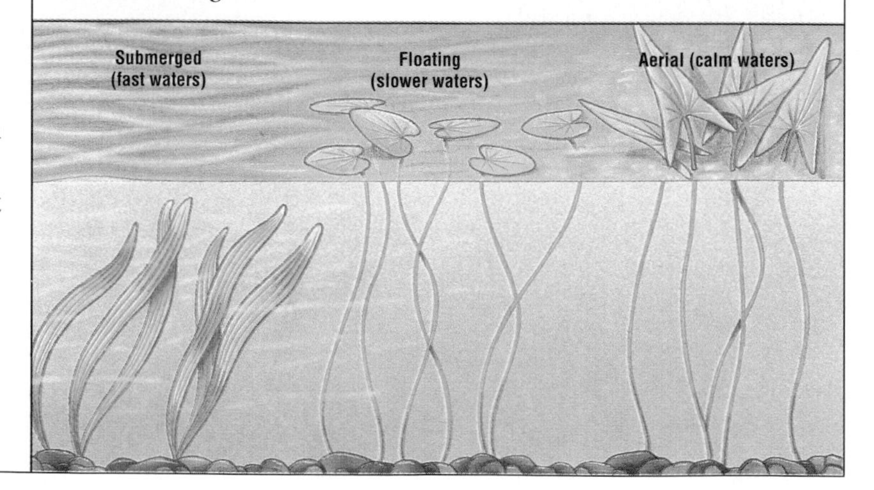

MARINE ECOSYSTEMS

AFTER READING THIS SECTION YOU SHOULD BE ABLE TO

❶ describe the characteristics of estuaries, coral reefs, oceans, and polar ecosystems.

❷ describe the adaptations of organisms to different marine environments.

The oceans of the world contain a wide variety of plant and animal communities. On land, the most important factors in determining what kinds of organisms live in an area are temperature and the amount of rainfall. In the oceans, lack of water is not a problem, so the types of organisms present depend on temperature and the amount of sunlight and nutrients available.

ESTUARIES

For a week each spring, huge horseshoe crabs, shown in Figure 4-46, crawl out of the ocean and onto the beaches of Delaware Bay. In the shallows along the shore, the crabs mate and lay billions of eggs. Shorebirds are waiting for them. Millions of migrating birds stop at the bay to gorge on the eggs.

Delaware Bay is an **estuary**—an ecosystem in which fresh water from rivers mixes with salt water from the ocean. Figure 4-47 illustrates how the waters mix in such a way that the estuary becomes a nutrient trap, where mineral-rich mud drops to the bottom. In shallow areas, marsh grass grows in the mud.

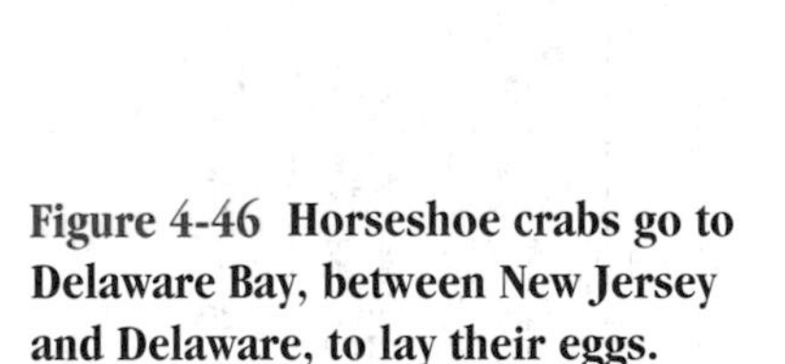

Figure 4-46 Horseshoe crabs go to Delaware Bay, between New Jersey and Delaware, to lay their eggs.

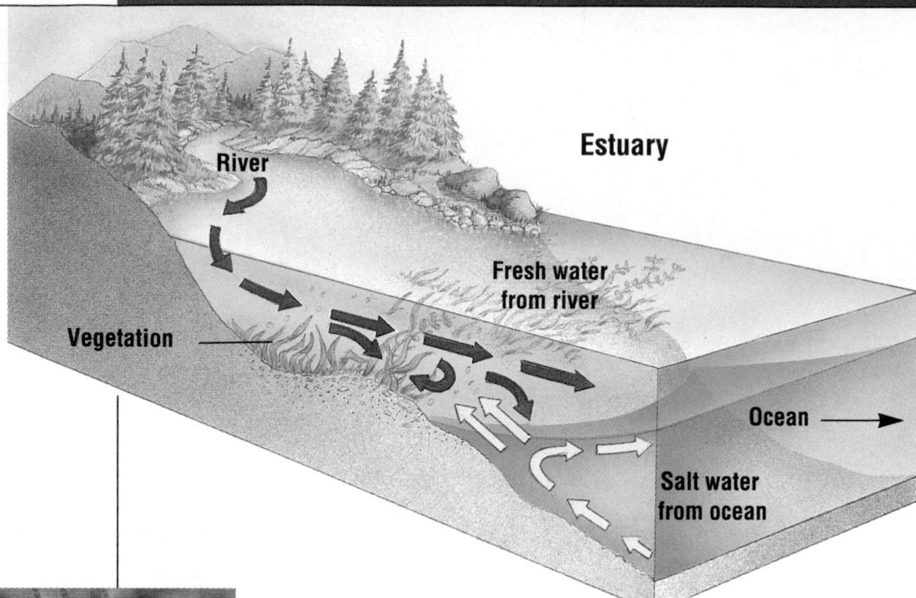

Figure 4-47 The mixing of fresh and salt water at the mouth of a river creates a nutrient trap.

Figure 4-48 The sticky, feathery appendages of filter feeders such as this barnacle wave through the water and trap tiny organisms.

Plant and Animal Adaptations Estuaries are among the richest ecosystems because they contain plenty of light and the nutrients that plants need for photosynthesis. Rivers supply nutrients washed from the land, and the water is shallow, so sunlight reaches all the way to the bottom. The light and nutrients support large populations of plants rooted in the mud, as well as phytoplankton and zooplankton. The plankton, in turn, provides food for larger animals such as fish. Dolphins, manatees, seals, and other mammals feed on fish and on plants in the estuary. Oysters, barnacles, and clams live anchored to marsh grass or in the bottom mud and feed by filtering algae and debris out of the water. All organisms that live in estuaries are adapted to tolerate variations in salinity because the salt content of the water varies as fresh and salt water mix.

Threats to Estuaries Estuaries provide protected harbors, access to the ocean, and connection with a river. As a result, many of the world's major ports are built on estuaries. Of the ten largest urban areas in the world, seven were built on estuaries (Tokyo, New York, Shanghai, Calcutta, Buenos Aires, Rio de Janeiro, and Bombay).

It is not surprising that estuaries have often been used as places to dump waste and then, as they fill in, as building sites. Much of this filling occurred in California, which now has plans to restore some of its estuary wetlands.

The pollutants that damage estuaries are the same ones that pollute lakes, rivers, and the oceans: sewage, industrial waste containing toxic chemicals, and agricultural runoff of soil, pesticides, and fertilizer. Estuaries can degrade most of these pollutants, but they cannot cope with the amounts produced by dense human populations.

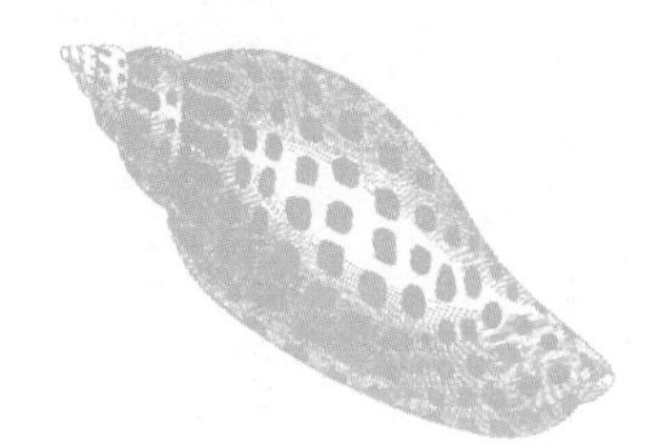

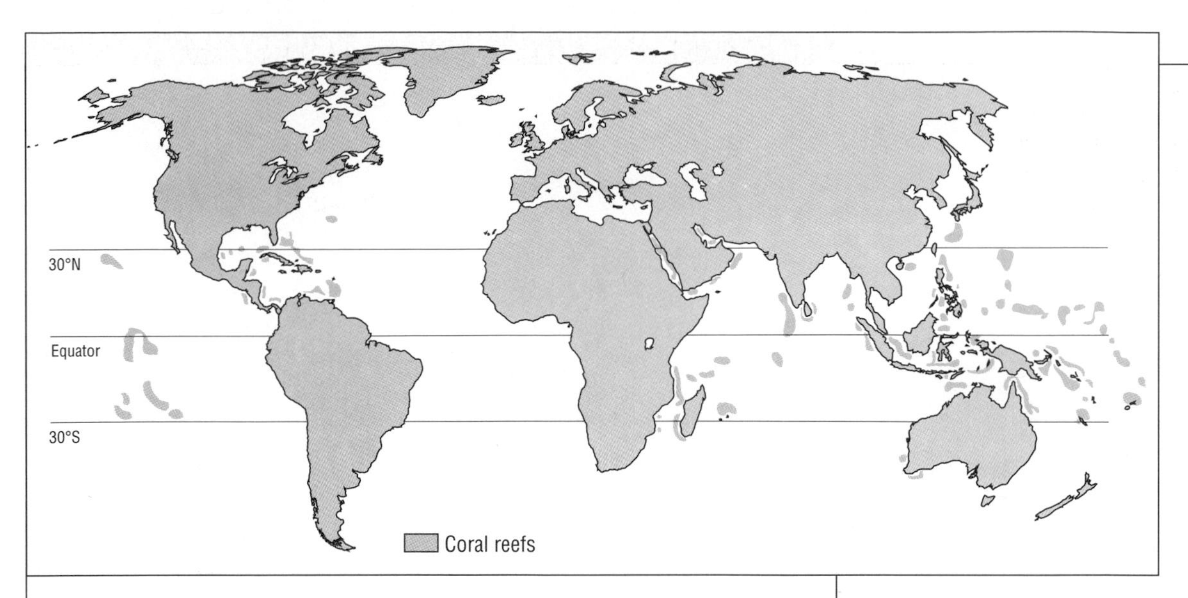

Figure 4-49 **Coral reefs are found in warm, shallow waters, where there is enough light for photosynthesis.**

Coral Reefs

Coral reefs are limestone islands in the sea that are built by coral animals. Thousands of species of plants and animals live in their cracks and crevices, making coral reefs among the most diverse ecosystems on Earth.

Corals can live only in warm salt water where there is enough light for photosynthesis, so coral reefs are found in shallow, tropical seas. Only the outer layer of a reef contains living corals, which build their rock homes with the help of photosynthetic algae.

Figure 4-50 **Coral reefs support a great diversity of species.**

Figure 4-51 A coral emerges from its protective structures to feed

Animal Adaptations Because of their convoluted shape, reefs provide habitats for a magnificent variety of tropical fish, as well as for snails, clams, and sponges. There is no shortage of food for predators such as moray eels, sharks, and groupers. The remarkable parrotfish has beaklike teeth that it uses to scrape algae and corals off reefs to eat.

Threats to Coral Reefs If the water surrounding a reef is too hot or too cold, or if fresh water drains into the water surrounding a reef, corals have difficulty producing limestone. If the water is too muddy, too polluted, or too high in nutrients, algae that live with the corals will either die or grow out of control and smother the corals.

Oil spills, sewage, and pesticide and silt runoff have been linked to coral-reef destruction. Furthermore, overfishing can devastate fish populations and upset the balance of a reef's ecosystem. Since coral reefs grow slowly, a reef may not be able to repair itself after chunks of coral are destroyed by careless divers, shipwrecks, ships dropping anchor, or people breaking off pieces of it for decorative items or building materials.

THE OCEAN

Ocean covers nearly three-quarters of Earth's surface. Although the ocean is vast, plants can grow only where there are nutrients and enough light for photosynthesis. As a result, much of the ocean's life is concentrated in shallow water around the edges of continents. Here, rivers wash nutrients from the land, and sunlight penetrates to the bottom, permitting the growth of seaweeds anchored to rocks and tiny floating plants in the phytoplankton near the surface. Invertebrates and fish that feed on these plants are also concentrated near the shore.

In the open ocean, sunlight penetrates only the surface layers, where phytoplankton grow if there are enough nutrients. As a

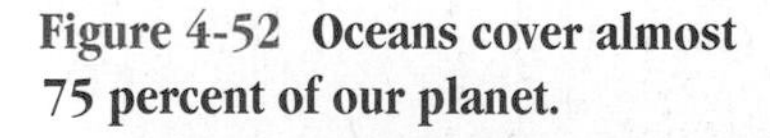

Figure 4-52 Oceans cover almost 75 percent of our planet.

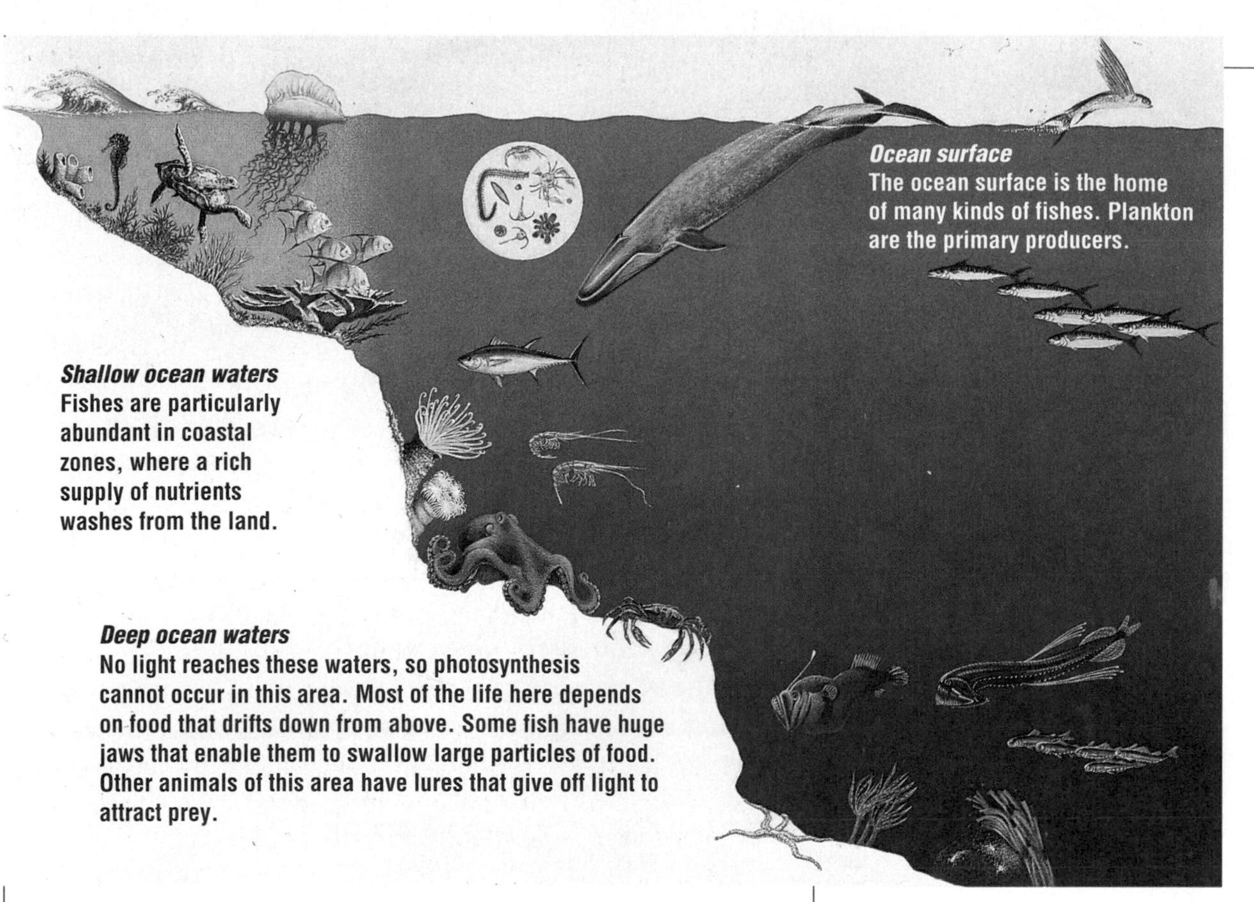

Figure 4-53 **The amount of sunlight determines which organisms can live in each layer of the ocean.**

result, the open ocean is one of the least productive of all ecosystems. (See Figure 4-54.) The depths of the ocean are perpetually dark, and most of the food consists of dead organisms that fall from the surface.

Plant Adaptations Flowering plants are absent from oceans, except around the edges. Food for herbivores in the open ocean is provided by the abundant phytoplankton. (See Figure 4-55.) These tiny cells all have buoyancy devices that prevent them from sinking

Figure 4-54 **The open ocean is among the least productive of all ecosystems.**

Ecosystem Productivity	
Ecosystem Type	**Average Net Primary Productivity** (grams of plant organic matter produced/sq. m/year)
Coral reef	2,500
Tropical rain forest	2,200
Estuary	1,500
Temperate forest	1,250
Tropical savanna	900
Taiga	800
Continental shelf	360
Lake	250
Open ocean	**125**
Tundra	90
Extreme desert	3

Figure 4-55 **Diatoms form one of the major groups of marine phytoplankton.**

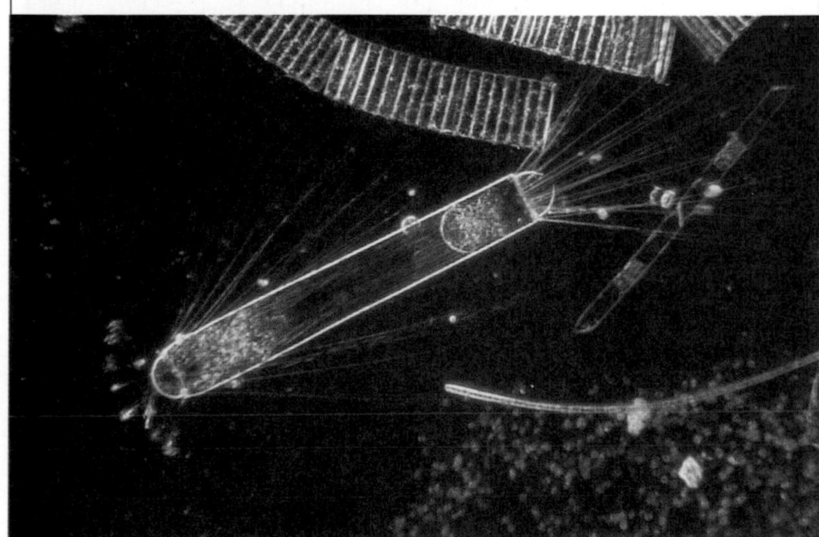

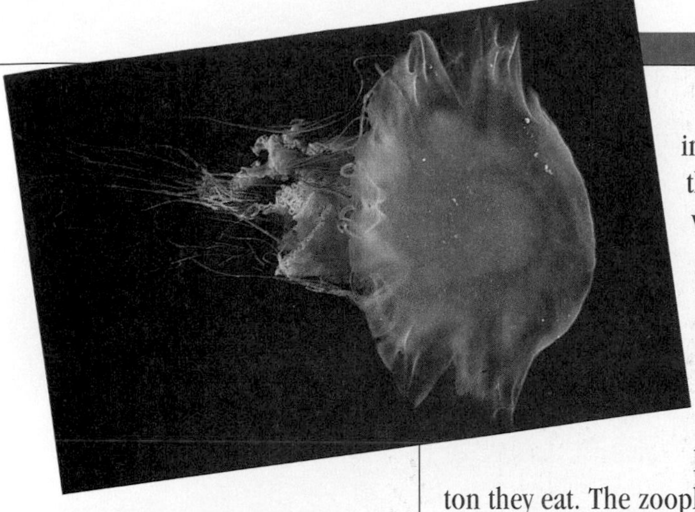

Figure 4-56 **The density of salt water permits many jelly-like animals, such as this red medusa jellyfish, to remain buoyant and move through the seas.**

into deep water that is too dark for photosynthesis. Some float by means of long spines or whiplike flagella. Others contain oil droplets that act as floats. When they die, the phytoplankton sink to the bottom, sometimes in vast numbers.

Animal Adaptations The sea's smallest herbivores are the zooplankton, living near the surface with the phytoplankton they eat. The zooplankton include jellyfish, tiny shrimp, and the larvae of fish and bottom-dwelling animals such as oysters and lobsters. Dozens of fish feed on the plankton, as do marine mammals such as seals and whales.

Many creatures of the open ocean have evolved sleek, tapered shapes for moving through dense water. Most fish that swim near the surface are tinged a silvery color, a protective camouflage for life in the open water with no place to hide. Many of them also have buoyancy devices that permit them to stay at one level in the water. Sharks have huge oily livers that act as floats. Many bony fish contain gas-filled swim bladders. Reptiles and mammals have lungs in place of swim bladders, and their lungs are used as floats as well as for breathing.

Water absorbs light, so sunlight penetrates only about 100 m (330 ft.) into the sea before it is all absorbed. Below this, there is no light for photosynthesis. In the depths of the ocean live decomposers and filter feeders and the organisms that eat them.

ECO-FACT

A jellyfish found in the Arctic Ocean, *Cyanea arctica*, can grow as large as 2.4 m (8 ft.) across and has tentacles 61 m (200 ft.) long.

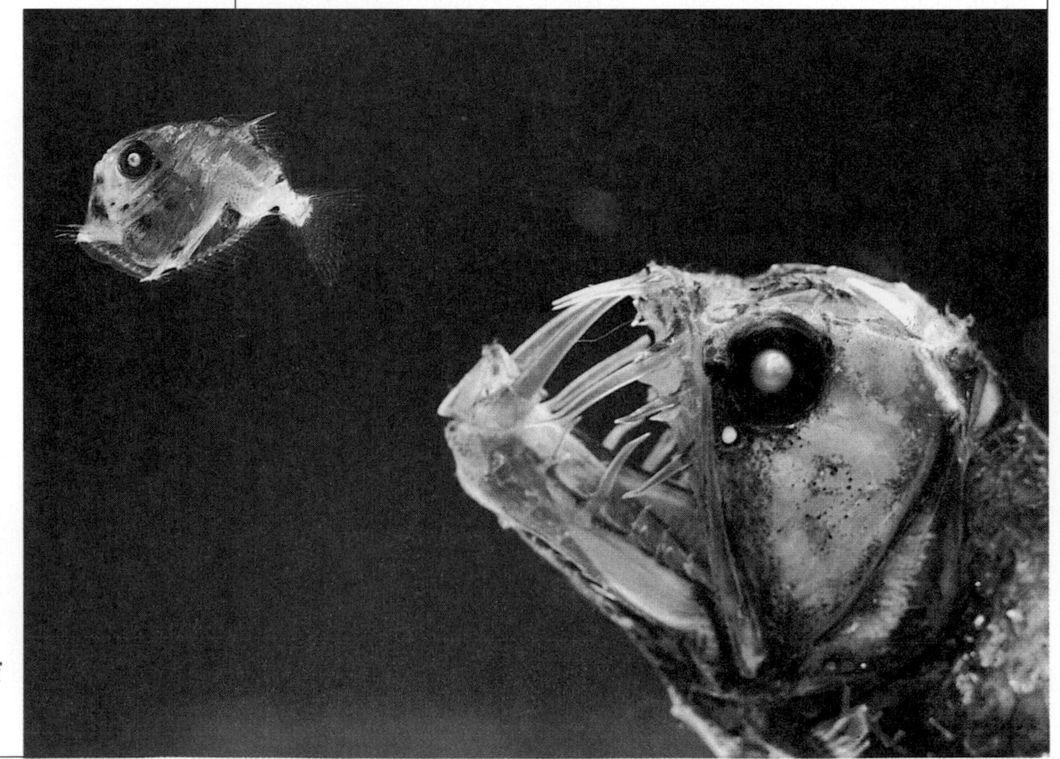

Figure 4-57 **The viper fish (right) is unmistakably a predator. This deep-sea fish lives at depths of 600 m or more.**

Figure 4-58 This flashlight fish in the Coral Sea communicates with light.

Also in the ocean depths, flashlight fish and other species use light for communication. Their bursts or blinks of light are produced by luminous bacteria incorporated into their bodies. Visibility is poor underwater, but sound carries extremely well over long distances. Whales communicate with beautiful, haunting "songs." Other marine mammals, such as dolphins, communicate by emitting clicks and calls.

Threats to the Oceans Although oceans are huge, they are becoming steadily more polluted. Most ocean pollution arises from activities on land. The pollutants are the same ones that cause problems on land. For instance, plant nutrients washing off the land as runoff from fertilized fields may cause blooms of algae, some of which are poisonous. Industrial waste and sewage discharged into rivers is the biggest source of coastal pollution in the United States. A modern addition to the list of substances that pollute the oceans is radioactive waste, particularly from nuclear power plants. Until 1970, much of this waste was dumped into deep water in drums and glass containers that have subsequently leaked.

Overfishing and certain fishing methods are destroying fishing grounds. Immense trawl nets entangle every living thing in the trawled area. Most of the catch is not used, and the dead fish are thrown back into the ocean. Marine mammals such as dolphins, which must breathe air, drown in the nets. Although it is against the law, ships may discard fishing lines into the ocean where they can strangle fish and seals. The toll is so vast that reduced reproduction is endangering many species.

Polar Ecosystems

The ice-covered polar caps at the North and South Poles can be considered marine ecosystems because nearly all food is provided by phytoplankton in the ocean. In other ways, the two poles are very different from each other. The South Pole lies on the continent

Figure 4-59 Humpback whales feed in the Arctic Ocean

Answers to Section Review can be found on page T146.

SECTION REVIEW

1. Coral reefs are especially sensitive to environmental disturbances. What are some of the factors that can damage coral reefs?
2. Why are estuaries particularly vulnerable to the effects of pollution? Why is this important to humans?

THINKING CRITICALLY

3. *Predicting Outcomes* Suppose the sea level were suddenly to rise by 100 m. What would happen to the world's coral reefs? Explain.
4. *Analyzing Processes* The parts of the ocean most remote from land are biological deserts. They contain far less life (in terms of mass per unit of surface) than most deserts on land. What do you think are some reasons for this?

of Antarctica, where it is covered with a permanent icecap that melts only around the edges of the continent in summer. The North Pole is not on land at all. It lies in the Arctic Ocean, much of which is frozen into a huge iceberg throughout the year, with smaller icebergs drifting around the edges.

The Arctic The Arctic Ocean is relatively shallow, so its waters are rich in nutrients from the surrounding landmasses, and it supports large populations of plankton. These in turn provide the food for a rich diversity of fish, which live in the open water and under the ice. In open waters, whales and ocean birds prey on the fish. The birds and many species of seals bear young on the ice. The seals keep holes open in the ice so that they can dive to catch fish. The seals and birds in turn provide food for the few humans who live in the Arctic and for the top predators, polar bears.

The Antarctic The Antarctic is the only continent never colonized by humans. It is governed by an international commission and is used mainly for research on the unusual animals that live there. Even during the summer, only a few plants grow at the rocky edges of the continent. As in the Arctic, nearly all the food comes from plankton in the ocean. This provides food for large numbers of fish, whales, and birds such as penguins, which cannot fly because their wings have evolved into paddles adapted to swimming.

Threats to Polar Ecosystems Both the Arctic and the Antarctic contain reserves of minerals, such as oil, whose extraction would disrupt these largely untouched ecosystems. Conservationists want the Antarctic to be made into a world wildlife refuge so that this unique ecosystem can be preserved. Meanwhile, the main threat to wildlife is an increase in tourism in recent years. The garbage left by tourists is difficult to dispose of in a climate so cold that nothing decays. Research stations and tour operators are working to solve this problem.

C H A P T E R 4

HIGHLIGHTS

SUMMARY

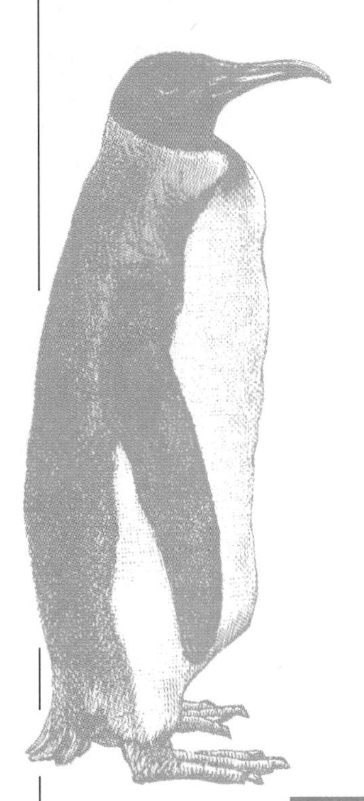

- Scientists classify the ecosystems of the world into classes called biomes. Biomes are named according to their plant life because the plants that grow in an area determine what other organisms live there.
- The major land biomes of the world include: tropical rain forests, temperate deciduous forests, taiga, savannas, desert, and tundra.
- In each biome, plants and animals have adapted to specific environmental conditions. These conditions, however, are threatened by human activities.
- Water ecosystems can be freshwater or marine. As in biomes, the plants and animals in water ecosystems are adapted to specific environmental conditions. These conditions are also threatened by human activities.
- Freshwater ecosystems include lakes and ponds, rivers and streams, and wetlands. The types of freshwater ecosystems are distinguished by the depth of the water, how fast the water moves, and the availability of mineral nutrients, sunlight, and oxygen.
- Marine ecosystems, which are identified by the presence of salt water, include estuaries, coral reefs, and oceans. The icecaps at the North and South Poles are also considered marine ecosystems because organisms living there obtain almost all their food from the ocean.

WATER ECOSYSTEMS
may be
freshwater
such as
lakes
rivers
wetlands
which include
swamps
marshes
marine
such as
estuaries
coral reefs

Sample EcoLog answers can be found on page T146.

EcoLog

Now that you've studied this chapter, revise your answers to the questions you answered at the beginning of the chapter, based on what you have learned.

❶ What are the general characteristics of the ecosystem in which you live? Are there ecosystems similar to yours in other parts of the world?

❷ Describe a plant or animal that lives in your area, and describe its survival strategy.

Vocabulary Terms

benthic zone (p. 100)
biome (p. 79)
canopy (p. 82)
coral reef (p. 107)
desert (p. 94)
drought-resistance (p. 96)
estivating (p. 96)
estuary (p. 105)
littoral zone (p. 100)
permafrost (p. 97)
taiga (p. 88)
temperate deciduous forest (p. 86)
temperate rain forest (p. 84)
temperate grasslands (p. 92)
tropical rain forest (p. 80)
tropical savanna (p. 90)
tundra (p. 97)
wetlands (p. 102)

C H A P T E R 4

REVIEW

Understanding Vocabulary

1. For each pair of terms, explain the difference in their meanings.
   a. savanna
   steppe
   b. biome
   ecosystem
   c. littoral zone
   benthic zone
   d. taiga
   tundra

Relating Concepts

2. Copy the unfinished concept map below onto a sheet of paper. Then complete the concept map by writing the correct word or phrase in each box containing a question mark.

Answers not indicated can be found on page T146.

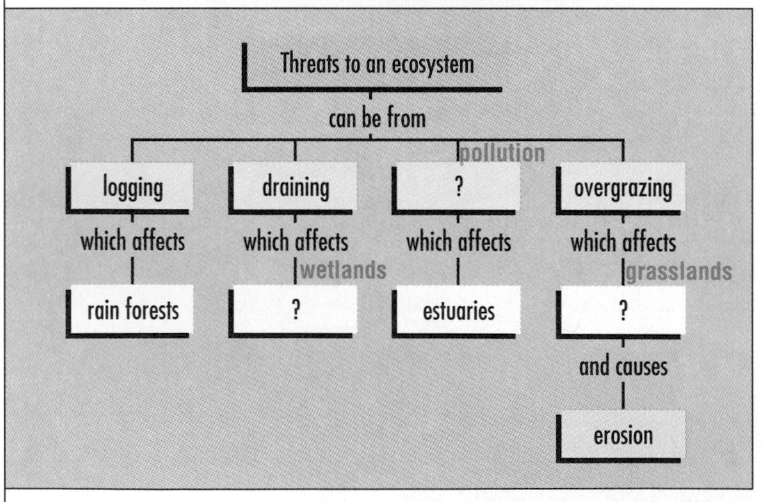

Understanding Concepts

Multiple Choice

3. Tropical rain forests cover approximately what percent of the Earth's surface?
   a. 7
   b. 20
   c. 70
   d. 80

4. Animal species of the tropical rain forest
   a. compete more fiercely for available resources than species native to other environments.
   b. specialize to avoid competition.
   c. have adaptations to cope with extreme climatic variations.
   d. never use camouflage.

5. Migration of animals in the savanna is largely a response to
   a. predation.
   b. altitude.
   c. rainfall.
   d. temperature.

6. Compared with temperate deciduous forests, temperate grasslands receive
   a. more rainfall.
   b. less rainfall.
   c. about the same amount of rainfall.
   d. sometimes more, sometimes less rainfall.

7. Spadefoot toads survive the dry conditions of the desert by
   a. migrating to seasonal watering holes.
   b. locating underground springs.
   c. burying themselves in the ground.
   d. drinking cactus juice.

8. The tundra would most likely be suitable to an animal that
   a. required nesting sites in tall trees.
   b. was coldblooded.
   c. had a green outer skin for camouflage.
   d. could migrate hundreds of kilometers each summer.

9. Wetlands are important to fishermen in the United States because
   a. wetlands are the easiest place to catch fish.
   b. wetlands breed insects eaten by fish.
   c. wetlands provide the most desirable species of fish.
   d. many of the fish caught each year use the wetlands for feeding and spawning.

10. Estuarine animals
   a. tend to produce few offspring.
   b. are usually found in unpolluted environments.
   c. must be adapted to varying levels of salinity.
   d. are adapted to cold-water conditions.

11. Bacteria cause eutrophication to occur in lakes that contain a large amount of plant nutrients by
   a. feeding on decaying plants and algae.
   b. reducing oxygen dissolved in the water.
   c. both **a** and **b**
   d. neither **a** nor **b**

12. Seals would MOST likely be found in which of the following?
   a. estuaries
   b. open ocean
   c. marshes
   d. benthic zone

13. Polar regions are considered marine ecosystems because
   a. they contain an enormous amount of frozen water.
   b. they are inhabited by few organisms.
   c. sunlight is limited.
   d. polar animals eat mostly phytoplankton from the sea.

Short Answer

14. Unlike depictions of jungles seen in movies, the floor of a tropical rain forest is largely devoid of vegetation. Why is this so?

15. When considering a grassland ecosystem, what is the relationship between root systems and erosion?

16. How is the practice of estivating similar to hibernating?

17. Fossilized mammoths have been found buried in the tundra. Explain why the tundra would be a good preserver of animal remains.

18. How does the phrase "best of both worlds" relate to an estuary?

Interpreting Graphics

19. **Examine the diagram below.** It shows how the relationship between temperature and rainfall affects the location of the biomes. According to the diagram, what type of climate would be found in the grasslands? in the tropical rain forest? in the tundra? Notice that there are two desert locations shown. Explain.

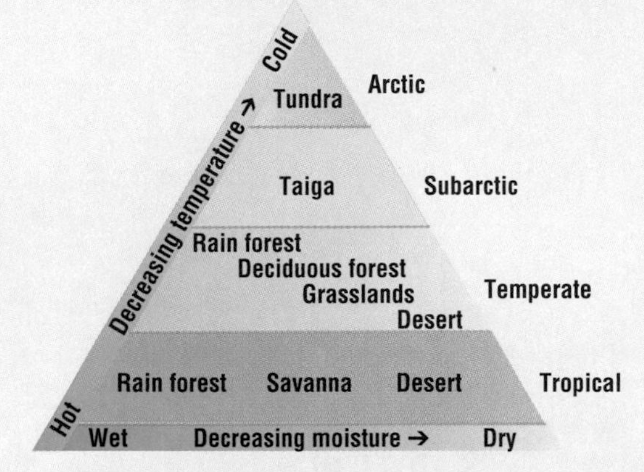

Thinking Critically

20. ***Analyzing Relationships*** Explain the relationship between the speed of a river and its oxygen content.

Themes in Science

21. ***Evolution*** Explain how corals and algae have evolved a beneficial relationship, and predict what might happen if a reef suddenly sank beneath the zone where light penetrates.

Cross-Discipline Connection

22. ***Geography*** Using a world map, identify locations of the various ecosystems you learned about in this chapter. Then mount photos or illustrations of various plants and animals that would be found in each of the ecosystems.

Discovery Through Reading

For an up-close look at a rain forest, read "The Secret Life of a Tree in the Rain Forest," by George Howe Colt, *Life,* June 1994, pp. 58–68. If you are interested in learning more about the way amphibians adapt to life in the desert, read "Frogs and Toads in Deserts," by Lon L. McClanahan, et al., *Scientific American,* March 1994, pp. 82–88.

PORTFOLIO ACTIVITY

Do a special project on the ecosystems found in your community. Using field guides of your area and other resources, find out what plants and animals make their home in the ecosystems you identify in your community. With the information you find, draw a food web that shows how the organisms in each ecosystem could be related.

CHAPTER 4

INVESTIGATION

IDENTIFY YOUR LOCAL BIOME

What sort of biome do you call home? Do you live in a temperate deciduous forest, a desert, or a prairie or other temperate grassland? In this Investigation, you will explore certain characteristics of the biome where you live. With the information you gather, you will be able to identify which biome you inhabit.

Answers to this Investigation can be found on page T147.

MATERIALS

- globe or map
- field guide to local flora and fauna
- binoculars (optional)
- notebook
- pencil or pen
- ruler
- graph paper (optional)

COLLECT INFORMATION FROM LOCAL RESOURCES

1. Using a globe or map, determine the latitude at which you live. Record this information in your notebook.
2. Consider the topography of the place where you live. To do this, study the contour lines on a map or surface variations on a globe. What clues do you find that might help identify your biome? For example, is your local area located in a rain shadow? Record your findings in your notebook.
3. Prepare a climatogram. A climatogram is a graph that shows average monthly values for two climatic factors—temperature and precipitation. Temperature is expressed in degrees Celsius and is plotted as a smooth curve. Precipitation values are given in centimeters and are plotted as a histogram. The climatogram for a city in Texas is shown below.

AUSTIN, TEXAS

PRECIPITATION (CM): 0, 5, 10, 15, 20, 25, 30, 35, 40, 45, 50

TEMPERATURE (°C): −60, −50, −40, −30, −20, −10, 0, 10, 20, 30, 40

J F M A M J J A S O N D

MONTHS

To make a climatogram of your area, first obtain monthly averages of precipitation and temperature from a local weather station. Make a data table, and record these values.

Next, draw the vertical and horizontal axes of your climatogram in your notebook or on graph paper. Then show the temperature scale along the right-hand vertical axis and the precipitation scale along the left-hand vertical axis. Show months of the year along the horizontal axis. Finally, plot your data.

PERFORM FIELD OBSERVATIONS

4. CAUTION: Before beginning your field study, review the safety guidelines on pages 409–412. Go outside and observe the plants growing in your area. Bring along a field guide, and respond to the following items in your notebook.
   - Sketch or describe as many plants as you can that are representative of the area. Using your field guide, identify each of these species.
   - Describe three or more adaptations that each plant has to the local climate.
   - Which of the plants you observed are native to your area? Which have been introduced by humans? Which of the introduced plants can survive on their own in local conditions? Which of the introduced plants require extensive human cultivation to remain alive?

5. Observe the animals of your biome. Also look for evidence that animals have left behind—footprints, scat, nests, dens or burrows, hair or feathers, scratches or urine markings, etc. Respond to the following items in your notebook.
   - Sketch or describe as many different animal species as possible. Identify each species with the help of your field guide.
   - Describe three or more adaptations that each animal has developed in order to survive in local climatic conditions.

ANALYZE YOUR FINDINGS

6. Compare your local climatogram to the biome climatograms shown in this chapter. Which biome has a climatogram most similar to your own?
7. Consider your latitude, topographical findings, and observations of local plants and animals. Synthesize all of this information with your climatogram, and determine which biome best matches the area in which you live. State this conclusion in your notebook.
8. Choose three adaptations you observed in the plants that grow naturally in your area. Explain in detail how each of these adaptations meets the conditions of your biome.
9. Name at least three adaptations you observed in local animals. Explain how each of these adaptations helps the animal meet specific conditions of your biome.
10. Does your local climatogram *exactly* match any of the seven major terrestrial climatograms shown in the chapter? Explain how any differences between your local biome and the one in the chapter that it most clearly matches might influence the adaptations of local animals and plants. Did the plants and animals you observed show any of these influences?

Which one of Earth's major biomes is represented in this neighborhood?

POINTS OF VIEW

The Future of the Arctic National Wildlife Refuge

During the 1970s, Congress passed the Alaska National Interests Land Conservation Act. The legislation gave Congress the responsibility for determining how Alaska's lands will be used. Included in this responsibility is the fate of the Arctic National Wildlife Refuge (ANWR). Oil company geologists believe that oil reserves underlie several areas of the northern Alaska coast, including the refuge. Debate has raged about whether Congress should maintain ANWR as a refuge or open it to oil exploration. Advocates of the refuge feel that the environmental cost of oil exploration would be too high. But those who favor oil exploration in the refuge believe its oil reserves must be tapped to meet U.S. oil consumption needs and to maintain economic security. Two points of view:

THE REFUGE SHOULD BE PROTECTED FROM OIL EXPLORATION

Conservationists and ecologists are concerned about the impact that oil exploration would have on animals that make their homes in the refuge. Oil exploration would occur on 1.5 million acres in the ANWR. This area includes the calving ground and grazing area for one of the last great herds of caribous in North America. Scientists studying the caribous believe that forcing the herd into other areas of the refuge would deprive the caribous of their main food source and expose caribou calves to increased predation.

In addition, migratory birds from all over the world travel to the refuge's tundra to nest and raise their young during the short arctic summers. Scientists believe that oil exploration would disrupt the nesting and feeding of these birds so much that they would be unable to finish rearing their young before the first freeze of early September.

Opponents of oil exploration in the refuge point to the environmental damage that has already been done in nearby Prudhoe Bay. When oil was found there, oil companies joined forces to extract the oil and build a pipeline across the state to tankers on the southern Alaska coast. Advocates of protection say that the use of this pipeline has exposed the fragile tundra ecosystem to toxic chemicals and destroyed natural habitats. They fear the same fate for the refuge if oil exploration is permitted there.

Advocates of protection also point out that nobody knows how much oil is really available there. And even if all the oil that could possibly be there *were* extracted, it would supply the United States for less than a year. Instead, conservationists contend that the American need for oil can be reduced by laws that require stricter energy conservation measures.

Migratory birds, such as these Canada geese, nest and raise their young on the refuge's tundra.

The Refuge Should Be Opened for Oil Exploration

Advocates of exploration in the refuge believe that U.S. demand for oil cannot be met by energy conservation. They insist that the United States must utilize every domestic source of oil available, including the ANWR. They point out that the United States depends too much on oil from other countries that control its price and availability. A significant amount of our oil is imported from the Middle East, a politically unstable area. Some of the countries in that region have been openly hostile to the United States. If those countries restrict or cut off sales of oil to the United States, our economy could be seriously affected. Those who favor exploration think economic security should take priority over environmental concerns.

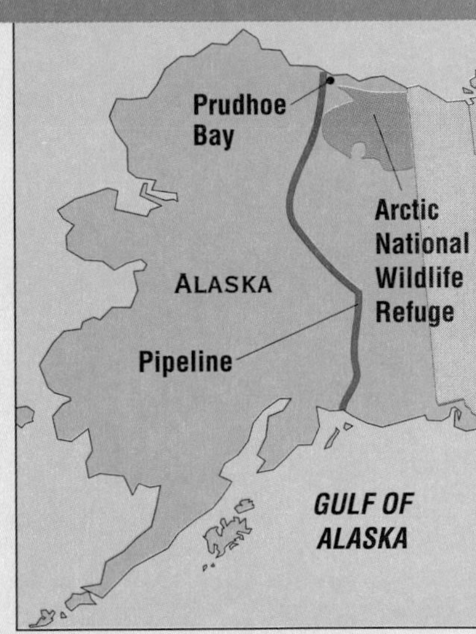

The Arctic National Wildlife Refuge is located in northeastern Alaska, conveniently close to the Prudhoe Bay complex and the oil pipeline.

The Prudhoe Bay oil production complex could be used to process oil found in the refuge.

Advocates of oil exploration in the refuge further stress that much of the oil in the Prudhoe Bay area has already been extracted, and production will soon begin to decline. The industrial complex that is already in place for the production of Prudhoe Bay oil could easily be used for the production of oil from the nearby refuge. New construction would be limited.

Government studies indicate a 19 percent chance of finding oil in the refuge, a percentage that the oil industry believes justifies exploration. Those who want to explore the refuge also suggest that oil companies can now extract oil with less environmental damage than was caused in Prudhoe Bay. They also believe that exploration of the refuge will not disrupt habitats as severely as predicted by environmentalists, pointing out that the caribou population actually increased in the area around Prudhoe Bay. To those who oppose the protection of the wildlife refuge, the economic benefits of oil exploration in this area far outweigh any remaining risks of environmental damage.

Analyze the Issue

1. ***Making Decisions*** Using the decision-making model presented in Chapter 1, decide whether oil exploration should be allowed in the refuge. Explain your reasoning.
2. ***Inferring Relationships*** Many factors can influence the population of a species. What other factors could have been responsible for the increased number of caribous in the Prudhoe Bay area? Explain your reasoning.

Answers to Analyze the Issue can be found on page T147.

This pipeline carries oil across the entire state of Alaska.

CHAPTER 5

WATER

"All is born of water; all is sustained by water."

JOHANN WOLFGANG VON GOETHE
GERMAN POET AND DRAMATIST

EcoLog

Before you read this chapter, take a few minutes to answer the following questions in your EcoLog.

1. Describe as best you can where the water in your home comes from.
2. Do you think the water you drink at home is pure or polluted? Explain your answer.

OUR WATER RESOURCES

AFTER READING THIS SECTION YOU SHOULD BE ABLE TO

1. explain why fresh water is a precious resource.
2. describe our main sources of fresh water.
3. explain why fresh water is often in short supply.

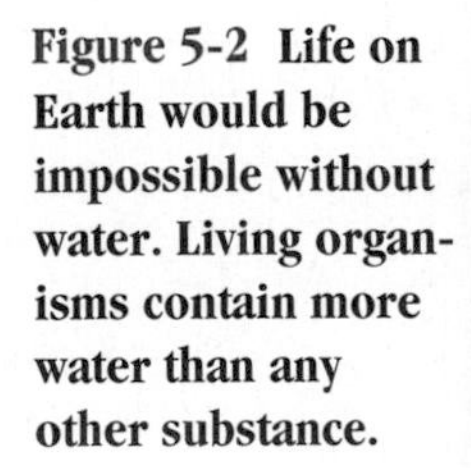

The next time you quench your thirst with some cold, clear water, think about this: that water may have been part of a rainstorm that pounded the Earth before life existed. It may have been part of a plant or fish that lived millions of years ago. Much of the water we drink today has been around since water first formed on Earth billions of years ago. When you drink water, you are actually sharing an ancient drink with the entire biosphere, both past and present.

Clean, fresh water is essential to life. People can survive for more than a month without food but can live for only a few days without water. Clean water is critical to human health. One of the main reasons people live longer today than they did 200 years ago is clean water: water to drink, water to bathe in, water to wash dishes and clothes, water to flush away sewage, and water to irrigate crops. Figure 5-1 shows how fresh water is used in the United States.

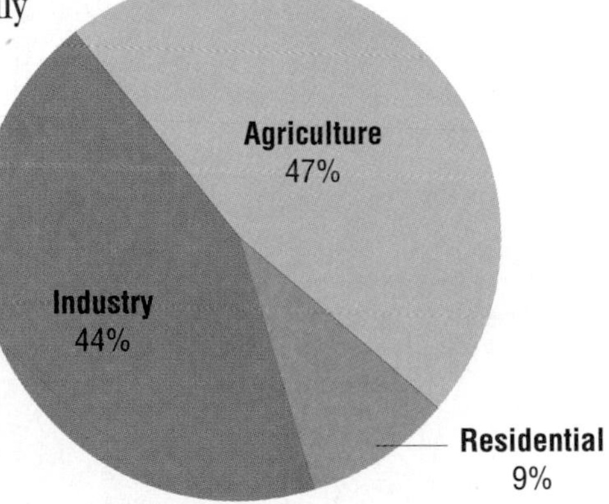

Figure 5-1 Agriculture and industry account for most of the water usage in the United States.

Figure 5-2 Life on Earth would be impossible without water. Living organisms contain more water than any other substance.

Water is a renewable resource. As you learned in Chapter 3, water is endlessly circulated by means of the water cycle. Water evaporates from the surface of the Earth, where it is held in oceans, lakes, rivers, soil, and living organisms. The water rises into the air, leaving behind salts, impurities, and other materials. Then the water vapor in the air condenses into clouds and falls as rain or snow. When rain or snow falls on land, it replenishes the land's fresh water.

Answer to caption:
Less than 1 percent

Figure 5-3 The diagram below shows how the Earth's water is distributed. How much of the fresh water is in lakes and rivers?

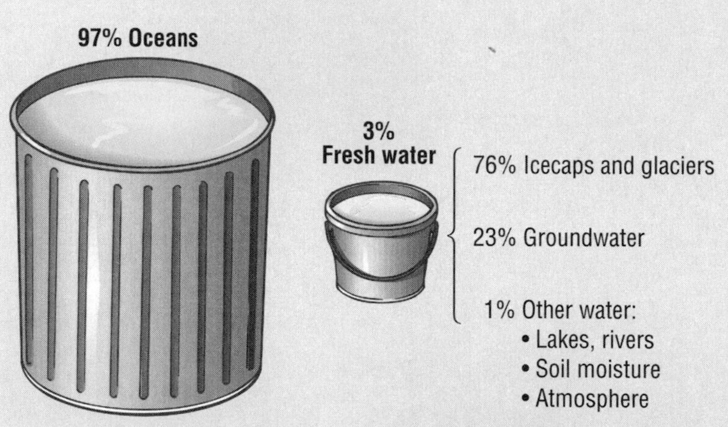

WATER, WATER EVERYWHERE, BUT . . .

The Earth is often called the water planet. Look at a globe or at photographs of the Earth taken from space, and you'll see why. Most of what you see is blue. About 70 percent of the Earth's surface is water. The Earth and its atmosphere contain an estimated 336 million cu. mi. of water, enough for every person on Earth to get about 1 trillion gal. each if the water were divided evenly. It would seem that a lack of water couldn't possibly be a concern on this planet.

However, if you look at the globe again, you can see one reason why water shortages are often a problem. As shown in Figure 5-3, fully 97 percent of the water on Earth is salt water and cannot easily be used for drinking. The remaining 3 percent is fresh water. But there's a catch there too. Well over half of the fresh water—about 76 percent, in fact—is frozen solid in the polar icecaps. Thus, only a tiny fraction of the Earth's water supply is available for our use. The water we require for all of our everyday needs comes from two sources: surface water and groundwater.

Figure 5-4 More than half of the world's fresh water is frozen solid in the polar icecaps. This photograph shows frozen fresh water on Ellesmere Island, Canada.

SURFACE WATER

Most large cities depend on surface water for their water supplies. **Surface water** is fresh water that is above ground in lakes, ponds, rivers, and streams. Throughout history, human societies have flourished where surface water was abundant. The water not only provided the supplies needed for life, but also allowed travel by boat.

You know that all water is part of the water cycle. But when you look at a huge river like the Mississippi, it may be difficult to believe that all of the water in the river fell to the Earth in the form of rain, sleet, or snow. Yet all rivers are the result of precipitation. The mighty Mississippi, the Columbia, the Colorado, the Missouri, and even the Amazon and Nile exist because of precipitation. In the case of the Mississippi, the river drains water that has fallen on thousands of square kilometers of land. As rain falls and snow melts, water drains into the Mississippi from mountaintops, hills, plateaus, and plains. The entire area of land that is drained by a river is known as its **watershed.** The watershed of the Mississippi River is shown in the map in Figure 5-5.

The amount of water that falls on a watershed varies from year to year and can have a significant effect on the amount of water that a river carries. As a result, it can be very dangerous for communities in some parts of the country to rely only on river water for water supplies. Lakes can provide more stable sources of water than rivers can, but lakes also depend on precipitation and the flow of water from rivers and streams for their water supply.

Diane Stout says you wouldn't believe the things she encounters along Oregon rivers. Find out more on pages 378–379.

Figure 5-5 The Mississippi River is the longest river in the United States. The map at the right shows its watershed.

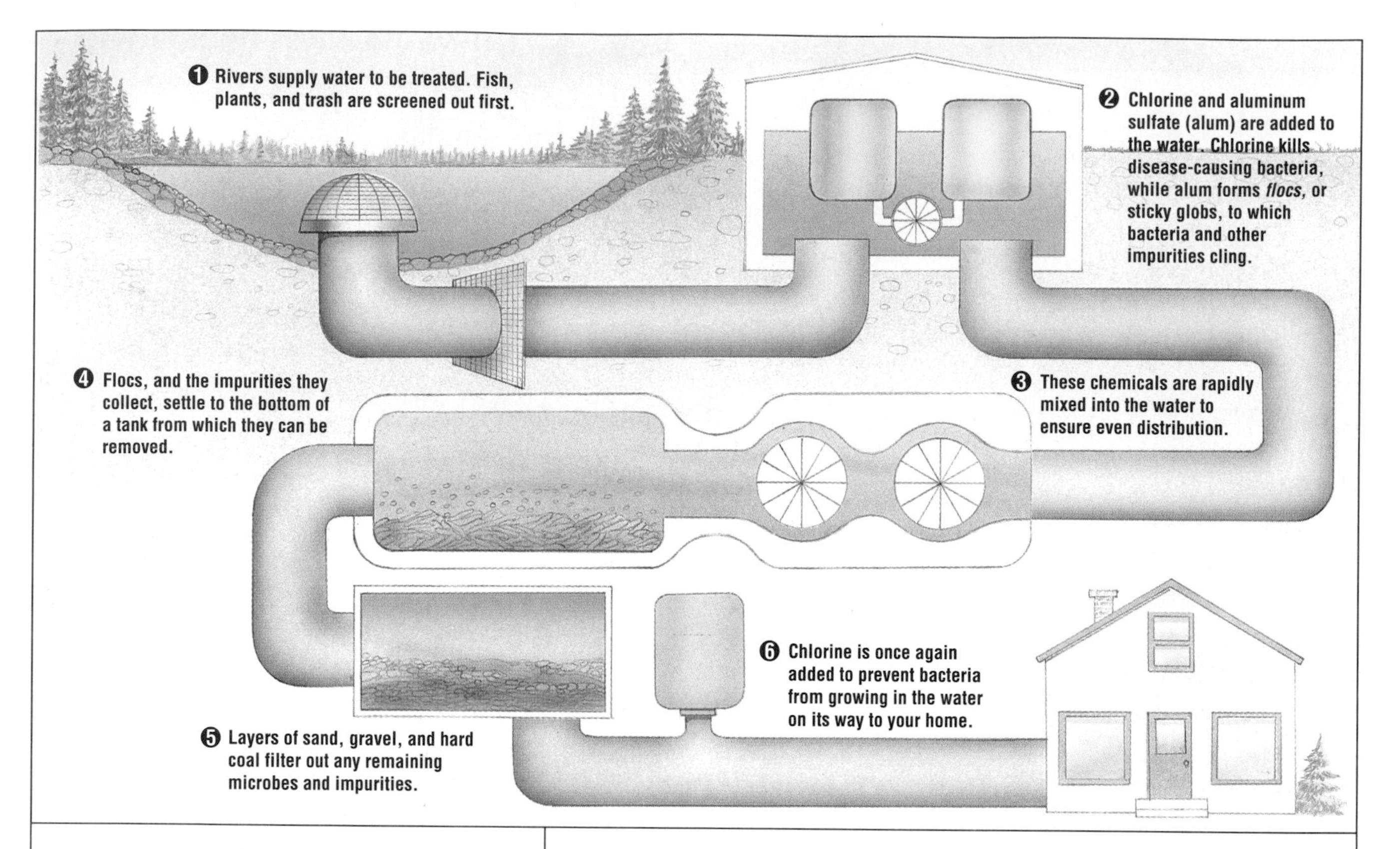

Figure 5-6 The diagram above shows what happens to water from a river before it reaches homes. Water from lakes and reservoirs undergoes the same process.

Rivers of Controversy River water is in such high demand that disputes arise over it. In the United States, for example, Arizona and California have fought over water rights to the Colorado River for a long time. The Colorado River is one of the largest bodies of water in the western United States. It flows for 2,335 km (1,450 mi.), passing through the Grand Canyon on its way to the Gulf of California. (See the map in Figure 5-7.) However, by the time the Colorado gets close to the gulf, very little water is left because so much water has been pumped out along the way.

Figure 5-7 As you can see in the map above, the Colorado River winds through the western United States. Because of heavy usage, the river trickles away in the deserts of Mexico, as shown in the photograph at right.

Disputes over rivers are not limited to the United States. Forty percent of the world's people rely on water that originates in another country, so conflicts over water rights are common. How much of a river's water does the country upstream have the right to use? How much water will the country downstream receive? Who will decide these issues and regulate international water rights? These are extremely difficult questions to answer.

Conflicts between countries are particularly common when dams are built on rivers. For example, Turkey is building a series of dams that will dramatically reduce the amount of water that flows into Syria and Iraq. One of the dams is shown in Figure 5-8. The problem is being negotiated, but with little success so far.

Figure 5-8 **This Turkish soldier is guarding one of the dams that will reduce the amount of water flowing into Syria and Iraq. Construction of the dams has caused conflict between Turkey and its neighbors.**

The problem of water rights is likely to get worse in the future. As the human population increases, the demand for fresh water will also increase. Many of the "shared" rivers flow through developing nations that want to build industries and irrigate their farmlands, which will require enormous amounts of water.

Dams A dam is a structure built across a river or stream that prevents most of the water from traveling downstream. The water that is prevented from flowing downstream collects behind the dam and forms a reservoir, an artificial lake. (See the dam and reservoir in Figure 5-9.) Water from reservoirs is used for drinking, manufacturing, and irrigation. When there is a drought, the reserves of water in the reservoir can be used to supply the water needs of the population. Dams can also provide flood control and electricity. Millions of people depend on dams for their survival.

However, dams are a mixed blessing. When a dam is built, the dry land behind the dam is completely covered with water, which destroys the existing ecosystem. In addition, ecosystems downstream are disrupted because they are accustomed to receiving a greater flow of water.

Figure 5-9 **This is the Lake Owyhee Dam in Oregon. Notice the reservoir of water that forms behind the dam.**

Because of environmental problems and the fact that dams have already been built at the most favorable sites, the era of building big dams in the United States is probably over. Still, all countries will probably continue to build some dams.

GROUNDWATER

Not all of the water that falls to Earth as precipitation drains off in rivers and streams. Some of the water soaks into the ground. Plants collect and use some of this water, but much of it seeps down through the soil. Water that seeps down in this way and is stored underground is called **groundwater.**

Large amounts of groundwater may be stored in underground rock formations called **aquifers.** (See Figure 5-10 on page 128.) Aquifers usually consist of rocks, sand, and gravel with a lot of air spaces in which water can accumulate. Occasionally, aquifers contain large areas of water without any rocks in them. Limestone, for instance, may dissolve to leave large caves full of water.

An aquifer constantly gains groundwater as it percolates down from the surface. Groundwater may take millions of years to collect—only a tiny amount reaches the aquifers each year.

THE OGALLALA AQUIFER:

AN UNDERGROUND TREASURE

Agriculture seemed a ridiculous prospect for settlers in the hot and dry Great Plains region of the United States. But when these settlers discovered that the enormous Ogallala Aquifer lay beneath the ground, the idea no longer seemed so absurd. Wells were dug, pipes were laid, and soon the Great Plains was bathed in an excess of water.

Decades passed, and landowners enjoyed a seemingly limitless water supply. But the Ogallala has begun to show its limits. Although the aquifer is still considered the largest underground water source in the United States, water is being withdrawn 10 to 40 times faster than it is being replenished. Most of the water is withdrawn to irrigate crops. In some places, the water level has dropped 30 m (100 ft.) since pumping began. Many people have abandoned their wells and moved on because digging any deeper was just too expensive.

In some areas, the Ogallala Aquifer flows onto the land's surface. These areas provide vital habitat for many animals, especially birds. Unfortunately, these wetlands are often the first to disappear when excessive crop irrigation lowers the water level in the aquifer.

Some farmers have begun to limit their crop irrigation to permit surface water levels to rise during bird migrations. Others have adopted water-saving practices such as drip irrigation systems or modified

The Ogallala Aquifer holds about 4 quadrillion liters of water, enough to fill Lake Huron (one of the five Great Lakes).

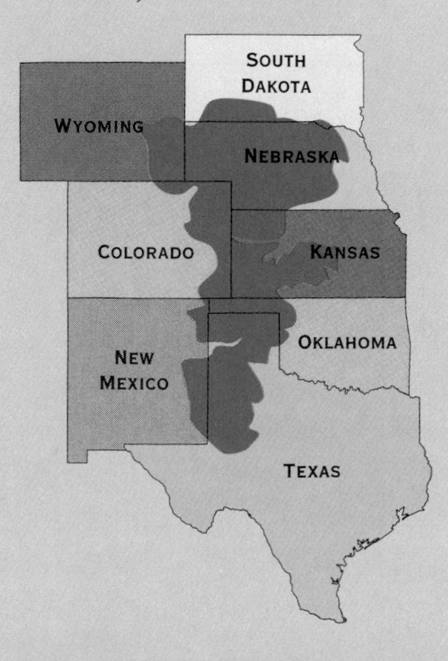

The area of land from which the groundwater originates is called its **recharge zone.**

Some large cities, as well as many rural communities and individual farms and ranches, depend on aquifers for their water needs. The United States has several huge aquifers that together supply millions of gallons of water for homes and agriculture. This resource is tapped by drilling a well into the ground until the hole reaches the groundwater. Sometimes the well must be drilled 30 m to 1,000 m deep to reach the groundwater. In other locations, the water is so close to the surface that it bubbles out of the ground as a spring. In either case, a pump is installed to force the water from the ground and through pipes to homes and other buildings.

Aquifers Are Running Low The problem with aquifers is that people are pumping out the water faster than it can be replaced naturally. Consequently, the water levels of many aquifers are beginning to drop rapidly. One aquifer that is being rapidly depleted is the largest aquifer in the United States, the Ogallala Aquifer, discussed in the Case Study below. Some communities that depend on aquifers are starting to investigate other sources of fresh water.

Answers to Thinking Critically can be found on page T147.

centerpivot sprinklers that deliver water directly to plant roots and reduce evaporation losses by as much as 98 percent. Still other farmers are turning to dry-land agriculture, which involves planting crops such as wheat or grain sorghum that need less water than more traditional corn or cotton crops.

For a farmer, reducing the amount of water used is not as simple as it may seem. Governmental policies often actually discourage such practices. For example, government money granted to grow corn may not be granted to grow crops that require less water. In addition, if a farmer uses less water than he or she has been allowed, the government may withdraw that person's future legal rights to that water. Thus, farmers are compelled to "use it or lose it."

Nevertheless, many farmers and other plains residents are recognizing the value of the Ogallala and are fighting to preserve it. They are pressuring politicians to change policies that encourage wasting water into policies that promote conserving it. These improved management practices may help save the Ogallala Aquifer.

These sandhill cranes are among the many birds that rely on the surface water generated by the Ogallala Aquifer.

THINKING CRITICALLY

1. *Applying Ideas* Why couldn't the Ogallala be replenished by watering its recharge zone?
2. *Expressing Viewpoints* Do you think residents of the Great Plains are the only citizens who should be responsible for conserving the Ogallala? Why or why not?

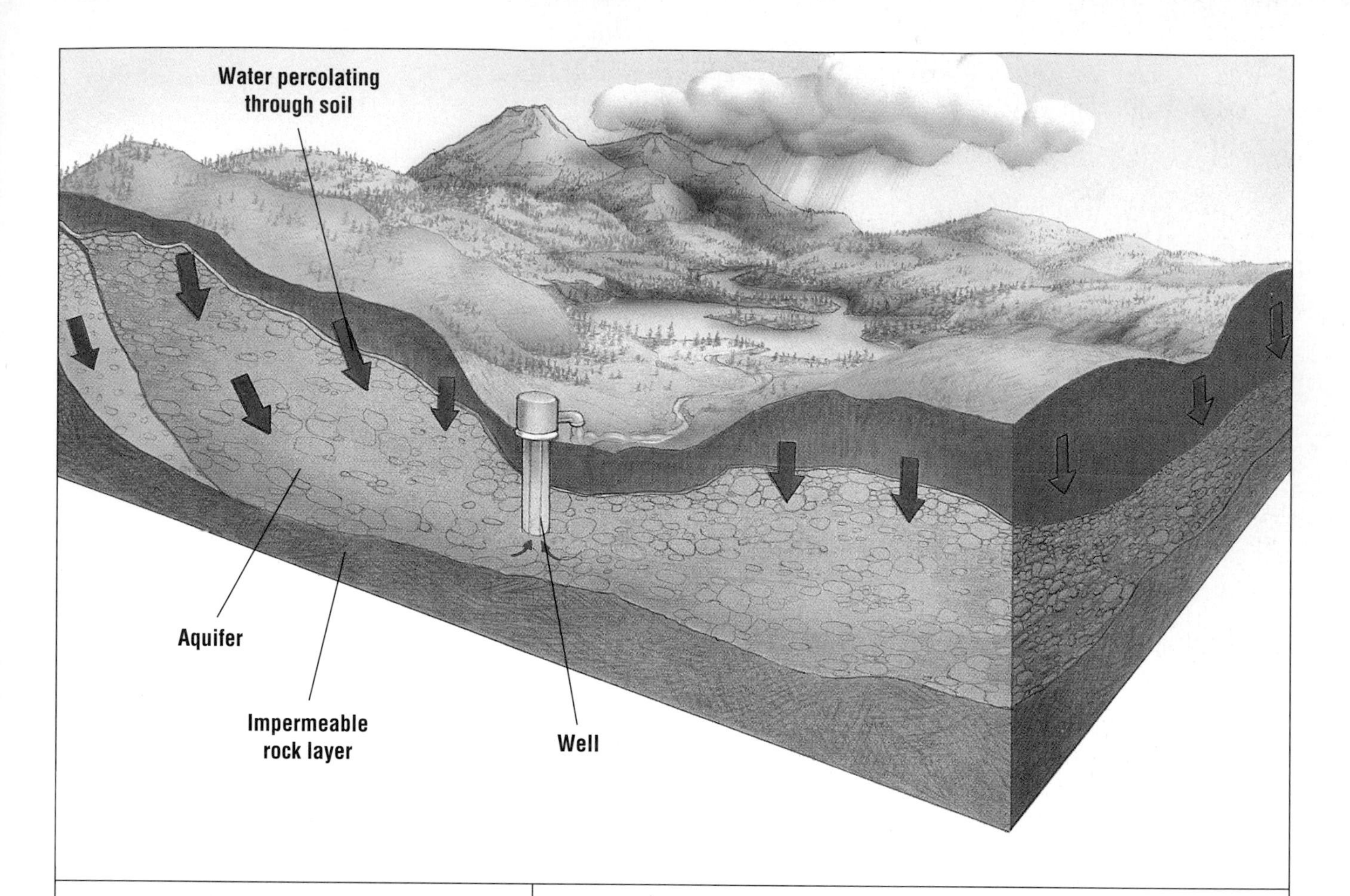

Figure 5-10 Aquifers are underground rock formations that hold water.

Solutions to Water Shortages

There are numerous potential solutions to water shortages, but it is probable that none of them alone is sufficient. We need to develop new sources of fresh water, to use less by practicing conservation, and to minimize pollution. Here are some of the relatively small-scale solutions that may one day add up to enough to prevent us from running out of fresh water.

Desalting the Sea Some coastal countries and communities are attempting to solve their water shortages by removing the salt from salt water, a process called **desalinization.** Salty water cannot be used for drinking and ruins the soil if used for irrigation. Nearly all the drinkable water in desert countries such as Saudi Arabia is produced by desalinization. Some California cities built desalinization plants after they experienced droughts.

The two main methods of desalinization are distillation and reverse osmosis. In distillation, heat is used to evaporate fresh water from salt water, leaving the salts behind. With reverse osmosis, pressure is used to push the water through a semipermeable membrane that will not permit the salts to pass.

Unfortunately, obtaining fresh water through desalinization is expensive. New technologies may someday reduce the cost, but today desalinization is still too expensive for many developing nations in dire need of fresh water.

Figure 5-11 **Most of the water from this automatic watering system is spilled into the street. Wasteful practices like this are becoming less common.**

Towing Water For years people have considered the possibility of towing water from one place to another to solve water shortage problems. One possibility is transporting icebergs. Saudi Arabia, where water costs more than oil, recently commissioned a series of experiments on towing icebergs. The flat-topped Antarctic icebergs would be easier to tow than the spiky Arctic ones, because they would be less likely to roll over. But even with the Antarctic icebergs, the potential problems are obvious: icebergs are hard to tow, and they melt rapidly as they get near the equator. Furthermore, how could the water be transported ashore once it arrived?

Alaska, which has more than 40 percent of the United States' fresh water, is experimenting with sending some of its water south in huge plastic bags and selling it. The idea is to float the bags down the Pacific coast to California, which is often short of water.

Did you know that every time you flush, you are probably sending far too much water down the drain? Don't worry; it's easy to fix. **See page 404.**

Water Conservation Water goes around and comes around, but its passage through the water cycle still takes time. The amount of usable fresh water that is available at a given time is limited, so people must do everything they can to use it wisely.

As more people realize that clean, fresh water is a precious and limited resource, they are doing more to conserve it. More people are installing low-flow faucets and shower heads and turning off the tap while they brush their teeth or shave.

Figure 5-12 **The average person uses about 295 L of water each day.**

Indoor Water Use

Typical Daily Water Use Per Person

Use	Approximate Percentage of Daily Use	Liters of Water
Toilets	39	115
Showers	33	100
Laundry	10	30
Washing dishes	5	15
Cooking and drinking	5	15
Brushing teeth	4	10
Cleaning	4	10

Figure 5-13 Planting native plants in yards instead of grass helps conserve water because the native plants do not need extra watering.

More people are watering their lawns at night to reduce water loss by evaporation. People in desert regions are replacing the Kentucky bluegrass in their lawns with native plants like the ones shown in Figure 5-13. Some people are even putting filled water bottles inside their toilet tanks to reduce the water used for each flush. See Figure 5-14 for more water-conservation ideas.

Can one person make a difference? When you multiply one by the millions of people who are trying to conserve water, it makes a big difference.

Figure 5-14 How could you save water?

What You Can Do to Conserve Water
1. Take shorter showers, and avoid taking baths unless you keep the water level low.
2. Install a low-flow shower head in your shower.
3. Install inexpensive low-flow faucet aerators in your water faucets at home.
4. Install a water-saving device in your toilet, or purchase a modern low-flow toilet. See page 404.
5. Don't let the water run while you are brushing your teeth.
6. Fill up the sink basin rather than letting the water run when you are shaving, washing your hands or face, or washing dishes.
7. Water your lawn in the evening to avoid losing much of the water to evaporation.

Answers to Section Review can be found on page T147.

Section Review

1. Why is fresh water considered a limited resource?
2. What is groundwater, and why should we take care not to use it too quickly?
3. What are some things you can do to help conserve the world's water supply? Give at least two examples.

Thinking Critically

4. *Making Decisions* How should disputes over water rights, such as the one over the dams in Turkey, be settled? Explain your reasoning.
5. *Inferring Relationships* What if the world's supply of fresh water ran almost completely dry tomorrow? How would this affect your life? How would this affect the world?

FRESHWATER POLLUTION

AFTER READING THIS SECTION YOU SHOULD BE ABLE TO

1. explain why groundwater pollution is hard to clean.
2. define and compare point and nonpoint pollution.
3. classify the kinds of water pollutants.
4. describe the impact of water pollution on people and the environment.

In 1969 the Cuyahoga River in Ohio was so full of industrial chemicals and other pollutants that it actually caught on fire.

In the last 20 years, developed countries have made enormous strides in cleaning up water supplies that had become heavily polluted. Despite this progress, we still have a long way to go. Some of the water is still dangerously polluted in the United States, and the lack of clean water restricts economic development in many areas.

Water pollution in less-developed countries remains a big problem. The former Soviet-bloc countries are probably the most polluted on Earth because industrial and agricultural pollution were not controlled by their governments. The Polish Academy of Sciences reports that more than half of Poland's water is too polluted even for industrial use. By 2000, it is estimated that Poland will have *no* water that is fit to drink.

In very poor countries, industry is not the cause of water pollution. Rather, the problem is that the population has outgrown the water supply. The only water available for drinking is often polluted with sewage and agricultural runoff, leading to sickness and even death from waterborne diseases.

Water pollution is the introduction of chemical, physical, or biological material into water that degrades the quality of the water and affects the organisms that drink it and live in it. Water pollution has two underlying causes: industrialization

Figure 5-15 Water pollution can create a public hazard.

Kinds of Pollutants
Pathogens Disease-causing organisms such as bacteria. Pollution occurs when human sewage is untreated or enters water through storm sewers, and when animal feces wash off the land into water.
Organic matter Biodegradable remains of animals and plants, including feces. These come primarily from nonpoint sources.
Organic chemicals Pesticides, fertilizers, plastics, detergents, gasoline and oil, and other materials made from fossil fuels such as petroleum. (Petroleum consists of hydrocarbons originally found in ancient plants.) Mostly nonpoint source pollution.
Inorganic chemicals Acids, salts, toxic metals; from both point and nonpoint sources.
Toxic chemicals Chemicals that are poisonous to living things, including heavy metals (lead, mercury, cadmium), and many industrial, and some household, chemicals.
Physical agents Heat and suspended solids such as soil.
Radioactive waste From power plants or nuclear processing and defense facilities.

Figure 5-16 **Which of these kinds of pollutants can come from homes?**

Answer to caption: All of them except radioactive waste

and the human population explosion. Both produce waste products that cannot be disposed of as fast as they are produced. Figure 5-16 describes some of the major kinds of water pollutants.

Point Pollution

You are walking along a riverbank when you see a rusty pipe jutting out of the bank above the water. The liquid spurting out of the pipe is turning the water red—or blue or green or gray. You turn around and see an industrial plant. The liquid pouring into and polluting the river obviously comes from the plant. You have discovered a source of point pollution. **Point pollution** is pollution that is discharged from a single source, such as a factory, a wastewater treatment plant, or an oil tanker. Figure 5-17 shows an example of point pollution.

Point pollution is relatively easy to regulate and control because it is easily identified and traced. But even when the source of point pollution is known, it is sometimes difficult to enforce cleanup, as the Case Study about the Pigeon River on pages 134–135 illustrates.

Figure 5-17 **This is an example of point pollution. The pollution in the river is coming from a single, easily identifiable source.**

Nonpoint Pollution

Have you ever noticed the black, oily spots on the pavement of parking lots and driveways? These spots are the result of small drops of oil dripping from automobile engines. These grimy spots of oil do not appear to be very dangerous. However, when it rains, the rainwater washes some of this oil into the storm sewers, which drain into waterways. Once there, the oil causes many environmental problems. This is an example of nonpoint pollution.

Figure 5-18 **This is an example of nonpoint pollution. When it rains, the oil on this street will wash into storm sewers, which drain into our waterways. The oil on streets and driveways all over the community will add up to an amount that can pollute a river or stream.**

Nonpoint pollution is pollution that comes from many sources rather than from a single specific site. Nonpoint pollution reaches bodies of water via streets and storm sewers. It can come from anywhere—homes, lawns, farms, highways, and almost any other land surface you can think of. Pesticides and fertilizers are washed off lawns and farmland; animal feces float away from farms, parks, and city streets. When it rains, any polluting material on city or suburban streets flows through storm sewers into waterways. Figure 5-20 on the next page shows just a few of the sources of nonpoint pollution from a single home. Because nonpoint pollution can enter bodies of water in many different ways, it is extremely difficult to regulate and control.

The accumulation of small amounts of pollution adds up to a huge pollution problem. The Environmental Protection Agency recently determined that 96 percent of the polluted bodies of water in the United States were contaminated by nonpoint sources. Controlling nonpoint pollution depends to a great extent on public awareness. Educating the public will probably be the most effective way of reducing this source of pollution.

Field Activity

Walk around your neighborhood and record potential sources of nonpoint pollution. Consult Figure 5-20 for examples. Tally the occurrences of each potential source of nonpoint pollution, and suggest ways that each can be prevented.

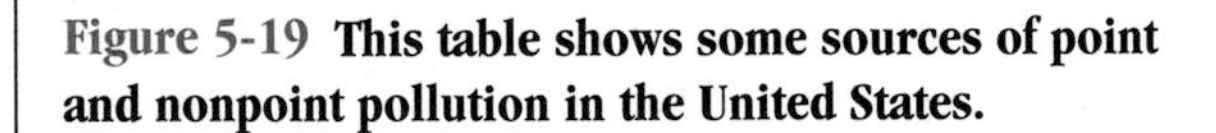

Figure 5-19 **This table shows some sources of point and nonpoint pollution in the United States.**

Where Does Water Pollution Come From?

Sources of Point Pollution	Sources of Nonpoint Pollution
• 23 million septic-tank systems	• Highway construction and maintenance: eroding soil and toxic chemicals
• 190,000 storage lagoons for polluted waste	• Storm-water runoff from city and suburban streets: oil, gasoline, dog feces, litter
• 9,000 municipal landfills	• Pesticides from the 112 million hectares of cropland treated with these substances each year
• About 2 million underground storage tanks containing pollutants such as gasoline	• 50 million tons of fertilizer applied to crops and lawns
• Thousands of public and industrial wastewater treatment plants	• 10 million tons of dry salt applied to highways for snow and ice control

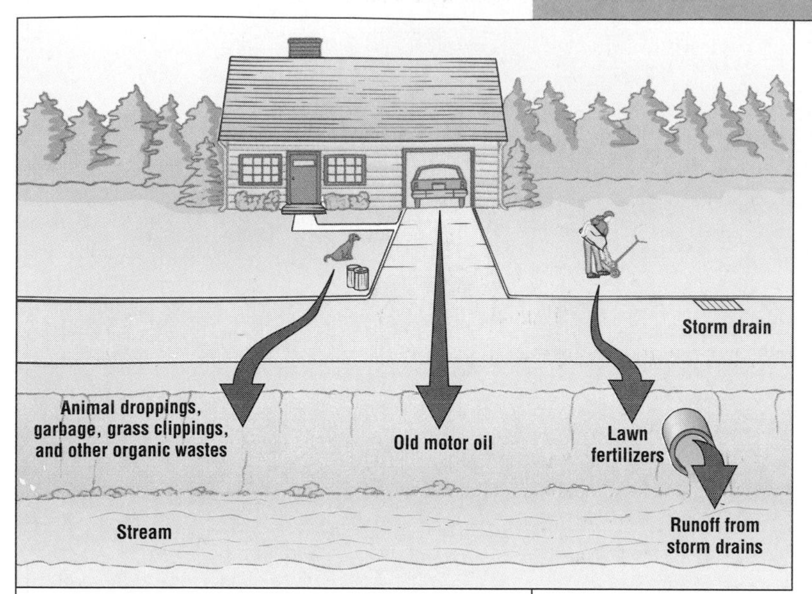

Figure 5-20 Some sources of nonpoint pollution from a single home

Wastewater Treatment Plants

Most of the wastewater from industrial and residential areas is collected in pipes and taken to a wastewater treatment plant, where it is treated before being returned to a river or lake. Figure 5-21 shows how a wastewater treatment plant works. Unfortunately, wastewater treatment plants may not be able to remove all of the harmful substances in waste. Most of the wastewater from homes is biodegradable and easy to purify, but industrial wastewater and storm runoff from streets and fields contain toxic substances that interfere with treatment.

The Fight Over Pigeon River

For 22 mi., from its source in Pisgah National Forest to the town of Canton, North Carolina, the Pigeon River is beautiful, sparkling clear, and alive with trout that attract vacationing fishermen. But the Pigeon River that flows out of Canton is dark brown, has a rank odor, and holds almost no aquatic life. The cause of this transformation is Canton's major industry, a paper mill that has sustained the local economy and unloaded wastes into the river for the past 80 years.

The river travels across the state line into Tennessee and toward Cocke County, where residents are angry about the pollution. Because of the river's condition in Cocke County, tourists do not vacation and fish in this otherwise beautiful area. Residents blame their hard economic times on this loss of potential income. They also blame the pollution for many illnesses and deaths in the county. A high rate of cancer deaths was reported in Hartford, a tiny town in Cocke County whose residents have always fished in the Pigeon River. When fish from this section of the river were tested, high levels of dioxin were found. Dioxin is a common byproduct of paper mills and a cause of cancer and birth defects in laboratory animals.

For years, residents of Cocke County have pleaded for stricter environmental controls on the paper mill. But the plant is so important to the economy of Canton that North Carolina officials were reluctant to call for any restrictions that would harm the plant's livelihood. In 1985, despite protests from residents of Tennessee, North Carolina approved a five-year extension of the paper mill's discharge permit with no changes to reduce the pollution.

Exasperated, Cocke County activists began to loudly express their concerns to state agencies, politicians, the media, and the Environmental Protection Agency (EPA). The EPA finally took away North Carolina's authority to issue the discharge permit and ordered the paper mill to clean up its act. Mill officials

The Pigeon River flows clean and sparkling into Canton, North Carolina, but flows out dark brown and polluted.

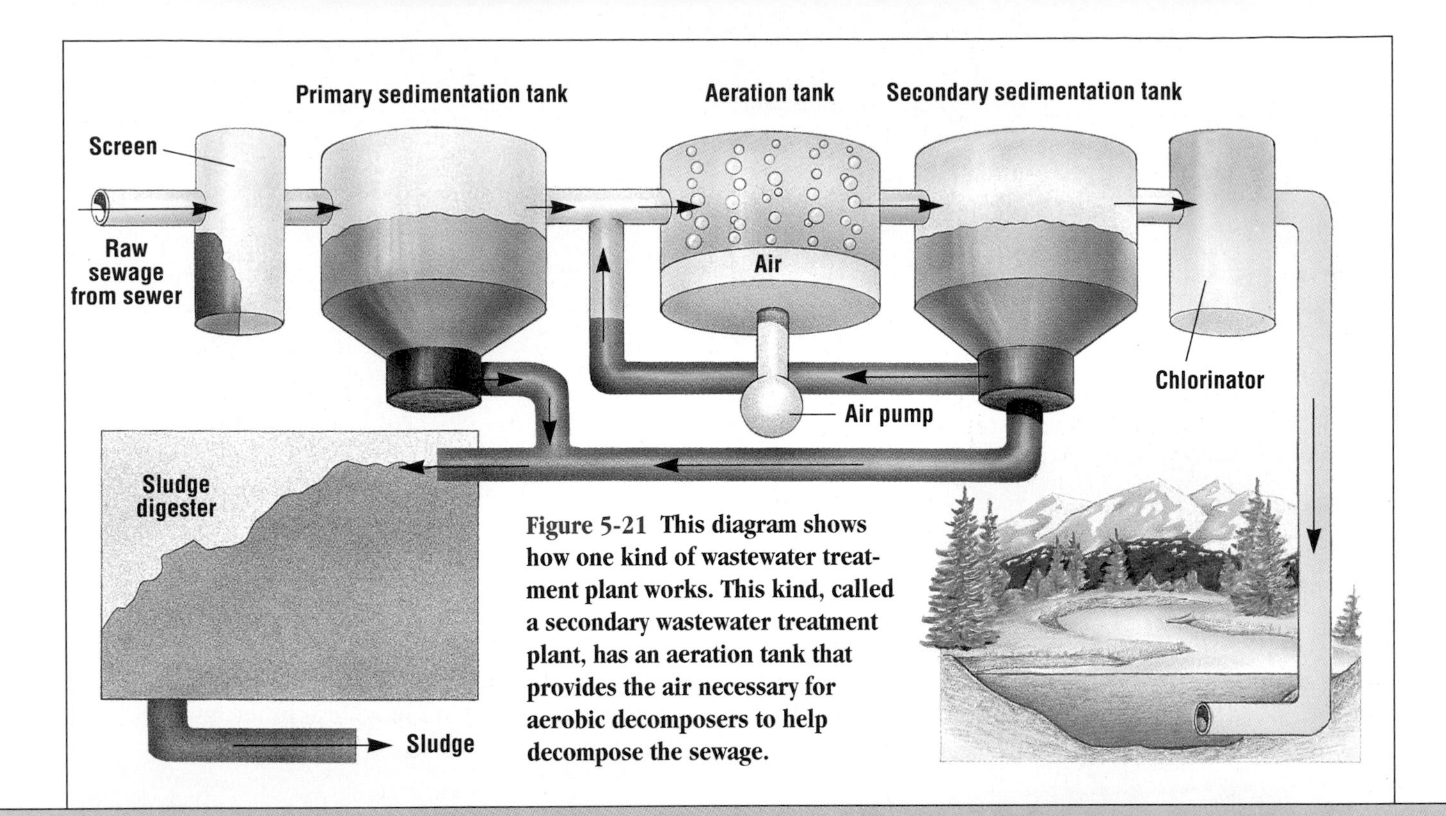

Figure 5-21 This diagram shows how one kind of wastewater treatment plant works. This kind, called a secondary wastewater treatment plant, has an aeration tank that provides the air necessary for aerobic decomposers to help decompose the sewage.

declared that the necessary changes would be too expensive—the mill would have to close if forced to comply. Panic swept through Canton as mill employees faced the prospect of losing their jobs, incomes, and possibly even their homes. Other Canton businesses also faced a frightening future because they depended on mill employees to buy their goods and services.

Thousands of mill supporters mobbed EPA public hearings in North Carolina to voice their concerns about their jobs and their community. Likewise, crowds of clean-river advocates attended hearings in Tennessee to vent their outrage about the pollution. A record number of letters about the Pigeon River debate—most in favor of the paper mill—were received at the EPA's regional office in Atlanta, Georgia.

In 1989, the paper mill agreed to spend hundreds of millions of dollars to treat its discharge and to modernize its operations. Fortunately, these changes did not affect the mill's work force as expected. For the past few years, tests for dioxin in the mill's discharge have been negative, although fish in the region still show unsafe levels of the toxin.

In Cocke County, the river's color is somewhat lighter and the smell isn't as strong, but the activists say that's not good enough. They continue to fight for a clean and safe river that one day might attract tourists to the area. Though weary from the struggle, the residents of Cocke County keep a watchful eye on Canton and the paper mill—especially since North Carolina recently regained its authority to issue the mill's discharge permit.

Unsafe levels of dioxin have been detected in fish taken from the Pigeon River. This sign warns people not to eat any fish caught there.

Answers to Thinking Critically can be found on page T147.

Tourists visit Canton, North Carolina, to fish in the clean Pigeon River. Residents of Cocke County, Tennessee, believe that pollution in their part of the river keeps tourists away.

THINKING CRITICALLY

1. *Analyzing Processes* What are some of the factors that make it difficult to enforce the cleanup of point pollution? Use the events reported in this Case Study to support your answer.
2. *Expressing Viewpoints* Which do you think is more important—protecting the environment or protecting people's jobs? Explain your reasoning.

Figure 5-22 **Materials that could become pollutants can sometimes be beneficial. This photograph shows sludge being applied as a fertilizer.**

If you look again at Figure 5-21, you'll see that one of the products of wastewater treatment is sludge, the solid materials left over after treatment. Sludge often contains dangerous concentrations of toxic chemicals, so it must usually be disposed of as hazardous waste. It is often incinerated and then buried in a secure landfill.

The volume of sludge that has to be disposed of every year is enormous and is an expensive burden to municipalities. Many communities, therefore, are attempting to reduce the toxicity of their sludge so that it can be disposed of in other ways. Sludge contains plant nutrients, so if it is free of toxic chemicals, it can be used as fertilizer, as shown in Figure 5-22. In one interesting process, sludge is combined with clay to make bricks that can be used in buildings. In the future, more industries will probably find innovative ways to use sludge.

Pathogens

Pathogens are disease-causing organisms such as bacteria, viruses, and parasitic worms. Pathogens can enter water supplies in untreated wastewater or animal feces. Cholera, hepatitis, and typhoid are among the many diseases that people can catch by drinking water that is polluted with pathogens.

Public water supplies are routinely monitored to check for the presence of pathogens. The most common method of testing to see if pathogens are in the water is to check for *Escherichia coli,* a common type of human intestinal bacteria. A test called the *fecal coliform test* is used to test for these bacteria.

How Water Pollution Affects Ecosystems

An entire ecosystem may suffer the effects of water pollution. Consider a river ecosystem. Soil tainted with toxic pesticides may wash into the river. This soil settles to the river bottom, and some of its toxins enter the bodies of tiny bottom-dwelling organisms. A hundred of these organisms may be eaten by one small fish. A hundred of these small fish may be eaten by a big fish. An eagle may eat 10 big fish. Each organism stores the toxins in its tissues, so at each step along the food chain, the amount of toxin passed on to the next "eater" increases. This is called **bioaccumulation** or biomagnification. The Case Study on pages 58–59 of Chapter 3 describes the effect of the pesticide DDT in an aquatic food chain at Long Island Sound.

Water pollution can also cause *immediate* damage to an ecosystem. In the past few years, there have been several spills of toxic chemicals directly into rivers or streams, killing every living thing for miles downstream from the spills.

Water pollution can be harmful to human health. And in fish, toxic chemicals can cause cancers and scale rot and fin rot, and they can accumulate in their tissues, making many fish too dangerous for humans to eat. In addition, some of the heavy metals and toxic chemicals that end up in waterways cause cancer or birth defects in humans, and others affect reproduction or damage the nervous system, liver, or kidneys.

Are there heavy metals in our drinking water? Not if Elizabeth Philip has anything to say about it!
Find out more on pages 148–149.

ARTIFICIAL EUTROPHICATION

Lakes and slow-moving streams can become eutrophic, which means that they contain an abundance of nutrients. This occurs naturally over a long period of time as organisms die and decompose, adding nutrients to the water. The process of decomposition uses up large amounts of oxygen dissolved in the water, which can affect the types of fish and other animals that can survive in the water. Plants take root in the nutrient-rich sediment at the bottom and start to fill the shallow waters. Eventually the body of water becomes a swamp or marsh. This process is an example of secondary succession, which you learned about in Chapter 3.

The natural process of eutrophication can be accelerated when inorganic plant nutrients such as phosphorus and nitrogen get into the water from sewage and fertilizer runoff. Eutrophication caused by humans in this way is called **artificial eutrophication.** Phosphorus, a plant nutrient contained in detergents, as well as animal wastes and fertilizers, causes the excessive growth of algae. The algae can form large mats, called *algal blooms,* which float on the water, as shown in Figure 5-23. As the algae die and decompose, large amounts of dissolved oxygen are used. Fish suffocate in the oxygen-depleted water.

In an effort to alleviate the problem, some states have banned phosphate detergents, which contain phosphorus. Others have limited the amount of phosphates in detergents.

Figure 5-23 **Algal blooms are caused by high levels of phosphates in the water.**

THERMAL POLLUTION

When excessive amounts of heat are added to a body of water, **thermal pollution** can result, as shown in Figure 5-24. Thermal pollution occurs when power plants and other industries located along lakes or rivers use the water in their cooling systems. Cool water from the river or lake is circulated around engines to absorb waste heat. The hot water is then returned to the lake or river, creating an unnatural warm or even hot area.

Thermal pollution can cause massive fish kills when the discharged water is too hot for the fish to tolerate. Furthermore, because hot water cannot hold as much oxygen as cool water can, aquatic organisms are deprived of oxygen and may suffocate. A constant influx of hot water into an aquatic ecosystem may totally disrupt it if the organisms are unable to adapt to the higher

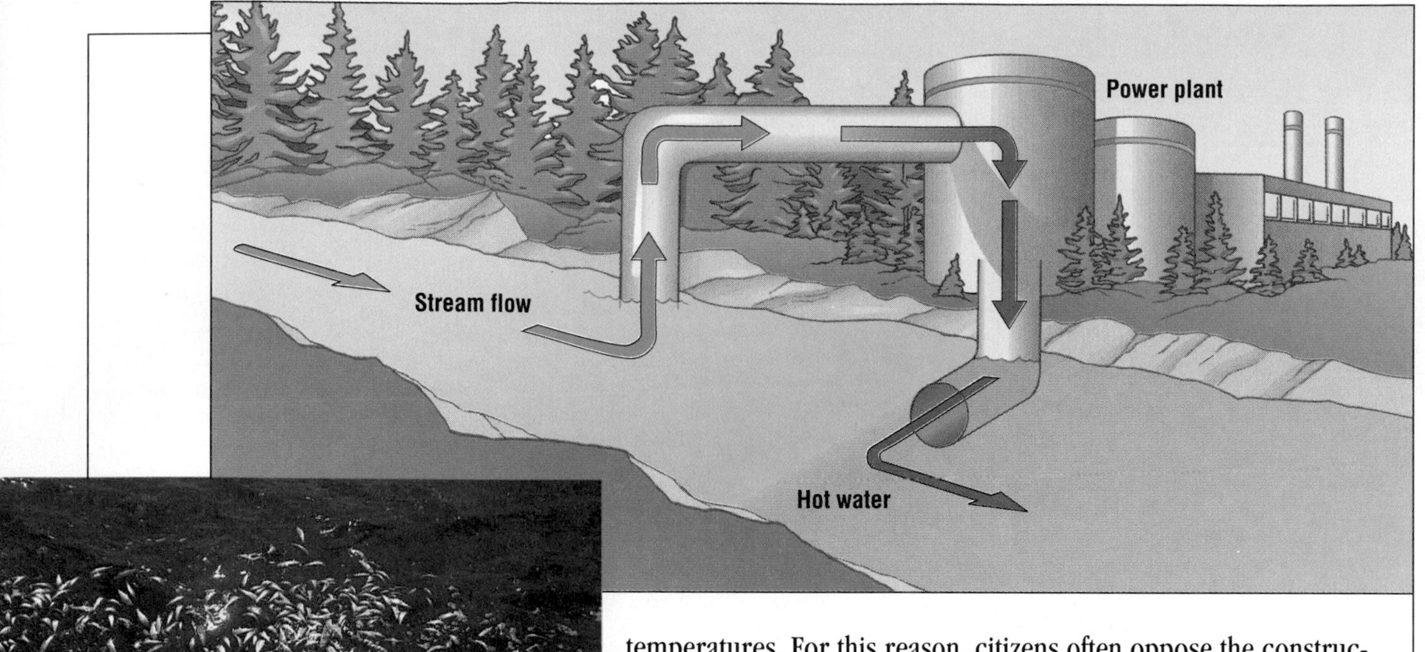

Figure 5-24 **The diagram above shows thermal pollution, and the photograph shows a massive fish kill that resulted from this type of pollution.**

Figure 5-25 **Some progress has been made in cleaning up our water, but much work remains to be done.**

temperatures. For this reason, citizens often oppose the construction of power plants on lakes and rivers or insist on cooling systems that reduce the temperature of the water before it is returned to the waterway.

Cleaning Up Water Pollution

In 1972 Congress passed the Clean Water Act (CWA). The stated purpose of the act was to "restore and maintain the chemical, physical, and biological integrity of the nation's waters." The goal was to make all surface waters clean enough for fishing and swimming. Other water-quality legislation is described in Figure 5-25. Since 1972 many states have passed their own, even stricter, water-quality standards. Together, federal and state regulations have had some positive effects on surface water. For instance, many toxic metals are now removed from wastewater, and many industrial wastes are treated prior to disposal.

Despite these successes in curbing point pollution, nonpoint pollution continues to be a problem that requires the cooperation of individuals and businesses throughout the nation. Progress has been made. Some agricultural wastes are channeled into lagoons, where pollutants are decomposed before the water is released into waterways. Similar waste-treatment processes have been adopted by some industries, including a few paper and

Water Quality Legislation in the United States

Here is a list of the main federal legislation designed to improve water quality in the United States. Many of the deadlines set by legislation in the 1970s, such as those for sewage treatment, have been extended because states and towns have not met them. Sometimes the deadlines were unrealistic, and sometimes money to implement changes has been lacking.

- **1972 Clean Water Act** (CWA). This is technically the **Water Pollution Control Act.** The act set a national goal of making all natural surface waters fit for fishing and swimming by 1983 and of no pollutant discharge into these waters by 1985. The Act required that metals be removed from wastewater beginning in the early 1980s.
- **1972 Marine Protection, Research, and Sanctuaries Act,** amended 1988. This act empowered the Environmental Protection Agency to control the dumping of sewage wastes and toxic chemicals in the ocean.
- **1975 Safe Water Drinking Act.** This act introduced programs to protect both groundwater and surface water from pollution.
- **1980 Comprehensive Environmental Response Compensation and Liability Act** (CERCLA). This is the Superfund Act, which makes owners, operators, and customers of hazardous waste sites responsible for their cleanup. It has reduced the pollution of water by toxic substances leached from hazardous waste dumps.

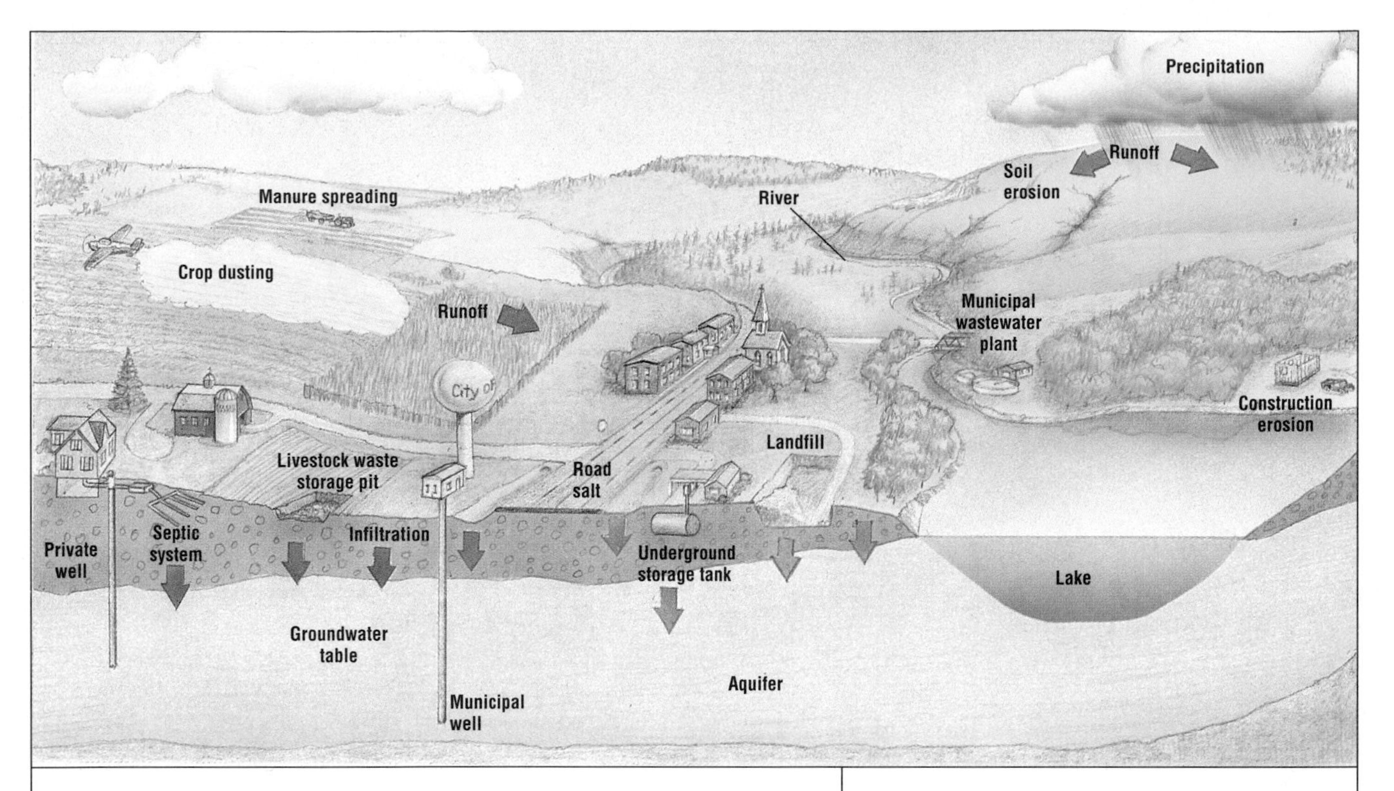

Figure 5-26 How pollutants get into groundwater

pulp mills. Some companies have even implemented innovative water-recycling processes that greatly reduce pollution.

The Special Problem of Groundwater Pollution

Groundwater pollution is likely to plague us for centuries to come. Pesticides, chemical fertilizers, and other agricultural chemicals are the primary pollutants that seep into groundwater. Leaky chemical storage tanks and industrial wastewater lagoons also contribute to the problem. Figure 5-26 illustrates how pollutants get into groundwater. The Environmental Protection Agency has detected at least 200 hazardous chemicals that can seep through the soil and into groundwater.

Unfortunately, even if groundwater pollution stopped tomorrow, the water would still be polluted generations from now. As you have learned, groundwater recharges very slowly. In addition, it is difficult to clean up water that is hidden in underground aquifers among sand grains and tiny rock fissures.

Bottled Water

Sales of bottled water have soared as more and more people decided their tap water was not fit to drink. Where does bottled water come from? Most people assume that it comes straight from a pure mountain stream. Much more often, however, bottled water is simply tap water that has been filtered and treated with various chemicals. Bottled-water plants are regulated by the government, but bottled water is not tested for pollutants as often as the public water supply is.

Answers to Section Review can be found on page T147.

Section Review

1. Why is point pollution easier to eliminate than nonpoint pollution?
2. What are pathogens, how do they pollute water, and what effect do they have on humans?
3. What are the main pollutants of groundwater? Why is groundwater so difficult to clean up?
4. How does nutrient-rich fertilizer cause fish kills?

Thinking Critically

5. *Interpreting Graphics* Draw a diagram showing how the bioaccumulation of a pollutant in an aquatic ecosystem may result in endangerment of a species.

OCEAN POLLUTION

AFTER READING THIS SECTION YOU SHOULD BE ABLE TO

❶ explain how and why the oceans are polluted and describe the effects of pollution on marine life.

❷ discuss the effects of polluted oceans on humans.

❸ explain how individuals can prevent ocean pollution.

"There isn't a clean spot in the Atlantic from Bermuda to the African coast," reported a sailor who made the journey. "A river of polystyrene cups and bits of plastic stretches across the ocean." If you walk along a beach, you can see some of these plastic cups, bottles, and bags that have washed up out of the ocean.

How much waste can the oceans absorb? How long will the waste take to decompose? Little research is being done on vital questions like these. And when we find the answers, often it is too late. However, we do know the answer to one question: Where does the ocean pollution come from?

How Pollutants Get Into Oceans

At least 85 percent of ocean pollution comes from the land. When pollutants enter rivers as runoff, the rivers may carry the polluted water to the ocean.

Figure 5-27 **Pollution on the beach in Oahu, Hawaii**

Pollutants are also dumped directly into the oceans. For example, scientists think that burns seen on the shells of lobsters and other shellfish are caused by sludge, an end product of wastewater treatment that has been dumped into the ocean. Oceangoing ships have also dumped wastewater and garbage overboard.

Accidental oil spills also contaminate ocean water. Disasters such as the 1989 *Exxon Valdez* oil spill in Prince William Sound, Alaska, make front-page news around the world. (See Figure 5-28.) However, such disasters are responsible for only about 5 percent of the oil polluting the oceans. Much of the oil pollution in the seas comes from less spectacular events. Small oil tankers also have accidents and leaks, for example, and there is often some spillage when tankers are loaded and unloaded. Offshore oil rigs sometimes leak petroleum as well.

Plastic is also a significant ocean pollutant because it does not break down easily. When plastic fishing lines are discarded into the sea, marine mammals can be strangled or disabled when they become entangled in them. Turtles may eat clear plastic bags that look like jellyfish and die from suffocation or blockage of the digestive system. Plastic six-pack rings end up around the necks of sea birds, strangling them, or around the bodies of fish, which die as the plastic cuts into their flesh.

FIELD ACTIVITY

Be part of a coastal cleanup. Every September, people from all over the world set aside one day to help clean up debris from beaches, riverbanks, or lakesides. You can join this international effort by writing to the Center for Marine Conservation, 1725 DeSales Street, NW, Washington, D.C. 20036.

PREVENTING OCEAN POLLUTION

There are laws that regulate or prohibit pollution of the seas. MARPOL (the International Convention for the Prevention of Pollution From Ships) prohibits the discharge

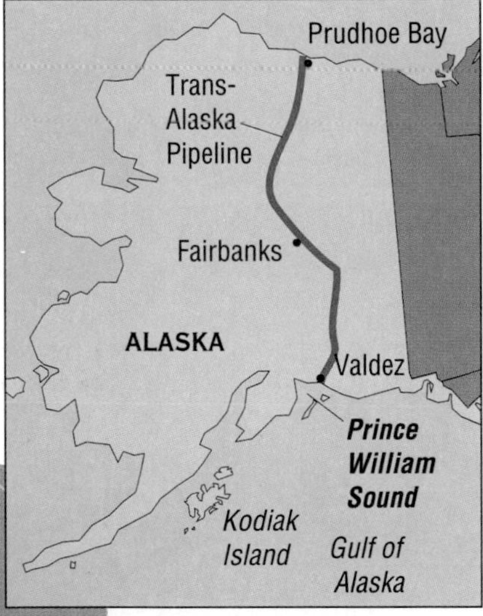

Figure 5-28 The sea otter (left) was one of the numerous marine animals covered in oil from the *Exxon Valdez* oil spill. The map above shows the location of the spill, Prince William Sound.

of oil and the disposal or abandonment of plastics in ocean or coastal waters.

The 1974 Helsinki Convention, supported by 120 nations, seeks to control land-based sources of ocean pollution such as toxic dumping, runoff, and discharging raw wastewater. So far, prohibitions against the dumping of toxins such as DDT, cadmium, and mercury have been enacted. Nations have also worked together to create 15 marine refuges to protect endangered marine species such as sea turtles and monk seals.

The United States has strengthened its laws against ocean dumping by enacting the Marine Protection, Research, and Sanctuaries Act. In addition, a new law will soon require all oil tankers arriving in United States waters to have double hulls as an added protection against oil spills. The Marine Mammal Protection Act prohibits any actions that could harm the many endangered marine mammals in our oceans.

However, as you might imagine, it is very difficult to monitor every ship on the ocean to ensure that none are discharging oil, throwing garbage overboard, or abandoning plastic fishing lines. To stop ocean pollution, individuals, businesses, and nations must first be convinced of the wisdom of obeying the laws and honoring the agreements.

Figure 5-29 **Left unaided, this elephant seal would strangle from the plastic ring around its neck.**

Who Owns the Oceans? Part of the problem of ocean pollution is uncertainty about who has jurisdiction over the oceans. Since time immemorial, international law has permitted nations to exercise complete control over their territorial waters, extending 3 mi. from the coast. The rest of the world's oceans were high seas, open to everyone. In the twentieth century, some countries claimed to extend their territorial waters to 12 mi., and sometimes to 200 mi., from their coasts.

In an attempt to clarify the situation, the Third United Nations Conference on the Law of the Sea met between the years 1973 and 1982. The conference resulted in the Law of the Sea Treaty.

The Law of the Sea Treaty states that the laws of a coastal nation extend to 22 km (12 nautical mi.) from its coastline. This area is called a nation's *territorial sea.* The area that extends 370 km (200 nautical mi.) from land is called a nation's *exclusive economic zone.* A nation has control over economic activity, environmental preservation, and research in this area. The rest of the world's oceans are designated as communal property to be controlled by the International Seabed Authority.

The final agreement was signed by 134 countries. Some of the most powerful developed nations did not sign the treaty, however, including the United States.

Answers to Section Review can be found on page T147.

Section Review

1. Where does most ocean pollution come from? What kind of pollution is it?
2. What are three major sources of oil pollution in the ocean?
3. Why is plastic considered an ocean pollutant? What are its sources, and how can it be eliminated?

Thinking Critically

4. *Inferring Relationships* What long-term effects on ocean life can be expected if ocean pollution is not reduced?
5. *Applying Ideas* What can individuals do in their own homes and communities to decrease ocean pollution? Give at least three examples.

HIGHLIGHTS

SUMMARY

- Only a tiny fraction of Earth's large water supply is suitable for drinking and other human activities. The tiny percentage of water used for everyday human needs comes from two sources: surface water and groundwater.
- Rivers, which collect water precipitated into watersheds, are an important source of surface water. Groundwater is precipitation that collects underground and is stored in rocky formations called aquifers. Water is pumped from aquifers to supply fresh water to agriculture, industry, and residences. As a result, aquifers are being depleted.
- Water may become polluted by chemical, physical, or biological material or by excess heat. Water pollution may be easily identified point pollution, or it may be nonpoint pollution, the source of which is often difficult to determine.
- Government actions, such as the Clean Water Act of 1972, have partially succeeded in curbing surface-water pollution. But even so, groundwater pollution will be with us for years to come. Ocean pollution is primarily due to runoff from the land, though pollutants are also sometimes dumped directly into the ocean.

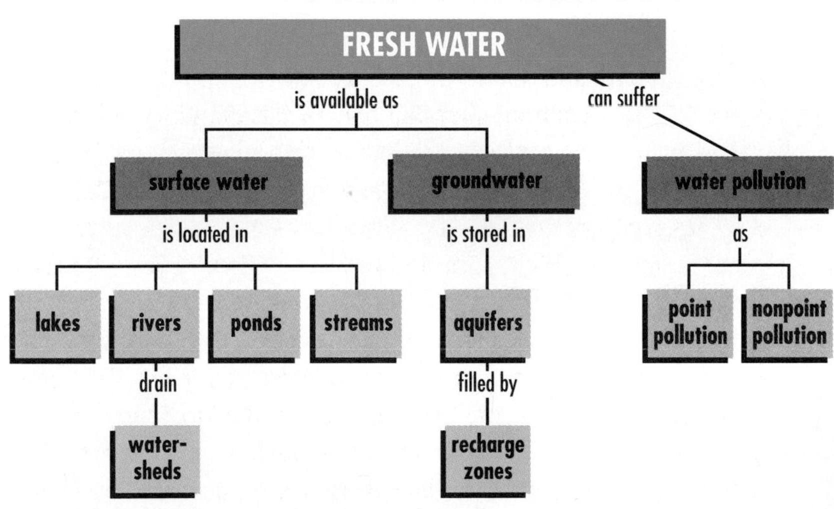

Sample Ecolog answers can be found on page T147.

Ecolog

Now that you've studied this chapter, revise your answers to the questions you answered at the beginning of the chapter, based on what you have learned.

1. Describe as best you can where the water in your home comes from.
2. Do you think the water you drink at home is pure or polluted? Explain your answer.

Vocabulary Terms

aquifer (p. 126)
artificial eutrophication (p. 137)
bioaccumulation (p. 136)
desalinization (p. 128)
groundwater (p. 126)
nonpoint pollution (p. 133)
point pollution (p. 132)
recharge zone (p. 127)
surface water (p. 123)
thermal pollution (p. 137)
water pollution (p. 131)
watershed (p. 123)

CHAPTER 5

REVIEW

Understanding Vocabulary

1. For each pair of terms, explain the difference in their meanings.
   a. surface water
      groundwater
   b. watershed
      recharge zone
   c. point pollution
      nonpoint pollution
   d. water pollution
      thermal pollution

Relating Concepts

2. Copy the unfinished concept map below onto a sheet of paper. Then complete the concept map by writing the correct word or phrase in each box containing a question mark.

Answers not indicated can be found on page T148.

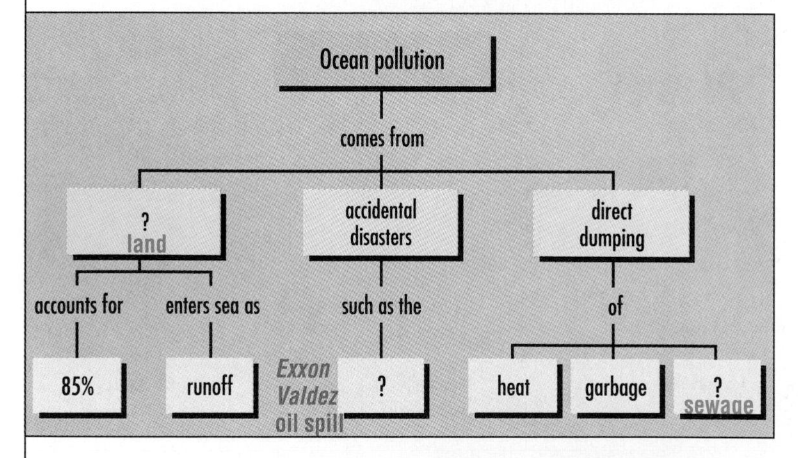

Understanding Concepts

Multiple Choice

3. The water in your drinking glass may have once been part of
   a. a primordial ocean.
   b. a dinosaur's bloodstream.
   c. water your neighbor used to water his lawn.
   (d.) all of the above

4. Lakes can provide more stable sources of water than rivers can because
   a. they tend to flow quickly.
   b. riverbeds keep changing direction.
   (c.) lakes do not depend directly on watersheds.
   d. rivers contain more salt than lakes do.

5. The era of building huge dams in the United States has probably come to a close MAINLY because
   a. smaller dams are more economical.
   b. we have all of the water and electricity we need.
   c. they are now too expensive to build.
   (d.) the environmental consequences are too great.

6. A result of pumping water from aquifers is that
   a. farms produce fewer crops.
   b. the recharge zone shrinks.
   (c.) the aquifer cannot be recharged fast enough.
   d. groundwater stops being collected.

7. Sludge is difficult to dispose of because
   a. it sinks to the bottom of wastewater-treatment plants.
   (b.) it often contains toxic or hazardous materials.
   c. there is so much of it.
   d. it is noncombustible.

8. Thermal pollution affects aquatic environments because it
   a. can make water too hot for fish to tolerate.
   b. reduces oxygen in the water.
   c. has been circulated around power-plant engines.
   (d.) both a and b

9. Severe oil-tanker spills account for what percentage of oil polluting the ocean?
   (a.) 5
   b. 10
   c. 25
   d. 65

10. Regulations that help protect the ocean environment from the dumping of garbage at sea include all of the following except the
  a. Marine Protection, Research, and Sanctuaries Act.
  b. Public Rangelands Improvement Act of 1978.
  c. International Convention for the Prevention of Pollution from Ships.
  d. Marine Mammal Protection Act.
11. Oil pollution is mostly caused by
  a. major oil spills such as the 1989 *Exxon Valdez* oil spill.
  b. the cumulative effect of small oil spills and leaks.
  c. decomposed plastic materials.
  d. intentional dumping of excess oil.

Short Answer

12. What difficulties would arise if the United States had to depend on water frozen in the polar icecaps for its water needs?
13. Which is more dangerous, point pollution or nonpoint pollution? Why?
14. What are the two methods commercially available for the desalinization of water? How do they work?
15. Describe how bioaccumulation might account for mercury poisoning in humans who eat fish caught near a factory.
16. Is the pollution of ocean water any more dangerous than pollution of fresh water? Explain your answer.

Interpreting Graphics

17. Examine the table below. It represents the water budget for a city in the southeastern United States. A water budget shows how much rainfall a city receives compared with how much its population uses. According to the table, in which month does the city receive the most rainfall? In which month does it experience the worst water shortage? Which months show a balanced budget? How might water availability be ensured during months with a deficit?

Rainfall (mm)	Jan	Feb	Mar	Apr	May	Jun	Jul	Aug	Sep	Oct	Nov	Dec
Credit	70	80	88	73	80	135	175	180	150	75	50	70
Debit	20	25	40	95	125	165	175	155	120	75	28	18
Surplus	50	55	48	0	0	0	0	25	30	0	22	52
Deficit	0	0	0	22	45	30	0	0	0	0	0	0

Thinking Critically

18. ***Communicating Ideas*** Describe the path of a nonpoint pollutant, such as engine oil, from a driveway to the ocean.

Themes in Science

19. ***Stability*** Do you think artificial eutrophication is more disturbing to the stability of a water ecosystem than natural eutrophication? Explain your reasoning.

Cross-Discipline Connection

20. ***Environmental Science and Mathematics*** If placing a container of water in your toilet tank reduces the amount of water for each flush by 2 L, how much water would be saved each day if all 250 million people in the United States placed a container of water in their toilet tanks? (Assume that each toilet is flushed five times per day per person.) How many gallons is that?

Discovery Through Reading

The Everglades National Park in southern Florida has endured numerous natural disasters, but can it survive the diversion of its fresh water? Find out in "The Everglades: Dying for Help," by Alan Maerson, *National Geographic,* April 1994, pp. 2–35.

PORTFOLIO ACTIVITY

Do a special project on where the water in your faucets at home comes from and goes to. Does it complete a cycle? Make a poster illustrating the source and destination of the water you use. You may want to work with several classmates and visit the source and destination that you discover.

CHAPTER 5 INVESTIGATION

HOW SAFE IS OUR GROUNDWATER?

In many parts of the United States, drinking water is supplied by underground wells. In fact, over 130 million Americans get their water from underground water supplies. Although this water varies in quality, it usually comes directly from the ground clean and safe to drink—at least it has in the past. In recent years there have been more and more reports of tainted wells and polluted groundwater.

Answers to this Investigation can be found on page T148.

MATERIALS

- 250 mL beakers (5)
- wax pencil
- glucose solution
- red food coloring
- graduated cylinder
- stirring rod
- glucose test paper
- 12 oz. plastic cups (4)
- thumbtack
- gravel
- sand
- soil
- metric ruler
- optional substances: motor or cooking oil, vinegar, fertilizer, detergent, antifreeze
- notebook
- pen or pencil

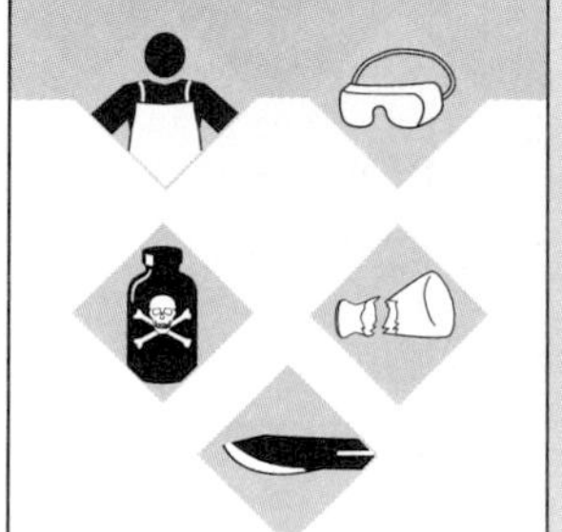

BACKGROUND

In this Investigation you will organize into small teams and explore how layers of earth act as a filter for groundwater supplies. You will make models of the Earth's natural filtration system and put them to the test to see how well they filter out various substances. Although most of the substances used in this exploration are not normally dangerous to human health, they will help you explore how safe our groundwater supplies are from surface contamination.

PREPARE

1. Label four beakers as follows: "contaminant: glucose," "contaminant: soil," "contaminant: food coloring," and "water" (control).
2. Fill these beakers two-thirds full with clean tap water. Then add to each beaker the contaminant listed on its label. (See the chart below for the amount of each contaminant.) Stir each mixture thoroughly. These mixtures will represent surface water.
3. Carefully observe each beaker and log your observations in your notebook. Create a data table similar to the one below to organize your observations.

Observations of Substances in Surface Water

Contaminant	Appearance	Measurements (if any)
Glucose (5 mL)		
Soil (5 mL)		
Food coloring (5 drops)		
Water (control)		

GROUND FILTRATION MODELS

4. Set the beakers aside and make four separate ground filtration systems. To do this, use a thumbtack to poke six holes in the bottoms of four plastic cups. Then fill each cup with layers of gravel, sand, and soil as shown on the facing page.

Make each model identical to the next. Be sure to leave approximately 2 cm of space between the surface of the soil and the top of the cup.

These cups of gravel, sand, and soil represent the earth through which surface water percolates on its way to underground water supplies. You have made four models so that you can test four different water mixtures.

Your ground filtration models should be layered as shown here.

TEST YOUR EARTH FILTERS

5. You are now going to pour each surface-water mixture through an earth filter model. But first predict how well the filters will clean each water sample. Write your predictions in your notebook.
6. Stir a water mixture in its beaker and immediately pour it through an earth filter into a clean beaker. Do the same for each water mixture. To reduce the number of beakers needed, clean the empty beakers as you proceed. Relabel the beakers to keep track of the contaminants.

ANALYZE YOUR RESULTS

7. Once each water mixture has been filtered, observe the resulting "groundwater." In your notebook, record your observations in a table similar to the one shown previously. This new table should be titled "Observations of Substances in Groundwater."
8. Test the glucose-water mixture for the presence of glucose. Is glucose still present after filtering? Can you see it?
9. Was the soil removed from the water by filtering? Was the food coloring removed? How do you know?
10. How accurate were your initial predictions? What conclusions can you draw from this filtration experiment?

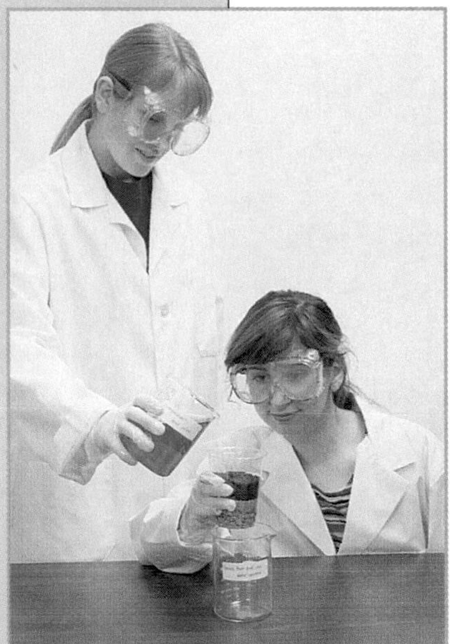

Pour each sample of contaminated surface water through an earth filter. Clean the "contaminant" beakers as you go along, and relabel them as "filtrate" beakers.

ON YOUR OWN

11. Choose a common substance from the materials list that hasn't been tested and that you think represents a real threat to our water quality. Or, you may want to pick your own substance. Predict what would happen if you mixed this substance in the water supply and then filtered it through the earth. Would the earth filter it out? What did you learn from the previous experiment that would lead you to your prediction?
12. Now test your prediction. Use the filter that was the control in the earlier experiment. This filter would not have been contaminated. How did your results compare with your prediction? CAUTION: Do not taste any of the substances you are testing. Dispose of all substances according to your teacher's instructions.
13. Compare your results with the results of other teams. What generalizations can you make about what substances will or will not be filtered by the natural percolation of surface water down through the earth? What precautions do you recommend for keeping groundwater clean?

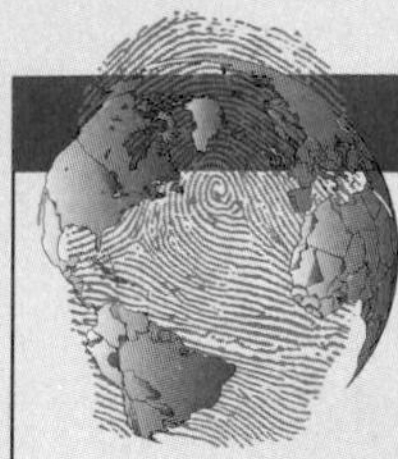

HIGH SCHOOL Chemist

Elizabeth Philip has received a long list of prestigious awards for her scientific work—at the age of only 17! What's all the attention about? Well, Elizabeth has developed an inexpensive and innovative way to remove heavy metals and other contaminants from drinking water. Her research could very well mean that you and other people have cleaner water to drink.

...even though young people don't have the experience or the resources our parents do, we can still have an impact on the environment.

Q: When did you become interested in water pollution?

Elizabeth: It all began one day as I stood in a grocery store checkout lane. Looking around the store, it occurred to me that lots of people were buying bottled water. I wondered why and decided to try to find out.

I discovered that the city's water supply had a high sodium content. I figured that this was the main reason people lugged home huge bottles of water instead of simply running their taps. But I decided to look into the subject a little further. That's when I came across the surprising statistic that one out of every four taps in the United States contains a significant amount of lead. Plus, I learned that even metals such as mercury, cadmium, and chromium can end up in the municipal water supply. By this point, I was *very* interested in water pollution and what could be done about it.

Elizabeth began her research after wondering why so many people purchased bottled water like this.

Elizabeth Philip presenting her science fair exhibit to judges

Q: Was anything being done about these water contaminants?

Elizabeth: Yes. Some people were purifying their water with various devices, but those instruments were expensive. It seemed that removing sodium and the more serious heavy metals such as lead, cadmium, and mercury was tough to do in a cost-effective manner with existing methods. In effect, I learned that we were using old technologies to solve new problems.

Q: What "new technology" did you think would be more appropriate?

Elizabeth: It seemed to me that there was probably some sort of living organism that could remove heavy metals from tap water. I decided to pursue the idea for a project that I could submit in my high school's science fair.

After considering algae, water hyacinths, and yeast, I decided to try using yeast to filter tap water. I found that baker's yeast, a common household product that costs less than $2 per pound, was able to remove 99 percent of the toxic trace metals in tap water. It turned out that yeast provided an ecologically safe and affordable alternative to current water-purification technologies.

Elizabeth's system for purifying water, shown above, could help provide cleaner drinking water for you and other people.

Q: What have you enjoyed the most about your research experience?

Elizabeth: Well, I got to present my findings at national and international science fair exhibitions. There I met lots of other students who were trying to find solutions to scientific and environmental problems.

It was nice to see people my age who were actually making a difference. I even met students from Germany and Sweden who face the same problems with pollution in the environment that we Americans do. We talked about what we could do together to help, since we all share the same air and water. It was great to realize that even though young people don't have the experience or the resources our parents do, we can still have a positive impact on the environment.

Q: What are your plans for the future?

Elizabeth: I plan to go to college, but I'm not sure what I'll study. I am sure that I'll continue researching water-purification technology. I think that's how I can help improve the environment. If I can say one thing to other students, it's that *you* can do something to make a difference. Just begin with something you care about, educate yourself, and then go from there.

Elizabeth Philip has been conducting experiments on water pollution since junior high school.

CHAPTER 6

AIR

"I thought I saw a blue jay this morning. But the smog was so bad that it turned out to be a cardinal holding its breath."

MICHAEL J. COHEN, PROFESSOR OF POLITICAL SCIENCE
BAR ILAN UNIVERSITY, ISRAEL

EcoLog

Before you read this chapter, take a few minutes to answer the following questions in your EcoLog.

1. What is the air quality like in your area?
2. If there is air pollution in your area, what do you think causes it?

WHAT CAUSES AIR POLLUTION?

AFTER READING THIS SECTION YOU SHOULD BE ABLE TO

❶ name the major causes of air pollution.

❷ distinguish between primary and secondary pollutants.

❸ explain how we could reduce air pollution.

❹ explain how a thermal inversion can make air pollution worse.

Students are sometimes told to stay home from school in Mexico City because the air pollution is so dangerous. At these times, some people wear dust masks when they go outside, and the hospitals are crowded with people suffering from respiratory problems. Birds occasionally drop out of the trees dead because they cannot escape the toxins in the air.

When substances that can cause harm to living things end up in the air, the result is **air pollution.** Figure 6-2 on the next page describes the major air pollutants. Substances that pollute the air can be in the form of solids, liquids, or gases.

Most air pollution is the result of human activities, but pollutants can come from natural sources as well. A volcano, for example, can spew clouds of particles and sulfur dioxide that can suffocate animals. Natural pollutants also include salt from the oceans and pollen and spores from plants. Natural pollutants do not often cause problems because they seldom become concentrated in particular areas.

Figure 6-1
Air pollution in Mexico City

Major Air Pollutants			
Pollutant	**Description**	**Sources**	**Effects**
Carbon monoxide (CO)	CO is an odorless, colorless, poisonous gas. It is produced by the incomplete burning of fossil fuels.	Cars, trucks, buses, small engines, and some industrial processes are the major sources of CO.	CO interferes with the blood's ability to carry oxygen, slowing reflexes and causing drowsiness. In high concentrations, CO can cause death. Headaches and stress on the heart can result from exposure to CO in cars stuck in heavy traffic. If inhaled by a pregnant woman, CO may hamper the growth and development of the fetus.
Nitrogen oxides (NO_x)	When combustion (burning) temperatures exceed 538°C (1,000°F), nitrogen and oxygen particles in the air combine to form nitrogen oxides.	NO_x come from burning fuels in vehicles, power plants, and industrial boilers.	NO_x can make the body vulnerable to respiratory infections, lung disease, and possibly cancer. NO_x contribute to the brownish haze often seen over congested areas and to acid rain. NO_x can cause metal corrosion and the fading and deterioration of fabrics.
Sulfur dioxide (SO_2)	SO_2 is produced by chemical interactions between sulfur and oxygen.	SO_2 comes largely from burning fossil fuels. It is released from petroleum refineries, smelters, paper mills, chemical plants, and coal-burning power plants.	SO_2 contributes to acid rain, which damages lakes, forests, metals, and stone. Some of the secondary pollutants that result from reactions with SO_2 can harm plant life and irritate the respiratory systems of humans and animals.
Volatile organic compounds (VOCs)	VOCs are organic chemicals that vaporize readily, producing toxic fumes. Some examples are gasoline, benzene, toluene, and xylene.	VOCs come from burning fuels and from solvents, paints, and glues. Cars are a major source of VOCs.	VOCs contribute to smog formation, and can cause serious health problems such as cancer. They may also harm plants.
Particulates	Particulates consist of smoke, ash, soot, dust, lead, and other particles from burning fuel.	Particulates come from industrial processes and motor vehicles that burn fossil fuel, burning wood, and dust from construction and agriculture.	Particulates can form clouds that reduce visibility and cause a variety of respiratory problems. Particulates have also been linked to cancer. They also corrode metals, erode buildings and sculptures, and soil fabrics.

Figure 6-2 Air pollutants can be solids, liquids, or gases.

A pollutant that is put directly into the air by human activity is called a **primary pollutant.** An example of a primary pollutant is soot from smoke. Figure 6-3 shows some sources of primary air pollutants. **Secondary pollutants** are formed when a primary pollutant comes into contact with other primary pollutants, or even with naturally occurring substances like water vapor, and a chemical reaction takes place. An example of a secondary pollutant

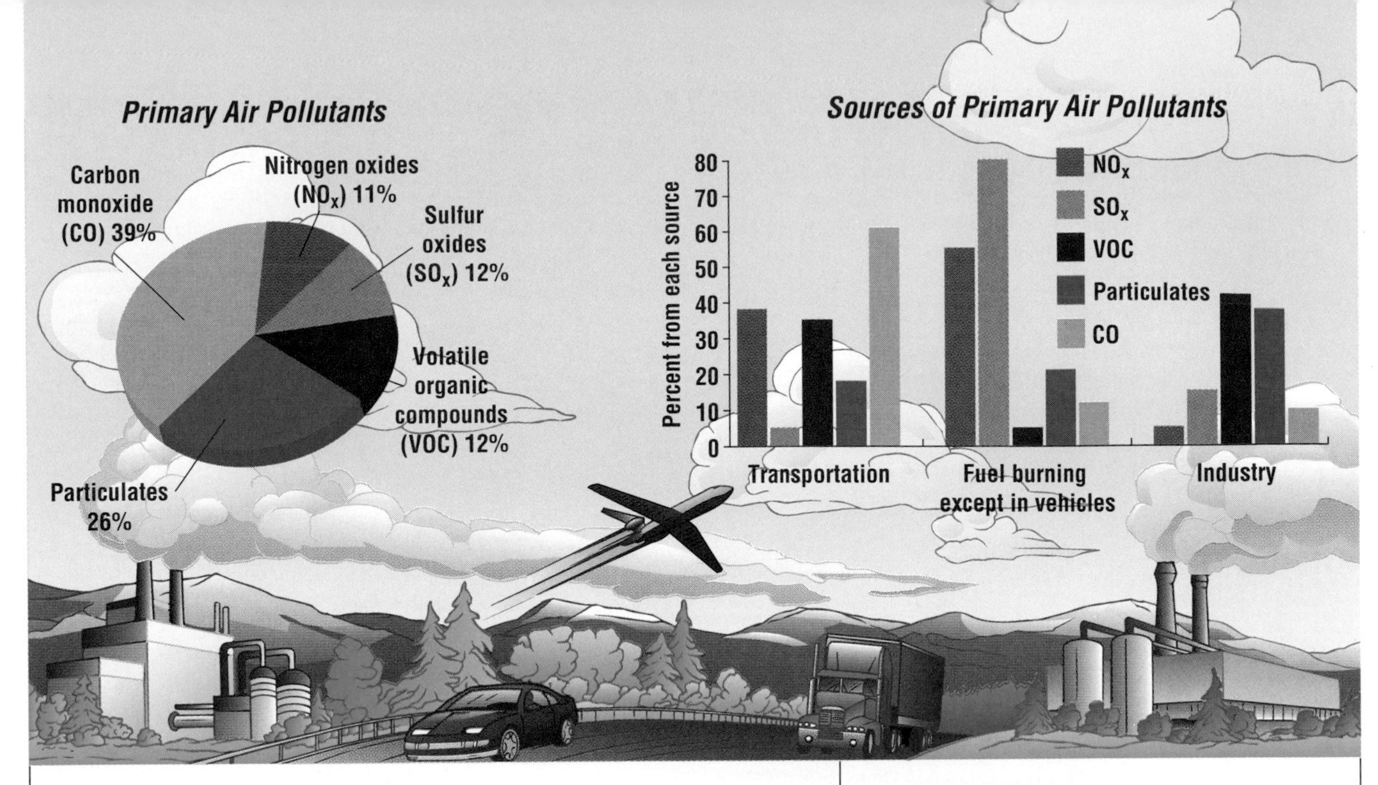

Figure 6-3 This illustration shows the primary air pollutants and their sources.

is ozone. Ozone is created as a result of a chemical reaction involving ultraviolet rays from the sun, the emissions from cars, trucks, and natural sources, and the oxygen gas in the atmosphere.

Air pollution is nothing new. Whenever something burns, pollutants enter the air. Two thousand years ago, Seneca, a Roman philosopher and writer, complained about the foul air in Rome. In 1273, England's King Edward I decreed that burning a particularly dirty kind of coal was illegal, and one man was even hanged for disobeying this medieval "clean air act."

The air-quality problem is much worse today because modern industrial societies burn such large amounts of fossil fuels. Most air pollution in urban areas comes from motor vehicles and industry.

Are electric cars the cure for air pollution? See pages 170–171 and decide for yourself.

Figure 6-4 This is a familiar scene to many Americans.

Air Pollution on Wheels

Fully one-third of our air pollution comes from gasoline burned in motor vehicles. In 1990, internal combustion engines propelled American vehicles over 2 trillion miles. Approximately 80 percent of that mileage was clocked by individual automobiles and the rest by trucks and buses. Figure 6-5 on the next page shows some of the effects automobile emissions have on the air.

Controlling Air Pollution From Vehicles

The Clean Air Act, passed in 1970 and strengthened in 1990, gives the Environmental Protection Agency (EPA) the authority to regulate automobile emissions. The EPA required the gradual elimination of lead in gasoline, and as a result, lead pollution has been reduced by 50 percent. In addition, catalytic converters clean exhaust gases of other pollutants before they exit a car's tailpipe.

One very effective way to reduce air pollution is to drive less. If people carpooled with just one other person, 15 million kg

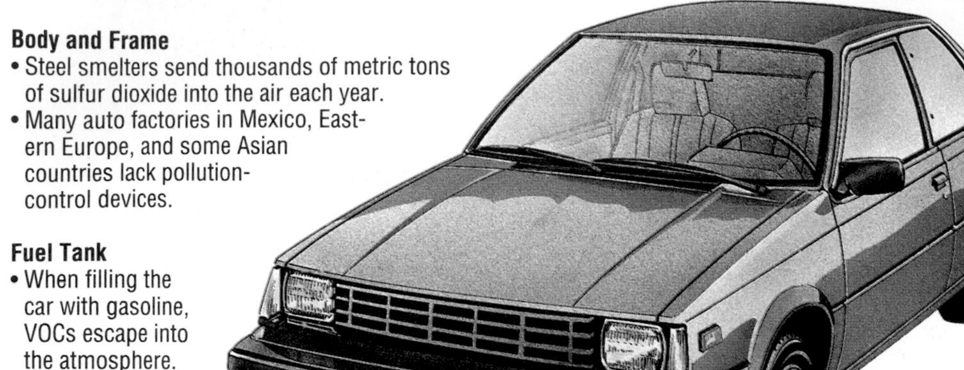

Figure 6-5 A car's contribution to air pollution

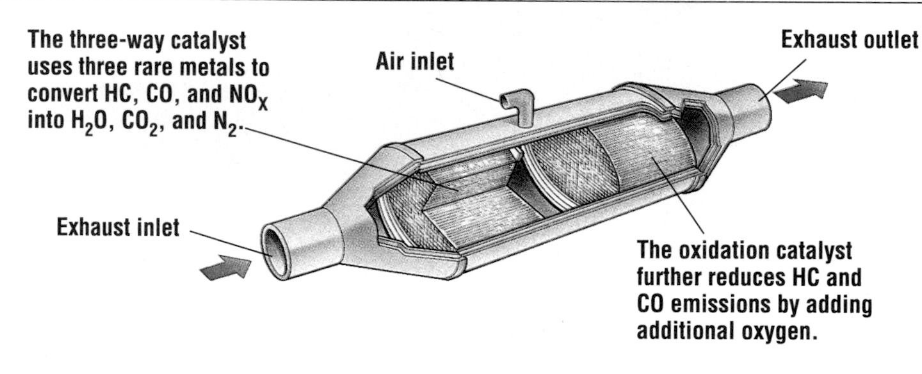

Figure 6-6 The catalyst material in a catalytic converter causes a chemical reaction that turns exhaust emissions into less harmful substances.

(33 million lb.) less carbon dioxide would be emitted into the air each year.

Manufacturers are building cars that run on fuels that cause less pollution. Some prototypes run on solar energy, some on batteries, and some on hydrogen or natural gas. In Brazil, 10 percent of the cars run on ethanol that is made from the remnants of sugarcane plants. Ethanol is a less-polluting hydrocarbon because it burns to form carbon dioxide and water. Corn farmers in the United States are eager to see ethanol engines in vehicles because ethanol can be made from fermented grain such as corn.

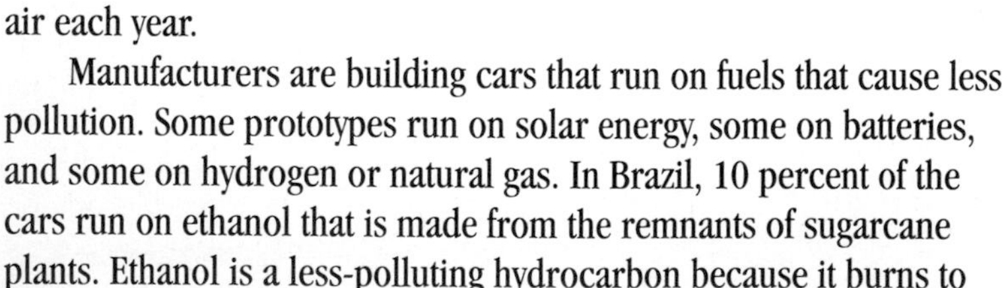

Figure 6-7 If more people carpooled, air pollution from motor vehicles could be reduced substantially.

Industrial Air Pollution

Many industries and power plants that generate electricity must burn fuel to get the energy they need. In most instances, this is a fossil fuel. Burning fossil fuels causes huge quantities of oxides to be released into the air. Electric power plants produce at least two-thirds of all sulfur dioxide and more than one-third of all nitrogen oxides that pollute the air.

Some industries also produce volatile organic compounds (VOCs), which form toxic fumes. The chemicals used by oil refineries, chemical manufacturing plants, dry-cleaning businesses, furniture refinishers, and auto-body shops significantly contribute to the VOCs in the air. Then when individuals use some of the products containing VOCs, more VOCs are added to the air.

Controlling Air Pollution From Industries The Clean Air Act requires many industries to use scrubbers or other pollution-control devices. Scrubbers remove some of the more noxious substances that would otherwise foul the air. A scrubber, shown in Figure 6-8, moves gases through a spray of water that dissolves many pollutants. Electrostatic precipitators are used in cement factories and coal-burning power plants to remove particulates from smokestacks. Electrostatic precipitators remove from the air 22 million metric tons (20 million tons) of ash generated by coal-burning power plants each year in the United States.

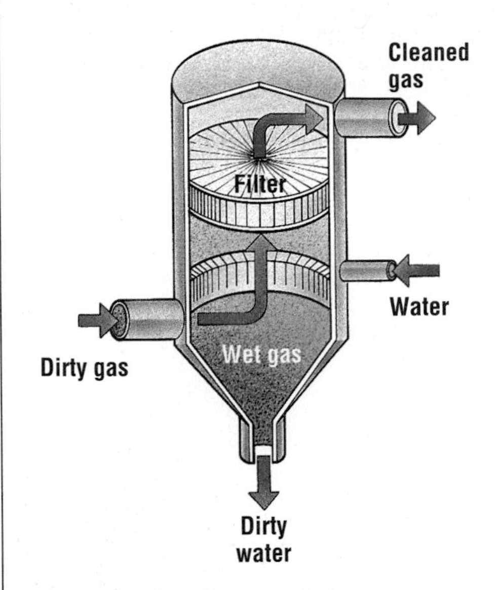

Figure 6-8 **Scrubbers work by spraying gases with water, removing many pollutants.**

THERMAL INVERSIONS

Air circulation in the atmosphere usually keeps air pollution from reaching dangerous levels. During the day, the sun heats the surface of the Earth and the air near it. The warm air rises through the cooler air above, carrying pollutants away from the Earth and into the atmosphere.

Sometimes, however, pollution is trapped near the Earth's surface by a thermal inversion. Normally, air temperatures decrease with height, but in a **thermal inversion,** the air above is warmer than the air below. Figure 6-9 shows how a thermal inversion traps

Figure 6-9 **The normal situation (top) and a thermal inversion (bottom)**

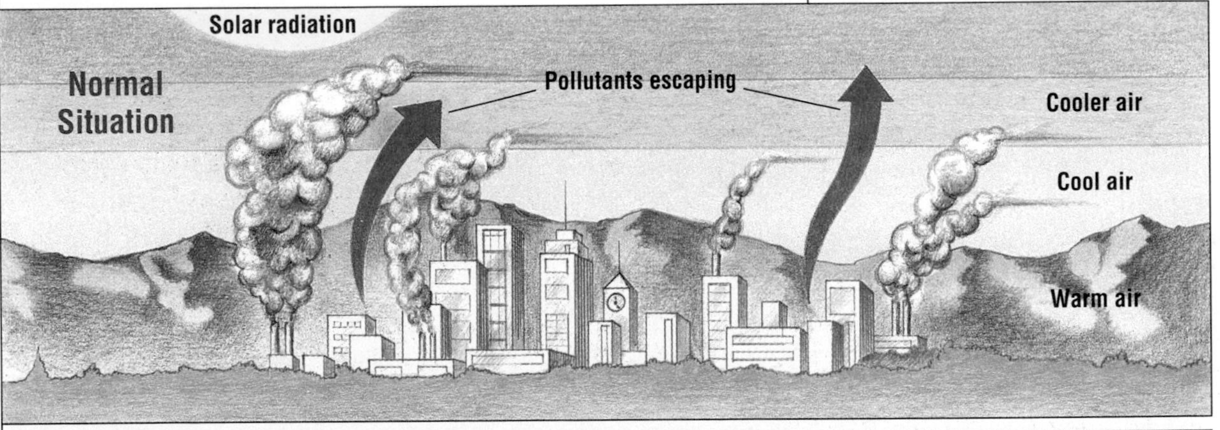

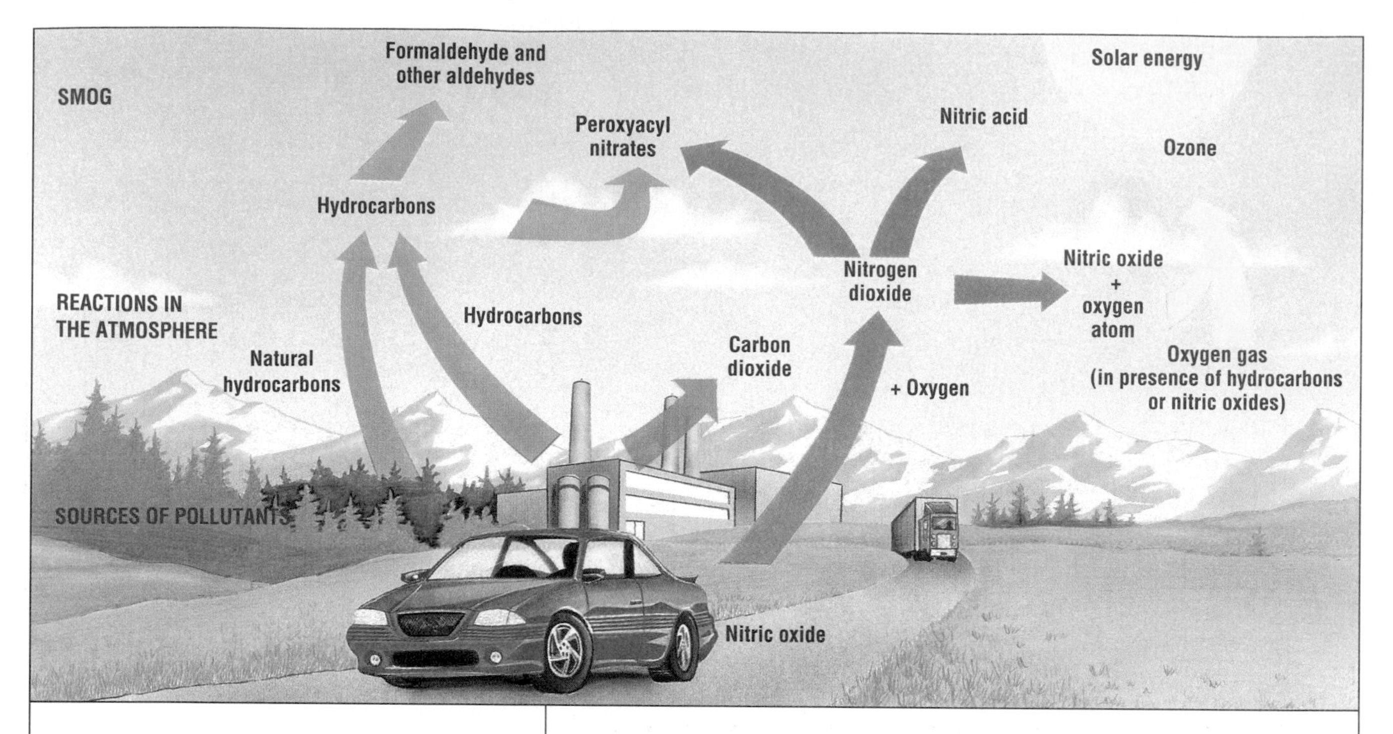

Figure 6-10 This illustration shows how smog is formed. Most of the reactions shown here take place only in the presence of sunlight.

Section Review answers on page T148.

Section Review

❶ Sulfur trioxide is a pollutant that is formed in a chemical reaction between oxygen in the air and sulfur dioxide emitted from smokestacks. Is sulfur trioxide a primary or a secondary pollutant? Explain your answer.

❷ How has the Clean Air Act reduced air pollution from automobiles? How has it reduced industrial air pollution?

❸ What are two factors that contribute to the air-pollution problem in Los Angeles?

❹ Air pollution was a serious problem in London 100 years ago, before automobiles were on the road. Offer an explanation for this.

Thinking Critically

❺ *Making Decisions* Think of several ways to reduce air pollution from motor vehicles, and then use the decision-making model to decide which is the best.

pollutants near the Earth's surface. The warmer air above keeps the cooler air at the surface from moving upward. In this way, pollutants are trapped below with the cooler air. If a city is located in a valley, it has a greater chance of experiencing thermal inversions. Los Angeles, which lies on the Pacific coast and is surrounded on the other three sides by mountains, has frequent thermal inversions that trap smog in the city.

Smog

When air pollution hangs over urban areas and reduces visibility, it is called **smog,** a term combining the words *smoke* and *fog.* As you can see in Figure 6-10, smog results from chemical reactions involving sunlight, nitrogen oxides, and hydrocarbons. Pollutants released by automobiles and industries are the main causes of smog. The cities that suffer most from smog have dense traffic and are located in dry, sunny areas. Los Angeles, Denver, and Mexico City are well-known examples of cities that have smog.

Figure 6-11 Denver has a serious problem with smog.

SECTION 6.2

EFFECTS ON HUMAN HEALTH

AFTER READING THIS SECTION YOU SHOULD BE ABLE TO

❶ describe some possible health effects of air pollution.

❷ explain what causes indoor air pollution and how it can be prevented.

Air pollution can cause serious health problems. People who are very young or very old and those with heart or lung problems are particularly susceptible to the effects of air pollutants. Decades of research have yielded convincing statistical evidence linking air pollution to disease. But because pollution aggravates existing diseases, no death certificates list the cause of death as "air pollution." Instead, diseases such as emphysema, asthma, and lung cancer are cited. The American Lung Association has estimated that Americans pay $93 billion a year in health costs to treat respiratory diseases attributed to air pollution.

Chronic Bronchitis and Asthma

If a person's lungs are perpetually irritated by pollutants, chronic bronchitis, a persistent inflammation of the bronchial linings, may result. Constant coughing weakens the bronchial tubes, and eventually breathing becomes difficult. Particulates, oxides, and acids of sulfur and nitrogen are the main irritants that contribute to chronic bronchitis.

These pollutants are also largely responsible for aggravating bronchial asthma, a condition in which the bronchial passages constrict and become blocked with mucus. Though most asthma is not caused by air pollution, pollutants may worsen an asthmatic condition.

Emphysema and Lung Cancer

Air pollution has been linked to emphysema. Emphysema results when the air sacs in the lungs lose their elasticity and can no longer push air out of the lungs. Thus, the exchange of oxygen and carbon dioxide in the lungs is impaired. People suffering from emphysema generally cannot perform the simplest physical tasks without gasping for breath.

Figure 6-12 **Air pollution can aggravate respiratory diseases.**

To investigate air pollution in your area, fill bowls with water, and leave them outdoors for several days. Bring them inside, and write a description of what you see in and on top of the water. You may wish to examine samples of the water under a microscope and draw what you see. Try to identify what you see under a microscope.

Answer to Field Activity:
Students may notice a thin film on the water's surface in addition to a layer of sediment on the bottom of the bowl. Suggest that students use light-colored bowls to improve visibility.

Lung cancer is another respiratory disease that is linked to air pollution. Although cigarette smoking is the major cause of lung cancer, automobile exhaust and particulates (especially asbestos, arsenic, beryllium, chromium, and nickel) also contribute to lung cancer. Nonsmoking city dwellers are three to four times more likely to develop lung cancer than nonsmoking people in rural areas.

Indoor Air Pollution

Air pollution is not limited to the outdoors. The air inside a building is sometimes worse than the air outside. In many homes, plastics and other industrial compounds are major sources of pollution. These compounds can be found in carpets, building materials, paints, and furniture, particularly when they are new. Figure 6-13 shows some indoor air pollutants.

In a building where the windows are usually open, pollutants will circulate and pass out of the building. But in a well-insulated building where the windows are tightly sealed, the pollutants continue to accumulate and reach higher levels than they could outdoors.

Buildings with particularly poor air quality are said to have **sick-building syndrome.** Sick-building syndrome is most common in places like the desert Southwest, Southern California, and

CASE STUDY

AIR POLLUTION IN POLAND

Poland may be the most polluted nation on Earth. The pollution in the southwestern part of Poland, known as Upper Silesia, is particularly bad. In the Upper Silesian town of Katowice, soot from smokestacks accumulates along the streets in small drifts. Children in the city routinely show high levels of lead in their blood. Human life expectancy is decreasing, infant mortality is increasing, and the incidence of leukemia is rising. By age 10, many children are being treated for chronic respiratory diseases.

In nearby Kraków, workers inside the Nova Huta steel-mill plant are usually coated with soot and ash. Work-related disabilities force most of them to retire in their fifties. Polish officials would like to close the polluting steel plant, but if this is done, how would all those workers make a living? As

Air pollution in southwestern Poland may be the worst in the world.

Figure 6-13 Some indoor air pollutants and their sources

Nitrogen oxides
From unvented gas stove, wood stove, kerosene heater

Asbestos
From pipe insulation

Fungi and bacteria
From dirty heating and air conditioning ducts

Tetrachloroethylene
From dry-cleaning fluid

Carbon monoxide
From faulty furnace, car left running

Methylene chloride
From paint strippers and thinners

Paradichlorobenzene
From moth-ball crystals, air fresheners

Radon-222
From uranium-containing rocks

Tobacco smoke
From cigarettes and pipes

Formaldehyde
From furniture, carpeting, particleboard, foam insulation

Gasoline
From car, lawn mower

Thinking Critically answers on page T148.

one mill worker said, "I'd rather have a sick, short life with a job than a longer, healthier life without work."

How did things get this bad? Poland, along with other Eastern European countries, experienced intense industrialization under communist rule. Production of goods was the most important consideration, with little thought given to environmental consequences. Most of the power plants had no pollution controls at all. Furthermore, the fuel that powered almost all of Poland's industries was a type of coal with an especially high sulfur content.

Things are beginning to change, however. Since the communist government fell in 1989, a free-market economy has begun to develop. Under the communist government, artificial price controls kept down the costs of coal, gasoline, and electricity. Now that there are no price controls, it is difficult for Poland's industries to make a profit. Consequently, factories in Poland are finding ways to use energy more efficiently, which usually means using less of it. Emissions of most pollutants are beginning to decrease.

A worker in the Nova Huta steel mill

THINKING CRITICALLY

1. *Identifying Relationships* The kind of coal used in Poland has a high sulfur content. Consult Figure 6-2, and explain why burning coal with a high sulfur content could be particularly harmful to human health.
2. *Expressing Viewpoints* Do you think that a free-market economy will solve Poland's pollution problems? Why or why not?

Figure 6-14 **This courthouse in Polk County, Florida, cost $37 million to build but had to be abandoned when many of the workers became sick.**

Answers to Section Review can be found on page T148.

Section Review

1. Why might it be difficult to establish a direct link between air pollution and health problems?
2. Why might asbestos be more likely to contribute to lung cancer in smokers than in nonsmokers?

Thinking Critically

3. *Applying Ideas* Your neighbor, who lives in a house built in 1949, keeps the house cool in the summer by opening windows and using lots of fans. He knows that you have been learning about sick-building syndrome at school and asks if that might be the cause of his recent respiratory problems. What would you tell him?

Florida, where buildings are tightly sealed to keep out the heat. In Florida, for example, a brand-new, tightly sealed county courthouse had to be abandoned when half of the people who worked there developed allergic reactions to fungi that were growing in the air-conditioning ducts, ceiling tiles, carpets, and furniture.

Air filters in air-conditioning systems can help relieve indoor pollution problems, but many pollutants cannot be removed by the filters. The best solution is to ventilate a building adequately by opening the windows and increasing the amount of fresh air.

Radon Gas Radon gas is invisible, tasteless, and odorless, and it is also radioactive. Radon is one of the elements produced by the radioactive decay of uranium, which occurs naturally in the Earth. Radon is most concentrated in porous soils overlying rocks that contain uranium. Radon can make its way into a house, where it adheres to dust particles. When people inhale the dust, radon enters the lungs.

In the lungs, radon decay releases alpha particles, which can destroy the genetic material in cells lining the air passages. Such damage can lead to cancer, especially among smokers.

Asbestos Asbestos is the name given to a combination of several fibers that contain silica and are valued for their strength and resistance to heat. Asbestos was widely used to reinforce cement and to make brake linings, vinyl floor tiles, residential siding, and garments to protect firefighters. It was used extensively in commercial buildings. Exposure to asbestos in the air is dangerous. The fibers can cut and scar the lungs, causing the disease asbestosis. Victims find it harder and harder to breathe and may eventually die of heart failure. Smokers exposed to asbestos are also more likely to develop lung cancer. The Asbestos Hazard and Emergency Response Act, passed in 1986, requires the removal of all exposed asbestos from school buildings and the overall phasing out of asbestos use in construction by 1997.

Figure 6-15 **All exposed asbestos must be removed from school buildings by 1997.**

SECTION 6.3

ACID PRECIPITATION

AFTER READING THIS SECTION YOU SHOULD BE ABLE TO

1. explain what causes acid precipitation.
2. explain how acid precipitation affects ecosystems.
3. describe ways that countries are working together to solve the problem of acid precipitation.

Imagine that you are hiking through the vast forests of New York's Adirondack Mountains. You come to a scenic lake and sit down to rest. You marvel at how clear the water is, so clear that you can see the bottom of the lake. But after a few minutes you become uneasy. Something is wrong. What is it? You realize that it is totally silent—there are no animals in or near the lake.

This lake, like thousands of others throughout the world, is a victim of acid precipitation, also known as acid rain. **Acid precipitation** is highly acidic precipitation (rain, sleet, or snow) that results from the burning of fossil fuels. When fossil fuels are burned, they releasc oxides of sulfur and nitrogen as byproducts. When the oxides combine with water in the atmosphere, they form sulfuric acid and nitric acid, which fall as precipitation. Figure 6-16 shows how these acids form.

Figure 6-16 **How acid precipitation forms.**

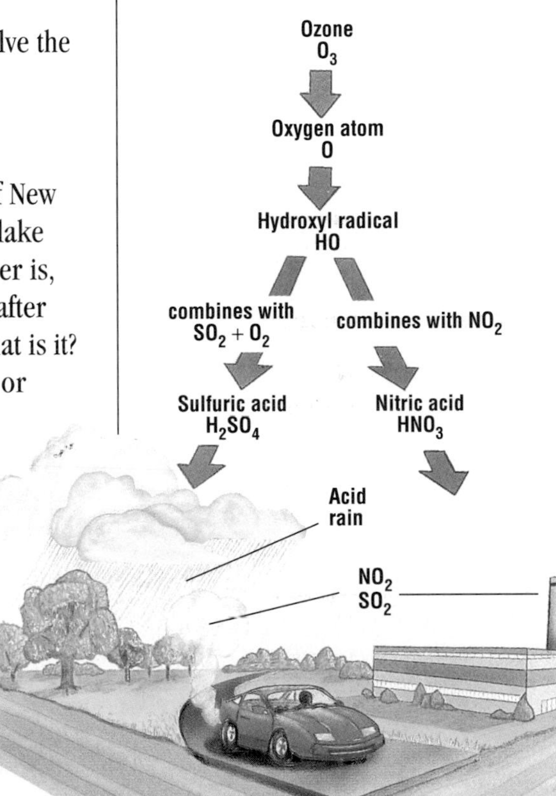

Figure 6-17 **Everything in this lake in the Adirondack Mountains has been killed by acid precipitation.**

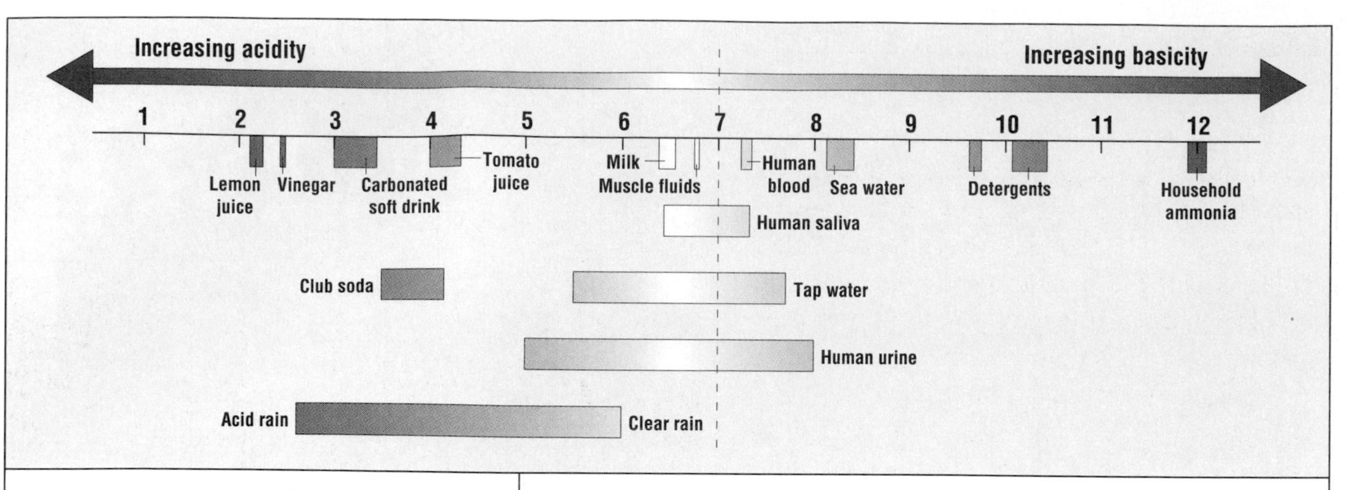

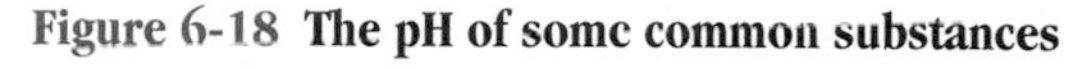
Figure 6-18 **The pH of some common substances**

Acid precipitation can kill living things, as it did in the Adirondack Mountains, and can even destroy entire ecosystems.

Precipitation is considered to be acid precipitation if it has a pH of less than 5. A pH number is a measure of how acidic or basic a substance is. A pH scale is shown in Figure 6-18. Each whole number on the pH scale indicates a tenfold change in acidity. In highly industrialized regions, acid precipitation can be extreme. The northeastern United States sometimes gets "soda-pop" rain with a pH of 3.5 to 4.0, similar to the pH of soft drinks.

Acid precipitation dissolves the calcium carbonate in common building materials such as concrete and limestone. Some of the world's most prized and historic monuments are being eaten away by acid precipitation. (See Figure 6-19.)

When it rains, place sterilized glass jars on windowsills or in open areas where the jars can collect the rain. Then test the pH of the rain with pH paper. Repeat this activity as often as possible, and graph the results. In the winter, you could test the pH of snow to see if the snow could cause acid shock when it melts in the spring.

How Acid Precipitation Affects Ecosystems

Animals are adapted to live in an environment with a particular pH range. If acid precipitation falls on a lake and changes the water's pH, it can kill aquatic plants, fish, and other animals. It is not only the pH of the water that kills fish. Acid precipitation causes aluminum to leach out of the soil surrounding a lake. The aluminum accumulates on the gills of fish, blocking the gills. Many fish slowly suffocate from the buildup of aluminum on their gills.

The effects of acid precipitation are worst in the spring, when acidic snow that accumulated all winter melts and rushes into lakes and other bodies of water. This sudden influx of acidic water causes **acid shock,** which can be so intense that entire populations of fish are wiped out. Acid shock also affects the reproduction of fish and amphibians. They produce fewer eggs, and these usually do not hatch. The young that do hatch are often defective and cannot reproduce.

Figure 6-19 **Acid precipitation has damaged this frieze on the Parthenon in Athens, Greece.**

To counteract the effects of acid precipitation on aquatic ecosystems, some states and countries spray tons of powdered lime on acidified lakes in the spring to help restore their natural pH. Because lime has a pH that is basic, it counteracts the acidic pH of the water. Unfortunately, we cannot spread enough lime to offset all acid damage to lakes.

Forest ecosystems are also affected by acid precipitation. Trees, like other organisms, can tolerate only specific pH ranges. If the water they take up through their roots is too acidic, they will die. Millions of hectares of forests in the northeastern United States and Canada are dying, partly due to acid precipitation. As the trees and other plants die, the animals they support die too. The ecologist shown in Figure 6-20 is attempting to determine the amount of acidity that plants can tolerate.

Figure 6-20 An ecologist monitors the effects of acid precipitation on plants.

International Conflict and Cooperation

One problem in controlling acid precipitation is that pollutants may be released in one geographical area and fall to the ground hundreds of miles away. In North America, for example, most of the acid precipitation in New England and southeastern Canada results from pollution produced in the midwestern and eastern United States.

Similarly, emissions from the highly industrialized Ruhr Valley in Germany have spread over most of the other nations in Europe. Weather patterns usually carry pollutants northward toward the Scandinavian countries, which have the most serious acid precipitation problems

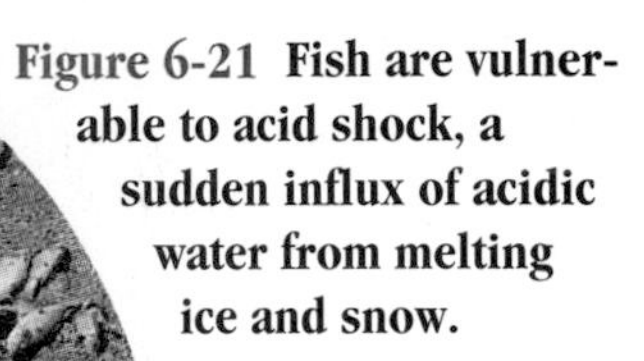

Figure 6-21 Fish are vulnerable to acid shock, a sudden influx of acidic water from melting ice and snow.

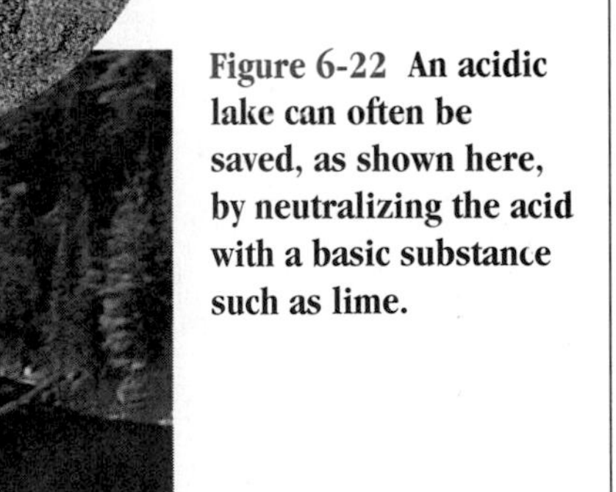

Figure 6-22 An acidic lake can often be saved, as shown here, by neutralizing the acid with a basic substance such as lime.

Figure 6-23 **This forest in Karkonoski National Park, Poland, suffers from the effects of acid precipitation.**

in Europe. Figure 6-24 shows which regions produce pollutants and which are affected by acid precipitation.

In 1985, the United Nations Helsinki Declaration was enacted, which requires countries to cut sulfur-oxide emissions by 30 percent over 10 years. The declaration was signed by 18 nations. The United States did not sign this agreement. In 1988, 27 nations signed the United Nations Sofia Protocol, which required a reduction in nitrogen-oxide emissions. The United States agreed to this protocol in 1989.

Later, the European Union, a coalition of European countries, mandated that refineries reduce the sulfur content in diesel fuel by more than 30 percent by 1994 and by another 80 percent by 1996. It is clear that even more international agreements such as these will be necessary to control the acid-precipitation problem.

Figure 6-24 **A global look at acid precipitation**

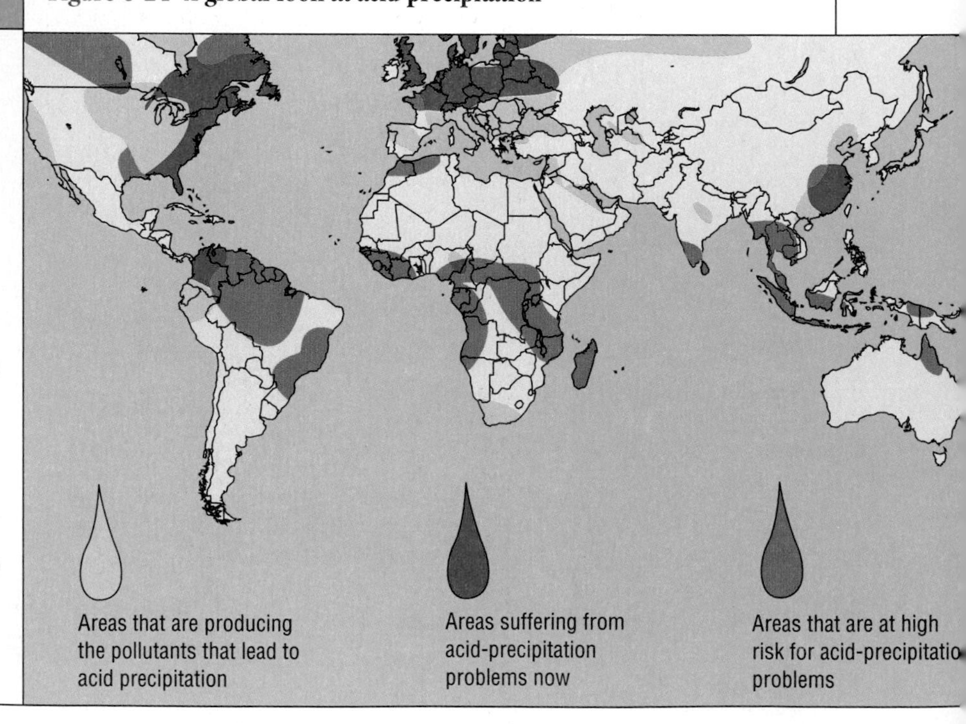

Answers to Section Review can be found on page T149.

Section Review

1. What human activities contribute to acid precipitation?
2. Why is international cooperation necessary to reduce acid precipitation?

Thinking Critically

3. *Inferring Relationships* Why might normal rainwater be more acidic than pure water?
4. *Interpreting Graphics* Lime is sprayed on lakes to counteract the effect of acid precipitation. Look at Figure 6-18, and predict whether lime has a pH above or below 7. Explain your reasoning.

HIGHLIGHTS

SUMMARY

- Some human activities release contaminants into the air that degrade its quality. This is called air pollution.
- Most air pollution is caused by the burning of fossil fuels, either by industries or by automobiles.
- Air pollution, which can occur both outdoors and indoors, can cause or aggravate a variety of health problems.
- One effect of air pollution is acid precipitation. Acid precipitation, which results when air pollutants react with moisture in the air, can cause widespread environmental damage.
- Acid precipitation is not just a local problem but a global one, and it will be solved only through international cooperation.

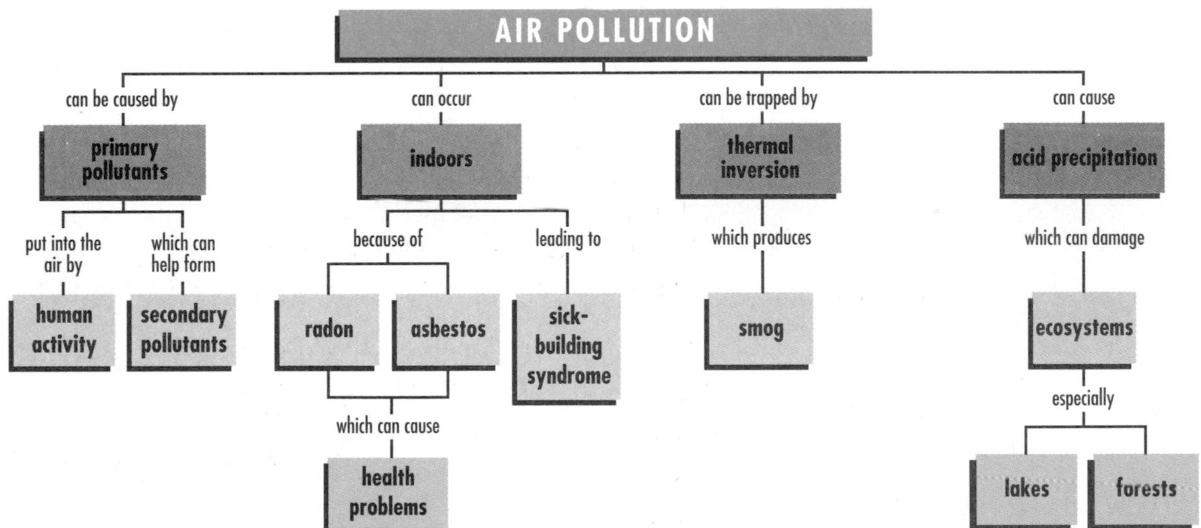

Sample EcoLog answers can be found on page T149.

EcoLog

Now that you've studied this chapter, revise your answers to the questions you answered at the beginning of the chapter, based on what you have learned.

❶ What is the air quality like in your area?

❷ If there is air pollution in your area, what causes it?

Vocabulary Terms

acid precipitation (p. 161)
acid shock (p. 162)
air pollution (p. 151)
primary pollutant (p. 152)
secondary pollutant (p. 152)
sick-building syndrome (p. 158)
smog (p. 156)
thermal inversion (p. 155)

CHAPTER 

REVIEW

Understanding Vocabulary

1. Use the following terms in a sentence to show that you know what they mean.
   - **a.** air pollution
   - **b.** acid shock
   - **c.** thermal inversion

Relating Concepts

2. Copy the unfinished concept map below onto a sheet of paper. Then complete the concept map by writing the correct word or phrase in each box containing a question mark.

Answers not indicated can be found on page T149.

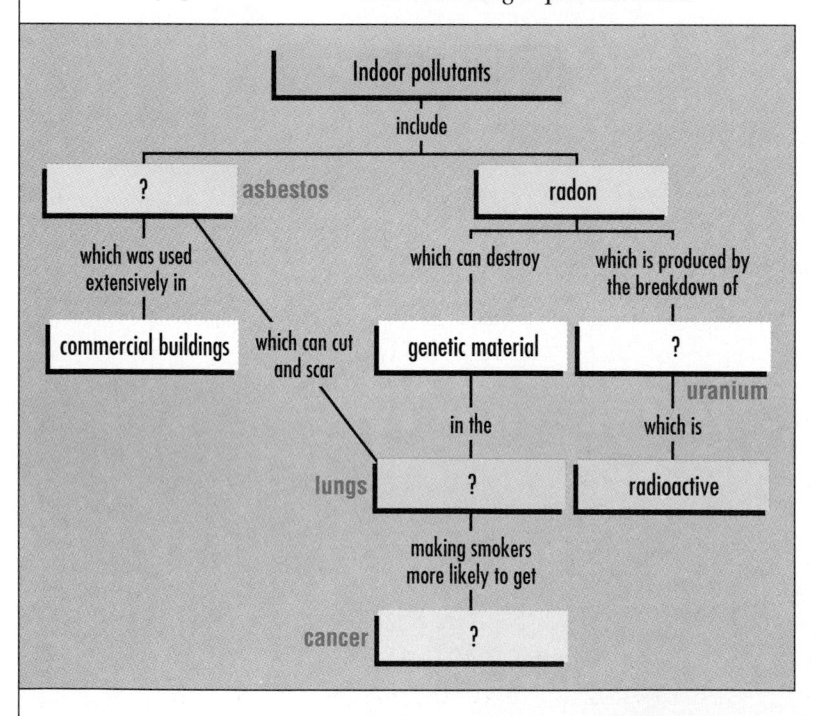

Understanding Concepts

Multiple Choice

3. Smog is an example of
   - **a.** a primary pollutant.
   - **b.** acid rain.
   - **c.** a secondary pollutant.
   - **d.** smoke.

4. Air pollution
   - **a.** originated in the past few years.
   - **b.** has occurred throughout history.
   - **c.** has grown much worse in recent times.
   - **d.** both b and c

5. Which of these disorders is NOT linked to air pollution?
   - **a.** emphysema
   - **b.** lung cancer
   - **c.** diabetes
   - **d.** chronic bronchitis

6. Sick-building syndrome affects what kinds of structures?
   - **a.** structures built near sewage treatment plants
   - **b.** tightly sealed, poorly ventilated structures
   - **c.** well-ventilated structures
   - **d.** hospitals

7. Which is an example of an indoor pollutant?
   - **a.** ozone
   - **b.** sulfur dioxide
   - **c.** radon gas
   - **d.** volcanic particles

8. Which type of air pollution would probably NOT be common in a city having a moist, cloudy climate and few traffic problems?
   - **a.** smog
   - **b.** acid rain
   - **c.** VOCs
   - **d.** none of the above

9. A thermal inversion is
   - **a.** like a sponge, soaking up pollutants.
   - **b.** like a giant fan, causing pollutants to disperse.
   - **c.** like a ceiling, preventing pollutants from moving upward.
   - **d.** something that happens only in Los Angeles.

10. Which is NOT an effect of acid precipitation on fish?
   a. suffocation
   b. increased food supply
   c. birth defects
   d. decreased egg production
11. Which is an example of a fuel that produces less pollution than gasoline?
   a. hydrogen
   b. ethanol
   c. natural gas
   d. all of the above
12. Precipitation is considered to be acidic if the pH
   a. is greater than the pH of clear rain.
   b. is greater than 5.
   c. is less than 5.
   d. cannot be measured.
13. Most of the acid precipitation in New England results from
   a. emissions from local industries.
   b. emissions from the heavy traffic in New York City.
   c. pollution produced in southeastern Canada.
   d. pollution produced in the midwestern and eastern United States.

Short Answer

14. Explain how carpooling can help control air pollution.
15. Agree or disagree with the following statement and explain your position: "All air pollution is the result of something being burned."
16. Radon poisoning is an example of "natural pollution." Explain what is meant by this term.
17. Explain how geography plays a role in the thermal inversions that occur frequently in Los Angeles.
18. Explain how spraying lime on acidified lakes helps restore their natural pH.

Interpreting Graphics

19. **Examine the map below.** Where does the worst acid rain seem to be concentrated? What might explain the strongly acidic rainfall in Birmingham? (It is not a large city.)

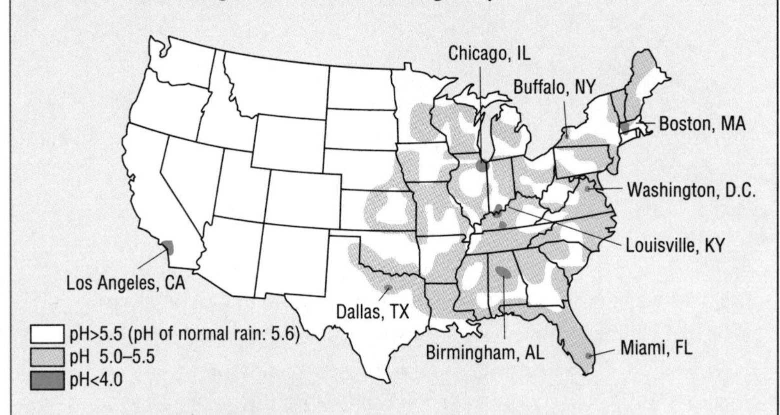

Thinking Critically

20. ***Analyzing Ideas*** Over the past 25 years, the pollution produced by automobiles has declined. Now, an average late-model car produces a fraction of the pollutants (except carbon dioxide) produced by an average 1970-model car. Given this trend, do you expect emissions to decline to zero in the near future? Explain your answer.

Themes in Science

21. ***Energy*** How might a switch from fossil fuels to alternative forms of energy affect the average adult life expectancy?

Cross-Discipline Connection

22. ***History*** In the past, some cities experienced "killer fogs," dense, smoky fogs that actually resulted in the deaths of some people. Find out more about these killer fogs, including which cities were affected by them, what caused them, and why they no longer occur.

Discovery Through Reading

For an interesting description of a naturally occurring indoor air pollutant, read "Radon: The 'Silent Killer'?" by Leonard A. Cole, and "Radon Risks Are Real," by Stephen Page, in *Garbage,* Spring 1994, pp. 22–31.

PORTFOLIO ACTIVITY

Develop a 30-second public service announcement on the seriousness of air pollution. It should be designed to persuade people to take some course of action to lessen this pollution. You may use either a storyboard or live-action format to create your presentation. If you use a live-action presentation, videotape the result.

CHAPTER 6 INVESTIGATION

How Does Acid Precipitation Affect Plants?

Acid precipitation, resulting from pollutants in the atmosphere, can damage buildings, erode statues, kill fish, and harm plants. Both crops grown for food and trees grown for timber or paper can be damaged by long-term exposure to acid precipitation. But what exactly happens to the plants? Does acid precipitation affect all plants the same way? Will the location or type of plant make a difference?

By studying rates of seed germination, you will observe one effect that acid precipitation can have on plants. Then you will devise your own experiment to explore how the effects of acid precipitation on seed germination may vary.

Answers to this Investigation can be found on page T149.

MATERIALS

- artificial acid precipitation
- 2 small beakers
- wax pencil
- graduated cylinder
- pH paper
- radish or mustard seeds
- 2 petri dishes
- cotton
- filter paper
- optional: acid solutions, other seeds, limestone chips and other rock fragments
- notebook
- pen or pencil

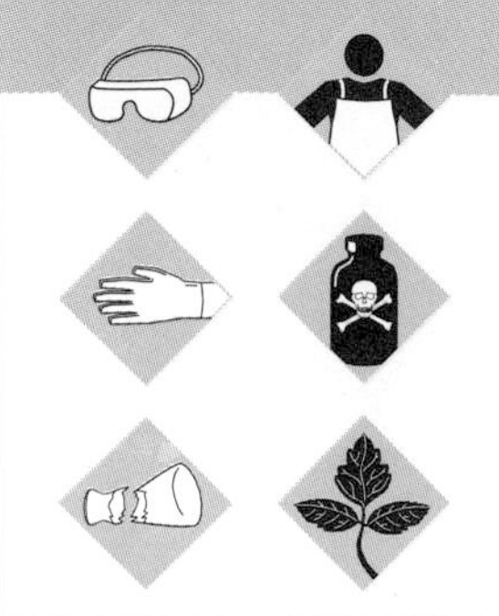

PREPARE

1. CAUTION: Acid may irritate your skin or damage your clothing. Wear goggles, gloves, and a lab apron during this Investigation.
2. Watch your teacher make artificial acid precipitation.

OBSERVE THE EFFECTS OF ACID ON SEEDS

3. Label one beaker "water" and one "acid precipitation." Put 25 mL of tap water in the "water" beaker and 25 mL of artificial acid precipitation in the "acid precipitation" beaker.
4. Use pH paper to test the liquid in each beaker. Copy the table below into your notebook, and record the pH values in the table.

Solution	Appearance of Seeds After Soaking	Total Number of Seeds	Number of Seeds That Germinated	Percentage of Seeds That Germinated
Water pH = ____				
Acid Precipitation pH = ____				

5. Put 50 seeds in each beaker. Leave them overnight.
6. After 24 hours, do the seeds look different? Record your observations in your data table.
7. Prepare two petri dishes by spreading a thin layer of cotton in the bottom of each one and covering the cotton with a circle of filter paper. Pour the liquid and seeds from the "water" beaker into one of the dishes and the liquid and seeds

from the "acid precipitation" beaker into the other dish. Label the dishes accordingly. Cover the petri dishes and leave them overnight.

8. Clean up your work area and wash your hands before leaving the lab.
9. After 24 hours, count the seeds in each dish that have germinated (begun to sprout). Record your results in your data table. Express the results as a percentage of the total number of seeds in each dish.
10. What was the effect of "acid precipitation" on the seeds? If a farmer planted 10,000 seeds in soil that had been exposed to acid precipitation, how many plants would he lose? Record your conclusions in your notebook.

After soaking the seeds overnight and observing any changes, pour them into petri dishes to germinate.

FORM A HYPOTHESIS

11. What factors do you think might influence how severely acid precipitation affects plants? For example:
    - Would the effects differ depending on the rain's level of acidity?
    - Are different types of seeds affected differently?
    - Will the type of rocks present make a difference? (In nature, some rocks and soils can help to neutralize acid precipitation.)

    Develop a hypothesis about one way that the effects of acid precipitation on plants might vary. Write your hypothesis in your notebook.

DESIGN AN EXPERIMENT

12. Think of an experiment to test your hypothesis. In addition to the materials you used in the first part of this Investigation, you may want to use some of the following materials:
    - a stronger or weaker "acid precipitation" solution
    - beans, corn, or other seeds
    - limestone chips and other rock fragments

    In your notebook, write down your experimental procedure. Remember to include a control. (You will need at least three or four beakers in your new experiment—what will be in them?)
13. Based on your hypothesis, what results do you expect to see from your experiment? Write your prediction in your notebook.

TEST YOUR PREDICTION

14. Carry out your experiment, recording your results in a table in your notebook.
15. Did your results agree with your prediction? Was your hypothesis supported?
16. In what ways did your experiment mimic the real-world effects of acid precipitation? In what ways was it different from a real-world situation? How might your experiment be modified to make it more realistic?

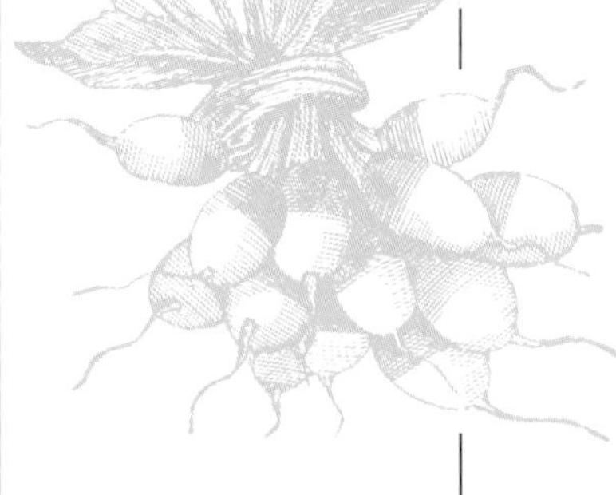

POINTS OF VIEW

Are Electric Cars the Cure for Air Pollution?

Automobile emissions are responsible for at least half of all urban air pollution and a quarter of all carbon dioxide released into the atmosphere. Therefore, the production of a car that emits no polluting gases in its exhaust is a significant accomplishment. The only such vehicle currently available is the electric car. Electric cars are powered by batteries, so they do not produce exhaust gases. Supporters believe that a switch to electric cars will reduce air pollution in this country. But critics believe that taxpayers will pay an unfair share for the switch, and the reduction in pollution won't be as great as promised. Two points of view:

ELECTRIC CARS WILL HELP REDUCE AIR POLLUTION

According to supporters, a switch to electric cars such as this one will reduce air pollution.

More cars are on the road every year, especially in big cities, and even the cleanest and most modern of these cars emit pollutants into the air. Supporters of a switch to electric cars believe the switch will be essential to a reduction of pollution in congested cities.

Some critics suggest that a switch to electric cars will simply move the source of pollution from a car's tailpipe to the power plant's smokestack, because most electricity is generated by burning coal. But in California, where electric cars would have the greatest impact, most electricity is produced by burning natural gas, which releases less air pollution than does burning coal. Nuclear plants and dams release no pollutants into the air when they generate electricity. Solar power and wind power are also emission-free ways to generate electricity. Supporters argue that a switch to electric cars will reduce air pollution immediately and that a further reduction will result when power plants convert to these cleaner sources of energy.

Electric cars are expensive now, but prices are expected to drop as more are produced and sold. Within 10 years of widespread introduction, electric cars should be comparable in price to conventional cars. And recharging a car to go about 120 mi. is expected to require less than a dollar's worth of electricity, much less than the price for an equivalent amount of gasoline.

To maximize an electric car's potential, a small gasoline engine could be added to recharge the batteries while the car is being driven. This would extend the car's range to 300 mi. or more. Such a hybrid electric car could use battery power for slower, stop-and-start driving in the polluted city and then switch to gas power for traveling on the open highway.

Advocates say that electric cars will reduce air pollution even more significantly when power plants begin using cleaner sources of energy, such as natural gas.

Electric Cars Won't Live Up to Their Promise

By 2003, 10 percent of all new cars sold in California must be zero-emission vehicles, meaning that they must produce no exhaust gases. Other state governments, including 12 in the Northeast, have adopted similar regulations. Some people believe that government officials have not considered all the consequences of a switch to electric cars.

Although the purchase price of an electric car is expected to drop, it's not just individual car buyers who will pay for electric cars. Government agencies are paying millions of dollars for the research and development of electric cars. A subsidy or rebate will probably be offered to car buyers—at several thousand dollars per car—to encourage them to buy electric cars. In addition, publicly owned electric companies will probably offer special low electric rates, free installation of recharging outlets, and rebates on expensive new batteries. Critics believe that taxpayers will end up paying for these rebates and incentives.

Critics say that electric cars will probably not replace older, poorly maintained cars, which may emit 100 times more pollutants than newer models.

This electric car is plugged into a recharging outlet. Some critics think that electric cars are too inconvenient because their batteries have to be recharged so often.

Electric cars are inconvenient because the batteries have to be recharged so often. They also have to be replaced every two or three years. The nation's landfills are already crowded with conventional car batteries, which contain acid and metals that may pollute groundwater. A switch to electric cars would aggravate this pollution problem because the batteries have to be replaced so often.

Finally, electric cars will likely replace the cleanest cars on the road, not the dirtiest. A new car may emit only a tenth of the pollution emitted by an older model. If an older car's pollution-control equipment does not work properly, it may emit a hundred times more pollution than a new car. But people who drive older, poorly maintained cars probably won't be able to afford expensive electric cars. Therefore, the worst offenders will stay on the road, continuing to pollute the air.

Analyze the Issue

1. ***Expressing Viewpoints*** In your opinion, are electric cars the best solution to the air pollution problem? Why or why not? What are some alternative solutions for reducing air pollution?

2. ***Making Decisions*** Would you be willing to drive an electric car? Do you think the benefits would outweigh the sacrifices? Explain your reasoning.

Answers to Analyze the Issue can be found on page T149.

CHAPTER 7

ATMOSPHERE AND CLIMATE

"The atmosphere is the key symbol of global interdependence."

MARGARET MEAD, AMERICAN ANTHROPOLOGIST

EcoLog

Before you read this chapter, take a few minutes to answer the following questions in your EcoLog.

1. Why do you think the warming of the Earth by gases such as carbon dioxide is called "the greenhouse effect"?
2. You learned in Chapter 6 that ozone is an air pollutant. Why, then, might scientists be concerned about the loss of ozone from the ozone layer?

THE ATMOSPHERE

AFTER READING THIS SECTION YOU SHOULD BE ABLE TO

1. explain how the atmosphere makes life possible on Earth.
2. explain how photosynthesis and respiration keep the amount of carbon dioxide in the air nearly constant.
3. describe how the atmosphere is structured in layers.

"When you look out across our atmosphere, it looks like the skin of an onion."

COL. CHARLES BOLDEN, SPACE SHUTTLE COMMANDER

The nearby planets Mars and Venus show no traces of living things. Why does our planet teem with life? One thing that makes life possible on Earth is its atmosphere. The **atmosphere** is a thin layer of gases, shown in the photograph in Figure 7-1, that surrounds the Earth. It extends from the surface of the Earth to hundreds of kilometers above the surface.

The atmosphere is 78 percent nitrogen and 21 percent oxygen. The remaining 1 percent is made up of carbon dioxide, helium, neon,

Figure 7-1 The Earth's atmosphere as seen from space

FIELD ACTIVITY

The Earth's circumference is 40,000 km (24,800 mi.), and the troposphere and stratosphere extend about 45 km (about 28 mi.) from the surface. You can get a visual image of how thin these layers are in relation to the Earth by following these instructions. On a paved area like a parking lot, use colored chalk to draw a circle with a diameter of 4 m (about 13 ft.) to represent the Earth. This is easy if you tie the chalk to a string 2 m (about 6.5 ft.) long and have one person hold the end of the string on the ground while another person moves around drawing the circle. Calculate how thick the combined troposphere and stratosphere need to be on your drawing to be in proportion to your "Earth." Then use white chalk to draw another circle around the first one to represent these two layers. Are they thinner than you guessed they would be?

Answer to Field Activity: The two layers form a band about 1.4 cm (.5 in.) wide. Most students will find this thinner than they expected.

argon, water vapor, and other substances. It is this mixture that we call "air." The most important of these substances for organisms are oxygen and carbon dioxide. As you have learned in previous chapters, oxygen is necessary for cellular respiration, and carbon dioxide is necessary for photosynthesis.

The Earth's atmosphere also protects living things from most of the sun's harmful ultraviolet radiation. At the same time, the atmosphere allows visible light to reach the Earth's surface, supplying energy and making photosynthesis possible. The atmosphere also reflects some infrared radiation back to the Earth, which warms the planet. Without the atmosphere, life on Earth would cease to exist.

How Photosynthesis Changed the Atmosphere

Living things played a very important role in forming the atmosphere we know today. The Earth's early atmosphere probably contained very little oxygen, unlike today's atmosphere. Then about 4 billion years ago, the first living things appeared. These early organisms began to change the Earth's atmosphere slowly but drastically.

The ancestors of green plants evolved photosynthesis, the process of making food from water and carbon dioxide, using sunlight for energy. When the process is complete, the leftover oxygen from the water is released into the air as oxygen gas. As these ancestors of plants multiplied, the amount of oxygen in the air began to increase. Oxygen now makes up about 21 percent of the gases in our atmosphere.

When organisms break down food molecules during cellular respiration, carbon dioxide is released into the atmosphere.

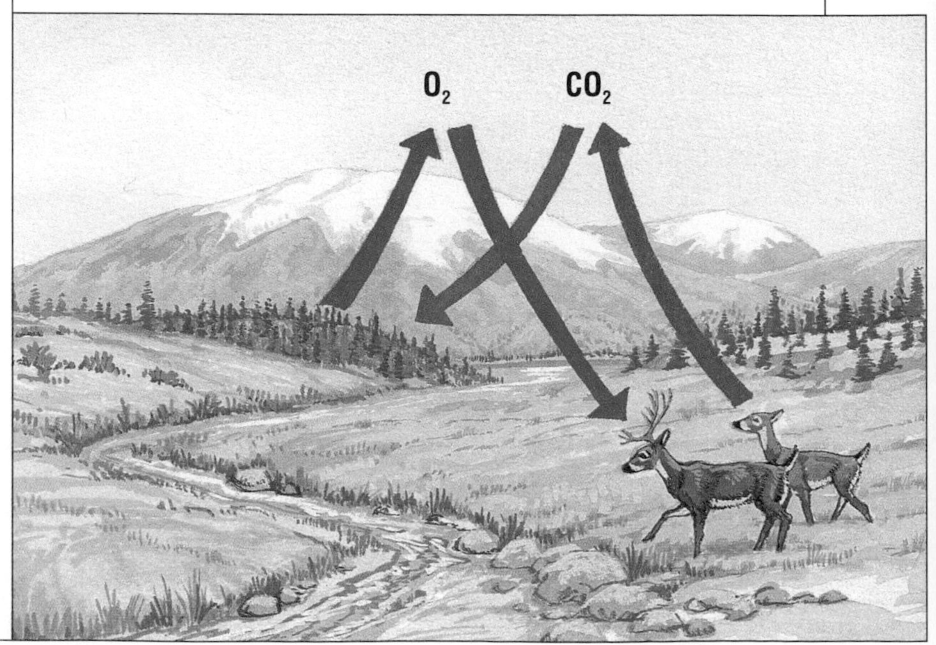

Figure 7-2 Photosynthesis and respiration affect the composition of the atmosphere.

Figure 7-2 shows how the balance between photosynthesis and respiration keeps the amount of carbon dioxide in the atmosphere nearly constant. Even though carbon dioxide makes up only .03 percent of the gases in the atmosphere, this small amount helps to keep the Earth within a temperature range that can support life.

The Five Layers

Scientists have found it useful to discuss the atmosphere in terms of five individual layers, which are shown in Figure 7-4. The layers become less dense the farther away from the Earth's surface they are.

The **troposphere** extends from the Earth's surface to about 10 km (about 6 mi.) above the surface. It contains nearly 90 percent of the atmosphere's gases. The air we breathe is part of the troposphere. The troposphere is also the layer in which most weather occurs, as air currents flow and swirl. The air currents may bring dry, icy air from the Arctic, for example, or hot, humid air from the Gulf of Mexico.

Above the troposphere is the **stratosphere,** which extends from 10 km to about 45 km (about 28 mi.) above the Earth. The air in the stratosphere is very thin. Winds blow, but without the swirling turbulence of the troposphere. Commercial airliners travel in the lower part of the stratosphere. The stratosphere contains the ozone layer, which protects us from harmful ultraviolet light from the sun. You will read about the ozone layer in Section 7.4. Beyond the stratosphere are the mesosphere, thermosphere, and exosphere. The gases in the exosphere become thinner and thinner until the exosphere merges with outer space.

Figure 7-4 The Earth's atmosphere has five layers. The closest layer is the troposphere, which is where weather occurs. The next layer, the stratosphere, contains the ozone layer.

Gases in the Atmosphere

Gas	Percentage of Total
Nitrogen	78
Oxygen	21
Carbon dioxide, helium, neon, argon, water vapor, and other gases	1

Figure 7-3 Nitrogen and oxygen make up 99 percent of the gases in the atmosphere.

Earth is the only place in the solar system where free oxygen exists in significant amounts.

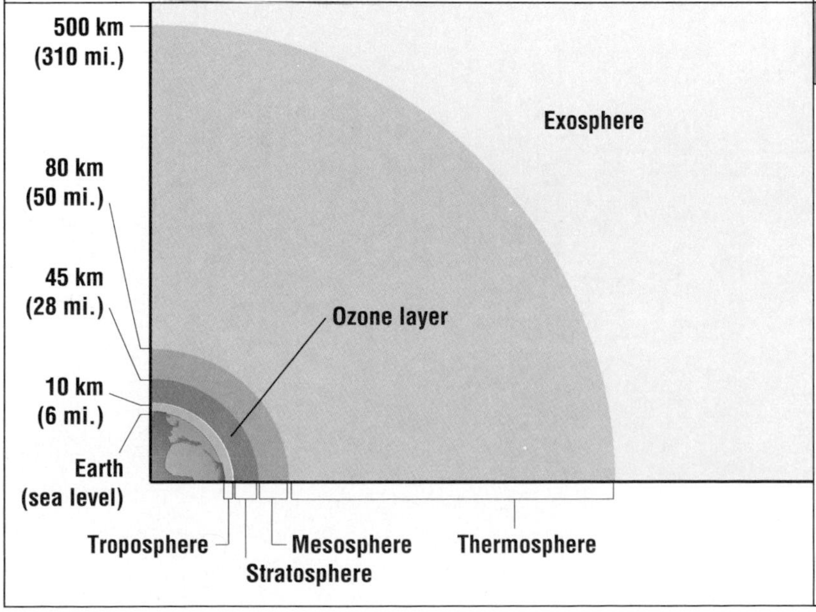

Answers to Section Review can be found on page T150.

Section Review

1. Name two characteristics of the atmosphere that make life possible on Earth.
2. What caused the amount of oxygen in the Earth's atmosphere to increase?
3. In which atmospheric layer do we live?

Thinking Critically

4. *Interpreting Graphics* Study the illustration in Figure 7-2. What process is represented by the arrow from the animal to carbon dioxide?

CLIMATE

AFTER READING THIS SECTION YOU SHOULD BE ABLE TO

❶ explain why different parts of the world have different climates.

❷ explain what causes the seasons.

You learned in Section 7.1 that weather occurs mostly in the troposphere, the atmospheric layer that touches the Earth's surface. But what *is* weather? **Weather** is simply what is happening in the atmosphere at a particular place at a particular moment. **Climate,** on the other hand, is the average weather in an area over a long period of time. To understand the difference between weather and climate, consider Seattle and Phoenix. These two cities have the same weather on a certain day if it is raining in both places. But their climates are quite different—Seattle is cool and moist, while Phoenix is hot and dry.

Important aspects of climate are temperature, humidity, winds, and precipitation (rain, snow, and sleet). Climate, particularly temperature and precipitation, determines what types of organisms live in a region.

What does the world look like 45 km (about 28 mi.) up? No one knows better than Richard Somerville. **See pages 384–385 for an interview with the climate researcher.**

WHAT DETERMINES CLIMATE?

Climate is determined by a variety of factors, including latitude, air circulation, ocean currents, and the local geography of an area. The most important of these factors is latitude.

Figure 7-5 Seattle, Washington (left) and Phoenix, Arizona (right) could have the same weather on a certain day, but the two cities have quite different climates.

Latitude

Latitude is the distance from the equator, measured in degrees north or south of the equator. The equator is defined as 0° (zero degrees). The most northerly latitude is the North Pole, at 90° north, while the most southerly latitude is the South Pole, at 90° south.

Latitude strongly influences climate because the amount of solar energy an area receives depends on its latitude. More solar energy falls on areas near the equator than on areas closer to the poles. Figure 7-6 shows why. At the equator, the sun is directly overhead, and its rays hit the Earth directly. The incoming solar energy is concentrated on a small area of the surface.

At higher latitudes (closer to the poles), the sun is lower in the sky. This reduces the amount of energy arriving at the surface. Sunlight hits the Earth at an oblique angle and spreads over a larger area of the surface than it does at the equator.

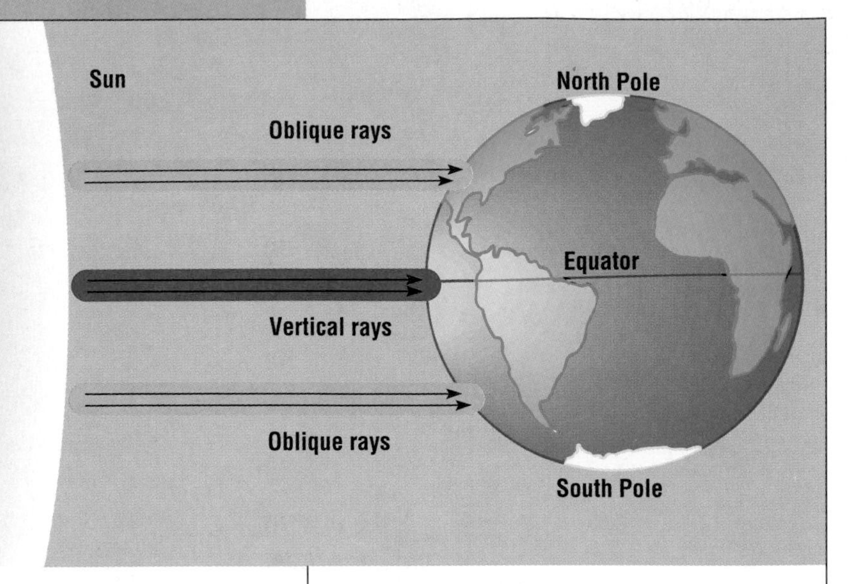

Figure 7-6 This diagram shows why latitude influences climate. At the equator, sunlight hits the Earth vertically. The sunlight is concentrated on a small surface area. Away from the equator, sunlight hits the Earth at an oblique angle and spreads over a larger surface area.

Atmospheric Circulation Patterns

Three important properties of air will help you understand how air circulation affects climate. First, cold air sinks and warms as it sinks. Second, warm air rises and cools as it rises. Third, warm air can hold more water vapor than cold air. This means that if warm air is cooled, the water vapor it contains will condense into liquid water, forming rain, snow, or fog.

Solar energy heats the ground, which warms the air above it. This warm air rises, and cooler air moves in to replace it. Heating of the atmosphere therefore causes wind, or the movement of air within the atmosphere. Because different latitudes receive different amounts of solar energy, the patterns of global circulation shown in Figure 7-7 result. This circulation pattern determines the global patterns of precipitation. For instance, the intense solar energy striking the Earth's surface at the equator causes the surface as well as the air above it to become very warm. This warm air can hold large amounts of water that evaporates from the oceans and land. As the air rises, however, it cools and loses some of its ability to hold water. Thus, it rains heavily at the equator; some areas receive over 450 cm (177 in.) of rain per year.

Answer to caption (below):
Solar energy warms the air closest to the Earth's surface and causes it to rise. The rising air is replaced by cooler air flowing in from farther north or south. The result is a circulation cell.

Figure 7-7 This diagram shows the major patterns of atmospheric circulation. How does solar energy affect these patterns?

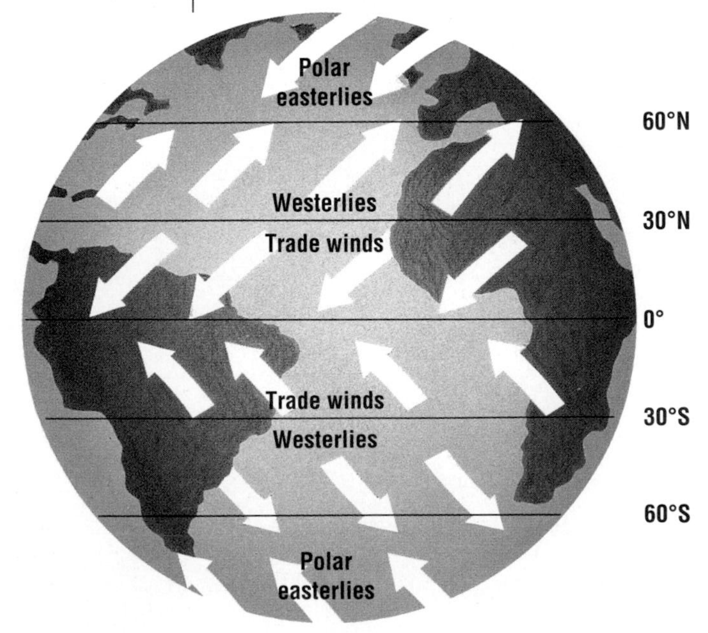

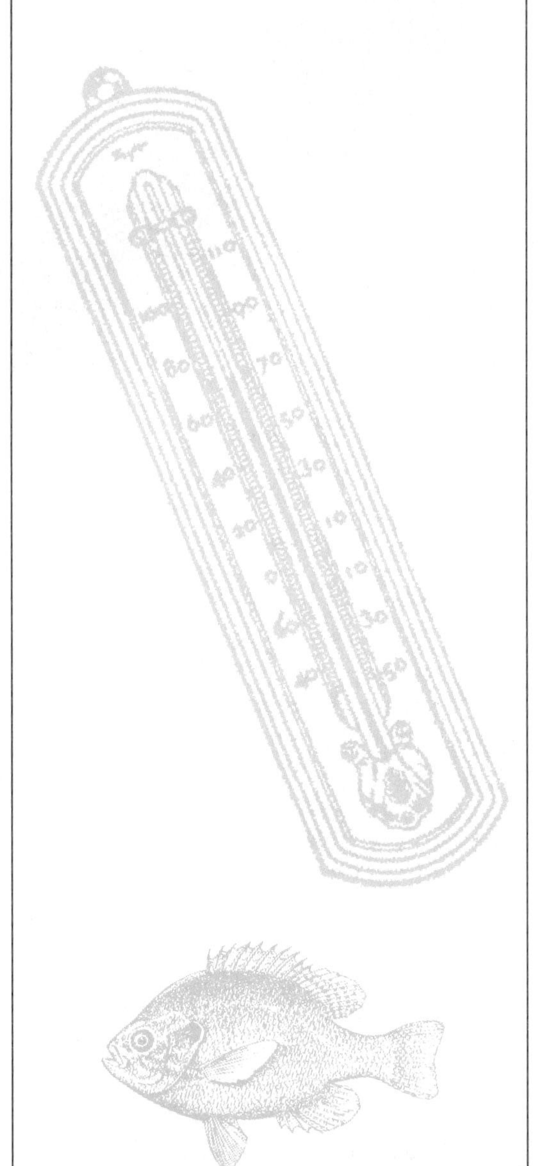

Cool air normally sinks, but the cool air over the equator cannot descend because hot air is moving up below it. So this cool air is forced away from the equator toward the poles. It sinks back to Earth at a latitude of about 30°, warming as it falls. As this warm, dry air moves across the surface, it causes water to evaporate from the land below, creating dry conditions. Most of the world's deserts, including those in southwestern North America, are located near 30°.

Air descending at 30° moves either toward the equator or toward the poles. Air going toward the poles warms while traveling near the surface, and then it rises again at about 60°. Cold, dry air descends at the poles, which are essentially very cold deserts. Their covering of ice and snow is precipitation that has accumulated over many years because it cannot melt or evaporate in the cold climate.

Ocean Circulation Patterns Ocean currents have a great effect on climate, because water holds large amounts of heat. The movement of surface ocean currents is caused largely by winds and the rotation of the Earth. You can see these major ocean currents in Figure 7-8. These currents redistribute warm and cool masses of water. For instance, the warm Gulf Stream moves along the east coast of the United States and then flows across the Atlantic Ocean, where it warms Western Europe.

Oceans tend to make climates more moderate, so coastal areas usually have warmer winters and cooler summers than inland areas. And since the ocean is the source of most of the water that falls as precipitation, coastal areas typically get more moisture than inland areas.

Local Geography Kilimanjaro, a mountain in Tanzania, is only about 3° south of the equator, but snow covers its peak year-round. Kilimanjaro illustrates the important effect that height above sea

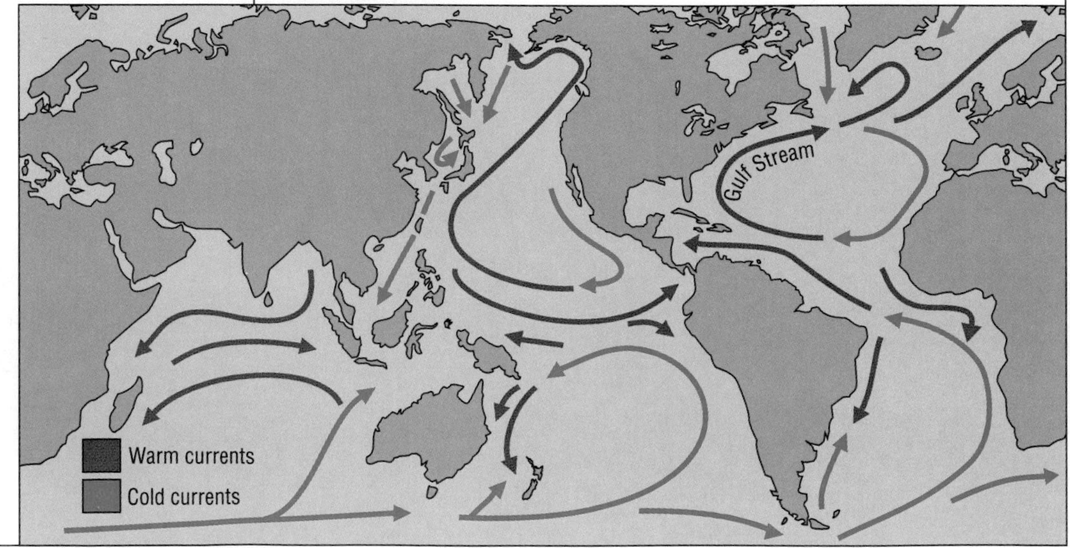

Figure 7-8 Surface ocean currents are caused mainly by winds and the rotation of the Earth. The currents redistribute warm and cold masses of water around the world.

level has on climate. Temperatures fall by about 6°C (about 11°F) for every 1,000 m (3,300 ft.) increase in elevation. Because of this temperature decrease, plants and animals living on high mountains often resemble those living in cold, northern climates.

Mountains and mountain ranges also influence the distribution of precipitation. For instance, consider the Sierra Nevada mountains of California. Air full of moisture blows from the Pacific Ocean in an eastward direction. When it reaches the Sierra Nevadas, it is deflected upward. This cools the air and causes it to release its moisture on the coastal side of the mountains. When the air crosses the mountains, it is dry. As it descends, it warms and draws up moisture from the surface. The Great Basin Desert of the western United States is the result. (See Figure 7-9.)

Figure 7-9 The coastal side of the Sierra Nevada mountains (above) receives a great deal of rain. Just 20 mi. east of the mountains, however, the land is a desert (below). What accounts for this difference? Answer on page T150.

Seasonal Changes in Climate

You know that temperature and precipitation change over the seasons. But do you know what causes the seasons? The seasons are the result of the Earth's orbit around the sun, as illustrated in Figure 7-10. The Earth is tilted at about 23° relative to the path of its orbit. This tilt means that the angle at which the sun's rays strike the Earth changes as the Earth moves around the sun.

During spring and summer in the Northern Hemisphere, the Northern Hemisphere tilts toward the sun and receives concentrated, direct sunlight. The Southern Hemisphere tilts away from the sun and receives less concentrated sunlight. During fall and winter in the Northern Hemisphere, the situation is reversed: the Southern Hemisphere is inclined toward the sun, while the Northern Hemisphere is tilted away.

The four seasons familiar to many people in the world do not occur in the tropics, which are the regions close to the equator. In the tropics, temperatures are high and constant throughout the year because most areas receive nearly direct sunlight year-round.

Answers to Section Review can be found on page T150.

Figure 7-10 The Earth's orbit around the sun causes seasonal changes.

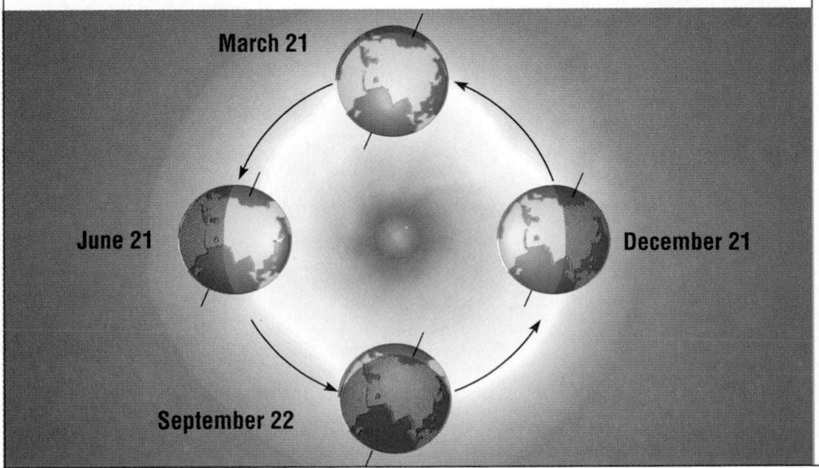

Section Review

1. Explain why the equator receives more direct sunlight than the poles receive.
2. Dublin, Ireland, and Moscow, Russia, are at nearly the same latitude. Yet Dublin is coastal and Moscow lies inland. How does this difference affect the climates of these two cities?

Thinking Critically

3. *Analyzing Processes* If the Earth were not tilted in its orbit, how would the climates and seasons be affected at the equator and in the temperate zones?
4. *Relating Concepts* Make a concept map showing what determines climate.

GREENHOUSE EARTH

AFTER READING THIS SECTION YOU SHOULD BE ABLE TO

1. explain why the Earth and its atmosphere are like a greenhouse.
2. explain why carbon dioxide levels in the atmosphere are rising.
3. explain why many scientists think that the Earth's climate will get warmer.
4. describe what a warmer Earth might be like.

Answer to caption:
In essence, certain trace gases in the atmosphere prevent heat from radiating back into space from the surface. This is similar to the effect seen in a greenhouse. The glass of the greenhouse allows sunlight to enter but prevents the heat produced by the absorption of sunlight from escaping.

Have you ever gotten into a car that has been sitting in the sun for a while with all its windows closed? Even if the day is cool, the air inside the car is much warmer than the air outside. On a hot summer day, opening the door to the car can seem like opening the door of a blast furnace.

The reason heat builds up inside the car is that the sun's energy streams into the car through the clear glass windows in the form of sunlight. The carpets and upholstery in the car absorb the light and change it into heat energy. Heat energy does not pass through glass as easily as light energy does. Sunlight keeps pouring into the car through the glass, but heat cannot get out. It just continues to build up, trapped inside the car. A greenhouse like the one shown in Figure 7-11 works the same way. By building a house of glass, gardeners can trap the sun's energy and grow delicate plants in the warm air inside the greenhouse even when there is snow on the ground outside.

Figure 7-11 How is the Earth like a greenhouse?

THE GREENHOUSE EFFECT

The Earth is similar to a greenhouse. We live inside "greenhouse Earth" like delicate plants, surrounded by the icy coldness of outer space. The Earth's atmosphere performs the same function as the glass in a greenhouse. As shown in Figure 7-12, sunlight streams through the atmosphere and heats the Earth. As heat rises, some of it escapes into space. The rest of the heat is trapped by gases in the troposphere and warms the air. This process is called the **greenhouse effect.**

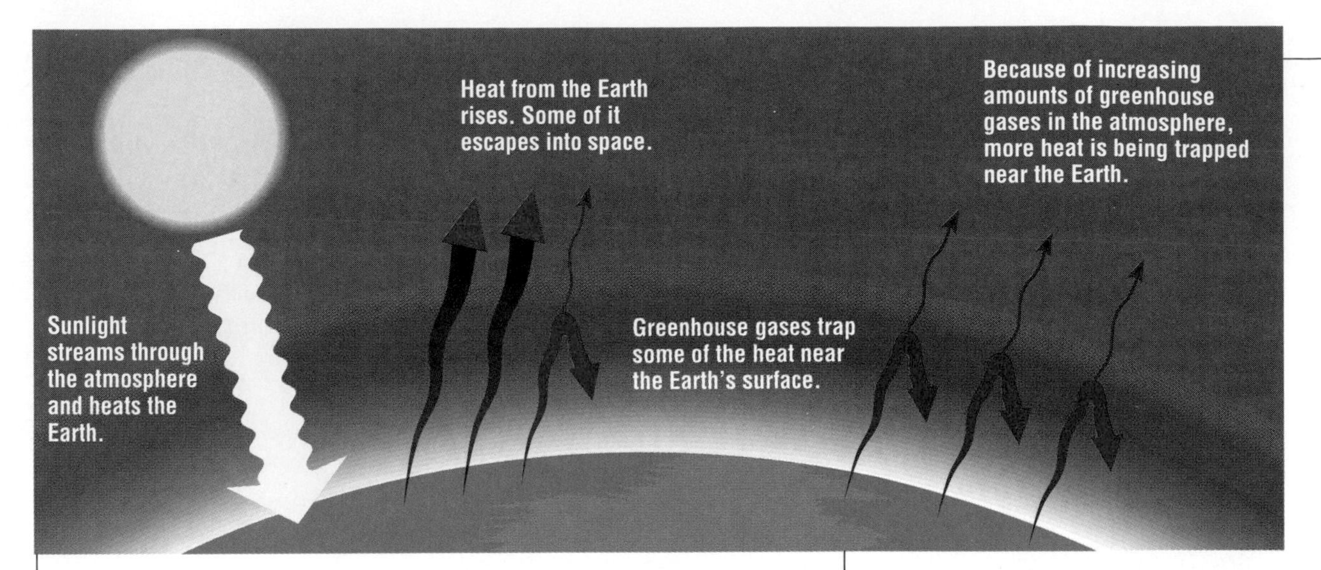

Figure 7-12 How the greenhouse effect works

Not every gas in our atmosphere traps heat in this way. The gases that do trap and radiate heat are called **greenhouse gases.** The major greenhouse gases are water vapor, carbon dioxide, chlorofluorocarbons (CFCs), methane, and nitrous oxide. After water vapor, carbon dioxide is the most abundant of the greenhouse gases.

More Carbon Dioxide in Our Atmosphere

In 1958 a geochemist named Charles Keeling installed an instrument at the top of a tall tower on the volcano Mauna Loa in Hawaii. An absolute perfectionist, Keeling wanted to measure the amount of carbon dioxide in the air very precisely, far away from forests and cities. In a forest, carbon dioxide levels rise and fall with the daily rhythms of photosynthesis. Near cities, carbon dioxide from traffic and industrial pollution raises the local concentration of the gas. The winds that blow steadily over Mauna Loa have come thousands of miles across the Pacific Ocean, far from forests and human activities, swirling and mixing as they traveled. Keeling reasoned that at Mauna Loa, the average carbon dioxide levels in the air could be measured for the entire Earth.

Keeling's first measurement, in March of 1958, was 314 parts per million of carbon dioxide in the air, or .0314 percent. The next month, the levels rose slightly. By summer the levels were falling, but in the winter they rose again. During the summer, growing plants use more carbon dioxide for photosynthesis than they release in respiration. This causes carbon dioxide levels in the air to decrease in the summer. In the winter, dying grasses and fallen leaves decay, releasing the carbon that was stored in them during the summer, and carbon dioxide levels rise.

After only a few years of measuring carbon dioxide levels, it became obvious that they were changing in ways other than just the seasonal fluctuations. Each year, the high carbon dioxide levels of winter were higher, and the summer levels did not fall as low. Figure 7-13 on the next page shows the carbon dioxide levels measured from 1958 to 1994. As you can see, the average

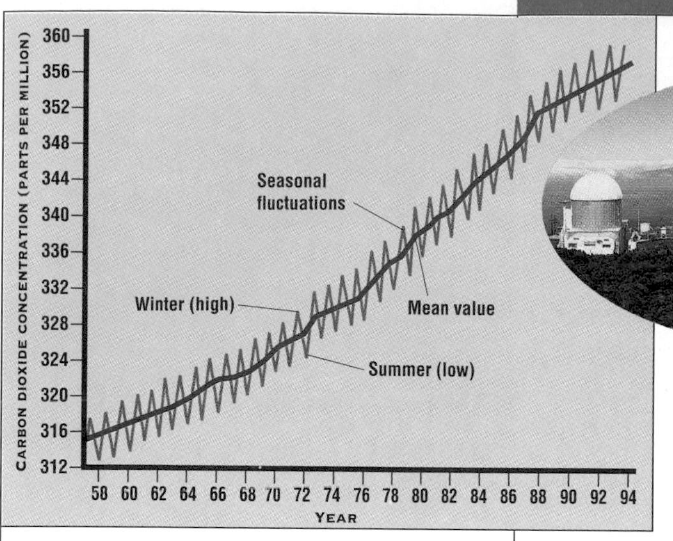

Figure 7-13 The graph shows the increase in carbon dioxide in the air since 1958. The photograph shows the Mauna Loa research station in Hawaii.

amount of carbon dioxide in the air has increased since 1958. By 1994, the average level of carbon dioxide was about 358 parts per million.

Where Is the Extra Carbon Dioxide Coming From?

During photosynthesis, a plant takes in carbon dioxide from the air. Some of the carbon in the carbon dioxide becomes part of the plant's body. This carbon is not returned to the air until the leaves fall or the plant dies and decays.

Some plants, however, never completely decay. Instead, they are covered by sand and silt. After millions of years underground,

Computer Models and Earth's Future Climate

Many Earth-science researchers think that the increase in greenhouse gases will result in a warmer Earth. But exactly how much warmer? Will sea levels rise? How much? How will the climate of Kansas or Hong Kong or Peru be different 50 years from now?

Scientists would like to be able to give definite answers, such as "The average world temperature will be 2.3°C warmer by the year 2050." But they are unable to make such precise predictions because climatic patterns are far too complicated, and too many variables must be taken into account.

Predictions about climatic change are based on computer models. These models are mathematical representations of how different variables affect the Earth's future climate. Scientists write equations representing the atmosphere and oceans, and enter data about prevailing winds, seasonal changes, levels of carbon dioxide, and many other variables. The computer models then predict how phenomena such as temperature, rainfall patterns, and sea levels will be affected.

However, the predictions may vary from one model to another, partly because different scientists may use different values for the variables. For example, one scientist might use one estimate of how much fossil fuel usage will increase in developing countries, and another scientist might use a slightly different estimate. The different estimates would result in different predictions of how much climatic change will occur and how rapidly it will occur.

The construction of climate models is also complicated by the Earth's own feedback processes. For example, as the Earth warms, more water will evaporate from the oceans, and more clouds will form. These clouds could reduce the amount of heat that reaches the ground, which would slow the warming trend. This is an example of a negative feedback process.

On the other hand, positive feedback could increase the rate of global warming. One effect of global warming is the partial melting of the polar icecaps, which is already happening in parts of Alaska. Ice reflects back most of the sunlight that shines on it, rather than changing it into heat energy as darker surfaces do. If the amount of ice near the poles decreases, global warming may speed up.

Computer models are becoming more reliable as more data are available and additional factors are taken into account. Scientists recently discovered, for instance, that warming at the Earth's surface might be reduced by sulfur compounds in air pollution because particles formed by the sulfur the plants become coal, oil, or natural gas. When these fossil fuels are burned, they release the stored carbon as carbon dioxide. Millions of tons of carbon dioxide are poured into the atmosphere each year from power plants that burn coal or oil and from cars that burn gasoline.

The burning of living plants also releases carbon dioxide. This increases the carbon dioxide in the air in two ways. First, a burning plant gives off carbon dioxide. Second, when a living plant is burned, there is one less plant to remove carbon dioxide from the air by photosynthesis. As millions of trees are burned in tropical rain forests to clear the land for farming, the amount of carbon dioxide in the atmosphere increases.

Greenhouse Gases and the Earth's Temperature

Since greenhouse gases trap heat near the Earth's surface, many scientists think that more greenhouse gases in the atmosphere will

Answers to Thinking Critically can be found on page T150.

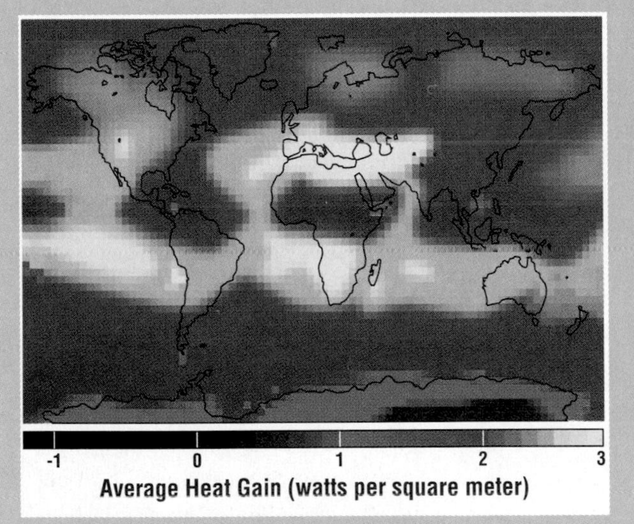

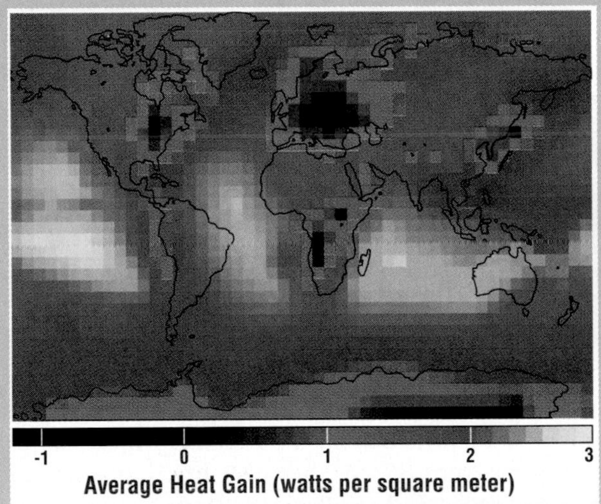

These maps were developed from computer models. The one on the left shows the effect of greenhouse gases on the warming of the Earth before sulfur pollution was factored into the model. The map on the right shows how the addition of the sulfur-pollution variable changed the results.

compounds reflect sunlight back into space. This discovery could explain something that had previously puzzled scientists. Temperatures had been predicted to increase faster in the Northern Hemisphere than in the Southern Hemisphere. However, temperatures have actually been rising faster in the Southern Hemisphere. Some scientists think that the northern areas have not warmed as fast as the southern areas because the sulfur pollution is worse in the industrialized areas of the Northern Hemisphere. The change to one model as a result of this discovery is illustrated in the computer-generated maps above.

Thinking Critically

1. *Identifying Relationships* What variables besides those mentioned in this Case Study might scientists consider as they construct their models of climatic change?
2. *Analyzing Conclusions* Given that scientists use different estimates when constructing models, should we be skeptical about their predictions of global warming? Explain your reasoning.

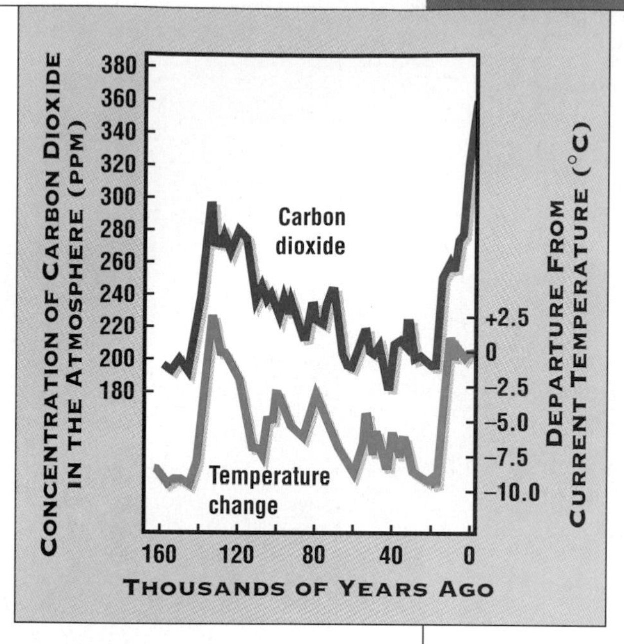

Figure 7-14 This graph shows the correlation between estimates of carbon dioxide levels in the troposphere and estimates of average surface temperatures of the Earth. The estimates were based on fossil evidence, rock strata, and samples of ice formed thousands of years ago.

result in a warmer Earth. A comparison of carbon dioxide levels in the atmosphere and average global temperatures for the past 160,000 years supports that view. As you can see in Figure 7-14, when carbon dioxide levels rose, temperatures also rose.

Today, we are releasing more carbon dioxide into the atmosphere than any other greenhouse gas. But we are also releasing other greenhouse gases—such as chlorofluorocarbons, methane, and nitrous oxide—in significant amounts. Figure 7-15 shows the sources of some major greenhouse gases.

Many scientists think that as a result of increasing greenhouse gases in our atmosphere, the average temperature of the Earth will increase by at least 2°C (about 4°F) by 2050. This predicted increase in temperature is called **global warming.**

A Warmer Earth

The Earth's climate has changed dramatically in the past as the great ice ages came and went. Those changes, however, occurred over hundreds or thousands of years. Scientists are not sure how quickly the Earth will warm or how severe the effects will be. Different computer models give different answers to these questions, as discussed in the Case Study on pages 182–183.

Weather Patterns If the Earth heats up significantly, the oceans will absorb more heat energy, which may make hurricanes and typhoons more common. Some scientists are concerned that global warming will also cause a change in ocean current patterns. Such a change could significantly affect the world's weather. Some regions might have more rain than normal, while others might have less. Severe flooding could occur in some regions at the same time that droughts devastate other regions.

Agriculture Disruptions in weather patterns could hit farmers especially hard. The American Midwest, which includes some of the most productive farmland in the world, is one of the regions that might get hotter and drier. With increasing average temperatures in North America, the weather patterns that are best for farming would shift northward.

Sea Levels As polar regions warm, more icebergs may break off glaciers and melt in the sea. Sea levels would then rise, not only from melting ice, but also because water expands as it warms. As a result of higher sea levels, some coastal areas might be covered with water.

Answer to caption (below): All of them

Figure 7-15 Which of these greenhouse gases are affected by human activity?

Some Major Greenhouse Gases

Greenhouse Gas	Main Sources
Carbon dioxide	Coal, oil, natural gas, deforestation
Chlorofluorocarbons (CFCs)	Foams, aerosols, refrigerants, solvents
Methane	Wetlands, rice, fossil fuels, livestock
Nitrous oxide	Fossil fuels, fertilizers, deforestation

How Much Carbon Dioxide Do You Send Into the Atmosphere?

If you want to figure your own monthly contribution of carbon dioxide to the atmosphere, begin by finding out how much electricity you use. Consult a recent family electric bill to find the number of kilowatt hours used in one month, and divide by the number of people in your household. If your house or water heater is heated with natural gas, do the same with the family gas bill. Using the odometer on the car, find out how many miles a month you drive or are driven somewhere. Then use the following figures for your calculations.

- Burning 1 gal. of gasoline produces 9 kg of carbon dioxide.
- Using one kilowatt hour of electricity from a coal-fired generating plant produces 1 kg of carbon dioxide.
- Burning 100 cubic feet of natural gas produces 5.5 kg of carbon dioxide.

When you have calculated your direct production of carbon dioxide, double it to account for the carbon dioxide produced in manufacturing the products you buy.

Figure 7-16 Perform the calculations shown at left to get a rough estimate of the amount of carbon dioxide you send into the atmosphere.

Slowing the Temperature Change

What can be done to slow global warming? The use of fossil fuels could be reduced so that less carbon dioxide is released into the atmosphere. Additionally, the Earth's existing forests could be preserved, and more trees could be planted. The trees would remove carbon dioxide from the atmosphere. The American Forestry Association has unveiled a new program called Global ReLeaf to encourage the planting of 100 million new trees. These trees would remove about 18 million tons of carbon, in the form of carbon dioxide, from the atmosphere.

Figure 7-17 This volunteer is part of Global ReLeaf, a program designed to remove carbon dioxide from the atmosphere by planting millions of trees.

Answers to Section Review can be found on page T150.

Section Review

1. Why is the air warmer inside a greenhouse than outside it?
2. What are the two main causes of the increase in carbon dioxide levels in the atmosphere?
3. What evidence convinces many scientists that global warming will occur?

Thinking Critically

4. *Analyzing Relationships* Some scientists predict that crop yields will increase in some parts of the world and decrease in others as a result of global warming. Name one factor that would account for this change.

THE OZONE SHIELD

AFTER READING THIS SECTION YOU SHOULD BE ABLE TO

1. explain how the ozone layer shields the Earth from much of the sun's harmful radiation.
2. explain how CFCs are damaging the ozone layer.
3. describe the damaging effects of excessive ultraviolet light.

The stratosphere contains the Earth's ozone shield. **Ozone** is a form of oxygen with molecules made of three oxygen atoms. Ozone in the stratosphere absorbs most of the ultraviolet (UV) light from the sun. UV light is very harmful to living things because it damages or destroys important biological molecules like DNA. By shielding the Earth's surface from most of the sun's ultraviolet radiation, the ozone in the stratosphere acts like a sunscreen for the Earth and its inhabitants.

Figure 7-18 **The CFC molecule in this illustration contains one chlorine atom, which can destroy 100,000 ozone molecules.**

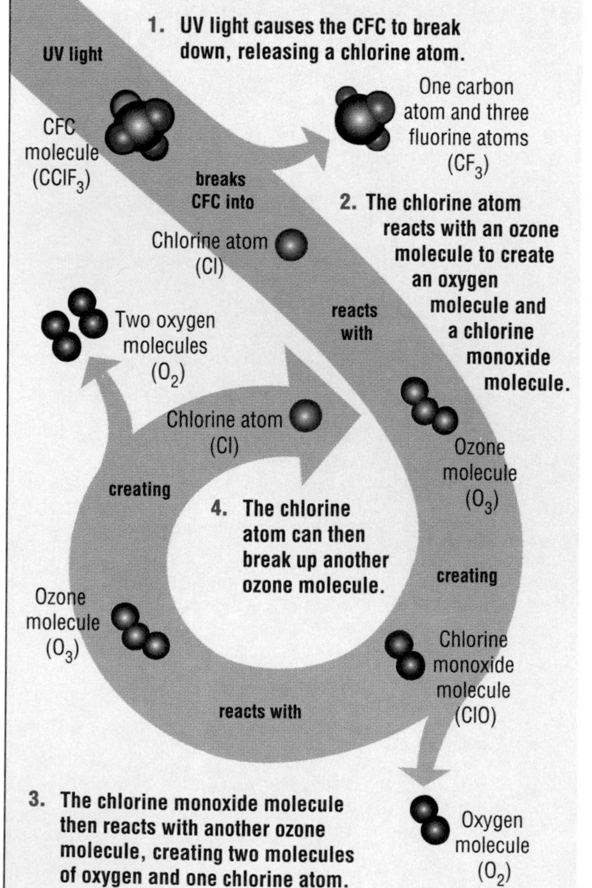

OZONE EATERS

During the 1970s, scientists began to worry that a class of human-made chemicals called **CFCs (chlorofluorocarbons)** might be damaging the ozone shield. For many years, CFCs were thought to be miracle chemicals. They were nonpoisonous and nonflammable, and they didn't corrode metals. CFCs quickly became popular as coolants in refrigerators and air conditioners. They were also used as a gassy "fizz" in making plastic foams such as Styrofoam and were used as a propellant in spray cans of everyday products such as deodorants, insecticides, and paint.

Breaking Apart CFCs At the Earth's surface, CFCs are chemically stable. They don't combine with other chemicals or break down into other substances. But CFC molecules can be broken apart high in the stratosphere, where UV radiation is absorbed. This is because UV radiation is a powerful energy source, powerful enough to break down CFC molecules.

Over a period of 10 to 20 years, CFC molecules released at the Earth's surface can make their way into the stratosphere. If you study Figure 7-18, you will see how the CFCs destroy ozone in the stratosphere. CFC molecules contain from one to four chlorine atoms, and it is estimated that a single chlorine atom can destroy 100,000 ozone molecules.

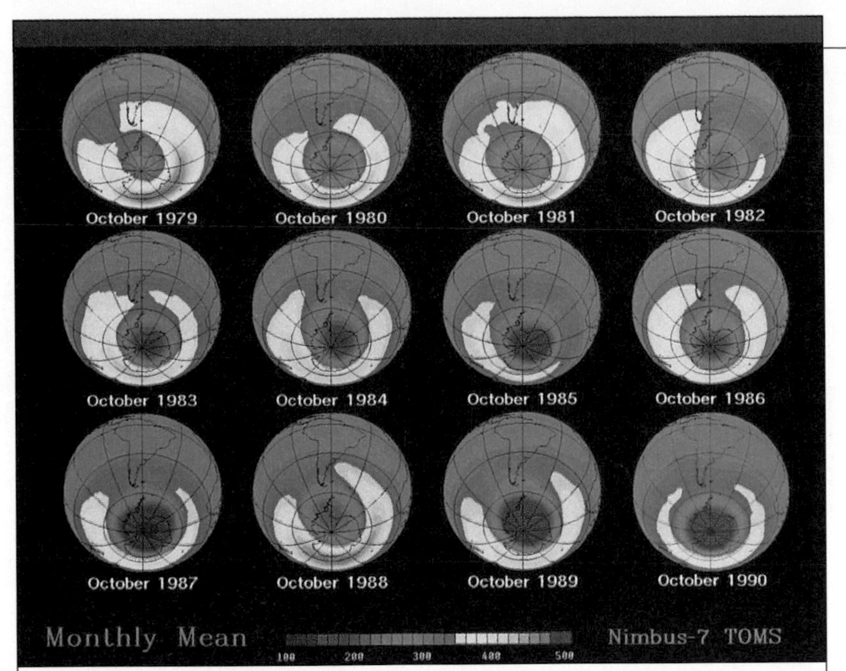

Figure 7-19 In this series of images (left) from the *Nimbus* 7 satellite, the ozone hole is identified by the purple color. An image taken in 1994 (below) by the Russian satellite *Meteor 3* indicates that the ozone hole (the area within the dark blue border) is still growing.

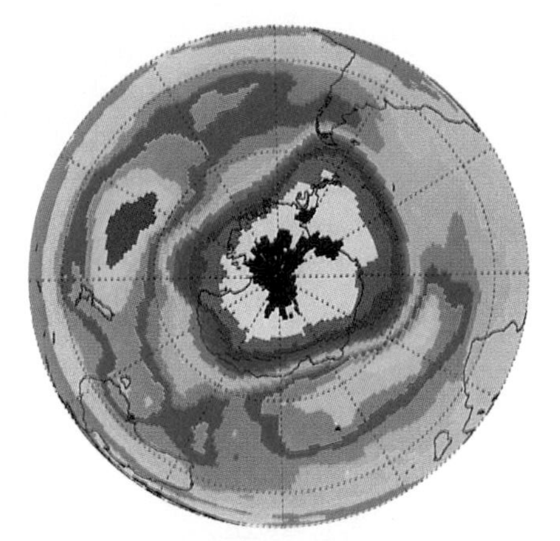

October 1994

THE OZONE HOLE

In 1985 an article in the British scientific journal *Nature* reported the results of studies by scientists working at Halley Bay, on the coast of Antarctica. The studies revealed that the ozone layer above the South Pole had thinned by 50 to 98 percent. This was the first news of the now-famous "ozone hole."

After the results from the Antarctic were published, NASA scientists reviewed data that had been sent back to Earth by the *Nimbus 7* weather satellite since its launch in 1978. They saw the first signs of ozone thinning in the data from 1979. Although the concentration of ozone fluctuates during the year, the data show a growing ozone hole, as shown in Figure 7-19. Scientists are seeing signs that ozone over the Arctic is also decreasing, but so far a large ozone hole like the one over Antarctica has not developed.

You read in Chapter 6 that ozone is being produced as air pollution, so you may be wondering why that ozone doesn't just float up to the stratosphere and repair the ozone hole. The answer is that ozone is very chemically reactive. Ozone produced by pollution breaks down or combines with other substances long before it can reach the stratosphere to replace ozone that is being destroyed.

Susan Solomon has braved polar climates at both ends of the Earth to study the ozone layer. Find out more about her on pages 194–195.

The Effects of Ozone Thinning

As the amount of ozone in the stratosphere decreases, more ultraviolet light is able to pass through the stratosphere and reach Earth's surface. (See Figure 7-20.) Because UV light damages DNA, it can cause mutations or the death of cells near the surface of the body. UV light is a major cause of skin cancer and cataracts. High levels of UV light can also kill one-celled organisms that live near the surface of the ocean. This would disrupt

Figure 7-20 Depletion of the ozone layer allows more ultraviolet (UV) radiation to reach the Earth's surface.

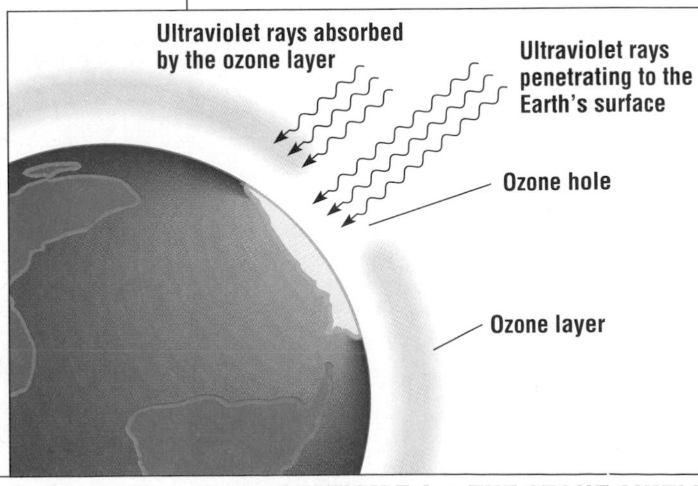

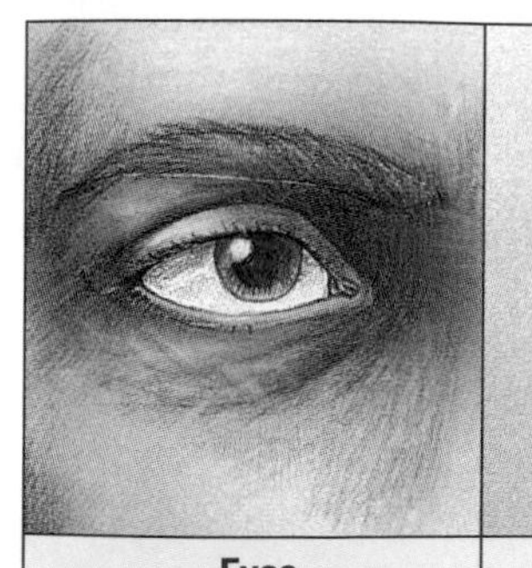	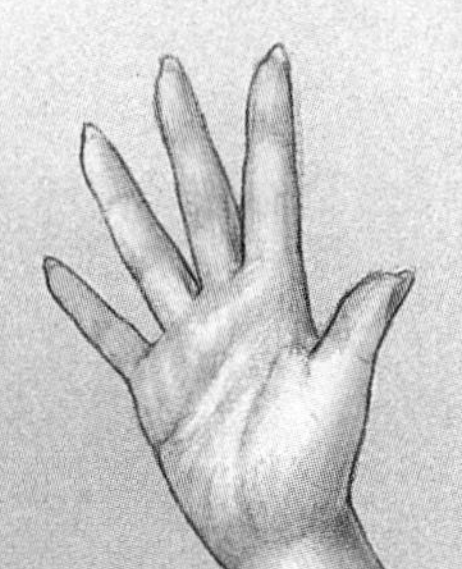			
Eyes Cataracts can develop, causing the lens to cloud up. Result: blurred vision or blindness.	**Skin** Faster aging, wrinkling, skin cancer.	**Immune system** Weaker immune response, making body more susceptible to most diseases.	**Crops** Interference with photosynthesis. Result: lower crop yields.	**Marine life** Kills phytoplankton near surface and disrupts ocean food chain.

Figure 7-21 This table shows some of the adverse effects of increased exposure to UV light as a result of depletion of the ozone layer.

ocean food chains and could reduce fish harvests. In addition, a reduction in the number of microscopic photosynthesizers would further increase the amount of carbon dioxide in the atmosphere. Other damaging effects of excessive UV light are shown in Figure 7-21.

Stopping the Ozone Eaters

In 1992 a conference on the problem was held in Copenhagen, Denmark. The 93 countries represented at the conference reached the following agreements.

- Industrialized countries agreed to eliminate CFCs completely by 1995. The United States pledged to ban by 2000 all substances that pose a significant danger to the ozone layer.
- Industrialized countries also agreed to set up a fund to help developing countries switch to substitutes for CFCs.
- Other substances that destroy ozone were also banned.

Substitutes for CFCs are now being developed, but ozone destruction will occur for years to come. For suggestions on how you can reduce the amount of CFCs released into the air, see Figure 7-22.

Figure 7-22 Individuals can make a difference in the amount of CFCs in the atmosphere.

What You Can Do to Reduce the Amount of CFCs Released Into the Air
1. Look for substitutes for plastic foams.
2. Keep car air conditioner in good repair so that it does not leak CFCs.
3. When having a car air conditioner serviced, make sure the shop captures and recycles the CFCs, as it is legally required to do.
4. When old air conditioners and refrigerators are discarded, find a repair shop that will capture the CFCs for recycling before the appliance is hauled off to the landfill. Call your local utility company for advice.

Answers to Section Review can be found on page T150.

Section Review

1. What is ozone? How does it act as a "sunscreen" for the Earth?
2. What are the main sources of CFCs in our atmosphere?

Thinking Critically

3. *Interpreting Graphics* Consult Figure 7-18 to answer the following question. Why is chlorine destructive to the ozone layer?
4. *Recognizing Relationships* If the ozone layer gets significantly thinner during your lifetime, what changes might you need to make to your lifestyle?

HIGHLIGHTS

SUMMARY

- The atmosphere is a thin layer of gases that surrounds the Earth. This mixture of gases is called "air."
- The oxygen in air is released by organisms that break apart water molecules during photosynthesis. The balance between photosynthesis and cellular respiration keeps the amount of carbon dioxide in the atmosphere constant.
- Climate is affected by latitude, wind and ocean currents, and local geography.
- Greenhouse gases trap heat in the troposphere. Called the greenhouse effect, this process keeps the Earth warm enough for life to exist.
- Burning fossil fuels and clearing forests are causing a rapid rise in carbon dioxide levels in our atmosphere. Some scientists predict that this will increase the greenhouse effect, causing changes in global weather patterns.
- Ozone in the stratosphere absorbs most of the UV light that comes to Earth. Ozone is destroyed by chlorine atoms released when chlorofluorocarbons (CFCs) are broken down.

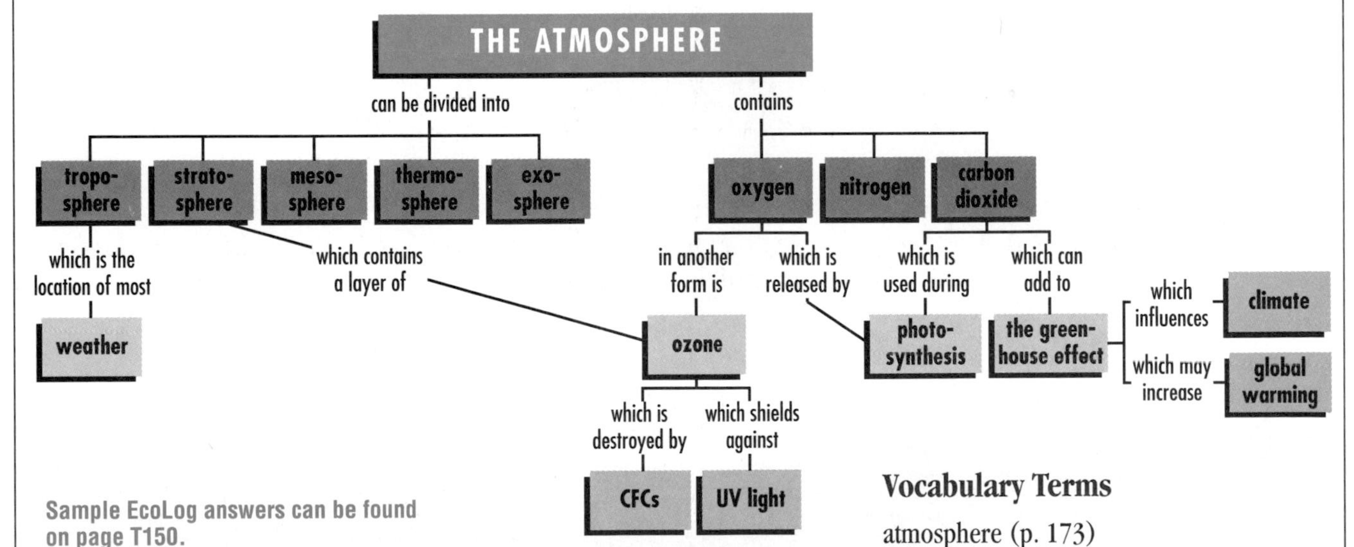

Sample EcoLog answers can be found on page T150.

EcoLog

Now that you've studied this chapter, revise your answers to the questions you answered at the beginning of the chapter, based on what you have learned.

❶ Why is the warming of the Earth by carbon dioxide and other gases called "the greenhouse effect"?

❷ Why are some scientists concerned about the loss of ozone from the ozone layer?

Vocabulary Terms

atmosphere (p. 173)
CFCs (chlorofluorocarbons) (p. 186)
climate (p. 176)
global warming (p. 184)
greenhouse effect (p. 180)
greenhouse gases (p. 181)
ozone (p. 186)
stratosphere (p. 175)
troposphere (p. 175)
weather (p. 176)

CHAPTER 7

REVIEW

Understanding Vocabulary

1. For each pair of terms, explain the differences in their meanings.
   a. weather
      climate
   b. atmosphere
      ozone
   c. greenhouse effect
      global warming
   d. troposphere
      stratosphere

Relating Concepts

Answers not indicated can be found on page T150.

2. Copy the unfinished concept map below onto a sheet of paper. Then complete the concept map by writing the correct word or phrase in each box containing a question mark.

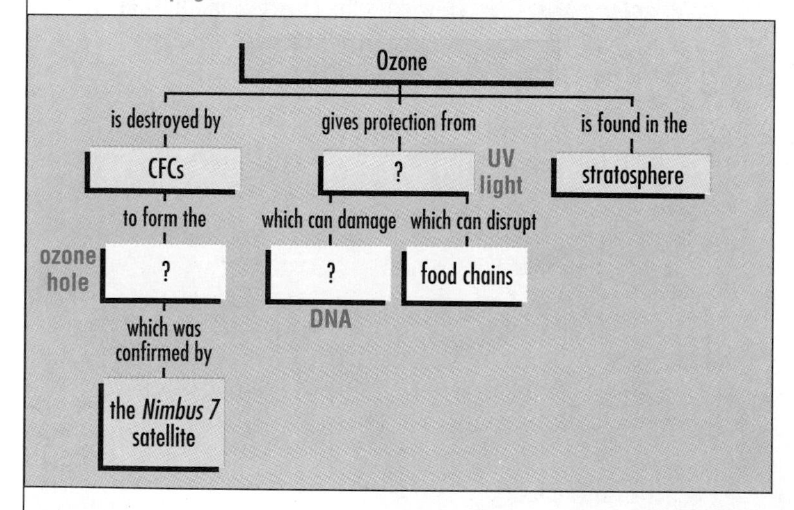

Understanding Concepts

Multiple Choice

3. The Earth's early atmosphere probably contained very little
   a. hydrogen.
   (b.) oxygen.
   c. methane.
   d. ammonia.

4. Today, 99% of the Earth's atmosphere is made up of
   a. hydrogen and oxygen.
   b. nitrogen and methane.
   c. oxygen and carbon dioxide.
   (d.) oxygen and nitrogen.

5. Carbon dioxide makes up about how much of the Earth's atmosphere?
   (a.) 0.03%
   b. 3.0%
   c. 10%
   d. 30%

6. Which is NOT a greenhouse gas?
   a. water vapor
   b. carbon dioxide
   c. methane
   (d.) nitrogen

7. At what time of year does carbon dioxide in the atmosphere decrease as a result of natural processes?
   (a.) summer
   b. fall
   c. winter
   d. The amount of carbon dioxide stays constant year-round.

8. How is climate related to weather?
   a. They are the same thing.
   (b.) Climate is the average weather in an area over a long period of time.
   c. Climate is simply the weather conditions at a particular time in a particular location.
   d. They have no relation to each other.

9. Which factor has the MOST influence on a region's climate?
   a. number of forests
   b. distance from the ocean
   c. number of mountain ranges
   (d.) distance from the equator

10. As a result of global warming, the agricultural regions of the midwestern United States are expected to
    a. become more like the tropics—hotter and wetter.
    b. become more like the southwestern United States—hotter and drier.
    c. become more like New England—cooler and wetter.
    d. change very little.
11. A chlorofluorocarbon molecule contains from 1 to 4 chlorine atoms. About how many ozone molecules can 1 chlorine atom destroy?
    a. 1
    b. 4
    c. 100,000
    d. 400,000
12. Which is NOT an adverse effect of high levels of UV light?
    a. increased incidence of skin cancer
    b. increased photosynthesis
    c. disruption of ocean food chains
    d. increased amount of atmospheric carbon dioxide

Short Answer

13. Explain how global warming could lead to both floods and droughts.
14. Why did Charles Keeling decide that Mauna Loa would be a good place to get accurate measurements of how much carbon dioxide is in the Earth's atmosphere?
15. Some scientists predict that sea levels will rise due to global warming. Explain why a warmer Earth might cause sea levels to rise.
16. How did NASA confirm reports of drastic thinning of the ozone layer over Antarctica?

Interpreting Graphics

17. **Examine the graph below.** It shows the average monthly temperature for two locations, both at the same latitude but in different parts of the United States. Which location has the smallest temperature range between summer and winter? What could cause this difference in climate between the two locations?

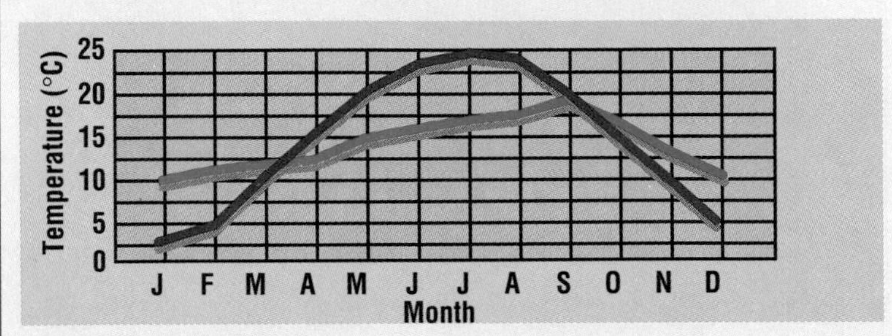

Thinking Critically

18. ***Making Predictions*** Over a long period of time, how might living things adapt to increased carbon dioxide levels and global warming? Do you think that most species will adapt, or are many species likely to become extinct? Explain your reasoning.

Themes in Science

19. ***Interacting Systems*** How does the clearing and burning of large regions of tropical rain forest increase the amount of carbon dioxide in the atmosphere?

Cross-Discipline Connection

20. ***Economics*** Insurance companies set some of their rates by estimating the number of destructive natural events, such as hurricanes and floods, that will occur in the next 20 years. Explain why insurance companies would be interested in knowing scientists' predictions about global warming for the next two decades.

Discovery Through Reading

For a discussion of the impact of global warming on public health issues, such as the incidence of heat-related illnesses and the availability of clean water, read "Global Warming: The Controversy Heats Up" in the January 1994 issue of *Current Health,* pp. 6–11.

PORTFOLIO ACTIVITY

Design a pamphlet explaining how car owners can help reduce the amount of CFCs that are released into the atmosphere by properly maintaining their air-conditioning systems. You might collect information for your pamphlet by calling auto-repair shops that service car air conditioners. Be sure to include an introduction to your pamphlet that explains why reducing CFCs is important. Distribute your pamphlet to your fellow students, and include it in your portfolio.

CHAPTER 7

INVESTIGATION

GLOBAL WARMING IN A JAR

How can we learn more about global warming? Since we only have one Earth, we can't experiment with it. However, we *can* experiment with a model. In this investigation, you will construct a simple "greenhouse" in a glass jar to model the warming of the Earth. Then you will choose a question about global warming that interests you and design and carry out your own experiment to answer it.

Answers to this Investigation can be found on page T151.

MATERIALS

- 2 large glass jars of the same size and shape
- 2 outdoor thermometers
- small pieces of thin cardboard
- tape
- clear plastic wrap
- rubber bands or string
- watch or clock
- graph paper
- straightedge or ruler
- notebook
- pen or pencil
- one or more of the following items: black and white latex paint, water, soil, sod or other plants, ice cubes

DEMONSTRATE THE GREENHOUSE EFFECT

1. In your notebook, make a table similar to the one shown.

Data Table for Greenhouse Model

Time	Temperature	
	Uncovered Jar (control)	Covered Jar ("greenhouse")
Before going outdoors		
When first placed in sun		
+ 2 minutes		
+ 4 minutes		
+ 6 minutes		
+ 8 minutes		
+ 10 minutes		

Which of these jars is the greenhouse model, and which is the control? What is the purpose of the control?

2. Tape thin pieces of cardboard over the bulbs of two weather thermometers so that the sun will not shine directly on them. Place one thermometer in each of two jars, taping them to the inside of the jars so that they are not touching the bottoms. Record the temperature inside each jar.
3. Cover the opening of one of the jars with plastic wrap, and secure it with rubber bands, string, or tape. This jar will be a model for "greenhouse Earth," while the uncovered jar will serve as a control.
4. Take the jars outdoors and place them in a bright sunny spot.
5. Every 2 minutes, read both thermometers. Always read the same thermometer first. Record the temperatures in your data table. Continue for 10 minutes or more, until the temperature increases noticeably in both jars.
6. Make a line graph that shows your results.

CHOOSE A VARIABLE TO TEST

7. As you read in Chapter 7, many scientists believe that the Earth will warm up as we add greenhouse gases to the atmosphere. However, not all parts of the Earth are expected to warm up by the same amount. Consider the following questions about how the warming effect may vary over different parts of the Earth:
   - Will the air heat up faster over snow-covered plains or newly plowed fields?
   - Which will heat up faster, air over the ocean or air over land?
   - Does the presence of plants affect the rate of warming?
   - How does the amount of moisture in the soil affect the rate of warming?
   - Does the temperature of the land affect the temperature of the air above it?

   Choose one of these issues to investigate. In your notebook, write a hypothesis about the issue you chose.

DESIGN AN EXPERIMENT

8. Think of an experiment you could do to test your hypothesis, using the same kind of jars you used to demonstrate the greenhouse effect.

   Choose materials from the following list for your experiment:
   - black and white latex paint
   - water
   - wet and dry soil
   - sod or other plants
   - ice cubes

   In your notebook, describe your proposed experiment. Remember to include a control jar as well as an experimental jar.
9. Based on your hypothesis, predict the results that you expect. Write your prediction in your notebook.

Which of these materials could be used to test your hypothesis about what affects the rate of global warming?

TEST YOUR PREDICTIONS

10. In your notebook, make a table for recording your data.
11. Set up and carry out your experiment, recording your results in your data table. If you do anything differently from the way you planned your experiment, note the changes that you made.
12. Graph your results.

ANALYZE YOUR RESULTS

13. Did your results agree with your predictions? What does this tell you about your hypothesis?
14. Compare your results with those of your classmates. Which variables had the greatest influence on the rate of warming in a greenhouse? Which had the least influence?
15. In what way did your experiment model the way in which global warming may occur at different rates in different parts of the world? What are some differences between your model and the real "greenhouse Earth"?

Making a DIFFERENCE

OZONE Scientist

Susan Solomon will not soon forget crawling across the roof of an Antarctic field station in windchill temperatures of –62°C (–80°F), moving heavy equipment and adjusting mirrors while the winds howled and whipped about her. Sounds like real adventure, right? It sure is! But it's just part of what Susan Solomon has done to establish herself as one of the world's leading authorities on ozone destruction.

Q: Is it true that you have traveled to the ends of the Earth to get information about the ozone layer?

Susan: Yes, I guess it is. My colleagues and I have studied the ozone hole in Antarctica, and we've measured and documented ozone chemistry above Greenland. But it's not all adventure. When I'm not visiting one of the poles, I run computer simulations of the atmosphere and study data at the National Oceanic and Atmospheric Administration (NOAA) in Boulder, Colorado.

Since our findings were announced, many of the world's countries have agreed to restrict or ban the use of CFCs.

Q: What is the significance of recent discoveries regarding the ozone hole?

Susan: Before British scientists discovered the ozone hole in Antarctica, no one was sure about ozone changes in the atmosphere. The popular belief was that in 100 years there might be 5 percent less ozone. So there were questions about whether it was a serious environmental problem. But when the British researchers released data that showed *50 percent* less ozone over Antarctica in 1985 than was present 20 years earlier—it raised our awareness that the problem was far more serious than previously thought.

Q: How have you contributed to "ozone science"?

Susan: Well, when the British data was first released, no one had much of an explanation about what was causing the destruction of the ozone layer. I thought about it a lot. Later that year, I sat in on a lecture

Susan Solomon has braved freezing polar temperatures to gather data about the ozone layer. "It's not so bad," she says. "After all, I grew up in Chicago!"

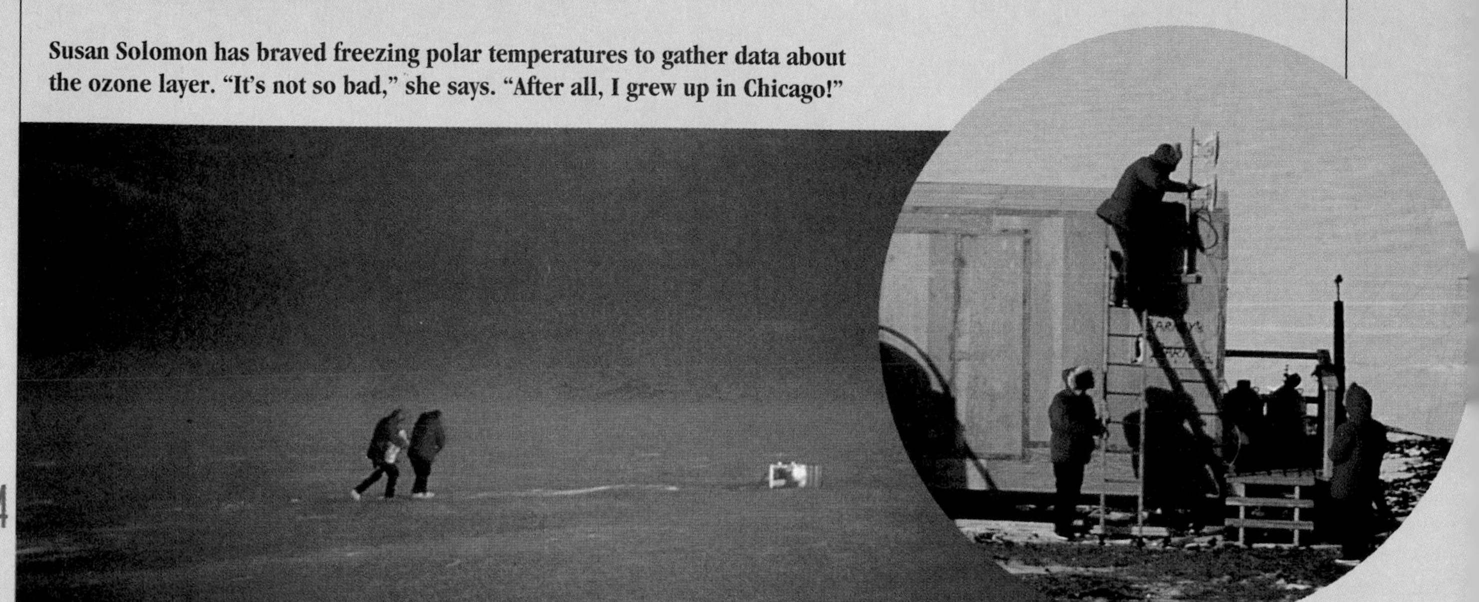

Clouds like these led Susan Solomon to make her important discoveries about the cause of ozone depletion.

about a type of cloud called polar stratospheric clouds. These are beautifully colored clouds that are known for their iridescence. While I was looking at these clouds, which are common in the Antarctic but rare elsewhere, it occurred to me that they may have something to do with ozone depletion. Perhaps they provide a surface for chemical reactions that activate reactive chlorine from CFCs (human-made chlorofluorocarbons). If so, once activated, the chlorine could contribute to reactions that destroy ozone.

Q: Did you get the chance to test your hypothesis?

Susan: Yes. The next year the National Science Foundation chose me to lead a group of 16 scientists for a nine-week expedition in Antarctica. We were the first scientific team from the United States sent to the Antarctic to study the ozone hole. Within one month we could see that unnaturally high levels of chlorine dioxide did occur in the stratosphere during ozone depletion. This discovery was very exciting because it seemed that we were on the right track. We kept collecting data that year and collected more data during a second trip the next year. Pretty soon, the evidence seemed to support my hypothesis that CFCs and ozone depletion are linked.

Q: How did this discovery make you feel?

Susan: On the one hand it's very exciting scientifically to be involved in something like this. On the other hand, sometimes I think it's a little depressing. It would be nice to be involved in something more positive, to bring people *good* news. So far, we've brought nothing but bad news. We were hoping that we wouldn't find the same ozone chemistry in the Arctic that we found in the Antarctic. Unfortunately, we did. We hope for a positive result for the planet, but we don't always get it.

Q: How has your research helped to make a difference in our world?

Susan: Since our findings and others were announced, many of the world's countries have agreed to restrict or ban the use of CFCs. As a result, the ozone hole will eventually go away, but it will take a very long time. So although most countries have slowed their use of CFCs in the last few years, CFCs from years past will still be hanging around in our atmosphere for the next 50 to 100 years. But I think our work has led in a small way to the realization that our actions do have consequences, and this realization should bring positive change.

For More Information

If you would like free information about the ozone layer and what you can do to protect it, contact the following organizations. **Environmental Defense Fund,** 257 Park Avenue South, New York, NY 10010. **Environmental Protection Agency,** Public Outreach, 401 M Street SW, Washington, D.C. 20460.

The ozone hole can be seen in this satellite image. It is the pale blue and black region immediately above Susan's shoulder.

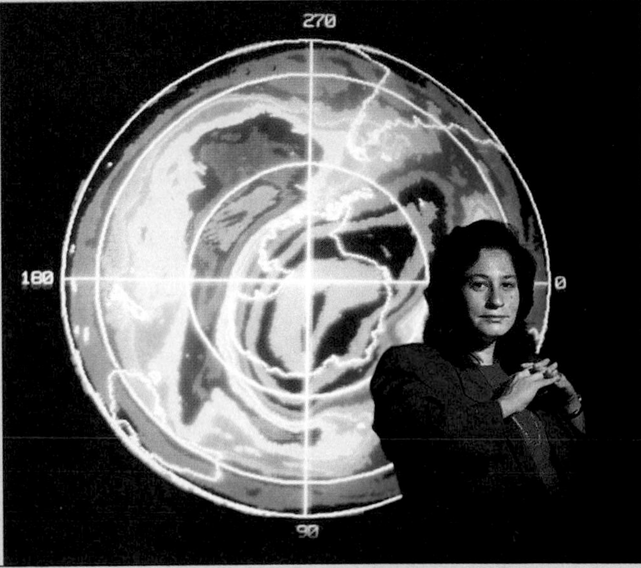

CHAPTER 8

LAND

"The size of the parcel of land matters less than the relationship of the people to it."

FRANCES MOORE LAPPÉ AND JOSEPH COLLINS
AMERICAN AUTHORS

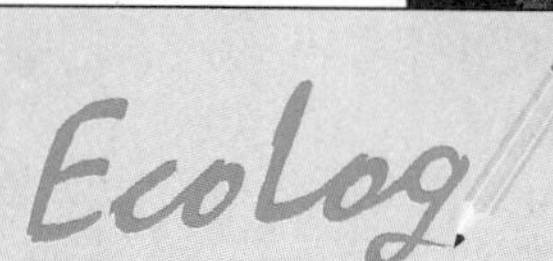

Before you read this chapter, take a few minutes to answer the following questions in your EcoLog.

1. Do you think your lifestyle affects the land in any way? Explain.
2. Do you think humans should plan how to use land, or simply let things work out as they will?

SECTION 8.1

THE CITY

AFTER READING THIS SECTION YOU SHOULD BE ABLE TO

❶ define suburban sprawl and explain why it is considered a problem.

❷ describe the urban crisis and explain what city planners are doing to relieve it.

In 1982 officials in California decided to find out just how land was being used in their state. Over a period of several years, careful research was conducted using maps, aerial photographs, field surveys, and a computerized mapping system. The results were startling—in just eight years (between 1984 and 1992) nearly 84,000 hectares (about 210,000 acres) of productive farmland, rangeland, and forests were converted into towns and cities.

Figure 8-1 Development at the edge of Danville, California, is replacing farmland and rangeland.

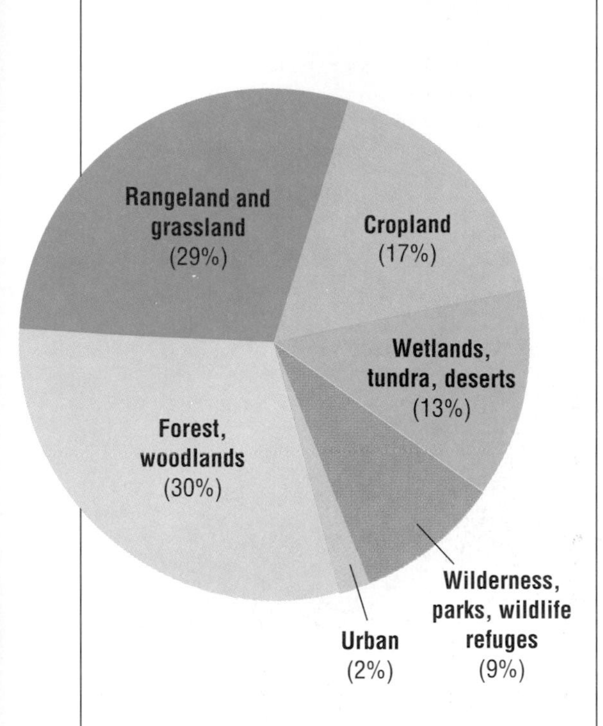

Figure 8-2 Land use in the United States

These findings have raised some serious concerns. How can California continue to be the nation's agricultural leader if its prime farmland is being transformed into cities? What if the same trend is occurring in other states and even in other countries? Obviously, if such trends continue, the world's agricultural land may be reduced to dangerous levels. Likewise, available forest lands and rangelands may be reduced to levels below what are required to serve the needs of our population.

The Urban-Rural Connection

The U.S. Census Bureau describes an urban area (or a city) as a place with at least 5,000 people per square mile. In Figure 8-2 you can see that urban lands actually take up only a small portion of this country's total land space. But almost 80 percent of Americans live in urban areas, and by the year 2000 about half of the world's population will probably live in cities.

Those urban dwellers draw heavily from the resources of their surrounding environment. In fact, cities and suburbs couldn't exist without the abundant resources of the nation's forests, croplands, rangelands, and other areas. So it is important that non-urban lands be cared for in a sustainable manner.

Figure 8-3 This pasture in Butte County, California, is an example of rural land. The cattle shown may eventually be used to produce beef or leather products that city dwellers could use.

URBANIZATION

Until about the 1850s, most people lived in villages and worked the surrounding land. Those people who were not farmers managed the forests, worked in local mines or mills, or manufactured the necessities of life for the village.

Today, modern machinery and methods have made it possible for farms to be operated by fewer people. In addition, efficient transportation networks have eliminated the need for manufacturers to locate near their customers. As a result, the number of jobs in the countryside has fallen, and people have moved to cities in search of work. The movement of people from rural areas to cities is known as **urbanization.** Urbanization occurred rapidly in developed countries between 1880 and 1950. Now it is occurring most rapidly in developing countries.

Figure 8-4 **A skyscraper could be viewed as a symbol of big-city life. It is only through advances in machinery and transportation that its existence is possible.**

The Urban Crisis

Not surprisingly, many countries have had trouble coping with their rapidly growing urban populations. People sometimes migrate to the cities faster than jobs become available for them and before an adequate infrastructure can be established to support them. **Infrastructure** is all of the things that a society builds for public use. Infrastructure includes roads, sewers, railroads, bridges, canals, fire and police stations, schools, libraries, hospitals, water mains, and power lines.

When more people live in a city than its infrastructure can support, living conditions deteriorate and increasing numbers of people become homeless. This problem has become so widespread, throughout the developed and developing world alike, that the term **urban crisis** was coined to describe it. According to the United Nations, the crisis is so bad that almost one-fourth of the world's city dwellers could be homeless by the year 2020.

Suburban Sprawl One consequence of the urban crisis is a phenomenon called suburban sprawl. **Suburban sprawl** is the expansion of a city's borders. Why do people move from cities to the suburbs? Generally, the suburbs offer more living space for less money, lower crime rates, and more privacy. With a car or two, a family can live in the suburbs and still work in the city. As of 1992, more Americans lived in suburbs than in cities and the countryside combined.

Each year suburbs spread over another 1 million hectares (2.5 million acres) of land in the United States. Most of the development that California officials recorded in their land-use study was due to suburban sprawl. The Los Angeles area, shown in Figure 8-5, has the largest and fastest-growing area of suburban sprawl in the world. To accommodate this growth, houses, schools, shopping centers, and roads are being built on agricultural and forest land.

Figure 8-5 Suburban sprawl around Los Angeles, California

LAND-USE PLANNING

When suburban sprawl began to take over the countryside surrounding Washington, D.C., in the 1960s, a commission was established to develop a growth plan for the region. The plan called for new cities about every 6.4 km (4 mi.) along a series of transportation corridors. These corridors would extend from the city like the spokes in a wheel, as seen in Figure 8-6.

The plan was called a "wedges-and-corridors" system because it integrated corridors of highly developed cities with wedges of non-urban, open spaces that allowed for recreation, agriculture, and the conservation of natural resources. In 1964 Montgomery County, Maryland, was the only county of those shown in Figure 8-6 to adopt the wedges-and-corridors plan. This plan helped county officials successfully anticipate and develop infrastructure for a population that grew dramatically in the next few decades.

As a result of the plan, Montgomery County has established itself as a leader in land-use planning. **Land-use planning** involves determining in advance where people will live, where they will locate their businesses, and where land will be protected for farming, wildlife, recreation, and other uses. Land-use planners also determine the best locations for roads, shopping malls, sewers, landfills, electrical lines, and other infrastructure.

Montgomery County had the insight to plan for future urban growth. But many other cities have been overwhelmed by rapid growth before they could develop land-use plans. What can be done to improve these cities? The use of mass transit, inner-city renovation, and open spaces are a few of the many possible solutions.

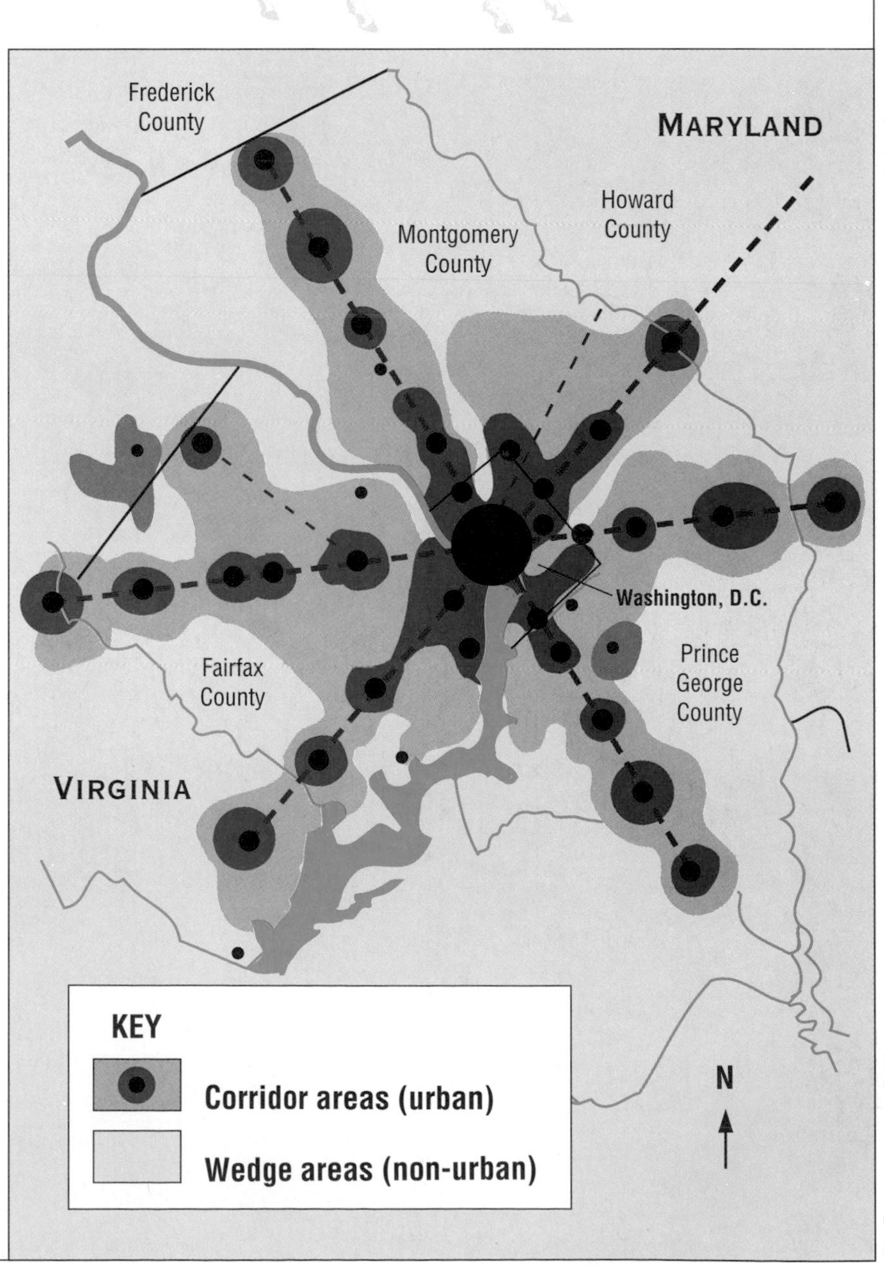

Figure 8-6 This plan integrates corridors of urban development with wedges of open spaces in an attempt to control suburban sprawl from the Washington, D.C., area. Montgomery County was the only county to actually follow through with this plan.

ECO-FACT

In Japan, the bullet train Nazomi reaches speeds of 269 km/h (168 mph). A similar train in France has a top speed of 510 km/h (319 mph).

Mass Transportation

Ask any urban dweller to name the number one annoyance of big-city life and the answer is likely to be "traffic." The roads of every major city in America are clogged with cars, most of which are carrying only one person. Because cars are so convenient, most Americans use them to travel everywhere. But this convenience has a cost. In 1993 the average American spent about $1,700 in gasoline and car-maintenance costs just to travel to and from work. This did not include the cost of parking, which can be steep in cities with limited parking space. By contrast, the average European spent half that amount to commute to work. Why the difference? Most Europeans use mass transit.

Mass transit is an economical, efficient alternative to the automobile. It is energy-efficient and reduces highway congestion, air pollution, and the loss of land to roadways and parking lots. Cities served by mass transit can be more compact, reducing suburban sprawl. Furthermore, studies have shown that mass-transit systems can help revitalize decaying urban areas. With an efficient mass-transit system, traffic congestion is usually reduced and a city becomes more livable.

Inner-City Renovation

As more people leave the cities for the suburbs, businesses follow them. Without the large sums of money that the original businesses were contributing to the community, many city areas fall into disrepair. However, cities are now attempting to renovate run-down areas. Some of these urban-renewal projects have been very successful. Figure 8-7 shows how Baltimore renovated its deteriorated waterfront.

Figure 8-7 **The successful renovation of this waterfront in Baltimore, Maryland, has encouraged other cities to invest in improving their urban centers.**

Figure 8-8 Open spaces help make urban areas more livable.

For the most part, the most successful urban-renewal projects have been those that use a combination of government grants, tax incentives, and other assistance to induce private citizens to renovate an area. A recent innovation of this type encourages residents in low-income housing projects to manage their own properties. These programs usually work well because, when planning, private citizens tend to understand the practical needs of their community better than a planner who is far removed from the community.

Open Spaces

As mentioned earlier, the "wedges" of open space that Montgomery County officials integrated into their land-use plans serve a valuable purpose. In fact, open spaces can help alleviate many of the problems experienced in crowded urban areas. Open spaces come in many forms, including parks, pools, stretches of bicycle and hiking trails, historical settings, gardens, and agricultural areas.

Open spaces have numerous benefits. They give people a place to escape the crowded, noisy conditions that can make big-city life so stressful. They can also be used for civic activities such as concerts, or simply for socializing. Open spaces serve valuable environmental functions as well. The greenery of open spaces absorbs carbon dioxide, produces oxygen, and filters out pollutants. Green spaces can even help keep a city cool in the summer.

Another important function of open spaces, especially those with vegetation, is to reduce drainage problems. Cement does not absorb water runoff from rain or melting snow, but open spaces with grasses and soil can absorb water. So by increasing the amount of open space in a city, flooding may be reduced.

Section Review answers on page T151.

Section Review

1. Explain why land that cities occupy is not a true reflection of the actual amount of land required to sustain them.
2. In the last 150 years, many people have moved from small rural communities to huge urban centers. Why?
3. What is the urban crisis? What might be done to relieve it?

Thinking Critically

4. *Making Decisions* In your opinion, should laws limit suburban sprawl? Why or why not?

SECTION 8.2

HOW WE USE LAND

AFTER READING THIS SECTION YOU SHOULD BE ABLE TO

❶ describe the usefulness of non-urban lands to humans.

❷ explain how logging, ranching, and mining activities affect the land.

❸ explain how lands can be logged, grazed, and mined sustainably.

As the human population grows, ever-increasing amounts of land and resources are needed to support it. Most of these resources come from the world's non-urban, or rural, lands.

Non-urban lands include forests from which we harvest timber for paper, furniture, and home construction. These lands also include rich grasslands that support livestock and can be used for farmland. Tremendous stores of mineral resources that power our engines and become part of our skyscrapers and electronic devices

Figure 8-9 This footbridge is part of the McKenzie River National Recreation Trail in Willamette National Forest, Oregon. Non-urban lands such as this can provide a surprising number of uses to humans. Unfortunately, humans sometimes overuse and degrade these lands.

also come from non-urban lands. But as our population grows, non-urban land areas are put under greater stress. Often, the usefulness of these areas is even destroyed by overuse.

Harvesting Trees

Trees are harvested to provide products we use every day, such as paper, furniture, and lumber and plywood for our homes. We all use enormous amounts of wood. The worldwide average is 1,800 cu. cm of wood per person each day. People in the United States use 3.5 times the per-capita average for the rest of the world. This is the equivalent of each person in the United States cutting down a 30-m-tall tree every year.

Harvested trees also provide firewood for many people. In fact, about 1.5 billion people in developing countries depend on firewood as their major source of fuel.

In some places, it's not the trees but the forest land that is valuable—the trees are removed to make way for farming or ranching enterprises. In developing countries, poor city dwellers sometimes move into forested regions, cut down the trees on a small parcel of land, and farm or ranch the area. By building roads into the forests and then letting people claim property rights to land that they clear, governments offer many people hope for a better future. This is similar to what happened in the United States during the settlement period.

Figure 8-10 Logging operations, such as this one in Washington, provide resources that are important in construction and manufacturing.

Deforestation Today forests around the world are being cleared at an alarming rate, and deforestation has become a serious environmental problem. **Deforestation** involves clearing trees from an area without replacing them. Today this situation is especially serious in tropical rain forests, which exist primarily in developing countries. Because the poor soil in tropical rain forests can usually support crops for only a short time, farmers must continuously move from one parcel of land to another, clearing additional forest with each move. The problem is made worse by the fact that human populations are growing so rapidly in developing countries.

There are many different methods of harvesting trees. Most methods cause damage to the forest, and some are more destructive than others. **Clear-cutting,** for example, is a process that involves removing all of the trees from a land area. The top illustration in Figure 8-11 shows a clear-cut area. Clear-cutting destroys wildlife

Each year 1.8 million hectares (4.5 million acres) of forest are cut down worldwide.

habitats, increases soil erosion, and diminishes the beauty of forests. Clear-cutting is popular because it is the least expensive way for timber companies to harvest trees. In addition, clear-cutting requires very little road building to harvest a large number of trees.

The main alternative to clear-cutting is **selective cutting,** shown in the bottom illustration of Figure 8-11. Selective cutting involves cutting only middle-aged or mature trees. These trees are removed from the forest individually or in small groups, and the rest of the trees are left alone. With time, the harvested trees are naturally replaced by reseeding. Selective cutting has less of an impact on the forest environment than any other method of tree harvesting. However, selective cutting requires more roads to harvest a given amount of timber than does clear-cutting or other cutting methods. Roads are the primary cause of soil erosion associated with logging. If not properly controlled, both clear-cutting and selective cutting can increase soil erosion, damage wildlife habitats, and cause other environmental problems.

Reforestation When trees die or are removed from a forest, reforestation helps restore the area to its original condition. **Reforestation** is the process of replacing trees that have died or been cut down. This can happen naturally when seeds fall from nearby trees, or it can happen when humans plant seeds or seedlings.

Sometimes when humans have harvested a large number of trees from an area, natural reforestation cannot occur. For example, if a very large area is clear-cut, there may be no way that the

Figure 8-11 Each of these methods of harvesting timber has benefits and drawbacks.

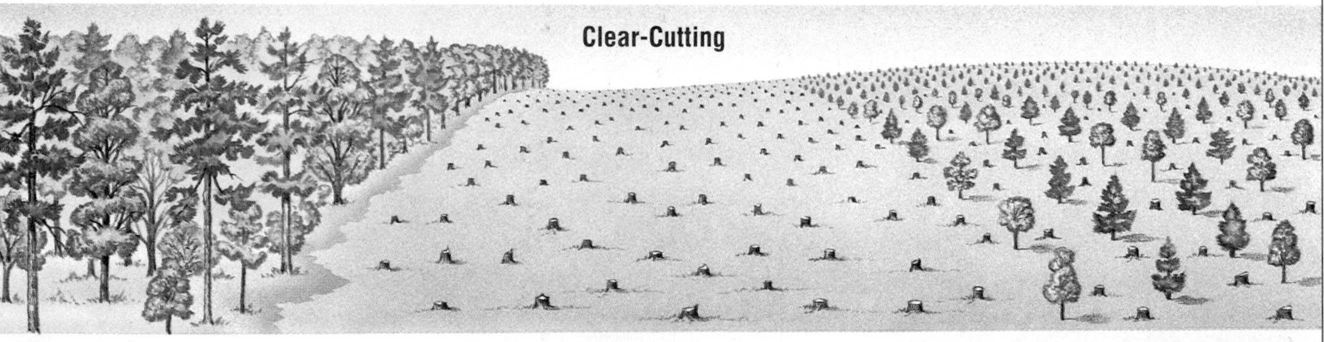

mature trees bordering the area can provide seed for the natural regeneration of the entire area. In addition, reforestation may fail because tree seedlings cannot grow where clear-cutting has caused soil erosion or where the seedlings will not be shaded by other trees.

Some governments require reforestation after timber is removed from public land. But, worldwide, more than 90 percent of all timber comes from forests that are not managed by the government. This means that some countries are still chopping down original forest land and not replanting the trees. Reforestation is required on public land in the United States, but this law is often not enforced, and most of the country's forest land is privately owned. Several states now require private landowners to reforest after timber harvest.

You can personally contribute to the reforestation effort simply by planting a tree. **Find out how on pages 398–399.**

Protecting Forests Many governments are currently working to improve reforestation efforts and promote less destructive harvesting methods. In addition, forest preserves established by government agencies, private conservation groups, and private citizens are appearing more frequently. Private organizations have also established effective tree-planting programs. These programs help guarantee the survival of forests. They also improve urban areas, where many trees are removed to make way for buildings and parking lots. The leader of one especially successful tree-planting organization is shown in Figure 8-12. It is important that these types of programs continue in order to protect forests and their many resources.

Figure 8-12 Wangari Maathai (below) is the founder of an organization in Nairobi, Kenya, called the Green Belt Movement. Her organization has helped over 50,000 people grow and plant more than 10 million tree seedlings.

In the arid West it takes about 31 hectares (77 acres) of grazing land to raise one cow, while in the rest of the United States it takes an average of about 7 hectares (17.5 acres) to raise one cow.

RANCHING

Rangelands support grasses and shrubs that are used by ranchers for grazing animals such as cattle, sheep, and goats. Along with farmland (discussed in Chapter 9), rangeland is essential for maintaining the world's food supply. Some experts predict that current trends in population growth will necessitate a 40 percent increase in food production from rangeland between the years 1977 and 2030.

Problems on the Range

Rangelands consist primarily of grasses and shrubs. Grasses are remarkably adaptable plants—they can live through droughts, freezes, fires, and years of animal grazing. This is because the growing point of grass is at the base of every leaf. So if the upper section of a leaf is damaged or cut, the leaf will grow back from its base to its original length. Shrubs are not as resistant to grazing as grasses are, but they also have numerous adaptations that help them survive grazing.

Another interesting feature of grass is its root system. As shown in Figure 8-13, the roots of grasses are fibrous, and they grow so densely that they form a matted tangle that extends several centimeters below the surface. This root system holds soil together, which prevents soil erosion.

Grass, even though it is adaptable, does have limits. When too many animals graze in an area for too long, they damage the grass beyond its ability to recover and much of it dies. This is called **overgrazing.**

When an area is overgrazed, much of the grass is destroyed and the fibrous root system decays. Often, vast fields of grass turn into mere patches, and plants that are less appealing to grazers take over. The roots of these plants are less efficient at protecting the soil from wind and water erosion. If a prolonged drought plagues the overgrazed area before other plants can take over, the land may become so degraded that it can never recover, like the land shown in Figure 8-14. This process, called *desertification,*

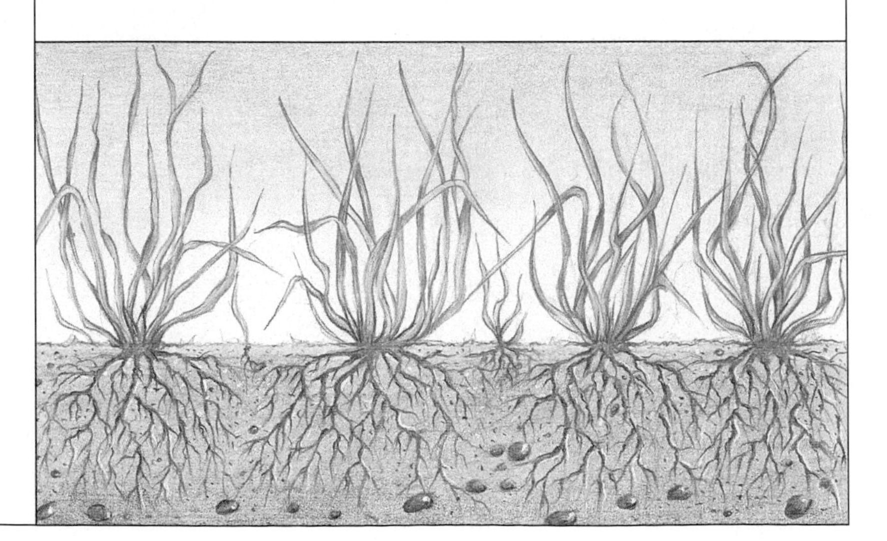

Figure 8-13 The fibrous root system of grass helps prevent soil erosion.

converts rangeland to desert. You will learn more about desertification in Chapter 9.

In some regions of the world, ranching has also contributed to the destruction of tropical forests. (See Figure 8-15.) In Brazil, an area of forest the size of South Carolina was cleared for cattle ranches between 1966 and 1978. In more recent years, however, government programs in Brazil have helped slow the rate of clearing.

Figure 8-14 Overgrazing converted this Australian ranchland into desert.

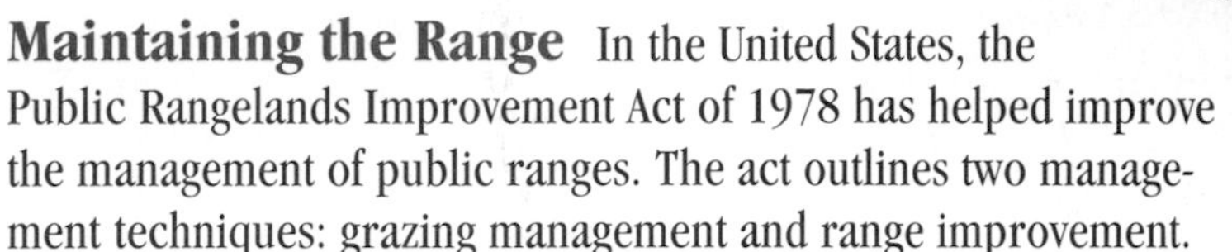

Maintaining the Range In the United States, the Public Rangelands Improvement Act of 1978 has helped improve the management of public ranges. The act outlines two management techniques: grazing management and range improvement.

Grazing management involves limiting animal herds to sizes the land can support. It also involves moving these herds in ways that will protect the plants that support them. Range improvement includes eliminating sagebrush and other weedy plants that invade overgrazed land, planting vegetation where soil is bare, fencing areas to let them recover from overgrazing, and digging enough small water holes to keep livestock from overgrazing the vegetation around a single watering area.

Figure 8-15 Rain forest in South America was cleared for this ranching operation.

Mining

A surprising amount of land is used for mining minerals. A **mineral** is a solid substance that is found in nature and consists of a single element or compound. Minerals include common substances such as salt, as well as rare ones such as gold and silver. Minerals that are useful to humans are called **mineral resources.** For instance, iron, copper, aluminum, and other metals are used in cars, stereos, refrigerators, and buildings. We build with concrete, brick, and glass, which are made from minerals. And many minerals go into paper, paint, plastics, chemicals, films, fertilizers, and a host of other products. Figure 8-16 shows some of the minerals that are important to industry.

Two common methods for extracting minerals from the Earth are open-pit mining and strip mining. In **open-pit mining,** machines are used to dig large holes in the ground and remove the **ore**, which is the mineral-containing rock. Sand, gravel, and building stone are also mined in this way. The copper mine shown in Figure 8-17 is an open-pit mine. In **strip mining,** huge bulldozers and other machines are used to clear away large strips of the Earth's surface. Phosphate rock, which is a raw material used to make fertilizers and phosphate chemicals, is often mined in this way.

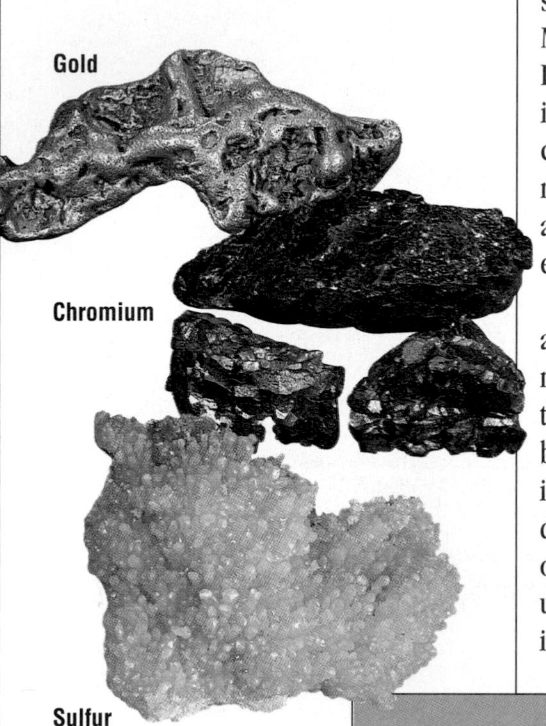

Some of the Most Important Minerals to Industry

Minerals	Some Uses
Diamonds	Industrial abrasives, jewelry
Chromium	Toolmaking, jet engines
Copper	Electrical wiring, alloys
Iron	Steel making
Aluminum	Structural material, packaging
Gold	Jewelry, money, electronics
Platinum	Electroplating, jewelry, cancer therapies, catalysts in pollution control and fertilizer synthesis
Silver	Jewelry, photography
Zinc	Metal alloys, electroplating
Sulfur	Papermaking, photography, food additives, paint, explosives, pesticides, pharmaceuticals, rayon
Graphite	Electronics, lubricants, nuclear reactor cores, pencil "lead"

Figure 8-16 This table identifies some of the most important minerals used in the manufacture of products.

Figure 8-17 **This open-pit copper mine in Bingham, Utah, is wider than 35 football fields and more than twice as deep as the Empire State Building is high! It is the largest human-made hole, and out of it comes 227,000 metric tons (250,000 tons) of copper each year.**

The Effects of Mining As you might imagine, extracting minerals from the Earth causes environmental damage. The most obvious damage is the disruption of the land surface and the ugly piles of waste materials left behind. When large areas of natural vegetation are cleared, wildlife habitat is lost, and the area's natural ecosystem is disrupted.

Such large-scale removal of vegetation and rock can cause land erosion and even landslides. In addition, toxic substances left behind at a mining site can pollute the air and water. It also requires a tremendous amount of energy to extract and process minerals.

Mines in the United States collectively produce more waste than all American cities and towns combined. Some of the waste that remains after mineral processing is dangerously toxic. The open-pit copper mine pictured in Figure 8-17 produces 1,200 metric tons of

Figure 8-18 **Mines such as this one in Colorado can leave the land polluted and heavily scarred.**

Figure 8-19 **Would you believe this area in West Virginia was once strip-mined for coal?**

waste material for every metric ton of copper metal produced. And every year the amount of waste increases as miners have to dig deeper and deeper to find the copper.

Responsible Mining One way to reduce damage from mining is to require mining companies to restore mined land to the condition it was in before mining began. This process is called **reclamation.** A successful reclamation project is shown in Figure 8-19. Environmental laws in the United States now require companies to reclaim mining sites on public land. Some states also have laws that require reclamation on private land.

Another way to reduce the destructive effects of mining is to reduce the need for more minerals. By recycling existing products made from minerals such as iron, copper, and aluminum, we not only save energy, water, and money, but also reduce the pollution caused by additional mining and processing operations.

Section Review answers on page T151.

Section Review

1. What is reforestation, and why is it important?
2. How can grazing-management and range-improvement techniques help sustain rangelands?
3. Name and describe two ways to decrease the impact of mining on the land.

Thinking Critically

4. *Identifying Relationships* It is difficult to provide an economic justification for the environmental protection of our nation's lands (such as providing wildlife habitat). How might this difficulty affect decisions about land use?

PUBLIC LAND IN THE UNITED STATES

AFTER READING THIS SECTION YOU SHOULD BE ABLE TO

❶ explain how public land is used in the United States.

❷ discuss the benefits and disadvantages of using public land for multiple uses.

In the early 1870s a group of explorers approached Congress with news of a magnificent expanse of land that they had seen in parts of the Wyoming and Montana territories. They expressed their concern that this land would be devastated by the logging, mining, farming, and development that had altered most of the land in the Northeast. They asked Congress to protect this land by setting it aside for the public to use and enjoy. Congress agreed, and the first national park—Yellowstone National Park—was born.

Today there are about 50 national parks in the United States, and the federal government has acquired and protected various other types of public land, as shown in Figure 8-21. The success

Figure 8-20 Scenic views in Yellowstone National Park, such as the one shown here, inspired the United States national park system.

ENVIRONMENTAL LAW

There are over 21 million hectares (52 million acres) of Native American lands that have had little or no environmental protection for many years. On pages 380–381, find out how Jana Walker is helping to protect those lands.

of national parks in the United States has encouraged many other nations to set aside public lands.

This Land Belongs to You and Me

Forty percent of all the land in the United States is publicly owned. Public lands are managed by federal, state, and local governments. The chief landowner is the federal government. As you can see in Figure 8-21, most of the land that the federal government owns is in the West and in Alaska. If you are a United States citizen, you are a partial owner of these lands.

Managing Our Public Land Because land resources are so valuable, the federal government allows most public land to be used in several ways, as shown in Figure 8-22. This approach, called multiple-use management, is designed to provide the greatest value for the greatest number of people. A national forest, for example, may be used for recreation, logging, and mining. This means that private individuals and corporations can often harvest

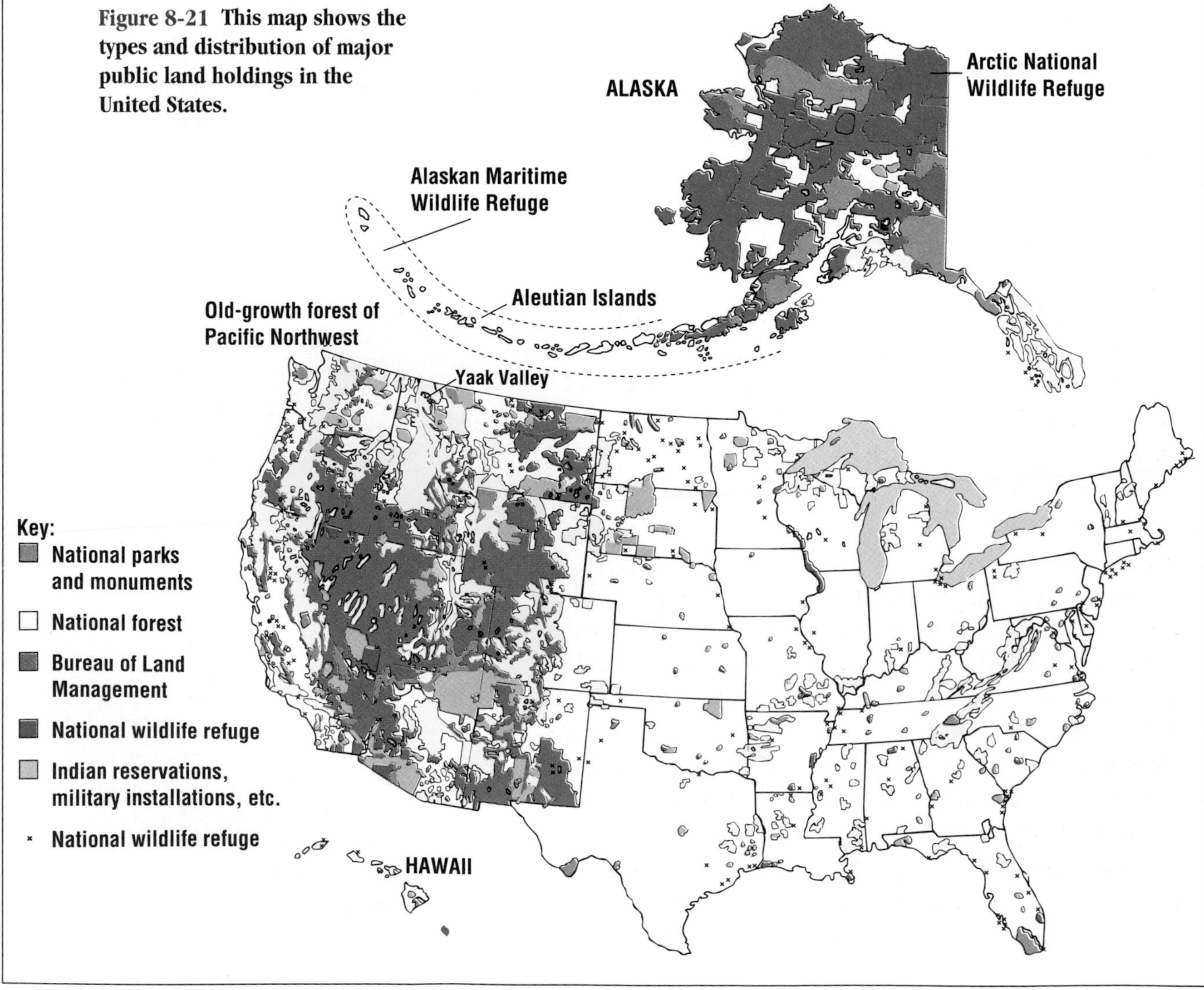

Figure 8-21 This map shows the types and distribution of major public land holdings in the United States.

Major U.S. Public Lands			
Major U.S. Public Lands	**Approximate Size**	**Administered by**	**How the Land Is Used**
National Parks and Monuments	29 million hectares (72 million acres)	National Parks System (Dept. of Interior)	Hiking, camping, boating, fishing, and other recreation; commemoration of historical sites. In some areas, hunting and mineral extraction are permitted.
National Wildlife Refuges	34 million hectares (84 million acres)	U.S. Fish and Wildlife Service (Dept. of Interior)	Wildlife conservation, recreational activities, breeding areas for commercial fish, education, and scientific research. On some refuges, hunting, fishing, mining for oil and gas, livestock grazing, and farming are permitted.
National Resource Lands	130 million hectares (321 million acres)	Bureau of Land Management (Dept. of Interior)	Recreation, wildlife conservation, and industry such as livestock grazing or mining
National Forests	76 million hectares (188 million acres)	U.S. Forest Service (Dept. of Agriculture)	Recreation and commercial uses such as logging, grazing, and mining. Sometimes the land is even leased for use as ski resorts or similar ventures.
Indian Reservations, Military Installations, etc.	29 million hectares (72 million acres)	Bureau of Indian Affairs, Department of Defense, and others	Indian reservations: recreation and commercial uses such as farming, logging, grazing, and mining. Military installations: firing ranges, troop maneuvers, dumping ground for old military vehicles

Figure 8-22 How some of America's lands are used

or extract resources that are found on public lands. In fact, corporations sometimes pay far less to use public lands than they would pay to use privately held land. See the Case Study on pages 216–217 for an interesting look at the Mining Act of 1872. This act still allows anyone to "stake a claim" on public lands.

The problem with multiple-use management is that different people have different ideas about what constitutes proper use of the land. As a result, multiple-use management has caused battles between developers and conservationists. Limited staffs and funding make it even more difficult for government agencies that control land use.

WILDERNESS

According to the Wilderness Act of 1964, **wilderness** is an area in which the land and the ecosystems it supports are protected from all development. Wilderness areas are found within several of the nation's public land systems. So far, 474 regions covering almost 13 million hectares (32 million acres) have been designated as wilderness. These areas are open to hiking, fishing, boating (without motors), and camping. Building roads or structures and using motorized equipment are not allowed in wilderness areas.

In addition to providing places where people can camp, hike, and fish, wilderness areas also have important ecological benefits. They provide habitats for wild plants and animals and serve as

Find a map and phone book for your local community, and determine how many land areas are available for public use. These will include public parks, forests, lakes, golf courses, etc. Visit or call as many of these areas as possible to determine how the land area is used and what activities are and are not allowed. Finally, record your opinions about how the land is managed, and note any suggestions you have for improving the site.

Figure 8-23 Clear-cutting is allowed in this national forest in Washington.

outdoor classrooms where people can learn more about the natural world and how it works.

Troubled Lands As our population grows, increasing numbers of people are using wilderness areas, parks, and wildlife refuges each year. Unfortunately, these people are leaving their mark on the land. Thousands of footsteps trample plants and erode soil from trails and campsites. Bathing and dish washing pollute streams and rivers, and trails of paper, aluminum, and plastic are left behind. Visitors drive for hours only to find in a so-called protected area the same noise and congestion that they had intended to escape.

When United States wilderness areas were first protected in 1964, ranching and mining were allowed to continue. In fact,

CASE STUDY

PUBLIC LANDS FOR THE ASKING

Back in 1872, President Ulysses S. Grant enacted a law to encourage mining during a period when our country was interested in promoting settlement of the West. At that time, the mining industry was made up mostly of independent prospectors and small companies. In addition to promoting settlement, the law also set standards for filing and holding claims. That law, called the Federal Mining Act of 1872, still governs the mining of hard-rock minerals (gold, silver, and other primarily metallic ores) on over 109 million hectares (270 million acres) of public lands.

The act gives anyone—including you—the right to stake a mining claim on public land if a "valuable deposit" of minerals is discovered. The land in question includes any federally owned land anywhere in the United States. Some public lands, however, such as certain national parks, are exempt from this law. Under this law, the federal government receives nothing for the gold or other metals removed from public lands. The miner also has the option to purchase the public lands from the government for a price that cannot exceed $5 per acre. This process is known as "patenting."

The Federal Mining Act contains no environmental protection provisions or land reclamation requirements. Although most individual states now have reclamation laws in place, standards vary widely from state to state. And since many of these reclamation laws are only 10 to 20 years old, the laws often do not apply to older mines.

This gravel pit is located in a national forest in Montana.

mineral claims could still be made in wilderness areas until 1983. Many mining companies filed claims at the last minute to keep their mining operations in wilderness areas busy long into the future. And ranch animals such as cattle can still be seen in some wilderness areas today.

In addition, the negative effects of nearby grazing, logging, oil and gas drilling, factories, power plants, and urban areas often affect these regions. Some scenic views in wilderness areas, such as that shown in Figure 8-24, are obscured by air pollution more frequently each year.

Figure 8-24 Pollution from nearby urban areas obscures the view of Grand Canyon National Park in Arizona.

Since 1872, over 1.4 million hectares (3.5 million acres) of public land have been sold to private owners. As of October 1992, mining companies, many of them foreign owned, have applied for patents to thousands of hectares of federal lands, which contain at least $91 billion worth of mineral deposits. Recent investigations by the General Accounting Office, the investigative arm of Congress, have shown many instances where these lands were patented for mining but were used instead for real-estate development.

It may seem clear that the Federal Mining Act of 1872 should be reformed. The situation is not that simple, however. Those in the mining industry argue that they need the special provisions of the act in order to remain profitable. They assert that patented land offers them the security they need to finance the opening of a mine and the purchasing of equipment. In addition, they maintain that the states have enacted sufficient environmental controls and thus the act need not be amended. Reforming the act, according to the mining industry, would lead to the loss of thousands of American mining jobs and might cause a shortage of affordable minerals and mineral-based products in the marketplace.

Cyanide seeping from an abandoned gold mine pollutes this Colorado stream.

THINKING CRITICALLY

1. *Analyzing Viewpoints* Some people suggest that many new jobs could be created by reclaiming abandoned mines and restoring the landscape to its original state. They propose that such programs could be funded by mineral royalties. Do you think this is a good solution to the possible loss of jobs associated with reforming the mining law?
2. *Expressing Viewpoints* Do you think the value of minerals in terms of jobs and resources outweighs the possible environmental damage that could result from mining activities? Why or why not?

Answers to Thinking Critically can be found on page T151.

"People Control" in Wilderness Areas To protect wilderness from damage, limits have been set in some areas for the number of people who can hike or camp at any one time. Certain areas have also been designated as off-limits to camping, and an increasing number of wilderness rangers now patrol vulnerable areas. In addition, volunteer programs are now active in many wilderness areas. Volunteers help pick up trash and help build trails. New education programs, such as the Forest Service's "Tread Lightly" program, are also being implemented to help people better understand and lessen their impact on the natural world.

Figure 8-25 These high school students are helping to alleviate problems associated with the overuse of New York City's Central Park.

Answers to Section Review can be found on page T152.

Section Review

1. What are the benefits and problems associated with using public land for multiple uses?
2. How do you benefit from the nation's public lands? Give at least three examples.

Thinking Critically

3. *Recognizing Relationships* Why do you suppose so many of the nation's public lands are in Western states?
4. *Expressing Viewpoints* Some people suggest that the best way to care for our nation's land is to designate more land as wilderness. Do you agree or disagree? Explain your reasoning.

CHAPTER 8

HIGHLIGHTS

SUMMARY

- Cities and surrounding suburbs have increased sharply in size in recent years as the population has grown and people have moved to the city from rural areas. Swelling urban populations threaten city infrastructures and burden the land.
- Many land-use practices cause damage to the land, reducing its value as a supplier of food and materials.
- The United States has millions of acres of public lands. "Multiple use" management allows these lands to be used for recreation, mining, and other purposes.
- Overuse and residual ranching and mining threaten the nation's wilderness lands.

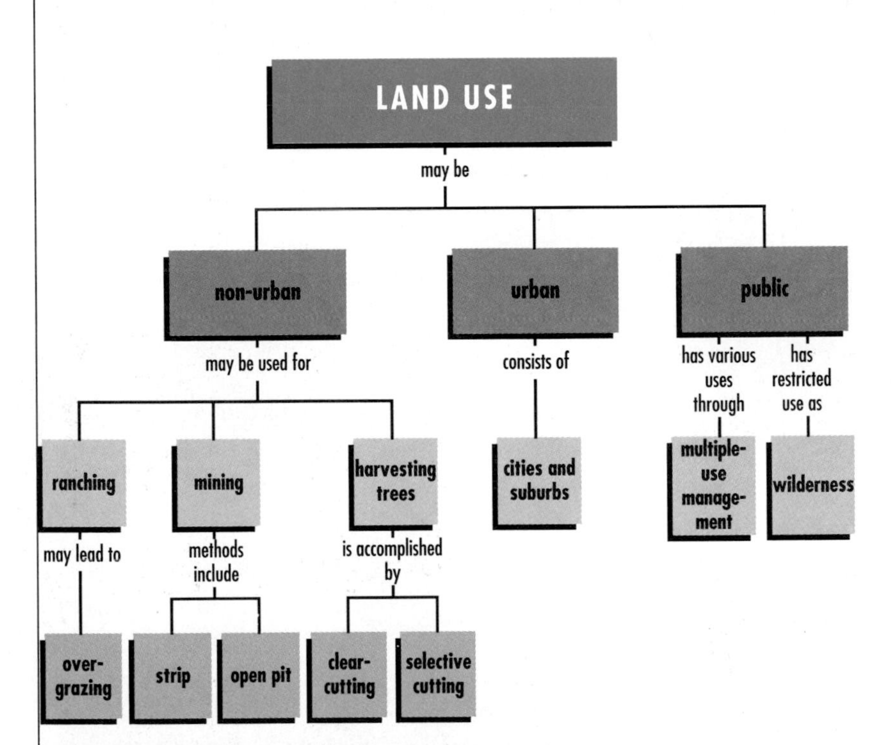

Sample EcoLog answers can be found on page T152.

Ecolog

Now that you've studied this chapter, revise your answers to the questions you answered at the beginning of the chapter, based on what you have learned.

❶ Do you think your lifestyle affects the land in any way? Explain.

❷ Do you think humans should carefully plan how to use the land, or simply let things work out as they will?

Vocabulary Terms

clear-cutting (p. 205)
deforestation (p. 205)
infrastructure (p. 199)
land-use planning (p. 201)
mineral (p. 210)
mineral resources (p. 210)
open-pit mining (p. 210)
ore (p. 210)
overgrazing (p. 208)
reclamation (p. 212)
reforestation (p. 206)
selective cutting (p. 206)
strip mining (p. 210)
suburban sprawl (p. 200)
urban crisis (p. 199)
urbanization (p. 199)
wilderness (p. 215)

CHAPTER 8

REVIEW

UNDERSTANDING VOCABULARY

1. Use the following terms in a sentence to show that you know what they mean.
   a. reclamation
   b. land-use planning
   c. suburban sprawl
   d. infrastructure

RELATING CONCEPTS

2. Copy the unfinished concept map below onto a sheet of paper. Then complete the concept map by writing the correct word or phrase in each box containing a question mark.

Answers not indicated can be found on page T152.

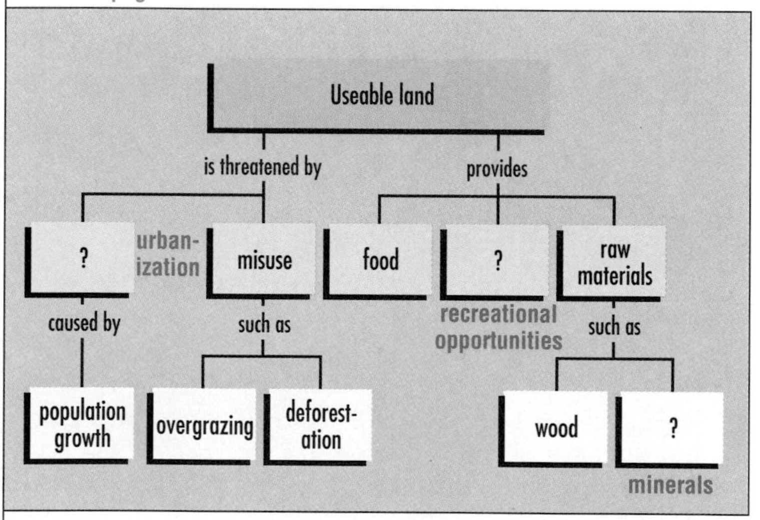

UNDERSTANDING CONCEPTS

Multiple Choice

3. Urbanization can best be described as
   a. the tendency for cities to develop large suburbs.
   b. the decline of the inner city as residents move to the suburbs.
   c. the revitalization of downtown areas that have fallen into decline.
   (d.) the growth of cities caused by the migration of people from rural areas to the cities.

4. What is the most common reason that people have moved to the city from the country?
   a. to escape war
   (b.) to search for jobs
   c. to seek higher environmental quality
   d. to find more spacious living accommodations

5. Which is NOT a part of the infrastructure?
   a. roads
   b. bridges
   (c.) restaurants
   d. power lines

6. What is a major advantage of mass transit over private vehicles?
   a. It reduces congestion on the roads.
   b. It is more convenient for most Americans.
   c. It uses energy more efficiently.
   (d.) both a and c

7. Selective cutting involves
   a. cutting all of the trees in a forest and then reseeding.
   b. cutting all of the trees of one species and leaving the rest.
   (c.) cutting the mature trees in a forest and leaving the rest.
   d. cutting the young trees in a forest and leaving the rest.

8. What is one advantage of clear-cutting?
   (a.) It is the cheapest method for harvesting trees.
   b. It enables the ecosystem to return rapidly to normal.
   c. It aerates the soil, allowing it to regenerate.
   d. It brings light to the heavily shaded plants of the forest floor.

9. Which choice best defines "wilderness"?
   a. any non-urban area
   b. any national park or monument
   (c.) any land protected from development and human impact
   d. any unspoiled area of more than 10,000 sq. km

10. One of Yellowstone's distinctions is that it was
    a. the first officially sanctioned wilderness area.
    (b.) the first national park.
    c. the first clear-cut area to be reclaimed.
    d. an experiment in land-use planning.

11. What is a possible result of overgrazing?
    a. Most of the grass dies off.
    b. The soil is eroded.
    c. The land is invaded by shrubs or cactuses.
    (d.) all of the above

12. A national forest may be used for mining, logging, and recreation because
    a. private companies regulate public land.
    (b.) the federal government allows most public land to be used in several ways.
    c. the federal government has no policy governing the usage of public land.
    d. private companies and conservationists typically agrec to share public land.

Short Answer

13. How does suburban sprawl contribute to the waste of resources?

14. Briefly summarize the causes of urbanization.

15. How does urbanization stress the infrastructure?

16. Why do environmentalists prefer the process of selective cutting to clear-cutting?

17. What are some reasons why land use should be managed, as opposed to unsupervised?

Interpreting Graphics

18. **Examine the graph below.** What percentage of total available land is used intensively by people? What percentage, more or less, serves the needs of the cities? What categories, if any, experience little human influence?

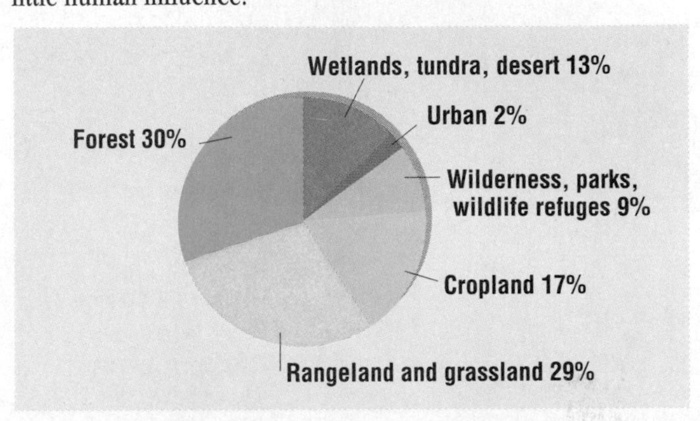

Thinking Critically

19. ***Analyzing Ideas*** Some experts suggest that we are "loving to death" our national parks and preserves. Because of this, they propose restricting access to the parks. Analyze the soundness of this proposal. Do you support it? Why or why not?

Themes in Science

20. ***Patterns of Change*** How has the increase in human population affected the land over the last 200 years?

21. ***Stability*** How could a recycling program in your community help restore stability in a deforested area?

Cross-Discipline Connection

22. ***Biology*** Attitudes toward the land differ not only from culture to culture but also from time to time. In America, popular attitudes toward the land have changed over the past 150 years. Do a little reading to find out about this subject. Cite, for example, the lyrics of songs or the writings of contemporary authors as evidence. Then summarize your findings.

Discovery Through Reading

For a closer look at the impact of tourists on a national park, read "Tranquility, Tourism—and Trouble: The Great Smokies Have It All," by David Nevin, *Smithsonian,* August 1993, pp. 20–30.

PORTFOLIO ACTIVITY

Do a special project about the difference between land-use traditions in the United States and those in other countries. For example, you might examine the land-use traditions of Latin America, Europe, or Africa. Prepare a skit with two or three other classmates that illustrates these traditions, and present it to your class.

CHAPTER 8

INVESTIGATION

MINING FOR PEANUTS

In what way do you depend on minerals? For starters, consider what the American Geological Institute has stated: "Without manganese, chromium, platinum, and cobalt, there can be no automobiles, no airplanes, no jet engines, no satellites, and no sophisticated weapons—not even home appliances." The photographs on the next page show other uses of minerals. Mining underground minerals has its costs, but it sure has its benefits too!

To a great extent, mining companies determine how much environmental damage is done as a result of locating, mining, and processing minerals. In this Investigation, you and your fellow students will operate a mining company. You will make a model mining site and then mine for peanuts that will represent mineral ore. How efficient and Earth-friendly will your mining company be?

MATERIALS

- plastic pan, 30 cm × 35 cm × 15 cm (12" × 14" × 6"), or shoe box
- metric ruler
- clean sand
- soil
- 5 to 10 peanuts
- grass clippings, twigs, small rocks, etc.
- index card
- paper clip
- craft stick
- watch or clock
- notebook
- pencil

Answers to this Investigation can be found on page T152.

CREATE A MODEL SITE

1. Place a layer of sand about 5 cm deep on the bottom of your pan.
2. Distribute 5 to 10 peanuts on the sand, either clustered together or randomly spread apart. In your notebook, record how many peanuts you deposited. Also make a sketch of how they were distributed.
3. Cover the peanuts with another layer of sand, to a total depth of about 10 cm.
4. Cover the sand with a layer of soil 1 cm deep.
5. Simulate a natural landscape by covering the top of the soil with features such as trees, grass, rocks, and perhaps a lake. Use common materials to represent these features. You might use grass clippings, twigs, small rocks, etc.
6. Make another sketch to show the features of your landscape.
7. Trade mining sites with another student group. Do not tell the other students how many peanuts are buried or how they are distributed.

MINE FOR ORE

8. Imagine that you and the other students in your group operate a mining company. Choose a name for your mining company, and use an index card to make a sign to put at the mining site.
9. You will have 5 minutes to extract as many peanuts, or "ore," as you can from the simulated landscape. Your goal is to extract as much ore as possible, while causing the least amount of environmental damage. Your group should work as a team to locate, dig, and process the ore. Use a straightened paper clip to probe for the ore, use a craft stick to dig for the ore, and use your fingers to separate the shells from the ore. Caution: the shells from the peanuts represent hazardous wastes from mining and processing the ore. Do not use your teeth to remove the peanuts from the shells! The shells, or tailings, must not be deposited back in the landscape. Record how many peanuts you extracted.

EVALUATE THE SITES

10. Retrieve your original landscape, and inspect the environmental damage. Compare the site with your sketches. Consider the following criteria and any others that you think are important.

- How badly was the land damaged by mining?
- How many peanuts did the mining company unearth? How many were buried?
- How did the number and distribution of peanuts affect the results of the mining operation?
- How much waste was produced?

RECLAIM THE LAND

11. Try to reclaim the damaged site. You will have 5 minutes to restore the land to its original condition.

12. At the end of the 5 minutes, evaluate the difficulty of restoring the site. How closely does the reclaimed site resemble the original site?

ANALYZE YOUR RESULTS

13. How do you think this simulation compares to actual mining operations?

14. Based on your experience, how expensive do you think reclamation activities are compared with extracting the ore?

15. Did all of the mining companies in your class recover the same amount of minerals? If not, what factors accounted for the differences?

16. What guidelines would you give to new miners prior to this activity so that they do less environmental harm?

This mining operation relies on heavy equipment to access mineral ore.

a

b

c

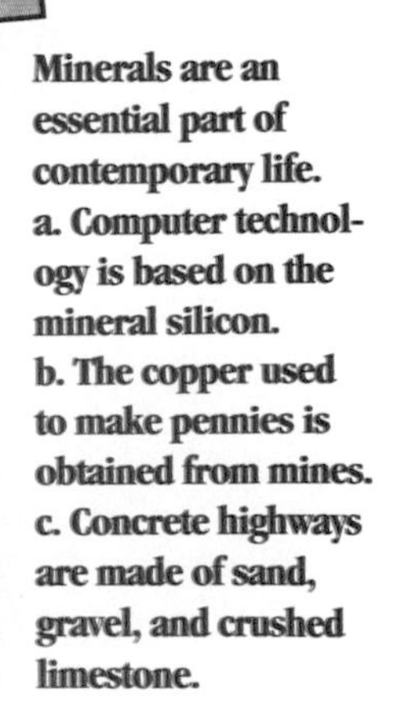

Minerals are an essential part of contemporary life.
a. Computer technology is based on the mineral silicon.
b. The copper used to make pennies is obtained from mines.
c. Concrete highways are made of sand, gravel, and crushed limestone.

Making a DIFFERENCE

BIG-CITY FARMER

You may not think that *city* and *farmer* fit well together, but in parts of New York City these words go hand in hand. Over the past 15 years, many vacant lots have been converted by urban farmers into productive miniature farms—right in the middle of the city. These small farms, although framed by an urban jungle of sidewalks and apartment buildings, are flourishing oases of vegetables, fruits, and flowers.

People give me a lot of attention because of my gardening success, but I'm just doing what my father taught me to do!

This is possible largely because of a program called Operation GreenThumb (OGT). OGT is sponsored by New York City's Department of General Services. The operation enables nonprofit organizations, such as a

Otis Butler (above) oversees two urban gardens in the South Bronx of New York City. Mr. Butler's gardens, and gardens like the one shown at right, in Philadelphia, provide urban residents with an unexpected, but welcome, source of beauty and fresh produce.

block association, to lease a city-owned vacant lot for $1 per year to create a community vegetable or flower garden.

Otis Butler first learned to farm as a child in the Bahamas. When Mr. Butler moved to New York City, he never thought he'd farm again—certainly not in the city. But in the late 1970s, a friend of Mr. Butler's told him about a vacant lot she owned. She kept the lot clean and planted flowers and vegetables there. She often told Mr. Butler how much she enjoyed the garden, and once or twice he visited the lot himself. That was enough to convince Mr. Butler that he'd like to begin farming or gardening again. His friend told him about Operation Green Thumb, and in 1980, he and his 20-member block association leased a large vacant lot in the South Bronx. After the city bulldozed all of the weeds and debris out of the lot, Mr. Butler once again dug his hands into the dirt to plant fruit trees, flowers, and vegetables.

Today, at 80 years old, Mr. Butler oversees two award-winning gardens. "Our gardens make a big difference in our lives," says Mr. Butler. "They give us someplace nice to visit and enjoy. Plus, sometimes it's tough to get fresh vegetables in the city. With the gardens, we can grow anything we want and then freeze things so that they can be eaten any time of the year. Squash, corn, okra, beans, tomatoes, sweet potatoes—you name it, and we probably grow it. I'm always eating fresh vegetables and fruit. I can even make my own grape juice!"

Mr. Butler grows a variety of fruits, vegetables, and flowers on a lot that was once a neighborhood eyesore, with unsightly weeds and discarded trash.

Other members of the community also benefit from the gardens. Every Thursday, the block association opens its gardens to neighborhood residents. On that day, the block association passes out baskets and allows the visitors to harvest fruits and vegetables. "Many of the people who visit the lots on Thursdays can't garden themselves or can't afford fresh fruits and vegetables. This way we can provide from 30 to 40 families with fresh, healthy produce," says Mr. Butler.

Since 1978, Operation GreenThumb has sponsored the transformation of thousands of vacant lots into community gardens. OGT provides new gardeners with valuable information about how to design and maintain a garden, as well as providing tools, fencing materials, lumber, soil, seeds and bulbs, ornamental and fruit trees, shrubs, and continued technical assistance.

This is a valuable service to many New York City residents. Vacant lots are often neighborhood eyesores that are filled with trash and rats. They also provide cover for people who sell drugs or are involved in other illegal activities. With a GreenThumb garden, a community group can gain some control over their neighborhood. They can transform an unsafe and unsightly area into one that the community can be proud of and benefit from.

In addition to his abilities as a leader, Mr. Butler's farming techniques have brought him much admiration and recognition as a master gardener and teacher. According to Mr. Butler, all of the credit goes to his father, who taught him how to farm back in the Bahamas. "People give me a lot of attention because of my gardening success, but I'm just doing what my father taught me to do!"

For More Information

If you are interested in participating in a community gardening project, contact the following organization for more information: American Community Gardening Association, c/o Philadelphia Green, 325 Walnut Street, Philadelphia, PA 19106.

CHAPTER 9

FOOD

"In simplest terms, agriculture is an effort by man to move beyond the limits set by nature."

LESTER R. BROWN
PRESIDENT, WORLDWATCH INSTITUTE

Before you read this chapter, take a few minutes to answer the following questions in your EcoLog.

1. Why do you think it is so difficult to provide adequate food for all of the world's people?
2. Besides using chemical pesticides, how might farmers keep insects from destroying crops?

FEEDING THE PEOPLE OF THE WORLD

AFTER READING THIS SECTION YOU SHOULD BE ABLE TO

1. explain why providing adequate food for all of the world's people is so difficult.
2. describe the advantages and disadvantages of the green revolution.

Ethiopia, 1985: Thousands of people were starving. Lack of rain combined with soil degradation and war had caused a major crop failure for the second time in 15 years. Near the end of the year, the rains finally came and washed millions of tons of soil into the rivers. That year, another 15,000 sq. km (6,000 sq. mi.) were added to the North African desert.

Events like those in Ethiopia present a frightening picture of the problems associated with feeding the people of the world. Many people go hungry, but efforts to produce more food sometimes cause environmental damage, which in turn makes food production more difficult. In this chapter, you will learn why it is so difficult to feed all of the world's people, and about efforts to increase food production.

Figure 9-1 Harvesting rice in China

Major Nutrients in Human Foods		
Nutrient	**Function**	**Major Food Sources**
Carbohydrates	Supplies energy	Bread, cereals, potatoes, beans
Proteins	Used to build and maintain the body	Meat, fish, poultry, eggs, dairy products, beans
Lipids (fats and oils)	Used to build membranes and some hormones. May also be stored and used later for energy.	Butter, margarine, vegetable oils

Figure 9-2 **This table shows the three main types of nutrients. Humans also need vitamins and minerals to stay healthy.**

WHAT PEOPLE EAT

Human beings must consume organic molecules produced by other living things in order to survive. The main types of organic molecules in all foods are carbohydrates, proteins, and lipids. In addition to these major nutrients, we need smaller amounts of vitamins and minerals to help our bodies function properly. The human body uses food both as a source of energy and as a source of materials for building body parts. Figure 9-2 describes how the body uses each of the major nutrients and lists some common foods that contain them.

When people do not get enough to eat, they can become sick and even die. When starving people die, it is usually as a result of diseases that their bodies cannot fight. Underfed people lose resistance to diseases that normally do not threaten the lives of well-fed people.

Even if people consume enough calories in the food they eat, they may still suffer from **malnutrition,** a condition caused by not consuming enough necessary nutrients. For example, lack of protein in the diet can cause serious illnesses, especially in young children. These illnesses can cause brain damage if they are not cured quickly. The child shown in Figure 9-3 is receiving food that may prevent conditions caused by malnutrition.

WHY PEOPLE GO HUNGRY

Why are people starving? Part of the answer is obvious—there is not enough food to go around. Because the world's population is increasing rapidly, more food is needed each year. As shown in Figure 9-4, world food production has been increasing for several decades, but the amount of food per person is no longer increasing. In other words, food production is not increasing as fast as the human population is. If everyone in the world today received an equal share of all the food produced, no one would have enough to stay strong and healthy.

However, the world's food is not divided equally. In many parts of the world, wealthy people have an abundance of food, while poor people have much less than they need. We often think of malnutrition as something that happens only in developing countries in Africa and Asia. But even in the United States many poor people suffer from malnutrition.

Starvation also occurs when food is available but cannot be transported to the people who need it. Transportation can be a problem when economic and political troubles cause trucking

Figure 9-3 **A relief worker in Bangladesh gives food to a starving baby.**

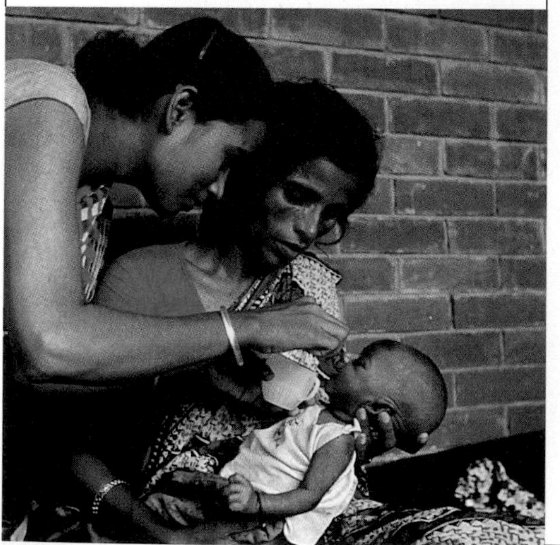

and railroad lines to break down. When there is war in a country, food donated by other countries may not reach the people who need it.

Droughts, or periods when rainfall is less than average, can contribute to starvation by causing crop failure. People can survive a crop failure if there is enough food saved from previous seasons and if there are effective systems for distributing that food. But if a drought is combined with war or inefficient transportation systems, the result can be famine. A **famine** is a food shortage so widespread that it causes malnutrition in many people.

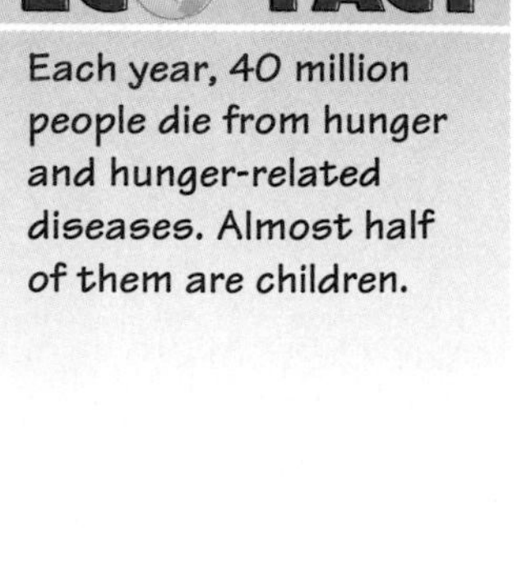

The Green Revolution

In an effort to increase crop yields, new varieties of grain were introduced in Asian and Latin American countries between 1950 and 1970. These new varieties had a much greater **yield,** or amount produced per unit area, than the old varieties. The introduction of the new grains, along with new farming techniques, were together called the "green revolution."

The green revolution allowed far more people to be adequately fed than was possible before. However, there were problems. The new varieties of grain did not grow well without the right kinds of machinery, fertilizers, and pesticides. (See Figure 9-5 on the next page.) Many **subsistence farmers**—those who grow only enough food to feed their families—could not afford the necessary equipment and chemicals. Therefore, most of the green revolution's increase in food production was from large farms that grow food to be sold. These farms are now producing about as much food as they can be expected to produce.

Another drawback of the green revolution was that the use of large amounts of pesticides and fertilizers can pollute the

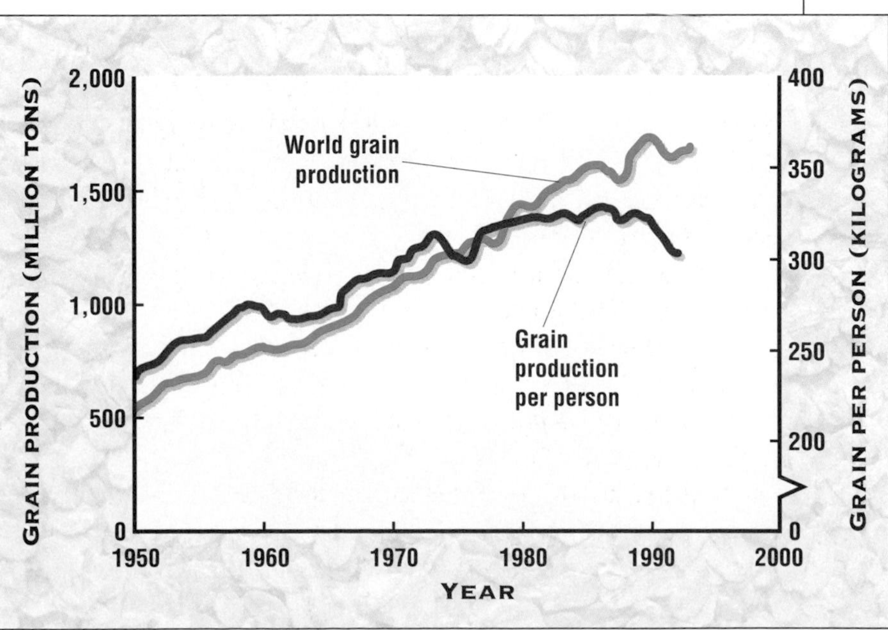

Figure 9-4 Grains make up most of the world's food. Since 1975, the amount of grain produced for each person on Earth has dropped slightly, indicating that world food production is not keeping up with population growth.

Figure 9-5 **Many of the new varieties of grain introduced during the green revolution require more water and fertilizer than older varieties. This corn crop in Guatemala is growing poorly because farmers cannot afford the crop's requirements.**

environment. Also, the necessary farm machinery consumes a great deal of energy, as does the manufacture of fertilizers and pesticides.

Political Changes

As Figure 9-6 suggests, making peace in a war-torn country is an important step toward supplying its people with enough food in the long run. Once the war ends, donated food can be distributed more effectively, and people can begin to work the land once more. They can also learn more sustainable farming techniques.

Answers to Section Review can be found on page T152.

Figure 9-6 **In Somalia in 1993, soldiers had to make sure that warring factions didn't take food that was meant for starving people in the countryside.**

Section Review

1. Give two reasons why people might get too little to eat even in a country with plenty of food.
2. Describe one benefit and one drawback of the green revolution.

Thinking Critically

3. *Interpreting Graphics* Study the graph in Figure 9-4. World grain production was at an all-time high in the mid-1990s. Why was the amount of grain *per person* declining?

SECTION 9.2

AGRICULTURE AND SOIL

AFTER READING THIS SECTION YOU SHOULD BE ABLE TO

1. describe fertile soil.
2. describe methods to prevent soil erosion.
3. explain how irrigation can cause salinization.
4. describe desertification and how it can be prevented.
5. compare low-input and conventional farming.

The Earth has a limited amount of **arable** land—land on which crops can be grown. This amount is decreasing every year. As shown in Figure 9-7, the amount of arable land in the world will have decreased by about one-fifth from 1985 to 2000. About 150 million hectares (about 371 million acres) of farmland will disappear under houses, mines, roads, factories, and power plants. Another 135 million hectares (about 334 million acres) will become unusable for farming because the soil will be damaged.

The shortage of fertile agricultural land threatens our ability to feed the human population. In this section, you will learn how food is produced and how we can ensure that we have enough land to grow the crops we need in the years to come.

Figure 9-7 **This graph shows the amount of the land farmed in 1985 that experts estimate will be lost from production by 2000.**

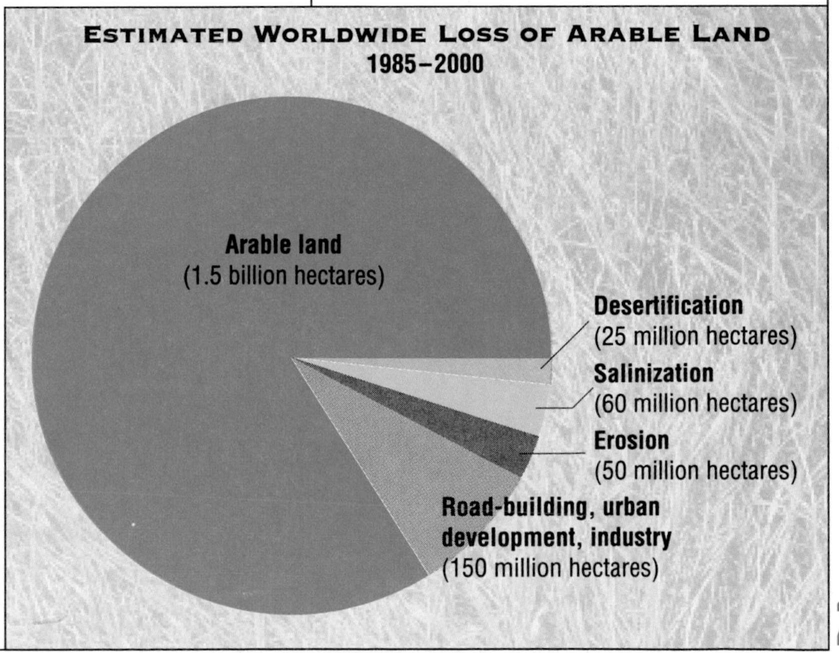

FARMING

Some of the basic methods of farming—plowing, fertilization, irrigation, and pest control—were begun by the earliest farmers. People or animals pulled plows through the soil. Plowing the soil (cutting through it and turning it over) helps crops grow by mixing up soil nutrients, loosening soil particles, and uprooting weeds. Organic fertilizers like manure were used to enrich the soil so that plants would grow strong and remain healthy. Fields were irrigated (supplied with water) by digging ditches for water to flow through. Weeds were removed by hand, and some plants were covered to keep out pests.

Figure 9-8 **Since ancient times, farmers have used human and animal labor to grow crops. This Indonesian farmer still uses oxen to plow his field.**

Some of these ancient methods are still used today, as you can see in Figure 9-8. On large farms in industrialized countries, however, new methods are also used. Machinery is used to plow the soil and harvest crops. (See Figure 9-9.) Synthetic fertilizers, which are produced in factories, are used instead of manure and plant wastes. A variety of overhead sprinklers and drip systems are used for irrigation. And many different synthetic chemicals are used to kill pests.

Fertile Soil

Soil in which most plants are healthy and able to grow rapidly is called **fertile soil.** Most plant roots grow in **topsoil,** the loose surface layer of soil. Fertile topsoil is composed of rock particles, water, air, and organic matter such as dead plants and animals and animal excrement. Living organisms also play an important role in fertile topsoil. Fungi, bacteria, and other microorganisms decompose the organic matter. Earthworms, insects, and other small animals help to break up the soil and let air into it. Figure 9-10 shows the large number of living organisms in fertile soil. As you can see in Figure 9-11, several other layers of soil lie under the topsoil. At the bottom is bedrock, which is solid rock.

Field Activity

Determine the living components of a soil sample from your area. Obtain the following materials: a glass jar, a funnel, a piece of gauze, a magnifying glass, and a small soil sample. Place the funnel, right side up, in the glass jar. Put the soil sample on a piece of gauze inside the funnel. Leave the jar under a lamp overnight. The light will force the living organisms in the soil down to the bottom of the jar. Use the magnifying glass to observe the kinds of organisms that inhabit the soil.

Figure 9-9 **In industrialized nations, machinery is used to do much of the work previously performed by humans and animals. This photograph shows the harvesting of lima beans in Quincy, Washington.**

Most soil is formed from bedrock. (See Figure 9-12 on the next page.) Temperature changes and moisture cause the bedrock to crack and break apart, creating smaller and smaller particles. These rock particles, combined with water, air, decaying organic matter, and the action of living organisms, form the soil. It may take thousands of years to form a few centimeters of soil.

Number of Organisms in Average Farm Soil	
Organisms	**Quantity**
Insects	670 million per hectare
All arthropods (including insects)	1.8 billion per hectare
Bacteria	1 billion per gram
Algae	100,000–800,000 per gram
Earthworms	1.8 million per hectare

Figure 9-10 Living organisms are crucial in maintaining the fertility of soil. The table at left shows the astounding number of organisms in farm soil.

Figure 9-11 The diagram below shows the structure of soil. The number of soil layers and characteristics of the layers may be different in different types of soil.

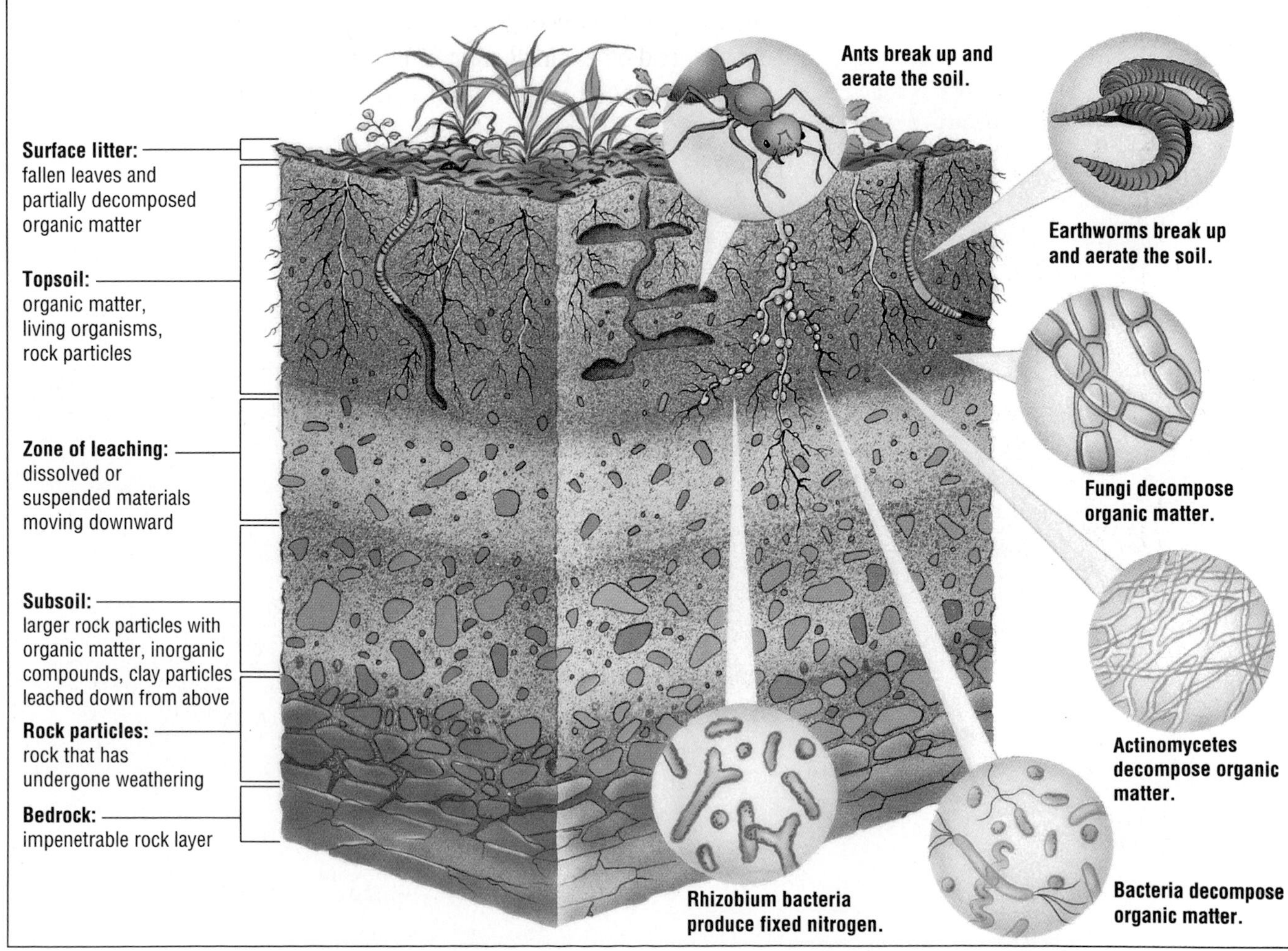

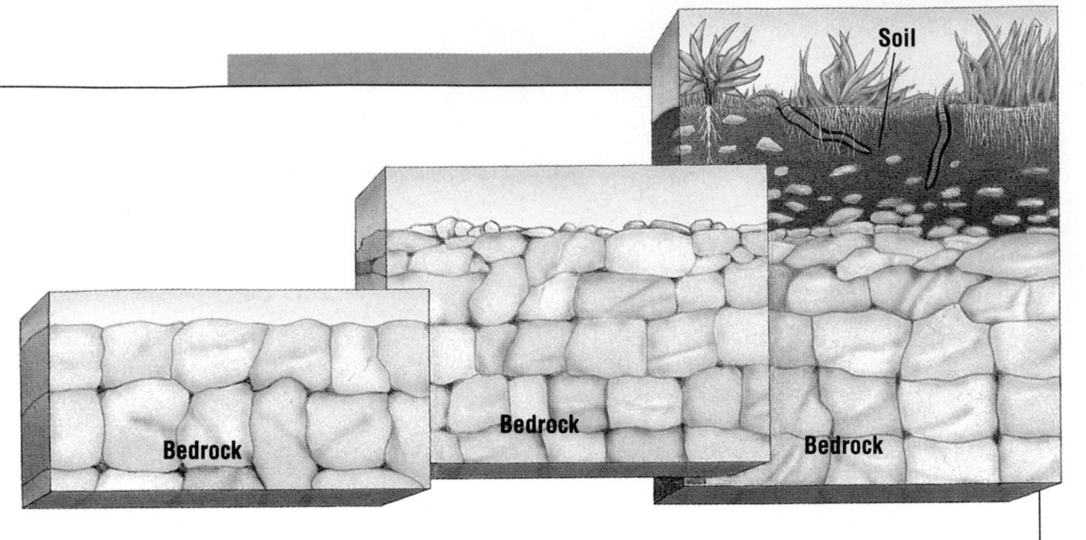

Figure 9-12 The formation of fertile soil can take thousands of years. These diagrams show how weathering causes the bedrock to crack into rock particles. Soil also contains decaying organic matter and living organisms.

Topsoil Erosion: A Global Problem

The soil that has taken so long to form is being lost to erosion at an alarming rate. **Erosion** is the wearing away of topsoil by wind and water. In the United States, about half of the topsoil has been lost to erosion in the past 200 years. Worldwide, it is estimated that about 11 percent of the soil has been eroded in the last 45 years. (See Figure 9-13.) Topsoil erosion is ranked as one of the most serious ecological problems we face. Without the valuable topsoil, crops cannot be grown to feed the world's people.

Certain farming practices can contribute to topsoil erosion. Plowing produces a loose surface layer of soil that is easily blown away by wind or washed away by rain. When a crop is harvested, soil fertility is decreased because the plants contain minerals that were once part of the soil.

Figure 9-13 This map shows the extent of soil erosion around the world. The worst soil erosion occurs in arid regions close to deserts.

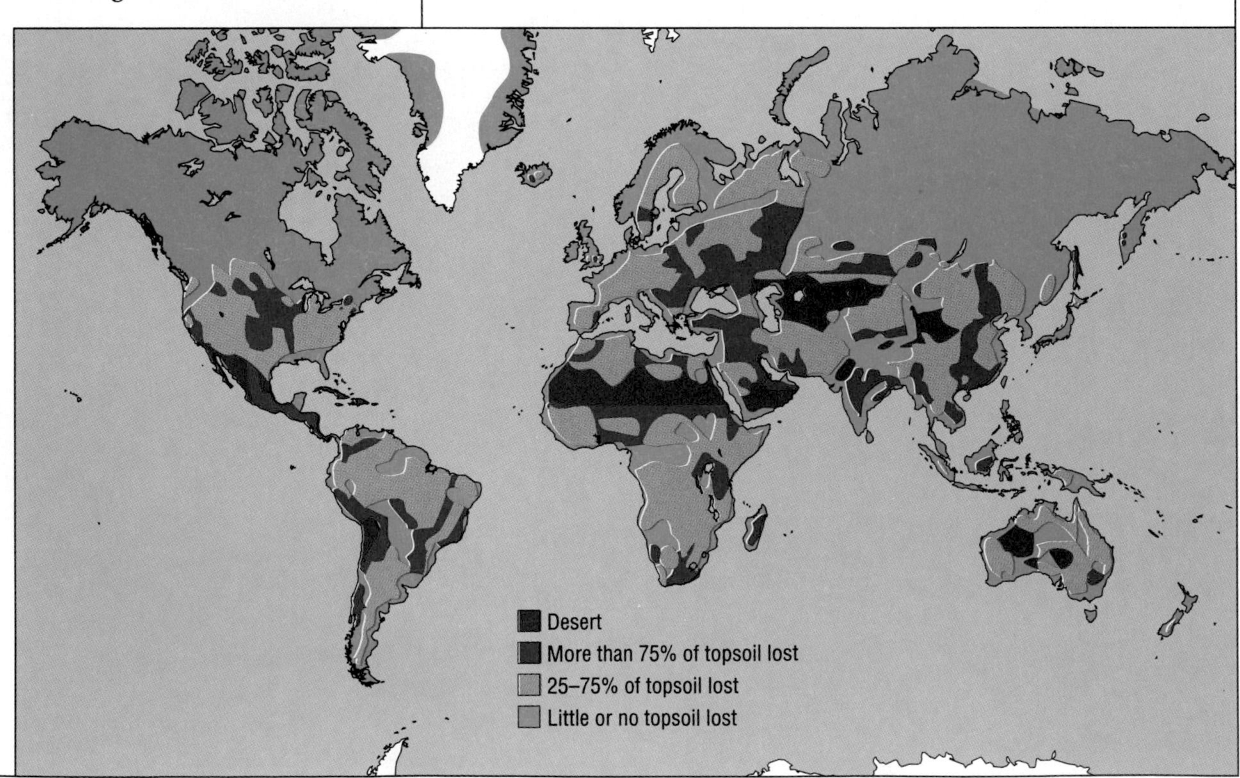

Clearing forests to produce lumber also contributes to soil erosion. Trees and shrubs absorb large quantities of water. When they are cut down, the amount of water running off the surface of the soil increases, and more soil is carried away. Sometimes the soil washes away slowly, and sometimes large amounts wash away in landslides.

Desertification

The loss of topsoil occurs all over the world, but it is especially severe in dry areas. Soil in these areas is easily destroyed because it is naturally thin. The land can easily deteriorate to the point where it becomes desert, a process called **desertification.** Worldwide, an area the size of Nebraska becomes desert each year.

An example of how desertification occurs is the Sahel region of northern Africa, shown in Figure 9-14. In the past, people in the drier part of the Sahel grazed animals, while those in the wetter part planted crops. The grazing animals were moved from place to place to find fresh grass and leaves. The cropland was planted for only four or five years at a time and then was allowed to lie fallow, or to "rest," for several years. These methods allowed the land to adequately support the people.

Today, an increased population has led to overuse of the land. Several crops a year may now be planted, and fallow periods may be shortened or eliminated, until the soil

Figure 9-14 **The map below shows the location of the Sahel region (in gold). The photograph shows how overuse of the land has led to desertification.**

finally loses its fertility. More and more animals are put out to graze. Trees and shrubs are cut for use as fuel or animal feed, until no plants are left to hold the soil in place or to trap any rain that falls. The topsoil in the Sahel is blowing away or being washed away by heavy rain. The land is becoming desert.

Soil Conservation

There are many ways of conserving topsoil and reducing erosion. One method is to pay attention to the slope of the land when planting. Contour plowing is plowing across the slope of a hill. The plow forms tiny ridges that help prevent the soil from washing down the hill. An even more effective technique is to leave strips of vegetation running across the hillside instead of plowing the entire slope, as shown in Figure 9-15. These strips catch soil and water that run down the hill. In very hilly areas, the best method of conserving topsoil may be to not farm the land at all, but to use it as forest or grassland.

Figure 9-15 **The ridges formed by plowing in curves around the hill prevent topsoil from washing away. This is contour plowing, a way of preventing erosion. Leaving strips of vegetation in place also reduces erosion.**

Using organic material instead of inorganic fertilizers can also help restore the soil. One type of organic material that can be used as fertilizer is compost. Compost is waste plant material that has been placed in piles and allowed to decompose. Animal manure also can be used as a fertilizer.

AQUACULTURE:
A Different Kind of Farming

Seafood provides about 40 percent of the animal protein consumed by people in developing nations. But overfishing is reducing the catches of fish and shellfish from the world's oceans. One solution to the problem is aquaculture.

Aquaculture—raising fish, shellfish, crustaceans, or seaweeds in artificial environments—is hardly a new idea. It probably began in China about 4,000 years ago. Today, China still leads the world in using aquaculture to produce freshwater fish.

There are a number of different types of aquaculture operations. The most common is known as a "fish farm." Farm operations generally consist of many individual ponds, each containing fish at a specific stage of development. Clean water is circulated through the ponds, bringing in oxygen and sweeping away carbon dioxide and fecal wastes. The fish grow to maturity in the ponds and are then harvested.

Another type of aquaculture operation is known as a "ranch." In this method, fish such as salmon are raised until they reach a certain age and then are released. The fish migrate downstream to the ocean, where they live until adulthood. At maturity, the fish return to their birthplace to reproduce. When they do so, they are captured and harvested.

Today, most of the catfish, crayfish, and rainbow trout and almost half of the shellfish consumed in the

Oyster farm in Washington State

Another method of soil conservation is to change the way the farmland is plowed, or tilled. In conventional farming, after a crop is harvested, the soil is plowed to turn it over and bury the remains of the harvested plants. Then it is raked before new seeds are planted. In **no-till farming,** shown in Figure 9-16, the seeds of the next crop are planted in slits that are cut into the soil, straight through the remains of the previous crop. The roots

Figure 9-16 **In no-till farming, a special machine cuts a slit in the ground to plant seeds (left). As the new crop grows, remains of the old crop slowly decay (right). In these photos, soybeans are planted among the remains of old rye plants.**

Answers to Thinking Critically can be found on page T153.

United States are products of aquaculture. Worldwide, about 10 percent of seafood comes from aquaculture.

Aquaculture is not without its drawbacks, however. If not managed properly, it can actually cause environmental problems. Large numbers of fish that are concentrated in a small space produce a great deal of waste matter, which can be a source of pollution. Also, large aquaculture operations can deplete local water supplies. One such operation in Texas used as much water as a city of 250,000 people, threatening the water supply of nearby San Antonio. Occasionally, sensitive wetlands are damaged or destroyed when large aquaculture operations are carved out of them. Fortunately, however, most aquaculture operations do not harm their surroundings.

Trout farm in Idaho

Aquaculture will probably continue to grow in importance as the world's population increases. While it cannot provide sustenance for the Earth's billions by itself, aquaculture may help to feed more people.

THINKING CRITICALLY

1. *Analyzing Processes* What are the advantages of aquaculture compared with traditional fishing?
2. *Inferring Relationships* Is aquaculture suitable for every part of the world? Why or why not?

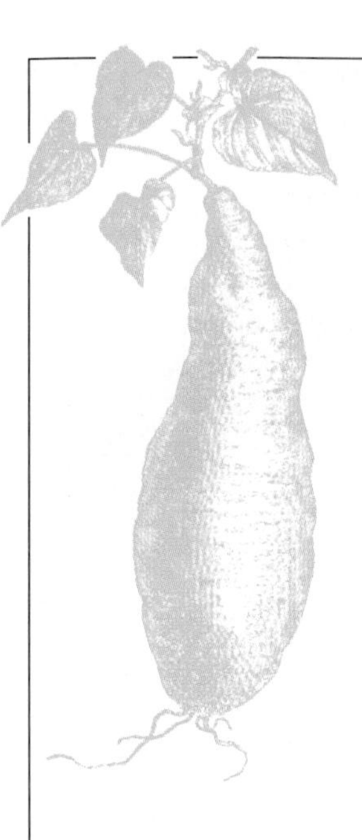

of the first crop hold the soil in place while the new crop develops. And as the remains of the first crop decay, organic matter is added to the soil. No-till farming uses less energy and saves time compared with conventional cultivation. It can reduce soil erosion to one-tenth of the amount caused by conventional methods.

Sustainable Agriculture

How can we continue to feed the world's population without continuing to deplete the world's resources? One answer is low-input farming. **Low-input farming** is farming without using a lot of energy, pesticides, fertilizer, and water.

One kind of low-input farming is organic farming, which is a method of growing plants without any synthetic pesticides or inorganic fertilizers. Organic farmers keep the soil moist and fertile by adding manure, compost, and other organic matter, by keeping the soil planted at all times to avoid erosion, and by alternating different crops to reduce pest populations. In addition to protecting the environment, organic farming saves money by reducing the need for water and expensive pesticides and fertilizer.

Another promising method of increasing food production is aquaculture, described in the Case Study on pages 236–237. Aquaculture is "fish farming," raising fish in artificial environments.

Preventing Salinization

The accumulation of salts in the soil is known as **salinization.** Although all soil naturally contains some salts, more salts are added when land is irrigated. This is because water for irrigation is taken from rivers or groundwater, which contain more salt than rainwater does. When land is irrigated, much of the water evaporates, leaving behind the salts it contained. Eventually, the soil may become so salty, like the land shown in Figure 9-17, that plants cannot grow in it.

Figure 9-17 **When irrigated land becomes salinized, it is difficult or impossible to farm.**

Another way that irrigation can cause salinization is by causing the level of groundwater to rise. Once the groundwater comes within a few meters of the surface, it is drawn up through the soil by capillary action, in much the way water is drawn up by a sponge. When it reaches the surface, the groundwater evaporates and leaves its salts in the soil.

Soil salinization can be slowed by careful irrigation methods. For example, irrigation canals can be lined to prevent water from seeping into the soil and raising the groundwater level. Another technique is to water the soil heavily to wash out salts before seeds are planted. This helps protect young seedlings, which are more easily harmed by salt than are older plants.

Planting salt-tolerant crops, which absorb salts from the soil, can also slow salinization. If the soil is too salty to grow crops, planting salt-tolerant trees, as shown in Figure 9-18, can help reclaim the land. Shade from the trees reduces the amount of water that evaporates from the soil surface. This slows the upward movement of groundwater. Tree roots penetrate the soil, allowing rainwater to sink in and wash out salts more rapidly. Fallen leaves improve the soil by adding organic matter. The water table can begin to fall within a year or two, although the complete restoration of salinized land takes decades.

Figure 9-18 **These Australian Girl Guides are planting trees to help reclaim former farmland that has become too salty to grow crops.**

Old and New Foods

Researchers are beginning to rediscover ancient plants and investigate new varieties of plants. They hope to find plants that are better adapted to different climates and that can produce high yields without large amounts of fertilizer, pesticides, and water. For example, some researchers are studying amaranth, shown in Figure 9-19. Amaranth was a sacred food of the Aztecs. It can survive with little water, and it produces a grainlike fruit that is rich in protein.

Other researchers are studying plants that have not been widely used for food in the past but that may be useful because they can grow on poor soil. For example, a plant called glasswort, also shown in Figure 9-19, grows naturally in salt marshes. It can be grown on saline soils and can even be irrigated with salt water.

Figure 9-19 **These two plants, amaranth (left) and glasswort (below), may provide food for the future.**

Answers to Section Review can be found on page T153.

Section Review

1. How do living organisms contribute to the fertility of the soil?
2. How does irrigation cause the soil to become salty?
3. Why are the world's deserts expanding?

Thinking Critically

4. *Relating Concepts* Create a concept map about soil conservation using the following terms: contour plowing, no-till farming, organic farming, careful irrigation, soil erosion, nutrient depletion, and salinization.
5. *Interpreting Graphics* Examine the map in Figure 9-13. Propose one possible reason for the great amount of soil erosion in Central America.

SECTION 9.3

PEST CONTROL

AFTER READING THIS SECTION YOU SHOULD BE ABLE TO

❶ explain why pest control is often necessary.

❷ explain how insects can become resistant to pesticides.

❸ describe alternatives to pesticides.

In North America, insects destroy about 10 percent of all crops. Crops in tropical climates suffer even greater insect damage because the insects grow and reproduce faster. In Kenya, for example, insects destroy 75 percent of the nation's crops. Worldwide, pests destroy about one-third of the world's potential food harvest. Clearly, pest damage must be controlled in order to feed the world's people.

As shown in Figure 9-20, insects are just one of several types of organisms that may be considered to be pests. A **pest** is any

Figure 9-20 Many kinds of organisms can be pests.

a. Tent caterpillars in cherry tree
b. Wasp on a grape
c. Weeds in a bean crop
d. Corn borer on an ear of corn
e. Hornworm eating a green tomato
f. Fungus on apricots
g. Brown rat in grain store

organism that occurs where you don't want it or in large enough numbers to cause damage.

Why are pests such a problem for cultivated plants? When plants grow in the wild, they are scattered among other plants, so insects and other pests have a hard time finding them. However, when they are grown as crops, they are planted in large fields containing only that one type of plant. The pests that feed on the plants are more likely to find them and do a lot of damage.

During the last 50 years, many new chemical **pesticides,** substances that kill pests, were invented. Some of the main types of pesticides are listed in Figure 9-21. The pesticides that kill insects, called insecticides, were so effective that farmers began to rely on them almost completely to protect their crops from insects.

Examples of Pesticides

Pesticide	Toxicity	Persistence
Chlorinated Hydrocarbons		Persistent (half-life 2–5 years)*
Aldrin	Extremely toxic	
DDT	Very toxic	
Lindane	Very toxic	
Organophosphates		Degradable (half-life 1–10 weeks)
Malathion	Moderately toxic	
Parathion	Extremely toxic	
Carbamates		Degradable (half-life 1 week)
Carbaryl	Moderately toxic	
Zectran	Extremely toxic	

**The half-life of a pesticide is the time it takes for half of the quantity applied to be broken down.*

Figure 9-21 **The table above shows some of the main types of pesticides developed in the last 50 years.**

Figure 9-22 **This field of pineapples in Hawaii is being sprayed with a pesticide.**

Drawbacks of Pesticides

Unfortunately, the new pesticides affect a lot more than just the pests that they were designed to kill. They can also harm people and wildlife.

Figure 9-23 This worker is wearing long sleeves, long pants, gloves, and a mask to protect himself while applying pesticides.

Health Concerns Many pesticides can cause people to get sick. For example, the San Joaquin Valley in California is an area of fruit and vegetable farms with very high pesticide use. Cancer rates among children there are eight times the national average. Workers in pesticide factories may also become ill. And people living near these factories, like the people of Bhopal, India, may be endangered by accidental chemical leaks. (See Figure 9-24.)

Pollution and Persistence The problem of pesticides harming people and wildlife is especially serious because many pesticides are persistent. Persistent pesticides do not break down rapidly into harmless chemicals when they enter the environment. As a result, they accumulate in the water and soil.

One of the best known and most persistent pesticides is DDT. This pesticide was used in the 1940s to kill the mosquitoes that cause malaria and the lice that spread typhus, saving millions of lives. However, because DDT is very persistent, it gradually accumulated in bodies of water. It was then absorbed by fish, which

Figure 9-24 This Union Carbide factory in Bhopal, India, made pesticides. In 1984 a cloud of deadly gas was accidentally released into the air. Thousands of people were killed or injured.

were eaten by birds. Organisms high on the food chain, like fish-eating birds, are likely to have high levels of DDT in their bodies because they eat many smaller organisms whose bodies contain low levels of DDT. The DDT caused the birds to lay eggs with shells so thin that they broke when the birds sat on them. Penguins, pelicans, peregrine falcons, and eagles were endangered and even wiped out in some areas as a result. DDT was eventually banned in many places, including the United States. Although it remains in the environment, it has gradually dispersed, so that most birds eat less of it. Many of the endangered birds have increased in number as a result.

Some of the most harmful pesticides have now been banned, but many of them will remain in the environment for years to come. In the United States, about 60 different pesticides have been found in groundwater, sometimes making wells unusable.

Did you know that you can get rid of household pests like ants and cockroaches without harmful chemicals? For more information, turn to pages 390–391.

Resistance You might think that the most effective way to get rid of pests is to spray often with large amounts of pesticide. However, in the long run this approach sometimes makes the pest problem worse. Pest populations evolve **resistance,** the ability to tolerate a particular pesticide.

How does resistance evolve? A few insects out of many may happen to have a gene that protects them from a pesticide. (For example, they may be able to break down the pesticide into harmless substances.) These few insects survive spraying with the pesticide while most of the others are killed. The surviving insects pass on the gene to their offspring. With continued pesticide spraying, the resistant insects continue to reproduce and increase in number while the nonresistant ones die off. Eventually, the entire insect population is resistant to the pesticide. As shown in Figure 9-25, 500 species of insects have developed resistance to pesticides since the 1940s.

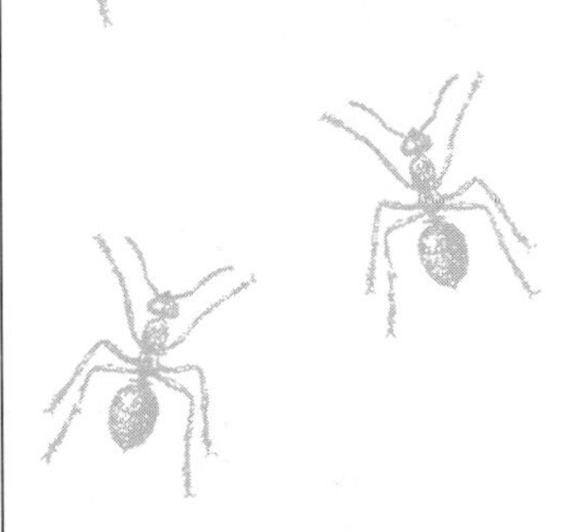

Figure 9-25 Pesticides have been widely used for about 50 years now. During that time, the number of insect species resistant to one or more insecticides has increased dramatically.

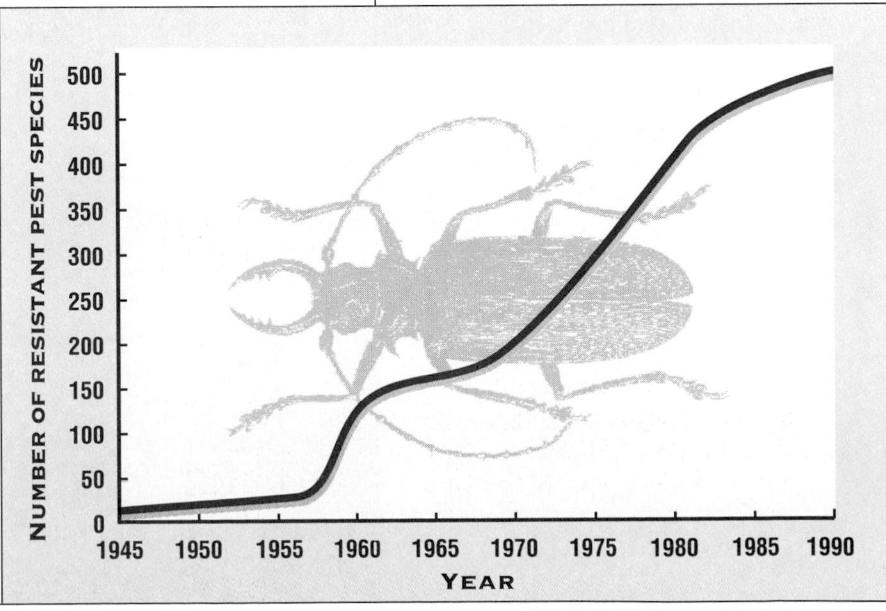

Biological Pest Control

Because resistance evolves rapidly, farmers and pest-control companies are using fewer pesticides. They are turning to **biological pest control,** using living organisms or naturally produced chemicals to control pests. Biological pest control methods generally do not harm organisms other than the pests they are designed to control, and resistance takes much longer to evolve.

Figure 9-26 **Gypsy moth caterpillars feed on the leaves of trees in the northeastern United States. Spraying with the bacterium *Bacillus thuringensis* can control outbreaks of the caterpillars.**

Answers to Section Review can be found on page T153.

Predators and Pathogens One type of biological control involves releasing a natural predator or parasite into the area where the pest lives. **Pathogens**—organisms that cause disease—can also be used to control pests biologically. One of the most common pest-control pathogens is the bacterium *Bacillus thuringensis,* often abbreviated *Bt.* This bacterium can kill the caterpillars of moths and butterflies, as well as other insect larvae. Each strain of *Bt* attacks the larvae of a specific insect. *Bt* is so successful at controlling gypsy moth caterpillars, shown in Figure 9-26, that it is one of the main methods of controlling this pest.

Chemicals From Plants Another type of biological pest control makes use of chemicals that some plants produce naturally to protect themselves from pests. Several of these chemicals have been extracted from plants, such as the chrysanthemums shown in Figure 9-27, and are now sold as pesticides. These products are biodegradable, meaning that they are broken down into harmless substances by bacteria and other decomposers.

Disrupting Insect Breeding Pheromones, chemicals produced by one organism that affect the behavior of another, can also be used in pest control. For example, female moths find mates by releasing pheromones that attract males from miles away. By treating crops with pheromones, farmers can confuse the male insects and interfere with mating. This method often can reduce the number of insects much more than pesticides can. Pheromones are delicate chemicals that are expensive to make and that must be handled with care, so they are not yet widely used. However, because they are so effective at controlling insects, they are likely to be used more in the future.

Another way to prevent insects from reproducing is to make it physically impossible for the males to reproduce. Male insects are treated with X rays to make them sterile and then are released. When they mate with females, the females produce eggs that do not develop.

Figure 9-27 **The photograph below shows painted daisies, a type of chrysanthemum that grows in Rwanda. Extracts from the plants are used as pesticides.**

Section Review

1. Why are pests more of a problem for cultivated plants than for wild plants?
2. Why are persistent pesticides dangerous?
3. How do insect populations evolve resistance to pesticides?

Thinking Critically

4. *Inferring Relationships* Are pheromones a type of pesticide? Explain your reasoning.
5. *Applying Ideas* What pesticides are used around your home? How could you reduce the amount of pesticides used at home?

CHAPTER 9

HIGHLIGHTS

SUMMARY

- Many people in the world today are malnourished or starving. People who do not get sufficient nourishment develop serious health problems and can even die.
- Malnutrition and starvation are caused partly by increased population and partly by human inefficiency or interference.
- New strains of food crops have resulted in tremendous increases in productivity, enabling more people to be fed than in the past.
- A limited amount of land is suitable for growing crops. Improper land-use practices cause that amount to decrease every year because of damage to the soil.
- In an effort to control crop-destroying pests, farmers typically apply large amounts of poisonous chemicals to crops.
- The act of controlling pests can result in their becoming resistant to the control.

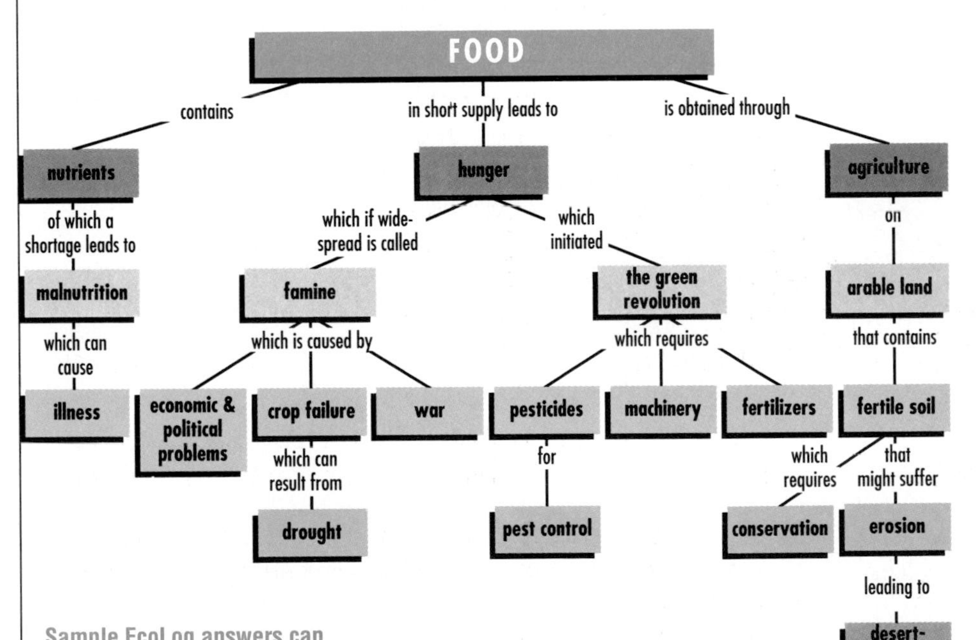

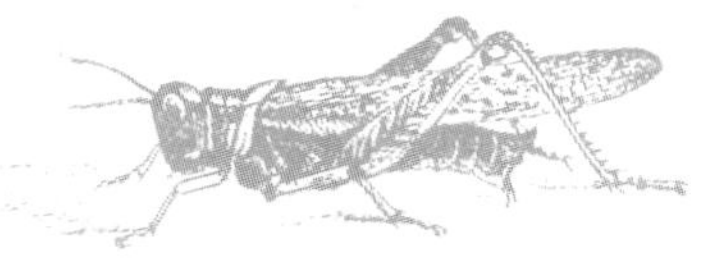

Vocabulary Terms

arable (p. 231)
biological pest control (p. 243)
desertification (p. 235)
droughts (p. 229)
erosion (p. 234)
famine (p. 229)
fertile soil (p. 232)
low-input farming (p. 238)
malnutrition (p. 228)
no-till farming (p. 237)
pathogens (p. 244)
pest (p. 240)
pesticides (p. 241)
resistance (p. 243)
salinization (p. 238)
subsistence farmers (p. 229)
topsoil (p. 232)
yield (p. 229)

Sample EcoLog answers can be found on page T153.

EcoLog

Now that you've studied this chapter, revise your answers to the questions you answered at the beginning of the chapter, based on what you have learned.

1. Why is it so difficult to provide adequate food for all of the world's people?
2. Besides using chemical pesticides, how do farmers keep insects from destroying crops?

REVIEW

UNDERSTANDING VOCABULARY

1. In your own words, explain the meaning of each of the following terms.
   a. malnutrition
   b. erosion
   c. desertification
   d. resistance

RELATING CONCEPTS

2. Copy the unfinished concept map below onto a sheet of paper. Then complete the concept map by writing the correct word or phrase in each box containing a question mark.

Answers not indicated can be found on page T153.

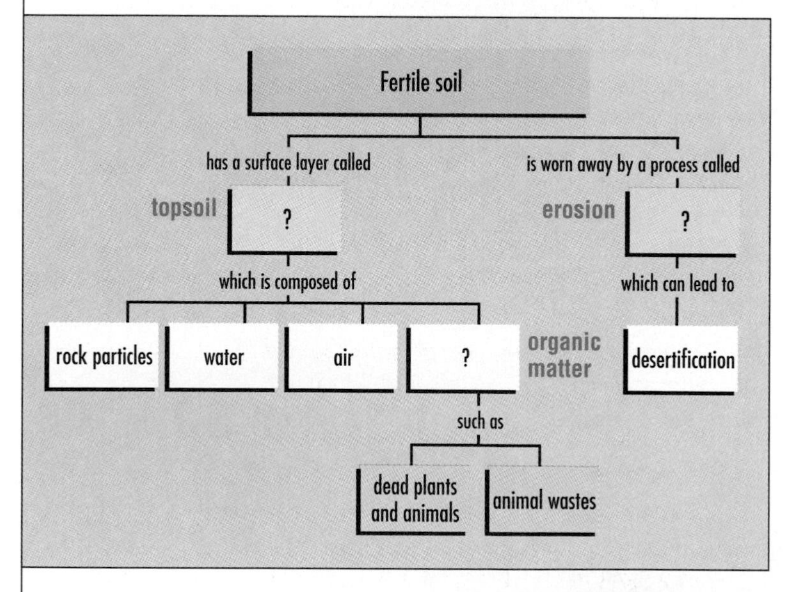

UNDERSTANDING CONCEPTS

Multiple Choice

3. The deterioration of the land until no plants are left is known as
   a. degradation.
   b. erosion.
   (c.) desertification.
   d. drought.

4. Consuming insufficient nutrients results in which condition?
   (a.) malnutrition
   b. drought
   c. desertification
   d. resistance

5. Famine can result if
   a. crops fail on a massive scale.
   b. war or inefficient transportation prevents food from reaching a population.
   c. food production increases faster than the human population grows.
   (d.) both a and b

6. The "green revolution" had what effect?
   a. Desertified areas were reclaimed.
   b. Subsistence farmers began making profits.
   (c.) The yields of many crops were greatly increased.
   d. Drought-resistant crops allowed farmers to operate in areas that were formerly too dry for farming.

7. Which is NOT a method of preventing soil degradation?
   a. contour plowing
   b. farming with organic methods
   (c.) irrigation
   d. no-till farming

8. Salinization
   a. is a method of ridding the soil of harmful pests.
   (b.) is the deposition of salt in the soil due to irrigation.
   c. destroys the soil forever.
   d. is a byproduct of desertification.

9. Pesticides can lose their effectiveness after a while because
   (a.) the pests develop resistance to the pesticide.
   b. the pesticides have a very short half-life.
   c. the insects learn to flee when the pesticide is applied.
   d. the pesticide breaks down rapidly in nature.

10. Growing crops on a plot of land tends to
    a. increase the fertility of the land because minerals and organic material are added to the soil.
    b. decrease the fertility of the land because minerals are removed from the soil.
    c. have no measurable effect on the soil.
    d. salinize the soil.
11. Which is NOT an example of biological pest control?
    a. pyrethrins
    b. chlorinated hydrocarbons
    c. *Bacillus thuringensis*
    d. pheromones
12. The impenetrable rock layer below the soil is called
    a. subsoil.
    b. the zone of leaching.
    c. rock particles.
    d. bedrock.
13. No-till farming
    a. requires less energy than conventional cultivation methods.
    b. causes severe soil erosion.
    c. is more time-consuming than conventional cultivation methods.
    d. all of the above

Short Answer

14. Why is soil erosion such a problem for farmers?
15. What are some of the dangers of excessive pesticide use?
16. What problems were caused by the persistence of DDT?
17. Explain how pathogens are used to control pests.
18. What are some of the drawbacks of irrigation?

Interpreting Graphics

19. Examine the graph below. It charts the yield of a farm over several years. In 1991 a pesticide was applied for the first time. It was reapplied in 1992 at the same dosage and in 1993 at twice the original dosage. In 1994 a new pesticide was applied. It was reapplied in 1995 at the same dosage. Provide an explanation for the pattern shown by the yield measurements.

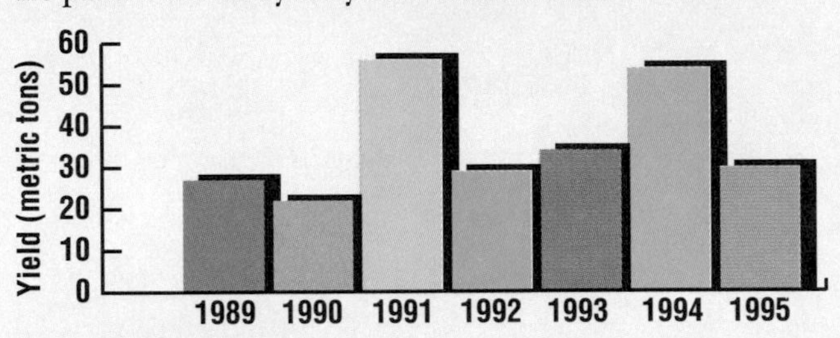

Thinking Critically

20. ***Recognizing Relationships*** What factors related to their reproduction ensure that insects inevitably gain resistance to the pesticides used against them?
21. ***Inferring Relationships*** Usable energy from sunlight is lost when moving higher up the food chain. In contrast, certain chemicals such as DDT become more concentrated as they move up the food chain. Explain this difference.

Themes in Science

22. ***Evolution*** What farming practices are actually causing insects to evolve at an accelerated rate? Explain.
23. ***Energy*** Humans are both primary and secondary consumers. Which do you think is more suitable for our large and growing population? Why?

Cross-Discipline Connection

24. ***History*** Although it now occurs more frequently and on a larger scale, human-caused environmental degradation has occurred throughout history. Research a historical example of environmental degradation, and write a short essay about it.

Discovery Through Reading

For an interesting look at the effects of excessive pesticide use on the environment, read *Silent Spring*, by Rachel Carson, Boston: Houghton Mifflin, 1962.

PORTFOLIO ACTIVITY

Do a special project comparing organic and conventional farming. Which method generally produces greater yields? Which method better maintains the quality of the soil? Prepare a montage to illustrate the two methods and how they compare in various categories.

INVESTIGATION

The Case of the Failing Farm

You work for the Bucolic County Soil Conservation District. One day you are given the following letter from a worried local farmer, describing a problem that you have been assigned to research and solve.

Answers to this Investigation can be found on page T154.

Materials

- cafeteria tray or sheet cake pan
- sandy soil
- metric ruler
- small blocks
- twigs
- toothpicks
- fresh grass clippings
- dried peas or beans
- spray bottle filled with water
- fork
- narrow strip of carpet
- materials of your choice, to be blended with soil
- notebook
- pen or pencil

Manager
Soil Conservation District
Bucolic County, Michigan

Dear Sir or Madam,

I hope you can help me solve a mystery. We have been experiencing some very unusual events over the past two years on our family farm. We own 80 acres of land, consisting primarily of rolling hills. The soil is mainly a sandy loam, which provides very good drainage. In the past few years, we have had the following problems:

- Productivity is decreasing.
- Bare patches of ground are appearing on the tops and sides of the hills.
- The stream is becoming wider and less deep.
- Small dust storms are occurring in July.
- Gullies are forming on the hillsides of the pastureland.

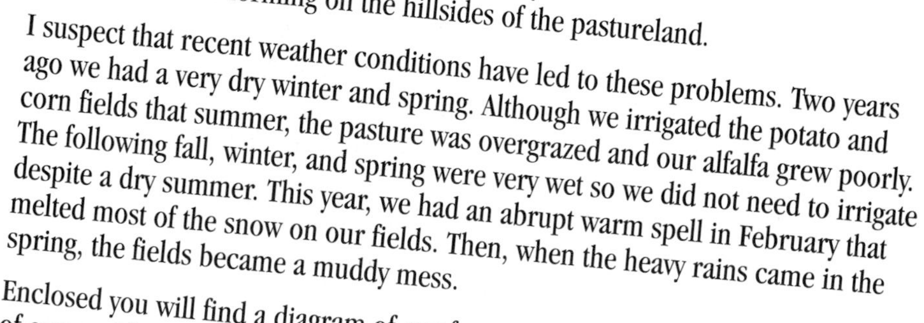

I suspect that recent weather conditions have led to these problems. Two years ago we had a very dry winter and spring. Although we irrigated the potato and corn fields that summer, the pasture was overgrazed and our alfalfa grew poorly. The following fall, winter, and spring were very wet so we did not need to irrigate despite a dry summer. This year, we had an abrupt warm spell in February that melted most of the snow on our fields. Then, when the heavy rains came in the spring, the fields became a muddy mess.

Enclosed you will find a diagram of our farm. I hope you can determine the cause of our problems and advise us on how to solve them.

Sincerely,

Francis Katawa

Francis Katawa

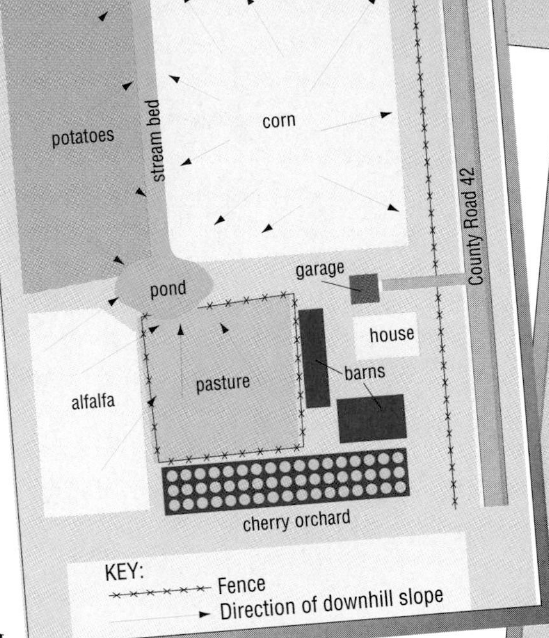

The Katawa farm

As you review the information that Mr. Katawa has given you, you realize that recent weather conditions have made the land very susceptible to soil erosion. You decide to make a model of the Katawa farm to observe the effects of erosion in more detail and to study ways of reducing erosion.

Construct a model similar to this one.

A MODEL FARM

1. Construct a model of the Katawa farm by filling a large tray with sandy soil, building the hilltops to a height of 5–10 cm, and making valleys and depressions where streams and a pond will form. Do *not* add water to your model at this time.
2. Add small blocks of wood, plastic, or metal to represent the house and farm buildings; add twigs to represent trees; and add toothpicks to represent fences.
3. Spread grass clippings over the areas that represent the pasture and alfalfa field. Stick blades of grass upright into the soil in rows to represent the corn in the cornfield. Place small dried peas or beans in rows to represent the potato plants in the potato field.

EROSION IN ACTION

4. Once your model is constructed, simulate precipitation using a spray bottle set on the "fine" setting. Continue spraying until water begins flowing in the valleys and accumulating in the pond area.
5. What happens as you spray more and more water onto the land? Write your observations in your notebook.
6. Why are the streams becoming wider and shallower?
7. What happens when you use a fork to punch small indentations into the pasture, simulating the tracks left by cows? (Be sure to stomp it up a bit!)
8. What happens if you cut the alfalfa (remove the clippings covering the field)?

MODEL EXPERIMENTS

9. Try the following modifications to the farm, and record their effects on erosion.
   - Plant the rows of corn and potatoes in a different direction.
   - Exchange the locations of the alfalfa crop and corn crop.
   - Add materials to the soil to reduce runoff. Use your imagination to choose materials.
   - Place a narrow strip of carpet in the stream bed to simulate planting the stream bed with grass.
10. What else could you do to your model to slow down the erosion? Try at least three ideas of your own. Note your changes and their results in your notebook.

RECOMMENDATIONS

11. What could Mr. Katawa do to help solve the erosion problems? List at least five different things he could do to prevent soil erosion and to improve the quality of the soil.

POINTS OF VIEW

GENETICALLY ENGINEERED FOODS

As the world population continues to grow, food production must try to keep pace with the increase. Genetic engineering provides one way to develop new foods. Biotechnologists develop desirable characteristics in an organism by altering its genes or inserting new genes into the organism's cells. For example, a gene that makes one plant species resistant to pests might be transferred to another plant species. The second plant species would then have the same resistance to pests.

In 1994 the first genetically modified whole food was offered for sale. It is a tomato called the Flavr Savr™ that softens slowly, allowing it to remain on the vine longer than most tomatoes. Biotechnologists developed the tomato by altering the gene that causes ripe tomatoes to soften. In clearing the new tomato for sale, the Food and Drug Administration (FDA) said it was as safe as other tomatoes. Here are two points of view on genetically engineered foods.

This hybrid squash was grown using traditional plant-breeding techniques.

THE BENEFITS OUTWEIGH THE RISKS

Scientists who support the manufacturing of genetically engineered foods view the process as simply an extension of previous plant-breeding techniques. Traditionally, farmers altered the genetic makeup of plants by crossbreeding different strains to combine the best traits of both plants. The direct manipulation of genes makes it possible to control genetic changes more precisely and efficiently than with traditional breeding methods.

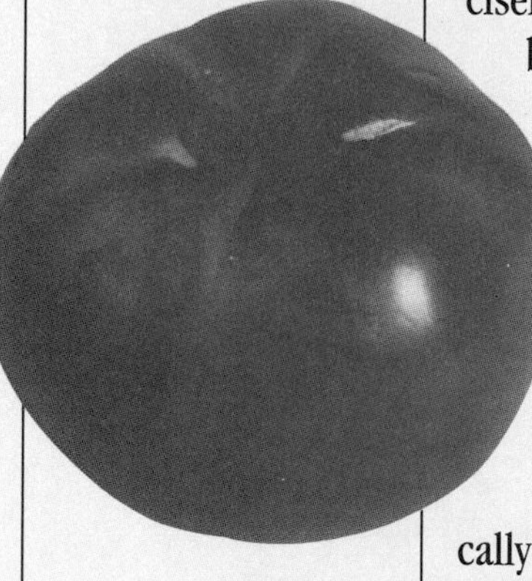

This tomato is just one of many genetically engineered food products.

Biotechnologists say that their new products are as safe for consumers as plants developed through crossbreeding. Why shouldn't genetically engineered foods sit beside others on grocery store shelves?

The benefits of creating genetically engineered fruits and vegetables include keeping produce fresh longer, adding nutrients, and creating more successful crops. For example, by inserting a gene that confers virus resistance in squash plants, scientists could boost the plants' resistance to viral infection. These resistant squash could produce five times the amount of squash per harvest as other squash. Or scientists could increase the amino acids in a food product to give it a higher nutritional value.

Crops could be developed to grow faster and have higher yields. To combat world hunger, scientists may be able to develop seeds that can grow well in areas with poor soil or water conditions. For more immediate relief, genetically engineered foods that would not spoil as quickly could be shipped to needy nations.

Proponents of genetic engineering say that the technology could provide fresher, more nutritious food for shipment to hungry people.

THE RISKS OUTWEIGH THE BENEFITS

Critics of genetically engineered foods want these foods labeled just as foods with additives such as saccharin are labeled.

Scientists in labs like this one refine the genetic-engineering techniques used to develop new foods.

Critics of genetically engineered foods believe that these products are significantly different from foods developed through traditional methods. Genetic engineering allows genes from any living organism, even animal or bacterial genes, to be placed into crops. Opponents are concerned about the safety of foods containing these "foreign" genes.

Another safety concern is the possibility of allergic reactions. Some foods, like peanuts and shellfish, cause allergic reactions in many people. Genes from these foods, if placed in an entirely different product, could cause allergic reactions in people who do not know they are eating food containing the foreign genes.

Other critics object because of religious or ethical reasons. Certain religions prohibit eating pork and other foods. People may object to the insertion of genes from pigs or other prohibited foods into foods they normally eat. Similarly, vegetarians might object to eating even one gene from an animal.

Some scientists are concerned that new, genetically engineered species may be accidentally introduced into the wild. Genetic engineering may give the new species an advantage over the existing wild species. If the new species thrives at the expense of the wild species, the wild species could become extinct.

The first genetically engineered foods were offered for sale in grocery stores in 1994.

ANALYZE THE ISSUE

1. ***Recognizing Relationships*** Some people propose that genetically engineered foods should have labels to identify the foreign genes used. How could such a measure decrease criticism about the safety of genetically engineered foods?

2. ***Making Decisions*** Use the decision-making model presented in Chapter 1 to decide whether you would buy genetically engineered foods at the grocery store.

Answers to Analyze the Issue can be found on page T154.

CHAPTER 10

BIODIVERSITY

". . . in wildness is the preservation of the world."

HENRY DAVID THOREAU
AMERICAN WRITER AND NATURALIST

EcoLog

Before you read this chapter, take a few minutes to answer the following questions in your EcoLog.

1. What would you guess the major cause of extinction is today?
2. Do you think humans should try to prevent the extinction of other species? Explain your reasoning.

BIODIVERSITY AT RISK

AFTER READING THIS SECTION YOU SHOULD BE ABLE TO

❶ explain how humans are causing extinctions of other species.

❷ explain why it is important that we preserve biodiversity.

The last dinosaurs died about 65 million years ago. Why? We don't know for sure, but one hypothesis is that a huge meteor crashed into the Earth. Another is that there was a dramatic increase in volcanic activity. In either case, it is possible that enough dust and ash were thrown into the atmosphere to block out much of the sunlight that normally reaches the Earth's surface. Without enough sunlight, many of the Earth's plants would have died. This would have led to the deaths of animals that ate plants and animals that fed on other animals. It is estimated that about half the species on Earth, including all the dinosaurs, became extinct during this time. As you learned in Chapter 2, a species is extinct when the last individual dies.

The dinosaurs' disappearance was part of a mass extinction, which is the extinction of many species during a relatively short period of time. Our planet has experienced several mass extinctions, as shown in Figure 10-1, each probably caused by a change in climate.

Answer to caption: The mass extinction most destructive to marine organisms was that which began just before the Triassic period.

Figure 10-1 The five mass extinctions indicated by ▼ were probably caused by changes in climate. This graph shows the effect on families (groups of related species) of marine organisms. Which mass extinction was the most destructive to marine organisms?

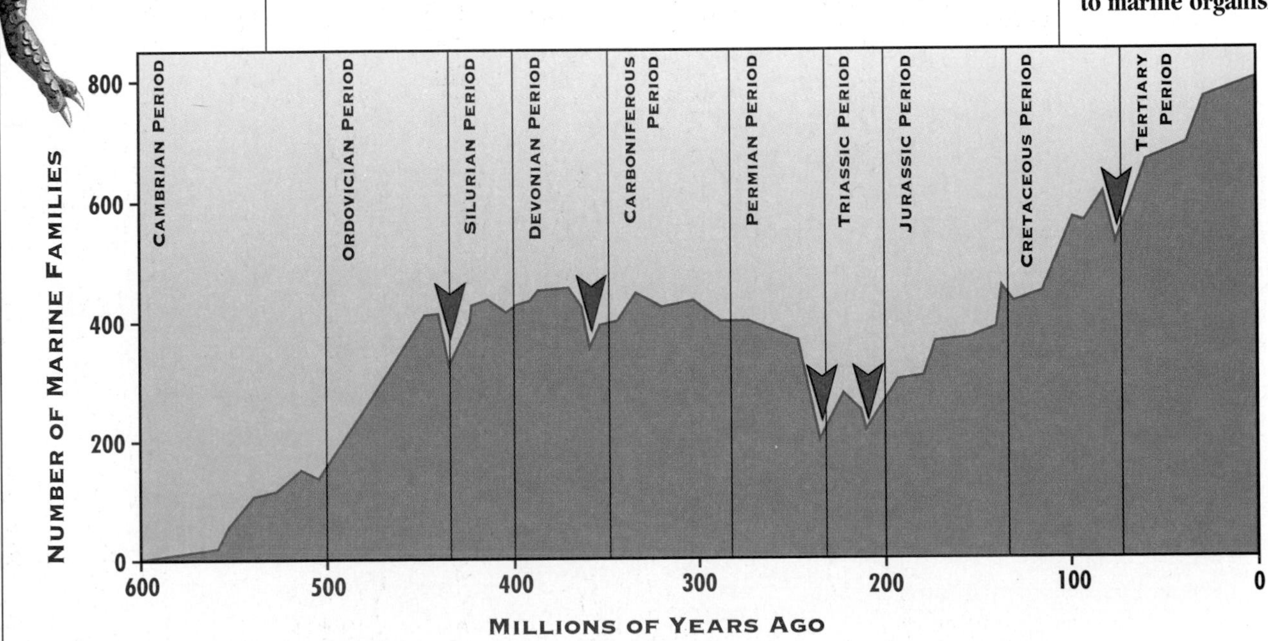

Discover the diversity of weeds and other plants in a small area. Yards, gardens, and vacant lots are good places to study. Mark off a section .5 m square. Using a field guide, identify every plant species. You may want to sketch some of the organisms. Then count the number of each kind of plant you identified, and record your results in a table.

Many scientists think we are living during another mass extinction. They fear that by the year 2100, 25 percent or more of all species of plants and animals that were on Earth in 1900 will have become extinct. If a mass extinction is underway, it is being caused not by a natural change in climate but by the actions of humans. Humans weren't responsible for the dinosaurs' extinction—that was millions of years before our time. But we are now causing the extinction of thousands of other living things.

A World Rich in Biodiversity

The term **biodiversity** refers to the number and variety of species on Earth. As shown in Figure 10-2, the number of species known to science is about 1.4 million, most of which are insects. The number of *known* species, however, by no means comes close to the *actual* number of species on Earth. Estimates of the actual number of species now living on our planet range from 10 million to 100 million.

In one study in Indonesia, British researchers counted the number of species of insects called hemipterans in a region of tropical rain forest. They found 1,690 species of hemipterans living on the ground there, but only 37 percent of these species had been known previously. This study and others point out that we do not know about many of the species alive today.

How Are Humans Causing Extinctions?

The human population of the world is increasing at a rate of about 260,000 people each *day*. Because the human population is growing so rapidly and changing the environment so dramatically, we are causing other species to become extinct at an accelerated rate.

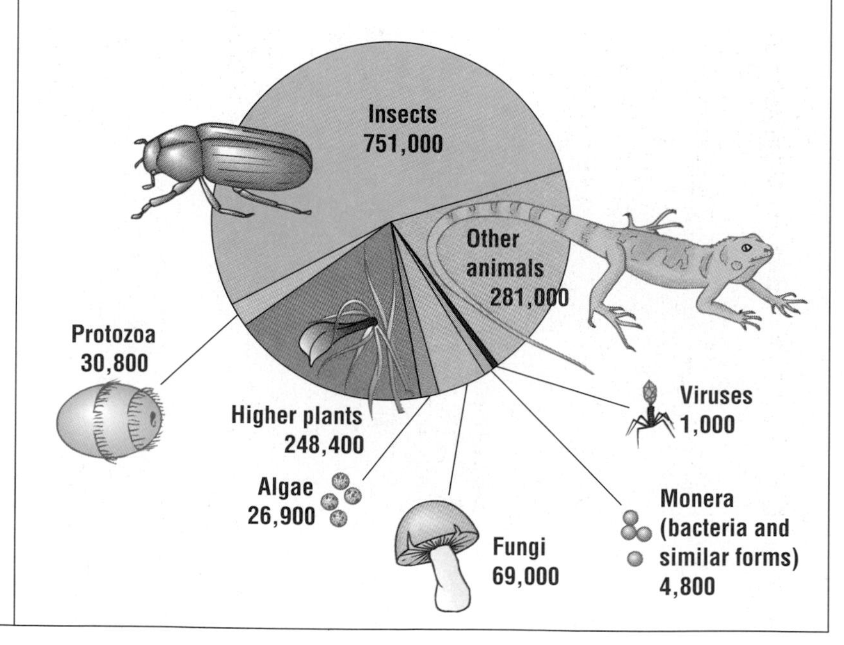

Figure 10-2 This pie chart shows the number of species now known to exist, divided into major groups. Scientists estimate, however, that the actual number of species far exceeds the 1.4 million shown here. Perhaps as many as 10 million to 100 million species exist, most of them undiscovered.

In fact, plant and animal species are disappearing at least 1,000 times faster than at any other time in the last 65 million years. Evidence seems to suggest that this dramatic increase in extinctions has largely taken place during the last century.

Why is the increase in the human population so devastating to other species? As humans take up more and more space on the planet, we destroy the habitats of other species. The most common cause of extinctions today is the destruction of habitats by humans. Unregulated hunting and the introduction of nonnative species also contribute to extinctions.

Habitat Destruction As human populations grow, we use more land to build homes and harvest resources. When we take over land, we destroy the habitats of other species. It is estimated that habitat loss causes almost 75 percent of the extinctions now occurring. As you read in previous chapters, organisms have evolved to survive in a particular place under particular environmental conditions. When their habitats are destroyed, they die.

One large mammal in grave danger because of habitat loss is the Florida panther, a kind of cougar. The Florida panther is shown in Figure 10-3. Two hundred years ago, cougars (also called mountain lions) ranged from Alaska to South America. Cougars require a large range and protective cover for feeding and resting, as well as an adequate source of large prey such as deer and elk. At one time the needs of the cougar were met in the forests of its original range. Today, much of the cougar's forest habitat has been destroyed or broken up into small patches by roads, canals, and fences. The only cougar population east of the Mississippi River is the Florida panther. Despite efforts to save the Florida panther, in 1994 only about 50 remained in the wild, making the Florida panther one of the most endangered animals in North America.

Figure 10-3 **Only about 50 Florida panthers remain in the wild. Almost all of the Florida panther's habitat was destroyed or fragmented by human development.**

Figure 10-4 **This Brazilian rain forest is being cleared by burning. Each minute, almost 2,000 hectares (5,000 acres) of rain-forest habitat are eliminated from the face of the Earth.**

Another animal endangered primarily by habitat loss is the whooping crane. The plight of the whooping crane is described in the Case Study below.

Most extinctions are occurring in tropical rain forests when the land is cleared for farming or cattle grazing. Biologists estimate that at least 50 percent of the world's species live in tropical rain forests, even though these forests cover only 7 percent of the Earth's land surface. Rain forests contain millions of species that have never been described, many of which may become extinct before we know much about them.

Hunting Unregulated hunting can also lead to species extinction. In the early 1900s, for example, 2 billion American passenger pigeons were legally hunted to extinction in the United States.

Similarly, the American buffalo (also called bison) was nearly

CASE STUDY

WHOOPING CRANES:

REGAINING LOST GROUND?

Whooping cranes spend the winter at Aransas National Wildlife Refuge in Texas.

Many years ago whooping cranes waded through wetlands and soared across prairies in Canada, the United States, and Mexico. The huge white birds, sometimes called whoopers, foraged for food while avoiding humans. But 200 years ago, a few early settlers wandered into the whoopers' habitat and discovered that the land was well suited for farming. A rush of settlers immediately descended upon the land—draining wetlands, uprooting vegetation, and cutting down trees to make room for crops and cattle. The birds were forced out.

As their habitat was destroyed, the whooping cranes kept moving their breeding grounds northward, until some of them ended up in a remote area of Canada. These were the lucky ones—beginning in 1922, the region was protected by the Canadian government as a national park.

From their breeding grounds in Canada, their annual journey south became especially long. They had to fly up to 4,000 km (2,500 mi.) to their winter home. The birds rested each night during the migration, occasionally staying for a week or so at a single spot. But when the whooping cranes stopped to rest, farmers shot them because they fed on crops. Hunters found them easy targets because of their large size and spectacular white hunted to extinction in the 1800s. When Europeans first came to North America, at least 60 million buffalo roamed the continent's vast prairies and forests. Early travelers across the Great Plains said there were so many buffalo that they covered the land like a dark sea. By 1906 only about 300 buffalo remained. Many of the buffalo had been killed just for their tongues (considered a delicacy) and hides. The rest of the animal was left to rot. Today, laws protect the buffalo, and their population has grown to about 130,000.

Legal hunting is no longer a major cause of extinction in countries that have wildlife laws. In the United States, hunting organizations and government agencies work together to make sure that only a certain percentage of game animals are killed each year.

In developing countries, however, where animal meat or cash for game can mean the difference between starvation and survival, hunting still threatens many species. In those countries, **poaching** (illegal hunting) can threaten animals with extinction. For example, African elephants were reduced in number when their habitats were

Answers to Thinking Critically can be found on page T154.

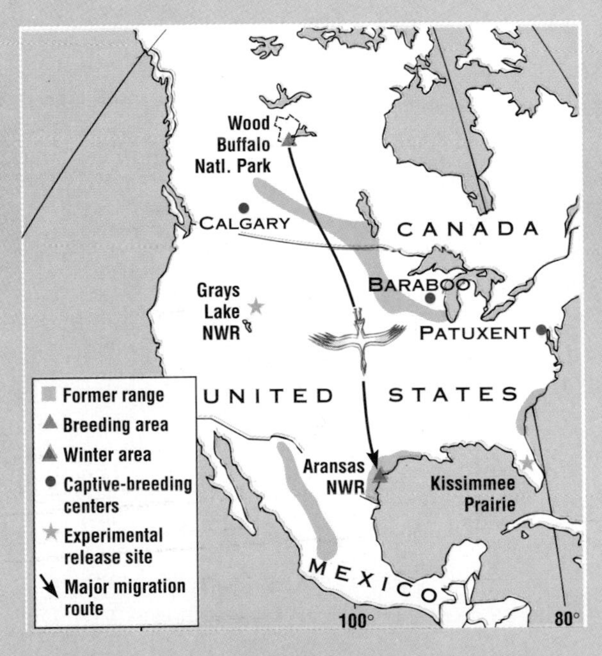

The natural range of the whooping cranes is now limited to a breeding area in Canada and a winter area in Texas.

plumage. Many whooping cranes didn't survive the long migrations.

Fortunately, in 1937 the Aransas National Wildlife Refuge was established in Texas to protect the whooping cranes and other wildlife. Without the protected breeding grounds in Canada and wintering grounds in Texas, the whooping crane would almost certainly be extinct today.

As it was, by the late 1930s the whooping crane was nearly extinct. Only two wild flocks remained—the flock that migrated between Canada and Texas, and another flock that lived permanently on the coast of Louisiana. The Texas flock had a population of only about 20, and the Louisiana flock numbered 13. By 1941, just 15 whooping cranes were left to spend their winters in Texas. All of the Louisiana cranes were dead by 1948.

Although the whooping crane is still considered one of the most endangered birds in North America, its population is slowly growing. As a result of habitat protection, captive-breeding efforts, and anti-poaching laws, there are now more than 230 whoopers.

THINKING CRITICALLY

1. *Applying Concepts* Do you think 250 whooping cranes make a large enough population to ensure the species' survival? Why or why not?

2. *Analyzing Viewpoints* Write a paragraph agreeing or disagreeing with the following statement. If a person owns land that is valuable habitat for the endangered whooping crane, that person should not be allowed to develop the land.

Figure 10-5 The melaleuca tree, imported from Australia, displaces the native plant species of Florida.

converted to farmland, and their numbers are still declining because of poaching. To meet the demand for elephant tusks, poachers killed half of all the elephants in Africa between 1979 and 1989.

Exotic Species An **exotic species** is a species that is not native to a particular region. Even such familiar organisms as cats and rats are considered to be exotic species when they are brought to regions where they have never lived before. Exotic species can threaten native species, which have no natural defenses against them.

In Florida, the exotic melaleuca tree, shown in Figure 10-5, displaces native plant species and is a serious threat to the Everglades wetlands. In fact, the tree was imported from Australia to Florida in the early 1900s in hopes that it would destroy the Everglades. Since melaleuca trees absorb a great deal of water, real-estate developers hoped that by planting them, the trees *would* dry up the wetlands and make them suitable for development. Now, the trees spread through Florida at a rate of up to 20 hectares (50 acres) a day. Researchers are searching for a way to check the spread of the melaleuca tree.

THE VALUE OF BIODIVERSITY

Extinction is a natural process. Throughout the history of life on Earth, species have appeared, flourished for a time, and then become extinct. Probably 99 percent of all the species that have ever lived are now extinct. Protecting species on the verge of extinction is difficult and expensive. So why should we work so hard to slow the rate of extinction?

Saving Species Preserves Ecosystems Species, along with abiotic (nonliving) factors, are what make up ecosystems. It is important to maintain healthy ecosystems because they ensure a healthy biosphere by regulating the flow of energy and the cycling of nutrients.

Each species has a role to play in its ecosystem, and each is dependent upon other species for survival. The ways that species depend on each other are not always obvious. If one species disappears from an ecosystem, the ecosystem changes. If another species disappears, the ecosystem changes a little more. At what point does the entire ecosystem collapse? How many threads can be pulled from a web of life before it completely unravels? We often don't know the answer until it's too late.

Some species are so important to the functioning of an ecosystem that they are called *keystone* species. One of the most famous examples of a keystone species is the sea otter. Sea otters live on the Pacific coast of North America, eating the plentiful shellfish and sea urchins in the offshore kelp beds. Sea otters were hunted for their fur throughout the 1800s until they were thought to be extinct. With the sea otters gone from the area, the sea urchins

Pacific coast sea otters lived in kelp beds, eating sea urchins.

In the 1800s, sea otters were hunted for their fur until they disappeared from the area.

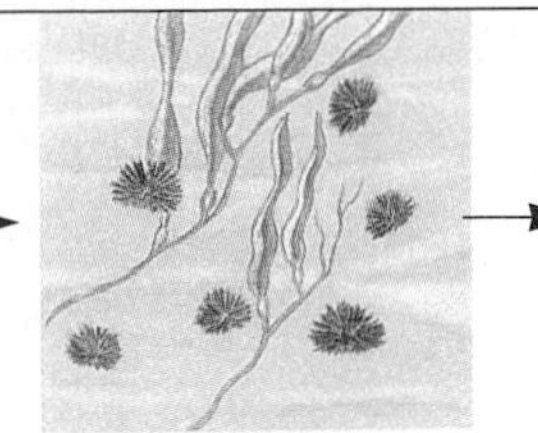

With their predators gone, the sea urchins multiplied and ate all of the kelp.

In 1937 a small group of surviving sea otters was discovered. With protection their numbers grew.

The otters, once again numerous, preyed on sea urchins. The kelp beds regenerated.

Figure 10-6 **When sea otters were hunted to near extinction on the Pacific coast of North America, their entire ecosystem was disrupted.**

multiplied. Since sea urchins eat kelp, the numerous sea urchins stripped the area of kelp. The kelp beds were bare. Fortunately, a small group of surviving sea otters was discovered in 1937 at Big Sur, California. With protection, the sea otter population grew, and the kelp beds again flourished. Figure 10-6 shows the sequence of events in the case of the sea otter.

Practical Uses of Species The mass extinction of species represents the loss of a potential gold mine of valuable products. For example, about 40 percent of all prescription drugs used in the United States were originally made from living things. Figure 10-7 lists just a few medicinal plants and the drugs made from them. One of them, vinblastine, is made from the rosy periwinkle of Madagascar. Vinblastine is a wonder drug used since the 1960s to treat childhood leukemia. Taxol, made from the bark of the Pacific yew tree, is used to treat certain kinds of cancer. L-dopa, used to treat Parkinson's disease, is derived from the velvet bean. Two of these valuable plants are shown in Figure 10-8.

Figure 10-7 **About 40 percent of all prescription drugs were derived from living things. This table shows just a few of the numerous plants used to make medicines.**

Medicines and Their Plant Sources

Medicine	Plant Source	Use
Bromelain	Pineapple	Controls tissue inflammation
Colchicine	Autumn crocus	Anticancer agent
D-tubocurarine	*Chondrodendron* and *Strychnos*	Surgical muscle relaxant
Digitoxin	Common foxglove	Cardiac stimulant
Ergonovine	Smut-of-rye or ergot	Control of hemorrhaging and migraine headaches
Glaziovine	*Ocotea glaziovii*	Antidepressant
Indicine N-oxide	*Heliotropium indicum*	Anticancer agent (leukemia)
L-dopa	Velvet bean	Parkinson's disease suppressant
Penicillin	Penicillium fungi	Antibiotic
Pilocarpine	*Pilocarpus*	Treats glaucoma and dry mouth
Quinine	Yellow cinchona	Antimalarial
Reserpine	Indian snakeroot	Reduces high blood pressure
Taxol	Pacific yew	Anticancer agent
Thymol	Common thyme	Cures fungal infection
Vinblastine, vincristine	Rosy periwinkle	Anticancer agent

Figure 10-8 **The rosy periwinkle (left) is the source of vinblastine, a drug used to treat leukemia. The bark of the Pacific yew tree (right) is used to make a cancer drug.**

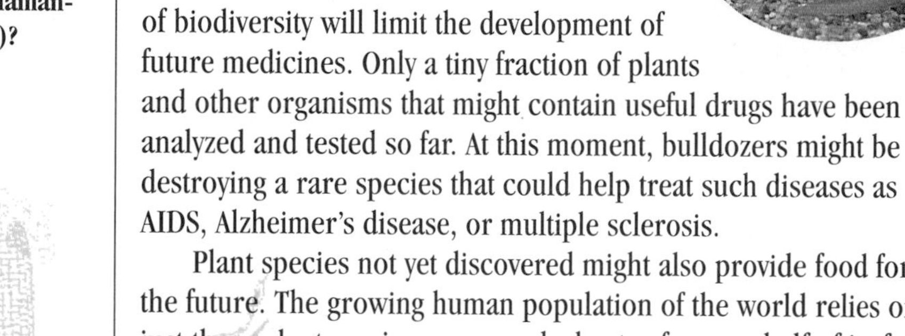

Figure 10-9 Should we be as concerned about protecting a salamander (right) as whales (above)?

Answer to caption: Answers will vary. Encourage class discussion, and emphasize that there are no wrong answers to this question, only individual opinions.

Answers to Section Review can be found on page T154.

SECTION REVIEW

1. Explain how habitat destruction can cause the extinction of species.
2. Explain how protecting one species could preserve an entire ecosystem.
3. What are some practical disadvantages of losing species?

THINKING CRITICALLY

4. *Drawing Conclusions* Do you think humans should try to protect all plant and animal species from extinction? Why or why not?

Scientists are concerned that the loss of biodiversity will limit the development of future medicines. Only a tiny fraction of plants and other organisms that might contain useful drugs have been analyzed and tested so far. At this moment, bulldozers might be destroying a rare species that could help treat such diseases as AIDS, Alzheimer's disease, or multiple sclerosis.

Plant species not yet discovered might also provide food for the future. The growing human population of the world relies on just three plants—rice, corn, and wheat—for over half of its food. It can be disastrous to depend on so few plants for food, as was demonstrated in Ireland in the 1800s. Ireland relied on only a few varieties of potatoes for almost all of its food. None of the varieties were resistant to a fungus that quickly spread through the potato fields in the 1840s and destroyed most of the crop. As a result, 2 million people died of starvation.

Researchers are continually looking for wild plants that can become new food crops or that can be crossed with present food crops to produce new strains resistant to disease or insects. In the 1970s, one research team discovered a rare species of wild corn growing on a mountainside northwest of Puerto Vallarta, Mexico. Resistant to most viral diseases and nematode rootworms, it grows year after year without needing to be replanted. This wild corn is so rare that it was found growing only on a 120-hectare (300 acre) plot. Cows grazed on the plot and could have eventually wiped out the valuable corn.

Aesthetic Reasons When species become extinct, we lose objects of fascination, wonder, and beauty. Almost everyone enjoys contact with wild living things, whether by feeding squirrels in the park, driving past woods on the way home from work or school, or camping in wilderness areas. Time spent with other living things renews our sense of connection with nature and our appreciation of the world as a living system.

SECTION 10.2

PUBLIC POLICY

AFTER READING THIS SECTION YOU SHOULD BE ABLE TO

1. describe the main provisions of the Endangered Species Act.
2. define *endangered species* and *threatened species.*
3. discuss controversies about efforts to protect endangered species.
4. describe worldwide efforts to prevent extinctions.

Most nations have laws and regulations designed to prevent the extinction of species. Those in the United States are considered to be among the strongest. Even so, there is much controversy about how to enforce such laws and regulations and about how effective they are.

THE ENDANGERED SPECIES ACT

The United States Congress passed the **Endangered Species Act** in 1973 and has amended it several times since. This law is designed to protect any plant or animal species in danger of extinction. The four main provisions of the Endangered Species Act are summarized in Figure 10-10.

Under the first main provision, the U.S. Fish and Wildlife Service must compile a list of all endangered and threatened species. A species is considered to be an **endangered species** if its numbers have fallen so low that it is likely to become extinct in the near future if protective measures are not taken immediately. **Threatened species** are species that are likely to become endangered if they are not protected. As of October 1994, 722 plants and

Figure 10-10 **The Endangered Species Act is a law designed to protect plant and animal species in danger of extinction.**

Main Provisions of the Endangered Species Act	
1.	The U.S. Fish and Wildlife Service must compile a list of all endangered and threatened species.
2.	Endangered and threatened animal species may not be caught or killed. Endangered or threatened plants on federal land may not be uprooted. No part of an endangered or threatened species may be sold or traded.
3.	The federal government may not carry out any project that jeopardizes endangered species.
4.	The U.S. Fish and Wildlife Service must prepare a species recovery plan for each endangered and threatened species.

animals were listed as endangered, and 198 were listed as threatened. Figure 10-11 shows just a few of the endangered and threatened species in the United States. Many more species are being considered for listing but have not yet been reviewed.

The second main provision of the Endangered Species Act states that endangered or threatened animals may not be caught or killed. Plants that are threatened or endangered may not be uprooted if they are growing on federal land. None of the listed species or any of their parts, such as fur, feathers, or flowers, can

Figure 10-11 Are any of these species native to your area of the country?

be sold or traded. Anyone who violates this part of the law is subject to a substantial fine.

According to the third main provision of the Endangered Species Act, the federal government may not carry out any project that jeopardizes an endangered or threatened species. The Case Study on pages 264–265 describes a controversial case that was the direct result of this provision.

Under the fourth main provision of the Endangered Species Act, the U.S. Fish and Wildlife Service must prepare a recovery plan for each endangered and threatened species. The Fish and Wildlife Service has found that the most effective way to save most species is to protect their habitats. Simply putting a few individuals into zoos, aquariums, or greenhouses does very little to preserve a species. Why? For one thing, many animals will not breed in captivity, so their numbers continue to dwindle. Also, tiny captive populations are vulnerable to infectious diseases and genetic disorders caused by inbreeding.

Setting aside small plots of land for endangered or threatened species is usually not effective either. An endangered species living in a small area could be wiped out by a single grass fire. Some species require a large range to find adequate food, find a suitable mate, and rear their young. Therefore, protecting the habitats of endangered and threatened species may mean restricting human use of some areas.

Hawaii is home to 40 percent of the nation's endangered birds and 25 percent of the endangered plants.

Developers vs. Environmentalists? Plans to restrict human use of land to preserve species are sometimes controversial. Real-estate developers invest large sums of money in buying land and designing plans for houses and commercial buildings. Then they may discover that they cannot build on the land because it is the habitat of an endangered or threatened species.

The local community may have been looking forward to the jobs that the new development would create. The city government may have been counting on property and sales taxes to finance local improvements to sewer lines, schools, police forces, and fire protection. Many times, both the real-estate developers and the local residents are angry that all of these economic benefits are threatened by what appears to be an insignificant weed or bug.

Although battles between developers and environmentalists are widely publicized, in most cases compromises are eventually worked out. Usually neither the developers nor the environmentalists get everything they want, but both sides get something. There have been tens of thousands of these cases in the United States, and all but about 20 were resolved by some kind of compromise.

Worldwide Efforts to Prevent Extinctions

Several international organizations work to protect species on a worldwide basis. The World Wildlife Fund works to protect biodiversity, especially in tropical forests, by encouraging the sustainable use of resources. The Nature Conservancy currently manages a system of about 1,300 nature sanctuaries in the United States and other countries. Conservation International develops ecosystem conservation projects with partner organizations and local people in countries that are rich in biological diversity. Friends of the Earth is a group that lobbies governments and disseminates information to the public. Greenpeace International stages dramatic protests like the one in Figure 10-12 to help stop the destruction of rain forests and the killing of endangered animals.

At the governmental level, the International Union for the Conservation of Nature and Natural Resources (IUCN) is in the forefront of efforts to protect species and habitats. This organization is a collaboration of almost 200 governments and government agencies and over 300 private conservation organizations. The IUCN publishes data books that list species in danger of extinction around the world, advises governments of the most effective ways

CUT!

A filmmaker was just doing his job for Conservation International when he was surrounded by machine guns and accused of being a terrorist. See pages 372–373 for an interview with him.

CASE STUDY

THE SNAIL DARTERS AND THE DAM

One of the first and most famous tests of the Endangered Species Act was the case of the Tellico Dam in Tennessee during the 1970s. The dam—a federal project—was being built on the Little Tennessee River to provide flood control and to generate hydroelectric power. Opponents had fought the dam for years, arguing that it was too expensive, that it was not really needed, and that it destroyed valuable farmland, forests, recreation areas, and historic and archeological sites.

The project continued, however, until it was discovered that the habitat of a tiny fish called the snail darter would be destroyed by the dam. Opponents of the dam filed suit, and the courts ruled that the dam project was covered by the Endangered Species Act. Construction was halted even though the dam was 80 percent completed and had already cost taxpayers $50 million. Appeals went all the way to the U.S. Supreme Court. The court ruled that under the provisions of the Endangered Species Act, construction of the dam must be stopped, no matter what the cost.

However, Congress then passed a law that granted certain exemptions to the Endangered

Efforts to protect the tiny fish shown at the upper right almost prevented the completion of the Tellico Dam (above) in Tennessee.

Figure 10-12 **These protesters are blocking the path of a Japanese whaling ship. Do you think this is an effective way to protect species?**

Answer to caption:
Answers will vary. Encourage class discussion, and emphasize that there are no wrong answers to this question, only individual opinions.

to manage their natural resources, and works with groups like the World Wildlife Fund to sponsor field projects around the globe. The projects range from attempting to halt poaching in Uganda to preserving the habitat of sea turtles on South American beaches.

Prevention of Poaching One offshoot of the IUCN is an international treaty, the Convention on International Trade in Endangered Species (CITES). CITES became well known for

Answers to Thinking Critically can be found on page T154.

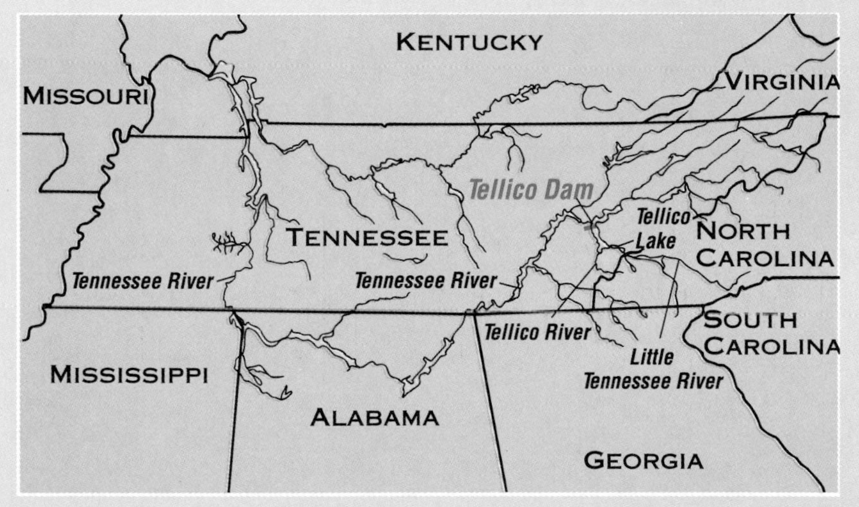

Tellico Dam is on the Little Tennessee River in eastern Tennessee.

Species Act. One way a project could be exempted is if an Endangered Species Review Committee decided that the economic benefits outweighed the potential harm to an endangered species. Supporters of the Tellico Dam hoped that the committee would exempt the project under this provision. But instead, the committee ruled that the dam project was economically infeasible and therefore could not be exempted.

Was this the end of the story? No. Congress then passed a bill stating that the Tellico Dam project did not have to comply with the Endangered Species Act. The dam was finally completed. The snail darters were transplanted to nearby streams and now appear to be reproducing. Their status has been upgraded from endangered to threatened.

THINKING CRITICALLY

1. *Making Decisions* Do you think the decision to halt the construction of the Tellico Dam was justified? Explain. Now answer the question again after considering the following scenarios.
   - Suppose the investment in the Tellico Dam had been $500 million, not $50 million.
   - Suppose that a type of beetle, rather than a fish, were endangered by the construction of the dam.
2. *Making Decisions* Should money ever be a consideration in cases where a species may be endangered by human activity? Explain.

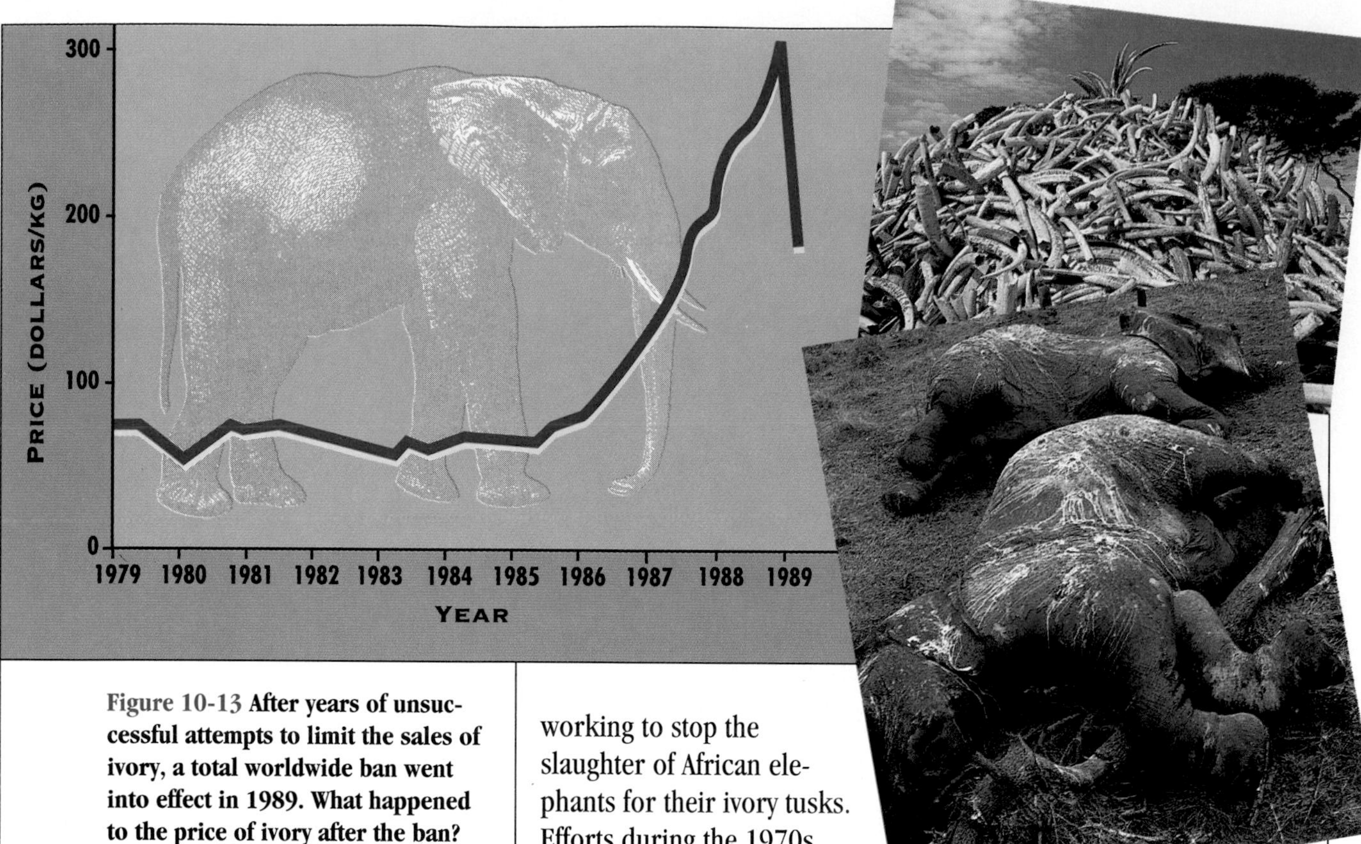

Figure 10-13 After years of unsuccessful attempts to limit the sales of ivory, a total worldwide ban went into effect in 1989. What happened to the price of ivory after the ban? Why do you think the price of ivory climbed before 1989?

Figure 10-14 Before a complete ban on the sale of ivory in 1989, scenes like these were common.

Answers to caption and Section Review can be found on page T154.

working to stop the slaughter of African elephants for their ivory tusks. Efforts during the 1970s and 1980s to limit the sale of ivory had little effect on the killing of elephants. In 1989, the members of CITES proposed a total worldwide ban on all sales, imports, and exports of ivory, hoping to put a stop to scenes like those in Figure 10-14.

Some conservationists were concerned that making ivory illegal would only *increase* the rate of poaching. They argued that illegal ivory, like illegal drugs, might become more expensive. Nevertheless, the sale of ivory was completely banned in 1989, and fears that the ban would increase poaching turned out to be unfounded. As you can see in Figure 10-13, the price of ivory fell steeply after the ban, and elephant poaching declined dramatically.

The Biodiversity Treaty One of the most ambitious efforts to tackle environmental issues on a worldwide scale was the United Nations Conference on Environment and Development, also known as the Earth Summit. More than 100 world leaders and 30,000 other participants met in Rio de Janeiro, Brazil, in 1992.

One of the agreements that came out of the Earth Summit was the Biodiversity Treaty, which encourages wealthier countries to give money to poorer countries for the protection of potentially valuable species. However, the treaty does not state exactly how this should be done. Arguing that it was too vague, former President George Bush decided not to sign the Biodiversity Treaty. One year later, however, newly elected President Bill Clinton signed the treaty.

SECTION REVIEW

1. What is the difference between an endangered species and a threatened species?
2. Why can't some species survive on small plots of land?

THINKING CRITICALLY

3. *Drawing Conclusions* Do you think it is reasonable for an expensive dam project to be stopped by the discovery of an endangered species that would be threatened by the dam? Why or why not?
4. *Inferring Relationships* Why do you think a complete ban of ivory sales was so much more effective in reducing poaching than the limited ban was?

THE FUTURE OF BIODIVERSITY

AFTER READING THIS SECTION YOU SHOULD BE ABLE TO

❶ describe how captive-breeding programs, botanical gardens, and germ-plasm banks help save species.

❷ explain the advantages of protecting entire ecosystems rather than individual species.

It is possible to slow the loss of species, but to do so we must develop new approaches to conservation and a sensitivity to human needs around the globe. In this section you will read about methods of saving individual species, as well as ways to protect entire ecosystems.

SAVING INDIVIDUAL SPECIES

In Section 10.2, you learned about attempts to save individual species through legislation such as the Endangered Species Act and anti-poaching laws. Additional ways to preserve individual species include captive-breeding programs, botanical gardens, and germ-plasm banks.

Captive-Breeding Programs Zoos and wild animal parks can increase the population of an endangered or threatened species by establishing captive-breeding programs. Such programs involve breeding animals under carefully managed circumstances.

One well-known example of a captive-breeding program involves the California condor, shown in Figure 10-15. Condors feed mainly on dead animals and need vast areas in which to search for food. Habitat loss and poaching almost brought about

Figure 10-15 California condors are slowly recovering, thanks to a captive-breeding program. The photograph at the left shows a release station where condors are being released into the wild.

the extinction of the species. In 1986, when there were only nine California condors left in the wild, a captive-breeding program was established. Early results are promising. After just a few years of breeding in captivity, a few condors have already been released into the wild. If all goes well, many more will be released in the next 10 years.

Botanical Gardens You may be surprised to know that botanical gardens like the one shown in Figure 10-16 are storehouses of genetic diversity. Botanical gardens worldwide house around 90,000 species of plants. Even so, the gardens don't have the space or the funds to preserve most of the world's rare and threatened plants.

Germ-Plasm Banks Germ-plasm banks store germ plasm for future use in case species become extinct. Germ plasm is the genetic material contained within the reproductive (germ) cells of organisms. Plants may be stored as seeds, and animals as frozen sperm and eggs. The germ plasm is stored in refrigerated and humidity-controlled environments that allow the genetic material to survive for many years.

Figure 10-16 **Botanical gardens are storehouses of genetic diversity. This botanical garden is in Queen Elizabeth Park, Vancouver, Canada.**

The Ecosystem Approach

Recently many conservationists have begun to concentrate on protecting entire ecosystems rather than individual species. There are two main reasons for this shift.

First, many more species are in danger of extinction than can possibly be placed on official lists. As you already read, biologists do not even know how many species actually exist on Earth today, let alone how many are endangered. By concentrating on entire ecosystems, we may be able to save most of the species in an ecosystem, rather than just the ones that are on an endangered species list.

Second, the health of the entire biosphere depends on the preservation of individual ecosystems. Forest ecosystems, for example, clean the air by collecting small particles of air pollution on the leaves and needles of trees. Wetland ecosystems provide filtration that helps clean water supplies. The loss of these and other ecosystems can have damaging consequences throughout the biosphere.

To protect biodiversity worldwide, some conservationists suggest that at least 10 percent of the Earth's land be set aside as protected preserves. Primary consideration, they argue, should

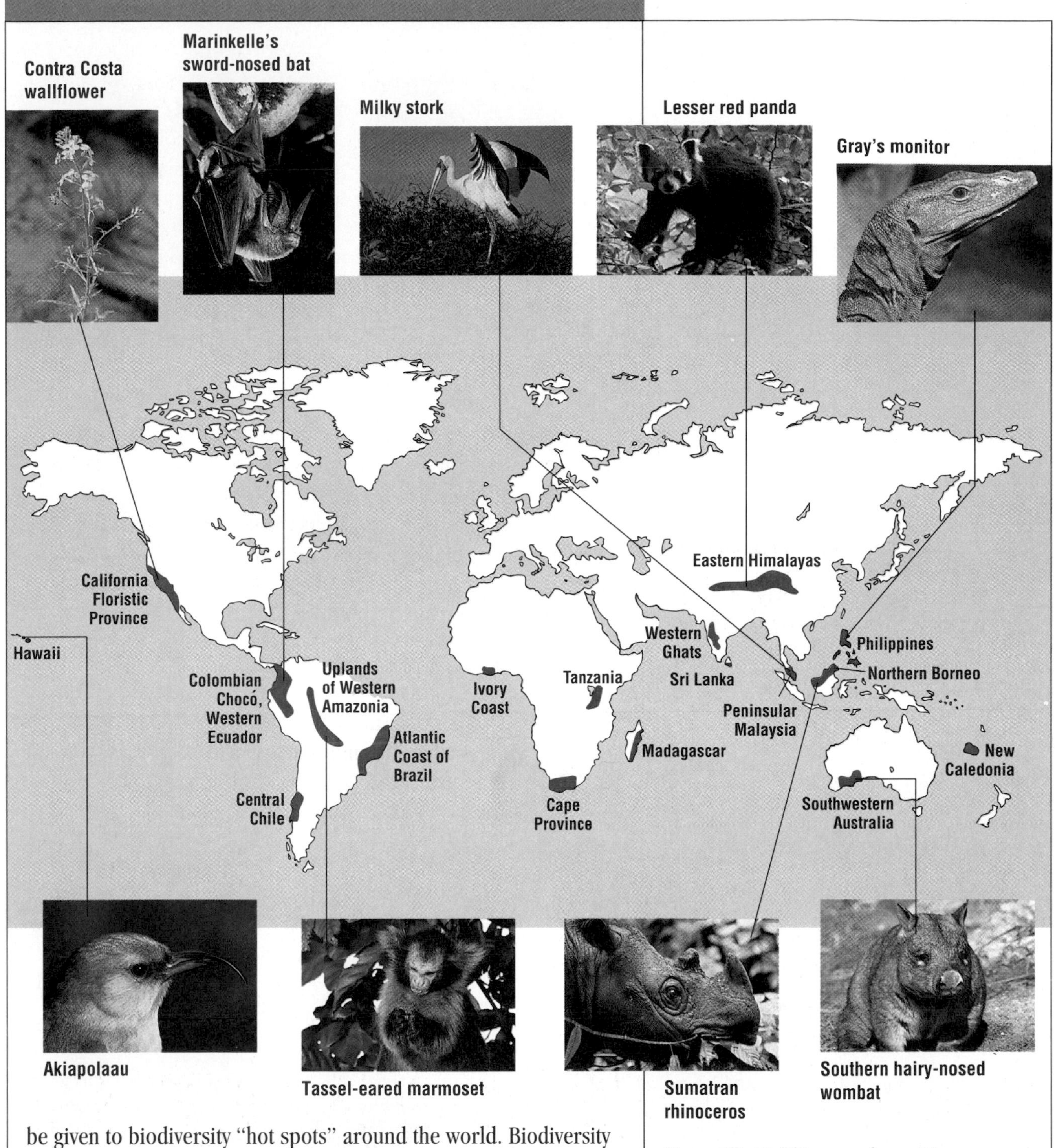

be given to biodiversity "hot spots" around the world. Biodiversity hot spots are regions that contain unusually large numbers of species. Eighteen of these hot spots are shown in Figure 10-17.

As part of the effort to preserve ecosystems, ecologists are working to answer such questions as, How large does a protected preserve have to be to maintain a diversity of species? Questions like this often take years of research to answer. One research project in Brazil is designed to determine the minimum amount of land necessary to sustain a rain-forest habitat. (See Figure 10-18.) Research is expected to continue into the next century. Unfortunately, decisions must often be made before there is sufficient data.

Figure 10-17 **This map shows 18 areas that contain large numbers of species, many of which are in danger of extinction. Each of these "hot spots" is home to species that are found nowhere else.**

Figure 10-18 Researchers in Brazil are trying to determine the minimum amount of land necessary to sustain a rain-forest habitat. This photograph shows a 1-hectare plot at the right and a 10-hectare plot in the center.

What About Human Needs?

As much as we may want to protect this planet's biodiversity, we can't ignore the needs of the human inhabitants of the world. In the developed countries of the Northern Hemisphere, the desire to protect biological resources often comes into conflict with the need to protect people's jobs. Such a situation arose in the states of Washington and Oregon in the late 1980s. The U.S. Fish and Wildlife Service designated millions of hectares of forest in these states as protected habitat for the northern spotted owl, which is threatened with extinction. Some argued that people in the timber industry would lose their jobs as a result.

In the poorer nations of the world, the conflict between human needs and biodiversity may even be a matter of life or death. It is difficult to blame a parent who shoots an endangered animal so that a child won't starve, or one who clears part of a rain forest to grow crops that will support a family.

We must find ways to meet human needs and manage living resources at the same time. Preserving these resources will perhaps be the greatest challenge of the twenty-first century.

Figure 10-19 Human needs must also be considered. This family cleared and burned land to plant a maize crop.

Should the habitat of an owl species be preserved even if it means that people might lose their jobs? See pages 276–277.

Answers to Section Review can be found on page T154.

Section Review

1. Which of the ways to save individual species may involve the preservation of only part of an organism?
2. Why are some conservationists beginning to concentrate on protecting entire ecosystems rather than individual species?

Thinking Critically

3. *Interpreting Graphics* The biodiversity hot spots shown in Figure 10-17 share several characteristics besides a great number of species. Look at the map, and name as many shared characteristics as you can.

HIGHLIGHTS

SUMMARY

- The term *biodiversity* refers to the number and variety of species on Earth. There are now an estimated 10 million, and perhaps as many as 100 million, species of living things.
- While extinction is a natural event, humans are accelerating the rate of extinction for many species.
- The major causes of extinction today are habitat destruction, poaching, and the introduction of exotic species.
- One reason for concern about the accelerated rate of extinction is that the disappearance of a single species can disrupt an entire ecosystem. In addition, many species have practical value to humans as medicines and foods.
- The Endangered Species Act is designed to protect plant and animal species that are in danger of extinction.
- The Biodiversity Treaty is an international agreement that encourages wealthier countries of the world to give money to poorer countries for the protection of potentially valuable species.

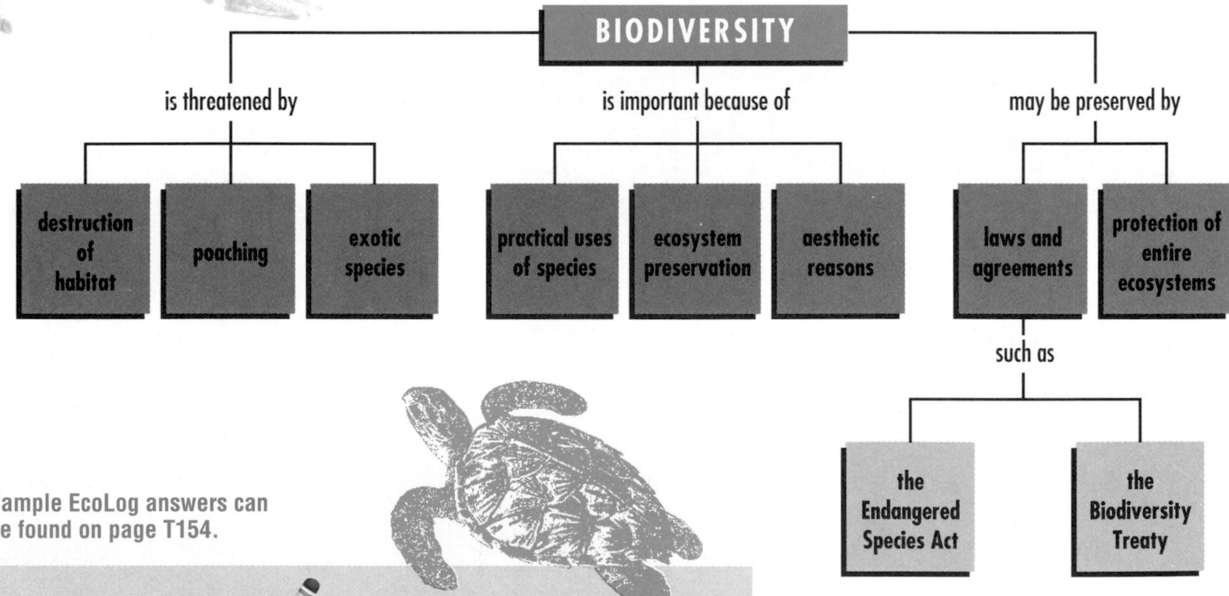

Sample EcoLog answers can be found on page T154.

EcoLog

Now that you've studied this chapter, revise your answers to the questions you answered at the beginning of the chapter, based on what you have learned.

❶ What is the major cause of extinction today?

❷ Do you think humans should try to prevent the extinction of other species? Explain your reasoning.

Vocabulary Terms

biodiversity (p. 254)
endangered species (p. 261)
Endangered Species Act (p. 261)
exotic species (p. 258)
poaching (p. 257)
threatened species (p. 261)

CHAPTER 10

REVIEW

UNDERSTANDING VOCABULARY

1. For each pair of terms, explain the differences in their meanings.
   a. hunting
      poaching
   b. native species
      exotic species
   c. endangered species
      threatened species

RELATING CONCEPTS

2. Copy the unfinished concept map below onto a sheet of paper. Then complete the concept map by writing the correct word or phrase in each box containing a question mark.

Answers not indicated can be found on page T155.

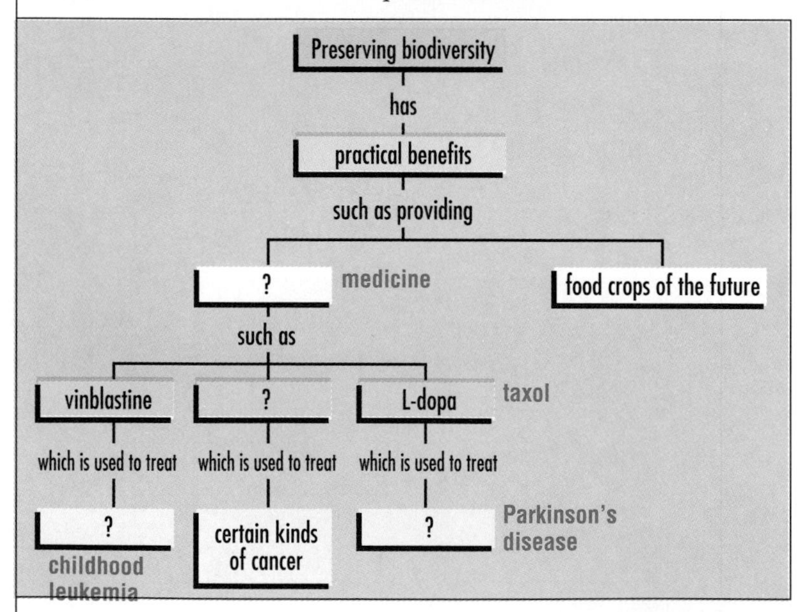

UNDERSTANDING CONCEPTS

Multiple Choice

3. A mass extinction is
   a. a rapid increase in biodiversity.
   b. caused by the introduction of exotic species.
   (c.) the extinction of many species in a short period of time.
   d. beneficial to the environment.

4. The term *biodiversity* refers to
   (a.) the variety of species on Earth.
   b. the large number of dinosaurs that became extinct.
   c. habitat destruction.
   d. the fact that 40 percent of prescription drugs come from living things.

5. Most of the species known to science today
   a. are large mammals.
   b. live in deserts.
   c. live in the richer countries of the world.
   (d.) are insects.

6. Some species are so important to the functioning of an ecosystem that they are called
   a. threatened species.
   (b.) keystone species.
   c. endangered species.
   d. extinct species.

7. When sea otters disappeared from the Pacific coast of North America,
   a. the area became overrun with kelp.
   b. the number of fish in the kelp beds increased.
   (c.) the number of sea urchins in the kelp beds increased.
   d. the area became overrun with brown seaweed.

8. Which is NOT true of the Endangered Species Act?
   (a.) Parts of an endangered animal, such as feathers or fur, may be traded or sold, but only if the animal is not killed.
   b. A species is considered endangered if it is expected to become extinct in the near future.
   c. The federal government cannot carry out a project that may jeopardize an endangered plant.
   d. A recovery plan is prepared for all threatened animals.

9. Because of efforts by the Convention on International Trade in Endangered Species (CITES),
   a. the poaching of elephants increased.
   b. the international trade of ivory was banned worldwide.
   c. the cost of ivory worldwide increased.
   d. a captive-breeding program for elephants was established.
10. By emphasizing the preservation of entire ecosystems,
   a. insect-resistant peaches can be developed.
   b. unknown species can be saved from extinction.
   c. the health of the biosphere will be jeopardized.
   d. biodiversity will be decreased.
11. Conflicts that arise in trying to preserve biodiversity include all of these EXCEPT
   a. short-term benefits versus long-term resources.
   b. job security versus the protection of biological resources.
   c. the cost of preserving animal species versus the cost of preserving plant species.
   d. balancing the needs of humans versus the needs of other species.

Short Answer

12. Why is legal hunting no longer a cause of extinctions in the United States today?
13. What are exotic species?
14. Explain why residents of a community may be unhappy when land is set aside for endangered species.
15. Describe three ways that preserving biodiversity can come into conflict with human needs.

Interpreting Graphics

16. The table below shows the number of vertebrate species on the Endangered Species List in 1992. Figure the percentage for each vertebrate group, and then construct a pie chart that represents the data.

Mammals	37
Birds	57
Reptiles	8
Amphibians	6
Fish	52

Thinking Critically

17. ***Comparing Processes*** How is the present pattern of extinction different from the pattern of extinction in the past?

Themes in Science

18. ***Evolution*** Contrast the process of natural selection discussed in Chapter 2 with the forces that are causing the present mass extinction.
19. ***Interacting Systems*** How might the loss of huge tracts of forested land be related to the global climatic changes discussed in Chapter 7?

Cross-Discipline Connection

20. ***Math*** There are 1,412,900 species known to exist. Consult Figure 10-2, and figure the percentage that are insects.
21. ***Geography*** Obtain a list of the plants and animals that are endangered in your state. Find out where those species reside, and mark the locations on a map of your state. Research the effects of habitat loss on species in your county and in surrounding counties.

Discovery Through Reading

For an interesting look at how humans have affected bird populations in the United States, read the article "Silence of the Songbirds" in *National Geographic,* June 1993, pp. 68–91.

PORTFOLIO ACTIVITY

Do a special project about one endangered species of your choice. Consider using one of the following media to inform your classmates about your chosen species or to persuade them of the importance of saving the species.

- **a poster**
- **an oral presentation**
- **a video**

INVESTIGATION

BACKYARD DIVERSITY

When we think about biodiversity, we usually think about endangered species, dwindling habitats, and international efforts to protect large ecosystems. The wildlife and rare plants of the rain forests are important to protect, but you don't have to travel that far to see habitat destruction and struggle for survival. You can watch the drama in your own backyard every day of the year. In this activity you will investigate the factors that affect diversity among the insects in your own area.

Answers to this Investigation can be found on page T155.

MATERIALS

- tape measure or meter stick
- 4 stakes
- string (about 10 m)
- hand lens
- field guide to insects (optional)
- notebook
- pen or pencil

BACKGROUND

You will survey two very different sites. One site will be an area that has been greatly affected by humans, and the other will be an area that has been less affected by humans. You are going to measure off a small area at each site and count the number and types of insects at each. You will also record the physical and biological features of each site.

First you will observe, gather information, and think about the factors that might affect insect diversity. Then you will propose a hypothesis and describe an experiment that would test it.

PREPARE

CAUTION: Before beginning your field study, review the safety guidelines on pages 409–412. Remember to approach all plants and animals with caution.

OBSERVE, RECORD, AND THINK

1. Choose two different sites for your survey. One should be an area that has been greatly affected by humans, such as a well-groomed lawn. The other should be an area that has been less affected by humans, such as a natural area or a vacant lot overgrown with weeds.
2. At Site 1, use the tape measure or meter stick, four stakes, and string to mark off a square plot 2 m × 2 m, as shown in the photograph at the upper right of the next page. This is your sample area.
3. In a table similar to Table A on the next page, record any features that you think might affect insects. Use the following list of features as a guide, adding any others you think of.
   - **Maintenance** Is the area maintained? If so, interview the person who maintains it, and find out how often the site is watered, fertilized, treated with pesticides, and mowed.
   - **Time Left Undisturbed** Estimate how long it has been since humans disturbed the site. For example, if the site is a well-tended lawn that was mowed one week ago, you would record "one week." If the site is a vacant lot that was mowed six months ago, you would write "six months."
   - **Sunlight Exposure** How much of the area is exposed to sunlight?
   - **Soil** Is the soil mostly sand, silt, or fine clay particles?

- **Rain** When was the last rain recorded for this area? How much rain was received?
- **Slope** Is the area flat or hilly?
- **Water Drainage** Is the water standing or pooling, or is the area well drained?
- **Vegetation Cover** How much of the soil is covered with vegetation? How much is exposed?

Table A

Feature	Site 1	Site 2
Maintenance	Well-maintained yard. Owner of home says she waters the yard once a week and mows it about twice a month. Doesn't treat with pesticides or fertilizer.	Vacant lot. City maintenance mows the grass about once every summer, more often if people complain. Pesticides also applied about once a year. Lot is not watered or fertilized.
Time Left Undisturbed	One week (mowed one week ago)	Six months (mowed and treated with pesticides six months ago)
etc.	etc.	etc.

Table B

	Site 1	Site 2
Number of Insects		
Number of Insect Types		

4. In your notebook, create a blank table similar to Table B shown above.
5. Use the hand lens to inspect the sample area, and count the number of insects you see, being careful not to disturb the soil or the organisms. Record the number of insects in Table B. Then count the *types* of insects, and record that number in Table B.
6. Follow exactly the same procedure at Site 2, starting with step 2.
7. Examine the two tables and think about what might be causing the difference (if any) between the two sites in the number and diversity of insects. In your notebook, jot down your hunch (your best guess) about what is causing the difference.

STATE A HYPOTHESIS

8. State your hunch in the form of a hypothesis.

DESIGN AN EXPERIMENT

9. Describe an experiment that could test your hypothesis.

Place stakes at the corners of a 2 m by 2 m square area, and loop string around the stakes to mark the edges of your site.

Record the number of insects and the number of types of insects in your sample area.

POINTS OF VIEW

OWLS VS. LOGGERS

The northern spotted owl became the center of controversy in 1990 when the U.S. Fish and Wildlife Service listed the bird as a threatened species. To save the owl, scientists and federal agencies developed a recovery plan that designated millions of hectares of old-growth forest in Washington and Oregon as protected habitat. Because spotted owls nest in dead trees, their habitat is limited to old-growth forests. The controversy stems from the fact that under the recovery plan trees cannot be cut down in the owls' protected habitat, so the loggers' jobs are threatened. Read the following points of view, and then analyze the issue for yourself.

PEOPLE AND JOBS SHOULD COME FIRST

Loggers, sawmill workers, and residents of communities that depend almost entirely on logging for their livelihoods argue that the recovery plan will result in the loss of tens of thousands of jobs. Towns and counties stand to lose millions of dollars from logging revenues that now fund schools and other essential community services.

Many logging families have lived in the area for generations. They say that if they lose their jobs, they will have to move away in search of other work, struggle to find one of the scarce jobs in the area, or depend on welfare to survive. Doing any of these things to save one species of bird seems absurd to the loggers.

As for saving the forest itself, they argue that when forests are logged and then replanted with young trees, they are healthier and grow faster than the old forests. They also point out that trees do not live forever even when they are not cut down. Eventually they will die from insects, disease, or fire. Why not make use of these resources rather than simply letting the trees fall to the ground and rot?

Furthermore, the loggers say, millions of hectares of old-growth forest will never be logged even without the plan. Some forest lands are already designated as parks and wilderness areas. Other forests are located in areas so inaccessible that it will never be profitable to cut them down.

Many loggers resent the fact that they are often portrayed as villains who destroy nature. They insist that they do not want to harm the forests. One of the reasons they want to keep their jobs is that they love going into the forest each day.

Logger at work in the Pacific Northwest

Members of a logging community protest the owl recovery plan.

Owls and Forests Should Come First

The northern spotted owl

Environmentalists argue that the spotted owl is a beautiful and gentle bird that deserves to live and to have its natural home protected. And more important, by protecting the owl, the old-growth forests will also be saved.

These forests, they insist, have a diversity of life that is not found in replanted, managed forests. If all of the old forests are logged, many species will become extinct because of habitat loss. It would take hundreds of years for the logged areas to recover the complexity of their ecosystems.

Environmentalists emphasize that these ecosystems do more than provide a home for plants and animals. They also reduce air pollution and soil erosion. And by reducing runoff from rains, the forests maintain water quality and protect the spawning grounds of fish.

Supporters of the recovery plan argue that many jobs have already been lost due to factors that have nothing to do with protecting the forest. Automation and advanced logging equipment produce more pulp and lumber with fewer workers. Also, logging companies in recent years have been shipping more lumber overseas for processing. This has led to a loss of jobs in the American lumber-processing industry.

Finally, they argue that change is coming to the logging industry no matter what. If logging is allowed to continue at the high rates of the past, all of the old trees will be gone in 50 years. Why not make adjustments now while there is still some old forest left?

Logs ready for processing

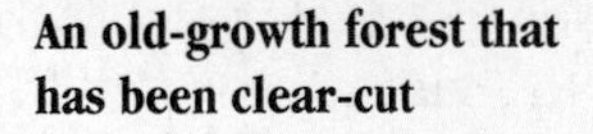
An old-growth forest that has been clear-cut

Analyze the Issue

1. ***Making Decisions***
   Based on the information you have, use the decision-making model presented in Chapter 1 to develop a recommendation for action in this case.
2. ***Analyzing Relationships***
   What additional information would be helpful in developing a recommendation? How might a scientist obtain this information?

Answers to Analyze the Issue can be found on page T155.

CHAPTER 11

ENERGY

"The law of conservation of energy tells us that we can't get something for nothing, but we refuse to believe it."

ISAAC ASIMOV, AMERICAN SCIENCE FICTION WRITER

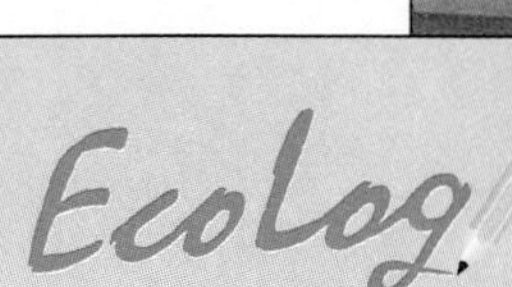

Before you read this chapter, take a few minutes to answer the following questions in your EcoLog.

1. You flip on a light switch, and the lights instantly come on. Where does the electricity come from?
2. What do you think our major source of energy will be 100 years from now? Explain your reasoning.

FOSSIL FUELS TO ELECTRICITY

AFTER READING THIS SECTION YOU SHOULD BE ABLE TO

❶ explain how fossil fuels are used to produce electricity.

❷ distinguish between renewable and nonrenewable resources.

❸ explain how our major sources of energy are dwindling.

It was a warm night in July. Twelve hundred feet above the streets of Manhattan in the World Trade Center restaurant, diners chatted quietly among themselves, stopping occasionally to admire the glittering New York City skyline. Then, without warning, all the lights in the restaurant went out. "Oh great," someone muttered. A waiter tripped in the darkness, sending a load of dishes crashing to the floor. "Hey, the elevator's out!" a woman shouted. "So is the rest of New York," replied the waiter, looking out the windows. The skyline was totally black against the night sky. All the lights of the city were out. And all over New York, people were stuck in elevators hundreds of feet in the air. A series of lightning strikes had knocked out several important power links to the city, which triggered automatic safety devices that shut down the entire power system.

Events like the 1977 blackout of New York City are newsworthy because they are so rare and because they underscore our dependence on electric energy. In this chapter you will learn how we get energy and how we use it to power our world. You'll also learn about some alternatives to our most important energy resources, which are dwindling rapidly.

Figure 11-1 The night the lights went out in New York City

ELECTRICITY: ENERGY ON DEMAND

Enormous amounts of electric energy—enough to meet the needs of thousands of people—can be transported through a wire the diameter of a quarter. You can use electricity to light up the night, cook dinner, heat a home, run a train, and much more. But what exactly is electricity?

Electricity is the flow of electrons, which are tiny charged particles that whirl around the nucleus of an atom. To generate electricity you simply have to move an electrically conductive material such as a copper wire through a magnetic field. The magnetic field puts the electrons in the atoms of the wire in motion, and presto!—you have an electric current. (See Figure 11-3.) In this example, the energy of the moving wire is converted into electricity. Of course, the amount of electricity produced in a single wire is very small. But imagine moving the equivalent of 100,000 wires through a magnetic field. This is what happens in a commercial electric generator.

An **electric generator** is simply a device for converting other forms of energy into electricity. The huge electric generators used in power plants require great amounts of energy for conversion into electricity. Where does this energy come from? Most power plants use fossil fuels.

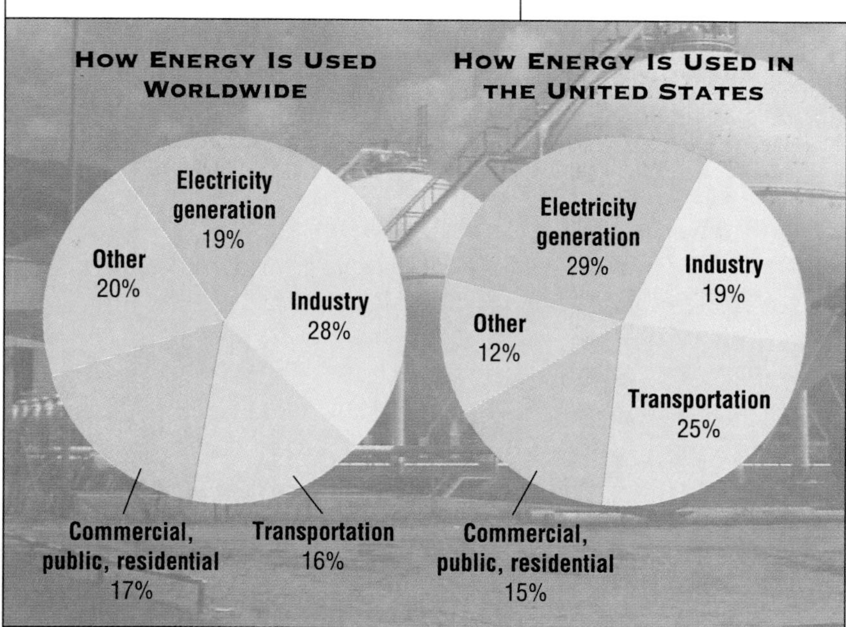

Figure 11-2 A comparison of energy usage worldwide and in the United States

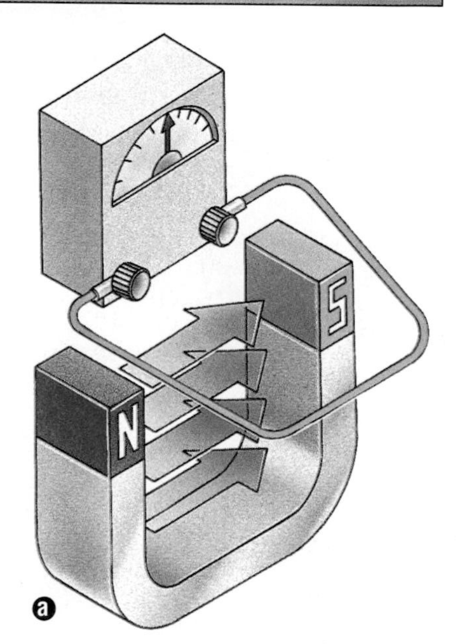

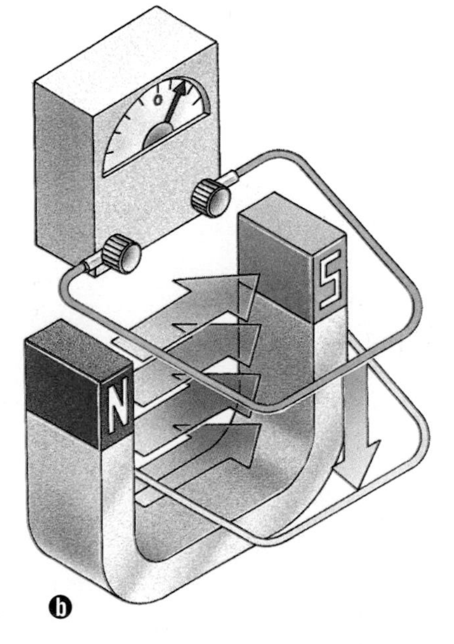

Figure 11-3 A very simple electric generator

a. The wire at rest in the magnetic field produces no electricity.
b. Moving the wire through the lines of magnetic force sets electrons in the wire in motion, generating electricity.

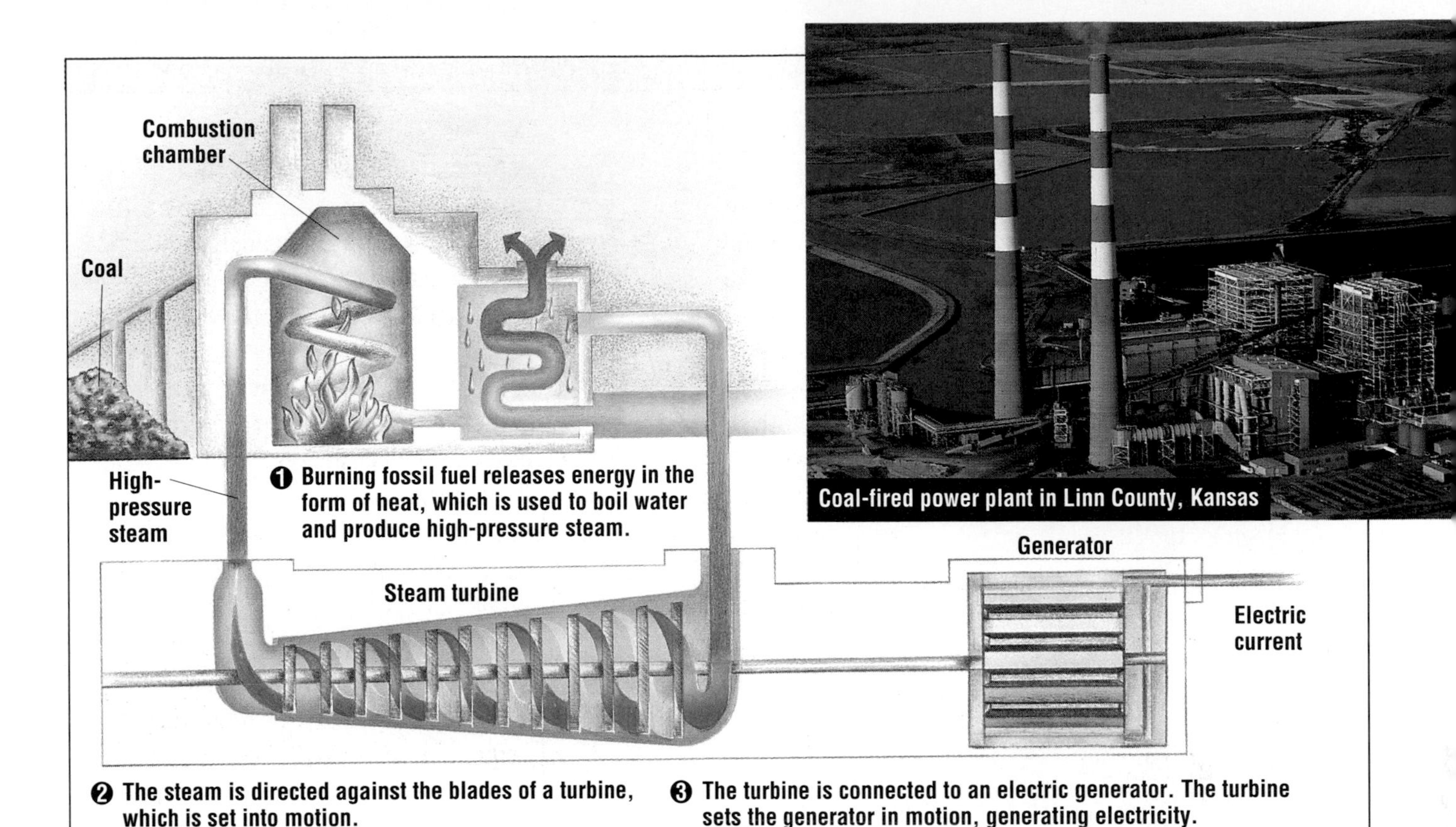

Figure 11-4 Power plants convert the energy in fossil fuels into electric energy.

Fossil Fuels

Power plants convert some form of energy, usually fossil fuels, into electric energy. (See Figure 11-4.) As you read in previous chapters, **fossil fuels**—coal, oil, and natural gas—are the remains of organisms that lived millions of years ago. (Figure 11-5 shows how a deposit of oil and natural gas is formed.) Fossil fuels are rich in

Figure 11-5 Fossil fuels take millions of years to form. This illustration shows the formation of an oil and gas deposit.

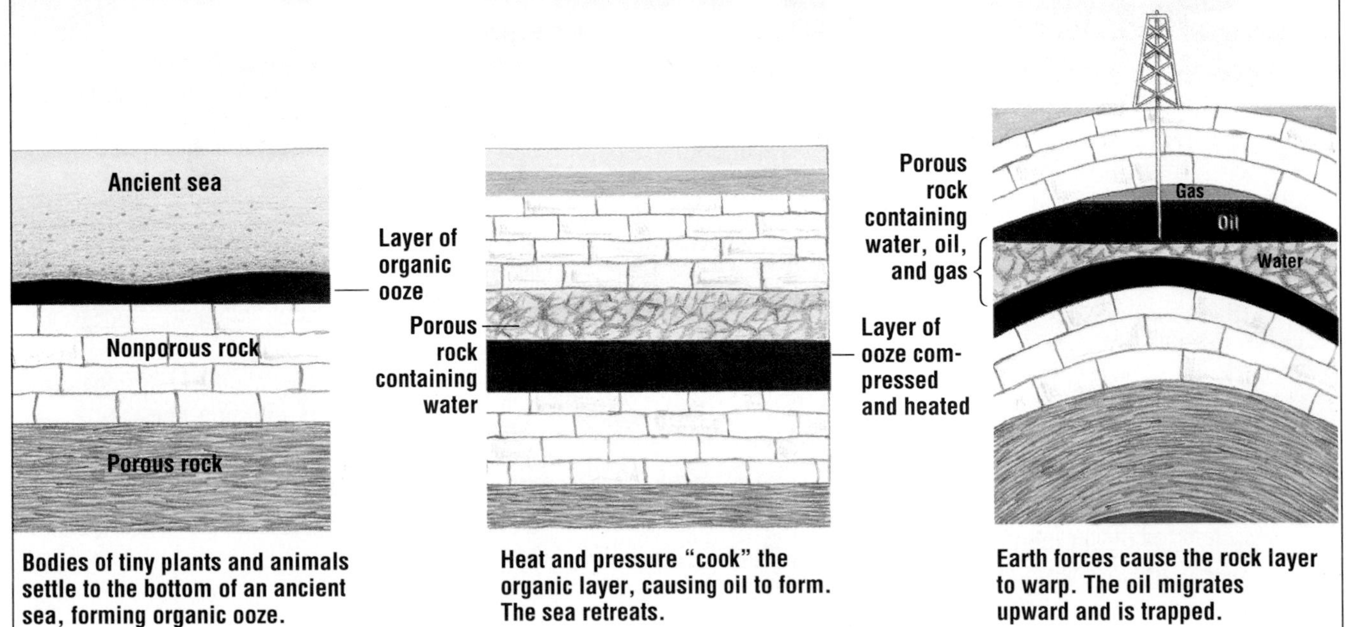

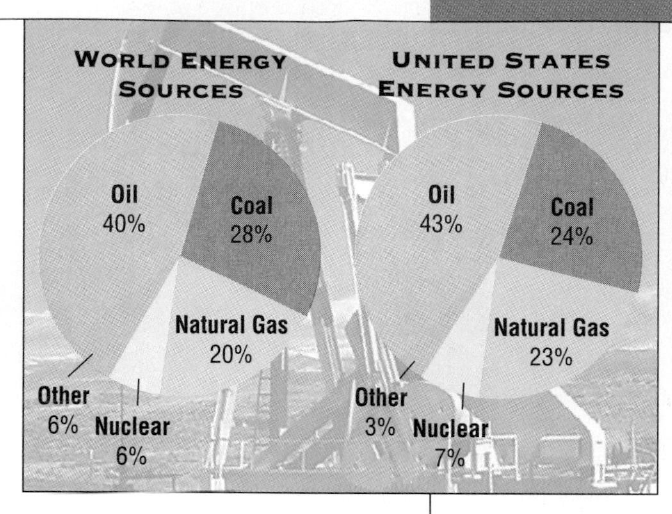

Figure 11-6 What percentage of the energy used in the United States comes from fossil fuels?

Answer to caption: About 90 percent

carbon compounds, which contain a lot of stored energy in their molecular bonds. Burning the fossil fuels releases this energy in the form of heat, which provides the energy to produce electricity. The electricity is then distributed to homes, businesses, and factories.

Fossil fuels are used not only to generate electricity, but also to operate the engines of planes, trains, trucks, cars, and buses. As you can see in Figure 11-6, fossil fuels are the world's main source of energy, accounting for almost 90 percent of the energy used.

Dwindling Supplies of Fossil Fuels Fossil fuel supplies are limited, and we are using these resources much faster than they can be replaced by nature. Much of the oil may be used up in your lifetime. Figure 11-7 shows approximately how long the oil will last, based on current estimates of global underground deposits. Only one fossil fuel—coal—exists in relative abundance. The United States, with about 22 percent of the world's known reserves, has enough coal to last about 400 years at the current rate of consumption.

In addition to being limited, fossil fuels have other drawbacks. The environment is threatened by exhaust from power plants and motor vehicles, by oil spills, and by strip mining for coal. And accidents in the oil-refining and coal-mining industries claim human lives every year.

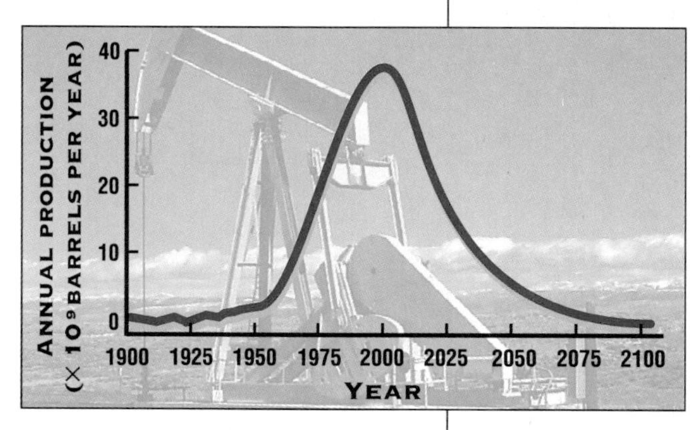

Figure 11-7 This graph shows past oil production and predicted production for the future.

Renewable and Nonrenewable Energy Resources Fossil fuels are being used quicker than they can be replaced naturally. As you read in Chapter 1, resources such as these are considered to be nonrenewable resources, meaning that they can be used up.

Renewable resources, on the other hand, are those resources that are continually produced. Some renewable energy resources, such as the wind and sunlight, are so abundant that they are considered to be inexhaustible.

While there is always the possibility that large, new deposits of fossil fuels will be found, such discoveries will simply prolong the inevitable. We are going to run out of fossil fuels. How can we avoid, or at least lessen the severity of, the approaching energy crunch? In the rest of this chapter, you will read about possible solutions to the energy problem.

Section Review answers on page T155.

SECTION REVIEW

1. Describe how fossil fuels are used to produce electricity.
2. What is the difference between renewable and nonrenewable resources?

THINKING CRITICALLY

3. *Interpreting Graphics* Examine Figure 11-7. What do you think accounts for the dramatic increase in the worldwide production of oil after 1950?

NUCLEAR ENERGY

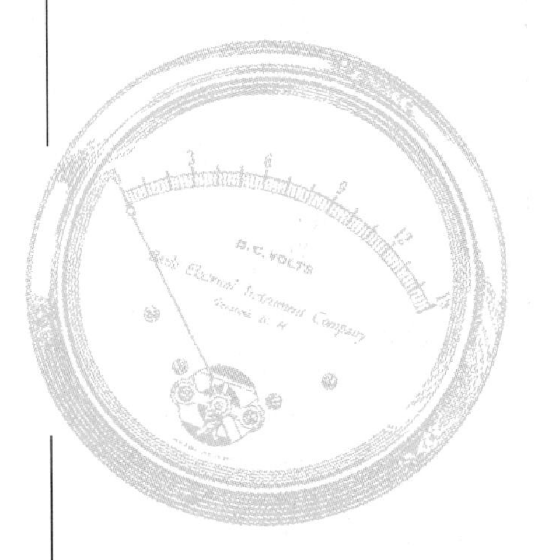

AFTER READING THIS SECTION YOU SHOULD BE ABLE TO

❶ define *atomic fission.*

❷ describe the operation of a nuclear power plant.

❸ explain how a nuclear power plant is similar to a plant that burns fossil fuels.

❹ explain the advantages and disadvantages of nuclear energy.

In the 1950s and 1960s, nuclear energy was seen as the wave of the future. It was predicted that a nationwide network of nuclear power plants would provide electricity "too cheap to meter." Many nuclear power plants were on the drawing boards. But in the 1970s and 1980s, almost 120 planned nuclear power plants were canceled, and about 40 partially constructed nuclear plants were abandoned. What happened? In this section, you'll learn why we don't get most of our electricity from nuclear power plants. You'll also learn how nuclear energy works.

HOW NUCLEAR ENERGY WORKS

Nuclear energy is the energy that exists within the nucleus of an atom. Powerful forces bind together the components of the atomic nucleus. When these subatomic bonds are broken apart in a process called **nuclear fission,** huge amounts of energy are released.

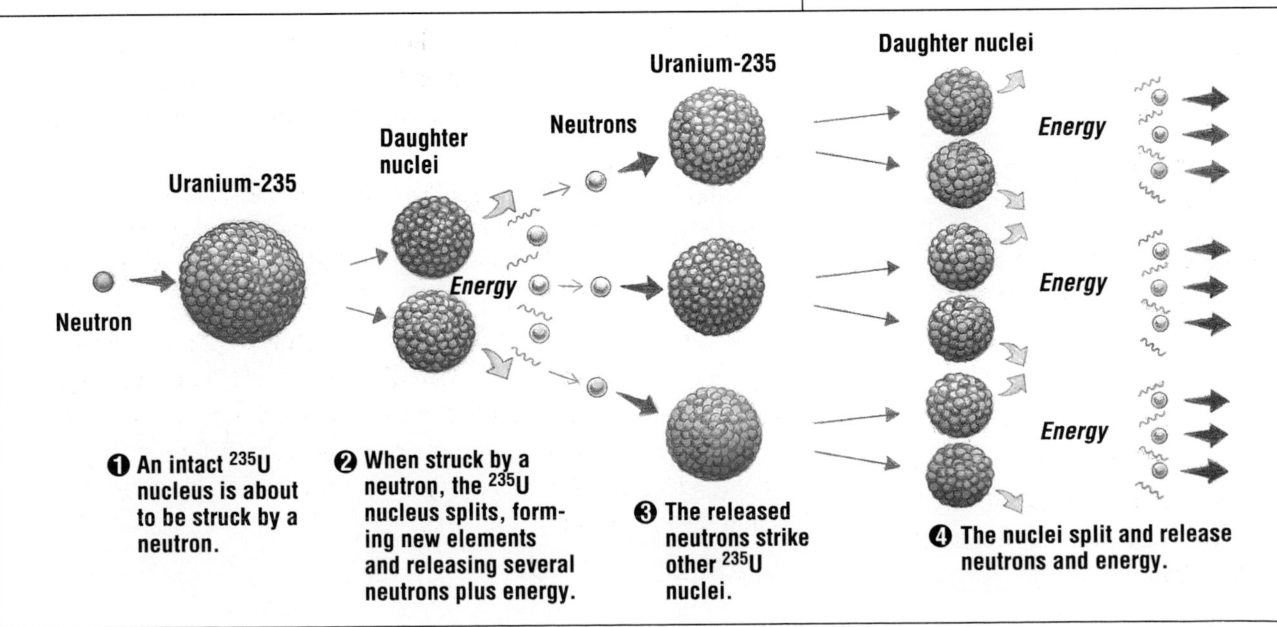

Figure 11-8 The process of nuclear fission

Figure 11-9 **The photograph below shows a nuclear power plant in France. The diagram shows how a nuclear power plant works.**

One example of nuclear fission is the explosion of an atomic bomb. An atomic explosion is an uncontrolled release of nuclear energy. Figure 11-8 on the previous page shows how this type of nuclear reaction takes place. If allowed to continue, the reaction quickly escalates out of control. It ceases only when all of the fissionable nuclear material is used up.

By contrast, the nuclear reactions in a nuclear power plant occur at a controlled, manageable pace and release energy slowly. Even so, a tremendous amount of heat is generated. This heat is used to boil water at very high temperatures, creating high-pressure steam that can be used to drive an electric generator. The diagram at the bottom of this page shows how a typical nuclear power plant works.

You can see in the diagram that the reactor is surrounded by a thick pressure vessel, which is a steel casing that surrounds the reactor and contains cooling fluid. The pressure vessel is designed to contain the fission products released in the event of a malfunction. The pressure vessel is completely enclosed in a containment building, which is usually a dome made of reinforced concrete several meters thick.

Advantages of Nuclear Energy

Nuclear materials are an extraordinarily rich energy source.

How a Nuclear Power Plant Works

1. Energy released by the nuclear reaction heats water in the pressurized primary circuit to a very high temperature.
2. The superheated water is pumped to a heat exchanger, which transfers the heat of the primary circuit to the secondary circuit.
3. Water in the secondary circuit flashes into high-pressure steam.
4. Steam is directed against a turbine, setting it in motion.
5. The turbine sets the generator in motion, generating electricity.

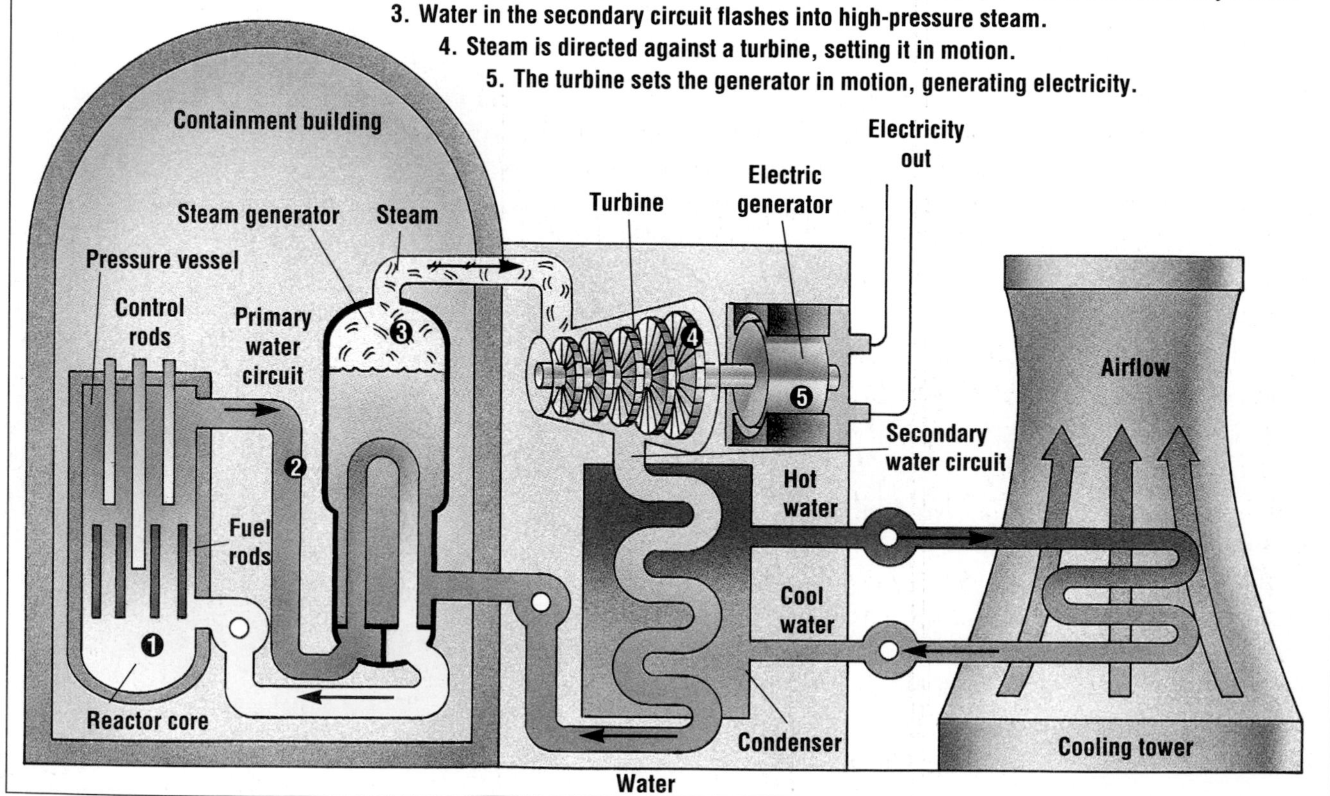

One gram of uranium-235 (^{235}U), the most common nuclear fuel, delivers about as much energy as 3.5 metric tons of coal. Well-designed reactors can run for years without having to be refueled or shut down for anything other than minor maintenance. Furthermore, nuclear energy does not produce gases such as carbon dioxide, which is an enormous advantage compared with fossil fuels. Many nations—especially those with meager fossil-fuel reserves—rely heavily on nuclear plants to meet their energy needs.

So Why Aren't We Using More Nuclear Energy?

The most serious disadvantage of nuclear energy is that it produces radioactive waste. At present, the United States has no facility for the permanent disposal of its commercial nuclear waste. Each nuclear power plant has its own temporary storage facilities.

Another disadvantage is that nuclear fuel is in relatively short supply. Estimates vary, but there may be only a 100- to 200-year supply of ^{235}U at current rates of consumption.

Nuclear energy is also extremely expensive. Nuclear power plants are very large and complex, with elaborate safety systems, so they are expensive to build. Minor problems in a nuclear facility can force weeks or months of costly downtime. All of these extra expenses are reflected in higher utility bills. It was mainly the expense of nuclear energy, along with the fear of nuclear accidents, that caused the worldwide slowdown in nuclear plant construction.

Safety Concerns The fission reaction creates radioactive products, many of which are very dangerous. If the reaction gets out of control, the enormous heat it generates may destroy the reactor building and spew the radioactive products into the air.

The former Soviet Union has experienced at least two major nuclear accidents. One occurred at a plant in the Ural Mountains in 1957, and the other occurred at Chernobyl in 1986. The details of the 1957 accident were kept secret, but the events at Chernobyl are well known.

Engineers at the Chernobyl nuclear power plant turned off most of the reactor's safety devices while they conducted an unauthorized test. This test caused massive explosions that demolished the reactor and the surrounding structure and blasted tons of hazardous materials high into the air. (See Figure 11-11.) Radioactivity contaminated thousands of square kilometers of land. Hundreds of firefighters and other workers died from radiation exposure, and thousands more may contract cancers as a result of their exposure to high levels of radiation.

The Chernobyl reactor was an obsolete type that, for safety reasons, is not used in the United States. In addition, the operators at Chernobyl violated basic safety guidelines.

ACTIVIST

Bob Geyer thinks the nuclear waste facility being built near his hometown might be dangerous. Find out what he's doing about it on pages 300–301.

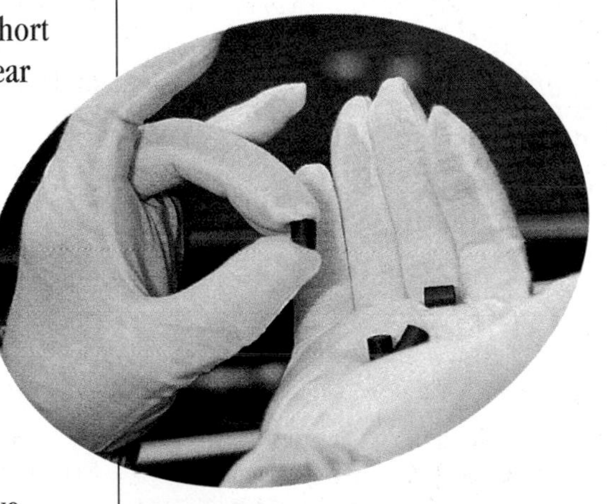

Figure 11-10 Each of these uranium fuel pellets contains the energy equivalent of about 1 metric ton of coal.

Figure 11-11 This is how the Chernobyl plant looked shortly after the accident in 1986. The arrow points to the damaged reactor.

In the United States, the most serious nuclear accident occurred in 1979 at the Three Mile Island nuclear power plant in Pennsylvania. Human error, combined with blocked valves and broken pumps, was responsible for this accident. Fortunately, only a small amount of radioactivity escaped the containment structure. Since the accident at Three Mile Island, the Nuclear Regulatory Commission has required more than 300 safety improvements to nuclear power plants.

Figure 11-12 **Experimental reactors such as this one at Princeton University have successfully achieved nuclear fusion for brief periods.**

Nuclear Fusion

One possibility as a future energy source is nuclear fusion. **Nuclear fusion** occurs when lightweight atomic nuclei combine to form a heavier nucleus, releasing huge amounts of energy in the process. This is basically the opposite of nuclear fission, in which the nucleus of an atom is split apart. Figure 11-13 shows how fusion works. Nuclear fusion is the process that powers the stars, including our sun, and is a potentially safer energy source than nuclear fission.

Unfortunately, although the potential of fusion is great, so is the technical difficulty of achieving it. For fusion to occur, the atomic nuclei must be heated to extremely high temperatures (100,000,000°C, or 180,000,000°F), be maintained at very high concentrations, and be properly confined. Achieving all three of these conditions simultaneously is extremely difficult. The problem is so complex that discovering a solution may take decades—or may never happen.

Figure 11-13 **During fusion, the nuclei of deuterium and tritium join to form helium, releasing huge amounts of energy.**

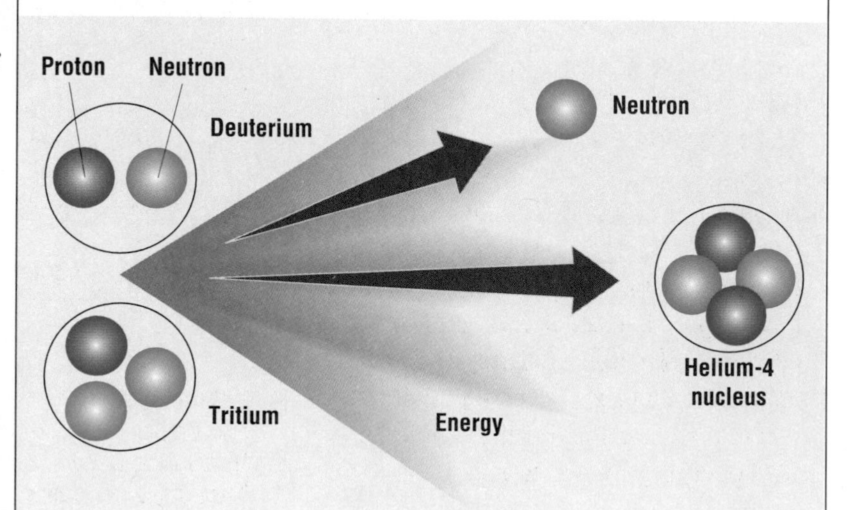

Section Review answers on page T155.

Section Review

1. How are a fossil-fuel-burning power plant and a nuclear power plant similar?
2. Name two advantages and two disadvantages of nuclear power plants.
3. Explain the difference between nuclear fission and nuclear fusion.

Thinking Critically

4. *Making Decisions* Suppose the world had an unlimited supply of either nuclear or fossil fuels, but not both. Use the decision-making model presented in Chapter 1 to decide which would be the better energy supply.

A SUSTAINABLE ENERGY FUTURE

AFTER READING THIS SECTION YOU SHOULD BE ABLE TO

1. describe methods of conserving energy.
2. describe several alternative energy sources.

Our fossil fuel supplies are dwindling, while our appetite for energy is increasing. The days when we could afford to be careless with energy are past. To achieve a sustainable energy future, we must develop new energy sources, make the most of the ones we already have, and use every form of energy as efficiently as possible.

The United States, with just over 4 percent of the Earth's population, uses 24 percent of all energy produced.

ENERGY CONSERVATION

The average American uses about twice as much energy as the average European, whose standard of living is comparable. Fortunately, there is much we can do to improve the situation. What comes to mind when you hear the word *conservation?* If you're like most people, you aren't exactly thrilled by the idea. Conservation doesn't sound like much fun. However, conservation doesn't simply mean making do with less. It means using materials and energy wisely.

Consider the ways you could conserve energy. You could turn the thermostat down in the winter and put on a sweater. In the summer, you could leave the air conditioner off as long as possible and use a fan. Another energy-saving strategy is to walk, bike, or ride a bus instead of driving. See Figure 11-14 for other energy-conservation tips.

Another way of conserving energy resources is to modify structures, systems, and vehicles so that they don't waste much energy. The Case Study on pages 288–289 focuses on a house that has high energy efficiency. The high-efficiency cars and appliances that have become popular in recent years are additional examples of this kind of energy conservation.

Figure 11-14 **Can you think of other energy-saving strategies besides those listed here?**

What You Can Do to Save Energy

Walk or ride a bicycle for short trips.
Carpool or use public transportation whenever possible.
Keep the engine in your car tuned.
Keep the thermostat on your water heater set at no more than 120°F.
Take short showers instead of baths.
Wash only full loads of clothes or dishes.
Close off and do not heat or cool unused rooms or closets.
Use fluorescent bulbs instead of incandescent ones.
Use natural lighting whenever possible.
Turn off unnecessary lights and appliances.
Buy only high-efficiency appliances.

In the winter:
Turn down the thermostat, and compensate by wearing a sweater.
Turn down the thermostat at night and when no one is home.
Caulk and weather-strip cracks.

In the summer:
Turn up the thermostat, and compensate with fans.
Turn off the air conditioner when no one will be home for four hours.
Close drapes on sunny days. Open them on cool days and at night.

Energy for the Future

What sources of energy will light our cities and power our vehicles in the future? You might imagine exotic, as-yet-unknown sources of energy. In reality, most of the future's energy sources are probably in use right now. Take a look at the following alternative energy sources.

Solar Energy

The energy from the sun is known as **solar energy.** Solar energy is clean, safe, and renewable. In fact, solar energy might be described as the ultimate renewable resource.

Solar energy is used in three basic ways: passive solar heating, active solar heating, and solar cells. The simplest and cheapest method is **passive solar heating,** in which sunlight is used to heat buildings directly without pumps or fans. As shown in Figure 11-15, houses with passive solar heating have energy-efficient windows that face south so that the house absorbs as much heat as possible from the sun. They are also built with large amounts of

ENERGY AUDIT

How energy efficient is your home? Find out how to check it out on page 389.

A Super-Efficient Structure

Imagine a home located deep in the Rocky Mountains, where winter temperatures can plunge to –40°F. The home has no furnace, yet it manages to stay comfortably warm even in the coldest weather. This home, built by energy experts Hunter and Amory Lovins in Snowmass, Colorado, is a prime example of a new generation of super-efficient structures.

Efficiency without sacrifice was the goal in designing the Lovins's home, which also houses the Rocky Mountain Institute (RMI), an energy-issues research organization. The structure uses one-tenth the electricity and one-half the water of comparable structures. The building cost somewhat more to build than comparable structures, but that extra cost was recovered through energy savings in only 10 months. Over 40 years, the savings in utility bills will pay for the entire structure.

Solar energy is the most important energy source for RMI. The abundant, south-facing windows let in plenty of sunshine. As a result, little daytime lighting is required. The artificial lighting that is necessary is provided by compact fluorescent lamps that draw only 18 W but provide as much light as a standard 75 W incandescent bulb. These lamps also last 10–15 times longer than ordinary bulbs. Sensors turn the lights off when a room is empty and back on when someone enters the room.

Much of the building's electricity is provided by solar panels. If it weren't for equipment such as

The Rocky Mountain Institute in Colorado

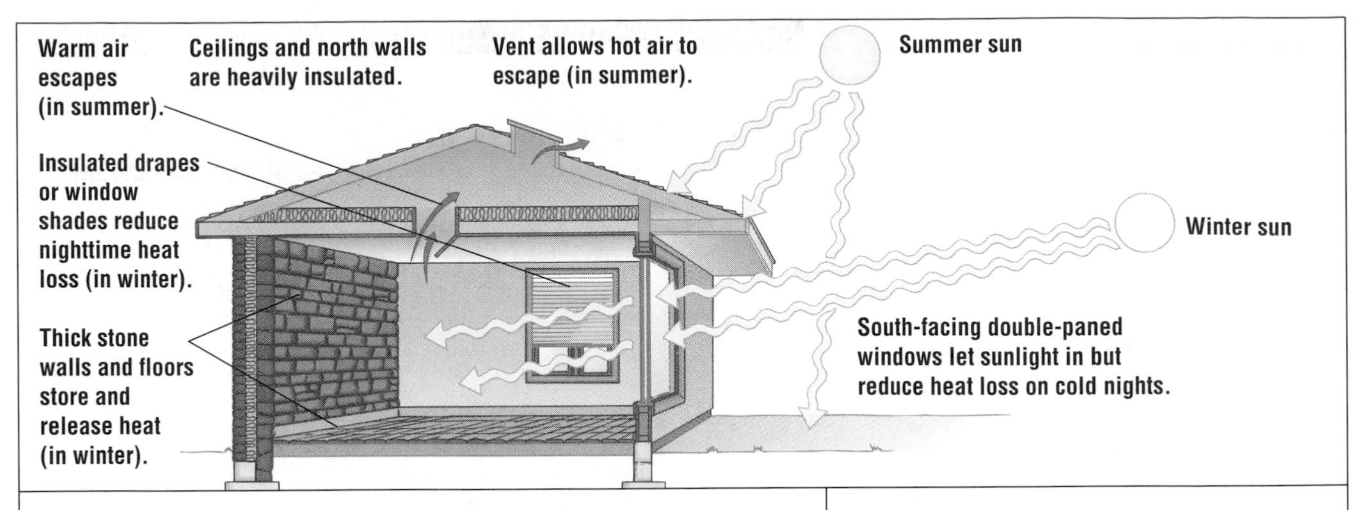

stone or concrete because these materials store heat better than wood or plaster. An efficient passive solar heating system can heat a house even in very cold weather without using any form of energy other than solar energy. The only requirement is reliable winter sunlight.

Structures heated by **active solar heating** have a series of solar collectors that absorb solar energy and convert it into heat.

Figure 11-15 In a passive solar house, double-paned windows facing south let in the winter sun but prevent heat from escaping. Thick stone walls and floors capture the heat of the sun and release it slowly. Thick insulation and heavy drapes over the windows reduce nighttime heat loss.

Answers to Thinking Critically can be found on page T156.

copiers and computers, the building might not require any outside electricity at all. RMI staffer Owen Bailey said, "When the copier is not running, we actually sell power back to the utility company."

RMI has some of the world's most efficient appliances. The state-of-the-art refrigerator uses 8 percent as much energy as a standard refrigerator, with no loss of cooling. A special solar-heated clothes dryer sits next to the high-efficiency washer.

Solar energy, plus the heat from appliances and human bodies, meets 99 percent of the heating needs. The other 1 percent is provided by two backup wood-burning stoves.

The walls and roof of RMI are superinsulated, greatly reducing heat loss. The walls and windows are airtight, eliminating another common cause of heat loss. During an extended period of cloudy winter weather (and therefore no solar heat input) the building loses only about 1°F per day. Nevertheless, the structure is well ventilated. It has specially designed air exchangers that use the exiting stale air to warm the incoming fresh air.

The RMI structure shows that conservation doesn't require discomfort. The building is comfortable, spacious, and livable. As Amory Lovins said, "The main thing the Institute demonstrates is that conservation—or as I prefer to call it, 'energy efficiency'—doesn't mean freezing in the dark."

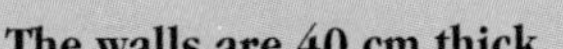

The walls are 40 cm thick.

This toilet uses up to 90 percent less water than ordinary toilets.

THINKING CRITICALLY

1. *Inferring Relationships* Many specially designed homes in Colorado are able to receive the majority of their heat from the sun. But in Canada and Alaska, where the climate is similar to the climate in Colorado, solar-heating systems are often inadequate. Review what you learned about latitude in Chapter 7, and offer an explanation for this.
2. *Analyzing Relationships* Currently, only about 1 percent of the homes built in this country are energy-efficient designs. What could be done to increase this percentage?

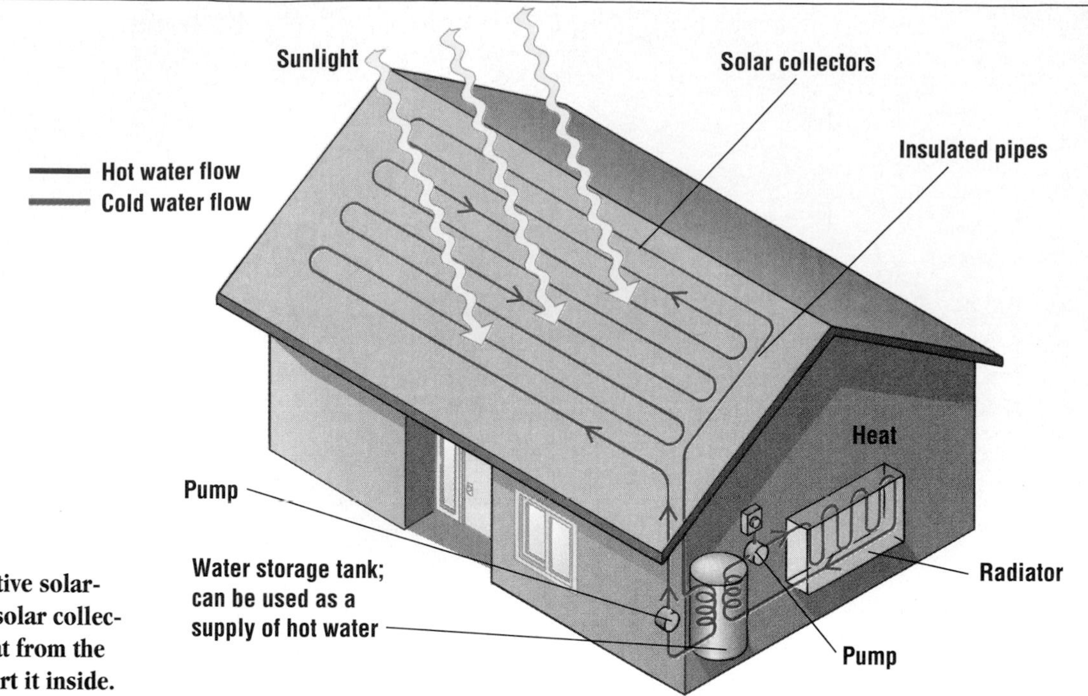

Figure 11-16 An active solar-heating system has solar collectors that absorb heat from the outside and transport it inside.

(See Figure 11-16.) These solar collectors are enclosed in an insulated box and are dark on the inside to better absorb the sun's energy. The heat is carried away from the box by water flowing through pipes or by air circulated by a fan. The heat is used to warm an insulated storage tank of water, and the air or water is then returned to the collection device. The storage tank provides hot water and heating for the building.

The third method for capturing solar energy is with solar cells, also called photovoltaics. **Solar cells** are devices that convert the sun's energy directly into electricity. Figure 11-17 shows how solar

Figure 11-17 Solar cells consist of thin wafers of semiconductors, which are substances that release electrons when struck by sunlight.

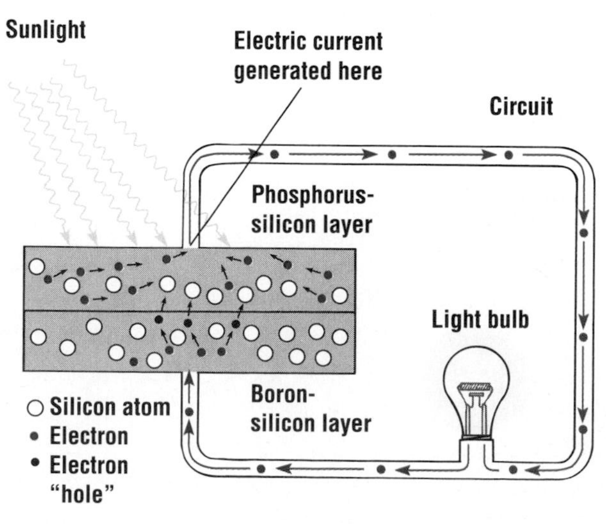

Sunlight falls on a semiconductor, causing it to release electrons. The electrons create a circuit that is completed when another semiconductor in the solar cell absorbs electrons and passes them on to the first semiconductor.

cells work. Until recently, solar cells were not a realistic option for most consumers because of their high cost and low efficiency. The manufacture of solar cells is a complicated, expensive process requiring the expenditure of much energy. However, great strides have been made in designing more efficient and cost-effective solar cells.

The main problem with solar-energy systems is that they require sunny weather to work. Few regions can count on extended periods of clear skies. Places that have the right kind of weather—deserts—are generally far from population centers. Additionally, there is only so much energy that can be extracted from a given amount of sunlit surface. To meet the electricity needs of even a small city would require covering hundreds of acres with expensive solar cells. Then, even in optimum weather, the system would work for only a few hours a day. Backup generators would be needed to supply electricity during "off hours."

Wind Energy

Wind is an indirect form of solar energy. The sun heats the Earth's surface unevenly, putting the atmosphere into turbulent motion. Tremendous energy is contained in the wind. This energy can be captured with a turbine, which is connected to an electric generator. A windmill is an example of a device that captures the energy of wind.

The technology of wind generators is well developed, and the cost of wind-generated electricity is lower than that produced by some other sources. The chief disadvantage of wind energy is that few regions have winds strong or consistent enough to make wind generators economical. And there are other disadvantages. The large commercial wind generators may not be attractive additions to the landscape. At full speed they can be very noisy. The blades of large wind generators can also interfere with microwave communications. While wind energy may never be a major source of energy, it is a practical alternative in some areas.

Figure 11-18 These wind generators produce millions of watts of electricity for the people of northern California.

Answer to Field Activity: Answers will vary depending on the materials used, but, in general, stone and metals will perform better than wood, plastic, and glass.

FIELD ACTIVITY

Find out what materials are best for converting sunlight to heat. Gather items made of different materials, such as aluminum, steel, copper, stone, wood, plastic, and glass. You can test any material you want. Tape a thermometer to each item, and put them outside in the sunlight. After 30 minutes, record the temperature of each item. (Be careful with very hot materials.) Make a list of the materials, showing the hottest materials at the top and the coolest at the bottom. Based on your research, which material do you think would make the best solar heater?

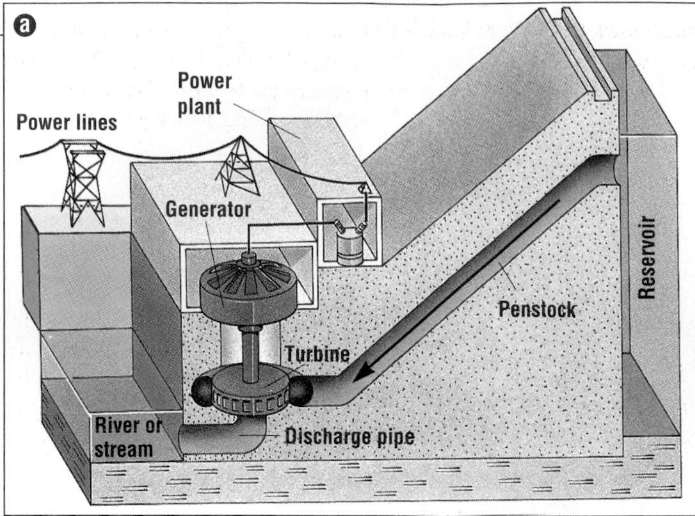

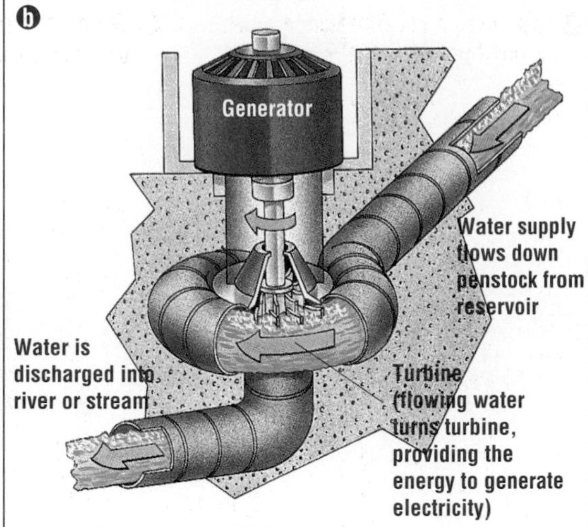

Figure 11-19 The diagrams above show how a hydroelectric dam works. The photograph below shows a dam in Oregon.
a. Cutaway view of a hydroelectric dam
b. Close-up of a generator

HYDROELECTRICITY

Thousands of years ago, humans learned to harness the energy of falling water. Today, water is used to generate electricity. To have enough energy to push a heavy turbine and generator, the water must fall a considerable distance. Dams provide this vertical distance. (See Figure 11-19.)

Power from water has several advantages. It is clean, renewable, and leaves no waste. There are indirect benefits as well. The reservoirs that form behind hydroelectric dams offer recreational opportunities and provide drinking water. The dams may also control flooding.

On the other hand, hydroelectricity has some disadvantages. When a dam is constructed, the water forms a reservoir that disrupts the river ecosystem and covers the existing land ecosystem. Furthermore, the cost of dam construction has soared. Finally, hydroelectricity is an option mainly in areas that have rivers and streams.

McNary Dam, a hydroelectric dam in Oregon

GEOTHERMAL ENERGY

The interior of the Earth is a very hot place. In some places, the heat lies very close to the surface. Such places are sources of **geothermal energy**—the heat inside the Earth. This heat can be used to drive electric generators, similar to the way in which fossil fuels and nuclear energy drive electric generators. Figure 11-20 shows one method of extracting geothermal energy from the Earth, called the *hot, dry rock method.*

The major problem with geothermal energy is that Earth processes replace the heat very slowly. In some locations, large-scale

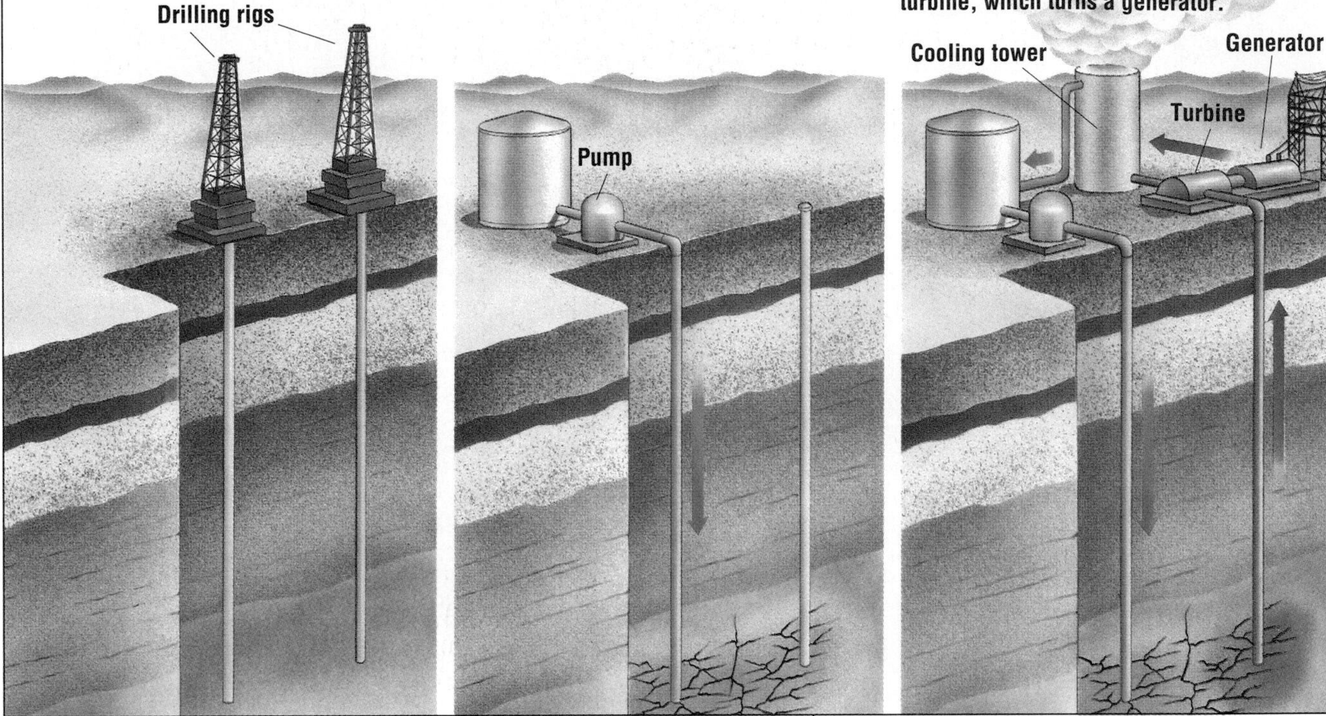

Figure 11-20 The hot, dry rock method of extracting geothermal energy from the Earth

geothermal operations can quickly use up the available geothermal energy. In New Zealand, for example, most electricity was once generated with geothermal energy, but excessive use of the resource depleted it. Another disadvantage is that geothermal energy can be tapped only in a few places, so its availability is limited.

ECO-FACT

Two billion people rely on wood as their primary energy source.

BIOMASS

Biomass is the organic matter in plants or plant products. Biomass is an important source of energy worldwide, especially in the form of wood. For much of the world, in fact, wood is the major source of energy. Industrialized nations use relatively little wood, but this is changing. In the United States, the use of wood for heating homes has increased in the last two decades as people have tried to conserve fossil fuels. Burning biomass for fuel, however, has the disadvantage of producing carbon dioxide and noxious smoke.

Waste to Energy An enormous amount of biomass exists in the form of waste material. Some industries have begun using this vast resource. The United States pulp and paper industry, for instance, generates more than half of the energy it uses from its own waste products. And the Hawaiian sugar industry burns waste

Figure 11-21 This biomass—litter from chicken coops—is burned in a British power plant to produce the steam that drives turbines.

Figure 11-22 Gasohol is a mixture of gasoline and alcohol.

Answers to Section Review can be found on page T156.

sugarcane to generate electricity. This industry meets its own energy needs and sells the excess electricity to the public.

Liquid Fuels From Biomass

Liquid fuels can be derived from certain types of biomass. For example, methanol, a type of alcohol, can be derived from wood. Ethanol, another type of alcohol, can be derived from corn, fruit, or agricultural waste. Cars and trucks can run quite well on either methanol or ethanol with little modification. As an added benefit, ethanol and methanol are much cleaner burning than either gasoline or diesel fuel. In some places the conversion from gasoline to alcohol is well underway. For example, 50 percent of the automotive fuel used in Brazil is alcohol.

Alcohol doesn't have to be used on its own as a fuel. Another option is to use a blend of gasoline and alcohol as fuel, a blend called *gasohol.* Typically, gasohol is a mixture of 10 percent alcohol and 90 percent gasoline. Gasohol is commonly used in some parts of the United States.

Hydrogen

Hydrogen is the most plentiful substance on the Earth's surface. Every molecule of water is made up of two hydrogen atoms and one oxygen atom. When separated from oxygen, hydrogen is a clean-burning gas that has great promise as a fuel for the future. The average internal-combustion engine can run on hydrogen gas without major modification. Hydrogen is also a nonpolluting fuel since the product of hydrogen combustion is water.

However, storing hydrogen is a problem. In its natural, gaseous state, hydrogen is simply not concentrated enough. Even a small engine would quickly use up the hydrogen contained in a fuel tank. One way to store usable amounts of hydrogen is to compress it to very high pressures. Another method is to chill it to extremely low temperatures, which causes it to liquefy.

Section Review

1. How could you conserve energy as you go about your daily activities?
2. What are some advantages of solar energy? What are some disadvantages?
3. How are solar energy and wind energy similar? How are they different?

Thinking Critically

4. *Relating Concepts* Make a concept map showing the advantages and disadvantages of hydroelectricity. Be sure to include at least the following terms: hydroelectricity, renewable, ecosystems, dams, reservoirs, waste, recreation, cost, and drinking water.
5. *Making Decisions* Which alternative energy source discussed in this section would be best for your region of the country? Why?

HIGHLIGHTS

SUMMARY

- Our society uses large amounts of energy, especially to generate electricity. Either directly or indirectly, most of this energy comes from fossil fuels.
- Fossil fuels are a limited, nonrenewable resource and are being used up at a very rapid pace.
- Nuclear energy is one possible alternative to fossil fuels. Nuclear materials are a rich source of energy. However, nuclear energy has many drawbacks that have kept it from being used more widely. The most serious drawback is the production of radioactive waste.
- There are a number of other alternatives to fossil fuels. All of these options have advantages and disadvantages. Most are applicable only to certain regions. Some of these technologies offer great promise if certain problems related to them can be solved.
- Our energy needs continue to grow. A sustainable energy future will require the efficient use of all forms of energy.

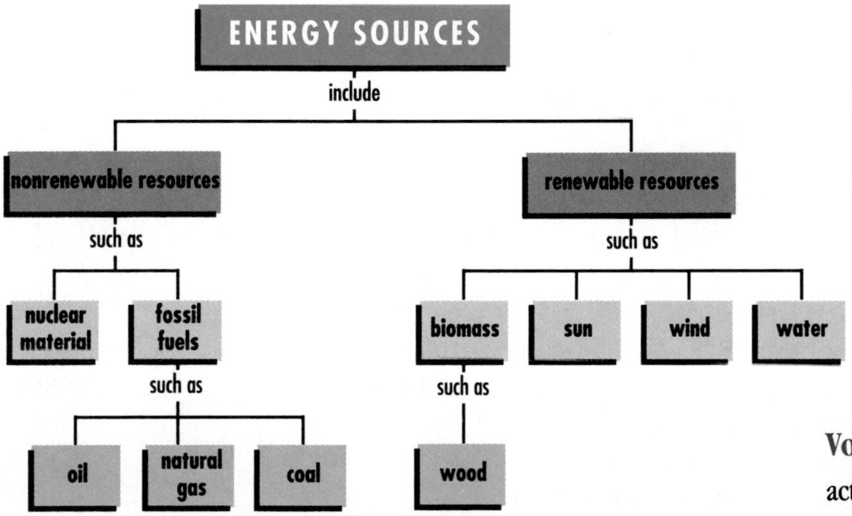

ENERGY SOURCES
include
nonrenewable resources
such as
nuclear material
fossil fuels
such as
oil
natural gas
coal
renewable resources
such as
biomass
such as
wood
sun
wind
water

Sample EcoLog answers can be found on page T156.

EcoLog

Now that you've studied this chapter, revise your answers to the questions you answered at the beginning of the chapter, based on what you have learned.

1. You flip on a light switch and the lights instantly come on. Where does the electricity come from?
2. What will our major source of energy be 100 years from now? Explain your reasoning.

Vocabulary Terms

active solar heating (p. 289)
electric generator (p. 280)
fossil fuels (p. 281)
geothermal energy (p. 292)
nuclear energy (p. 283)
nuclear fission (p. 283)
nuclear fusion (p. 286)
passive solar heating (p. 288)
solar cells (p. 290)
solar energy (p. 288)

CHAPTER 11

REVIEW

UNDERSTANDING VOCABULARY

1. For each of the following pairs of terms, describe the difference in their meaning.
   - **a.** nonrenewable resource
     renewable resource
   - **b.** nuclear fission
     nuclear fusion
   - **c.** active solar heating
     passive solar heating

RELATING CONCEPTS

2. Copy the unfinished concept map below onto a sheet of paper. Then complete the concept map by writing the correct word or phrase in each box containing a question mark.

Answers not indicated can be found on page T156.

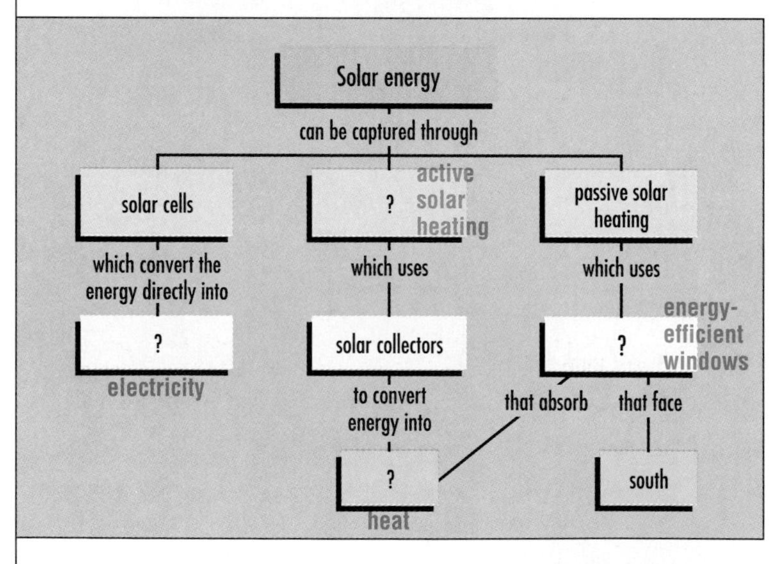

UNDERSTANDING CONCEPTS

Multiple Choice

3. Which is a reason for the widespread use of fossil fuels?
   - **(a.)** They contain a great deal of energy.
   - **b.** They are a renewable source of energy.
   - **c.** They are not harmful to the environment.
   - **d.** all of the above

4. Which fossil fuel will most likely last the longest?
   - **a.** oil
   - **(b.)** coal
   - **c.** natural gas
   - **d.** All will run out at about the same time.

5. Which is NOT true of geothermal energy?
   - **a.** It is effectively available in only a few locations.
   - **b.** Individual sources can be depleted quickly.
   - **c.** It can be used to operate electric generators.
   - **(d.)** When a source is depleted, the Earth replenishes it quickly.

6. The main reason for the worldwide slow-down in nuclear plant construction is that
   - **a.** we have already run out of nuclear fuel.
   - **b.** reactors require too much maintenance.
   - **(c.)** nuclear energy is extremely expensive.
   - **d.** reactors release too many greenhouse gases.

7. Solar energy
   - **a.** can meet the energy needs of most places cheaply and efficiently.
   - **(b.)** has much potential but also has many limitations.
   - **c.** can supply all of our energy needs.
   - **d.** is a nonrenewable resource.

8. The problem with nuclear fusion is that
   - **(a.)** there are many technical difficulties to solve before it becomes feasible.
   - **b.** it doesn't release much energy.
   - **c.** it is more dangerous than nuclear fission.
   - **d.** the process by which it works is unknown.

9. Why will wind probably never be a major source of energy?
   a. The wind contains very little energy.
   b. The technology to generate energy from the wind is too difficult.
   c. The winds in most regions are not strong or consistent enough to run a wind generator cost-effectively.
   d. Wind-generated electricity costs more than electricity generated by other sources.
10. Which is NOT a renewable energy source?
   a. solar energy
   b. fossil fuels
   c. biomass
   d. hydroelectricity
11. Which is an example of a DIRECT use of fossil fuels?
   a. an oil-fired furnace
   b. a nuclear reactor
   c. a wood-burning stove
   d. an electric generator
12. One gram of uranium-235 delivers the same amount of energy as how much coal?
   a. 1 g
   b. 3.5 g
   c. 1 metric ton
   d. 3.5 metric tons

Short Answer

13. Explain why geothermal energy is not a realistic option in all locations.
14. Wind and water are often called indirect forms of solar energy. Is this an accurate statement? Explain your reasoning.
15. In general, nuclear energy is used more widely in countries that have meager energy supplies. Explain this connection.
16. Could the sun be economically used as a source of energy in your area? Why or why not?
17. Hydroelectricity works well only in areas that have hilly or mountainous topography. Why is this so?

Interpreting Graphics

18. **Examine the graph below.** It shows the change in known oil reserves from 1980 to 1992. What could have caused this change? Did more oil form? How and why might a worldwide tax of $30 per barrel affect the search for reserves?

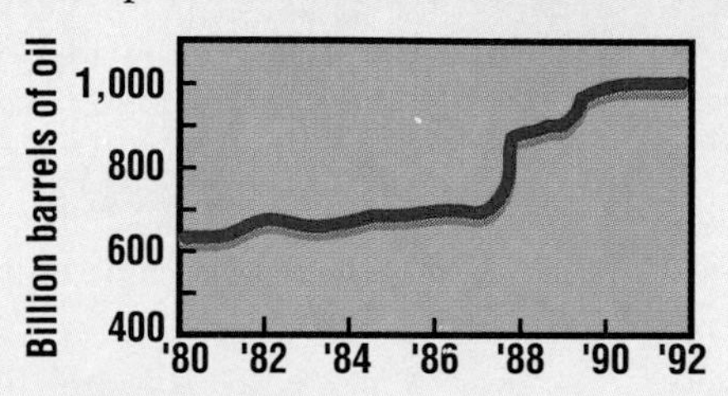

Thinking Critically

19. ***Analyzing Conclusions*** It has been said that geothermal energy is globally renewable but locally nonrenewable. Explain the reasoning behind this statement.
20. ***Inferring Relationships*** Consider that 100 barrels of oil require the equivalent of several barrels of oil in energy to process into usable form. What's left is called the net energy yield. What conclusion would you draw about an energy source that has a negative net energy yield? Explain.

Themes in Science

21. ***Energy*** Explain why fossil fuels can be considered a form of stored solar energy.
22. ***Patterns of Change*** How has our pattern of energy use evolved over time? How has our lifestyle changed in the process?

Cross-Discipline Connection

23. ***Physics*** Recall the discussion of net energy yield from Question 20 above. Using your knowledge of physical principles, do you think there could be an energy source with a net yield of greater than 100 percent? Explain.

Discovery Through Reading

How much progress is this nation making, in terms of using known technology, to improve the energy efficiency of buildings? For a status report and some strategies for conserving energy in your home, read "Keeping Warm and Staying Cool, Economically and Efficiently," by Daniel Yoon, et al., *Garbage,* Spring 1994, pp. 52–57.

PORTFOLIO ACTIVITY

Do a special project about electrical power in your community. How is it supplied now? How might it be supplied in the future? Does your community have potential as a site for solar energy? wind energy? hydroelectric energy? Take a local map and locate the power plants that supply your city, or locate potential areas where alternative sources might be developed. Prepare a display using your map and any other appropriate materials.

INVESTIGATION

SOLAR DESIGN

Have you ever returned to your car on a cold but sunny day to discover that its interior temperature was warmer than the outside air? The temperature in your car rose due to the greenhouse effect. Solar energy, energy from the sun, heated the interior surfaces of the car. Thermal energy built up inside the car because, unlike solar energy, thermal energy does not pass as readily through the car's glass windows.

Architects sometimes design homes to take advantage of the sun's heating ability. These solar homes have features that collect and store the energy from the sun, often eliminating the need for gas, oil, or electric heating. In this Investigation, you will experiment with models of passive solar homes to find out what features best utilize the sun's energy.

Answers to this Investigation can be found on page T156.

MATERIALS

- two identical cardboard boxes with lids
- knife or single-edged razor blade
- clear plastic wrap
- thermometer
- tape
- metric ruler
- watch or clock
- one or more of the following: cardboard, foam insulation, caulk, paint, foil, tiles, water containers, colored construction paper, rocks, soil, etc.
- notebook
- pen or pencil

MAKE A MODEL OF A PASSIVE SOLAR HOME

Imagine that you are an architect and that you and the other architects in your firm are designing the most effective passive solar home that is possible. Your group will first make a model of a passive solar home to serve as an experimental control. Then you will make a second model home, only this time you will add a feature to it to improve its efficiency.

1. Use a cardboard box to build a model of a passive solar home. To make your model, cut out a window that is 5 cm (2 in.) long on each side. Cover the window with plastic wrap secured with tape. Place a thermometer in the model so that it can be seen through the window.
2. Set your model in the sun, facing it in the direction that allows maximum sunlight to hit the window directly.
3. Record the temperature in the model at the start of this experiment and every 5 minutes thereafter for 30 minutes. Use a table similar to the one below to record your results.

	Time						
	Start	5 min.	10 min.	15 min.	20 min.	25 min.	30 min.
Temperature							

HOW CAN YOU IMPROVE YOUR HOME'S PERFORMANCE?

4. Working with your group, develop a hypothesis about a design feature that would best improve your model's passive solar heating capability. Consider the following design features in developing your hypothesis:
   - orientation of the home
   - location of windows
   - surface area of windows

- skylights
- roof overhang
- insulation
- interior and exterior surface colors and finishes
- heat-absorbing materials, including rocks, tile, water containers, and soil
- air leakage around windows and doors

Sample hypothesis:
The greater the surface area of windows on the side of the home facing the sun, the more solar energy the home will collect and the more rapidly it will heat up.

Passive solar homes use the sun's clean, plentiful energy for inexpensive heating.

PERFORM AN EXPERIMENT TO TEST YOUR HYPOTHESIS

5. After your teacher has approved your hypothesis, design and build a new model to test your hypothesis.
6. Test your model, and record your results in a table like the one on the facing page.

ANALYZE YOUR RESULTS

7. Did your second model reach a higher internal temperature than your first model? How do you account for any differences in the effectiveness of your two models?
8. Compare your results with the results of other groups. Were their designs more efficient? If so, what did they do differently? If you could redesign your model yet again, what changes would you make?
9. If a passive solar heating system adds $5,000 to conventional home construction costs, the maintenance cost of the system is $25 per year, and the estimated savings on heating costs are $475 per year, how long will it take for the solar system to pay for itself?

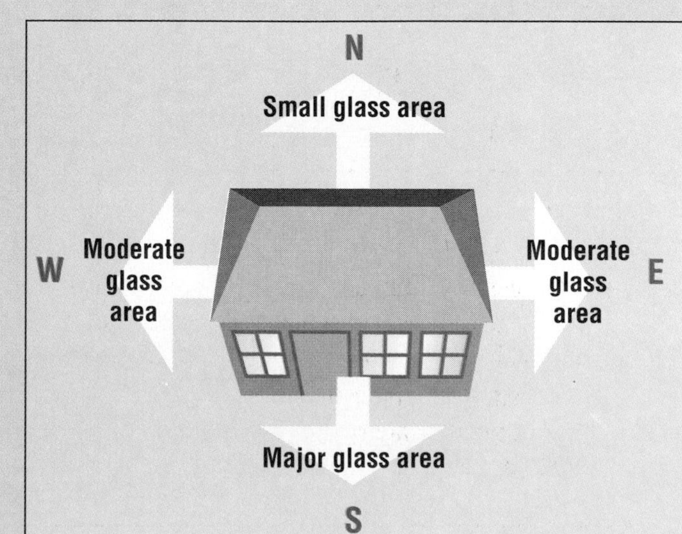

During the winter, the south side of a house receives almost three times more solar radiation than any other side. For this reason, passive solar windows usually face south.

Making a DIFFERENCE

ENVIRONMENTAL ACTIVIST

Bob Geyer is making a difference both on and off his job. As the planning manager for the mass-transit system in El Paso, Texas, Bob encourages his associates and the people of El Paso to take full advantage of alternative-fuel vehicles, car-pooling programs, and mass transit in general. In his spare time, Bob educates himself and others about issues and policies that affect the environment. When he learns of some action, activity, or company that he believes will threaten the environment, Bob doesn't hesitate—he acts.

If I saw something that bothered me, even back in high school or college, I would write a letter or speak out against it.

Q: What environmental issue is most important to you right now?

Bob: The Texas state government has selected Sierra Blanca, Texas, as a disposal site for low-level nuclear waste. This type of waste is still very toxic and could pose a serious threat to the people in the vicinity. And many of the people who live in Sierra Blanca are poor, undereducated, and basically ill prepared to fight against this sort of an environmental hazard.

Plus, the Sierra Blanca nuclear dump site is only 86 mi. from El Paso and 16 mi. from the Río Grande. I hate to think about the area becoming contaminated by nuclear waste. I grew up in El Paso, and I love it. In recent years I've seen the area plagued by all sorts of environmental problems—air pollution, uncontrolled development, and so on. I'll do whatever I can to get people to really think about the long-term consequences of actions such as the disposal of nuclear waste.

Q: But nuclear waste has to go somewhere, right?

Bob: Sure it does. It's not that I'm opposed to the storage of nuclear waste. And it's not that I just want it to go somewhere else. I just think the people in charge of finding disposal sites have an obligation to find the most appropriate sites. And "appropriateness" should be based mainly on physical and safety conditions. Plus, those in charge have the obligation to demonstrate to the community that the facility will be safe as long as it exists. Personally, I think that nuclear waste should be stored above ground and at the location where it is produced.

Bob Geyer points to the proposed nuclear dump site in Sierra Blanca, Texas.

Bob prefers that nuclear power plants store their spent fuel rods on-site, as this plant in Morris, Illinois, is doing.

Q: How did you become interested in the environment?

Bob: I guess I became an environmentalist as a kid when I went with my family on camping trips to the mountains of New Mexico. I really learned to appreciate nature during those trips, and I developed a real concern for what happened to the natural world. If I saw something that bothered me, even back in high school or college, I would write a letter or speak out against it. In the last year and a half or so, I really got going full steam into the environmental movement when I heard about the proposed nuclear dump site at Sierra Blanca.

Q: As an environmental activist, exactly what do you do?

Bob: I do what I can to raise people's consciousness about certain environmental issues. I write letters and organize and attend protests and marches. Recently I organized a visit from an author of a book about nuclear energy. She spoke to various groups of influential people and really stirred up the community. This sort of activity is good because it helps people think about the long-term consequences of certain activities, such as the health consequences of radiation dumping.

Q: Do you ever feel overwhelmed by the number and complexity of environmental problems?

Bob: No. I have a great deal of experience with overwhelming odds. My wheelchair has given me a lot of experience dealing with difficult problems. What I have learned is that by working harder, you can change people's perceptions, which is often the hardest step in solving a problem. For 25 years I have been changing people's perceptions about what a person who uses a wheelchair can accomplish.

Q: Do you mind talking about your disability?

Bob: Not at all. I was in an automobile accident when I was a junior in college. The accident left me with a broken neck and a damaged spinal cord, and I've been using a wheelchair ever since.

I've also been an activist ever since. It all began when I realized that it was impossible to continue at the university I was attending because there were so many physical barriers to a person using a wheelchair. I had to transfer to another university, where I earned a bachelor's and a master's degree. I also became involved in making buildings and buses and other things more accessible to people with disabilities. We have come a long way since then, but we still have a long way to go. But I don't give up easily—I never have. So whether you're talking about fighting for the rights of people with disabilities or fighting to improve the environment, I don't intend to become easily discouraged.

As an environmental activist, Bob reads about current environmental issues, writes letters to politicians and newspapers, and arranges for experts to speak to the El Paso community.

CHAPTER 12

WASTE

"Nothing can be forgotten, only left behind."

JOY HARJO
NATIVE AMERICAN POET

EcoLog

Before you read this chapter, take a few minutes to answer the following questions in your EcoLog.

1. Which one of the following materials do you think makes up the largest portion of household and business waste: plastic, glass, or paper?
2. What actions do you now take to limit the amount of waste you produce? What else could you do to limit the waste you produce?

SOLID WASTE: THE THROWAWAY SOCIETY

AFTER READING THIS SECTION YOU SHOULD BE ABLE TO

1. define *solid waste*.
2. explain how most municipal solid waste is disposed of.

It's lunchtime. You stop at a fast-food restaurant and buy a burger, fries, and a soda. Within minutes, the food is devoured, and you toss your trash into the nearest wastebasket. Figure 12-1 shows what might be in your trash: a paper bag, a polystyrene burger container, the cardboard carton that held the fries, a paper cup with a plastic cap, a plastic straw, a handful of paper napkins, and several little ketchup and mustard packets. Once it's gone, you probably don't give the trash a second thought. Why should you? It's gone. You threw it away. It's not your problem anymore. Wait a minute. Where exactly is *away*?

The trash from the wastebasket probably will be picked up by a collection service and taken to a landfill, where it will be dumped with thousands of tons of other trash and covered with a layer of dirt at the end of the day. That trash won't bother anyone anymore, will it? Maybe not, unless the landfill fills up next year and the city has nowhere to put the garbage anymore. Or what if rainwater runs down into the landfill, dissolves harmful chemicals such as paint thinner and nail-polish remover, and seeps into the groundwater? All of a sudden, *away* doesn't seem so simple anymore.

Now imagine multiplying the waste disposal problems that come with your lunch by the number of things that you and everyone else throw away each day. It adds up to enough waste to cause some

Answer to caption: Trash "goes" into garbage trucks and is hauled to landfills or incinerators.

Figure 12-1
Where does your trash go when you throw it away?

Figure 12-2 The barge *Mobro* (right) from Islip, New York, sailed up and down the East Coast for five months looking for a place to dump its load of garbage. The map above shows its route.

real crises. In 1987 the barge shown in Figure 12-2, loaded with garbage from the town of Islip, New York, sailed up and down the East Coast for more than five months in search of a dump. When no one would accept the garbage, it was finally incinerated (burned), and the ashes went back to Islip for burial.

Every year, Americans generate more than 4 billion tons of solid waste. **Solid waste** is any discarded material that is not a liquid or a gas. Solid waste includes everything from junk mail and coffee grounds to junked cars and the discarded trimmings from paper mills.

Many products today are designed to be used once and then thrown away. Partly as a result of this, the amount of waste each American produces each year has almost doubled since the 1960s. (See Figure 12-4.) Although Americans generate the most waste, other countries are also throwing away an increasing amount of trash. Every year, the amount of waste increases in every country.

Figure 12-3 Few people are happy to have a waste-disposal facility near their home. These people from the Bronx in New York City are protesting an incinerator in their neighborhood.

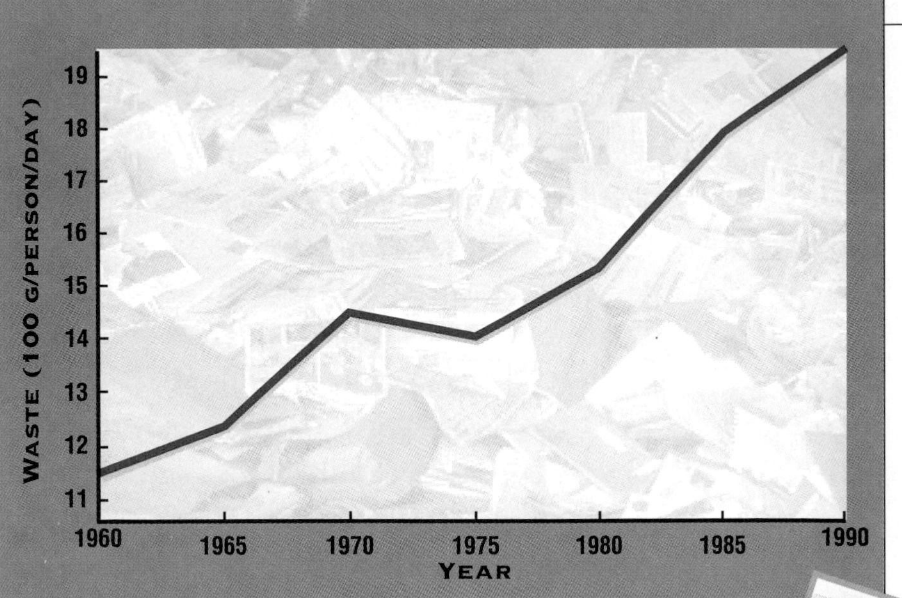

Figure 12-4 **This graph shows the amount of municipal solid waste produced per year in the United States, divided by the population. About how much solid waste did the average person produce per day in 1990?**

Answer to above caption: 1,950 g per day

Figure 12-5 **The increase in waste is tied to an increasing dependence on disposable products. The 1955 *Life* magazine article shown below promoted the new products.**

Throwaway Living

DISPOSABLE ITEMS CUT DOWN HOUSEHOLD CH

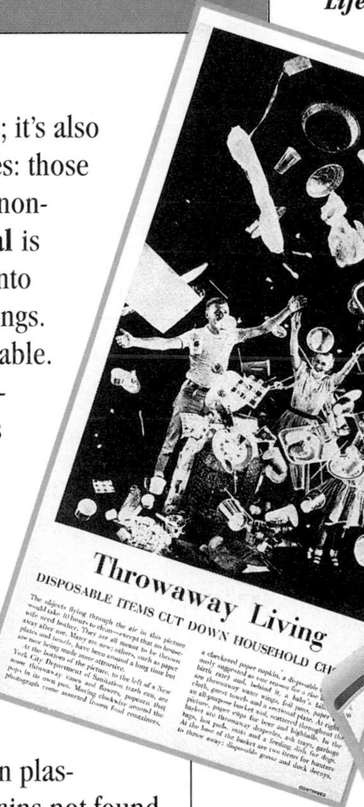

ALL WASTES ARE NOT EQUAL

It isn't just the amount of waste that causes problems; it's also the *kind* of waste. There are two basic kinds of wastes: those made of biodegradable materials and those made of nonbiodegradable materials. A **biodegradable material** is a material that can be broken down by living things into simpler chemicals that can be consumed by living things. Products made from natural materials are biodegradable. Examples of biodegradable products include newspapers, paper bags, cotton fibers, and leather. Products made from many synthetic materials are not biodegradable. A synthetic material is a material that is made in a laboratory. Some examples of synthetic materials are polyester, nylon, and plastic.

Plastics illustrate how nonbiodegradable materials can cause problems. Plastics are made from petrochemicals (oil). They consist mainly of carbon and hydrogen, the same elements that make up most molecules found in living things. But in plastics, these elements are put together in molecular chains not found in nature. Over millions of years of life on Earth, microorganisms have evolved ways to break down nearly all biological molecules. But because plastics do not occur in nature, microorganisms have not developed ways to break down their molecular structures. Thus, when we throw away plastics, they may last for hundreds of years.

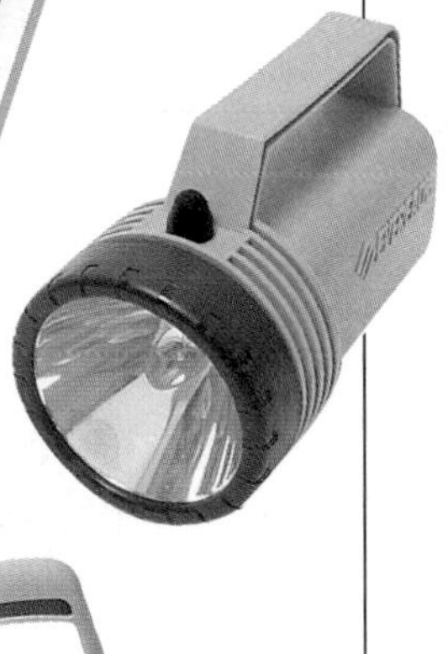

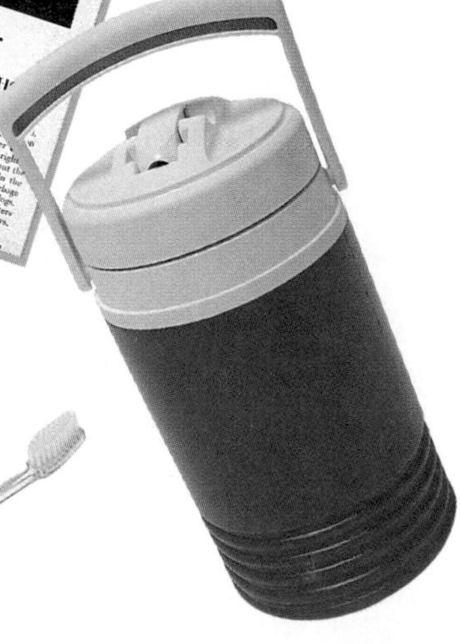

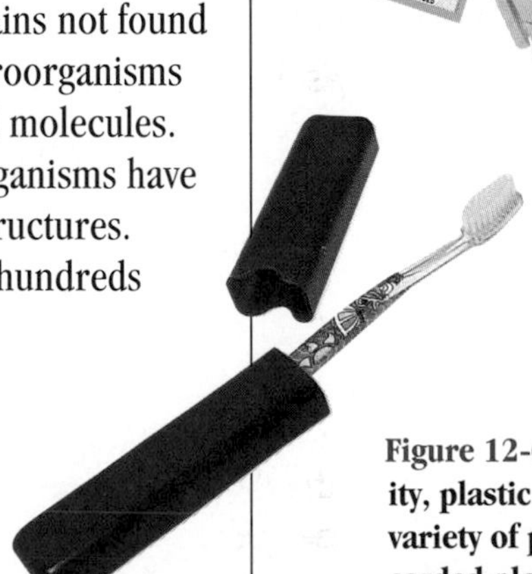

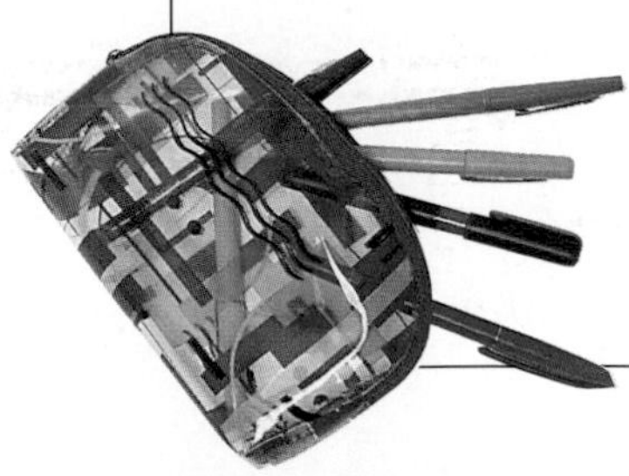

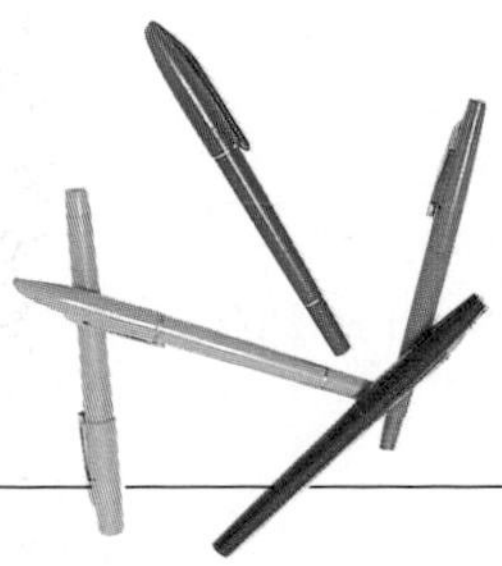

Figure 12-6 **Because of its versatility, plastic is used for an incredible variety of products. As a result, discarded plastic products make up a growing portion of our solid waste.**

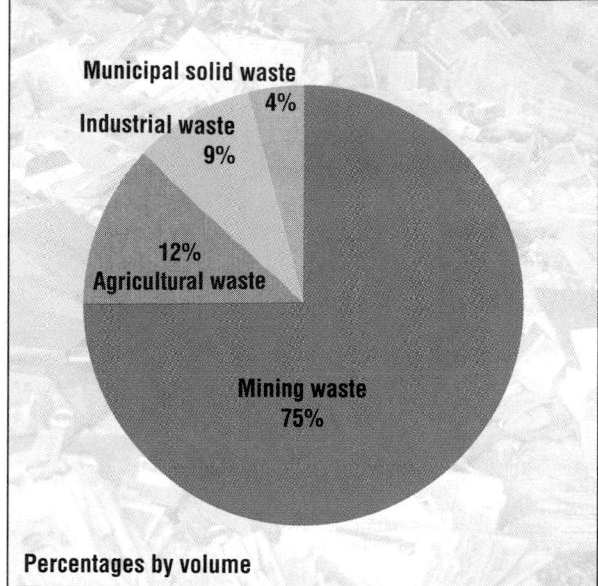

Figure 12-7 This pie chart shows the composition of all solid waste generated in the United States. Municipal solid waste is mainly waste collected from homes and businesses.

What We Throw Away

As shown in Figure 12-7, about 87 percent of the solid waste produced in the United States is from mining and agriculture. About 9 percent of the country's solid waste is produced by industries. Much of this waste contains hazardous chemicals and must be specially treated so that it will not contaminate the air, water, or soil. Hazardous waste will be discussed in Section 12-3.

The remaining 4 percent is **municipal solid waste**, the trash produced by households and businesses. Figure 12-8 shows the composition of municipal solid waste in the United States. As you can see, paper makes up a large part of our waste, partly because almost everything we buy comes in at least one layer of paper or cardboard packaging.

These figures may make you think that municipal waste is relatively unimportant. After all, it is only 4 percent of the total solid waste. Nevertheless, municipal waste is still a huge amount of waste—more than 180 million tons each year in the United States. That's enough to fill a bumper-to-bumper convoy of garbage trucks that would stretch around the Earth almost six times. Furthermore, the amount of municipal waste that we produce is growing much faster than the amount of mining or agricultural waste. If every American reduced the amount of waste he or she produced by just a small amount, our total waste would be reduced by millions of pounds.

Answer to caption (below):
At least 81 percent

Figure 12-8 Paper makes up the largest portion of municipal solid waste. How much of the waste shown in this chart could be recycled?

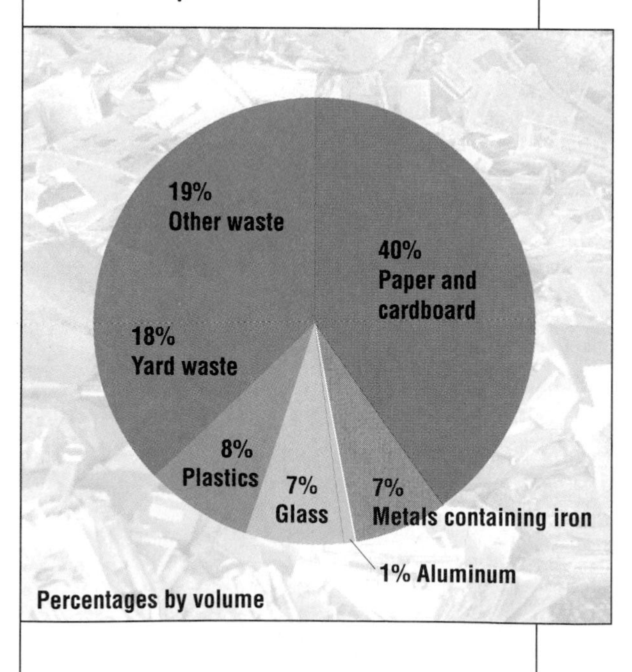

Where Our Trash Goes

Most trash is sent to landfills, and a small amount is incinerated. As you can see in Figure 12-9, very little of our waste is recycled or converted into energy.

Landfills Eighty percent of our waste ends up in landfills. A **landfill** is a waste-disposal facility where wastes are put in the ground and covered each day with a layer of dirt or plastic, or both. A modern landfill is pictured in Figure 12-10.

Where Our Trash Goes	
Waste-Disposal Method	**Percentage of Waste by Volume**
Stored in landfills	80
Recycled	11
Converted to energy	6
Incinerated	3

Figure 12-9 Though more people are recycling, most of our waste still goes to landfills.

Landfills are safer than the open dumps of the past, which produced obnoxious smells and provided breeding grounds for flies and rats. Even so, many problems remain.

One problem with landfills is leachate. **Leachate** is water that contains toxic chemicals dissolved from wastes in a landfill. Leachate is formed when water seeps down through a landfill, dissolving chemicals from decomposing garbage along the way. Leachate may contain chemicals from paints, pesticides, cleansers, cans, batteries, and appliances. This poisonous "chemical soup" sometimes flows into groundwater supplies, making water from nearby wells unfit to drink.

Another problem with landfills is methane. As waste decomposes deep in the landfill, where there is no oxygen, it produces methane, a highly flammable gas. The methane may seep through the ground into basements of homes up to 300 m (1,000 ft.) from a landfill. If the methane is ignited by a spark, it could cause deadly explosions.

The Resource Conservation and Recovery Act (RCRA), passed in 1976 and amended and updated in 1984, requires that new landfills be built with safeguards to reduce pollution problems. New landfills must be lined with clay or a plastic liner and must have systems

Figure 12-10 New landfills like the one in the photograph are lined with clay or plastic and have a system for collecting and treating leachate. The landfill shown in the diagram generates electricity by burning the methane produced by decomposing garbage.

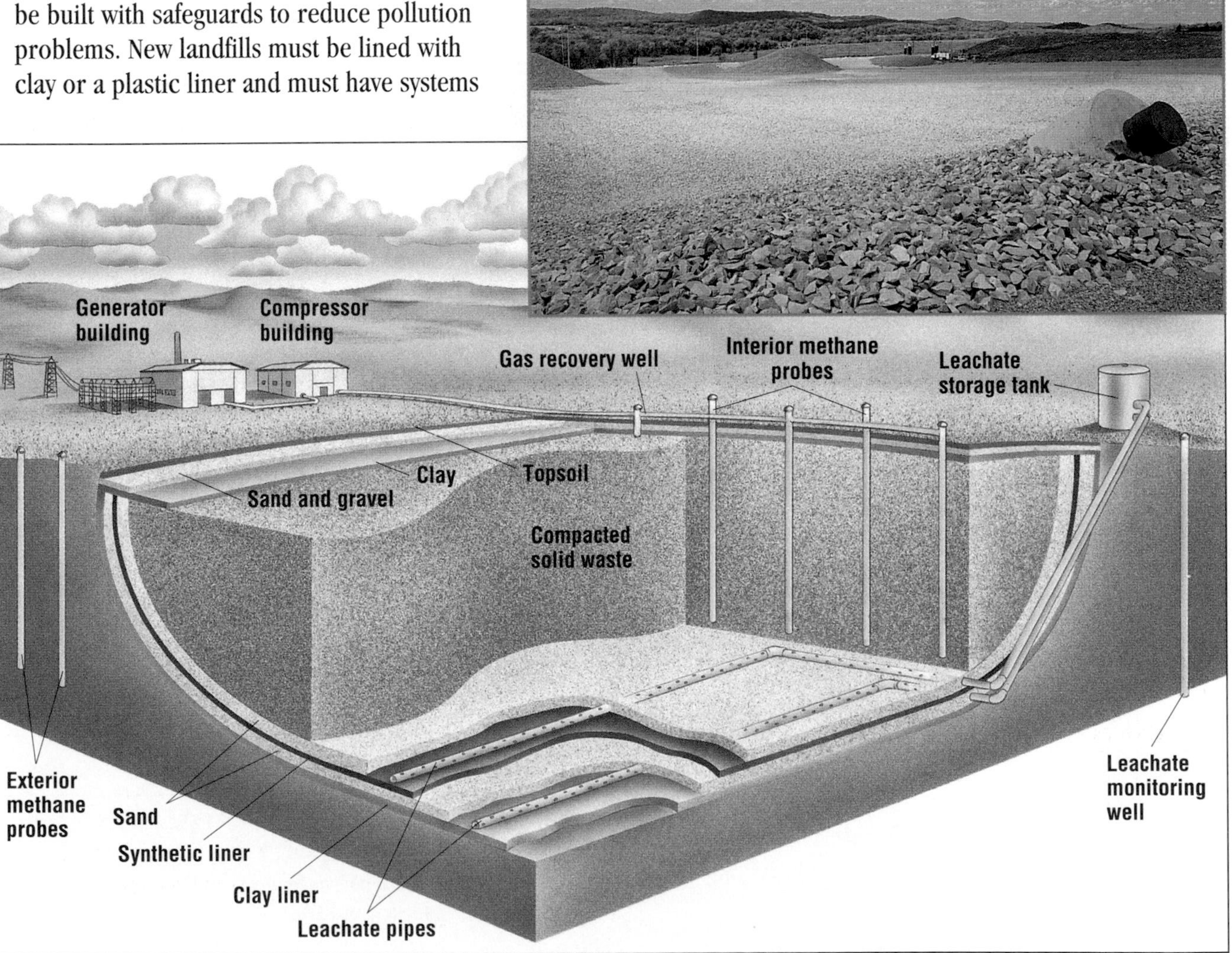

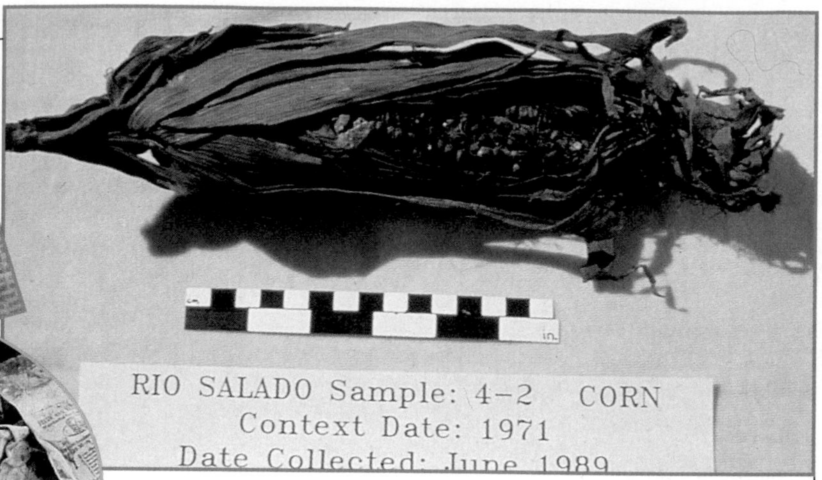

Figure 12-11 **Biodegradable items don't necessarily degrade in modern landfills. All the items shown in the three photographs above were put in a Tempe, Arizona, landfill in 1971.**

for treating leachate. Vent pipes may be installed to carry methane out of the landfill, where it may be released into the air or burned to produce energy.

Adding these safeguards to landfills increases the cost of building them. In addition, it is getting harder to find acceptable places to build landfills. The landfills must be close enough to the city producing the waste, yet far enough from residents who object to having a landfill near their homes. Any solution is likely to be expensive, either because of the legal fees a city must pay to fight residents' objections or because of the cost of transporting garbage to a distant site. Because of the expense and hassle of building new landfills, most municipal solid waste is still being dumped in old, polluting landfills. But this cannot go on much longer. Half of the country's landfills are scheduled to fill up and close within the next five to ten years. The U.S. Environmental Protection Agency (EPA) estimates that by the year 2000 more than 25 states will be out of landfill space. (See Figure 12-12.)

LANDFILL MANAGER

Being a landfill manager may sound like a dirty job, but it's a vitally important career. Find out more on pages 376–377.

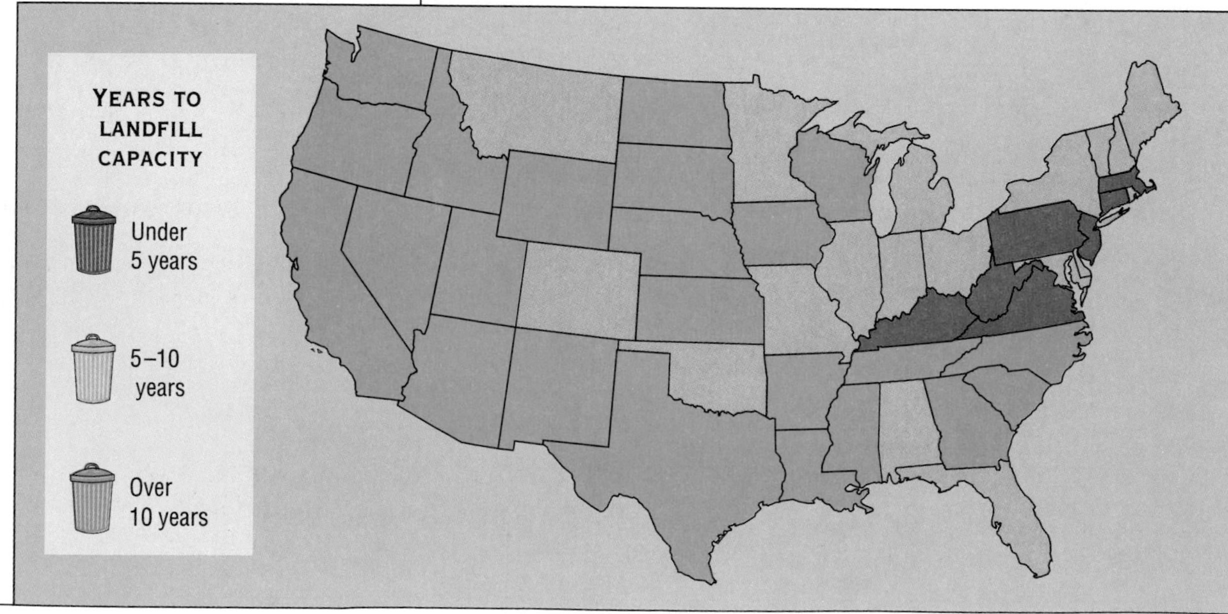

Figure 12-12 **As this map shows, we are running out of places to bury our garbage.**

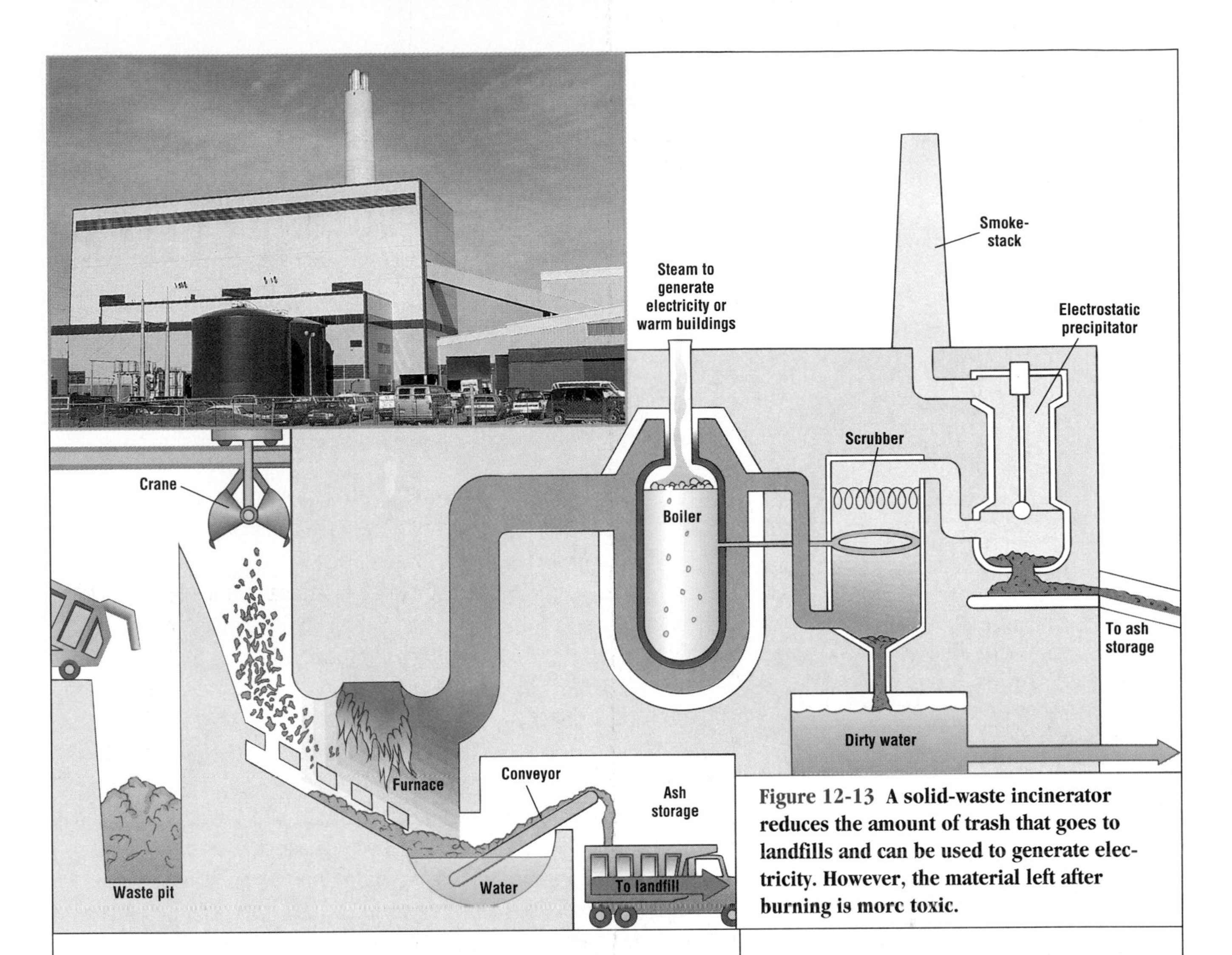

Figure 12-13 A solid-waste incinerator reduces the amount of trash that goes to landfills and can be used to generate electricity. However, the material left after burning is more toxic.

Even if building new "state-of-the-art" landfills were easy, some problems would remain. The liners at the bottom of the landfills can't last forever, so eventually even the best-designed landfills can leak wastes into water supplies. And if we continued to bury our garbage this way for hundreds of years, the Earth would become overcrowded with closed landfills!

Incinerators One option for reducing the amount of solid waste sent to landfills is to burn it in incinerators, as shown in Figure 12-13. Some incinerators have special equipment that uses the heat from burning waste to produce steam. The steam can be used to generate electricity. However, waste that is burned does not disappear. Some of it ends up in the air as polluting gases, and the rest is converted into ash that must be disposed of in a landfill. The amount of space needed in the landfill is reduced, but the material to be buried is more toxic. And even modern incinerators with specialized air-pollution control devices release small amounts of poisonous gases and particles of toxic heavy metals into the air. What's more, because incinerators need a certain amount of waste to keep operating, a community that builds an incinerator may not be as motivated to recycle or reduce its waste.

Answers to Section Review can be found on page T156.

Section Review

1. Why do nonbiodegradable wastes cause waste-management problems?
2. What is the difference between solid waste and municipal solid waste?
3. Describe the structure of a landfill.
4. What is one advantage of incinerating solid waste? What is one disadvantage?

Thinking Critically

5. *Identifying Relationships* Name one nonbiodegradable product you use and a biodegradable substitute for it.

SOLID WASTE: OPTIONS FOR THE FUTURE

AFTER READING THIS SECTION YOU SHOULD BE ABLE TO

- describe three ways to reduce the amount of waste that goes to landfills and incinerators.

If landfills and incinerators are expensive and polluting, what are some other options? This section examines three options for dealing with solid waste problems—producing less waste, recycling, and changing the materials used in products.

Figure 12-14 **One way to reduce the amount of waste you produce is to avoid overpackaged products like this one.**

PRODUCING LESS WASTE

If we produce less waste, we will reduce the expense and difficulty of collecting and disposing of it. Many ideas for reducing waste are simple common sense, like using both sides of a sheet of paper and not using unneeded bags, napkins, or utensils at stores and restaurants. You can also reduce the amount of trash you have to throw away by refusing to buy products with unnecessary packaging, like the product shown in Figure 12-14. By choosing products with as little packaging as possible, you also help influence manufacturers to create products with little packaging. Families with babies can reduce the amount of solid waste they produce by using washable cloth diapers instead of disposable ones.

Beverage manufacturers could use the kind of refillable bottles that were used in the past. Until about 1965 most bottled beverages were sold in bottles that were designed to be returned to stores when empty. They were then collected, washed, and refilled at bottling plants. Manufacturers could also reduce waste and conserve resources by redesigning products to use less material. A return to sturdy products that last longer and are designed to be easily repairable would both save resources and reduce waste disposal problems.

RECYCLING

In addition to reducing waste, we need to find ways to make the best possible use of all the materials we discard—to recycle as much as possible. Making products from recycled materials usually saves energy, water, and other resources. For example, it takes 95 percent less energy to produce aluminum from recycled aluminum than from ore. It takes about 75 percent less energy to make steel from scrap than from ore. And it takes about 70 percent less energy to make paper from recycled paper than from trees.

Figure 12-15 Many cities now have recycling programs in which newspapers, cans, and bottles are picked up at the curb.

When most people talk about recycling, they think of only the first step—bringing their bottles, cans, and newspapers to a recycling center or putting them at the curb in specially marked containers for city trucks to collect. However, as shown in Figure 12-16, recycling actually involves a series of activities, all of which must occur for recycling to work. First, the discarded materials must be collected and sorted by type. Next, each type of material must be taken to a facility where it can be cleaned and made ready to be used again. For example, glass is sorted by color and is crushed, and paper is sorted by type and made into a pulp with water. Then the materials are used to manufacture new products. Finally, buyers for these new products must be found.

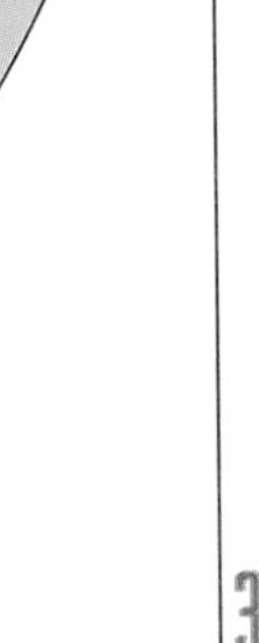

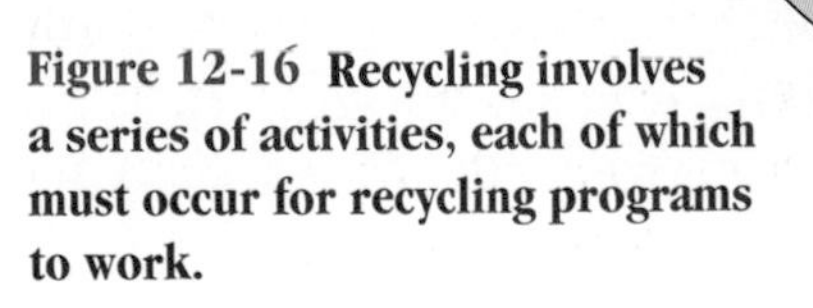

Figure 12-16 Recycling involves a series of activities, each of which must occur for recycling programs to work.

If Americans recycled just half of the newspapers that are thrown out, we could reduce our garbage by 6 million tons each year.

"Buy Recycled" In the 1980s, so many cities and towns started recycling programs that many of them ended up with more recyclable materials, especially newspapers, than they could sell. The best solution to this problem starts with more people buying products made from recycled materials. Increasing the demand for these products encourages manufacturers to build facilities to make them. When such facilities are built, it becomes easier for communities to sell the materials that they collect from residents.

Composting Yard waste often makes up nearly a fifth of a community's municipal solid waste. None of this waste really has to go to a landfill. Because it is biodegradable, yard waste can be allowed to decompose naturally in a compost pile. Many people also put fruit and vegetable trimmings and table scraps in their compost piles. The warm, moist, dark conditions inside a large pile of biodegradable material allow bacteria to grow and break down the waste rapidly. Eventually it becomes **compost,** a dark

PAPER OR PLASTIC?

Does the question "paper or plastic?" sound familiar? If you've ever stood in the checkout line of a grocery store, it probably does. Almost every grocery store today offers a choice between either paper or plastic bags for sacking grocery items. Many people make their choice based on convenience. But what is the best choice for someone who is concerned about the environment?

On the surface, it may seem that paper is the better choice. Paper comes from a renewable resource—trees—and is biodegradable. Plastic, on the other hand, comes from petroleum, which is usually considered a nonrenewable resource. In addition, plastic bags are not biodegradable.

Upon closer examination, however, the decision may not be as simple as it seems. Removing large numbers of trees from forests in order to manufacture paper can disrupt woodland ecosystems. Plus, a tremendous amount of energy is required to convert trees into pulp and then manufacture paper from the pulp.

In order to make the best decision about which product is better for the environment, the following factors should be considered.

- How much raw material, energy, and water is needed to manufacture each bag?
- What waste products will result from the manufacture of each bag, and what effect will those wastes have on water, the atmosphere, and land?
- Can recycled materials be used in the manufacture of the bag? If so, to what degree will the use of recycled materials reduce the amount of raw materials, energy, and water used and wastes produced in making the bag?
- How will the bag be decomposed, and what will the environmental impact be if it is incorrectly disposed of?

brown, crumbly material made from decomposed vegetable and animal materials. Compost is rich in the nutrients that help plants grow.

Some cities collect yard waste from homes and compost it at a large, central facility. Although most municipal (city-run) composting in the United States is limited to yard wastes, several European cities also collect and compost food wastes in municipal facilities. Composting can also be an effective way of handling waste from food-processing plants and restaurants, manure from animal feedlots, and municipal sewage sludge. If all such biodegradable wastes were composted, the amount of solid waste going to landfills could be greatly reduced.

Conserve landfill space and make your own rich soil at the same time. **Find out how on pages 394–395.**

Changing the Materials We Use

Much waste could be eliminated by simply changing the materials used to package products. Single-serving drink boxes, for example, are made of a combination of foil, cardboard, and plastic. The

Answers to Thinking Critically can be found on page T157.

Although several studies have analyzed these factors, most have been conducted by parties with a vested interest, such as plastic or paper manufacturing companies. As you might expect, the studies done by plastic manufacturers conclude that plastic bags have the least environmental impact, while studies done by paper producers conclude that paper bags have the least environmental impact. Often, these researchers fail to study all of the important factors listed previously.

But one thing is certain: the plastic-versus-paper debate has caused both industries to improve the way their products affect the environment. For example, paper bags recently outsold plastic bags because they were considered stronger, better for reusing or recycling, and less harmful in a landfill.

Then, new technology allowed the plastics industry to gain a larger market share. By incorporating recycled plastic into the bags, manufacturers improved the image of plastic bags.

Paper bag manufacturers responded to this increased competition by improving the quality of their recycled bags. This action again increased their sales.

The tug-of-war between the manufacturers of plastic and paper bags continues. And environmentally conscious people are still wondering which is better. Right now there seems to be no single right answer. However, the following are environmentally sound options.

A reusable canvas shopping bag may be the best response to the paper-or-plastic question.

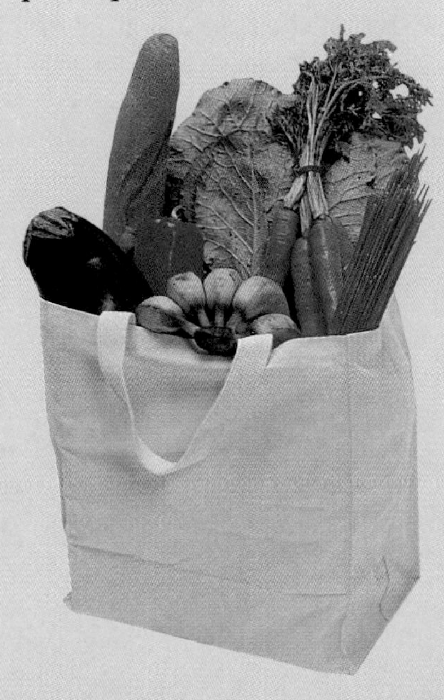

1. Carry your groceries in bags brought from home (paper, plastic, or canvas bags).
2. Choose a recycled bag, if available.
3. Choose the bag you are most likely to reuse in the future.
4. If you have only one or two small items, tell the checker you don't need a bag at all.

Thinking Critically

1. *Identifying Relationships* Explain how environmentally conscious shoppers have helped improve paper and plastic bag manufacturing in this country.
2. *Recognizing Relationships* Why should a person care which bag he or she is given at the grocery store?

Some Household Recyclables		
Product	**Primary Material**	**Recycled Products**
Newspapers	Newsprint	Newsprint, cardboard, egg cartons, building materials
Telephone books	Paper	Building materials
Magazines and catalogs	Clay-coated paper	Building materials
Aluminum beverage cans	Aluminum	Beverage cans, lawn chairs, siding, cookware
Steel food and beverage cans	Steel	Food and beverage cans, automobile parts, tools, construction materials
Glass jars and bottles	Glass	Glass jars and bottles
Plastic beverage containers	Plastic	Non-food containers, insulation, carpet yarn, textiles, fiberfill, scouring pads
Plastic milk containers	Plastic	Non-food containers, toys, crates, plastic lumber

Figure 12-17 What happens to the products you recycle? They come back in the forms shown in the right-hand column.

drink boxes are hard to recycle because there is no good way to separate the three different components. More of our waste could be recycled if such products were simply eliminated and all drinks came in recyclable glass or aluminum containers.

Degradable Plastics? As you read earlier, most plastics are not biodegradable. In order to make plastic products more appealing to people who are concerned about the environment, several companies have developed new kinds of plastics that they call "degradable." One type, called photodegradable plastic, is made so that when it is left in the sun for many weeks, it becomes weak and brittle and eventually breaks into pieces. Another type, usually called biodegradable by the manufacturers, is made by blending cornstarch and a special chemical agent into ordinary plastic. When this plastic is buried, the bacteria in the soil eat the cornstarch, leaving the plastic weakened because it is full of microscopic holes. The chemical agent then gradually causes the long plastic molecules to break into shorter molecules. These two effects combine to cause the plastic to eventually fall apart into small pieces.

The main problem with these "degradable" plastics is that although they break apart, what's left is just smaller pieces of plastic. In other words, the plastic doesn't go away, it just gets spread

What You Can Do to Help With Waste Problems
1. Take your recyclable materials to a recycling center or put them out for curb-side collection.
2. Buy products made from recycled materials whenever possible.
3. At the grocery store, use durable cloth shopping bags instead of paper or plastic, or bring your old paper or plastic ones back and use them again.
4. Ask for "no bag, please" when buying just one or two items at a store.
5. When shopping, look for items sold with little or no packaging. Choose foods packaged in large quantities instead of many individually wrapped servings, and avoid products that have multiple layers of packaging (like a plastic-wrapped dish of noodles inside a plastic-wrapped cardboard box).
6. Don't take extra paper napkins, plastic utensils, or condiment packets when buying takeout food.
7. At meals, use washable cups, dishes, utensils, and cloth napkins instead of throwaways.
8. Keep a reusable cup or mug at school or work for beverages.
9. Use both sides of a sheet of paper when writing or photocopying.
10. Save unused pages from class notebooks for doing homework assignments or other writing.
11. Put dead leaves, weeds, and fruit and vegetable trimmings in a backyard compost pile instead of in the trash.
12. When mowing the lawn, leave grass clippings on the lawn to decompose naturally and add nutrients to the soil.

Figure 12-18 Individuals can make a difference. Is there anything you could do to help with waste problems?

around. Photodegradable plastic may help reduce the harm caused by plastic litter, because after many months of sitting on the ground or in the water it probably will break into pieces too small to strangle or choke most animals. However, neither type does anything to reduce the amount of waste in landfills.

A few companies are developing new plastic-like materials from substances produced by living things, such as starch or a compound made by bacteria. (See Figure 12-19.) These new materials are truly biodegradable, but very few products have been made from them so far.

Figure 12-19 Packing material made from cornstarch is biodegradable.

Answers to Section Review can be found on page T157.

Section Review

1. Name four things you could do each day to produce less waste.
2. What is the main drawback of photodegradable plastics? What is one potential benefit?

Thinking Critically

3. *Analyzing Relationships* Describe how composting is a form of recycling.

HAZARDOUS WASTE

AFTER READING THIS SECTION YOU SHOULD BE ABLE TO

1. define *hazardous waste.*
2. explain how most hazardous waste is disposed of in the United States.
3. explain the two best ways to deal with the hazardous-waste problem.

Many of the products we use today, from laundry soap to computers, are produced in modern factories using thousands of chemicals. Some of these chemicals are ingredients that become part of the products themselves, while others are used as cleansers or to make chemical reactions happen. Large quantities of the chemicals are often left as waste. Many of these chemicals are classified as **hazardous wastes**, which are wastes that are toxic or highly corrosive or that explode easily. Hazardous wastes may be solids, liquids, or gases. Some examples of hazardous wastes include the following substances.

- some dyes, cleansers, and solvents
- PCBs (polychlorinated biphenyls) used in insulating material, plastics, solvents, lubricants, and sealants
- toxic heavy metals such as lead, mercury, cadmium, and zinc
- pesticides (discussed in Chapter 9)
- radioactive wastes (discussed in Chapter 11)

The methods used to dispose of hazardous wastes often are not as carefully thought out as the manufacturing processes that produced them. One case of careless hazardous-waste disposal with horrifying results occurred at Love Canal in Niagara Falls, New York. The Case Study on pages 318–319 describes that situation.

The events at Love Canal shocked people into paying more attention to how hazardous wastes were being handled throughout the country. In thousands of other places, improperly stored or disposed of wastes—like those shown in Figure 12-20—were leaking into the air, soil, and groundwater. Federal laws were passed to provide a system for cleaning up old waste sites and regulating future waste disposal.

Figure 12-20 A hazardous-waste site near Texas City, Texas

Laws Governing Hazardous Waste Disposal

The Resource Conservation and Recovery Act (RCRA) requires producers of hazardous waste to keep records of how their wastes are handled from the time they are made to the time they are placed in an approved disposal facility. If the wastes ever cause a problem in the future, the producer is legally responsible for cleaning them up. RCRA also requires all hazardous-waste treatment and disposal facilities to be built and operated according to standards that are designed to prevent them from polluting the environment.

The Superfund Act Because the safe disposal of hazardous wastes is expensive, companies that produce hazardous waste may be tempted to dump it illegally to save money. The Comprehensive Environmental Response, Compensation, and Liability Act of 1980, more commonly known as the Superfund Act, tried to discourage this by giving the Environmental Protection Agency (EPA) the right to sue the owners of hazardous-waste sites to make them pay for the cleanup. It also created a fund of money to pay for cleaning up abandoned hazardous-waste sites.

Cleaning up improperly disposed of wastes is difficult and extremely expensive. At Love Canal alone, $275 million was spent to put a clay cap on the site, install a drain system and treatment plant to handle the leaking wastes, and relocate the residents. Now, nearly 20 years after Love Canal was evacuated, many waste sites are still waiting to be cleaned up, as shown in Figure 12-21. Of the nearly 1,300 approved or proposed Superfund sites, cleanup has been completed at only 75. Even more discouraging is that about three-quarters of the Superfund money has been spent on legal fees and administrative costs.

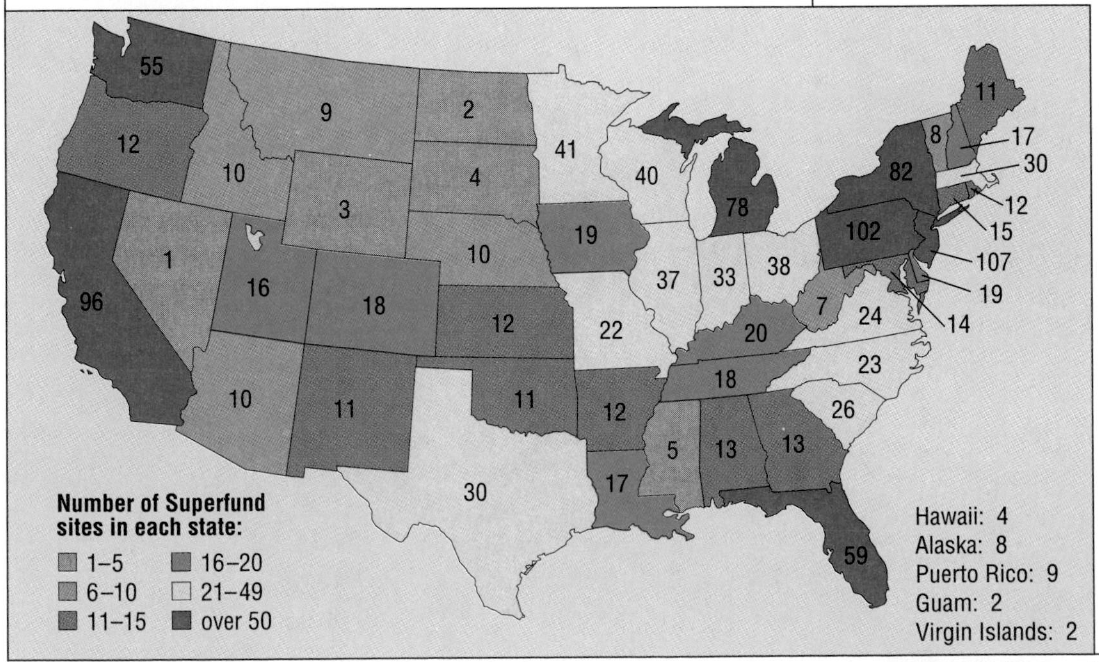

Figure 12-21 This map shows the number of approved and proposed Superfund sites as of the spring of 1995. These sites represent some of the nation's most hazardous areas.

How Can We Manage Hazardous Wastes?

Each year, the United States produces about 275 million tons of hazardous waste, and this amount is growing by about 3 percent each year. Unfortunately, the disposal techniques used today cannot guarantee that hazardous waste will not eventually pollute our air, food, or water.

Produce Less The best way to handle hazardous waste is to produce less of it. The U.S. Office of Technology Assessment estimates that the amount of hazardous waste produced in the United States could be cut in half. In recent years, many manufacturers have discovered that they can redesign manufacturing methods to produce less hazardous waste or none at all. For example, some manufacturers who used chemicals to clean metal parts have discovered that they can use tiny plastic beads instead. The beads, which "sandblast" the parts clean, can be reused several times and are not hazardous when discarded. Often, such techniques end up saving the manufacturers money in material costs as well as waste-disposal costs.

Love Canal: A Toxic Nightmare

To someone who has never heard of it, Love Canal may sound like a pleasant place for a picnic. But in the minds of those familiar with the abandoned canal site in Niagara Falls, New York, the area is synonymous with chemicals, disease, and loss.

It all began in 1942 when a chemical company purchased the area as a dump for toxic wastes. Over the next 11 years, the company buried almost 20,000 metric tons (22,000 tons) of hazardous chemicals in the canal. The chemicals were usually contained in steel barrels. At the time, disposing of chemical wastes in this way was legal. It was thought that the thick clay that lined the canal would prevent the wastes from escaping into the surrounding soil.

By 1953 the dump was full. It was covered with a cap of clay and soil and sold to the school board of Niagara Falls. The school board, ignoring warnings from the chemical company, built an elementary school on top of the canal. In addition, homes were built nearby, and roads and sewer lines were constructed across the site, which disturbed its clay cap and occasionally exposed barrels of waste. The new homes drew many new residents, who were not warned about the nearby hazardous-waste dump.

By the late 1950s, problems started occurring. Children playing near the school were burned by chemicals they encountered. In

Families living near the chemical dump were evacuated in 1978.

Reuse The second-best way to deal with hazardous waste is to find a way to reuse it. In the United States, about 20 programs have been set up to help industries get in touch with other companies that can use the materials that they normally throw away. For example, one company that has to throw away a cleaning solvent after one use can sell it to another company whose product (such as steel castings) is not harmed by small amounts of contamination in the solvent.

Conversion Into Nonhazardous Substances Some types of wastes can be treated chemically to make them less hazardous. For example, lime (a base) can be added to acids to neutralize them, converting them into salts that are less harmful to the environment. Cyanides can be combined with oxygen to form carbon dioxide and nitrogen. In other cases, wastes can be treated biologically. Sludge from petroleum refineries, for example, may be spread on the soil and left to decay into less harmful substances.

Incineration Some toxic substances are disposed of by burning, often in specially designed incinerators. If properly designed

Answers to Thinking Critically can be found on page T157.

the sixties and seventies the chemical leaks became more obvious. Residents near the former dump often noticed a strong odor in the air, especially after a rain. Puddles of chemicals appeared in their backyards. Thick, black sludge oozed into their basements. Worst of all, health problems such as asthma, dizziness, blurred vision, seizures, miscarriages, stillbirths, and birth defects were becoming more common among the residents.

Local, state, and federal officials began to take notice of the problems at Love Canal in the mid-seventies. Water-, soil-, and air-quality tests showed chemical contamination. In 1978 the governor of New York ordered the 235 families living closest to the chemical dump to evacuate. The state purchased their homes and paid for their relocation. In 1980 an additional 792 families were moved at taxpayers' expense.

Who was responsible for the Love Canal mess? Who should pay the $275 million in cleanup and relocation costs? The chemical company has denied all responsibility. Nevertheless, the company paid compensation to Love Canal residents in 1984. The amounts received ranged from $2,000 to $400,000. And in 1988, a federal judge ruled that the chemical company was indeed responsible for its wastes and would have to pay the cleanup bill.

Love Canal is now being redeveloped.

THINKING CRITICALLY

1. *Evaluating Viewpoints* Who do you think should be held responsible for the Love Canal disaster? Why?
2. *Comprehending Processes* Use the Love Canal situation to demonstrate why we can never really throw anything away.

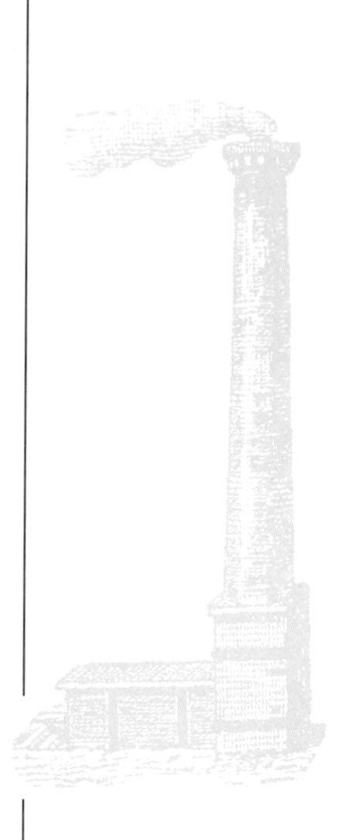

and maintained, incinerators can be a safe means of waste disposal, but they face several problems. Incineration is generally the most expensive form of waste disposal. Incinerators need pollution-control devices and careful monitoring to avoid releasing hazardous gases and particles into the air. And even if perfectly safe incinerators existed, they would not solve all hazardous-waste disposal problems, because not all wastes are burnable.

Land Disposal Most hazardous waste produced in the United States is disposed of by some form of land disposal. One such method, illustrated in Figure 12-22, is **deep-well injection**, in which wastes are pumped deep into the ground, where they are absorbed into a dry layer of rock below the level of groundwater. Another common land-disposal facility is a **surface impoundment**, which is basically a pond with a sealed bottom. The wastes settle to the bottom of the pond, while water evaporates and leaves room to add more wastes. Hazardous wastes in concentrated or solid form are often put in drums and buried in landfills similar to those used for ordinary solid waste but with extra precautions taken to prevent leakage.

In theory, all of these facilities, if properly designed and built, provide safe ways to dispose of hazardous wastes. In practice, however, it is possible for any of them to develop leaks, resulting in contamination of the air, soil, or groundwater. Contamination seems especially likely if you consider that the wastes may be left in these sites for hundreds of years—plenty of time for something to go wrong.

Although we talk about waste disposal, putting wastes into land-disposal facilities is really just long-term *storage,* because the wastes do not go away. Disposal of radioactive wastes from nuclear reactors is an especially difficult storage problem. The only way to make the wastes nonradioactive is to let them sit for thousands of years,

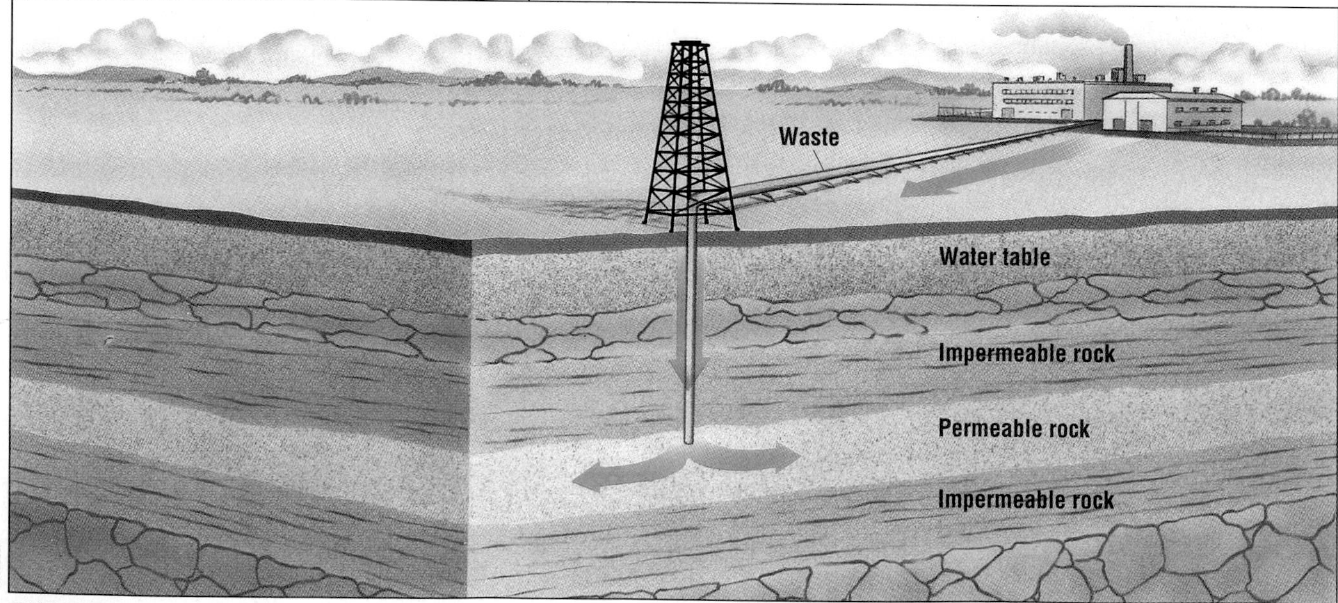

Figure 12-22 One method of hazardous-waste disposal is deep-well injection, in which hazardous wastes are pumped deep into the ground below the groundwater level.

allowing the radioactivity to decrease over time. Therefore, engineers search for disposal sites that probably will not be damaged by movements of the Earth or by groundwater for thousands of years.

Dumping Waste on Your Neighbors

Until recently, only local laws regulated waste disposal in the United States. Companies would often get rid of hazardous waste by sending it to landfills in other states, especially the poorer southern states. In the 1980s, as southern populations grew, these states began to refuse hazardous wastes from other states. Now, hazardous waste is a worldwide problem. Industrialized countries sometimes ship hazardous wastes to developing countries, where laws governing the disposal of these wastes are less stringent. Some international agreements to control hazardous substances are now being made, and more such agreements are needed.

Should radioactive waste travel across the country to a storage site deep underground? Turn to pages 328–329 for two points of view.

Hazardous Waste at Home

We usually think of hazardous-waste management as a problem that only big industries have to face. However, many hazardous chemicals, including house paint, pesticides, and batteries (which contain heavy metals) are used in homes, schools, and small businesses. People often do not realize that hazardous materials poured down the drain or put in the trash end up in sewage sludge and in solid-waste landfills, contaminating the environment.

To deal with this problem, more and more cities around the country have begun to provide collection for household hazardous wastes. Some collect materials only once or twice a year, while others have permanent facilities where residents can drop off their wastes. Trained workers sort the wastes, sending some materials for recycling and packing others into drums for disposal. Used batteries and motor oil are commonly recycled. Paint may be blended and used for city park maintenance or graffiti cleanup work, as shown in Figure 12-23.

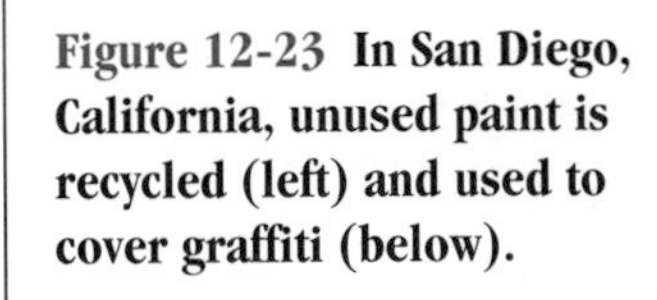

Figure 12-23 **In San Diego, California, unused paint is recycled (left) and used to cover graffiti (below).**

FIELD ACTIVITY

Find out how hazardous wastes are stored and disposed of in your own home. Prepare an inventory of hazardous household chemicals. Check for hazardous materials among cleansers, auto supplies, garden supplies, and wherever chemicals are stored in your home. Determine the proper disposal method for each product. Clearly mark each container with the safest disposal method for it.

Section Review answers on page T157.

SECTION REVIEW

1. Define *hazardous waste.*
2. What is the most common way to dispose of hazardous waste in the United States? What are two dangers of this method of disposal?
3. Explain how the United States could produce less hazardous waste.

THINKING CRITICALLY

4. *Evaluating Ideas* Suppose that a surface-impoundment site for hazardous waste is planned for your community. Would you oppose locating the site in your community? Explain your answer.
5. *Applying Ideas* Have you ever disposed of something that is considered to be a household hazardous waste? Describe the hazardous waste. How could you safely dispose of it next time?

Figure 12-24 What can you do with your old, dirty motor oil? Take it to a service station or to a municipal oil-collection site like the one shown here.

An important benefit of such programs is that they make people more aware of which household wastes are hazardous, encouraging them to handle wastes properly and to look for ways to use less hazardous materials. For example, individuals can look for safer substitutes for toxic home chemicals, reduce the use of pesticides, and buy paints in the smallest quantities possible to avoid having to throw away leftovers.

What About Motor Oil? If you've ever changed the oil in your car yourself, you've probably wondered what to do with the old, dirty oil. Just pour it on the ground? Throw it in the trash? It may surprise you to find out that American backyard mechanics throw away 450 million L (about 119 million gal.) of used motor oil every year. This doesn't even include the oil discarded by service stations and auto-repair shops. The 1989 *Exxon Valdez* oil tanker spill polluted the environment with less than one-tenth of the amount that individuals in the United States throw away each year.

Across the country, discarded oil is going down the drain, into landfills, and into the ground, where it can seep into groundwater or waterways. Just 2 L of old oil, which is less than half the amount in a car, can make 1 million L of fresh water undrinkable.

So what can people do with the oil? One option is to take it to a service station, where it will be turned in for recycling. Some cities have municipal oil-collection sites, and some even have pickup services for used oil. The cities recycle the used oil turned in by citizens. If you don't know what services your community provides, you can call your local city hall and find out.

HIGHLIGHTS

SUMMARY

- Each year, Americans generate over 4 billion tons of solid waste that must be disposed of in one way or another.
- Synthetic materials, which include non-biodegradable plastics, are a major cause of disposal problems.
- Though municipal solid waste makes up only a small fraction of the total solid waste generated, it is still a significant amount.
- Waste can be stored in landfills or incinerated. As landfills become full, however, the pressure to find other options increases.
- To reduce waste problems, we can begin by producing less waste, recycling used products, buying products made from recycled materials, and developing more biodegradable materials.
- Hazardous wastes are produced by many manufacturing processes. The wastes must be carefully disposed of or disastrous results may occur.
- Home activities also can involve hazardous wastes. Proper disposal at community collection sites is important to preserving the environment.

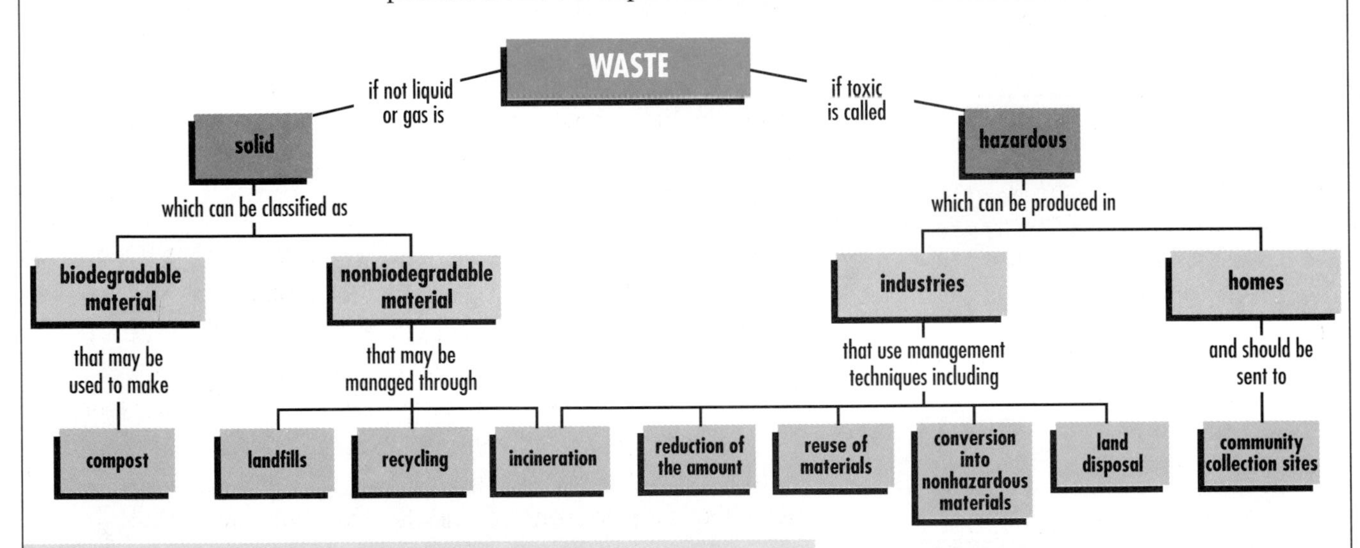

Ecolog

Sample EcoLog answers can be found on page T157.

Now that you've studied this chapter, revise your answers to the questions you answered at the beginning of the chapter, based on what you have learned.

❶ Which of the following materials makes up the largest portion of household and business waste: plastic, glass, or paper?

❷ What actions do you now take to limit the amount of waste you produce? What else could you do to limit the waste you produce?

Vocabulary Terms

biodegradable material (p. 305)
compost (p. 312)
deep-well injection (p. 320)
hazardous wastes (p. 316)
landfill (p. 306)
leachate (p. 307)
municipal solid waste (p. 306)
solid waste (p. 304)
surface impoundment (p. 320)

CHAPTER 12

REVIEW

UNDERSTANDING VOCABULARY

1. For each pair of terms, explain the differences in their meanings.
   - **a.** biodegradable material
     synthetic material
   - **b.** solid waste
     municipal solid waste
   - **c.** leachate
     hazardous waste

RELATING CONCEPTS

2. Copy the unfinished concept map below onto a sheet of paper. Then complete the concept map by writing the correct word or phrase in each box containing a question mark.

Answers not indicated can be found on page T157.

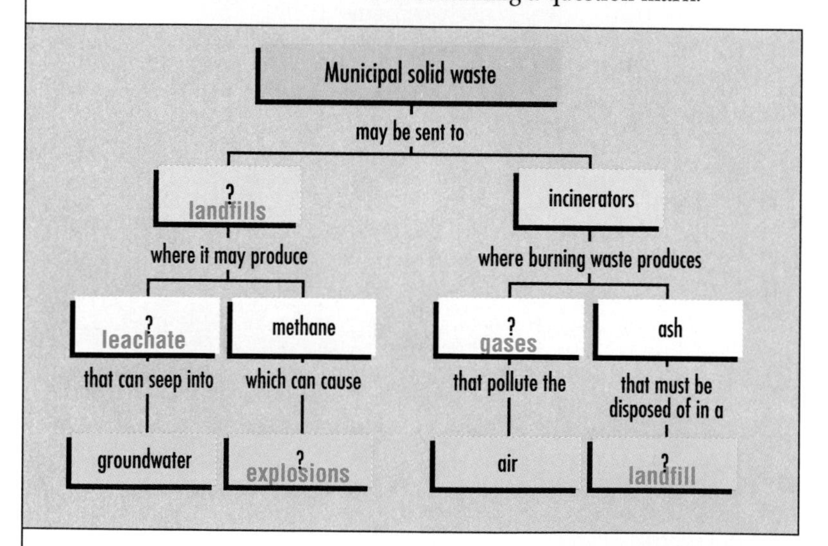

UNDERSTANDING CONCEPTS

Multiple Choice

3. Solid waste includes all of the following EXCEPT
   - **a.** newspaper and soda bottles.
   - **b.** food scraps and yard clippings.
   - **(c.)** ozone.
   - **d.** junk mail and milk cartons.

4. If your shirt is made of 50% cotton and 50% polyester, what part is biodegradable?
   - **(a.)** cotton
   - **b.** polyester
   - **c.** both
   - **d.** neither

5. Microorganisms are unable to break down plastics because
   - **a.** plastics are too strong.
   - **b.** plastics are too abundant.
   - **c.** plastics are made of elements not found in any other substance.
   - **(d.)** plastics do not occur in nature.

6. Municipal solid waste is approximately what percentage of all solid waste?
   - **(a.)** 4%
   - **b.** 20%
   - **c.** 60%
   - **d.** 90%

7. Leachate is a substance that
   - **a.** is produced in a compost pile.
   - **b.** is a byproduct of bacterial digestion.
   - **c.** leeches have for dinner.
   - **(d.)** contains dissolved toxic chemicals.

8. Which is NOT a benefit of incinerating waste?
   - **a.** It reduces the amount of material sent to the landfill.
   - **b.** It produces energy in the form of heat.
   - **c.** It can be used to produce electricity.
   - **(d.)** It destroys all of the toxic materials in the waste.

9. Manufacturers could reduce waste and conserve resources by making products that
   - **a.** use more materials.
   - **(b.)** are more durable.
   - **c.** are difficult to repair.
   - **d.** are disposable.

10. What is the BEST solution to decreasing an overabundance of recyclable materials?
   a. building more recycling plants
   b. limiting the amount of recyclable materials that can be collected
   c. increasing the demand for products made from recycled materials
   d. putting the excess in landfills
11. Examples of hazardous wastes include all of the following EXCEPT
   a. battery acid.
   b. paint.
   c. yard waste.
   d. motor oil.
12. Composting
   a. increases the amount of materials put in landfills.
   b. is a method of making bricks out of sludge.
   c. is a natural way to decompose biodegradable materials.
   d. all of the above
13. Which is the BEST way to manage hazardous wastes?
   a. incineration
   b. land disposal
   c. reduction of the amount produced
   d. conversion to nonhazardous substances

Short Answer

14. What is the second largest source of solid waste in terms of volume?
15. How do plastic liners and layers of clay help protect the environment around a landfill?
16. What does compost consist of, and what are its benefits?
17. How does the Superfund Act allow the federal government to ensure proper disposal of hazardous waste?
18. Explain how the statement "You dump it, you drink it!" could apply to used motor oil.

Interpreting Graphics

19. **Consult Figure 12-4 to answer the following questions.** During what five-year period did production of municipal solid waste actually *decrease?* An economic recession was occuring during this time. How might the reduction in waste have been related to the recession? Conversely, why did the amount of waste increase again after the recession was over?
20. **Consult Figure 12-8 to answer the following question.** If every home in the United States had a compost pile, by what percentage (at least) could the amount of municipal solid waste be reduced?

Thinking Critically

21. ***Communicating Ideas*** Some people feel that "new is better." How would you explain to them that recycled products are often better than the original product?
22. ***Evaluating Ideas*** Because plastics cannot be completely broken down, some people might call for a complete ban on plastics. How would such a ban affect both the environment and society?

Themes in Science

23. ***Scale and Structure*** Considering that the average American throws away 1.5 kg (3.3 lb.) of solid waste per day, how long would it take you to personally fill a dump truck that had an 18,000 lb. capacity?
24. ***Interacting Systems*** How do a person's current shopping habits affect the quality of the environment 100 years in the future?

Cross-Discipline Connection

25. ***Social Studies*** Using an almanac, determine which five states have the most hazardous-waste sites and which five have the fewest sites. Based on this information, what factors do you think might account for the number of hazardous-waste sites located in a state?

Discovery Through Reading

For an intriguing look at solid waste and what it says about our society, read the book *Rubbish! The Archaeology of Garbage,* by William Rathje and Cullen Murphy, New York: HarperCollins, 1992.

PORTFOLIO ACTIVITY

Do a special project about recycling in your community. Determine what types of materials are collected, where they go for processing, how they are reduced, who buys recycled materials, and what products are made from them. To communicate your results, prepare a display using photographs, diagrams, written abstracts, models, or any combination of these or other materials.

CHAPTER 

INVESTIGATION

DOES FAST FOOD HAVE TO BE WASTEFUL?

You work for a consulting firm that specializes in environmental issues. One day, your boss gives you the following letter and memo describing a new assignment:

Answers to this Investigation can be found on page T158.

MATERIALS

- notebook
- pen or pencil
- calculator (optional)

Joe Benedetti, President
Environmental Consulting Organization

Dear Mr. Benedetti:

I am planning an exciting business venture—fast-food restaurants with a difference! My new EarthFood® restaurants will serve fresh, healthy meals in a way that creates the least possible impact on our environment.

Here is where I need your help: I want to design my restaurants and serving methods so as to minimize solid waste from the start. I would like you to conduct a preliminary study of the type and quantity of solid waste generated by other fast-food restaurants. Based on your findings, I would like you to suggest how my operations could produce less waste that must be sent to landfills or incinerators.

If I am pleased with the results of your preliminary study, I may wish to hire your firm to prepare detailed plans for EarthFood's entire waste-handling system. I look forward to your reply.

Sincerely,

Angela Fairweather
President,
EarthFood® Enterprises

INTEROFFICE MEMO

To: All ECO Staff
From: Joe Benedetti, President
Subject: New Solid-Waste Contract

The attached letter describes a new job we have just accepted. I have informed Ms. Fairweather that for our preliminary study, we will survey only the waste generated after customers receive their food. Wastes generated in food preparation areas may be the subject of a separate study. I'm counting on all of you to do an exemplary job on this study so that we can land the big contract that Ms. Fairweather hinted might come our way.

Your boss, Mr. Benedetti, asks you to conduct your study as follows:

How much solid waste do you think is generated by a typical fast-food restaurant?

COLLECT DATA

1. Choose a fast-food restaurant to study. Arrange to spend one hour at the restaurant with two to four of your co-workers to gather data.
2. Throughout this study, be as quiet and unobtrusive as possible, and avoid aggravating restaurant management or customers.
3. At the restaurant, observe how food is packaged as it is served. To get a firsthand look, one or more of your team members may want to buy a meal. Make notes about the type of packaging used, paying particular attention to the following questions:
   - What materials are the packages made from (e.g., paper, cardboard, plastic, metal, or a combination material)?
   - How many layers of packaging are used for each food item? (For example, is a burger wrapped in just paper or in a wrapper inside a box?)
   - Do servers provide utensils, condiments, and napkins, or do customers help themselves? How are the condiments packaged?
4. Choose a centrally located trash container, and observe it for 30 minutes. Count the number of people who deposit trash in the container in that time. Make notes about how frequently you see items other than empty packaging (such as food waste, unused napkins, or condiment packets) go into the trash container.
5. Estimate the total number of restaurant customers per day as follows: Begin with the number of people who visited your selected trash can in half an hour. Multiply this number by two, and then multiply by the number of hours per day the restaurant is open. Multiply the result by the number of trash cans in the restaurant. Finally, assume that each person deposited trash for an average of two customers, and multiply by two to obtain the total number of customers.
6. Estimate what waste items (cups, straws, napkins, etc.) are thrown away by the average customer. Based on this estimate, how many of each item would be thrown away each day?
7. Are separate recycling containers provided for customers? If so, do signs on the containers make it clear which waste items are supposed to go into the recycling containers and which go into the trash? Are customers following the instructions?

REPORT TO YOUR CLIENT

8. Summarize your findings in a report to Ms. Fairweather. Include a description of the types and amount of waste generated at the restaurant you visited. Based on your findings, suggest ways in which the amount of trash could be reduced. Be sure to include the following:
   - ways to avoid producing certain types of waste altogether
   - ways to replace disposable items with reusable ones
   - discussion of what waste items could be recycled or composted
   - ways to replace nonrecyclable, noncompostable items with items that can be recycled or composted

Before throwing away your trash, think about how much waste could be avoided if the restaurant were equipped for recycling and composting.

POINTS OF VIEW

SHOULD *Nuclear Waste* BE STORED AT YUCCA MOUNTAIN?

Yucca Mountain in Nevada has been chosen as the site for the nation's first storage site for high-level radioactive waste. High-level radioactive waste includes solids, liquids, and gases that contain a high concentration of radioactive isotopes that take thousands of years to decay. The idea is to seal 77,000 tons of radioactive waste in steel canisters and store them in a maze of underground tunnels designed to last 10,000 years. Yucca Mountain is scheduled to receive its first shipment of nuclear waste by 2015.

Construction of the facility has already begun. But the debate continues about whether it would be safer to store radioactive wastes at Yucca Mountain, or keep them where they are now—in temporary storage facilities at each nuclear power plant.

FOR THE YUCCA MOUNTAIN SITE

Supporters of the Yucca Mountain storage facility think that this isolated spot in Nevada is a suitable place for permanent nuclear-waste disposal. Opponents of the site disagree.

Those who support construction of the facility point out that there are two major advantages to the plan. First, Yucca Mountain is located in a remote region, far from large populations of people. Second, the climate is extremely dry. A dry climate means that rainfall is unlikely to cause the water table to rise and come in contact with the stored nuclear waste.

Many opponents of the site worry that changes in the climate might cause the water table to rise. They say groundwater could then reach the stored nuclear waste and become contaminated. However, supporters of the site claim that these concerns are unfounded. They point to three scientific studies. All three studies determined that no significant rise or fall of the water table has occurred in the past.

The operators of nuclear power plants around the country are anxious for the Yucca Mountain facility to be completed as planned. Currently, each power plant stores its nuclear waste near the plant. Many of these storage sites have been in use for decades now and are approaching their maximum capacity.

Some people believe that storing wastes in one main location will be safer than storing them at the individual power plants. Many of those storage facilities are already leaking. In addition, some of the nuclear waste is contained in pools of water rather than in underground containers. These conditions worry some people, who fear that the hazardous wastes could leak into neighborhoods around the country.

Spent fuel rods stored underwater at a nuclear power plant

Against the Yucca Mountain Site

Perhaps the fiercest outcry against the Yucca Mountain site comes from residents of Nevada. They fear that if tons of highly toxic waste are stored in one place, some of it might eventually leak out. Since some of this waste is so toxic that just a tiny speck could be lethal, even a small leak could result in a major environmental disaster.

Some people are concerned that the radioactive waste might leak into the groundwater underneath the facility. How might the radioactivity enter the water? The waste containers are expected to last 500 to 1,000 years. After that, the cavern that holds the containers will have to remain leakproof for 10,000 years. Opponents of the plan say that nobody can guarantee that the cavern will be leakproof for that long.

If radioactive waste leaked out of the facility, it could contaminate the water in wells, springs, and streams. In time, the contamination could spread farther and farther from the site and into the biosphere.

Some people also worry that transporting nuclear waste across vast distances to Yucca Mountain is riskier than leaving the material near the facilities where it is produced. Any accident along the way could release radioactivity into the environment.

Most opponents of the Yucca Mountain site agree that current methods of storing nuclear waste are dangerous and should be improved. They suggest that by transferring the waste to solid steel and concrete containers, the waste could be safely stored at each nuclear power facility for 75 to 100 years. By that time, they suggest, more will be known about how to store the wastes safely for thousands of years.

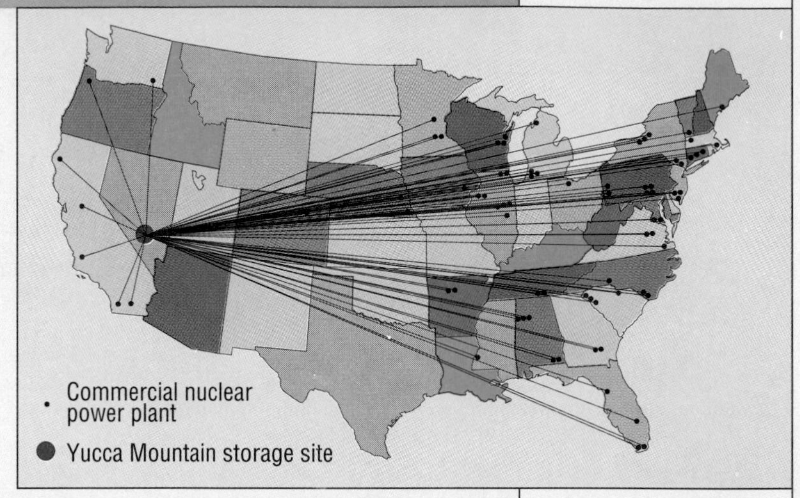

This map shows the nuclear power plants around the country that are possible sources of nuclear waste for the Yucca Mountain facility.

Preliminary plan for the Yucca Mountain nuclear-waste storage facility

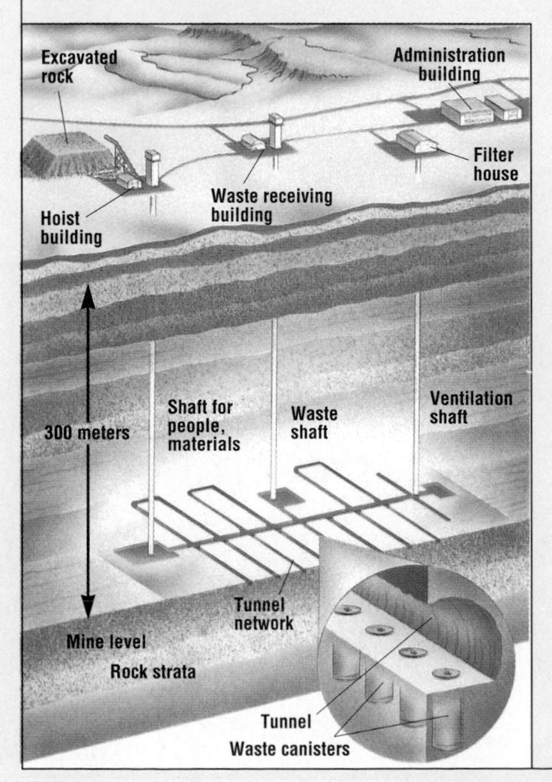

Analyze the Issue

1. ***Making Decisions*** Use the decision-making model presented in Chapter 1 to develop a recommendation for action in this case.
2. ***Analyzing Relationships*** Is there a nuclear power facility near your community? If so, did that affect your answer to question 1? Explain.

Answers to Analyze the Issue can be found on page T158.

CHAPTER 13

POPULATION GROWTH

SECTION 13.1

HOW POPULATIONS CHANGE IN SIZE

SECTION 13.2

A GROWING HUMAN POPULATION

SECTION 13.3

PROBLEMS OF OVERPOPULATION

"People are everywhere. Some people say there are too many of us, but no one wants to leave."

CHARLES M. SCHULZ
AMERICAN CARTOONIST

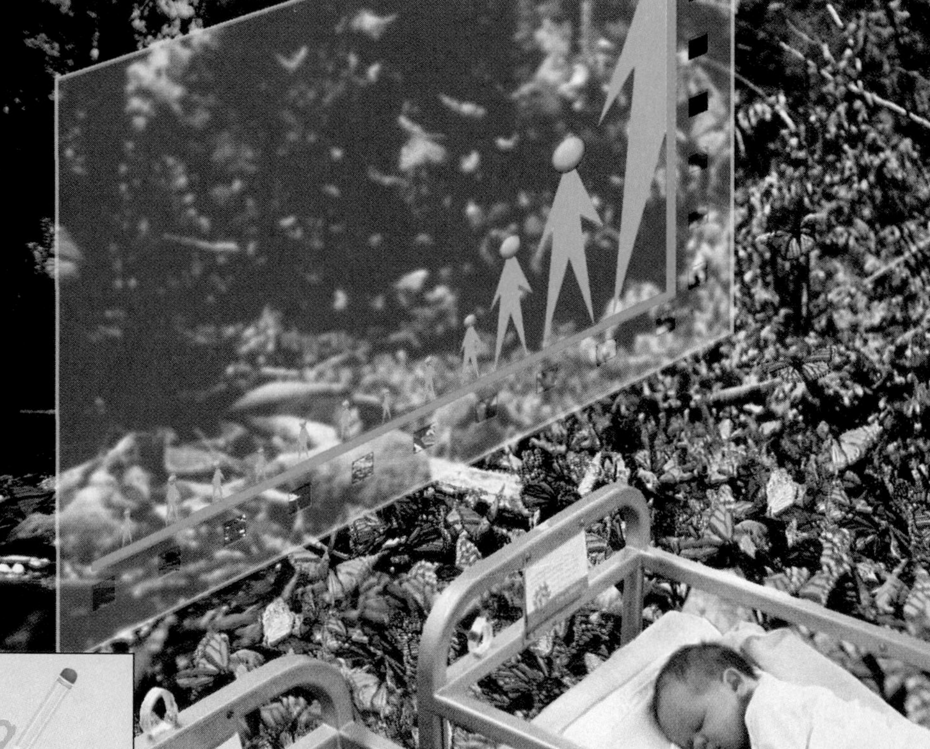

EcoLog

Before you read this chapter, take a few minutes to answer the following questions in your EcoLog.

1. Could there ever be too many people on Earth? Why or why not?
2. How could overcrowding in another part of the world affect you?

SECTION 13.1

HOW POPULATIONS CHANGE IN SIZE

AFTER READING THIS SECTION YOU SHOULD BE ABLE TO

1. describe the factors that affect a population's size.
2. explain why populations grow.
3. explain what limits a population's growth.

The Earth's human population has reached nearly 6 billion and is climbing fast. It is estimated that there will be 10 billion people alive by the middle of the twenty-first century. The rapid growth of the human population is a fundamental cause of many environmental problems. To understand human population growth, you need to know something about the properties of populations of organisms in general.

As you read in Chapter 2, a population is a group of individuals of the same species living in a particular place. Birth and immigration (moving in) add individuals to a population. Death and emigration (moving out) subtract individuals. A population grows when the number of individuals added to the population is greater than the number of individuals subtracted, that is, when

$$\begin{array}{c}\text{number of births}\\+\\\text{number of immigrants}\end{array} > \begin{array}{c}\text{number of deaths}\\+\\\text{number of emigrants.}\end{array}$$

Figure 13-1
It is estimated that the human population is growing at a rate of 94 million people per year.

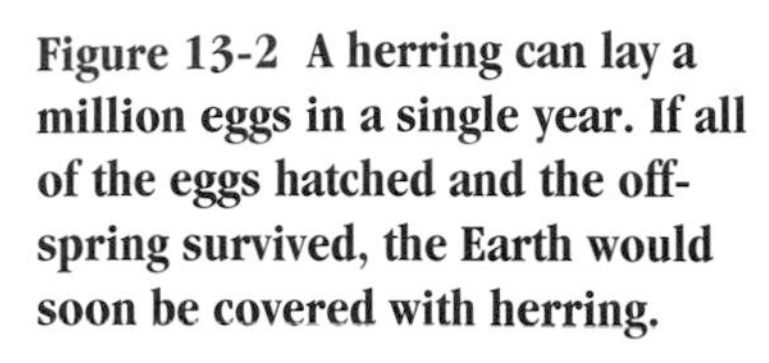

On the other hand, a population will shrink if

$$\text{number of deaths} + \text{number of emigrants} > \text{number of births} + \text{number of immigrants}.$$

How Fast Can a Population Grow?

Most organisms produce many more offspring than can survive to grow up and reproduce. For example, a herring may lay a million eggs a year. If all of the eggs hatched and survived, the entire surface of the Earth would be knee-deep in herring in just a short time. It is obvious that this is not what happens in normal

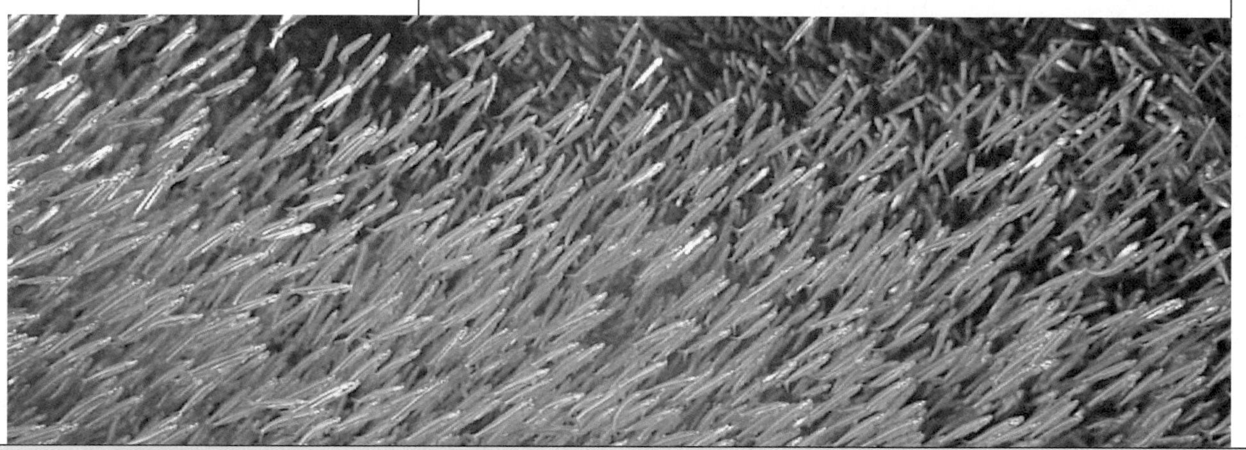

Figure 13-2 A herring can lay a million eggs in a single year. If all of the eggs hatched and the offspring survived, the Earth would soon be covered with herring.

Overshooting the Carrying Capacity

Carrying capacity is the maximum number of individuals an environment can support for a long period of time. When a population grows larger than the carrying capacity of the land, it uses up its resources faster than they are replenished. Environmental degradation is one obvious effect of exceeding the carrying capacity. An ecosystem in which starving animals are scrambling for food is an ecological disaster area, as the animals try to tear every last edible leaf from the trees and every blade of grass from the soil. The result is that the habitat's carrying capacity is reduced, and the population usually shrinks substantially through emigration or starvation. The ecosystem and its carrying capacity may recover, but it may take hundreds of years.

A spectacular and well-studied example of a population exceeding

Reindeer were introduced to Saint Paul Island, off the coast of Alaska, in 1911.

populations. However, in order to understand how populations work, we must first look at what would happen if there were no limits on the growth of a population.

Each population has a characteristic maximum growth rate. This maximum growth rate, also known as **biotic potential,** is the rate at which a population would grow if every new individual survived to adulthood and reproduced at its maximum capacity. Given these conditions, any population will grow exponentially, meaning that a larger number of individuals is added to the population in each generation, as shown in Figure 13-3. (The curved line showing exponential growth is called a J-curve because of its J shape.)

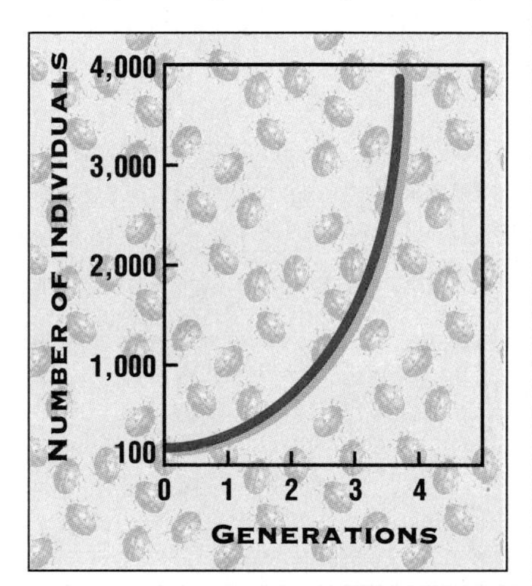

Figure 13-3 If there are no limits to the growth of a population, the population will grow exponentially. Each generation will contain a larger number of individuals than the one before it. This kind of growth curve is called a J-curve.

The biotic potential for the human population is estimated to be about 6 percent per year. If the human population were growing at this pace, it would double in size

carrying capacity occurred in the Pribilof Islands off the coast of Alaska in the first half of the twentieth century. In 1911, 25 reindeer were introduced to Saint Paul Island, a 41 sq. mi. island that is part of the Pribilofs. The reindeer were introduced to replace the native caribou, which had been hunted to extinction. Scientists monitored the size of the introduced population for the next 39 years.

The reindeer found the small island very suitable. As you can see in the graph below, the reindeer population soared. There were 2,046 individuals in 1938—that translates to an average annual growth rate of 16 percent.

The population of reindeer on Saint Paul Island between 1910 and 1950

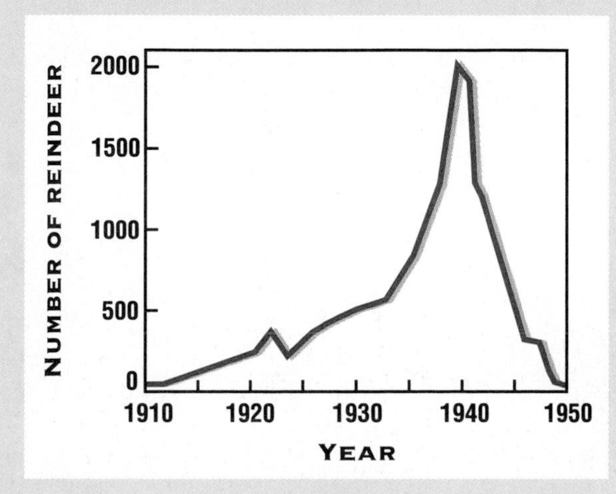

The large population quickly overgrazed and trampled the island's food supply of lichens (the multicolored, flaky growths you see on trees and rocks). Lichens grow slowly in the cold climate of the Pribilofs, so the reindeer's food supply could not recover from overconsumption. Short of food and unable to leave the island, the reindeer starved. By 1950, only eight reindeer were left on Saint Paul Island. The reindeer population did not die out, however. It began to grow again after 1950 and reached a stable size of about 250.

Answers to Thinking Critically can be found on page T158.

THINKING CRITICALLY

1. *Interpreting Graphics* Compare the graph at the left with Figure 13-3. How are the two graphs similar? How are they different?
2. *Inferring Relationships* Suppose predators had also been introduced to Saint Paul Island. How might their presence have changed the outcome of the reindeer introduction? Explain your reasoning.

To understand exponential growth, calculate the following. Suppose you have a flask containing one bacterium that divides into two at the end of one hour. If at the end of each succeeding hour, each bacterium divides, how many bacteria will be present at the end of 24 hours? Graph your results.

Answer to Field Activity: 2^{24} or 16,777,216

in just over 11 years. And the biotic potential for humans is small compared with the rates for many other organisms. For rats, the biotic potential is 1.5 percent *per day*—a population of rats living in a warehouse could double in size in only 47 days. Bacteria are probably the champion reproducers. Given perfect conditions, the growth rate of certain kinds of bacteria is 250 percent *per hour!* That means each bacterium divides more than once an hour.

What Limits Population Growth?

Despite the potential of populations to grow rapidly, most populations *don't* change in size very much from year to year. In most habitats, the rare species remain rare and the common species remain common. If you could count the number of spiders, houseflies, and weeds in your neighborhood, you would probably find that they show up in much the same numbers year after year. Given that populations have the capacity for explosive increase, why is it so rare?

To answer this question, consider that every habitat, no matter how rich, contains limited supplies of important resources like food, shelter, mineral nutrients, water, and

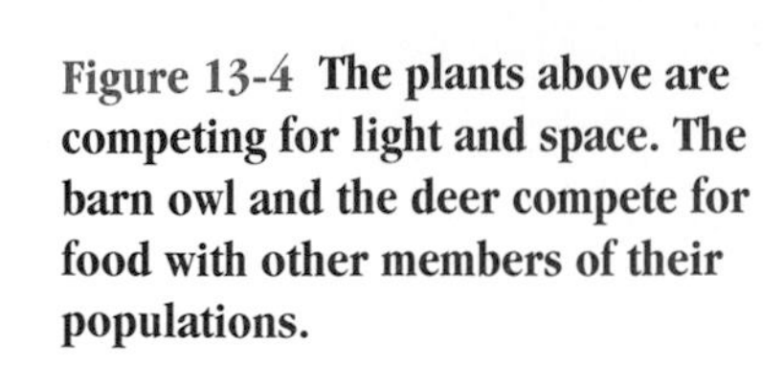

Figure 13-4 **The plants above are competing for light and space. The barn owl and the deer compete for food with other members of their populations.**

light. When a population grows rapidly, it consumes these resources at an increased rate. The resources that limit the growth of a certain population are known as **limiting resources.** A limiting resource for animals might be the availability of a food source, water, or nesting sites. In plants, the supply of water, sunlight, or a particular mineral nutrient is usually the limiting resource.

If the population increases, more and more individuals must use the limiting resources. It becomes harder for each individual to survive. Members of the population begin to compete with each other and with members of other populations in the ecosystem for the resources. As competition intensifies, the death rate rises and the birth rate declines. Eventually, the population reaches a size at which it ceases to grow.

The size at which the population ceases to grow is known as the carrying capacity. The **carrying capacity** is the maximum number of individuals an environment can support for a long period of time. A population may increase beyond the carrying capacity, but it cannot stay at the increased size for very long. Figure 13-5 shows a graph of a population increasing to its carrying capacity and then falling. (The curved line that shows a population reaching its carrying capacity and then falling below it is called an S-curve.)

Figure 13-5 This graph shows what happens when a population exceeds its carrying capacity. When rabbits were first introduced into Australia, their population exploded. They became so numerous that they couldn't find food, and disease spread throughout the population. A population crash resulted, followed by a series of ups and downs that eventually stabilized.

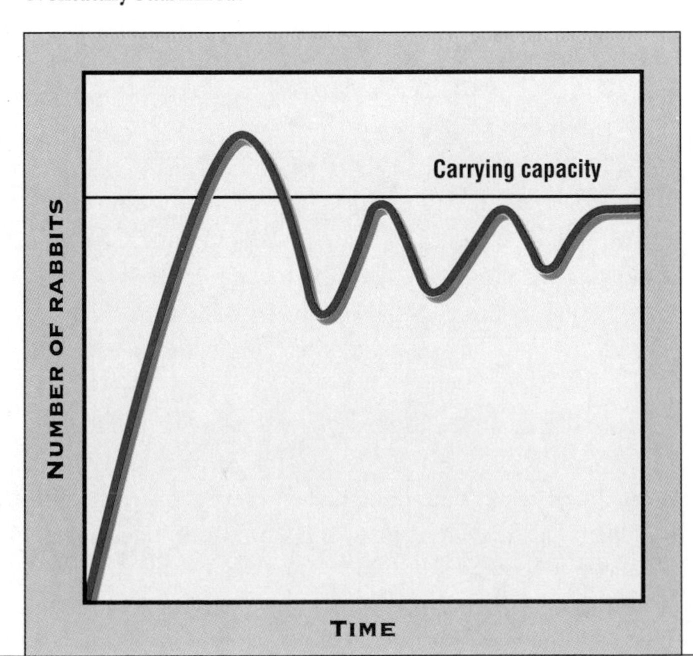

Answers to Section Review can be found on page T158.

SECTION REVIEW

1. What factors limit the growth of populations?
2. Why can't a population increase indefinitely?
3. What factors could cause the carrying capacity of the environment to increase or decrease?

THINKING CRITICALLY

4. *Relating Concepts* Make a concept map using the following terms: population growth, population shrinkage, birth, death, immigration, and emigration.
5. *Analyzing Processes* What are some reasons the human population does not reach its biotic potential?

SECTION 13.2

A GROWING HUMAN POPULATION

AFTER READING THIS SECTION YOU SHOULD BE ABLE TO

1. describe how the size of the human population has changed.
2. identify the factors that led to changes in the human population.
3. describe the stages of population growth.

Remember that carrying capacity is the maximum number of individuals an environment can support for a long period of time. A population has reached its carrying capacity in a particular ecosystem when it is using up some necessary resource as fast as the ecosystem can produce it.

Like a single ecosystem, the entire Earth also has a carrying capacity. However, no one knows the exact carrying capacity of the Earth for the human population. A recent United Nations study showed that approximately 30 billion people could be fed, using technology now available. Most demographers expect the world's population to stabilize long before it reaches 30 billion, probably somewhere between 8 and 20 billion. But food is just one consideration. What would happen to water and air resources, biodiversity, and human health if the population grew to 30 billion?

Though we can only estimate the future effects of population growth, we have been able to trace past population growth. Studying the patterns of the past may help us plan for the future.

FROM HUNTING AND GATHERING TO AGRICULTURE

For about 99 percent of our history, all humans were **hunter-gatherers,** people who obtain their food by hunting, fishing, and gathering wild plant foods. Today, very few people survive in this way. Most hunter-gatherer groups moved around from place to place, following game animals and plant foods that were available only in certain places at certain times of the year.

Hunter-gatherer populations remain small. Most hunter-gatherers live in groups of 25 to 50 people, spread widely across the landscape. It is estimated that there were only about 1 million individuals on Earth when all people lived by hunting and gathering. (This may seem like a lot until you consider that the population of Detroit is now about 1 million.)

Then about 10,000 years ago, people began to raise animals and grow crops instead of relying on wild animals and plants for their subsistence. This change from hunting and gathering to agriculture had such dramatic results that it is often called the **agricultural revolution.** Agriculture probably did not originate in just one area. People probably began raising animals and cultivating plants in many different places at about the same time. But once it started, agriculture spread all over the face of the globe.

Agriculture allowed people to produce more food than they could by gathering wild plants. In other words, farming increased the carrying capacity of the land. Agriculture also led to food storing, because, unlike hunters and gatherers, agriculturists stayed in one place year-round. Farmers stored food supplies to help them survive cold winters or dry summers.

The human population began to grow, partly because more food was available. Also, larger families are advantageous to people who practice agriculture. Children can help out on a farm beginning when they are very young. A family with many healthy children could plant more crops and raise more animals than could a small family.

At first the population grew slowly, as you can see in Figure 13-7. It took almost 10,000 years—until about A.D. 1800—for the human population to reach 1 billion.

Figure 13-6 A large family was advantageous to farm families because the children could help with the work. The more children a family had, the more food they could produce.

ECO-FACT

The Earth's human population is increasing by 179 people each minute, 10,730 each hour, 257,534 each day, and 94 million each year.

Figure 13-7 After growing slowly for thousands of years, the human population began a period of rapid growth in the 1800s.

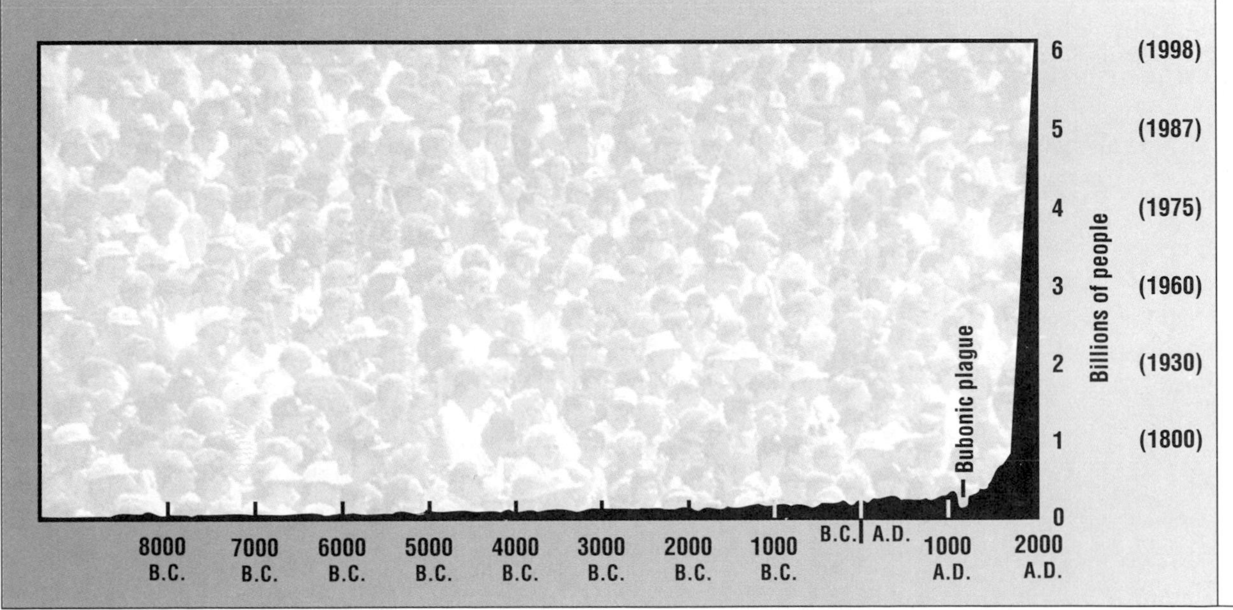

Number of Years to Add 1 Billion People		
	Year	Years to add
First billion	1800	all of human history
Second	1930	130
Third	1960	30
Fourth	1975	15
Fifth	1987	12
	Projected	
Sixth	1998	11
Seventh	2008	10
Eighth	2019	11

Figure 13-8 What is the projected population size for 2008?

Answer to Caption: 7 billion

THE POPULATION CONTINUES TO GROW

After thousands of years of slow growth, the population doubled between 1800 and 1930, as shown in Figure 13-8. By 1975 the human population of Earth was 4 billion, and just 12 years later it had reached 5 billion. Why did the human population grow so quickly?

The answer is quite simple: people started living longer. Because of improvements in sanitation, nutrition, and medical care, more people survived childhood and grew up to have children of their own.

THE DEMOGRAPHIC TRANSITION

As we enter the twenty-first century, the human population is still growing rapidly. However, it may surprise you to learn that population growth has slowed because birth rates have dropped in most parts of the world. Birth rates have decreased in the developed countries and in some developing countries in Latin America and Asia.

THE STORIES BEHIND THE STATISTICS

Demographers often use mathematical formulas to predict population changes, but demography actually has more to do with understanding human behavior. Many complicated and interrelated factors influence people's choices about how many children to have and when to have them.

Take the people in the United States, for example. Hidden behind the demographic data found in census reports are the reflections of war and peace, economic boom and bust, and social change. Between the lines of the dry statistics are the stories of how the lives of average Americans have been affected by the changes of the past century.

As America moved from the nineteenth into the twentieth century, the numbers reflect a phenomenal rise in the number of people coming to America from other countries. Almost 9 million immigrants came to America between 1900 and 1910, accounting for more than 50 percent of the growth in the population.

After the stock market crash of 1929, the United States was plunged into the Great Depression. Unemployment skyrocketed and millions of once-comfortable families found themselves in poverty. Many young people, convinced that the economy would soon improve, decided to wait a few years before marrying. And married couples

Since the first Europeans arrived, immigrants from all over the world have been anxious to come to the United States to live.

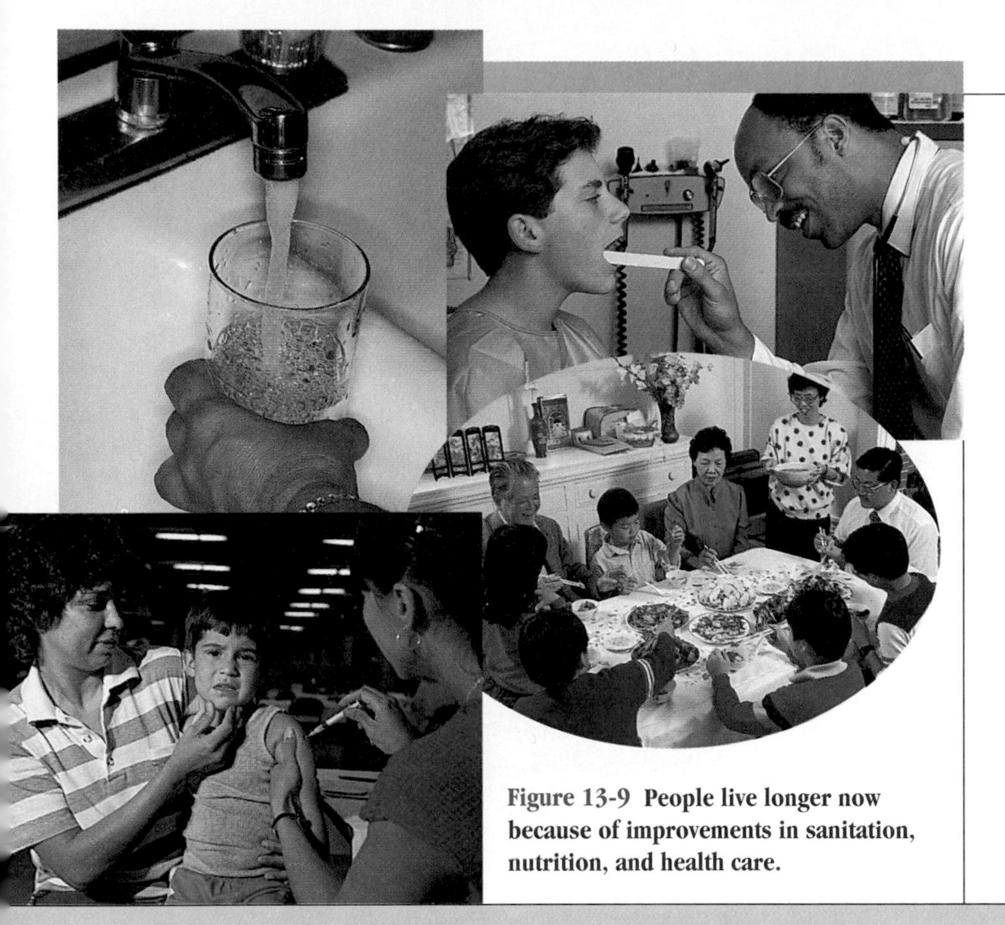

Figure 13-9 **People live longer now because of improvements in sanitation, nutrition, and health care.**

FIELD ACTIVITY

At your local library, gather population figures for your town or city dating from the present back to at least 100 years ago. Chart the change in population in five- or ten-year intervals. Predict how the population of your area will change in the next 10, 20, and 50 years. Identify environmental factors that might be affected by your predicted change in population.

Sample answer on page T158.

decided that it was not a good time to have another child. Consequently, the birth rate fell drastically.

In 1941 Pearl Harbor was bombed and the United States entered World War II. Millions of young men left their homes to fight. People left at home made sacrifices and devoted their energies to contributing to the war effort. Birth rates therefore remained low.

For a generation that had grown up during the depression and survived a world war, the 1950s must have seemed the ideal time to have a family. The GI bill covered college expenses for many, and well-paying jobs were plentiful. People were much better off financially than their parents had been. Young married couples had babies—lots of babies. By 1958, the average woman had almost four children, twice as many as the average in the late 1930s. In the 1950s and 1960s, communities struggled to build schools fast enough to educate the influx of baby boomers—children born between 1945 and 1965.

However, when they reached adulthood, the baby boomers had fewer children, and many put off having children until they were in their late twenties or early thirties. Twenty-five percent fewer children were born in the 1970s than demographers had expected.

Predicting demographic trends is as difficult as predicting human behavior. A family choosing to abandon the city rat race and move to a small town, or a young woman waiting until after college to marry and have children—these are the stories behind the statistics. It's hard to predict the exact patterns of America's future population, but we know the pattern will be formed from the individual choices of millions of human beings. To understand population, you have to understand people.

After World War II, the United States birth rate soared. By 1958, the average woman had almost four children.

Answers to Thinking Critically can be found on page T158.

THINKING CRITICALLY

1. *Inferring Relationships* People from other countries still immigrate to the United States each year. What direct and indirect effects does this have on the population?
2. *Expressing Ideas* What do you think will be the main factors influencing the number of children born to people of your generation?

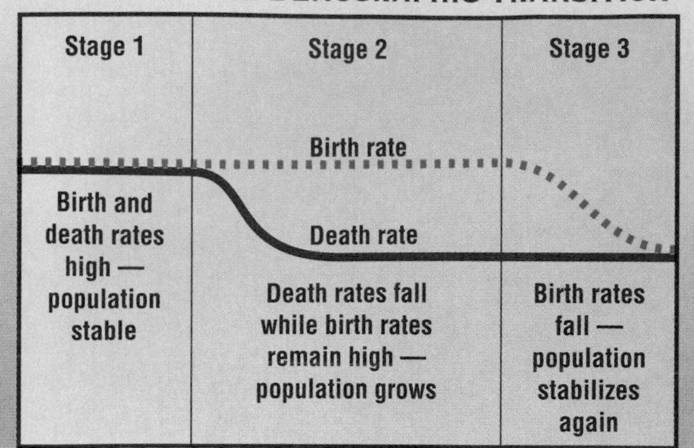

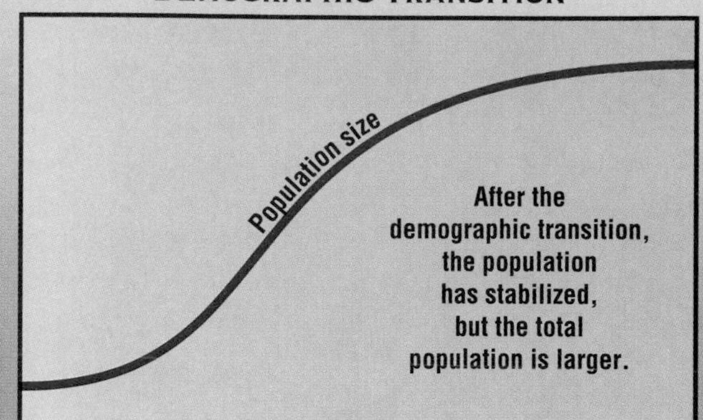

Figure 13-10 According to the theory of demographic transition, population grows rapidly during the time when death rates fall and birth rates remain high. Eventually, birth rates fall and population growth levels off.

Frank Notestein, a scientist in the field of demography—the study of human population patterns—outlined a theory to explain how this happened. Notestein's theory, known as the theory of demographic transition, states that the economic and social progress of the industrial revolution affects population growth in three stages. As illustrated in Figure 13-10, the three stages of the demographic transition are as follows:

Stage 1: Birth and death rates are both high in preindustrial societies, and the population grows slowly, if at all.

Stage 2: When health care improves, the population size increases. Birth rates continue to be high, and people are living longer. The population increases by about 3 percent each year, which means that it doubles every 25 years and is 20 times its original size by the end of 100 years.

Stage 3: Birth rates fall until they roughly equal death rates, and population growth slows down and stops.

Why do birth rates drop during the third stage? There are many reasons. Parents begin to realize that their children will most likely survive to adulthood, so they begin to want only as many children as they can adequately care for. People who live in the city begin to see each child not as another worker but as an expensive responsibility.

Figure 13-11 Some countries are still in the second stage of the demographic transition. Death rates have fallen, but birth rates remain high. The population continues to grow faster with each generation.

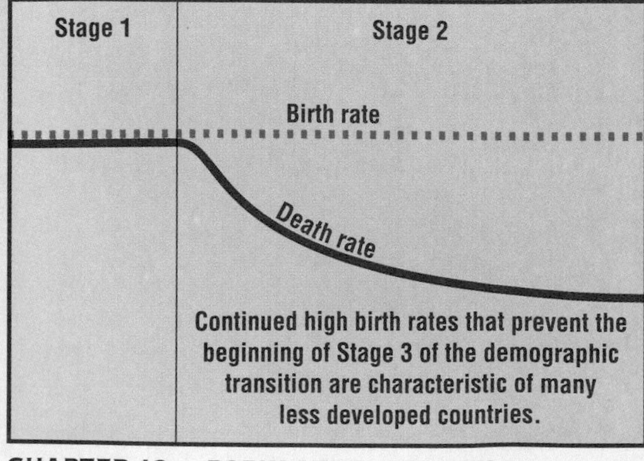

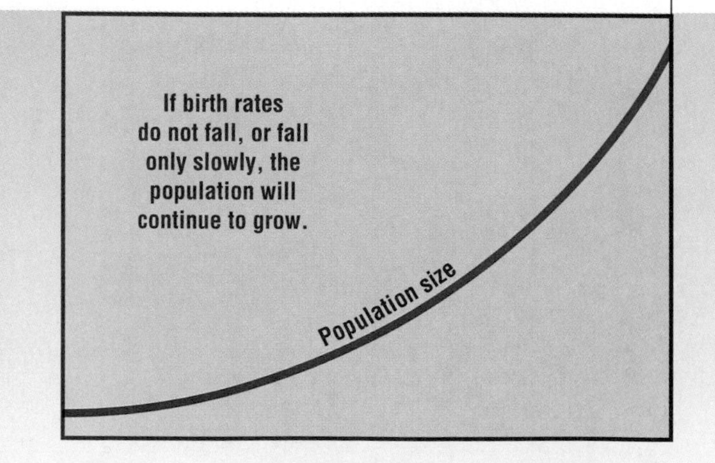

And people who live on farms need fewer children to help out because machines now do much of the work. Furthermore, as a country becomes more affluent, there are greater opportunities, especially for women. People marry at a later age and start having children at a later age. And finally, the availability of reliable methods of birth control allows parents to intentionally limit the number of children they have.

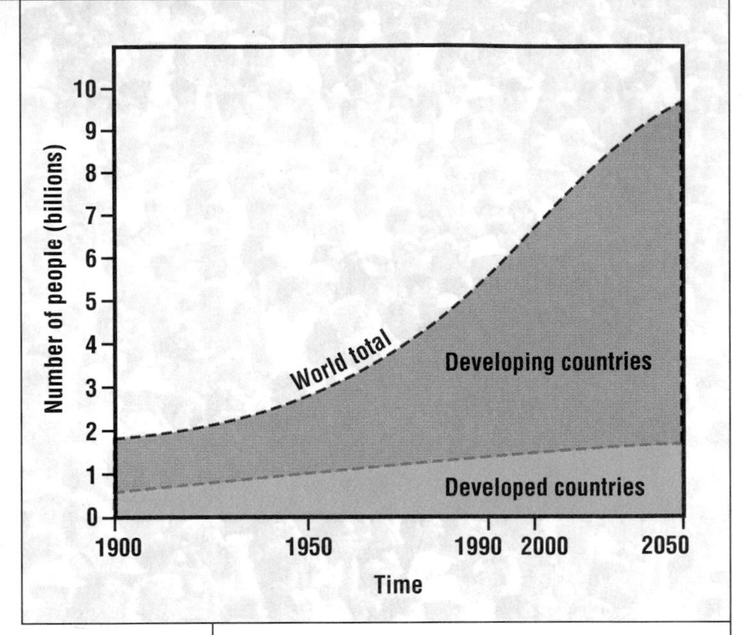

Figure 13-12 This graph shows a demographically divided world. Most of the world population increase since 1950 has been in developing countries.

Answers to caption (at left) and Section Review can be found on page T158.

A Demographically Divided World

Some countries are still in the second stage of the demographic transition. Figure 13-11 shows the population patterns of these countries. The world is now divided into countries that have completed the demographic transition and countries that are still in the second stage. (See Figure 13-12.) In many developing countries, the average number of births per woman is still quite high. In other developing countries, the number of children per woman has dropped, and yet the population is still growing. Why is this so? These countries have a high proportion of young people, as shown in Figure 13-13. As the large number of young people become parents, the number of children born each year stays high, even though the number per woman is decreasing. Consequently, most of the increase in human population is occurring in the developing countries.

Figure 13-13 These are age-structure histograms, which show the proportion of people of each age level and sex in a population. The developing countries have a larger proportion of young people. How will the large number of young people influence future population growth?

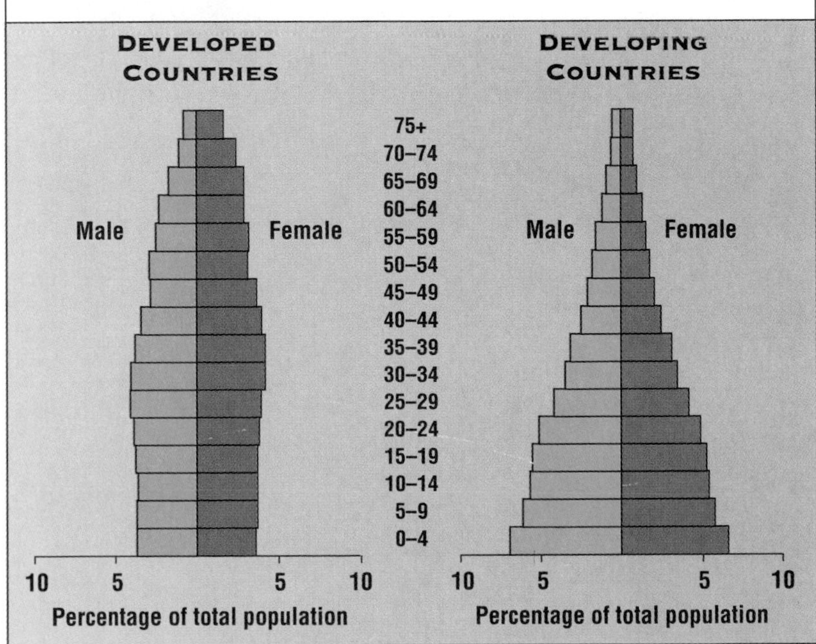

Section Review

1. Why was it advantageous for people who lived on farms to have many children?

2. It took thousands of years for the human population to reach 1 billion but only about 130 years for it to reach 2 billion. What are the main reasons for the change in growth rate?

Thinking Critically

3. *Analyzing Relationships* The average life expectancy of a person in the United States was 47 years in 1900. Yet most adults lived well past that age. Explain this apparent contradiction.

4. *Inferring Relationships* How might an improvement in a country's educational opportunities lead to a decrease in its birth rate?

PROBLEMS OF OVERPOPULATION

AFTER READING THIS SECTION YOU SHOULD BE ABLE TO

- describe the problems stemming from overpopulation.

The rapid population growth you read about in Section 13.2 is overwhelming the ability of some countries to provide for their people. In this section, you'll read about the consequences of overpopulation, including a shortage of fuelwood, contaminated water, urban problems, and the displacement of people from their homes.

A Shortage of Fuelwood

The United Nations and the World Bank have determined that fuelwood is the limiting resource in many developing countries. Often there is enough food to support more people, but there is not enough wood. Having enough fuelwood can be a matter of life and death. Although some foods can be eaten raw, many foods release

Figure 13-14 Having enough wood to cook food and boil water is essential for survival in many parts of the world. These young Tanzanian women are cooking millet.

more of their nutrients when cooked. And some foods are completely inedible if they are not cooked. Without the ability to cook their food, many people suffer from malnutrition or even starvation.

A supply of wood also ensures that a person can boil water, which is extremely important in developing countries. People in developed countries can usually feel safe drinking water straight from the tap. In some parts of the world, however, a person who drinks unboiled water will almost certainly suffer from waterborne parasites or other diseases.

WATER THAT KILLS

Each year, at least 25 million people, most of them children, die of diseases contracted from dirty water. In many developing nations, the local water supply is used not only for drinking and washing, but also for sewage disposal. As a result, the water supply becomes a breeding ground for organisms that cause dysentery, typhoid, and cholera.

Lima, Peru, offers an example of the relationship between population growth and waterborne diseases. In 1991 that South American city was the site of the first cholera epidemic the Western Hemisphere had experienced in more than 75 years. The cholera epidemic was directly linked to Lima's exploding population, which has grown from less than 3 million in 1970 to about 7 million today. More than half of Lima's population is housed in shantytowns, like the one shown in Figure 13-15. People in shantytowns live in houses made from plastic, packing cases, and other

FIELD ACTIVITY

Obtain three large planters, such as large plastic milk jugs, and 70 germinated bean seeds. Place the same amount of soil in each jug. In the first jug, plant 10 germinated bean seeds spaced at regular intervals. In the second jug, plant 20 bean seeds. Finally, plant 40 bean seeds in the third jug. Keep all three jugs in sunlight, and water the seedlings as needed. After three to four weeks, compare the growth of bean plants in each jug. Determine the effects of overpopulation on the growth and development of the bean plants.

Answer to Field Activity: Students should find that overpopulation typically prevents or slows the growth of the bean plants.

Figure 13-15 This photograph shows a shantytown in Lima, Peru. When cities grow faster than adequate housing and clean water can be provided, outbreaks of disease become more common.

garbage. They have no running water and no sewage system. The bacteria that cause cholera found a perfect environment in the shantytowns of Lima.

To stop diseases such as cholera, Peru must supply clean water and hygienic sewage disposal to all its people. Unfortunately, this is almost impossible because the country's population continues to grow by 2 percent each year, and its urban population is growing even faster.

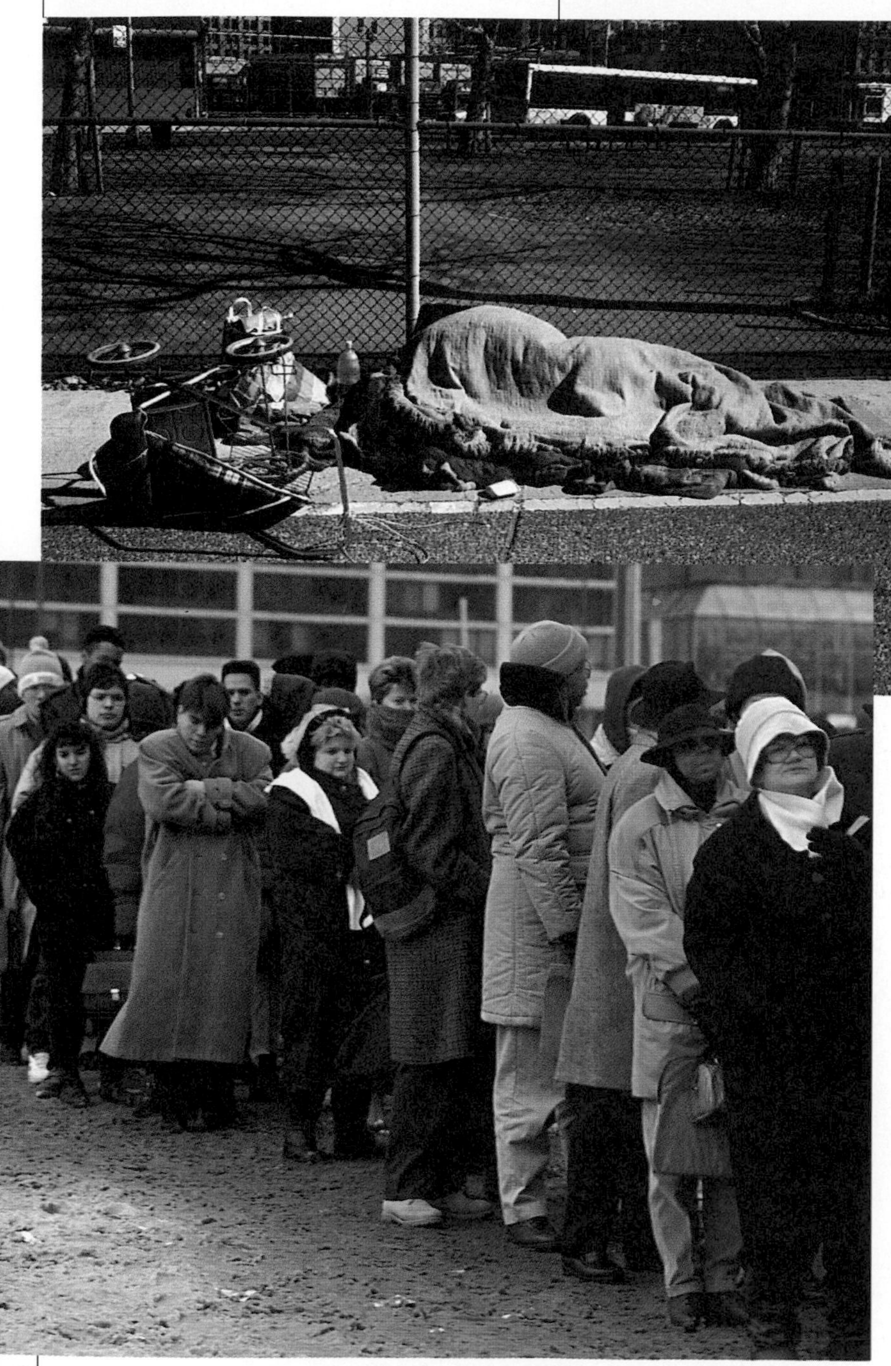

Figure 13-16 The United Nations estimates that by the year 2020, almost one-fourth of the people living in large cities will be homeless.

Figure 13-17 Rapid population growth means fewer jobs are available. These people are in line to apply for a single job.

The Urban Crisis

As a country's population grows, more and more people choose to live in cities. They go there seeking work or educational opportunities. In some countries, a shortage of fertile land means there are few jobs in the countryside, and this pushes even more people into the cities.

The flood of new residents then often overwhelms the cities, and people end up on the streets or in shantytowns. The United Nations projects that by 2020 almost one-fourth of the world's city dwellers will be homeless.

Social Unrest

Rapid population growth in a nation can lead to conflicts between groups of people competing for scarce resources. One ethnic group may dominate the government and monopolize most of the available jobs—a situation that can erupt into civil war.

Population growth can also cause disputes between two or more nations. For instance, rivers often flow through more than one nation, and disputes about water rights are common if water is scarce. Rapid population growth has also been linked to food riots in Egypt and problems caused by illegal immigration into the United States and Europe.

Environmental Refugees

When we think of refugees, we usually think of people who are fleeing war zones. However, environmental damage also creates refugees. After the nuclear power plant explosion at Chernobyl, for example, 10,000 residents had to leave the city because it was contaminated by radiation. These people are **environmental refugees,** people driven from their homes by severe environmental damage.

The millions who leave their homes in Somalia, Kenya, and Ethiopia in search of food are also environmental refugees. Even though the immediate cause of their plight might be drought or civil war, that is not the whole story. Overpopulation has resulted in intensive farming in those countries, which has destroyed the fertility of the soil. Most of the refugees can never return to their homes because the carrying capacity of the land will take hundreds of years to recover—if it ever does.

When many refugees resettle in another country, they may overwhelm the ability of the host country to provide education, health care, and jobs. As the world's population grows and food becomes more scarce, the number of environmental refugees will probably increase. The flood of environmental refugees in the 1990s is a good argument for helping other countries solve their environmental problems and reduce their population growth.

POPULATION GROWTH

Should population growth be limited? **Turn to pages 352–353 and analyze the issue yourself.**

Figure 13-18 People crowd a refugee camp in Zaire

Answers to Section Review can be found on page T158.

Section Review

1. Explain why the human population of the Earth is still growing rapidly even though birth rates have begun to fall slightly.
2. How is overpopulation contributing to environmental degradation?
3. What are some of the social problems caused by overpopulation?

Thinking Critically

4. *Relating Concepts* Create a concept map describing the problems of overpopulation. Include: a shortage of fuelwood, contaminated water, urban problems, displacement of people, and an example of each.
5. *Making Decisions* One person says there are too many people in the world, while another disagrees. What criteria would you use to decide whether the human population is too large?

Solving the Problems of Overpopulation

Slowing the growth of the human population is considered vital to future peace and prosperity. A country with a stable population can better provide services like education and health care to its citizens. In addition, environmental damage to natural resources can be slowed, and eventually some of the damage may be repaired.

The benefits of slower population growth extend to individual families as well. A study in Kenya showed that family members were healthier in families in which a woman had her first child after the age of 20 and had no more than three children, spaced two to three years apart. Children born a few years apart are also more likely to get a good education.

When people take measures to ensure that they will have the size of family they want, they are practicing family planning. An important component of family planning is birth control, a means of preventing births. People might practice birth control if they want to wait awhile before having another child, or if they do not want children at all.

A worldwide survey found that half of all women in developing countries said they want no more children. But half of those women who said they want no more children do not practice birth control. Why is this? Many people do not practice birth control because it is not accepted within their culture or because of their own religious beliefs. Others may not know how to prevent pregnancy. In addition, certain methods of birth control may be too expensive or available only through medical clinics, which are few and far between in many developing countries.

Overconsumption in Developed Nations

As you learned in Chapter 1, the population crisis is worst in developing countries, but the consumption crisis is most severe in the developed nations of the world. Developed countries have lowered their birth rates, but they are using up much more than their share of the Earth's resources.

To achieve a sustainable future, the problems of overpopulation *and* overconsumption must be solved. This will require effort on the parts of both the developing and developed countries.

CHAPTER 13

HIGHLIGHTS

SUMMARY

- Each population has a characteristic biotic potential. The biotic potential of any population represents the greatest growth possible if all members of that population reproduce at maximum capacity.
- If there are no limits to the growth of a population, it will grow exponentially. Factors such as scarce resources and disease prevent populations from growing at their biotic potential.
- Carrying capacity is the maximum population size that a habitat can support indefinitely. A population that exceeds carrying capacity for long periods degrades its environment and reduces future carrying capacity.
- The transition from hunting and gathering to farming initiated the agricultural revolution, which allowed the human population to increase. Industrialization has led to further population increases by lowering the death rate.
- Today, the human population is growing at different rates in different parts of the world. The theory of demographic transition predicts that human populations in developing countries will grow rapidly with increased industrialization, but will later stabilize as birth rates fall.

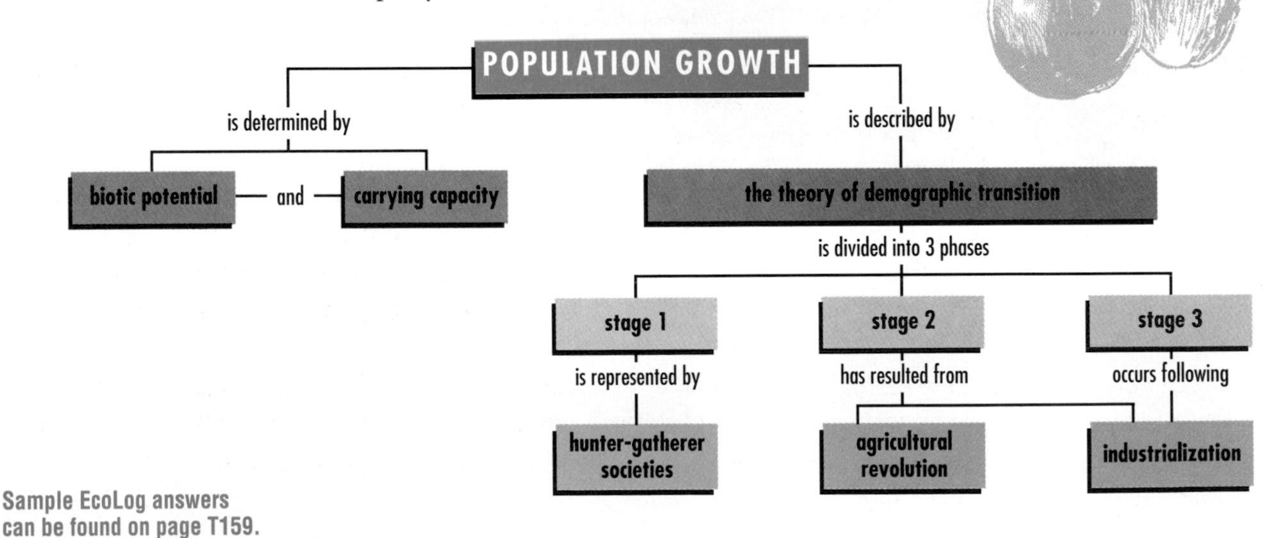

Sample EcoLog answers can be found on page T159.

EcoLog

Now that you've studied this chapter, revise your answers to the questions you answered at the beginning of the chapter, based on what you have learned.

1. Could there ever be too many people on Earth? Why or why not?
2. How could overcrowding in another part of the world affect you?

Vocabulary Terms

agricultural revolution (p. 337)
biotic potential (p. 333)
carrying capacity (p. 335)
environmental refugees (p. 345)
hunter-gatherers (p. 336)
limiting resources (p. 335)

REVIEW

UNDERSTANDING VOCABULARY

1. For each pair of terms, explain the differences in their meanings.
   a. carrying capacity
      biotic potential
   b. environmental refugees
      hunter-gatherers
   c. demographic transition
      agricultural revolution

RELATING CONCEPTS

2. Copy the unfinished concept map below onto a sheet of paper. Then complete the concept map by writing the correct word or phrase in each box containing a question mark.

Answers not indicated can be found on page T159.

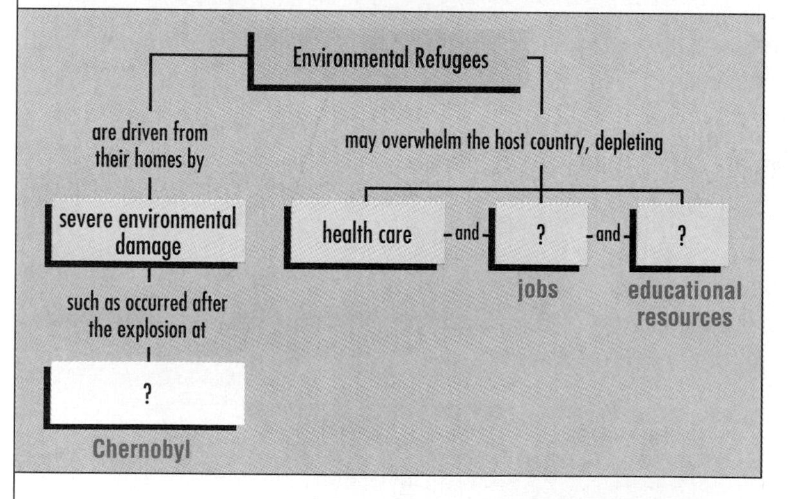

UNDERSTANDING CONCEPTS

Multiple Choice

3. Which of the following would cause a population to grow quickly?
   a. Many offspring are produced from each pregnancy.
   b. Reproduction usually occurs more than once in life.
   c. Members of the population begin to reproduce at a young age.
   (d.) Any of these factors would cause a population to grow quickly.
4. A population is most likely to grow explosively
   a. if the species is a very common one in the area.
   (b.) when a species moves into a suitable habitat where it has never lived before.
   c. after a natural disaster such as a flood.
   d. if there are few individuals of that species in the area to begin with.
5. A population will degrade its environment if
   (a.) the population stays above carrying capacity for a long period of time.
   b. it must share resources with many other species.
   c. it moves frequently from one habitat to another.
   d. it has a high biotic potential.
6. It took 130 years for the Earth's human population to double from 1 billion to 2 billion. How long did it take for the population to double again to 4 billion?
   a. 100 years
   b. 75 years
   (c.) 45 years
   d. 25 years
7. What was the most important factor leading to the explosive growth of the human population during the last 200 years?
   a. an increase in birth rates
   (b.) a decrease in death rates
   c. birth rates increased while death rates decreased
   d. a decrease in the number of people living in poverty
8. Because birth rates have begun to fall, demographers expect that the human population of the Earth will
   a. soon stabilize at the level it is today—about 5.6 billion.
   b. begin to decrease until it is only half the current population by the end of the twenty-first century.

c. continue to increase for a short time and then decrease to current levels by the end of the twenty-first century.

d. stabilize somewhere between 8 and 20 billion.

9. The cholera outbreak in Lima, Peru, was caused by
   a. the inability to provide clean water for such a rapidly growing urban population.
   b. inadequate medical care.
   c. widespread malnutrition.
   d. a population explosion of rats and disease-carrying insects.

10. Refugees from Somalia are considered environmental refugees because
   a. they are fleeing war.
   b. they are searching for food.
   c. they are leaving an environmentally damaged homeland.
   d. they are depleting the health-care system of their host country.

Short Answer

11. Explain how established species may be affected when a new species moves into an ecosystem.

12. If a population exceeds the carrying capacity of its habitat, what are the likely results?

13. How did our ancestors' lives change as a result of the agricultural revolution?

14. According to the theory of demographic transition, why do birth rates fall during the third stage?

15. Give two reasons why rapid population growth in rural areas results in the explosive growth of big cities in many countries.

Interpreting Graphics

16. **Examine the table below.** It shows the numbers of people in different age groups in two populations that are now the same size. Make a bar graph of this data. From your graph, predict which population will grow the fastest. Explain the reason for your answer.

	0–10	11–20	21–30	31–40	41–50	51–60	61–70	71–80	81+
Population 1	500	700	1100	1800	1600	1100	600	200	100
Population 2	300	200	500	800	1200	1500	1300	1200	700

Thinking Critically

17. ***Predicting Outcomes*** Describe what you think life would be like for the average person over the next 50 years in a country that is now entering the third stage of demographic transition.

Themes in Science

18. ***Patterns of Change*** Explain why populations grow exponentially under ideal conditions.

Cross-Discipline Connection

19. ***Health*** Viruses are the cause of many infectious diseases. The common cold, for instance, can be caused by an assortment of different viruses. These viruses can be passed along in many different ways. For example, when someone with a cold coughs, the tiny droplets that are coughed out carry the cold viruses. These droplets, which float easily on air currents, can infect someone else if they are inhaled. Why do you think viral diseases spread more easily in an overpopulated area? What could be done to help slow the spread of such viruses?

Discovery Through Reading

To learn more about how population growth affects people's lives, read the article "Population, Plenty, and Poverty" in *National Geographic,* December 1988, pages 914–945.

PORTFOLIO ACTIVITY

At the library, find the *Demographic Yearbook/Annuaire démographique* published by the United Nations Department of Economic and Social Development, Statistical Division. Obtain copies for every other year going back 20 years (10 in all). For one continent, compile total population figures for each country, and then use the data to make a graph for each country. Using an encyclopedia, add current average yearly income for each country to its graph. Then mount all of your graphs on a poster, and share the results of your research with your class.

CHAPTER 13

INVESTIGATION

What Causes a Population Explosion?

Can you predict the future? If you were a demographer, you might be asked to do so. Demographers must answer questions about how a population is changing, how it is likely to change in the future, and what factors will affect the rate of change. For example, both the number of children per family and the age at which people have children affect the rate of population growth. But which has a greater effect?

To explore this question, you will use age-structure histograms, like those shown below. These graphs show the distribution of age groups in a population at one point in time. The number of people in each age group is represented by a horizontal bar, with the bar on the bottom representing the youngest group.

Answers to this Investigation can be found on page T159.

Materials

- notebook
- graph paper
- pen or pencil
- straightedge or ruler
- colored pencils or markers
- calculator

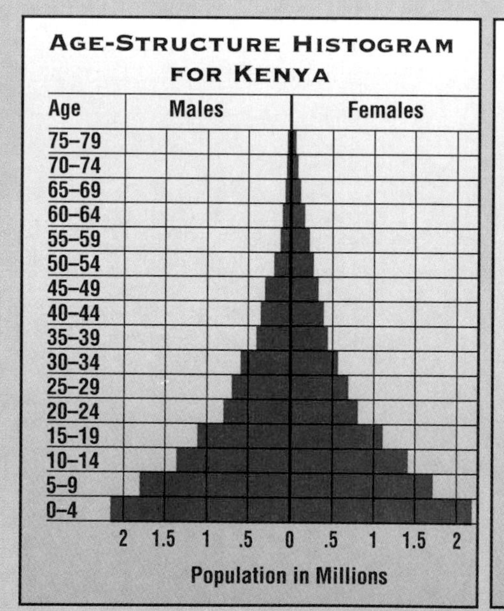

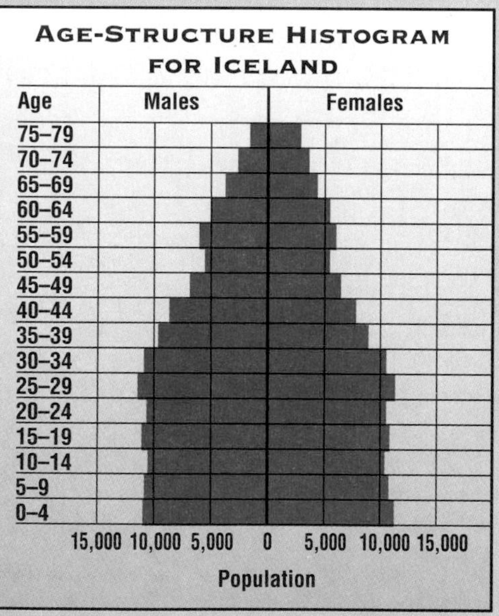

The age-structure histogram for Kenya, on the left, is typical of a rapidly growing population. Because there are so many young people, the number of women of child-bearing age will continue to increase each year, and so will the number of babies born. The age-structure histogram for Iceland, on the right, shows that about the same number of children have been born each year for the past 35 years. If current trends continue, the number of women of child-bearing age will stay about the same, and the population of Iceland will grow very little.

BACKGROUND

In this Investigation you will calculate future population trends for an imaginary city. To compare how different factors may affect population growth, different groups of students will do the calculation using different assumptions. For simplicity, assume the following:

- In each age group, exactly half the population is male and half is female.
- Each woman has all of her children during a five-year period of her life.

- Everyone who is born lives to the age of 80 and then dies.
- No one moves into or out of the city.

PREDICT POPULATION CHANGES

1. Your teacher will divide the class into four groups. Each group will project population growth using different assumptions, as follows:
   - Group A will assume that each woman gives birth to five children when she is between 15 and 19 years old.
   - Group B will assume that each woman gives birth to five children when she is between 25 and 29 years old.
   - Group C will assume that each woman gives birth to two children when she is between 15 and 19 years old.
   - Group D will assume that each woman gives birth to two children when she is between 25 and 29 years old.
2. The table at right shows the 1995 population of our imaginary city. Use the data in the table to make an age-structure histogram for the city.
3. Make a table similar to the one shown, adding columns for the years 2000, 2005, and every fifth year until the year 2050.
4. Calculate the number of 0- to 4-year-olds in the year 2000. To do this, first determine how many women will have children between 1995 and 2000. (Hint: Remember that half the population in each age group is female.) Multiply that number by the number of children that each woman has.
5. Fill in the rest of the column for the year 2000, calculating the number of people in each age group older than 5 by "moving up" the population from 1995. For example, the number of 5- to 9-year-olds in 2000 will equal the number of 0- to 4-year-olds in 1995, the number of 10- to 14-year-olds in 2000 will equal the number of 5- to 9-year-olds in 1995, and so on.
6. Repeat the process described in steps 4 and 5 ten times to complete the table through the year 2050.
7. Add up the total population for each year. Plot these figures on a line graph.
8. Make an age-structure histogram for the 2050 population from your table.

Age Group	1995 Population
75–79	500
70–74	600
65–69	700
60–64	800
55–59	900
50–54	1,000
45–49	1,250
40–44	1,500
35–39	2,000
30–34	2,500
25–29	3,000
20–24	4,000
15–19	5,000
10–14	6,500
5–9	8,000
0–4	10,000
Total	48,250

INTERPRET YOUR PREDICTIONS

9. Compare your graphs with those made by the other three groups. Which factor had a bigger effect on population growth: the number of children each woman had, or the age at which women had children?
10. Did any of the graphs show no growth in population? Explain these results.

Most families in Kenya have many children, contributing to rapid population growth.

Families in Iceland, whose population is stable, tend to be small.

POINTS OF VIEW

Should POPULATION GROWTH Be Limited?

In 1650 the human population was growing by 0.3 percent each year, doubling every 250 years. The Industrial Revolution raised standards of living, and medical advances increased life spans and reduced infant deaths. By 1970 there were 3.6 billion people, and the population was growing by 2.1 percent each year, doubling every 30 years. Today, the world's population is nearly 6 billion and could reach 10 to 14 billion in the twenty-first century. Some people believe that population growth must be limited to avoid exceeding the Earth's carrying capacity. But other people object to population control for cultural, political, and economic reasons. Two points of view:

HUMAN POPULATION GROWTH MUST BE LIMITED

Population growth is contributing to a rapid depletion of the Earth's resources. For example, overfishing has reduced the annual catch in some marine fisheries.

Like all organisms, humans depend on the Earth for food, water, shelter, and space. As the human population grows, people use more and more of the Earth's resources to fulfill these needs. Supporters of population control believe this has serious consequences for the Earth and for people.

Some people believe that uncontrolled population growth has stretched the Earth's resources to their limits. Most of the land suitable for farming has already been cultivated; little more land remains that could be used to grow more food. Forests around the world are being destroyed at an alarming rate to provide lumber and to clear land for agriculture. Overfishing has seriously depleted the number of fish in some marine fisheries. And water disputes will erupt as more and more rivers are diverted for irrigation and dams, thereby depleting the water supply of people in other areas.

Advocates of population control argue that the poorest countries in the world are those with the fastest-growing populations. Controlling population growth would promote economic development in these countries, leading to a better standard of living for the citizens. If population growth continues unchecked, people in these countries will never have enough food, jobs, or education to go around.

Already, many people around the world do not get enough food or clean water to thrive, and starvation is a significant cause of death in some developing countries. Although population growth is only one factor in the world's food supply problems, the countries with the highest birth rates also experience the worst food shortages. Supporters of population control say that limiting growth will help ease food supply problems, allowing these countries to work toward a brighter economic future.

Almost every bit of available space is taken in this area of Jodhpur in northwestern India. Some people fear that if population growth is not limited, we may exceed the Earth's carrying capacity.

No Limits Should Be Placed on Human Population Growth

Population control is often opposed because of cultural values. In some societies, a higher value is placed on male children than on female children, so parents strive to have as many sons as possible. Such social pressure keeps birth rates high. Also, many countries lack social programs such as unemployment compensation or retirement plans. Parents rely on children to contribute to family earnings and support them when they grow old. More children can earn more money and provide more support.

Religious values also may influence opinions about population control. Studies have shown differences in the birth rates of various religions. Followers of some faiths think that birth control is immoral because they believe only God can determine the number of children that are meant to be born.

Other people believe population-control policies violate personal rights. These opponents point to a population-control measure that was used in China in the 1980s. The Chinese government offered couples incentives such as cash and good medical care if they pledged to have only one child. Penalties were given and the incentives taken away if a second child was born. Birth rates declined, but critics claim that the program inhibited freedom of choice. They believe that civil rights will suffer further if these policies are adopted elsewhere.

Some people oppose population control because they fear that a decrease in their country's population will slow economic growth and weaken the country's cultural identity. Other opponents of population control suggest that people are the world's greatest resource—more people will contribute more technological advances in agriculture and resource recycling that will lead to a higher standard of living for everyone, no matter how large the population.

This young boy is selling pastries on a street in Ho Chi Minh City, Vietnam. Children in some countries contribute to family earnings, so more children can mean a higher family income.

This agricultural scientist studies new strains of food crops. Some people believe that population control is unnecessary because technological advances, such as agricultural improvements, will solve the problems of a growing population.

Analyze the Issue

1. ***Making Decisions*** Using the decision-making model presented in Chapter 1, decide whether population growth should be limited. Explain your reasoning.
2. ***Analyzing Relationships*** China now uses education instead of penalties and incentives to try to lower the birth rate. Do you think this will be a more successful method? Why or why not?

Answers to Analyze the Issue can be found on page T159.

CHAPTER 14

TOWARD A SUSTAINABLE FUTURE

"Act, act, act. You can't just watch."

ANGELES SERRANO
COMMUNITY ACTIVIST, THE PHILIPPINES

SECTION 14.1

INTERNATIONAL COOPERATION

SECTION 14.2

ENVIRONMENTAL POLICIES IN THE UNITED STATES

EcoLog

Before you read this chapter, take a few minutes to answer the following questions in your EcoLog.

1. What do you think the world will be like in 50 years? What could you do now to contribute to the kind of future you want?
2. At which level of government—national, state, or local—can individual citizens have the most influence on environmental policies? Why?

INTERNATIONAL COOPERATION

AFTER READING THIS SECTION YOU SHOULD BE ABLE TO

1. describe the results of the Earth Summit.
2. describe international agreements relating to the environment.

As you learned in Chapter 1, environmental science is not a pure science. It has a practical goal. That goal is to develop ways of living that permit the human species and other species to survive and prosper into the future. This is what we call a sustainable future. A sustainable future can be achieved only if we can find ways to preserve and expand our environmental resources.

The solutions to our environmental problems require both individual and group action. Throughout this book, you have read about ways that individuals can help solve environmental problems. But what about group action? In this chapter you will learn how people can work together to help create a sustainable future. Cooperation is necessary at international, national, and local levels. The Case Study on pages 356–357 describes one example of international cooperation—the effort to regulate whale hunting.

Figure 14-1 **International cooperation is required to save species such as the orangutan.**

THE EARTH SUMMIT: PROMISE AND PROBLEMS

Rio de Janeiro, Brazil, was the site of the 1992 Earth Summit, sponsored by the United Nations Conference on Environment and Development. For 12 days, hundreds of government officials (including 118 heads of state) worked together to hammer out international agreements. Thousands of environmentalists

from around the world also gathered to share ideas and to influence the governmental proceedings. More than 9,000 reporters—the largest gathering of media in history—covered the proceedings.

The Earth Summit yielded some positive results, described on page 358. Agenda 21 was passed, which is a blueprint for protecting the environment and promoting sustainable development. Agenda 21 includes an agreement by industrialized nations to help developing countries become industrialized without destroying the environment. Participants also agreed to the Earth Charter, which set forth 27 broad principles for preserving ecosystems and helping developing nations. However, both the Earth Charter and Agenda 21 are nonbinding, which means that they cannot be enforced. Nations are very protective of their sovereignty, their right to manage their own affairs. They don't want to be in a position that permits other countries to tell them what to do.

CASE STUDY

A WHALE OF AN INTERNATIONAL CONTROVERSY

One species of whale, the blue whale, is the largest animal ever known to inhabit the Earth—bigger than even the dinosaurs. Blue whales reach lengths of up to 27.4 m (90 ft.) and weights as great as 136,000 kg (150 tons). Unfortunately, these enormous whales and other whale species have been hunted almost to the point of extinction. A look at the history of agreements between nations to regulate whale hunting illustrates both the problems and successes of international cooperation.

In 1949 the International Whaling Commission (IWC) voted to limit commercial whaling vessels to operations within a nation's territorial waters. However, France objected and invoked the IWC's "opting out" provision to ignore the ruling. Under this provision, any member nation can opt out of an IWC decision within 90 days. France was the first among many nations to use this loophole, which hampers the IWC's ability to enforce its own regulations. Over the years, member nations have frequently opted out of IWC attempts to protect threatened whale species.

The 1949 agreement also established a method of calculating the permitted quota of whales that could be killed. However, this method did not control whaling by species—a nation could "harvest" its quota by

The blue whale, the largest animal in the world, is also among the most hunted by whalers.

Figure 14-2 The Earth Summit in Rio de Janeiro, Brazil

a. Indigenous people from a Brazilian rain forest leave a meeting
b. World leaders confer

killing any species of whale, even one with an already low population. Consequently, blue, fin, humpback, and sei whales were hunted nearly to extinction. Then in 1960 the IWC suspended quotas entirely. The whaling free-for-all that ensued resulted in the highest whale catch in the history of the IWC.

The IWC reestablished a quota in 1967. And in 1972, whale protection was strengthened when the IWC instituted a program in which observers from member nations monitored the catch on other member nations' whaling ships.

Meanwhile, the public was urging the IWC nations to "save the whales." In response, the IWC restricted whaling substantially in 1977 and passed a resolution urging nations to stop importing whale products. In 1982 the IWC enacted a total ban on whaling, beginning in 1984. Norway, Japan, and the former Soviet Union opted out of this agreement.

In 1993 Norway violated the worldwide ban by harpooning 157 minke whales, claiming that their numbers (now estimated at over 900,000) could sustain limited hunting. Norway's action has generated international debate. Some scientists believe that a controlled annual harvest will not hurt the population, but others fear that whaling of any kind will lead other whaling nations to resume the hunt. Critics further point out that Norway's action weakens the consensus among IWC countries, already fragile after decades of controversy.

In 1994 the IWC created a permanent whale sanctuary in Antarctica. Even if the current worldwide ban on whaling is lifted, the waters of Antarctica—the largest feeding ground for whales in the world—will remain off-limits to commercial whaling. Norway and Japan, with the world's largest whaling industries, strongly opposed the measure.

An Icelandic whaler with harpooned fin whales

THINKING CRITICALLY

1. *Analyzing Processes* What measures could international organizations take to enforce international agreements (without starting wars)?
2. *Analyzing Processes* When reading about the Earth Summit, you learned that nations want to protect their sovereignty and that this influenced the agreements reached at the summit. How has the desire to protect sovereignty affected efforts to protect whales?

Answers to Thinking Critically can be found on page T159.

Some International Environmental Agreements	
Earth Summit 1992 Agreements	
Agenda 21	Blueprint for protecting the environment and promoting sustainable development. Industrialized nations agreed to provide money to help poorer countries become industrialized without destroying the environment.
Biodiversity Treaty	Agreement to inventory and protect endangered and threatened species. Rich nations agreed to compensate poor nations for use of plants and animals in making products.
Earth Charter (also called the Rio Declaration)	Sets forth principles for preserving ecosystems and helping developing nations.
Global Warming Convention	Agreement to reduce emissions of greenhouse gases.
Statement on Forest Principles	A broad statement of principles about protection of the world's forests.
Other International Agreements	
Antarctic Treaty and Convention	Agreement that Antarctica be used solely for peaceful purposes, that all nations cooperate in scientific research there, and that the Antarctic be used "in the interest of all mankind." Mining is banned for 50 years, and strict environmental protections are enforced.
CITES (The Convention on International Trade in Endangered Species)	Classifies endangered and threatened species worldwide and monitors international trade in these species. The ban on the ivory trade to save the elephant is among its most recent successes.
Law of the Sea	Provides for new treaties on controlling ocean pollution from runoff (farm and city), ocean dumping, transport of hazardous materials, oil exploration, mining, and air pollution. Signing nations agree that deep-sea resources such as minerals are "the common heritage of mankind."
MARPOL (The Protocol of 1978 Relating to the International Convention for the Prevention of Pollution From Ships)	Agreement to decrease ocean pollution due to shipping, such as releases of oil and disposal of waste. MARPOL applies to both commercial and recreational vessels.
Migratory Species: Convention on the Conservation of Migratory Species of Wild Animals	Protects wild animal species that migrate across international borders.
Montreal Protocol on Substances That Deplete the Ozone Layer	Agreement to completely eliminate the manufacture and use of substances such as CFCs that destroy the protective atmospheric layer of ozone.
Ocean Dumping: The Convention on the Prevention of Marine Pollution by Dumping of Wastes and Other Matter	Regulates disposal and dumping of materials at sea, promotes regional agreements, and establishes a procedure for settling international disputes.
RAMSAR: The Convention on Wetlands of International Importance, Especially as Waterfowl Habitat	Seeks to eliminate encroachment on wetlands and prevent their destruction. Lack of funding by signing nations has compromised the success of this treaty.

Figure 14-3 Nations have managed to overcome their differences and sign these environmental agreements.

Figure 14-4 Issues faced by poor and wealthy countries are often quite different. To the people of developed countries, recycling may represent savings of energy and materials. To people of developing countries, recycling may be necessary for survival.

Consequently, binding agreements did not fare as well at the Earth Summit. A treaty to limit emissions of greenhouse gases was adopted, but only after the other nations agreed to the United States' request that no timetable be included in its provisions. The Biodiversity Treaty, a binding agreement designed to decrease the rate of extinctions, was also signed. President George Bush did not sign the Biodiversity Treaty, but his successor, President Bill Clinton, subsequently joined 150 other nations by signing the treaty. The Earth Summit was not an unqualified success, but at least it focused attention on global environmental issues and paved the way for future agreements.

Wealthy and Poor Nations

One theme that emerged during the Earth Summit was the contrasting needs of the wealthy and poor nations of the world. The wealthy nations asked the poor countries to curb their rapid rates of population growth and take measures to preserve natural resources. The poor countries emphasized that their primary concerns were feeding their people and surviving as nations. They said they want to preserve their resources but simply cannot afford to do so. The poor countries suggested that the wealthy nations should contribute money to help them. The poor countries also pointed out that overconsumption in the developed nations is a major cause of environmental problems.

Our most urgent environmental problems—overpopulation, pollution, depletion of resources, and loss of natural habitats and biodiversity—require international cooperation. But international cooperation is difficult because of the countries' different needs. Yet nations at the Earth Summit overcame some of their differences and managed to adopt environmental agreements.

Answers to Section Review can be found on page T160.

Section Review

1. In your opinion, what was the most significant result of the Earth Summit? Explain your reasoning.
2. Imagine that you are on the planning committee for an Earth Summit that will take place this year. What three environmental issues would you recommend for discussion? Why would you recommend those three?

Thinking Critically

3. *Analyzing Processes* How could the wealthy countries and the poor countries cooperate to preserve resources and provide a decent standard of living for everyone?

ENVIRONMENTAL POLICIES IN THE UNITED STATES

What on Earth is an Earthship? Find out on pages 370–371.

AFTER READING THIS SECTION YOU SHOULD BE ABLE TO

❶ explain how environmental impact statements are prepared.

❷ describe the role of local government in environmental conservation.

❸ give examples of how citizens can influence environmental decisions at all levels of government.

The United States has been among the leaders in environmental protection. The world's first national park, Yellowstone, was created in 1872. Today 9 percent of the United States is protected in national parks and wilderness areas. The first Earth Day, in 1970, can be credited with initiating widespread environmental awareness in the United States. (See Figure 14-5.) In the same year, the Environmental Protection Agency was created. Today, there are numerous government departments and agencies charged with implementing hundreds of environmental laws and regulations. Figure 14-6 lists some of these federal bodies and the laws they enforce.

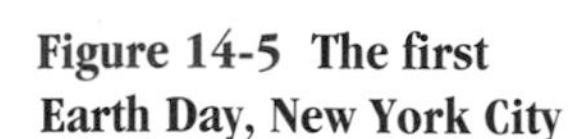

Figure 14-5 The first Earth Day, New York City

United States Departments and Agencies, and Environmental Laws They Enforce

Department of Agriculture

Soil Conservation Service
- Soil and Water Conservation Act

U.S. Forest Service—controls logging of national forests while protecting wildlife and forest ecosystems
- Forest Reserves Management Act
- National Forests Management Act

Department of Commerce

National Oceanic and Atmospheric Administration

National Marine Fisheries Service
- Marine Mammal Protection Act

Environmental Protection Agency—responsible for research, setting guidelines, and enforcing regulations on numerous environmental issues
- National Environmental Protection Act
- Water Quality Act
- Federal Water Pollution Control Acts
- Safe Drinking Water Act
- Clean Water Act
- Clean Air Act
- Solid Waste Disposal Act
- Resource Conservation and Recovery Act
- Waste Reduction Act
- Toxic Substances Control Act
- Superfund (Comprehensive Environmental Response, Compensation, and Liability Act)
- Federal Insecticide, Fungicide, and Rodenticide Control Act

Department of Energy
- National Energy Act
- Public Utilities Regulatory Policies Act
- Electric Consumers Protection Act

Department of the Interior
- Wild and Scenic Rivers Act (with Department of Agriculture)

U.S. Fish and Wildlife Service
- Endangered Species Act
- National Wildlife Refuge System Act
- Alaska National Interest Lands Conservation Act (with Bureau of Land Management and U.S. Forest Service)
- Species Conservation Act
- Fish and Wildlife Improvement Act
- Fish and Wildlife Conservation Act

Bureau of Land Management—responsible for wilderness and other land management acts on land it controls
- Federal Land Policy and Management Act
- Taylor Grazing Act

National Park Service—manages nation's national parks

Bureau of Mines, Land Management, and Reclamation
- Surface Mining Control and Reclamation Act

Department of Justice—handles lawsuits involving federal environmental laws

Department of Transportation—responsible for jet noise, mass transit, pollution via transportation
- Quiet Communities Act

Environmental Impact Statements

One of the laws listed in Figure 14-6 is the National Environmental Protection Act. This act led to the requirement that nearly all federal agencies file an **environmental impact statement (EIS)** for any proposed legislation or project that would have a significant effect on the quality of the environment. Environmental impact statements are also known as environmental assessments. Proposals for the construction of dams, highways, airports, and any other projects the federal government controls or subsidizes must be evaluated by an EIS.

An EIS states the need for a project, its environmental impact, and how such an impact can be minimized. After an EIS for a project is drafted, it must be available to the public for at least 90 days before the project begins. The public must have at least 45 days to comment on the project, and citizens' comments must be taken into account when the final EIS is written. Hundreds of plans for projects have been modified as a result of public and government reaction to these environmental assessments.

Environmental impact statements often do not work as their designers intended, however. An individual, business, or government agency planning a project may hire its own environmental

Figure 14-6 This table identifies federal departments, agencies, and environmental laws. Which department or agency has the major responsibility for preventing water pollution?

Answer to caption: Environmental Protection Agency

In a 1992 poll of registered voters, 22 percent said they had voted for a political candidate based on his or her environmental record.

Figure 14-7 Plans for construction of Two Forks Dam on the Platte River, southwest of Denver, Colorado, were modified as a result of an environmental impact statement. The Platte River is a crucial stop for migrating cranes, including these sandhill cranes.

consulting firm to prepare the EIS. Because it is paid by the interested party, the consulting firm may be biased in presenting the data. However, independent environmental organizations may carefully review crucial EISs and challenge the conclusions in court.

Influencing National Policy

How can you influence national environmental policies? One way is to write or telephone your U.S. representative and senators and let them know what you think about proposed policies and laws. Another way to influence environmental policy is to join organizations that lobby Congress. Environmental organizations with large memberships can hire lobbyists to make sure that their points of view are heard.

It is sometimes difficult to keep up with proposed environmental laws. Citizens too often find out about environmental legislation after it has already been signed into law. One way to keep up with proposed legislation is to consult the *Weekly Bulletin,* published by the Environmental and Energy Study Institute. The bulletin lists all upcoming environmental and energy legislation in Congress. The same organization publishes the *Briefing Book on Environmental Issues and Legislation,* an annual summary of environmental legislation with quarterly updates. These publications may be found in the library.

State Environmental Policies

The federal government may pass laws that set certain environmental standards, but these are generally considered minimum standards. Individual states are usually free to pass laws that set higher standards. For example, air pollution in California is so bad that the state has set very stringent air quality standards that far exceed federal guidelines.

Most states have one or more departments responsible for enforcing environmental laws. These departments are often required to hold public hearings on proposed projects that may affect the environment in a community. You can influence state policy by expressing your views at these hearings and by contacting your state representative or the appropriate department in your state capital.

Field Activity

Write a letter to your senator or representative about an environmental issue. Choose an issue you care about, and research it thoroughly. Go to the library and read all you can about the issue in books, magazines, and newspapers. In your letter, first state your position on the issue, and then offer facts to support your position. Be polite and respectful, and make sure to include your return address at the top of the letter.

LOCAL COMMUNITIES

Local governments have a significant impact on the environment. In most states, local governments and planning and zoning boards determine the type of development that will be allowed in their area. Consequently, local governments make decisions about issues such as the preservation or destruction of habitat, and human population density. Local governments also regulate recycling, sewage treatment, water quality, and other important concerns.

Figure 14-8 Anyone can express an opinion on environmental issues at state and local public hearings.

Government at the local level is often more responsive to citizen input than any other level of government—especially when citizens organize around an issue. People who are active in local government can have an immediate, tangible impact on their communities.

One common obstacle to local influence, however, is lack of coordination among neighboring communities. For example, your community may decide to limit commercial development. But if a neighboring community permits construction of a huge mall, traffic may become intolerable as shoppers drive through your town to the mall. In some states, different communities are trying to work together to preserve the environment. Communities along the Hudson River in New York, for instance, are cooperating to create a "greenway" of open space that will stretch hundreds of miles along the river. (See Figure 14-10 on the next page.)

Do you want to contribute to a healthier environment? Why not start in the grocery store? Find out how on pages 392–393.

Figure 14-9 Your opinion can make a difference in whether your community looks like the one shown at left or the one shown above.

Figure 14-10 Because hundreds of communities and planning boards along the Hudson River worked together, the Hudson Valley Greenway will improve the quality of life for all residents of the region.

Answers to Section Review can be found on page T160.

Section Review

1. What influence can an environmental impact statement have on the construction of a federal project? How can citizens influence an environmental impact statement?
2. At what level of government can individual citizens have the most influence? Explain your reasoning.

Thinking Critically

3. *Relating Concepts* Make a concept map showing how local governments can affect the environment. Include at least the following terms: local governments, habitat, population density, recycling, sewage treatment, and water quality.
4. *Expressing Viewpoints* Do you think people alive today have any obligation to future generations? Explain your answer.

Into the Future

You have learned that environmental problems can be very complex. Complex problems usually require complex solutions, often based on compromises. Making compromises sounds reasonable—until you realize that such compromises may require a change in your lifestyle. But if we do not make difficult decisions now about how to use the Earth's resources, we may have no control over our future.

Education is probably the most important single step toward a better future. People educated about the environment know that solutions can be found to problems such as pollution, water shortages, and overconsumption of energy. They understand environmental discussions in the media and educate their elected representatives through letters and testimony at public hearings.

Perhaps none of us alone can save the world, but the actions of individuals and small groups can add up to the kind of progress that makes a difference. Your actions, along with the actions of others, can go a long way toward improving the future of our planet. It is still within our power to build a sustainable future, a future in which the environment is not destroyed and all people have a decent standard of living.

CHAPTER 14

HIGHLIGHTS

SUMMARY

- The goal of environmental science is to attain a sustainable future by solving environmental problems through individual and group action.
- The 1992 Earth Summit, attended by representatives from most of the world's nations, was intended to develop world-wide environmental policy and foster solutions to current environmental problems.
- The United States government has established a set of environmental rules and regulations to govern environmental quality. For example, all major federally funded projects must be accompanied by an Environmental Impact Statement.
- States and communities can set their own environmental standards. They are usually more rigorous than federal standards.
- Citizens can influence environmental policy at all levels. The biggest barrier to influencing policy is lack of coordination.
- Much damage has been done to the environment, but much progress has also been made. With concerted action, a sustainable future is within reach.

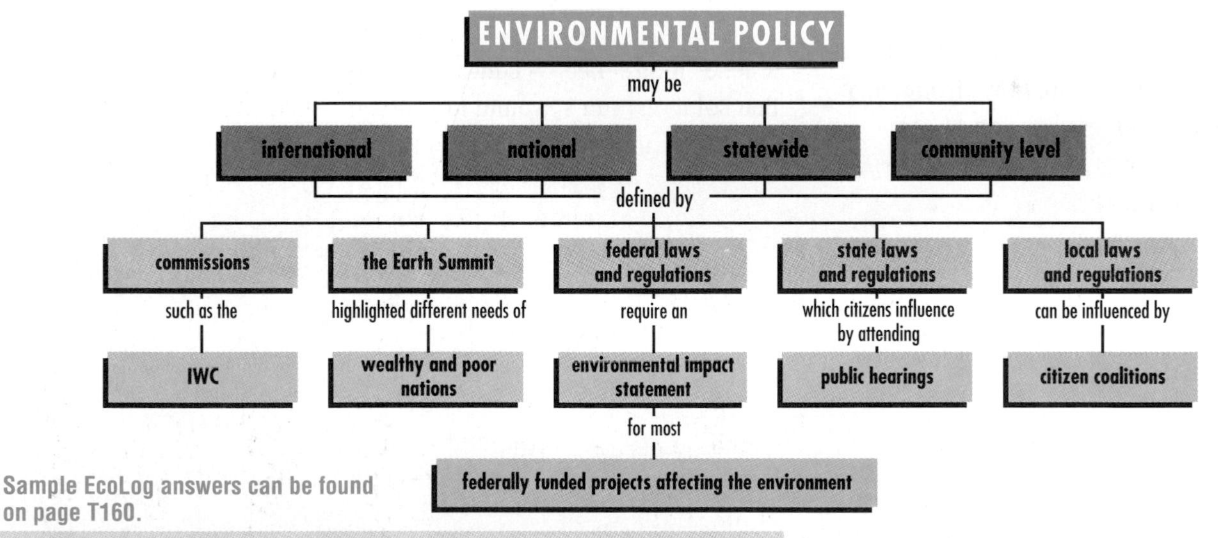

Sample EcoLog answers can be found on page T160.

EcoLog

Now that you've studied this chapter, revise your answers to the questions you answered at the beginning of the chapter, based on what you have learned.

❶ What do you think the world will be like in 50 years? What could you do now to contribute to the kind of future you want?

❷ At which level of government—national, state, or local—can individual citizens have the most influence on environmental policies? Why?

Vocabulary Terms

Environmental Impact Statement (EIS) (p. 361)

CHAPTER 14

REVIEW

Understanding Vocabulary

1. Explain what an environmental impact statement is and how it works.

Relating Concepts

2. Copy the unfinished concept map below onto a sheet of paper. Then complete the concept map by writing the correct word or phrase in each box containing a question mark.

Answers not indicated can be found on page T160.

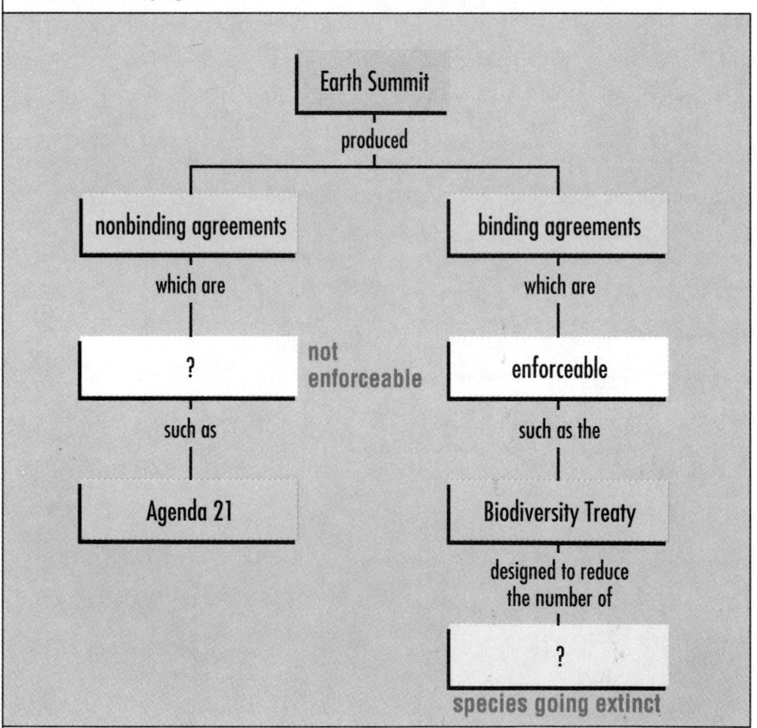

Understanding Concepts

Multiple Choice

3. The Earth Summit was
   - **a.** designed to force all nations of the world to submit to an international environmental policy.
   - **(b.)** intended to help diverse nations develop common environmental policy guidelines.
   - **c.** a corporate convention.
   - **d.** a complete failure as an instrument of policy.

4. At the Earth Summit, rich and poor nations
   - **a.** agreed that their needs were exactly the same.
   - **b.** agreed to disagree.
   - **(c.)** realized that their needs were different.
   - **d.** could not reach agreement on any matters of substance.

5. Most of the agreements drawn up at the Earth Summit
   - **a.** became international law.
   - **(b.)** were nonbinding.
   - **c.** will be administered by the United Nations.
   - **d.** will expire after five years.

6. The function of an environmental impact statement is
   - **(a.)** to clarify the effect that a project would have on the environment.
   - **b.** to generate a record of the ongoing impact to the environment of existing structures.
   - **c.** to satisfy international legal requirements.
   - **d.** to limit development to a bare minimum.

7. State and local environmental regulations
   - **a.** are generally less strict than federal standards.
   - **(b.)** are generally more strict than federal standards.
   - **c.** may be more or less strict than federal standards.
   - **d.** are nonexistent; all environmental regulation takes place at the federal level.

8. Environmental impact statements
   - **a.** must be filled out by anyone carrying out any project that has any effect on the environment.
   - **b.** are filled out for both the state and federal governments.
   - **(c.)** are filled out for nearly all federally funded projects that have significant effects on the environment.
   - **d.** are no longer required.

9. Air pollution standards set by California are
   a. much more stringent than federal requirements.
   b. much less stringent than federal requirements.
   c. nonbinding.
   d. a model for the rest of the nation.
10. Local governments do not regulate
   a. recycling.
   b. sewage treatment.
   c. water quality.
   d. the filing of environmental impact statements.
11. One of the best ways to help improve the future of the environment is through
   a. force.
   b. education.
   c. government regulation.
   d. world conferences.
12. Making environmental compromises can mean
   a. getting everything you want.
   b. allowing half the rain forest to be destroyed.
   c. a change in your lifestyle.
   d. making sure your point of view is heard.

Short Answer

13. Why are gatherings such as the Earth Summit necessary?
14. Why are state environmental regulations no less stringent than federal regulations?
15. Why might there be fundamental disagreements between rich and poor nations regarding environmental policy?
16. What are some ways in which citizens can influence environmental policy?

Interpreting Graphics

17. **Examine the graphs below.** They show that federal money is not spent exactly as citizens think it should be spent. Propose an explanation for this.

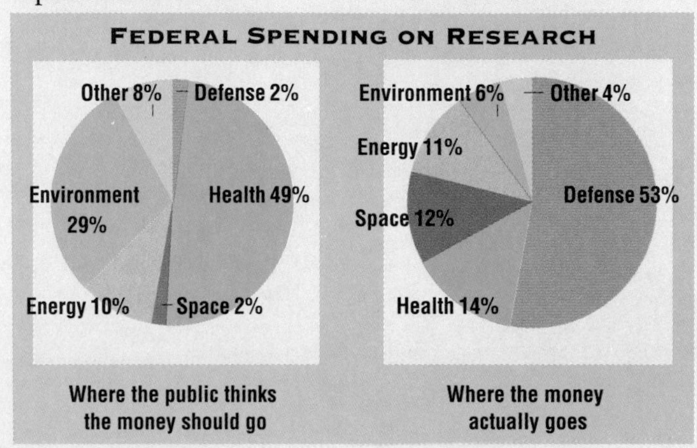

Thinking Critically

18. ***Communicating Ideas*** One major barrier to achieving significant environmental reform is lack of enforcement. Nations cannot be held responsible in any real way for their environmental abuses because they won't give up sovereignty to a higher governing body. How might you convince skeptical nations to give up sovereignty in this matter?

Themes in Science

19. ***Scale and Structure*** A favorite expression of the environmental movement is "Think globally, act locally." Explain how action on a local scale can end up having global impact.

Cross-Discipline Connection

20. ***Biology and History*** The developed nations have urged developing nations to take better care of their resources. Developing nations have responded that they are just doing what the developed nations did in the past. How valid is this claim? Do a little research to find out.

Discovery Through Reading

Read about one woman's fight to clean up air pollution from a waste incinerator in "When Citizens Slapp Back," by Jessica Speart, *National Wildlife,* June–July 1994, pp. 12–15.

PORTFOLIO ACTIVITY

Write a proposal for the creation of a new governing body that would have the responsibility of monitoring and enforcing a uniform environmental policy for all of the world's nations. Describe the agency's responsibilities and characteristics, and the skills needed by its officials to do the job. Describe also the methods the agency would use to monitor and enforce environmental policy.

CHAPTER 14

INVESTIGATION

BE AN ENVIRONMENTAL JOURNALIST

In almost all cases, governmental agencies take public opinion into account before passing environmental legislation. The debates are reported by the media in newspapers and magazines as well as on television and radio. In addition, concerned individuals and groups often try to win support for their side of the debate through advertisements of their own. Sometimes the truth gets stretched a little, but usually there are valid arguments on each side of an environmental issue. An environmental journalist investigates these issues thoroughly and reports on what he or she has learned.

Answers to this Investigation can be found on page T160.

MATERIALS

- back issues of newspapers and magazines
- tape
- scissors
- notebook
- pen or pencil

BACKGROUND

In this activity, you will find out what environmental legislation is currently being considered by your federal, state, county, or local government. After picking a topic that concerns you, you will research the progress of the pending legislation, taking care to identify biases in your sources of information. By investigating both sides of the issue, you will be able to write an educated and informative report on what you have found.

BE AN INVESTIGATIVE REPORTER

1. Look through several newspapers, magazines, and other publications to find out what environmental issues are being considered by legislators at various levels of government. If a computerized database is available at your school or local library, use it to quickly locate articles by topic or key word. When you find a relevant article, summarize it briefly in your notebook, stating what the issue is about, who is involved in the controversy, and where you read about it.

The general public depends on articles from magazines and newspapers to learn about environmental issues and legislation.

2. Choose an issue that is important to you, and spend two or three days researching it thoroughly. You may be able to read current newspaper and magazine articles to learn more, but you will probably have to visit your school or local library to find back issues for further information. The library may also have books or other publications that relate to the environmental issue you choose.

ANALYZE YOUR FINDINGS

3. When you find a source, answer the following questions about it:
   - Who is the author, and what type of source is it (magazine, pamphlet, etc.)?
   - Does the source present both sides of the issue? If not, which side is the author arguing for?
   - What are the main points that the source makes?
   - Does any of the information conflict with other sources that you have read?
   - What logical flaws do you find in the arguments people give?
   - What biases (either information omitted or information presented in a misleading way) do you find in the research sources? Explain how the way an issue is reported may affect public opinion about it.

Environmental legislation often dictates how land may be used, whether for commercial and residential development, for recreation, or as a wilderness refuge.

LET THE WORLD KNOW

4. Write a newspaper article about your issue. Include information about the people and organizations that are involved, the legislation, and the various opinions expressed by the public. Write your report as if it were a feature article in a newspaper, using headlines and, if possible, pictures with captions. At the end of your article you may present your own opinion, but in the rest of the article, you should try to be as unbiased as possible.
5. Let your classmates read your article. Discuss the issues with them, and answer any questions they may have. Are their opinions the same as yours?

EXPLORE FURTHER

6. The League of Conservation Voters tracks the voting record of members of Congress on all federal environmental legislation. If you are interested, you may contact the League to get a summary record of how your representative and senators voted on various environmental laws. If possible, you may want to telephone your representative and find out his or her opinions on specific environmental issues. Contact the League of Conservation Voters by calling 202-785-8683 or by writing to the League at 1707 L Street, NW, Suite 550, Washington, D.C. 20036.

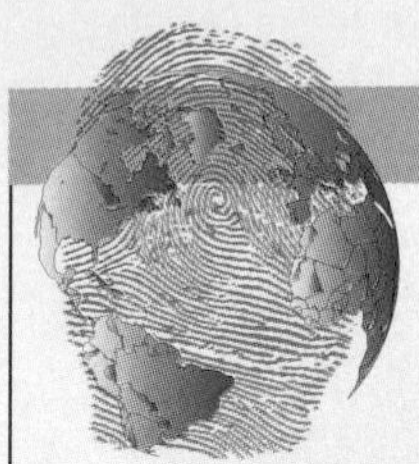

Making a DIFFERENCE

building toward THE FUTURE

To Michael Reynolds, a house is not just a home, and old tires and empty soda cans are not just trash. For 25 years, this Taos, New Mexico, architect has been designing and building energy-efficient houses out of automobile tires, cans, and other discarded items. These houses, which Michael now calls "Earthships," not only provide a comfortable, affordable place for people to live but also contribute to a sustainable future for our planet.

To Michael, an Earthship is not just a home—it's also a lifestyle.

In 1970 a television report about the growing number of beverage cans littering the streets and fields of America started Michael thinking about ways that trash could be used to build houses. Through many years of experimentation, he found that sturdy walls could be built by packing soil into old tires, stacking the tires like bricks, and covering them with cement or adobe, a heavy clay often used in buildings in the Southwest. Michael had this design tested by structural engineers to ensure that the walls would meet or surpass any existing building code requirements. One engineer even commented that the design could be used to construct dams!

The tire-stack design is used for three of the outside walls of an Earthship. These walls are approximately 1 m (3 ft.) thick, and this large mass causes the walls to act like a battery, storing energy from the sun and releasing it when needed. Also, the base of the Earthship is built below the frost line (the deepest level to which the ground freezes). Below this line, the ground maintains a constant temperature—around 15°C (59°F)—and walls anchored below the frost line

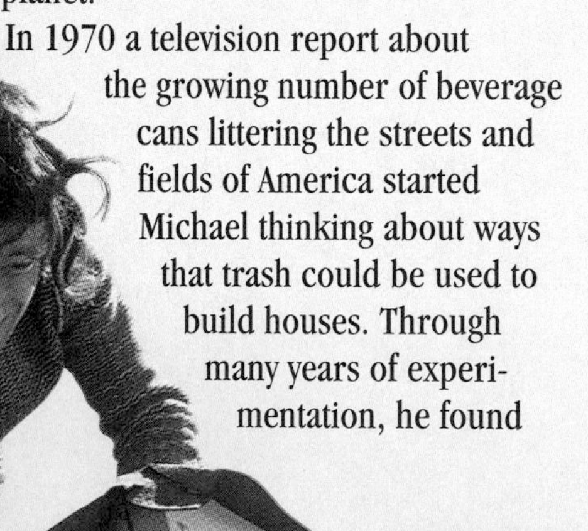

Michael Reynolds uses discarded materials, such as used soda cans, to construct environmentally friendly houses.

Earthships often look more like natural land formations than like houses.

usually stay at that temperature, too. The fourth wall, which faces south, is constructed completely of glass to capture as much sunlight as possible. In the winter, the tire-stack walls hold in the sun's warmth; in the summer, cool air enters through windows in the front while warm air escapes through a skylight in the back.

Even the soil that is excavated for the site is used to build the house. Some of the soil is pounded into the tires to construct the walls, and the remaining soil is piled against the outside of these walls and on top of the roof (constructed of beams) for further insulation. The most suitable location for this design is a south-facing slope, where the Earthship can simply be "carved" out of the hill. Often, Earthships look more like natural formations of land than houses.

Inside, walls between rooms are constructed by embedding empty beverage cans into mortar or mud. If these walls are covered with cement and then painted with latex paint or some other durable finish, they look just like walls constructed with conventional materials. Other inside surfaces, including stairs and even bathtubs, can be built using the beverage-can technique. Because the cans are so lightweight, this method can even be used to create dynamic interior structures such as arches and domes.

Earthships are typically built to obtain electricity from photovoltaic cells that convert sunlight to electricity. All household water is supplied by rainwater that is collected on the roof. Wastewater from sinks, tubs, and the laundry room is recycled to nourish plants in the greenhouse, which can provide a sustainable source of food. With these features, Earthship inhabitants use fewer of the Earth's resources and often have *no* utility bills.

Because Earthships are kinder to the environment and less expensive to buy and maintain, more and more people are choosing them instead of conventional homes. Over 100 Earthships have already been built, most of them in the Southwest. However, the design can be used anywhere. In fact, Earthships have been built in Florida, Vermont, Canada, Mexico, Bolivia, and Japan. Wetter environments simply require that the house be built entirely above the ground and use more cans and tires.

More tires in the design certainly wouldn't be a problem. According to the Environmental Protection Agency, more than 240 million tires are discarded in this country every year. But most landfills do not accept tires because of their tendency to rise to the surface even when the landfill is covered over. Tire dealers usually pay to have used tires hauled away to stockpile areas, where they sit indefinitely. Earthships provide one way to diminish the stockpiles.

A greenhouse, built along the Earthship's southern glass wall, can provide residents with a sustainable food source.

The tire-stack design of the outer walls accounts for much of the Earthship's energy efficiency. These tire stacks will be covered with cement or adobe for a finished exterior.

Michael enthusiastically shares his Earthship concept with others. He has published manuals with complete instructions for building an Earthship, and he leads seminars for prospective builders. Architecture students often study with him and learn his techniques. And he has established three self-sustaining communities of Earthships in the mountains near Taos. To Michael, an Earthship is not just a home—it's also a lifestyle. His dedication to designing Earth-friendly homes is a result of his commitment to "reducing the stress involved in living on the Earth, for both humans and the planet."

ENVIRONMENTAL FILMMAKER

Haroldo Castro considers himself a "citizen of the planet." It's easy to see why: he was born in Italy to a Brazilian father and a French mother, he was educated in France, he speaks five languages, and he has visited more than 80 countries. Furthermore, Haroldo has devoted his life to improving the planet's well-being. He has accomplished this by taking photographs, writing books and articles, and producing award-winning video documentaries. Haroldo works for Conservation International (CI), an environmental organization that establishes partnerships with countries all over the world to develop and implement ecosystem conservation projects.

The military thought we were terrorist guerrillas and surrounded us with machine guns. I think you might call that a crisis situation!

▶ What do you do at CI?

Haroldo: I am the International Communications Project Director. What I do is make documentaries and take photographs of CI's conservation projects. These videos and photos are designed to teach people how to better interact with their local environment. Most of our work is done in countries with tropical rain forests, such as those in Latin America, Asia, and Africa.

▶ Can you describe one of your documentaries?

Haroldo: Sure. We made a documentary in Guatemala about products that local people can sustainably harvest from the northern tropical forests.

After one year of production, we completed a half-hour documentary called *Between Two Futures.* CI then distributed the video to government officials, environmental organizations, university professors, and teachers. We also encouraged its broadcast on TV channels in Guatemala and other Latin American countries.

The film has been a real success story. I think our ability to be culturally sensitive to the Guatemalan people contributed in large part to the film's success. Each of us who worked on the project had a Latin American background.

Haroldo Castro (left) directing a video crew in Rio de Janeiro, Brazil

Haroldo Castro filming slash-and-burn agriculture (left)

We worked closely with the Guatemalan people; we had a Guatemalan narrator and we used only Guatemalan music. If you are trying to deliver an important message to people of a different culture, it's important to step into their shoes and deliver it from their point of view.

▶ What is your educational background and experience?

Haroldo: Although I do have a degree in economics, my best education and training has definitely come from traveling and other real-life experiences. I learn by studying the diverse cultures around the world.

Once I spent two years traveling around Latin America by van; another time I drove from Europe to India in six months. These experiences are my education. When my friends say that it is necessary to have a master's or doctor's degree to gain respect, I respond by saying that I have a Ph.D. in "Travelology." That's a degree I think my real-world experience on the road has earned me.

▶ Do you ever have to deal with crisis situations?

Haroldo: [laughter] If there is *not* a crisis when I'm traveling, I'm worried—it usually means there will be a disaster later! Anyone who travels a lot has to deal with crises, such as getting sick on local food or getting robbed. I've had equipment stolen from Lebanon to Peru!

I would like to tell you a story. Two years ago we were working in a remote rain-forest region of Mexico for 10 days. When we were ready to leave, we boarded a small plane and set out for the nearest commercial airport, only to learn that it had been closed. We were forced to go to a nearby military airport instead.

When we landed and began to unload our large boxes of equipment, the military personnel got very nervous. We looked pretty grungy—unshaven and covered with mud. It was obvious that we'd been in the rain forest awhile. They thought we were terrorist guerrillas and surrounded us with machine guns. For three hours we pleaded our case, and finally they let us go. I think you might call that a crisis situation!

▶ If a high school student were to ask you what he or she could do to help the environment, what would your answer be?

Haroldo: I would say . . . Learn all you can, appreciate the world around you, and follow your passion. If you like photography, go out and take pictures of things that leave you with good and bad impressions. If you like gardening, start experimenting with seedlings. Whatever your interest, my advice is, *just go for it!*

For Haroldo, capturing on film images like this Guatemalan girl holding a hummingbird allows him to recall rich travel experiences.

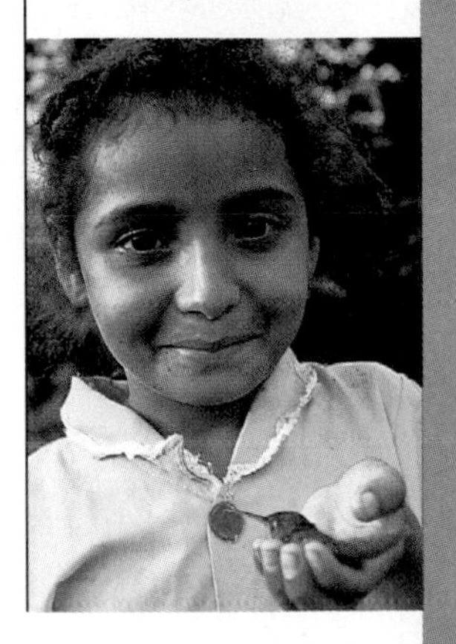

MORE ON THIS CAREER

Many government offices, publishers, and environmental organizations have in-house communications departments for producing films or photographs. Have a librarian help you make a list of such places, and then call them for more information and for possible volunteer or internship ideas.

While you're at the library, look through *The Guide to International Film and Video Festivals* for any mention of environmental film festivals in your area. Haroldo recommends attending a film festival if at all possible. "Doing so," he says, "would give you the invaluable opportunity to see some of the best films produced and to talk to the people who made them." If you can't find the guide or would like further information, contact the **Association of Independent Video Filmmakers** at 625 Broadway, 9th Floor, New York City, NY 10012.

RESEARCH WILDLIFE BIOLOGIST

Many people imagine wildlife biologists wrestling large game animals to the ground, slapping radio collars around their necks, and then creeping through the forest for weeks on end to study the creatures. According to Mariko Yamasaki, research wildlife biologist for the U.S. Department of Agriculture, Forest Service, there's a lot more to wildlife biology than that. To Mariko, "nature is fascinating on many, many levels, from the tiniest ant all the way up to charismatic animals such as bears and wolves. We have to get away from the notion that animals with feathers or fur and big brown eyes are more important than slimy, scaly creatures with beady eyes. All organisms have a role—we must be sure that their contribution to the big picture is recognized."

To Mariko Yamasaki, every creature, no matter how small or seemingly insignificant, has an important role in this biosphere. Below, she is searching for salamanders in the wild. Above, she and an associate are examining a salamander under a microscope.

▶ What is your educational background and experience?

Mariko: My background is basically a long and colorful stringing together of different experiences. I have bachelor's degrees in anthropology and zoology and a master's degree in natural resources (specific to wildlife). By the time I got out of school in the late 1970s, I came up against a surprising attitude—people in my home state really couldn't conceive of having female biologists supervising in the field. So I looked outside my home state. I ended up studying bald eagles for the Bureau of Land Management out West. This sort of snowballed into a permanent appointment in Washington as a wildlife biologist for the Bureau of Land Management. Today I work at the Northeastern Forest Experiment Station, where I do research in forested lands that cover a 200 mi. radius including parts of Maine and New Hampshire.

▶ What organisms are you studying in the field right now?

Mariko: I'm studying small mammals, such as mice, shrews, voles, and squirrels. My colleagues and I also study insectivorous bats, migratory birds, and terrestrial salamanders. These are animals that we know something about, such as their basic biology, but we don't know how they respond to forest management. We're looking at these critters to get a sense of how they fit into the bigger picture.

▶ What types of questions are you trying to answer about these animals?

Mariko: One question my colleagues and I are trying to answer right now is how terrestrial salamanders respond to "even-aged management" of northern hardwoods. Even-aged management involves harvesting a large area

We have to get away from the notion that animals with feathers or fur and big brown eyes are more important than slimy, scaly creatures with beady eyes.

of trees whose ages are within 20 years of each other. Foresters often use even-aged management because it is an efficient means of harvesting large amounts of timber at one time. My hypothesis is that when a large area of trees has been harvested, the ground temperature might change because the area is suddenly exposed to direct sunlight. This might affect the population and distribution of terrestrial salamanders in a negative way. I use the data I gather to make recommendations to forest managers about how they can manage tracts of forest to best support the needs of salamanders and other wildlife.

▶Do you work with other people a lot?

Mariko: There's an old stereotype that a wildlife biologist leads a solitary life studying nature. This simply isn't true—it's important to know how to work with people and how to understand and deal with a variety of viewpoints. There is rarely a day that I sit alone in my office. But I will say that a wildlife biologist does have some control over the matter—generally, you can work with people as much or as little as you want.

▶Do you ever have to deal with crisis situations?

Mariko: Not really, but I do see a lot of controversy, particularly related to wildlife and the use of natural resources. My work has often become the object of heated debate. Some people will support my findings wholeheartedly while others call them worthless. There are any number of ways of dealing with this kind of pressure. I've found that it's real important to get my information together and analyze it as thoroughly as possible so that I can really stand behind what I'm saying. It's also important to realize that everyone is entitled to an opinion.

Mariko wants to know if the way in which trees are harvested from forested areas like this one affects the survival of terrestrial salamanders.

▶What are the most interesting or exciting aspects of your work?

Mariko: Oh heavens! Being out and observing the natural world. Being able to test hypotheses. Being up real early on a bird survey. It's never the same twice. I also enjoy discovering something new—there's nothing any more special than that. There's a lot out there! The scale of things to observe and study is mind-boggling.

▶What advice might you give to someone who is searching for a career?

Mariko: I think it's important to do something you are really interested in. My career, just like anybody else's, is not always a bed of roses. But if you really care about what you do, you can get beyond the problems and complications inherent to any job. It's also important to think that you've got something to contribute. I think that I can help contribute to the way people view wildlife, and that's important to me.

This group of community leaders, politicians, and scientists is discussing how best to use the natural resources of a forested region in Maine.

MORE ON THIS CAREER

If you are interested in learning more about a career in wildlife biology, contact **The Wildlife Society,** 5410 Grosvenor Lane, Bethesda, MD 20814-2197, or the **American Institute of Biological Sciences,** Office of Career Service, 730 11th Street NW, Washington, D.C. 20001.

Landfill Manager

Do you think careers associated with waste are dirty, uninspiring, and of little value? After listening to James Bailey describe his work for a few minutes, it's obvious that this common impression is far from accurate. James manages two landfills in Atlanta, Georgia, and he and his team of employees are dedicated to protecting the area's environment and the quality of life for Atlanta's citizens.

I often tell students that what I do for a living is play in the dirt—which is properly disposed-of waste!

Above, James Bailey is doing management planning in his Atlanta office. Below, he is inspecting a landfill site to make sure that waste has been converted to a harmless dirtlike substance.

What does your daily work involve?

James: I often tell students that what I do for a living is play in the dirt—which is properly disposed-of waste! You see, it's my job to make sure that waste is converted into a harmless dirtlike substance and that it does not contain dangerous levels of pollutants. To do this I review a lot of reports, go out to the landfill sites and examine actual samples taken from the deposited waste, and make sure that the sites I manage do not get overloaded.

What inspired you to enter the field of waste management?

James: In August of 1965, I was a kid when riots broke out in the Watts neighborhood of Los Angeles, California. The damage and destruction caused by the riots really made an impression on me. I remember thinking that if people just had a nice place to live, such destruction would not be happening.

Since I really enjoyed building things at the time, I started to think about how great it would be to build a city. I remember thinking about how I could work with architects and other people to plan and build cities that would instill pride and inspiration in the people who lived there.

So how did those ideals lead to your current position?

James: Well, I never forgot that goal. After a tour of duty in the U.S. Army, I went to school at Clark-Atlanta University to get a degree in urban planning. Then I went to Georgia Institute of Technology to pursue a master's degree in city planning with an emphasis on industry and the environment. After that, I got my current job, and my career began.

Now I feel that I can help build a name for landfills—a name of respect and quality.

I'm also gaining a lot of firsthand knowledge about what goes into the development of a city. You see, it's impossible to manage the waste of a city without knowing what types of businesses and housing developments exist in the city, how they are distributed, and how they function. In addition, it takes a lot of our time and energy to manage the waste produced by a city and to look for ways to *minimize* it in order to maintain a balance with the environment.

▶Can you give us an example of a specific city-planning situation you've been faced with?

James: Sure! As you probably know, the Olympic games were held in Atlanta in 1996. In order to prepare for this huge event I had to consider many issues, such as: How many restroom facilities will be needed for the games? Where will they be needed? What will happen to the waste from those facilities each day? Who will be responsible for handling the waste materials? What kind of strains will be placed on our current system of waste management? I met with city planners and Olympic officials to discuss these questions, and together we developed an effective plan for handling these problems.

▶Have you encountered much controversy about whether landfilling is a safe method for disposing of waste?

James: Oh, yes! Many people, including government officials, politicians, business people, community organizers, and individuals, are concerned about the safety of landfills for individuals and for the environment as a whole.

James's work helps keep the city of Atlanta beautiful.

▶What do you do to reassure those who disapprove of landfilling?

James: I show them our facilities and demonstrate our elaborate system of safety checks and balances. By doing that, I can reassure people that our waste facilities are safe for local residents and the environment as a whole. You'd be surprised how many individual residents, school teachers, students, and politicians support our efforts. I've even gotten letters of approval from citizens and politicians who oppose the whole concept of landfilling waste. We've also had a few citizens bake cookies for us! And I'll never forget one letter I got from a third-grade student who had recently toured our facilities. The student wrote, "Thank you. I want to be an environmentalist when I grow up."

▶What personal qualities do you consider most important for a person in your career?

James: Technical aptitude, broad-based education, common sense, open-mindedness, and the ability to define options and consider better ways of doing things. In addition, the person must have an appreciation for the environment. Finally, the person must show a passion and dedication for his or her work. That passion should not be fueled by a desire for instant rewards, but instead driven by the knowledge that what he or she is doing is making a real, long-term difference in the world.

James helps reduce the fears and concerns of individuals by giving them tours of the landfills he manages.

MORE ON THIS CAREER

For more information on this career, contact one of the following organizations.

Environmental Protection Agency, 401 M Street SW, Washington, D.C. 20460

Air and Waste Management Association, P.O. Box 2861, Pittsburgh, PA 15230

FISH AND WILDLIFE TROOPER

Someone's been killing deer out of season in an Oregon forest. Diane Stout, trooper and law enforcement officer for the Oregon State Police, Fish and Wildlife Division, wants to catch the offender. So she sets up a decoy—an artificial deer—and hides in some nearby bushes. After waiting in silence for some time, she hears something. A large man with a 30.06-caliber rifle appears in the distance and slowly raises his gun toward the decoy. Diane watches silently. Suddenly, though, he becomes nervous. He jerks down the gun and takes off in a full sprint. Diane leaps out of the bushes and pursues. But as she nears the offender, he turns and aims his rifle at her. Suddenly Diane's life is in danger . . .

Diane Stout stakes out a poacher in this Oregon forest

I may find some offenders, but they're almost always quick to comply. But every once in a while I'll run into an angry person who will challenge me.

Is this sort of thrilling adventure something you encounter every day?

Diane: No, but I'm never really sure what's going to happen on any given day. Some days are really exciting. Other days I'll spend all my time patrolling and not have a single problem. I may find some offenders, but they're almost always quick to comply. But every once in a while I'll run into an angry person who will challenge me.

We have kind of a unique situation here in Oregon in that the Oregon State Police has a Fish and Wildlife law enforcement division. So I'm a state trooper who has the authority to enforce all Oregon state laws. However, my primary job responsibility is to enforce fish and wildlife and environmental protection laws. As a state trooper, I am also required to provide assistance in any kind of accident. This is a lot of responsibility—I have to be well-schooled in first aid, firearms, and emergency medicine, as well as many laws and regulations.

So what does your daily work involve?

Diane: Well, I begin each day by determining where to direct my time. I consider what hunting seasons are in progress and what problems we are likely to have from people who are hunting out of season or who are hunting incorrectly. I also consider any new or seasonal fishing regulations that are in effect, and any problems we may have with regard to those. In addition, I communicate with wildlife biologists who monitor game populations and habitat conditions. They let me know if they've seen any problems or have any concerns that I could check out. Also, we sometimes get complaints from residents about people who are poaching, dumping hazardous substances, or harming the environment in some other way. So, because my priorities are always different, each day (or night, if I do a night patrol) is unique.

▶ What is the connection between your job and the environment?
Diane: The areas I patrol are full of valuable natural resources—trees, lakes, rivers, wildlife. People use these resources in a variety of ways. Whether the main interest is fishing or camping or hunting or boating or whatever, I want those resources available for people to enjoy for a long time. So my job is to help keep the areas safe and clean, and the ecosystems healthy.

▶ Does seeing the "bad side" of people all the time ever get to you?
Diane: Yes, of course. People litter and dump hazardous wastes and abuse all sorts of hunting and fishing regulations. Sometimes their offenses are shocking, and depressing! But by informing people of the laws and enforcing the laws, I think that I'm actually doing something to improve the situation. So I feel really, really good about what I do for a living. Besides, there are a lot of terrific people out there too, so it's not just the bad ones that I run into.

▶ What are the most frustrating aspects of your job?
Diane: Limited funding. There are only four officers patrolling two counties. I have Polk County, Oregon, which includes a 60 mi. radius. It's a big area and I often wish I could be in several places at once. We definitely need more enforcement, but the funds just aren't available.

This is especially true in situations that require investigation. For example, when I

Part of Diane's job involves informing people of new fishing and hunting regulations.

find barrels of chemicals dumped along a rural river, it's seldom obvious who left them there. So I have to conduct an investigation that may involve court orders and search warrants and trials. Undertaking an investigation of this sort takes a lot of time. Knowing that you don't have the time to really do it right is frustrating.

▶ What is your educational background and experience?
Diane: I have three years of college in industrial drafting, one year of training as a medical assistant, and five years of police reserve experience.

▶ Do you work with other people?
Diane: No, not on a regular basis. If there's a case to investigate or prosecute, I may work with a lawyer or another officer. But you do have to like people to be effective in this job because so much time is spent talking to people about what they're doing!

MORE ON THIS CAREER

If you are interested in becoming a law enforcement officer in environmental protection, contact the United States Department of the Interior, **U.S. Fish and Wildlife Service,** Department of Labor, Washington, D.C. 20240; the United States Department of the Interior, **National Park Service,** Department of Labor, Washington, D.C. 20240; or the **American Park Rangers Association,** 3801 Biscayne Boulevard, Miami, FL 33137.

ENVIRONMENTAL LAWYER

Jana L. Walker used to be a nurse, but now she's showing her concern for individuals and their safety in a different way. She owns her own law practice that focuses on environmental protection and Native American issues.

Jana is a member of the Cherokee nation. She supports the "Great Law" of the Six Nations Iroquois Confederacy: "In our every deliberation, we must consider the impact of our decisions on the next seven generations." According to Jana, the Great Law is particularly relevant to environmental issues. This is because, she says, it reflects the need to establish laws to protect our natural resources for future generations *now,* before lands and waters are permanently damaged and species are driven to extinction.

Jana L. Walker, shown here next to the Río Grande in New Mexico, is working to improve the quality of river water running through tribal lands.

▶ What inspired you to change from nursing to law?

Jana: Well, I'd always been interested in law, but I guess I never thought I'd be able to do it. But after seven years of nursing, I was really ready for a more independent career. So I made getting through law school my goal. Now I know that it's never too late to get additional education or to fulfill a personal goal.

After law school, I worked at a couple of different law firms—one large and one small. Then I decided that what I really wanted to do was practice law on my own. So I started a solo law practice to focus on Indian and environmental law issues. These issues are very important to me as an inhabitant of the planet, as an attorney, and as an Indian.

▶ What is the relationship between environmental law and the Indian nations?

Jana: Well, tribal lands have suffered from many environmental problems. You see, although the first federal environmental laws were enacted several decades ago, those laws did not address Indian tribes and reservations. And the tribes lacked the money to start these programs on their own. As a result, there are now over 53 million acres of tribal lands that have had little or no environmental protection for many years. So the environmental movement that took off in other parts of the country during the 1970s is only now reaching many Indian lands.

▶ What kinds of environmental problems do you encounter?

Jana: The problems range from leaking underground storage tanks to acid rain to radioactive contamination to water pollution to illegal, or "wildcat," dumping of trash.

▶ Those are tough problems. What can you do about them?

Jana: I help tribes set up regulatory programs to protect the wildlife, land, air, and water resources of the reservations. I also review environmental bills that could affect Indian lands to determine whether they would have a positive or negative impact. Then I lobby for those bills that would help tribal programs. It's an awful lot of reading and writing—definitely not what a television lawyer does!

The environmental movement that took off in other parts of the country during the 1970s is only now reaching many Indian lands.

The work Jana does in her New Mexico office/home is improving the quality of life for many tribal people.

▶ What are the most frustrating aspects of your work?

Jana: It's frustrating to see a tribe begin to move forward with environmental regulation and then have its efforts challenged by a neighboring community. For example, I know of a case in which the tribe wanted to establish water quality standards for a large river that ran through its reservation. The river is listed as one of the 10 most endangered rivers in America because of severe pollution. The Environmental Protection Agency approved of the tribal standards. Then officials for a large city upriver learned that the new standards would limit their use of the river for municipal waste discharge. So the city planners disputed the EPA's approval of the new standards. As a result, the improvements in water quality are again delayed for the tribal people as well as for other communities downstream from this city. This is frustrating! Persistence and the ability to cooperate with government authorities are necessary tools in such a situation.

▶ What personal qualities do you think are most important in your field?

Jana: Determination and self-motivation are musts. It's a long haul getting a law degree. Then, once you're a lawyer, the law is constantly changing. That means you must be willing to continue to learn and to study these changes. Creativity is also essential because many times a law may not directly address your client's problem or need. As a result, you often have to weave together several legal theories to address a particular situation.

▶ What message would you like to send to high school students today?

Jana: I'd like to emphasize that protecting the Earth is everybody's job. But before we can tackle the work, we must become aware of the environment and how we fit into this world. And often that's not something you can learn from a book. It's only after we become truly conscious of nature and the environment that we can begin to see how our actions affect it and what steps must be taken to protect the Earth. So my advice is, go out and enjoy the natural world, and develop a real appreciation for it!

Jana hopes that actions taken because of her work and the work of others will increase awareness about, and help solve, serious environmental problems such as water pollution.

MORE ON THIS CAREER

If you're interested in learning more about a lawyer's work, check with your high school guidance counselor. You may be able to get a part-time or summer job in a law office. Or contact the **American Bar Association**, Law Student Division, 750 North Lake Shore Drive, Chicago, IL 60611, (312) 988-5264 for more information.

Environmental Educator

As a child who watched *The Undersea World of Jacques Cousteau* on public television every chance she got, Niki Espy dreamed of one day studying aquatic mammals for a living. She went to college with the intent of continuing on to graduate school to do behavioral studies in marine biology. But while pursuing a bachelor's degree in biology, she got a job interning as a naturalist.

Seven years later, Niki is still a park naturalist and environmental educator. She works at the Wehr Nature Center, which is a Milwaukee County Park in Franklin, Wisconsin. Niki enjoys it so much that her dream of becoming a marine biologist is—perhaps temporarily, perhaps permanently—on hold. According to Niki, "What hooked me is that I can teach, do research, learn, work with kids and animals, be outdoors a lot, and talk. These are all of my favorite things!"

▶ What does your daily work involve?

Niki: It varies so much—I'll just tell you about some of the stuff I do. First of all, the Wehr Nature Center is a beautiful 220-acre facility with prairie, oak savanna, woodland, and wetland ecosystems—including a terrific lake. What I do is educate people about these resources. I conduct hikes and give nature talks to students from local schools; I tell preschoolers stories and lead them through activities; and I give workshops for teachers and community leaders.

Each day I talk to people who visit the nature center. I show them the animals that we keep here. Sometimes I'll have a conversation with a visitor to explain, for example, "why that snake isn't nasty!" I also help take care of the animals, which means doing yucky stuff like cleaning out cages, and doing interesting stuff like helping animals that are injured or sick.

I also plan public events such as Earth Day activities and assist with Halloween nature walks. And I have several ongoing projects that I work on when I have time, such as adding to and maintaining the center's insect collection and establishing a library of multicultural resources for the center.

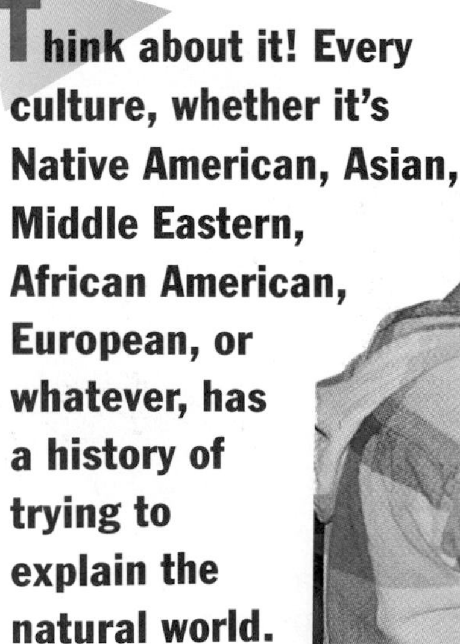

Think about it! Every culture, whether it's Native American, Asian, Middle Eastern, African American, European, or whatever, has a history of trying to explain the natural world.

Here Niki Espy (second from left) tells a group of students about "Foxy" the fox snake. She hopes her work will help eliminate misconceptions and increase respect for wildlife.

▶ How does your work relate to environmental issues?

Niki: Many of our programs have the general goal of helping visitors become more appreciative of natural resources. And some programs are related to a specific environmental topic, such as recycling. On Earth Day, for instance, I had a group of toddlers make some recycled paper, create an Earth Day greeting card out of it, and then send it to President Clinton. We just got a letter back from him—it's pretty cool!

I also helped develop a biodiversity studies program. It involves taking older students out to examine one of the natural ecosystems here. The students measure the physical and biological characteristics of the ecosystem and look for problems such as pollution. We then go back to the center, tabulate the information, and think about plans for improving any problem situations.

▶ You mentioned a multicultural library. What are your goals in terms of multicultural education?

Niki: I guess the way I see it is that the environment links together all cultures. Think about it! Every culture, whether it's Native American, Asian, Middle Eastern, African American, European, or whatever, has a history of trying to explain the natural world. Their stories, poems, and traditions show a variety of viewpoints that stem from different circumstances. I think I can help people appreciate the variety of cultures in our diverse society today by sharing those stories.

▶ Do you think that all cultural groups share a concern for environmental issues?

Niki: Yes, I think everyone cares. But I also think that environmental issues often take a back seat to more pressing social issues such as crime, unemployment, soaring school dropout rates, and drug abuse. So the attitude may be, "Yeah, this environmentalism stuff is kinda neat, but who can worry about that right now?"

▶ As an environmental educator, what can you do about this attitude?

Niki: What I try to do is sneak in there and teach people something about our natural world without their knowing that they're being taught. I try to spark their interest. That way maybe I can change their perspective and open their eyes a little bit.

▶ What is your most memorable work experience?

Niki: One experience I can think of actually happened before I started working here, but it really affected me. It was when I went with a class to look for eagles. We'd gotten up at 5:30 A.M. and headed out to a nearby river. By the time we got there we were really excited about seeing the eagles. An hour later, when we still hadn't seen any, we were beginning to wonder why we'd gotten up so early. We started piling back into the van to return to the school when someone suddenly yelled "Here they come!" To see a bald eagle—that was just, like, WOW! We were closer than half a mile from the eagles. They're so big and so grand. Even as they were flying in you could just see their huge, white feathers. It was *humbling.* You need to have that kind of thing to humble you.

I've also seen the birth of triplet fawns in the wild. Whooo, that was neat. Sometimes you can go weeks without having any remarkable experiences because you may be chained to your desk doing paperwork or something, and then there are other times when it's just constant awe. It's great!

For more information about a career as an environmental educator or as a naturalist, contact the **Environmental Careers Organization,** 68 Harrison Avenue, 5th Floor, Boston, MA 02111; or contact your state's **Department of Natural Resources.**

CLIMATE RESEARCHER

One summer when Dr. Richard Somerville was just a child, he built a weather station in his backyard. His creation grew out of a fascination for the great power of weather—a phenomenon that affects everyone every day. So with instruments made out of coffee cans, balloons, and rubber bands, Richard began keeping track of daily weather conditions and questioning how the world's weather systems worked. As time went on, he began to question more than just the weather—he looked at clouds, oceans, and the world of living things as well. These pursuits led Richard to the prestigious Scripps Institution of Oceanography. Today he is a professor of meteorology and the director of the climate research division of the Scripps Institution, which is part of the University of California at San Diego.

What exactly is meteorology?
Richard: Simply put, it is the science of the atmosphere—especially the study of weather and weather forecasting.

What most appeals to you about your job?
Richard: Probably the most exciting aspect of any scientist's work involves those few, rare "Eureka!" moments when you realize that you've discovered something that no one else on Earth knows about. That's quite a feeling. It's also rewarding to know that you're adding to the knowledge of others, transferring important pieces of information to important people who can use that information to improve this world.

What does your research involve?
Richard: Well, I do research on the greenhouse effect, on climate changes in general, and on the effects of long-range climate changes. I also study El Niño events and Indian monsoons. I see how these events and phenomena affect people—such as people involved in agriculture. The climate *really* affects the way people live!

I'm also researching whether the activities of humans are affecting the atmosphere. For

There are two general classes of technology that are most important to my work: satellites and computers.

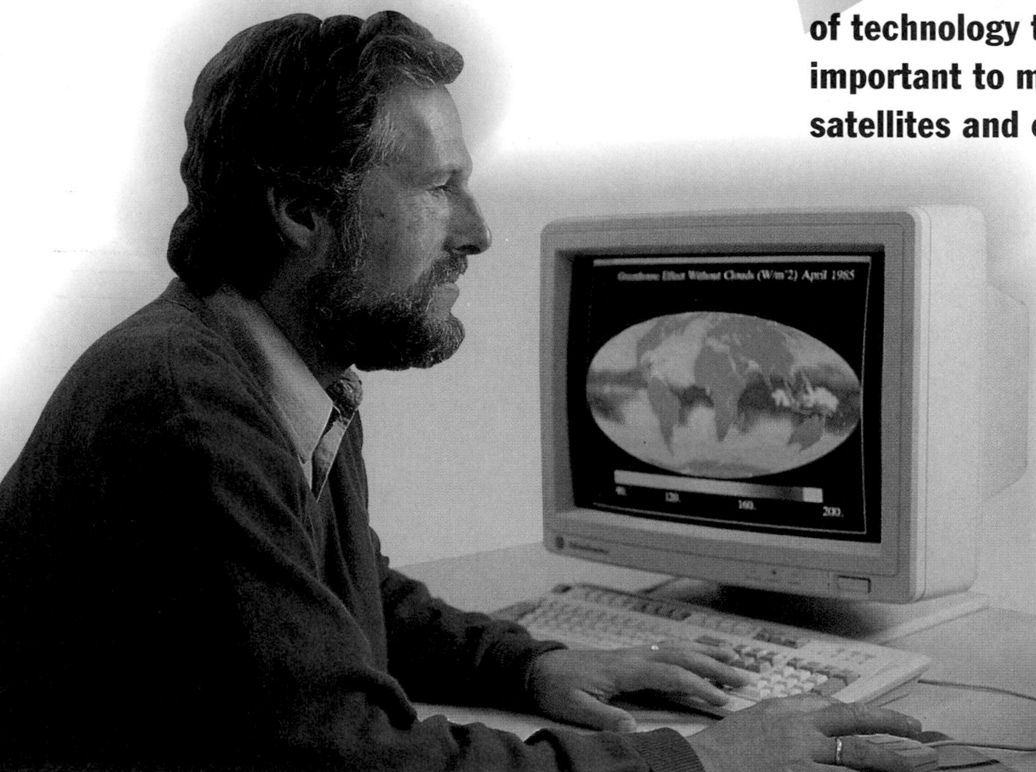

This computer simulation of increasing global temperatures allows Richard to study the possible repercussions of a changing climate on our planet.

example, 5 billion people use more and more energy by burning coal, oil, natural gas, and wood each year. When all of these substances are burned, they add carbon dioxide to the atmosphere. So I study the atmosphere to see how much the added carbon dioxide is intensifying the greenhouse effect. Then I try to determine how those changes will affect humans. You see, the more we know about the atmosphere, the better we can predict what will happen next.

▶ How is your research data used?
Richard: Many of my findings can affect public policy. For instance, How should the energy of the world be generated? I can help policymakers explore this question by providing them with data about the effects of fuels such as coal, oil, and gas on the atmosphere. Then I can recommend that they establish policies to reduce human reliance on those fuel sources. I can also encourage the use of resources such as solar, wind, and hydroelectric power.

▶ What tools do you use to obtain your data?
Richard: There are two general classes of technology that are most important to my work: satellites and computers. Together these two items have virtually revolutionized this field by hugely expanding what we've been able to observe and understand. Satellites, for example, can provide us with a whole different perspective of our world. The photographs generated by a satellite allow us to look at global temperatures as well as specific weather and sea conditions. Data are also collected on clouds, soil, and vegetation. By analyzing these observations we can monitor changing conditions and identify possible problem areas.

Computers help us make sense of the data. Computer equipment in the satellites helps to answer our questions and helps us to better visualize the data. Personal computers help us record and summarize our findings. Then we have "super computers," which can simulate the motions of the atmosphere and the ocean, and thereby help us to answer questions and make predictions.

We also have access to ships and airplanes that are loaded with highly specialized equipment. These research platforms can be sent to specific areas of the world to gather more information about a situation or condition.

▶ What are the most frustrating aspects of your job?
Richard: Other demands that limit the time I spend doing research. There's a large fraction of time and energy that must be spent making research possible—you have to find money, so you spend lots of time writing proposals and doing other administrative work.

▶ What school subjects turned out to be the most important for your career?
Richard: You might be surprised. Math and science classes are essential, but in retrospect I value my English courses the most. Scientists are writers—the final products of their research are shown in published papers.

▶ What personal qualities do you think are most essential for a successful person in your field?
Richard: There is an enormous variety of scientists—some are sloppy, some are organized, some like to work alone, some in teams. One thing all good scientists have in common, though, is dedication—they all want to do science above anything else. I think Thomas Edison's famous quotation, "Genius is 1 percent inspiration and 99 percent perspiration," is really on the mark. Not everyone can be born a genius, but anyone who is really dedicated can have a good career in science.

This scientist uses state-of-the-art equipment to gather information about changes in ocean temperatures over time.

MORE ON THIS CAREER

If you are interested in learning more about a career in meteorology, contact the **American Meteorological Society,** 45 Beacon Street, Boston, MA 02108.

ANY JOB CAN BE ENVIRONMENTAL

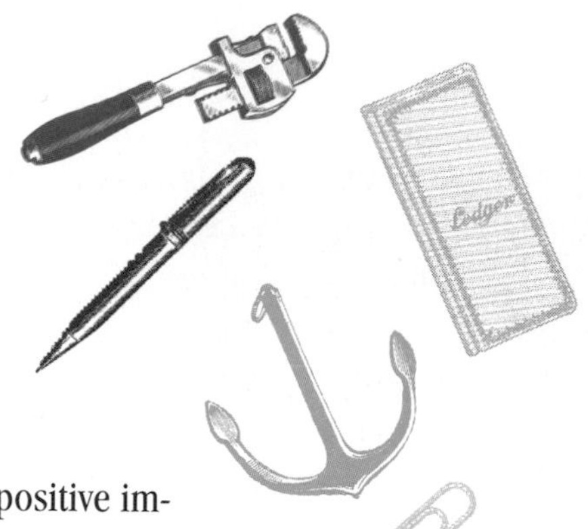

You don't have to be in an environmentally oriented field to make a positive impact on the environment. The following four people are excellent examples of how you can make a difference through your career even if it's not your job to do so.

Gun Denhart is the cofounder and chief executive officer of Hanna Andersson in Portland, Oregon. Hanna Andersson is a company that specializes in selling baby clothing and children's clothing through a mail-order catalog service.

As a parent, Gun knows that children outgrow their clothing very quickly. So she instituted a program called "Hanna-downs." The program allows customers to send outgrown clothes that were originally purchased from Hanna Andersson back to the company for recycling. Participating customers receive credit toward a purchase of new clothing that is equal to 20 percent of what they originally spent on the returned items. Hanna Andersson then donates the clothing to children in need. Over the years, Hanna Andersson's customers have sent back over 240,000 pieces of clothing, and Hanna Andersson has issued more than $1 million in credits. These clothes could have ended up in a landfill but have instead clothed kids all over the world.

This clothing will be recycled thanks to an innovative program designed by Gun Denhart.

How many natural resources does it take to run a luxury hotel? *A lot,* according to **Lewis Ware.** As director of housekeeping for the Boston Park Plaza Hotel & Towers in Boston, Massachusetts, Lewis has spent the last several years implementing an aggressive natural resource conservation program for his department.

Lewis's program began several years ago when the hotel's owners asked the company's managers to integrate environmental action into their daily business operations. Lewis was quick to respond—he had conserved energy, water, and other resources at home for years and was thrilled to learn that he now had full support to implement such programs at work.

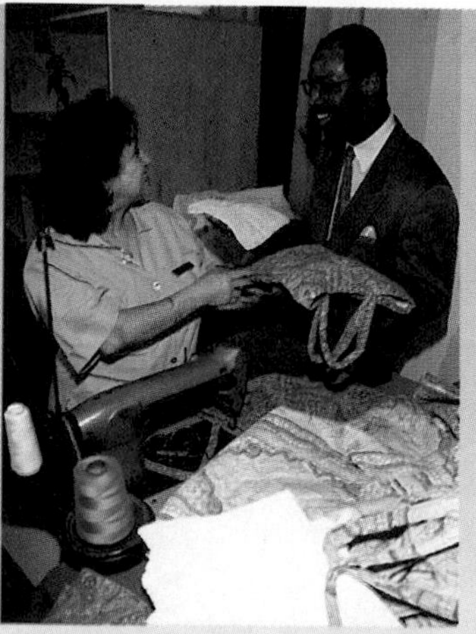
One of Lewis's innovative programs includes turning old table linens into chefs' aprons and taking worn bedspreads, pillows, and extra soap to homeless shelters.

Lewis has implemented many positive changes in the housekeeping department. Some of them include replacing harmful chemical cleaners with less harmful ones, replacing 2 million 1 oz. plastic shampoo and conditioner containers with elegant refillable ones, purchasing soap and shampoo products made from all-natural ingredients, and setting up a program to recycle in-room wastes such as bottles, cans, newspapers, and telephone books.

Thanks to the efforts of dedicated individuals like Lewis Ware, the Boston Park Plaza Hotel has earned much national recognition. This recognition includes the prestigious President's Environment and Conservation Challenge Gold Medal for 1992.

Jeff Hahn began his career as a legislative analyst for the Austin, Texas, office of Motorola Corporation. His job was to make sure that the laws passed in the state of Texas did not interfere with the company's ability to conduct business. Motorola is a company that manufactures electronic devices such as computer chips, cellular telephones, digital pagers, and two-way radios.

Jeff Hahn teaches local school children about their environment

Manufacturing computer chips requires the use of a number of hazardous chemicals that must be properly handled and disposed of after use. Jeff knew the public was concerned about the way companies manage the chemicals they use during manufacturing, so he began to look for ways to tell people how Motorola takes care of its waste. Soon he had taken this task one step further by telling people in the community how they could use the same methods to manage hazardous chemicals in their own homes.

Today, Jeff spends most of his time creating programs to educate Motorola employees and members of the community about how to improve the local environment. For example, Jeff wrote the *Discovery Pack,* a book filled with experiments and projects that teach kids how they can help protect the environment. The book has already been distributed to all Austin elementary schools, and plans are underway to deliver the *Discovery Pack* to other Motorola facilities around the nation and the world.

Jeff says he never thought that he'd be so heavily involved in promoting the environment, but his career has simply "taken off" into environmental issues, and he couldn't be happier about it!

Debbie Aguirre is the president of Tierra Pacifica Corporation, a construction and real-estate services company in Irvine, California. Debbie has been in the construction business for nearly 20 years. In that time, she's noticed that there are often excessive amounts of waste and many non-energy-efficient practices involved in construction projects. About 10 years ago Debbie decided that construction didn't have to be so wasteful, and she began thinking of ways to cut back on construction waste.

Today, Debbie's company is located in an energy-efficient facility—complete with a recycling and paper-reduction program. But perhaps even more significant is Debbie's corporate waste-management program. This program involves meeting with clients before starting construction to discuss how Tierra Pacifica can incorporate recycled products into the project, reduce construction waste, and make the construction process as energy efficient as possible.

Many of the materials used at this construction site are made from recycled materials, thanks to Debbie Aguirre's innovative thinking.

For example, Tierra Pacifica recently used several recycled materials to construct a new parking facility next to a local hospital. The materials included a type of asphalt base made with recycled ground glass; wheel stops and benches were made from recycled plastic.

According to Debbie, "Construction is a field not known for its environmental sensitivity. We would like to become a model for other construction companies so that they might take a more active stance in improving their relationship with the environment."

EcoSkills

What do bat houses, toilets, and shopping bags have in common? Well, not much unless you are developing skills for improving your relationship with the environment. This EcoSkills section contains plans, tips, and information for doing such things as building a home for bats, planting a tree, making a water-saving device for a toilet, and changing your shopping habits. All of the projects are inexpensive and fun to do. And they all help reduce your impact on the environment or help improve your natural surroundings, not to mention the fact that several of the projects can actually save your household money. So, roll up your sleeves and develop your EcoSkills!

Contents

BOOSTING YOUR HOME'S

ENERGY EFFICIENCY

Many people don't realize the impact that energy production has on the environment.

No matter what kind of energy plant serves your area, the production of that energy carries with it certain environmental risks. For example, when we burn coal to create electricity, many pollutants are released into the air—substances that may cause environmental problems such as global warming and acid rain. The more energy each of us uses, the more we contribute to these problems. So it makes environmental sense to conserve energy. Conservation is also a good way to save money—just a few energy-saving measures can substantially lower an energy bill.

Could the energy efficiency of your home be improved? Perform the following energy audit to find out.

The Wind Test

One day when it's windy outside, fasten a sheet of tissue paper onto a hanger with a piece of tape as shown below. Next, hold the hanger in front of a window at the point where the window meets the wall. Hold the hanger still. If the paper moves, you've found a draft. Note the location of the draft in your EcoLog. Check all around the window, making comments about the drafts you find. Then examine all of the other windows, doors, electrical outlets, plumbing pipes, and baseboards that are on the outer walls of your home. Note every place that the tissue moves.

These drafts of air that you've discovered can add 20 to 35 percent to your heating and cooling bills. Fortunately, you can seal these air leaks with weatherstripping and caulk. Weatherstripping is for moving parts such as doors and window frames. Caulk is for sealing cracks along joints and edges. These materials are relatively inexpensive, can be found at any hardware store, and can save 7 to 20 percent on your heating and cooling bills.

This simple device could help you improve the energy efficiency of your home.

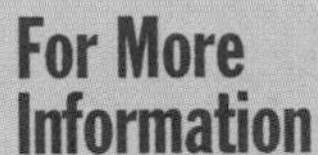

For More Information

Your local electric company can probably send you a packet of energy- and cost-saving ideas. In addition, your city may sponsor thorough in-house energy audits, as well as rebates and loans for improving the energy efficiency of your home. Contact your city's electric utilities conservation department for more information.

Even the tidiest of homes can be bugged by insect pests. If this happens to your home, fight back—naturally!

Eliminating Pests NATURALLY

A huge cockroach is crawling across your floor. How will you get rid of it? Don't reach for an expensive store-bought chemical that could possibly contaminate the local water supply or even harm someone in your household. Instead, try a natural remedy!

Cockroaches Make a roach trap by putting honey in the bottom of a jar and setting it upright where the pests are most likely to visit. The sweet smell of the honey will lure roaches into the jar, but the stickiness of the substance will make it impossible for them to escape. You could also line the cracks where you think roaches are entering your home with bay leaves. The smell of bay leaves repels roaches. Prevent roaches from entering your home by keeping all food covered and stored and by cleaning dirty dishes. Seal cracks in walls, baseboards, and ducts with caulk so that roaches and other pests can't get in.

Ants Sealing cracks with caulk will also help keep ants out of your home. In the meantime, squeeze fresh lemon or lime juice into the holes or cracks. Then leave the peels where you've seen ants. Scatter mint around your shelves and cabinets, or pour a line of cream of tartar, red chili pepper, salt, paprika, dried peppermint, or talcum powder where ants enter your home. These substances either repel or kill the pests. Another effective

remedy for ridding your home of ants or cockroaches is to sprinkle a mixture of equal parts of boric acid and confectioners' sugar in dry areas where ants and cockroaches are found. The pests will eat the sugar and then die from the effects of the boric acid. **Caution:** *If ingested, boric acid is acutely toxic to pets and small children. Use boric acid only in areas that are out of reach of kids and pets.*

Ticks and Fleas If your pet has a problem with ticks or fleas, try feeding the animal brewer's yeast or vitamin B. Also wash your pet regularly with soap and water; then dry the animal and spray an herbal mixture of rosemary and water onto its coat. (You can make the mixture by steeping $\frac{1}{2}$ cup of fresh or dried rosemary in one quart of boiling water. Let the liquid cool, pour it into a pump bottle, and then spray it onto your animal's coat.)

You can control the ticks and fleas in your yard by sprinkling the grass with diatomaceous earth, which is available at many nurseries. Diatomaceous earth consists of tiny glasslike skeletons of diatoms (a type of single-celled algae). These organisms scratch the outer layer of an insect's body as it crawls along the ground. Bacteria then enter the insect's body through the open wounds, and the insect dies of disease. **Caution:** *Diatomaceous earth can be harmful to your lungs if inhaled. Wear a protective mask when spreading the substance.*

You can help reduce the number of ticks and fleas that bother your pet by bathing it frequently and spraying an herbal mixture on its coat.

For More Information

Your city's environmental and conservation services department (if you have one) may have some other remedies for pests and some recipes for nontoxic household cleaners. Also, check your local bookstore or library for books on natural pesticides, organic gardening, and chemical-free homes.

On your next few shopping trips, think about the products you choose. If you're like most Americans, you'll probably be amazed at how many wasteful shopping habits you have.

Environmental Shopping

Try to count how many products you've used today. It's probably not as easy as you think. In the first few minutes of your day, you may have used a dozen products.

Each of those products, and their packaging, is made from valuable resources. More often than not, once those resources are used, they're tossed in a trash can and eventually hauled to the local landfill.

You can cut back on the amount of waste you send to the landfill and conserve resources in the process. On your next few shopping trips, think about the products you choose. If you're like most Americans, you'll probably be amazed at how many wasteful shopping habits you have. But after a while you'll begin to know instinctively which products are best for you and the environment.

Your Personal Shopping Guide

Read the following information and think of a way to reproduce it so that you (and other members of your household) have it handy when you set out on a shopping trip. For example, you may want to copy the questions and answers on the side of a brown paper bag. That way you'll have a shopper's guide, and you'll need one less sack at the checkout stand. Another option is to write your guidelines on the back of an old grocery receipt and then adhere the receipt to the refrigerator with a magnet so that it will be handy for the other shoppers in your household. The options are limitless, so be creative and try to incorporate recycled items into your design.

Review Figure 12-18, shown on page 315. Is there anything from the list you could add to your personal shopping guide?

Do I really need this product? Can I use something I already have?	Borrow or rent products you don't use often.
Is this a "throwaway" item that is designed to be used once or twice and then thrown away?	Avoid using disposable products whenever possible. Nondisposable alternatives may be more expensive initially, but in the long run they often save you money.
Does this product have more packaging than it really needs?	Look for alternatives with less packaging or wrapping. Purchase products in bulk or in a larger size so that, in the long run, you use less packaging (and save money!). Buy fresh vegetables and fruit instead of frozen or canned products.
Was this product's container or packaging made with recycled materials?	Choose products that have recycled paper, aluminum, glass, plastic, or other recycled materials in their packaging.
Is this product's container or packaging made from cardboard, aluminum, glass, or another material that I can easily recycle?	Find out which materials you can conveniently recycle, and then buy those sorts of containers. Also, think of ways to reuse old containers rather than throwing them out.
Does this product have bleaches, dyes, or fragrances added to it? Does it contain phosphates? Is it made from a petroleum-based synthetic fabric, such as polyester?	Phosphates and many other chemicals can pollute water sources. Look for natural, organic, and phosphate-free alternatives. When purchasing clothing, choose cotton or wool over synthetic fabrics.
Does the company that makes this product have a good environmental record?	You may have to do a little research to answer this one. Try the references listed below.
Although this product has a "green" label, is it really good for me and the environment?	Don't be deceived by advertising and product labeling; carefully examine the contents of a product before you purchase it.
Do I really need a shopping bag to carry home the items I'm purchasing? If so, will I be more likely to recycle or reuse a plastic shopping bag or a paper one?	If you purchase just one or two items, tell the grocer that you don't need a bag to carry the products. For more items, bring old paper or plastic sacks with you when you go to the store, or use a canvas bag that will last through many trips.
How much energy do I spend getting to the store?	Ride your bike or walk to the store, if possible. If not, condense several short trips for one or two items into one longer trip for a bigger supply of items.

For More Information

The Green Consumer Supermarket Guide: Brand-Name Products That Don't Cost the Earth, by Joel Makower, with John Elkington and Julia Hailes. New York: Viking-Penguin Books, 1991.

The Green Supermarket Shopping Guide, by John F. Wasik. New York: Warner Books, 1993.

By making your own compost heap, you can reduce the amount of waste you send to the local landfill and create excellent natural fertilizer for your garden.

MAKING YOUR OWN COMPOST HEAP

Why on Earth would you want to pile a bunch of garbage in your yard and let it rot? Crazy as the idea may sound, it's actually a very good one—copied straight from nature itself.

Compost is the natural product of the Earth's organic decaying process. When a dead organism decomposes, it returns nutrients to the soil. A compost heap is a collection of organic materials such as leaves, grass, and fruit peelings that will decompose over time to create rich, fertile soil. By making your own compost heap, you can reduce the amount of waste you send to the local landfill and create excellent natural fertilizer for your garden.

There are many opinions on how to construct the best compost heap—it can be as basic or as fancy as you like. Either way, composting is easy, and it's almost impossible to foul up the process.

A compost heap can be placed just about anywhere in the yard. Either a sunny or a shady spot will be fine. You will want to keep it out of the way of normal activity, however.

Many people choose a spot on a concrete slab or a grassy area and then simply pile

This is an easy and effective way to make your own compost heap.

Anatomy of a Compost Heap

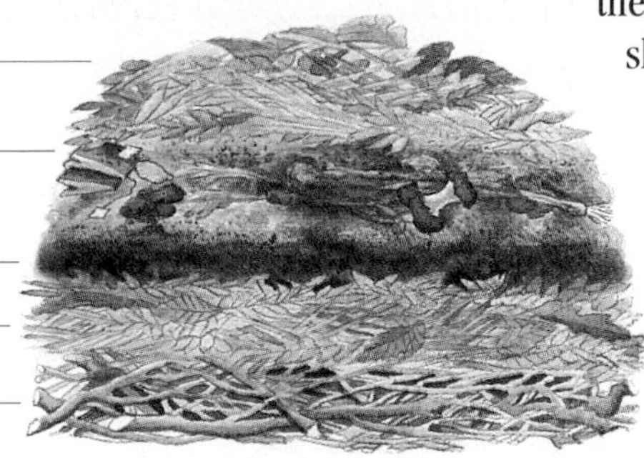

their materials there. See the photo shown on page 394. This method is easy and effective.

A compost heap contains a mishmash of many different organic materials. Most of your heap will probably consist of grass clippings and leaves. You can also add raw vegetables, other uncooked food scraps, coffee grounds, tea bags, cotton, dust, discarded plants, and weeds. Avoid adding pet manure, cooked foods, and meat of any kind. If you add raw food wastes, cover them with leaves to keep away flies and to prevent an unpleasant odor.

Your heap will begin to decompose through the action of microorganisms. It's a good idea to shovel a couple of scoops of soil from your yard into the heap. The microorganisms in the soil will immediately begin decomposing the items in the heap.

Turn the heap at least once a month to keep it well aerated and active. Once the organic matter has broken down to the point that no single item is recognizable, it's ready to work into your garden's soil. The entire process can take anywhere from two months to one year, depending on the kinds of materials being decomposed and how often the heap is turned. Composting is more of an art than a science, so be prepared to experiment!

Compost Container

If you choose to contain your compost pile, you will be able to add more materials to a smaller area. You can buy a ready-made container from a hardware store, or you can build one yourself.

If you decide to build one, you may wish to use metal stakes and chicken wire to create a container like the one shown at right. Keep in mind, however, that as long as the container allows air to get in and out, the type of container you choose is limited only by your imagination!

You can build this container for your compost heap with a few materials from your local hardware store.

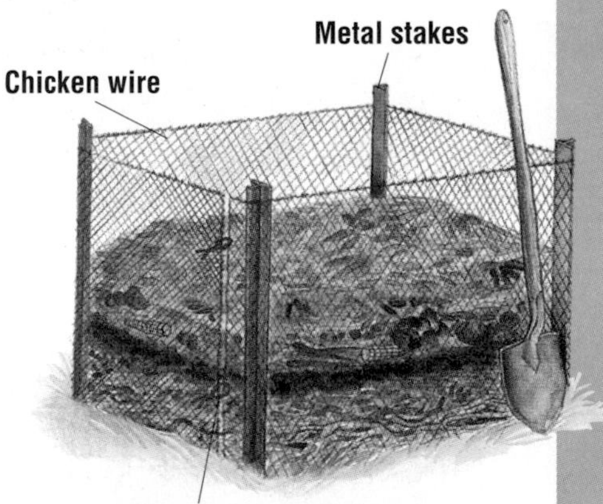

Loose wire can be twisted around two sections of chicken wire to create a "door" for easy turning.

For More Information

Consult your library for a manual on composting. You might find one of these helpful.

Let it Rot! The Home Gardener's Guide to Composting, by Stu Campbell. Pownal, VT: Storey Communications, 1990.

Rodale Book of Composting, edited by Grace Gershuny and Deborah L. Martin. Emmaus, PA: Rodale Press, 1992.

To attract wildlife to your home, you simply need to provide native plants and the sorts of water sources and shelters naturally available to the wildlife in your area.

Creating a Wildlife Garden

Manicured lawns and nonnative vegetation are not part of a natural ecosystem. Although these have been standard in urban and suburban neighborhoods for years, they usually require pesticides, fertilizers, water, and attention just to survive. In addition, they often exclude wildlife by removing some of their natural sources of food, water, and shelter.

To attract wildlife to your home, you simply need to provide native plants and the sorts of water sources and shelters naturally available to the wildlife in your area.

Plants

Plants are probably the most crucial element of your wildlife garden. Whether you have a lot of space for planting a wildflower meadow, a balcony on which you can create a container garden full of native plants, or a few windows to which you can attach boxes full of bright and cheery wildflowers, you will need a variety of native plants. Check with a local nursery, library, or bookstore for recommendations.

Water

People often overlook the need all animals have for water. Although some animals obtain enough water from the foods they eat, most require additional water for drinking and bathing.

Water sources are easy to provide. Many people purchase hanging or standing birdbaths from a nursery or hardware store. Others create ponds. You can make a simple pond by setting an old trash-can lid upside down in a corner of your yard and filling it with water. Surround your water source with plants, rocks, and other items so that the wildlife can find cover if necessary. In addition, make sure your pond or birdbath is at least partially shallow so that no animal is in danger of drowning, and keep the water clean.

Food and Shelter

Many different kinds of birdhouses and feeders are available at nature stores, hardware shops, and nurseries. Most of these can be hung on a balcony, and some can even be adhered to a window. Or, you could make your own birdhouse or feeder. A milk jug with a large hole cut in its side that is filled with seed and hung from a tree or balcony is an excellent way to feed many birds. If you would like to attract bats to your yard, see pages 400–403 for directions on how to make (or purchase) a bat house.

Woodpiles, rock piles, and brush piles are valuable sources of shelter for wildlife such as lizards and toads that might not frequent your backyard habitat otherwise. The most successful pile is one that incorporates different-sized spaces among the various components. You can make your pile attractive by planting vines in and around it.

Caution: *A shelter like the one described above may attract poisonous snakes. Find out if any live in your area, and if so, you may want to refrain from making a shelter pile.*

For More Information

Consult your library or bookstore for books on gardening with plants native to your area, gardening for the wildlife in your area, and Xeriscape techniques. You might find these books helpful.

Noah's Garden: Restoring the Ecology of Our Own Back Yards, by Sara Bonnett Stein. Boston: Houghton Mifflin, 1993.

Your Backyard Wildlife Garden: How to Attract and Identify Wildlife in Your Yard, by Marcus Schneck. Emmaus, PA: Rodale Press, 1992.

No matter where you live, you'll undoubtedly find a tree a welcome addition to your neighborhood.

Planting a Tree

In many towns and cities across America today, more trees are dying or being cut down than are being planted. If you live in an urban environment, you can help reverse this disturbing trend by planting a tree.

No matter where you live, you'll undoubtedly find the tree a welcome addition to your neighborhood. If you don't have a yard or can't plant a tree in your yard, consider getting a small tree that you could put in a pot on your patio or balcony. Or, you could participate in a community tree-planting project. Check with a library, nursery, or environmental organization to find out if there are any tree-planting projects in your area.

Finding the Right Tree

There are countless varieties of trees that you can choose from. The following guidelines should help you find the tree that best suits your needs.

Go Native Native trees usually don't require fertilizers, pesticides, or excessive watering. So choose a native tree, and save yourself considerable time, money, and effort.

Long-Living or Quick-Growing? Most people want a tree that will grow quickly to its full height. Unfortunately, many fast-growing trees have a long list of serious problems. As a result, they often do not live as long as slower-growing varieties. So choose a medium- to slow-growing tree, and look forward to watching it mature over the years. (Many slow-growing trees can live over 100 years!)

Follow the Sun Think about the effect the tree will have on its surroundings. For example, by planting a tree within 4.5 m (15 ft.) of your home's south or west side, the tree can shade the house in the summer. If the tree is deciduous (sheds its leaves in the fall), sunlight can filter through the branches in the winter and help keep your home warm.

Watch Those Roots and Branches Keep in mind that roots can seriously damage sidewalks, driveways, and sewage systems and that branches can damage shingles, windows, and house siding. Therefore, keep trees that will have large root systems or branches an appropriate distance from anything they might damage.

Cost Trees sold in plastic 1–5 gal. containers range from $3 on sale to $50 or more for a rare or nonnative species.

Seeking Advice The nursery is an obvious place to get advice about trees you could plant, but nearby college agricultural departments, university extension offices, U.S. Forest Service offices, and city government offices can also help you. Sometimes they even have trees for sale, often at their cost.

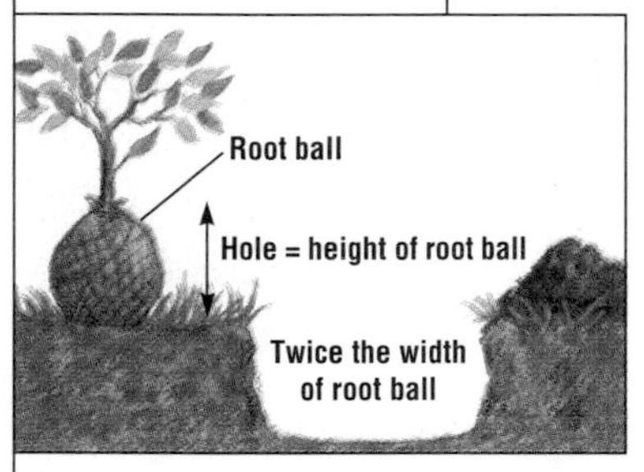

Planting Your Tree

The following directions will help you give your tree a good start.

1. Dig a square hole *exactly* the same depth as the root ball and approximately twice as wide. Use a ruler or measuring tape to make sure the hole is the right depth. The sides of the hole do not have to be smooth. In fact, it is better if they are rough and jagged, as shown in the diagram at left.

2. Now put the tree in the hole, and refill the hole with the soil you took out. You may want to mix it with some organic material such as bark mulch or finished compost *(well-decomposed* organic matter). Do not use pure manure, sand, bark, or peat moss. Compact the soil slightly, but not too firmly. Water the tree slowly and deeply to help the soil settle into the hole.

3. It's a good idea to cover the area around your newly planted tree with about 1 in. of compost and 3 in. of mulch (shredded hardwood bark, wood chips, or hay). The compost will serve as a natural fertilizer that gives the tree necessary nutrients, and the mulch will help the soil retain moisture.

Caring for Your Tree

The first couple of years of the tree's life are especially important. Don't just plant the tree and forget about it! Water the tree thoroughly when the top 3–4 in. of soil dry out or if the tree's leaves start to wilt.

For More Information

Global ReLeaf Forests: American Forests, P.O. Box 2000, Washington, D.C. 20013, (800) 368-5748

Tree People: 12601 Mulholland Drive, Beverly Hills, CA 90210-1332, (818) 753-4649 (Ask about their Campus Forester program and their book called *The Simple Act of Planting a Tree.*)

Give a bat a home and keep annoying insects, such as mosquitoes, at bay.

Building a Bat House

If mosquitoes bother you, you should love bats. A single little brown bat is capable of eating 600 mosquitoes in just one hour. Other bats pollinate flowers and disperse the seeds of the fruit they eat, helping many plants to reproduce. Bats are a vital part of nearly every ecosystem on Earth.

Despite the importance of bats, people have feared and persecuted them for centuries, incorrectly believing them to be vicious, bloodsucking creatures that attack humans and transmit diseases. The fact is, bats are gentle and useful creatures, and the chance of catching a disease from one is no greater than the chance of catching a disease from any other wild animal. If we simply avoid handling bats, we have nothing to fear.

Unfortunately, many bat species in the world today are threatened with extinction. Nearly 40 percent of those living in the United States are on the endangered list or are official candidates for it. You can do your part to help these unique creatures by building a bat house.

Bat houses come in many shapes and sizes. This activity provides plans for building the popular (white) house shown at right. For other options, see "For More Information" on page 403.

Getting Started

To build a bat house for 30 or more bats, collect the following materials. Because you will be working with sharp tools, exercise extreme caution.

Materials

- one 8' piece of 1" × 8" nontreated lumber (for the front and back pieces)*
- one 5' piece of 2" × 2" nontreated lumber (for the ceiling and sides)*
- one 5' piece of 1" × 4" nontreated lumber (for the roof and the posting board for mounting the bat house)*
- one piece of $15\frac{1}{2}$" × 21" fiberglass window screening (Do not use metal screening.)
- nails (approximately sixteen 8d galvanized nails and thirty 6d galvanized nails)
- safety goggles
- staple gun and staples, or approximately 16 upholstery tacks
- metal utility knife
- hammer
- saw
- measuring tape
- silicone caulk (may or may not be necessary)
- small can of exterior latex paint or varnish, light colored or clear if you live in a warm southern climate, dark if you live in a colder northern climate
- ladder (to mount the bat house)

* ***Note:** The lumber sizes given are the sizes you will need to ask for at a lumber company. The* actual meas-urements *of those boards are slightly smaller, however. The measurements given in the directions and drawings for this activity reflect the* actual measurements *of those boards. (For example, a 1" × 8" actually measures $\frac{3}{4}$" × $7\frac{3}{4}$".)*

1. Cut the 5' piece of 1" × 4" lumber into two pieces, one piece $16\frac{1}{2}$" long (roof) and the other 3' long (posting board).
   Caution: *Exercise extreme caution while sawing.*

2. Cut the 1" × 8" × 8' board into six pieces, each measuring $15\frac{3}{4}$" in length. Next, cut one of these pieces lengthwise to make a strip measuring 1" wide; then cut the strip to $12\frac{3}{4}$" long. Be as accurate as possible when sawing.

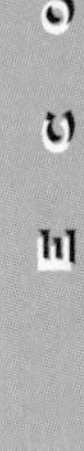

3. Using 6d nails, nail three of the pieces cut in Step 2 (note: all three should be identical in size) to the posting board as shown in Diagram A. Fit the three boards together as tightly as possible.
4. Using the staple gun or upholstery tacks, tightly secure the fiberglass screening to the three boards you assembled in Step 3. (Use the side without the posting board.)
5. Cut the 2" × 2" × 5' piece into two pieces measuring $21\frac{3}{4}$" long (sides) and one piece measuring $12\frac{3}{4}$" long (ceiling).
6. Construct the internal frame by nailing the sides, ceiling, and entrance restriction (this is the $12\frac{3}{4}$" piece you cut off of one of the six identical boards in Step 1) together with 8d nails, as shown in Diagram B.
7. Nail the back pieces to the internal frame, using 6d nails.
8. Using 6d nails, secure the front three pieces to the internal frame, allowing a gap of $\frac{1}{2}$" for ventilation between the bottom and middle pieces. (Note: the bottom piece is smaller than the other two.) See Diagram B.
9. Nail the roof to the top of the structure, allowing the excess to extend over the front of the bat house. See Diagram C.
10. Be sure the house is as draft-free as possible. If there are gaps between the pieces in the top two-thirds of the house, seal those spaces with silicone caulk. Gaps in the lower third of the house will allow ventilation for the bats.
11. Paint or varnish the exterior of the bat house.

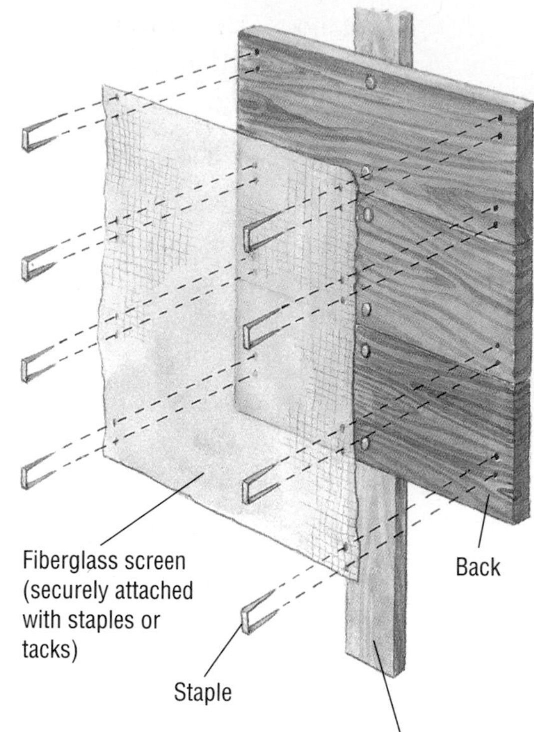

Diagram A

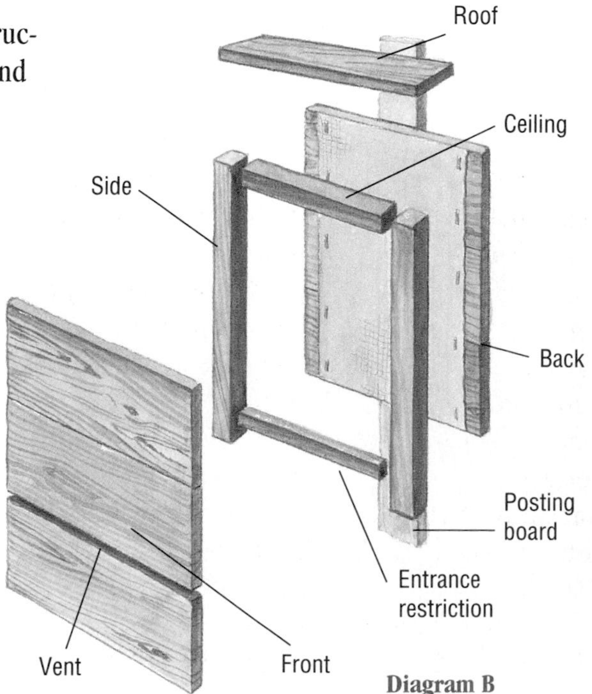

Diagram B

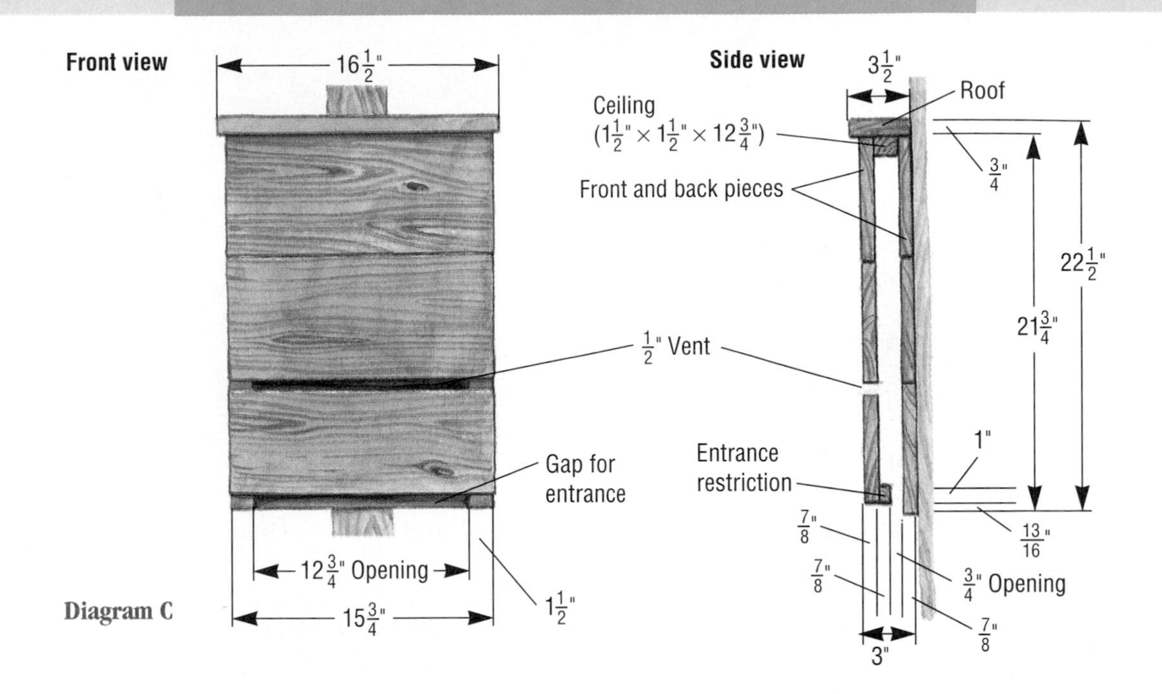

Diagram C

Mounting Your Bat House

Mount your bat house at least 12 ft. high (ideally, 15–20 ft. or more) on a sturdy pole or on the side of a building away from lights. You can also put the house in a tree as long as the tree's leaves do not shade the bat house too heavily in the summer. If possible, the bat house should be within a mile or so of a water source such as a lake, river, or creek. The amount of sunlight the bat house receives will influence whether bats will inhabit it. All bat houses should receive at least four hours of sunlight per day.

If you live in the southern United States, the bat house will need protection from the sun's heat. Position the house so that it receives sunlight in the morning but not in the afternoon. Paint the house white or light brown to reflect the sun's rays.

If you live in the northern part of the country, paint the bat house black or dark brown to retain the sun's heat. Position the house so that it will receive 4–6 hours of sunlight if it is black or 6–12 hours of sunlight if it is painted dark brown. If you live in the central United States you can paint the house any color, but make sure it receives at least 4 hours of sunlight.

Check your bat house regularly for signs of habitation. If the house is occupied, you can look at the bats with a flashlight, but only for brief periods. If bats don't immediately take up residence, don't get discouraged. It might take a while for them to find your house, especially in the winter (when many species hibernate). Be prepared to experiment with different colors of paint and different locations.

For More Information

For more information about how to attract bats to your house, how to build a more elaborate bat house, or to purchase a ready-made bat house, contact:

Bat Conservation International
P.O. Box 162603
Austin, Texas 78716
(512) 327-9721

FLUSHING LESS WATER

A typical American uses over 100 gal. of water before he or she even leaves for work or school in the morning, and much of that water is wasted.

Many Americans are beginning to change their wasteful practices, however. One simple and inexpensive way you can waste less water is by making a water displacement device for your toilet's tank. This device takes up space in the tank so that less water is required to fill it with every flush. It only takes about 10 minutes to make, and with it you can save 1–2 gal. of water every time you flush. This may not sound like much, but it adds up quickly. Most toilets use 5–7 gal. of water with every flush. If a toilet is flushed an average of eight times per day, it uses around 52 gal. of water per day, or 18,980 gal. per year. If you can save $1\frac{1}{2}$ gal. of water with every flush, you'll save 4,380 gal. of water each year. If just 250 other people take similar measures, over 1 million gal. of water could be saved each year.

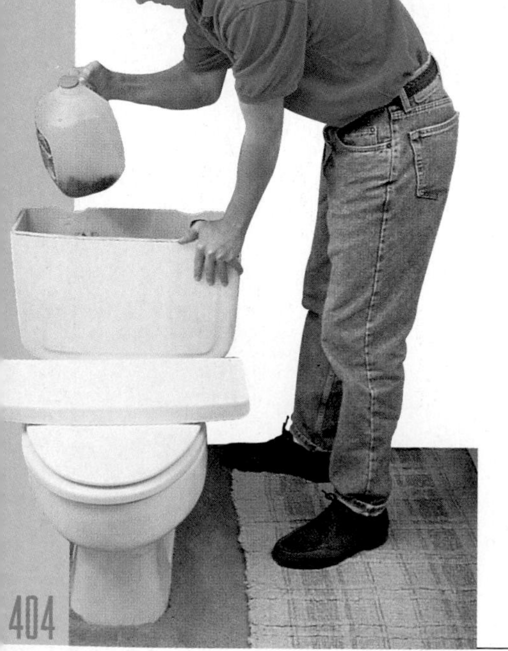

Making a Quick and Easy Water Displacer

1. Remove the label from a plastic container. (Milk jugs, juice bottles, and dishwashing soap bottles work well. Be prepared to experiment with different-sized containers.) Drop a few rocks into the container to weigh it down, fill the container with water, and put the lid back on.
2. Place the container in the toilet tank as shown at left.
3. Be certain that the container doesn't interfere with the flushing mechanism inside the tank.
4. Experiment with different containers. Your goal is to use the largest container that the tank will hold while still maintaining an effective flush.

One Final Important Note

The more water you save, the less you pay for. No matter which water-saving device you install, your water bill should be noticeably lower.

For More EcoSkills Projects . . .

The following books offer more suggestions for how you can work toward a better environment.

Embracing the Earth: Choices for Environmentally Sound Living, by D. Mark Harris. Chicago: Noble Press, 1990.

Environmental Vacations: Volunteer Projects to Save the Planet, 2nd ed., by Stephanie Ocko. Santa Fe, NM: John Muir Publications, 1992.

The Green Lifestyle Handbook, edited by Jeremy Rifkin. New York: Henry Holt and Company, 1990.

Heloise: Hints for a Healthy Planet, by Heloise. New York: Putnam Publishing Group, 1990.

How to Make the World a Better Place: A Beginner's Guide to Doing Good, by Jeffrey Hollender. New York: William Morrow and Company, 1990.

How to Save Your Neighborhood, City, or Town: The Sierra Club Guide to Community Organizing, by Maritza Pick. San Francisco: Sierra Club Books, 1993.

Our Earth, Ourselves, by Ruth Caplan and the staff of Environmental Action. New York: Bantam Books: 1990.

The following books, which are published by Earthworks Press of Berkeley, CA, may also be helpful.

50 Simple Things You Can Do to Save the Earth, 1989.

50 More Things You Can Do to Save the Earth, 1991.

The Recycler's Handbook, 1990.

The Student Environmental Action Guide, 1991.

Vote for the Earth: The League of Conservation Voters' Election Guide, 1992.

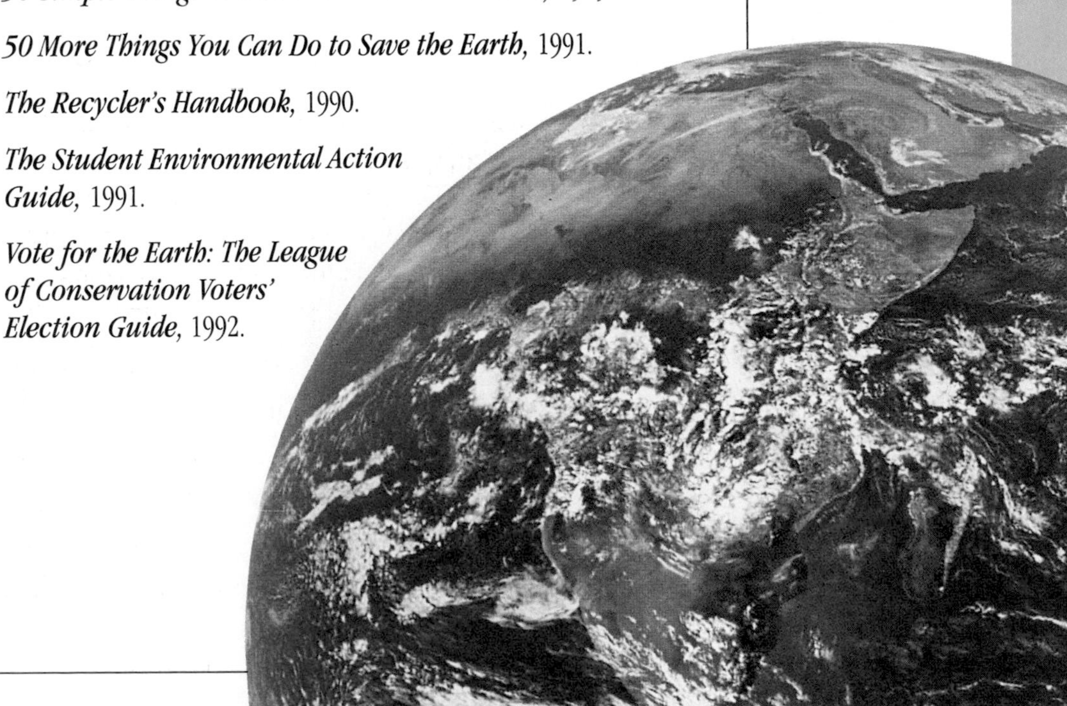

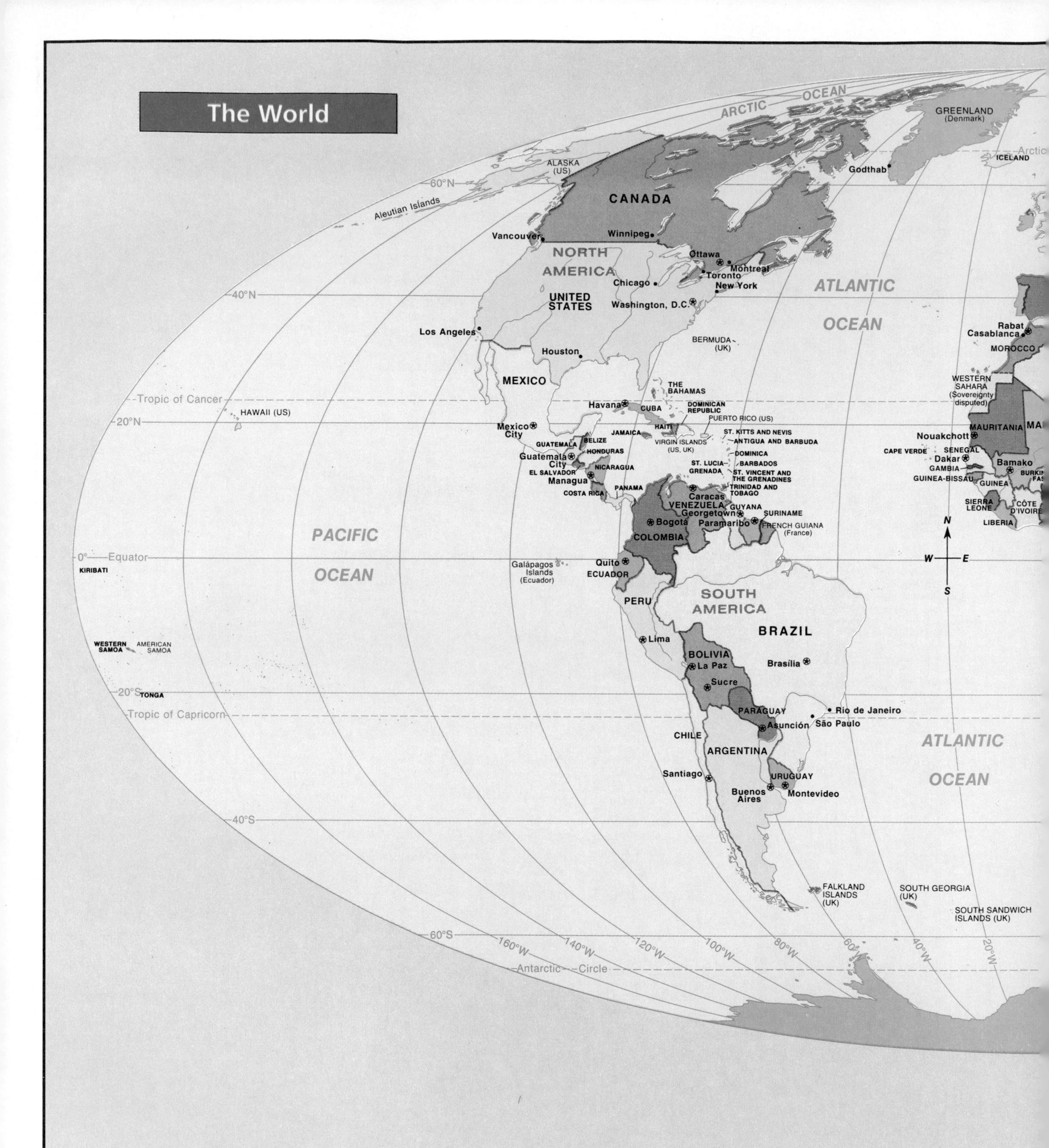
The World
ARCTIC OCEAN
GREENLAND (Denmark)
ICELAND
Godthab
ALASKA (US)
Aleutian Islands
CANADA
Vancouver
Winnipeg
NORTH AMERICA
Ottawa
Montreal
Toronto
Chicago
New York
UNITED STATES
Washington, D.C.
ATLANTIC OCEAN
Los Angeles
Houston
BERMUDA (UK)
Rabat
Casablanca
MOROCCO
MEXICO
THE BAHAMAS
WESTERN SAHARA (Sovereignty disputed)
Tropic of Cancer
Havana
CUBA
DOMINICAN REPUBLIC
HAWAII (US)
PUERTO RICO (US)
Mexico City
JAMAICA
HAITI
ST. KITTS AND NEVIS
ANTIGUA AND BARBUDA
MAURITANIA
Nouakchott
GUATEMALA
BELIZE
HONDURAS
VIRGIN ISLANDS (US, UK)
DOMINICA
CAPE VERDE
SENEGAL
Dakar
Guatemala City
ST. LUCIA
BARBADOS
GAMBIA
Bamako
EL SALVADOR
NICARAGUA
GRENADA
ST. VINCENT AND THE GRENADINES
GUINEA-BISSAU
GUINEA
Managua
PANAMA
TRINIDAD AND TOBAGO
COSTA RICA
Caracas
VENEZUELA
GUYANA
SIERRA LEONE
CÔTE D'IVOIRE
Georgetown
SURINAME
LIBERIA
PACIFIC OCEAN
Bogotá
Paramaribo
FRENCH GUIANA (France)
COLOMBIA
N
W
E
S
Equator
KIRIBATI
Galápagos Islands (Ecuador)
Quito
ECUADOR
SOUTH AMERICA
PERU
BRAZIL
Lima
WESTERN SAMOA
AMERICAN SAMOA
BOLIVIA
La Paz
Brasília
Sucre
TONGA
Tropic of Capricorn
PARAGUAY
Rio de Janeiro
Asunción
São Paulo
CHILE
ARGENTINA
Santiago
URUGUAY
Buenos Aires
Montevideo
FALKLAND ISLANDS (UK)
SOUTH GEORGIA (UK)
SOUTH SANDWICH ISLANDS (UK)
60°N
40°N
20°N
0°
20°S
40°S
60°S
160°W
140°W
120°W
100°W
80°W
60°W
40°W
20°W
Antarctic Circle
National capitals
Other cities
SCALE: at Equator
0 500 1,000 1,500 2,000 Miles
0 1,000 2,000 Kilometers
Projection: Mollweide

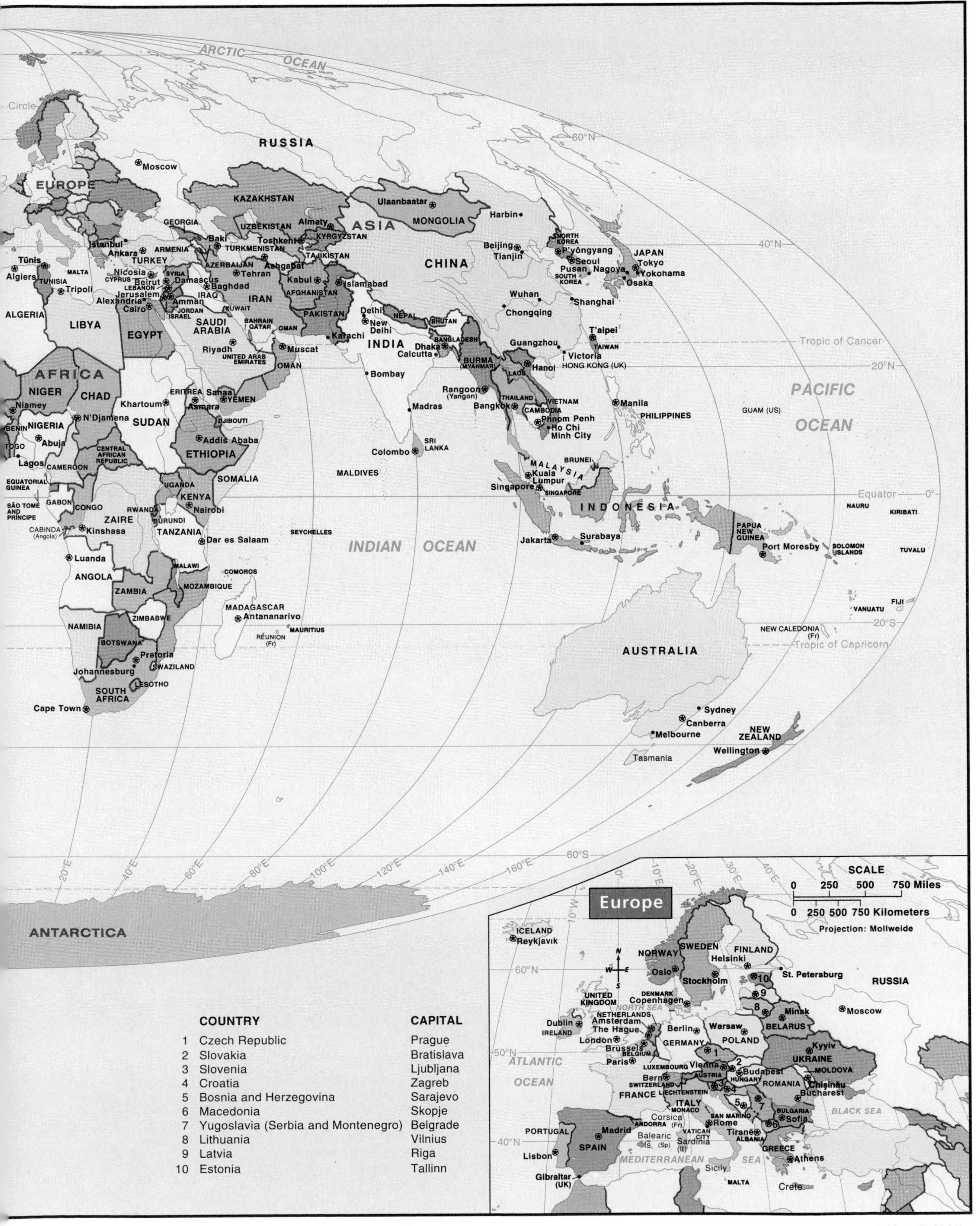
ARCTIC OCEAN
Circle
RUSSIA
Moscow
EUROPE
KAZAKHSTAN
Ulaanbaatar
MONGOLIA
Harbin
GEORGIA
UZBEKISTAN
Almaty
ASIA
KYRGYZSTAN
Baki
Toshkent
NORTH KOREA
İstanbul
Ankara
ARMENIA
TURKMENISTAN
TAJIKISTAN
Beijing
Tianjin
P'yŏngyang
JAPAN
TURKEY
AZERBAIJAN
Ashgabat
CHINA
Seoul
Tokyo
Tūnis
MALTA
Nicosia
CYPRUS
SYRIA
Tehran
Pusan
Nagoya
Yokohama
Algiers
TUNISIA
Tripoli
Beirut
LEBANON
Damascus
Baghdad
Kabul
Islamabad
SOUTH KOREA
Ōsaka
Jerusalem
IRAQ
AFGHANISTAN
Alexandria
Amman
IRAN
Wuhan
Cairo
ISRAEL
JORDAN
KUWAIT
Shanghai
ALGERIA
LIBYA
EGYPT
SAUDI ARABIA
BAHRAIN
QATAR
OMAN
PAKISTAN
Delhi
New Delhi
NEPAL
BHUTAN
Chongqing
Karachi
T'aipei
Riyadh
UNITED ARAB EMIRATES
Muscat
INDIA
Dhaka
BANGLADESH
Calcutta
Guangzhou
Victoria
TAIWAN
HONG KONG (UK)
Tropic of Cancer
AFRICA
OMAN
Bombay
BURMA (MYANMAR)
LAOS
Hanoi
20°N
NIGER
CHAD
ERITREA
Sanaa
YEMEN
Rangoon (Yangon)
THAILAND
Bangkok
VIETNAM
PACIFIC OCEAN
Niamey
Khartoum
Asmara
Madras
CAMBODIA
Manila
GUAM (US)
BENIN
NIGERIA
N'Djamena
SUDAN
DJIBOUTI
Phnom Penh
Ho Chi Minh City
PHILIPPINES
TOGO
Abuja
Addis Ababa
Colombo
SRI LANKA
Lagos
CENTRAL AFRICAN REPUBLIC
ETHIOPIA
BRUNEI
CAMEROON
MALDIVES
MALAYSIA
EQUATORIAL GUINEA
SOMALIA
Kuala Lumpur
UGANDA
Singapore
SINGAPORE
KENYA
Equator
0°
SÃO TOMÉ AND PRÍNCIPE
GABON
CONGO
RWANDA
Nairobi
INDONESIA
NAURU
KIRIBATI
BURUNDI
ZAIRE
CABINDA (Angola)
Kinshasa
TANZANIA
SEYCHELLES
PAPUA NEW GUINEA
Dar es Salaam
Jakarta
Surabaya
INDIAN OCEAN
Port Moresby
SOLOMON ISLANDS
TUVALU
Luanda
ANGOLA
MALAWI
COMOROS
ZAMBIA
MOZAMBIQUE
FIJI
MADAGASCAR
Antananarivo
VANUATU
ZIMBABWE
NAMIBIA
MAURITIUS
NEW CALEDONIA (Fr)
20°S
RÉUNION (Fr)
BOTSWANA
Tropic of Capricorn
AUSTRALIA
Pretoria
Johannesburg
SWAZILAND
LESOTHO
SOUTH AFRICA
Cape Town
Sydney
Canberra
NEW ZEALAND
Melbourne
Wellington
Tasmania
60°N
40°N
60°S
20°E
40°E
60°E
80°E
100°E
120°E
140°E
160°E
ANTARCTICA
Europe
SCALE
0 250 500 750 Miles
0 250 500 750 Kilometers
Projection: Mollweide
ICELAND
Reykjavík
NORWAY
SWEDEN
FINLAND
Helsinki
Oslo
Stockholm
St. Petersburg
RUSSIA
UNITED KINGDOM
DENMARK
Copenhagen
NORTH SEA
Minsk
Moscow
NETHERLANDS
Amsterdam
The Hague
Dublin
IRELAND
Berlin
Warsaw
BELARUS
London
GERMANY
POLAND
Brussels
BELGIUM
Kyyiv
Paris
UKRAINE
ATLANTIC OCEAN
LUXEMBOURG
Vienna
Budapest
MOLDOVA
Bern
SWITZERLAND
AUSTRIA
HUNGARY
ROMANIA
Chişinău
FRANCE
LIECHTENSTEIN
Bucharest
ITALY
MONACO
Corsica (Fr)
SAN MARINO
BULGARIA
Sofia
BLACK SEA
PORTUGAL
Madrid
ANDORRA
Rome
VATICAN CITY
Tiranë
ALBANIA
SPAIN
Balearic Is. (Sp)
Sardinia (It)
GREECE
Lisbon
MEDITERRANEAN SEA
Athens
Sicily
Gibraltar (UK)
MALTA
Crete
50°N
60°N
40°N
10°W
0°
10°E
20°E
30°E
40°E
COUNTRY
CAPITAL
1 Czech Republic Prague
2 Slovakia Bratislava
3 Slovenia Ljubljana
4 Croatia Zagreb
5 Bosnia and Herzegovina Sarajevo
6 Macedonia Skopje
7 Yugoslavia (Serbia and Montenegro) Belgrade
8 Lithuania Vilnius
9 Latvia Riga
10 Estonia Tallinn

THE METRIC SYSTEM IS USED FOR MAKING MEASUREMENTS IN SCIENCE. THE OFFICIAL NAME OF THIS SYSTEM IS THE SYSTÈME INTERNATIONAL D'UNITÉS, OR INTERNATIONAL SYSTEM OF MEASUREMENTS (SI).

SI Conversions

SI Units	From SI to English	From English to SI
Length		
kilometer (km) = 1,000 m	1 km = 0.62 mile	1 mile = 1.609 km
meter (m) = 100 cm	1 m = 3.28 feet	1 foot = 0.305 m
centimeter (cm) = 0.01 m	1 cm = 0.394 inch	1 inch = 2.54 cm
millimeter (mm) = 0.001 m	1 mm = 0.039 inch	
micrometer (µm) = 0.000 001 m		
nanometer (nm) = 0.000 000 001 m		
Area		
square kilometer (km^2) = 100 hectares	1 km^2 = 0.3861 square mile	1 square mile = 2.590 km^2
hectare (ha) = 10,000 m^2	1 ha = 2.471 acres	1 acre = 0.4047 ha
square meter (m^2) = 10,000 cm^2	1 m^2 = 10.765 square feet	1 square foot = 0.0929 m^2
square centimeter (cm^2) = 100 mm^2	1 cm^2 = 0.155 square inch	1 square inch = 6.4516 cm^2
Volume		
liter (L) = 1,000 mL = 1 dm^3	1 L = 1.06 fluid quarts	1 fluid quart = 0.946 L
milliliter (mL) = 0.001 L = 1 cm^3	1 mL = 0.034 fluid ounce	1 fluid ounce = 29.57 mL
microliter (µL) = 0.000 001 L		
Mass		
kilogram (kg) = 1,000 g	1 kg = 2.205 pounds	1 pound = 0.4536 kg
gram (g) = 1,000 mg	1 g = 0.0353 ounce	1 ounce = 28.35 g
milligram (mg) = 0.001 g		
microgram (µg) = 0.000 001 g		

Temperature

°F 0 20 40 60 80 100 120 140 160 180 200 220

°C −20 −10 0 10 20 30 40 50 60 70 80 90 100

Freezing point of water

Room temperature

Normal human body temperature

The top of the thermometer is marked off in degrees Fahrenheit (°F). To read the corresponding temperature in degrees Celsius (°C), look at the bottom side of the thermometer. For example, 50°F is the same temperature as 10°C. You may also use the formulas at the right for conversions.

Conversion of Fahrenheit to Celsius:

$°C = \frac{5}{9}(°F - 32)$

Conversion of Celsius to Fahrenheit:

$°F = (\frac{9}{5}°C) + 32$

GENERAL GUIDELINES FOR LABORATORY SAFETY

In the laboratory, you can engage in hands-on explorations, test your scientific ideas, and build practical laboratory skills. However, the laboratory can be a safe place or dangerous place, depending on your knowledge of and adherence to safe laboratory practices. Read and follow the basic safety guidelines described below.

Before You Begin an Investigation...

- Be prepared. Study assigned Investigations before class. Resolve any questions about procedures before starting work.
- Keep your work area uncluttered. Store books, backpacks, jackets, or other items you do not need out of the way.
- Arrange the materials you are using for an Investigation in an orderly fashion on your work surface. Keep laboratory materials away from the edge of the work surface.
- Tie back long hair and remove dangling jewelry. Roll up sleeves and secure loose clothing.
- Do not wear contact lenses while performing an experiment that involves chemicals. If you must wear them by a doctor's order, inform your teacher before beginning such an experiment.
- Avoid wearing sandals or open-toed shoes in the laboratory, because they will not protect your feet if any chemical, glassware, or other object is dropped on them.
- Know the location of the nearest phone. Find out where emergency telephone numbers, such as the number for the nearest poison control center, can be found.
- Find out where laboratory safety equipment (such as eyewash stations and fire extinguishers) is stored, and know how to operate this equipment.
- Know the fire evacuation routes established by your school.
- Before you begin the experiment, review the supplies you will be using and the safety issues you should be concerned about. Be on the alert for the safety symbols shown here. The symbols indicate particular safety concerns.

While You Are Working...

- Approach all laboratory work with a mature and serious attitude. Most accidents are caused by carelessness or horseplay. Decrease your risk by concentrating on your work and staying alert.
- Never perform an experiment not authorized by your teacher.
- Never work alone in the laboratory.
- Wear safety goggles and a lab apron when you are working with chemicals, hot liquids, lab burners, hot plates, or apparatus that could break or shatter.
- Wear protective gloves when working with toxic or irritating chemicals or preserved specimens and when handling plants, animals, or other items as directed by your teacher.
- Never look directly at the sun through any optical device or use direct sunlight to illuminate a microscope. The focused light can seriously damage your eyes.

Wear lab apron

Wear safety goggles

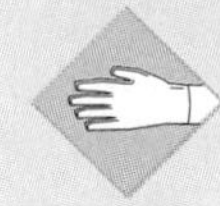
Wear gloves

Sharp/pointed object

Electrical hazard

Dangerous chemical/poison

Flame/heat

Glassware

Plants

Live animals

Biohazard

- When heating substances in a test tube, always point the test tube *away* from yourself and others.
- Keep your hands away from the sharp or pointed ends of scalpels, scissors, and other sharp instruments.
- Observe all of the safety symbols that accompany the procedural steps of Investigations. Employ specific safety practices that are called for.
- Never put anything in your mouth and never touch or taste substances in the laboratory unless your teacher instructs you to do so.
- If your teacher instructs you to smell a chemical in the laboratory, follow the correct procedure. The correct method is to gently fan your hand over the substance, waving its vapors toward your nose. Do not put your nose directly over the substance.
- Never eat, drink, chew gum, or apply cosmetics in the laboratory. Do not store food or beverages in the lab area.
- Report any accident, chemical spill, or unsafe incident to your teacher immediately.
- Check labels on containers of chemicals to be certain you are using the right material.
- When diluting an acid or base with water, always add the acid or base to water. Do NOT add water to the acid or base.
- Dispose of chemicals according to your teacher's instructions.
- Never return unused chemicals to the containers you obtained them from. Do not put any object into a bottle containing a laboratory chemical.

Finishing Up

- Dispose of materials and wash used glassware and instruments according to your teacher's instructions.
- Clean tables and sinks.
- Put away all equipment and supplies.
- Make sure all water faucets, gas jets, burners, and electrical appliances are turned off.
- Return all laboratory materials and equipment to their proper places.
- Wash your hands thoroughly with soap after completing an Investigation.

Emergency Procedures

Don't panic. In a common laboratory emergency follow these instructions.

In the event of a fire, alert the teacher and leave the laboratory immediately.

If your clothes catch fire, STOP, DROP, and ROLL! The quickest way to smother a fire is to stop immediately, drop to the floor, and roll.

If your lab partner's clothes or hair catches fire, grab the nearest fire blanket and use it to extinguish the flames. Inform your teacher.

If a chemical gets into your eyes or on your face, wash immediately with plenty of water for at least 15 minutes. Flush under each eyelid, and have a classmate notify the teacher.

If a chemical spills on your skin or clothing, wash it off immediately with plenty of water, and notify your teacher.

If a chemical spills on the floor, do not clean it up yourself. Keeping your classmates away from the area, alert your teacher immediately.

If you receive a cut, even if it is just a small one, notify your teacher.

Safety With Animals in the Laboratory

Observing and experimenting with animals can vastly enrich your understanding of environmental science. Yet you must use extreme caution to assure your own safety as well as the safety and comfort of animals you work with. General rules are as follows.

- Do not touch any animal unless your teacher specifically gives you permission.
- Do not tease or disturb any animal unnecessarily.
- Do not bring any animal into the laboratory without your teacher's permission.
- Wear gloves or other appropriate protective gear when working with animals.
- Wash your hands after touching any animal.
- Inform your teacher immediately if you are scratched, bitten, stung, or otherwise harmed by an animal.
- Always follow your teacher's instructions regarding the care of laboratory animals. Ask questions if you do not clearly understand what you are supposed to do.
- Keep each laboratory animal in a suitable, escape-proof container in a location where the animal will not be frequently disturbed. Animal containers should provide adequate ventilation, warmth, and light.
- Keep the container clean. Clean cages of small birds and mammals daily.
- Provide water at all times.
- Feed animals regularly, according to their individual needs.
- If you are responsible for the care or feeding of animals, arrange for necessary care on weekends, holidays, and during vacations.
- No study that involves inflicting pain on a vertebrate animal should ever be conducted.
- Vertebrate animals must not be exposed to excessive noise, exhausting exercise, overcrowding, or other distressing stimuli.
- When an animal must be removed from the laboratory, your teacher will provide a suitable method.

Safety With Plants in the Laboratory

Plants, as living organisms, deserve care and safe keeping. Some guidelines for their proper care are given below. On the other hand, many plants or plant parts present a safety hazard to you. Some plants or plant parts are poisonous to the point of fatality, depending on the weight of the person and the amount of plant material ingested. A common-sense approach is to take the following precautions with *all* plants.

- Never place any part of any plant in your mouth unless instructed to do so by your teacher. Seeds obtained from commercial growers can be particularly dangerous because such seeds may be coated with hormones, fungicides, or insecticides.

- Do not rub sap or juice of fruits on your skin or into an open wound.
- Never inhale or expose your skin or eyes to the smoke of any burning plant or plant parts.
- Do not bring unknown wild or cultivated plants into the laboratory.
- Do not eat, drink, or apply cosmetics after handling plants, without first scrubbing your hands.
- Provide adequate light and water and appropriate soil and temperature for plants growing in the laboratory.
- If you are responsible for plants, make necessary arrangements for their care on weekends, holidays, and during vacations.

Safe and Successful Fieldwork

Environmental scientists conduct much of their research in the field. For environmental scientists—and environmental science students like you—there are three important issues to consider when working in the field. One issue is your personal safety. Another issue is the successful completion of the scientific work you set out to do. The third consideration is protection of the environment you have come to study. The following guidelines will help you achieve these three goals.

- Dress in a manner that will keep you comfortable, warm, and dry. Wear long pants rather than shorts or a skirt. Wear sturdy shoes with closed toes. Do not wear sandals or heels. Wear waterproof shoes if you will be working in wetlands.
- Bring rain gear if there is any possibility of rain.
- Bring sunglasses, sunscreen, and insect repellent as needed.
- Do not go alone beyond where you can be seen or heard; travel with a partner at all times.
- Do not approach wild mammals, snakes, snapping turtles, or other animals that may sting, bite, scratch, or otherwise cause injury.
- Do not touch any animal in the wild without specific permission from your teacher.
- Find out whether there are likely to be poisonous plants or dangerous animals where you will be going. Learn how to identify any hazardous species.
- Do not pick wildflowers or touch plants or plant parts unless your teacher gives you permission. Do not eat wild plants.
- Report any hazard or injury to your teacher immediately.
- Be sure you understand the purpose of your field trip and any assignments you have been given. Bring all needed school supplies, and keep them organized in a binder, backpack, or other container.
- Be aware of the impact you are having on the environments you visit. Just walking over fragile areas can harm them, so stay on trails unless your teacher gives you permission to do otherwise.
- Sketching, photographing, and writing field notes are generally more appropriate than collecting specimens for observation. Collecting from a field site may be permitted in certain cases, but always obtain your teacher's permission first.
- Do not leave garbage behind at the field site. Strive to leave natural areas just as you found them.

CHEMISTRY REVIEW

Atoms and Elements

Every object in the universe is made up of particles of some kind of matter. Matter is anything that takes up space and has mass. All matter is made up of elements. An element is a substance that cannot be separated into simpler components by ordinary chemical means. This is because each element consists of only one kind of atom. An atom is the smallest unit of an element that has all of the properties of that element.

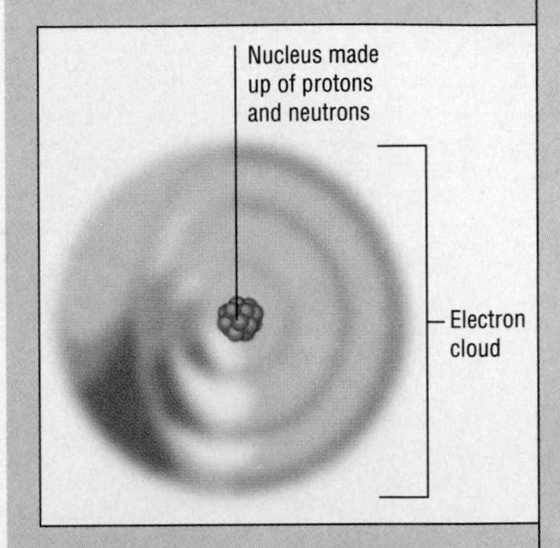

The nucleus of the atom contains the protons and neutrons. The protons give the nucleus a positive charge. The negatively charged electrons are in the electron cloud surrounding the nucleus.

Atomic Structure

Atoms are made up of small particles called subatomic particles. The three major types of subatomic particles are **electrons, protons,** and **neutrons.** Electrons have a negative electrical charge, protons have a positive charge, and neutrons have no electrical charge. The protons and neutrons are packed close to one another to form the **nucleus.** The protons give the nucleus a positive charge. The electrons of an atom move in a region around the nucleus known as an **electron cloud.** The negatively charged electrons are attracted to the positively charged nucleus. An atom may have several energy levels in which electrons are located.

Atomic Number

To help in the identification of elements, scientists have assigned an **atomic number** to each kind of atom. The atomic number is equal to the number of protons in the atom. Atoms with the same number of protons are all the same kind of element. In an uncharged, or electrically neutral, atom there are an equal number of protons and electrons. Therefore, the atomic number also equals the number of electrons in an uncharged atom. The number of neutrons, however, can vary for a given element. Atoms of the same element that have different numbers of neutrons are called **isotopes.**

Periodic Table of the Elements

A periodic table of the elements is shown on the next page. In a periodic table, the elements are arranged in order of increasing atomic number. Each element in the table is found in a separate box. As you go from left to right, each element has one more electron and one more proton than the element to its left. Each horizontal row of the table is called a **period.** Changes in chemical properties across a period correspond to changes in the elements' electron arrangements. Each vertical column of the table, known as a **group,** lists elements with similar properties. The elements in a group have similar chemical properties because they have the same number of electrons in their outer energy level. For example, the elements helium, neon, argon, krypton, xenon, and radon all have similar properties and are known as the noble gases.

Molecules and Compounds

When the atoms of two or more elements are joined chemically, the resulting substance is called a **compound.** A compound is a new substance with properties different from those of the elements that compose it. For example, water (H_2O) is a compound formed when atoms of hydrogen (H) and oxygen (O) combine. The smallest complete unit of a compound that has all of the properties of that compound is called a **molecule.**

A chemical formula indicates what elements a compound contains. It also indicates the relative number of atoms of each element present. The chemical formula for water is H_2O, which indicates that each water molecule consists of

The Periodic Table of the Elements

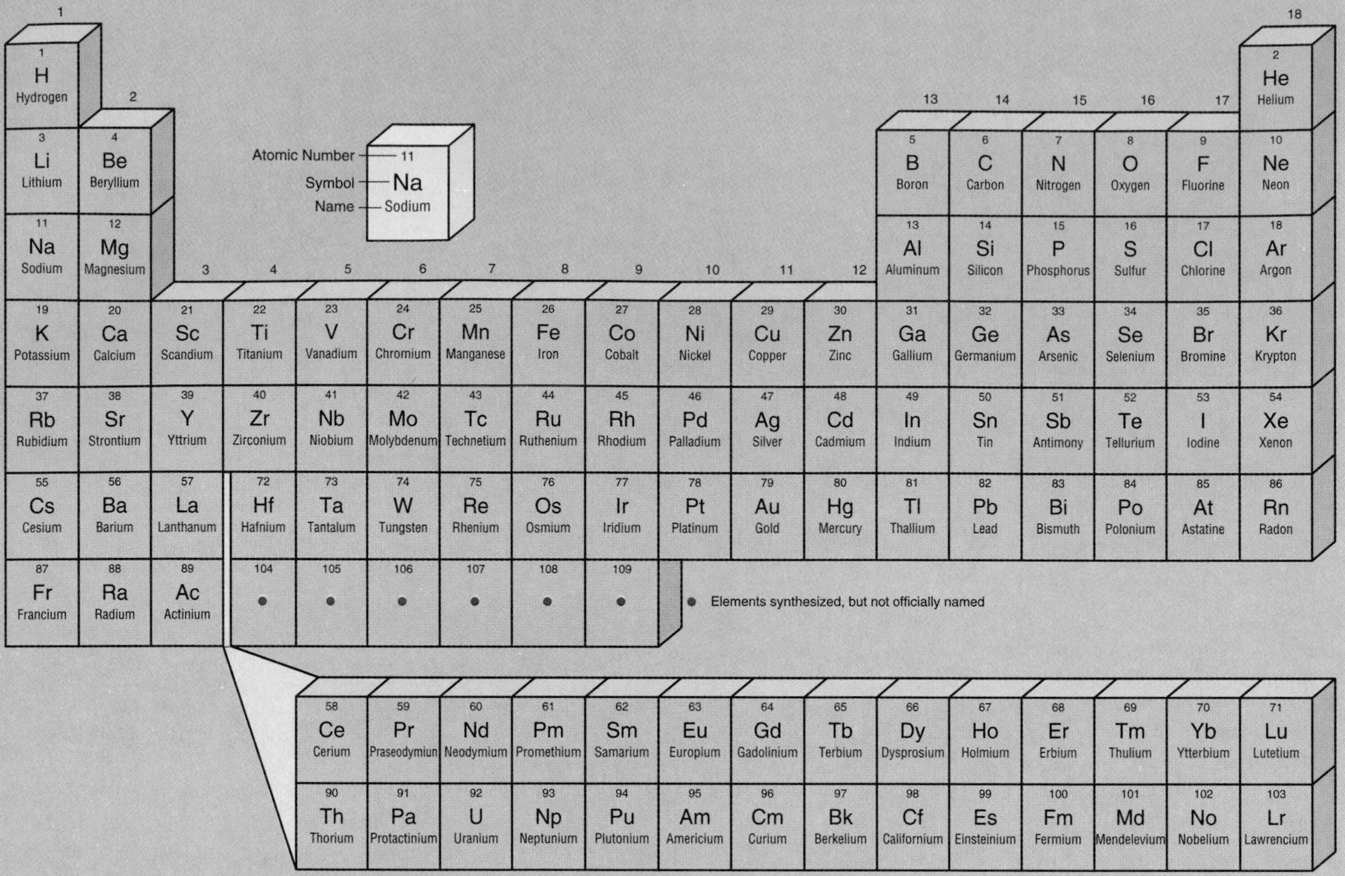

two atoms of hydrogen and one atom of oxygen. The subscript number is used after the symbol for an element to indicate how many atoms of that element are in a single molecule of the compound.

Chemical Equations

A chemical reaction occurs when a chemical change takes place. (In a chemical change, new substances with new properties are formed.) A chemical equation is a useful way of describing a chemical reaction by means of chemical formulas. The equation indicates what substances react and what the products are. For example, when carbon and oxygen combine, they can form carbon dioxide. The equation for this reaction is as follows: $C + O_2 \rightarrow CO_2$

Acids, Bases, and pH

An ion is an atom or group of atoms that has an electrical charge because it has lost or gained one or more electrons. When an acid such as hydrochloric acid (HCl) is mixed with water, it separates into ions. An **acid** is a compound that produces hydrogen ions (H^+) in water. The hydrogen ions then combine with a water molecule to form a hydronium ion (H_3O^+). A solution that contains hydronium ions is an acidic solution. A **base**, on the other hand, is a substance that produces hydroxide ions (OH^-) in water.

To determine whether a solution is acidic or basic, scientists use pH. **pH** is a measure of how many hydronium ions are in solution. The pH scale ranges from 0 to 14. The middle point, pH = 7, is neutral, neither acidic nor basic. Acids have a pH of less than 7; bases have a pH of more than 7. The lower the number, the stronger the acid. The higher the number, the stronger the base. A pH scale is shown in Figure 6-18 on page 162.

GLOSSARY

A

abiotic factors (ay by OT ik) nonliving parts of an ecosystem (35)

acid precipitation (pree sip uh TAY shun) highly acidic rain, sleet, or snow that results from the release of oxides of sulfur and nitrogen into the air from burning fossil fuels (161)

acid shock sudden influx of acidic water caused by melting acidic snows that rush into lakes and streams, killing large numbers of fish and amphibians and affecting the offspring of others (162)

active solar heating (SOH lur) system by which buildings are heated by collection devices that absorb solar energy and convert it into heat (289)

adaptation (ad up TAY shun) an inherited trait that increases an organism's chance of survival and reproduction in a certain environment (46)

aesthetic (es THET ik) relating to something that is beautiful or pleasing (20)

Agenda 21 a program adopted at Earth Summit that is a blueprint for protecting the environment and promoting sustainable development (356)

age-structure histogram illustration that depicts in comparative ways the age and sex of a population (341)

agricultural revolution change from a hunting and gathering society to an agricultural society that began about 10,000 years ago (337)

air pollution (puh LOO shun) condition in which the air contains substances harmful to living things (151)

algal bloom (AL gul) excessive algae growth that forms large, floating mats (137)

amaranth (AM uh ranth) plant that can survive with little water, producing a grain-like fruit rich in protein (239)

applied science study and activity that uses information provided by pure science to solve problems; examples are engineering and medicine (13)

aquaculture (AK wuh kul chur) cultivation of fish for human use or consumption (238)

aquifer (AK wuh fur) an underground rock formation that contains water(126)

arable land (AR uh bul) land on which crops can be grown (231)

artificial eutrophication (yoo trof ih KAY shun) introduction of inorganic plant nutrients into a body of water through sewage and fertilizer runoff (137)

asbestos (as BES tus) mineral that separates into long, thread-like fibers, frequently used for insulation (160)

atmosphere (AT muhs fir) the thin layer of gases that surround the Earth (173)

atomic explosion (uh TOM ik) uncontrolled release of nuclear energy (284)

autotroph (AW toh troaf) self-feeder organism that obtains its nutrients by synthesis from the environment, such as green plants (56)

B

bar graph a graph in which parallel bars are used to compare data (18)

bedrock solid rock that lies below the layers of soil (232)

benthic zone (BEN thik) bottom of a body of water, inhabited by decomposers, insect larvae, and clams (100)

bioaccumulation (by oh uh kyoom yoo LAY shun) accumulation of larger and larger amounts of a toxin within the tissues of organisms at each successive trophic level (136)

biodegradable material (by oh di GRAY duh bul) a material that can be broken down by living things into simpler chemicals (305)

biodiversity (by oh duh VUR suh tee) term used to indicate the number and diversity of species on Earth (254)

Biodiversity Treaty (by oh duh VUR suh tee) agreement resulting from the Earth Summit that encourages wealthier countries to give money to poorer countries for the protection of potentially valuable species (266)

biological pest control (by uh LAJ i kul) pest control using living organisms or naturally produced chemicals (243)

biomass (BY oh mas) organic matter in plants or plant products (293)

biome (BY ohm) regions that have distinctive climates and organisms and contain many separate but similar ecosystems (79)

biosphere (BY oh sfir) the layer around the Earth in which life occurs naturally, extending to about 8 km above the Earth to the deepest part of the ocean, which is about 8 km deep (9)

biotic factors (by OT ik) the living parts of an ecosystem (35)

biotic potential (by OT ik poh TEN shul) rate at which a population would grow if every new individual survived to adulthood and reproduced at its maximum capacity; the characteristic maximum growth rate for a population (333)

bronchial asthma (BRON kee ul AZ muh) chronic disorder of the bronchial tubes characterized by wheezing and difficulty in breathing (157)

C

canopy (KAN uh pee) in a forest, the covering of tall trees whose intertwining branches absorb a great amount of sunlight and shade the area beneath (82)

captive-breeding program breeding endangered or threatened species in zoos and wild animal parks (267)

carbohydrate (kar boh HY drayt) organic compound, such as starch or sugar, made of carbon, hydrogen, and oxygen in which the ratio of hydrogen to oxygen is 2:1 (228)

carbon cycle process in an ecosystem in which producers take in carbon dioxide from the atmosphere during photosynthesis and consumers, having eaten producers, release carbon into the atmosphere as carbon dioxide (63)

carnivore (KAR nuh voar) consumer that eats only other consumers (57)

carrying capacity the maximum number of individuals an environment can support for a long period of time (335)

catalytic converter (kat uh LIT ik) device attached to exhaust system of automobiles to clean exhaust gases before they leave the car (153)

cellular respiration (SEL you lur res puh RAY shun) process of breaking down food to yield energy (57)

CFCs (chlorofluorocarbons) (KLOR oh floor uh kar bunz) human-made chemical compounds restricted in use because they destroy ozone (186)

chronic bronchitis (KRON ik bron KYT is) recurring inflammation of the mucous lining of the bronchial tubes (157)

Clean Air Act act passed by U.S. Congress that gave EPA authority to regulate automobile emissions (153)

Clean Water Act act passed by U.S. Congress to restore and maintain the chemical, physical, and biological integrity of the nation's waters (138)

clear-cutting method of harvesting trees in which all of the trees are removed from a land area, as opposed to selective cutting (205)

climate (KLY mut) the average weather in an area over a long period of time (176)

climatogram (kly MAT oh gram) graph that shows monthly variations in temperature and precipitation (116)

climax community (KLY maks) final, stable community that forms when land is left undisturbed (67)

coevolution (koh ev uh LOO shun) two or more species evolving in response to each other (46)

commensalism (kuh MEN sul iz um) relationship between two species in which one species benefits and the other is neither harmed nor helped (42)

community (kuh MYOO nuh tee) a group of interacting populations of different species (37)

competition (kom puh TISH un) the relationship between species that attempt to use the same limited resource (40)

compost (KOM pohst) mixture of decomposing vegetable refuse, manure, and plants used as fertilizer and soil conditioner (312)

consumer (kun SOOM ur) organism that gets its energy by eating other organisms; heterotroph (56)

consumption crisis situation in which natural resources are being used up, wasted, or polluted faster than they can be renewed, replaced, or cleaned up (11)

contour plowing (KON toor) plowing across the slope of a hill to prevent soil from being washed away (236)

control constant factor used in an experiment to test a hypothesis (16)

Convention on International Trade in Endangered Species (CITES) offshoot of the IUCN that is responsible for limiting the sale of ivory and the killing of elephants (265)

coral reef limestone islands in the sea built by coral animals (107)

D

data (DAYT uh) information observed or gathered from which conclusions can be drawn (17)

DDT powerful insecticide, use of which is limited by law because of its potential damage to the environment (58)

decomposer (dee kom POHZ ur) consumer that gets its food by breaking down dead organisms, causing them to rot (57)

deep-well injection method of hazardous-waste disposal in which wastes are pumped deep into the ground below the level of the groundwater (320)

deforestation (dee foar est AY shun) clearing trees from an area without replacing them (205)

demographer (di MOG ruh fur) one who is involved with

the statistical science dealing with the distribution, density, and vital statistics of human populations (338)

demographic transition (dem uh GRAF ik tran ZISH un) theory that states that the economic and social progress of the industrial revolution affects population in three stages (338)

desalinization (dee sal uh nuh ZAY shun) process in which salt is removed from salt water, as from the oceans, rendering the water fit for drinking and cooking (128)

desert biome that receives less than 10 in. of precipitation a year and occurs between 15° and 30° north and south latitude around the Tropic of Cancer and the Tropic of Capricorn (94)

desertification (di zurt uh fi KAY shun) deterioration of land to the point that it becomes desert (235)

developed countries highly industrialized countries with high incomes and high standards of living (11)

developing countries less industrialized countries in which the average income and standard of living are low (11)

distillation (dis tuh LAY shun) process of heating a liquid and cooling and condensing the resultant vapor to separate substances and produce a more refined liquid (128)

drought (DROUT) period when rainfall is less than average, causing crop failure (229)

drought-resistance (DROUT) characteristic of a plant that allows it to survive in areas of light or sporadic rainfall (96)

E

Earth Summit conference dealing with the environment and development sponsored by the United Nations in 1992 (355)

ecology (ee KOL uh jee) the study of how living things are related to each other—how they interact and depend on each other (14)

ecosystem (EK oh sis tum) all living organisms in a certain area as well as their physical environment (34)

ecosystem approach (EK oh sis tum) conservation method that concentrates on protecting an entire ecosystem (268)

electric generator device that converts other forms of energy, such as fossil fuels, wind, and flowing water, into electric energy (280)

electricity flow of electrons, the tiny particles that whirl around the nucleus of an atom (280)

emigration (em i GRAY shun) act of organisms leaving one area to locate in another (331)

emphysema (em fuh SEE muh) swelling of the lung tissue due to the permanent loss of elasticity or destruction of the alveoli (157)

endangered species (SPEE sheez) a species whose numbers have fallen so low that it is likely to become extinct in the near future (261)

Endangered Species Act (SPEE sheez) 1973 act of U.S. Congress designed to protect any plant or animal species in danger of extinction (261)

energy (EN uhr jee) the quantity that describes the capacity to do work (55)

energy conservation (EN uhr jee kon sur VAY shun) use of energy efficiently without necessarily reducing effectiveness (287)

energy pyramid (EN uhr jee PIR uh mid) diagram in the form of a pyramid that shows how energy is lost from one trophic level to the next (61)

environment (en VY run munt) the surroundings of an organism that affect its life and development (4)

environmental impact statement (en vy run MEN tul IM pakt) an assessment of the effect that a proposed project or law will have on the environment (361)

environmental refugees (en vy ron MEN tul ref yoo JEEZ) people driven from their homes by severe environmental damage (345)

environmental science (en vy run MEN tul) study of how humans interact with the environment (4)

erosion (ee ROH zhun) the wearing away of topsoil by wind or water (234)

estivating (ES tuh vayt ing) practice of lying dormant underground during the summer (96)

estuary (ES tyoo er ee) aquatic ecosystem in which fresh water from rivers mixes with salt water from the ocean, forming a nutrient trap (105)

eutrophication (yoo trof ih KAY shun) process that increases the amounts of nutrients, especially nitrogen and phosphorus, in a marine or aquatic ecosystem (101)

evolution (ev uh LOO shun) change in the genetic characteristics of a population from one generation to the next (44)

exosphere (EKS oh sfir) outermost portion of the Earth's atmosphere (175)

exotic species (eg ZOT ik SPEE sheez) species that is not native to a particular region (258)

experiment activity designed to test a hypothesis under controlled conditions (16)

extinction (ek STINK shun) the irreversible disappearance of a population or a species (46)

F

family planning practice of people who take measures to ensure that they will have the size of family they want (346)

famine (FAM in) widespread food shortage (229)

Federal Mining Act act of U.S. Congress to regulate the filing and settling of land claims in the West (216)

fertile soil (FURT ul) soil in which most plants are healthy and able to grow rapidly (232)

fire-maintained communities those communities where natural fires caused by lightning make possible a secondary succession (68)

food chain the sequence in which energy is transferred from one organism to the next as each organism eats and is eaten by another (59)

food web all of the feeding relationships in an ecosystem (60)

fossil fuel (FOS ul) organic substance such as coal, oil, and natural gas, found underground in deposits formed from the remains of organisms that lived millions of years ago (281)

fresh water water that contains little salt (122)

G

gasohol (GAS uh hol) blend of gasoline and alcohol for motor fuel (294)

genetic disorder (juh NET ik) disruptive disease causing a condition in an organism that injures its health or prevents normal functions (263)

genetic engineering (juh NET ik) scientific activity that develops desirable characteristics in an organism by altering its genes or inserting new genes in the organism's cell (250)

geothermal energy (jee oh THUR mul) energy that is drawn from heat within the Earth and used to drive electric generators (292)

germ plasm (PLAZ um) genetic material contained within the reproductive cells of organisms (268)

greenhouse effect warming effect on the air caused by heat rising from the surface of the Earth and being trapped by gases in the troposphere (180)

greenhouse gases the gases in the atmosphere that trap and radiate heat (181)

green revolution the introduction of new farming techniques and new varieties of crops, especially grains, in Asia and Latin America (229)

groundwater water that seeps down through the soil and is stored underground (126)

H

habitat (HAB i tat) place where an organism lives (38)

hazardous wastes (HAZ uhr dus) wastes that are toxic or highly corrosive or that explode easily (316)

heavy metals dense metals, such as lead and mercury, which can be toxic to organisms (321)

herbivore (HUR buh voar) consumer that eats only producers (57)

heterotroph (HET uhr oh trof) consumer organism that gets its energy from eating other organisms (56)

high-level radioactive waste extremely hazardous radioactive waste from nuclear plants (328)

host (HOHST) organism from which a parasite takes its nourishment (41)

hunter-gatherers people who obtain their food by hunting, fishing, and gathering wild plant foods (336)

hydrocarbon (hy droh KAR bun) compound, such as methane, that is composed of hydrogen and carbon and vaporizes at normal temperatures (156)

hydroelectricity (hy droh ee lek TRIS i tee) electricity produced by converting the energy of moving water (292)

hypothesis (hy POTH uh sis) a testable explanation for a specific problem or question, based on what has already been learned (15)

I

immigration (im uh GRAY shun) act of organisms moving into one area from another (331)

incinerator (in SIN uhr ayt uhr) plant used for burning solid waste material (309)

infectious disease (in FEK shus) disease that can be transmitted to another member of the species (263)

infrastructure (in fruh STRUK chur) the necessary structures such as roads, buildings, bridges, and sewers that a society builds for public use (199)

International Union for the Conservation of Nature and Natural Resources (IUCN) a collaboration of almost 200 governments and over 300 conservation organizations to protect species and habitats (264)

International Whaling Commission (IWC) an international agency whose purpose is to limit whaling practices for the preservation of various species (356)

irrigation (ir uh GAY shun) process of supplying water to an area by artificial means, such as ditches and sprinklers (125)

J

J-curve curved line showing exponential growth (333)

K

keystone species species of great importance to an ecosystem, the loss of which might cause the complete collapse of the ecosystem (258)

krill small, shrimplike crustaceans, the main food of whales (59, 60)

L

landfill waste-disposal facility where wastes are put in the ground and covered each day with a layer of dirt, plastic, or both (306)

land-use planning determining the location of homes, businesses, and protected areas before an area is developed (201)

latitude (LAT uh tood) distance measured in degrees north and south of the equator, which is 0° latitude (177)

Law of the Sea Treaty United Nations treaty that states the laws of a coastal nation extend to 22 km (12 nautical mi.) from its coastline (142)

leachate (LEECH ayt) water that contains toxic chemicals dissolved from wastes in a landfill (307)

lichen (LY kun) composed of a particular fungus and a particular alga growing in a symbiotic relationship and forming a dual organism (69)

limiting resources those resources that limit the growth of a certain population (335)

line graph graph in which data points are plotted and connected with lines to show relationships (18)

lipid (LIP id) organic compound that contains fats and other substances that living organisms use to get energy (228)

littoral zone (LIT uh rul) the shallow-water area near the shores of lakes and ponds where sunlight reaches the bottom (100)

Love Canal location of burial of tons of hazardous chemicals from industrial waste (316, 318–319)

low-input farming farming that does not use a lot of energy, pesticides, fertilizer, and water (238)

M

malaria (muh LER ee uh) infectious disease carried by mosquitoes (242)

malnutrition (mal noo TRISH un) poor health condition caused by not consuming enough necessary nutrients (228)

marsh land covered with water; contains non-woody plants (102)

mass extinction extinction of many species during a relatively short period of time (253)

mass transit transportation facilities to move large numbers of people at a time (202)

melaleuca tree (mel uh LOOK uh) tree imported from Australia to Florida that is threatening the existence of native plants (258)

mesosphere (MES oh sfir) atmospheric zone above the stratosphere (175)

methane (METH ayn) colorless, odorless, flammable gas present in natural gas and formed by the decomposition of plant matter (307)

mineral solid substance found in nature that consists of a single element or compound (210)

mineral resources minerals that have economic value and are useful in some way (210)

multiple-use management term used to indicate variety of ways in which land can be used for the greatest good for the greatest number of people (214)

municipal solid waste (myoo NIS uh pul) trash produced by households and businesses (306)

mutualism (MYOO choo ul iz um) a relationship between two species in which both benefit (41)

N

native species as opposed to exotic species, native species is one original to a particular region (258)

natural resource any natural substance that living things use, such as sunlight, soil, water, plants, and animals (5)

natural selection term used to describe the unequal survival and reproduction of organisms that results from the presence or absence of particular inherited traits (43)

niche (NICH) the way of life of an organism (38)

nitrogen cycle (NY truh jun) process by which atmospheric nitrogen is converted into compounds for use by plants and animals, eventually returned by decay (64)

nitrogen-fixing bacteria (NY truh jun bak TIR ee uh) bacteria that convert nitrogen gas from the atmosphere into a form that plants can use (64)

nonpoint pollution (puh LOO shun) pollution that comes from many sources rather than a single specific site, such as pollution that reaches a body of water from streets and storm sewers (133)

nonrenewable resources resources that can be used up faster than they can be replenished naturally, such as coal, oil, and natural gas (6, 282)

non-urban land rural land, such as forest, grassland, and farmland, that is not densely populated and without large infrastructures (204)

no-till farming procedure in which the seeds of the next crop are planted in slits cut into the soil through the remains of the previous crop (237)

nuclear energy (NOO klee ur) energy that exists within the nucleus of an atom (283)

nuclear fission (NOO klee ur FISH un) process in which subatomic bonds that bind the components of the atomic nucleus are broken apart, releasing huge amounts of energy (283)

nuclear fusion (NOO klee ur FYOO shun) the process in which lightweight atomic nuclei combine to form a heavier nucleus, releasing huge amounts of energy; basically the opposite of nuclear fission (286)

O

observation use of our senses to report the characteristics of properties and phenomena (14)

old-field succession succession that occurs after farmland is abandoned (67)

omnivore (OM ni voar) consumer that eats both plants and animals (57)

open-pit mining method of mining in which large holes are dug in the ground to remove materials such as ore, sand, gravel, and building stone (210)

open space areas designed or purposely left undeveloped for the relief of urban areas (203)

Operation GreenThumb (OGT) program sponsored by New York City's Department of General Services enabling nonprofit organizations to create community vegetable and flower gardens (224)

ore rock that contains minerals (210)

organic farming (oar GAN ik) method of growing plants without synthetic pesticides or fertilizers (238)

organism (OR guh niz um) an individual living thing (37)

overgrazing damage to a grassland caused by too many animals eating in a limited area so that the grass cannot recover (208)

overpopulation condition in which an area cannot support its human population with its available resources, or in which the population, because of growth, suffers problems that affect its general welfare (342)

ozone (OH zohn) form of oxygen with molecules made of three oxygen atoms (186)

P

pampas (POM puhz) large, treeless plains of Argentina and other parts of South America (92)

parasite (PAR uh syt) organism that lives in or on another organism and feeds on it without immediately killing it (41)

parasitism (PAR uh syt iz um) the relationship between a parasite and its host (41)

particulate (par TIK you lit) very small, separate particles, as in soot and ash (152)

passive solar heating (PAS iv SOH lur) system in which sunlight is used to heat buildings directly without pumps or fans (288)

pathogens (PATH uh juns) disease-causing organisms, such as bacteria, viruses, and parasites (136)

perennial grass (puhr EN ee ul) tall grass plants that live several years (as contrasted to annuals) and have a tendency to shade the ground, limiting growth of smaller plants (68)

permafrost (PUR muh froast) permanently frozen soil a few inches below the active soil in tundra biomes (97)

pest any organism that is not wanted or exists in large enough numbers to cause damage (240)

pesticides (PES tuh sydz) substances that kill pests (241)

pheromone (FER uh mohn) chemical substance secreted externally by certain animals that conveys information to and produces certain responses in other animals of the same species (244)

pH number measure of the acidity or basicity of a substance (162)

photodegradable plastic (foht oh dee GRAY duh bul) plastic that will decompose into smaller pieces under certain kinds of radiant energy, especially ultraviolet light (314)

photosynthesis (foht oh SIN thuh sis) biological synthesis of chemical compounds in the presence of light to produce organic substances such as sugar (55)

pioneers (py uh NIR) first organisms to colonize any newly available area and start the process of succession (67)

poaching illegal hunting (257)

point pollution (puh LOO shun) pollution discharged from a single source, such as from a factory or wastewater treatment plant (132)

pollution (puh LOO shun) the contamination of the air, water, or soil (8)

population (pop you LAY shun) a group of individuals of the same species living in a particular place (37)

population crisis (pop you LAY shun) situation in which the number of people grows so quickly that a region cannot support them (11)

prairie (PRER ee) large area of level or slightly rolling grasslands (92)

precipitation (pree sip uh TAY shun) the process in which water in the atmosphere returns to the surface of the Earth in the form of rain, sleet, or snow (63)

predation (pree DAY shun) the act of killing and eating another organism (39)

predator (PRED uh tur) organism that kills and eats another organism (39)

prediction (pree DIK shun) statement about what one expects will happen (15)

pressure vessel steel casing containing cooling fluid that surrounds the reactor in a nuclear power plant (284)

prey (PRAY) organism that is killed and eaten by a predator (39)

primary pollutant (PRY mer ee puh LOOT nt) pollutant put directly into the air by human activity, such as soot from smoke (152)

primary succession (PRY mer ee suk SESH un) succession that occurs in areas where no ecosystem has existed previously (69)

producer (proh DOOS ur) organism that makes its own food; autotroph (56)

protein (PROH teen) group of organic compounds used for food by living organisms and contained in meats and dairy products (228)

pure science study and activity that seek answers to questions about how the world works; examples are biology and physics (13)

R

radon gas (RAY don) gas produced naturally in the Earth by the decay of uranium (160)

rangeland grassland used for grazing animals (208)

reactor (ree AK tur) main element of nuclear power plant from which a steady, manageable amount of energy is released (284)

recharge zone area of land on the Earth's surface from which groundwater originates (127)

reclamation (REK luh may shun) the process of restoring land to the condition it was in before mining operations began (212)

recycling (ree SY kling) reusing discarded material (311)

reforestation (ree foar is STAY shun) process of replacing trees that have died or been cut down (206)

remora (REM uhr uh) small fish that cling to larger fish and other bodies by means of a sucking disc (42)

renewable resources resources in great abundance that are continually produced, such as the wind and sunlight (6, 282)

reservoir (REZ uhr vwor) artificial lake used to store water, control drainage, and provide recreation (125)

resistance (ri ZIS tuns) the ability of a pest population to tolerate a particular pesticide (243)

resource anything that is ready for use or can be drawn on for use by an organism (40)

Resource Conservation and Recovery Act (RCRA) act of U.S. Congress that requires new landfills to be built with safeguards to reduce pollution problems (307)

resource depletion (dee PLEE shun) exhaustion of a natural resource, such as the extraction of oil from the earth or the absence of nutrients from soil that has been overused (5)

reverse osmosis (os MOH sis) desalinization process in which pressure is used to push water through a semipermeable membrane that will not permit salts to pass (128)

rhizoids (RY zoidz) rootlike structures of mosses and ferns that attach the plant to rocks and substratum (104)

rhizome (RY zohm) creeping stem lying below soil but differing from a root in having leaves and shoots and producing roots from its undersurface (87)

S

salinization (sal uh ni ZAY shun) the accumulation of salts in the soil (238)

science systematized knowledge derived from observation, study, and experimentation; also the activity of specialists to add to the body of this knowledge (14)

scientific methods the systematic analysis of the natural world whereby scientists follow methods to collect data by observing, hypothesizing and predicting, experimenting, organizing and interpreting, and reporting (14)

S-curve curved line that shows a population reaching its carrying capacity and then falling below it (335)

secondary pollutant (SEK un der ee puh LOOT nt) pollutant that forms when a primary pollutant or a naturally occurring substance such as water comes into contact with other primary pollutants and a chemical reaction takes place (152)

secondary succession (SEK un der ee suk SESH un) pattern of change in an area where an ecosystem has previously existed (67)

selective cutting method of harvesting only middle-aged or mature trees individually or in small groups (206)

sick-building syndrome condition of buildings with particularly poor air quality, frequently caused by sealed windows and poor air circulation (158)

sludge (SLUJ) solid material left over after wastewater treatment (136)

smog air pollution over urban areas that reduces visibility; combination of the words *smoke* and *fog* (156)

solar cells (SOH lur) devices that convert the sun's energy directly into electricity (290)

solar energy (SOH lur) energy from the sun (288)

solid waste any discarded material that is not a liquid or gas (304)

specialists organisms that are adapted to exploit a special resource in a particular way to avoid competition (82)

species (SPEE sheez) a group of organisms that are able to reproduce together and that resemble each other in appearance, behavior, and internal structure (37)

steppe (STEP) large, grassy plains area with few trees in southeast Europe and Asia (92)

stratosphere (STRAT uh sfir) one of the five layers of the atmosphere, immediately above the troposphere and extending from 10 km to about 45 km above the Earth's surface (175)

strip mining method of mining in which huge machines clear away large strips of the Earth's surface, as in phosphate mining (210)

subsistence farmers (sub SIS tuns) farmers who grow only enough food to feed their families (229)

suburban sprawl (suh BUR bun) expansion of a city's borders primarily for residential development (200)

succession (suk SESH un) the regular pattern of changes over time in the types of species in a community (66)

Superfund Act act passed by the U.S. Congress to discourage illegal dumping of hazardous wastes and to pay for the cleanup of abandoned waste sites (317)

surface impoundment (SUR fis im POUND ment) a pond with a sealed bottom that serves as a disposal facility in which wastes settle to the bottom (320)

surface water fresh water found above ground in lakes, ponds, rivers, and streams (123)

sustainable world world in which human populations can continue to exist indefinitely with a high standard of living and health (12)

swamp land covered with water; contains woody plants or shrubs (103)

synthetic fertilizer (sin THET ik) fertilizer produced in factories and used in place of manure and plant wastes (232)

T

taiga (TY guh) biome dominated by conifers and characterized by harsh winters, that occurs in an area just below the Arctic Circle; also called northern coniferous forest (88)

temperate deciduous forest (dee SIJ oo us) forest in an area of extreme seasonal variation, in which trees drop their leaves each fall (86)

temperate grassland biome occurring in semi-arid interiors of continents; examples are the prairies of North America, the steppes of Russia and Ukraine, and the pampas of South America (92)

temperate rain forest cool, humid biome where tree branches are draped with mosses, tree trunks are covered with lichens, and the forest floor is covered with ferns (84)

thermal inversion (THUR mul in VUR shun) atmospheric condition in which the air above is warmer than the air below, sometimes trapping pollutants near the Earth's surface (155)

thermal pollution (THUR mul puh LOO shun) addition of excessive amounts of heat to a body of water, such as in runoff from industrial cooling systems (137)

thermosphere (THIR moh sfir) atmospheric zone above the mesosphere (175)

threatened species (SPEE sheez) species likely to become endangered if protective measures are not taken immediately (261)

topsoil loose surface layer of soil (232)

trophic level (TROF ik) a step in the transfer of energy through an ecosystem; the level of a food chain that an organism occupies (60)

tropical rain forest warm, wet biome that occurs in a belt around the Earth near the equator and contains the greatest diversity of organisms on Earth (80)

tropical savanna (suh VAN uh) plain or grassland characterized by scattered trees (90)

troposphere (TROH poh sfir) one of the five layers of the atmosphere, extending from the Earth's surface to about 10 km above the surface (175)

tundra (TUN druh) biome without trees, where grasses and tough shrubs grow in the frozen soil; extends from the Arctic Circle to the North Pole (97)

typhus (TY fus) infectious disease carried by lice (242)

U

ultraviolet (UV) light (ul truh VY oh lit) harmful light from the sun (186)

understory shrubs and plants that grow beneath the main canopy of a forest (82)

urban area land area with dense population and including housing and infrastructure necessary for citizens' well-being (198)

urban crisis (UR bun) condition in which more people live in a city than its infrastructure can support (199)

urbanization (ur bun ih ZAY shun) movement of people from rural areas to cities (199)

urban-renewal project program to rebuild a worn-down urban area by providing improved facilities (202)

V

value what a person considers to be important, as when making a decision (20)

variable (VER ee uh bul) changed or changing factor used to test a hypothesis in an experiment (16)

VOCs (volatile organic compounds) chemical compounds that form toxic fumes (155)

W

wastewater treatment plant structure that filters out contaminants from waste water (134)

water cycle the continual process in which water circulates between the atmosphere and the Earth (62)

water pollution (puh LOO shun) introduction of foreign substances into water that degrade its quality, limit its use, and affect organisms living in it or drinking it (131)

watershed entire area of land that drains into a river (123)

weather conditions in the atmosphere at a particular place and particular moment (176)

wetland area of land covered by water for at least part of the year (102)

wilderness area designated area in its natural state where the land and the ecosystems it supports are protected (215)

wind energy (EN uhr jee) wind, an indirect form of solar energy, that can be captured to generate electricity (291)

Y

yield (YEELD) amount of crops produced per unit area (229)

INDEX

Note: Page numbers in italics indicate illustrations.

D

E

F

G

N

O

P

R

S

Abbreviations used: (t) top, (c) center, (b) bottom, (l) left, (r) right, (bckgd) background

Cover Photo Credits

Front cover: sand dunes, Tony Stone Worldwide; sumach leaves, Andreas Feininger; Silvaculture, Index Stock; wind turbines, Glen Allison, Tony Stone Worldwide; impalas, Frans Lanting, Minden Pictures; wave, Superstock; trees, FPG International; sunflower, John Cancalosi, Natural Selection; suburbia, Superstock; apples, Superstock; flasks, Four By Five

Back Cover Photo

Andreas Feininger

Title Page

John Cancalosi, Natural Selection

Table Of Contents

Page iv(t), Surgio Purtell/FOCA; iv(b), Heather Angel/Biofotos; v(t), Laurance B. Aiuppy/Aiuppy Photography; v(b), Thomas C. Boyden; vi(t), HRW photo by Sam Dudgeon; vii(t), courtesy Cliff Lerner; vii(b), Simon Fraser/Science Photo Library/Photo Researchers; viii(t), Dr. Jeffrey Kiehl, Climate and Global Dynamics, National Center for Atmospheric Research; viii(b), Peter Gridley/FPG; ix(t), William E. Ferguson; ix(b), Ray Richardson/Animals Animals; x(t), HRW photo by Sam Dudgeon; x(c), Cameramann International; x(b), HRW photo by Sam Dudgeon; xi(t), Prof. Rathje/The Garbage Project; xi(b), Jean-Paul Ferrero/Auscape International; xii(t), Contact Press Images; xii(c), Ron Sherman; xii(b), Stephen Dalton/Oxford Scientific Films/Animals Animals; xiii(t), Michelle Bridwell/Frontera Fotos; xiii(ct), Richard T. Bryant; xiii(cb), Art Wolfe.

To The Student

Page xiv(tl), John Cancalosi/Peter Arnold; xiv(c), Bob Wolf; xiv(bl), HRW photo by Sam Dudgeon; xiv(br), Karen Allen; xv(tl), Ron Garrison/Zoological Society of San Diego; xv(r), Gerhard Gescheidle/Peter Arnold.

Chapter 1

Page 2(bckgd), Luiz C. Marigo/Peter Arnold; 2(tl), HRW photo by Russell Dian; 2(c), Doug Cheeseman/Peter Arnold; 2(b), Harvey Lloyd/Peter Arnold; 2(cr), Jason Lauré; 2-3(c), Altitude/Y. Arthus-B./Peter Arnold; 3(l), Julie Robinson; 3(r), courtesy of Cliff Lerner; 4, Peter Frank/Tony Stone Images; 5, Art Wolfe/Tony Stone Images; 7(t), Renee Lynn/Davis/Lynn Photography; 7(c), Wolfgang Kaehler; 7(b), Visuals Unlimited/W. A. Banaszewski; 8(t), Robert Dawson/F-Stock; 8(c), Karen Allen; 9(tl), Jacques Jangoux/Tony Stone Images; 9(tr), Ron Sherman/Tony Stone Images; 9(b), Francis and Donna Caldwell/Affordable Photo Stock; 10(t), Elizabeth Harris/Tony Stone Images; 10(b), Paula Lerner/Tony Stone Images; 12(t), Renee Lynn/Davis/Lynn Photography; 12(bl), HRW photo by Sam Dudgeon; 12(br), Chromosohm/Sohm/AllStock; 13(t), Peter Yates/SABA; 13(bl), Paul S. Conklin; 13(br), Bruce Forster/Tony Stone Images; 14(t), Brownie Harris/Tony Stone Images; 14(bl), HRW photo by John Langford; 14(br), Terry Vine/Tony Stone Images; 15,16, courtesy of Cliff Lerner; 17, HRW Photo by Sam Dudgeon; 18(t), Surgio Purtell/FOCA; 18(b), HRW photo by Stock Editions; 21, Chad Ehlers/Tony Stone Images; 22(t), Charles Mauzy/AllStock; 22(b), Michelle Bridwell/Frontera Fotos; 23, 29, HRW photos by Sam Dudgeon; 30-31 (all photos), Karen Allen.

Chapter 2

Page 35(clockwise from top), George O. Miller, F. Stuart Westmorland/AllStock, George O. Miller, Arthur C. Smith III/Grant Heilman Photography, Robert Landau/Westlight; 37(tl), Larry Mulvehill/Science Source/Photo Researchers; 37(tr), Frans Lanting/Minden Pictures; 37(bl), Lynn M. Stone; 37(br), George O. Miller/TexaStock; 38(tl), Heather Angel/Biofotos; 38(tr), Beverly Joubert/Courtesy, National Geographic Society; 38(bl), Tom Brakefield; 38(br), Y. Arthus-Bertrand/Peter Arnold; 39(clockwise from top), Alan Blank/AllStock, Comstock, Harry M. Walker, W. Perry Conway, W. Perry Conway; 40(tl), Gary Braasch; 40(bl), Greg Brant/Texas Department of Agriculture; 40(tr), Kenneth Garrett; 40(br), William H. Allen, Jr.; 41(tl,tr), Heather Angel/Biofotos; 41(bl,br), Runk/Schoenberger/Grant Heilman Photography; 42(t), Patti Murray/Animals Animals; 42(b), Mark Stouffer/Animals Animals; 43(b), John Cancalosi/Peter Arnold; 44(both), Kim Taylor/Bruce Coleman, Ltd.; 46, Heather Angel/Biofotos; 51(t), HRW photo by Sam Dudgeon; 51(b), David Dvorak, Jr.; 52,53, Lincoln Brower.

Chapter 3

Page 55(t), Fred Atwood; 55(c), Jim Brandenburg/Minden Pictures; 55(b), Philippe Giraud/Sygma; 56(t, left to right), NASA, Doug Wilson/Westlight, David R. Frazier Photolibrary, Tom and Pat Leeson/Photo Researchers; 56(b), Richard A. Lutz/Institute of Marine and Coastal Sciences, Rutgers,The State University; 57(clockwise from top), Michael Fairchild/Peter Arnold, courtesy Minister of the Environment of Quebec, HRW photo by Russell Dian, Hans Pfletschinger/Peter Arnold; 59, David R. Frazier Photolibrary; 64, Runk/Schoenberger/Grant Heilman; 66(l), Zimberoff/Sygma; 66(r), William E. Ferguson; 67, Grant Heilman/Grant Heilman Photography; 69, Ovis-Jeff Vanuga; 69(t), Superstock; 69(b), Ovis-Jeff Vanuga; 70, David R. Frazier Photolibrary; 74, HRW photo by Sam Dudgeon; 76-77, Dan Grandmaison/Courtesy International Wolf Center; 77(t), Art Wolfe.

Chapter 4

Page 80-81 (background), Thomas C. Boyden; 80(l,r), Thomas C. Boyden; 80(c), Robert and Linda Mitchell; 81(tl,bl), Frans Lanting/Minden Pictures; 81(tr,c), Thomas C. Boyden; 82(l), Merlin D. Tuttle/Bat Conservation International; 82(r), Whit Bronaugh; 83(t), Thomas C. Boyden; 83(b), Alberto Venzago/Magnum; 84, Gary Braasch; 85, Renee Lynn/Photo Researchers; 86(all), Antman/The Image Works; 87(tl), Carl R. Sams II/Peter Arnold; 87(tr), John Cancalosi/Peter Arnold; 87(b), S. J. Krasemann/Peter Arnold; 88, Ron Levy; 88(inset),Visuals Unlimited/Peter K. Ziminski; 89(clockwise from top right), Daniel J. Cox/DJC & Associates, S. J. Krasemann/Peter Arnold, S. J. Krasemann/Peter Arnold, Art Wolfe, Art Wolfe; 91(t), Tim Laman/TheWildlife Collection/Hillstrom Stock Photos; 91(cr), Ric Ergenbright; 91(bl,br), Lynn M. Stone; 92(tl), Laurence Parent; 92(bl), Joel Bennett/Peter Arnold; 92(br), Jeff Gnass; 93(l), Garry D. McMichael/Photo Researchers; 93(r), Thomas A. Wiewandt; 94-5(bckgd), Jon Mark Stewart/Biological Photo Service; 94(bl), Dan Porges/Peter Arnold; 94(r), Fred Bruemmer/Peter Arnold; 95(r), Frans Lanting/Minden Pictures; 95(l), 96(both), Thomas A. Wiewandt; 97, Anthony Bannister/NHPA; 98(t), Jo Overholt/AlaskaStock Images;

98(b), S. J. Krasemann/Peter Arnold; 99(t), George Herben/AlaskaStock Images; 99(b), John W. Warden/West Stock; 101(t), Gary Braasch; 101(cl), Jack Dermid/Bruce Coleman, Ltd.; 101(cr), Doug Wechsler; 101(bl), Robert & Linda Mitchell; 101(br), Visuals Unlimited/Science Visuals; 102, Doug Wechsler; 103, Doug Wechsler; 104(both), Gary Braasch; 105, Fred Bavendam; 106, Fred Bavendam/Peter Arnold; 107, Norbert Wu; 108(t), Norbert Wu; 108(b), Galen Rowell/Hillstrom Stock Photos; 109, Flip Nicklin/Minden Pictures; 110(t), Neil G. McDaniel/Photo Researchers; 110(b), Norbert Wu/Peter Arnold; 111, Visuals Unlimited/Alex Kerstitch; 112, Kim Heacox/DRK Photo; 117, Glen Allison/Tony Stone Images; 118(tl), Johnny Johnson/AlaskaStock Images; 118(bl), Art Wolfe/Tony Stone Images; 118-9, Ken Graham/AllStock; 119(t), Clyde H. Smith/FPG.

CHAPTER 5

Page 121, SI/J-P. Nova/Nawrocki Stock Photos; 122(b), Stephen J. Krasemann/DRK Photo; 123, Georg Gester/Comstock; 124, Jim Richardson/Westlight; 125(t), Coskun Aral/Sipa Press; 125(b), C. C. Lockwood/DRK Photo; 127, Dwight B. Miller; 129, Richard A. Feeny; 130, courtesy City of Austin Environmental & Conservation Services; 131, Jerry L. Ferrara /Photo Researchers; 132, Jeremy Walker/Tony Stone Images; 133, HRW photo by Sam Dudgeon; 135(tr), W. Campbell/Sygma; 135(bl), Marianne Austin-McDermon; 136, Visuals Unlimited/R. F. Ashley; 137, Barbara Van Cleve/Tony Stone Images; 138, Leo de Wys; 140, Warren Bolster/Tony Stone Images; 141, Jeff Schultz/Alaska-Stock Images; 142, Frank S. Balthis; 147, 148(t), HRW photos by Sam Dudgeon; 148(b), 149(t), courtesy Elizabeth Philip; 149(b), Bill Meeks.

CHAPTER 6

Page 150(tl), Werner H. Muller/Peter Arnold; 150(tr), Kevin Schafer/Peter Arnold; 150(cl), HRW photo by Helena Kolda; 150(cr), Ray Pfortner/Peter Arnold; 150(bc)HRW photo by Russell Dian; 150(br), Bill and Jan Moeller/The Stock Market; 151, Photri; 153, Ken Biggs/Tony Stone Images; 154, HRW photo by Sam Dudgeon; 156, Ted Spiegel/Black Star; 157, David R. Frazier Photolibrary; 158, Christopher Pillitz/Matrix; 159, G. Gianni Giansanti/Sygma; 160(t), HRW photo by Michael Landes Images; 160(bl), Thomas Ives/The Stock Market; 160(br), Jim Pick-erell/Tony Stone Images; 161, Buddy Mays/Travel Stock/Hillstrom Photos; 162, Ted Spiegel/Black Star; 163(t), Argonne National Laboratory/U. S.Department of Energy ;163(c), David R. Frazier Photolibrary; 163(b), Bill Weed-mark/Panographics; 164, Simon Fraser/Science Photo Library/Photo Researchers; 169, HRW photo by Sam Dudgeon; 170(t), 94 Steve Winter/Black Star; 170(b), Fred Hirschmann/AllStock; 171(t), Michelle Bridwell/Frontera Fotos; 171(b), Spencer Grant/Photo Researchers.

CHAPTER 7

Page 173(both), NASA; 176(l), Doug Wilson/Westlight; 176(r), Robert Landau/Westlight; 179(both), W. Kleck/Terraphotographics/Biological Photo Service; 180, Randall Hyman; 182, Hank Morgan/Photo Researchers; 183, Dr. Jeffrey Kiehl, National Center for Atmospheric Research, Climate and Global Dynamics Division; 185(t), David R. Frazier Photolibrary; 185(b), Rob Badger; 187, NASA; 188(b), Millman/Sudmeier/Rocky Mountain Institute; 192, 193, HRW photos by Sam Dudgeon; 194(both), R. Sanders/Courtesy Susan Solomon; 195(tl), courtesy Susan Soloman; 195(br), 1990 Louis Psihoyos/Matrix International, Inc.

CHAPTER 8

Page 196(tl), John Kieffer/Peter Arnold; 196(c), Jim Wark/Peter Arnold; 196(tr), HRW photo by Yoav Levy; 196(bl), Steve Allen/Peter Arnold; 196(bc), H. R. Bramaz/Peter Arnold; 196-197(bc), Scott T. Smith; 197, Visuals Unlimited/L. Linkhart; 198, Jeff Gnass Photography; 199, HRW photo by Yoav Levy; 200, Peter Gridley /FPG; 201, courtesy Maryland-National Park and Planning Commission; 202, Chromosohm/Joe Sohm/AllStock; 203(l), Dennie Cody/FPG; 203(r), HRW photo by Russell Dian; 204, Jon Gnass; 205, Bruce Forster/AllStock; 207(t), William Campbell/Time Magazine; 209(b), Visuals Unlimited/John D. Cunningham; 207(b), Jason Lauré; 209(t), Visuals Unlimited/J. Alcock; 210, E.R. Degginger; 211(t), R. Caton/FPG; 211(b), Rob Badger; 212, Visuals Unlimited/C. P. Hickman; 213, Fernando Bueno/The Image Bank; 216(t) Gary Braasch/AllStock; 216(b), Rob Badger; 217(t), Appel Color Photogra-phy; 217(b), Rob Badger; 218, Stephen Kline /Bruce Coleman, Inc.; 223(t), Bob Glaze/Artstreet; 223(c), HRW photo by A. Sirdofsky; 223(bl), HRW photo by Sam Dudgeon; 223(br), HRW photo by Dennis Fagan; 224(l), Photos Courtesy of Otis Butler; 224(r), Sam Abell/National Geographic Society; 225(c), Photo Courtesy of Otis Butler.

CHAPTER 9

Page 226(clockwise from top left), David R. Frazier Photolibrary, John R. MacGregor/Peter Arnold, Walter H. Hodge/Peter Arnold, Wyman Meinzer/Peter Arnold, G. J. James/Biological Photo Service, David R. Frazier Photolibrary, John Cancalosi/Peter Arnold, Walter H. Hodge/Peter Arnold; 227, Robb Kendrick/Aurora; 228, A. Alain Nogues/Sygma; 230(t), Chris Bryant/Tony Stone Images; 230(b), Chris Rainier/J. B. Pictures; 232(t), Bernard Pierre Wolf/Photo Researchers; 232(b), David R. Frazier Photolibrary; 235, Robert E. Ford/Terraphotographics/Biologocial Photo Service; 236(t), Larry Lefever/Grant Heilman Photography; 236(b), Doug Plummer/Photo Researchers; 237(tl), Robert J. Bennett/Photri; 237(tr), Grant Heilman/Grant Heilman Photography; 237(b), David R. Frazier Photolibrary; 238, Mark Gibson; 239(t), Courtesy of Girl Guides Association of Victoria, Australia; 239(bl), David Cavagnaro/Peter Arnold; 239(br), Alan Bonicatti/Liaison International; 240(clockwise from top), Robert Borneman/Photo Researchers, Tony Stone Images, William E. Ferguson, Andrew Henley/Biofotos, William E. Ferguson, Tony Stone Images, Randall Hyman; 241, John Zoiner; 242(t), Runk/Schoenberger/Grant Heilman Photography; 242(b), Laurie Sparham/Network/Matrix; 244(t), Grant Heilman/Grant Heilman Photography; 244(b), Gerry Ellis/Ellis Nature Photography; 249, HRW photo by Sam Dudgeon; 250(t), Jim Strawser/Grant Heilman Photography; 250(bl), Rob Badger; 250-251(bc), Patrick Robert/Sygma; 251(tl), Sidney/Monkmeyer Press; 251(tr), USDA/Science Source/Photo Researchers; 251(bc), Rob Badger.

CHAPTER 10

Page 252-3, Courtesy Audubon Society and General Electric; 255, Art Wolfe/AllStock; 256(t), David Hiser/Photogra-phers/Aspen; 256(b), Texas Highways Magazine; 257, Ray Richardson/Animals Animals/Earth Scenes; 258, Heather Angel/Biofotos; 259(bl), Patti Murray/Animals Animals/Earth Scenes; 259(br), William E. Ferguson; 260(t), Flip Nick-lin/Minden Pictures; 260(inset), Laurence Parent; 264(t), Richard T. Bryant; 264(b), Tennessee Valley Authority; 265,

Baker/Greenpeace; 266(tr), Louise Gubb/J. B. Pictures; 266(br), Robert Caputo/Aurora; 267, David Clendenen/U.S. Fish and Wildlife Service/Condor Research Center; 268, Cameramann International; 269(t, left to right), Dan Suzio/Photo Researchers, Merlin D. Tuttle/Bat Conservation International, Dr. M. P. Kahl/Photo Researchers, Toni Angermayer/Photo Researchers, Tom McHugh/Photo Researchers; 269(b, left to right), Jack Jeffrey, Luiz C. Marigo/Peter Arnold, Frans Lanting/Minden Pictures, John Cancalosi/Tom Stack & Associates; 270(t), Richard O. Bierregaard, Jr.; 270(b), Franz Lanting/Minden Pictures/Bruce Coleman Limited; 275, HRW photos by Sam Dudgeon; 276(tl), Dan Lamont/Matrix; 276(bl), Bill Gabriel/Biographics; 276(br), Gary Braasch/AllStock; 277(tl), Tim Davis/Davis/Lynn Photography; 277(tr), Walter Hodges/AllStock; 277(bl), David J. Cross.

CHAPTER 11

Page 279, Bob Gomel/LIFE Magazine, © Time, Inc.; 280(c), Energy Technology Visual Collection, Department of Energy; 280(c), Stock Editions; 280, HRW Photo; 281, Jeri Gleiter/Peter Arnold; 282, HRW Photo; 284, David R. Frazier Photolibrary; 285(t), Courtesy Westinghouse Nuclear Fuel Division; 285(b), Reuters/Tass/Bettmann Newsphotos; 286, Plasma Physics Laboratory, Princeton University; 288-289, Robert Millman/The Rocky Mountain Institute; 290, David Hurn/Magnum; 291, Cameramann International; 292, David R. Frazier Photolibrary; 293, James King-Holmes/Science Photo Library/Photo Researchers; 294, David R. Frazier Photolibrary; 299, Arthur Tress/Photo Researchers; 300, Paul Ferris; 301(t), Cameramann International; 301(b), Paul Ferris.

CHAPTER 12

Page 302(tl), Phil Degginger; 302(tr), Visuals Unlimited/Bob Newman; 302(bl), Steve Niedorf/The Image Bank; 302(br), C. A. Wilkinson/FPG; 303, Michelle Bridwell/Frontera Fotos; 304(t), Mark Sands/Sipa Press; 304(b), James Lukoski/Black Star; 305(c), Peter Stackpole/LIFE Magazine © Time Warner, Inc.; 305(b), HRW photo by Sam Dudgeon; 307, Ray Pfortner/Peter Arnold; 308, Professor Rathje/The Garbage Project; 309, Visuals Unlimited/John Sohlden; 310, HRW photo by Sam Dudgeon; 311(b, clockwise from top), Walter Bibikow/The Image Bank, Paul Merideth/Tony Stone Images, Jose Azel/Aurora, Tom Tracy/The Stock Market, HRW photo by Sam Dudgeon; 312, 313, HRW photo by Sam Dudgeon; 315, HRW photo by Eric Beggs; 316, John Nordell/J. B. Pictures; 318, Michel Philippot/Sygma; 319, James Cavanaugh; 321(both), Jose Azel/National Geographic Society; 322, HRW photo by Sam Dudgeon; 327(both), Michelle Bridwell/Frontera Fotos; 328(t), Sander/Gamma-Liaison; 328(b), Cameramann International.

CHAPTER 13

Page 330(clockwise from tr), Frans Lanting/Minden Pictures, Lincoln Brower, Harvey Lloyd/Peter Arnold, James H. Karales/Peter Arnold; 331, Bo Zaunders/The Stock Market; 332(t), Al Grotell; 332(b), Robert E. Barber; 334(l), Kevin Vandivier/TexaStock; 334(c), Joe McDonald/Joe McDonald Wildlife Photography; 334(r), Daniel J. Cox/DJC & Associates; 337(t), James Karales/© LOOK Magazine; 338(l), Lewis W. Hine/George Eastman House; 338(r), The Bettmann Archive; 339(tl), Michael Sullivan/TexaStock; 339(tr), Blair Seitz/Photo Researchers; 339(cl), Kennedy/TexaStock; 339(cr), Lawrence Migdale/Tony Stone Images; 339(b), Robert J. Bennett/FPG; 342, Bernard P. Wolfe/Photo Researchers; 343, Michael S. Yamashita/Westlight; 344(t), Christopher Morris/Black Star; 344(b), Alan Hawes/Sygma; 345, Roger Job/Gamma Liaison; 351(t), Wendy Stone/Odyssey/Chicago; 351(b), Randall Hymann; 352(t), Jim Nilsen/AllStock; 352(b), David Ball/AllStock; 353(t), Jeffrey Aaronson/Network Aspen; 353(b), Lowell Georgia/Science Source/Photo Researchers.

CHAPTER 14

Page 354(t), John B. Hyde/AlaskaStock Images; 354(cl), Ray Pfortner/Peter Arnold; 354(cr), John T. Morrison/Photonica; 354(bl), HRW photo by Eric Beggs; 354(br), NASA; 355, Peter Charlesworth/J. B. Pictures; 356(tl), Najlah Feanny/SABA; 356(tr), AP/Wide World Photos; 356(b), SuperStock; 357, R. Sorensen & J. Olsen/NHPA; 359(l), Robert Caputo/Aurora; 359(r), Steve Rubin/J. B. Pictures; 360, Archive Photos; 362, Tom & Pat Leeson/Photo Researchers; 363(t), Terry Farmer/Tony Stone Images; 363(c), HRW photo by Sam Dudgeon; 363(bl), Dan Lamont/Matrix; 363(br), David Young-Wolff/PhotoEdit; 364(t), David Forbert/SuperStock; 364(b), NASA; 368, HRW photo by Sam Dudgeon; 368(l), Darnin Farrell Images; 368(r), Chesapeake Bay Foundation/both courtesy the Cousteau Society; 369, Alan L. Detrick/Photo Researchers; 370(l), Louie Psihoyos/Contact Press Images, Inc.; 370(r), Solar Survival Architecture, Taos; 371(t), Pamela Freund/Solar Survival Architecture, Taos; 371(b), Solar Survival Architecture, Taos.

ENVIRONMENTAL CAREERS

Page 372, Flavia Castro; 373, 374, Ken Dudzik; 375(t), Calex S. MacLean/Landslides, Boston; 375(b), Ken Dudzik; 376, 377, Ron Sherman; 378, 379 George Ostertag; 380, Courtesy Jana Walker; 381, Jonathan A. Meyers; 382(t), Ted M. Conde, Park Artist, Milwaukee County/Photo courtesy Niki Espy; 382(b), R. Selvakumar; 383, John Herron/AllStock; 384, Michael Newman/Photo Edit; 385, HRW photo by Russell Dian; 386(t), Photo courtesy Hanna Anderson; 386(b), Mark Selick/Fay Photo Service, Inc.; 387(t), HRW photo by Michelle Bridwell; 387(b), Courtesy Tierra Pacifica Corporation, CA.

ECOSKILLS

Page 389, HRW photo by Sam Dudgeon; 391, Bob Wolf; 394, HRW photo by Sam Dudgeon; 396, Hans Reinhard/Bruce Coleman, Ltd.; 397(t), Paul S. Conklin; 397(b), Mae Scanlon; 397(inset), George H. Harrison/Grant Heilman Photography; 398, Skjold/Photri, Inc.; 400(tl), Tony Tilford/Oxford Scientific Films/Animals Animals; 400(tc), Merlin D. Tuttle/Bat Conservation International; 400(tr), Merlin Tuttle/Bat Conservation International; 400(b), Donna Hensley/Bat Conservation International; 400(inset), HRW photo by Sam Dudgeon; 401(t), Stephen Dalton/Oxford Scientific Films/Animals Animals; 401(b), HRW photo by Sam Dudgeon; 402, Stephen Dalton/Oxford Scientific Films/Animals Animals; 403(t), Stephen Dalton/Oxford Scientific Films/Animals Animals; 403(b), HRW photo by Sam Dudgeon; 404(t), Ken Cole/Animals Animals/Earth Scenes; 404(b), HRW photo by John Langford.